The Life of Vertebrates

THE LIFE OF VERTEBRATES

J. Z. YOUNG

Emeritus Professor of Anatomy, University College London

THIRD EDITION

CLARENDON PRESS · OXFORD   1981

Oxford University Press, Walton Street, Oxford OX2 6DP

OXFORD LONDON GLASGOW
NEW YORK TORONTO MELBOURNE WELLINGTON
KUALA LUMPUR SINGAPORE JAKARTA HONG KONG TOKYO
DELHI BOMBAY CALCUTTA MADRAS KARACHI
NAIROBI DAR ES SALAAM CAPE TOWN

© J. Z. Young, 1981

Published in the United States by Oxford University Press, New York.

FIRST EDITION 1950
SECOND EDITION 1962
THIRD EDITION 1981

British Library Cataloguing in Publication Data

Young, J. Z.
  The life of vertebrates. – 3rd ed.
  1. Vertebrates
  I. Title
  596   QL605

ISBN 0–19–857172–0
ISBN 0–19–857173–9 Pbk

Composition in Times by
Filmtype Services Limited
Scarborough, North Yorkshire
Printed by Butler & Tanner Ltd
Frome and London

Preface to the Third Edition

In the last twenty years there has been an immense addition of knowledge on nearly all aspects of vertebrate life, especially of physiology and ecology. In this new edition we have introduced much new detail, especially about the nervous and endocrine systems and behaviour. It has been found possible to incorporate many new facts within the original framework of the book and this has confirmed my belief that the best way to study animals is to look at all aspects of their lives. To understand the special features of the carpus or the cortex of an animal you need to know how it lives.

Fortunately there have not been great changes in knowledge of gross anatomy and we have kept the descriptions of skeletons and dissections of a few types, and the beautiful drawings made by the late Miss E.R. Turlington. The facts set out in these sections of the book are hard to obtain elsewhere. We have also retained the systematic classifications; even though controversial they provide a framework for which many workers may be grateful. We have not always followed recent cladistic revisions since we believe that classifications that emphasize grades of organization are useful, at least for the beginner.

The whole book is organized around the theme that mechanisms of homeostasis have become increasingly more complex during vertebrate evolution, allowing life to continue under conditions not possible before. Discussion of this involves questions of value about the aims of life and the meanings of 'higher' and 'lower'. Such topics are often avoided by scientists but no honest treatment of living organisms can avoid them. I believe that emphasis on the pervasive tendency to self-maintenance (homeostasis), and its progressive evolution, allows us to organize our study of the life of vertebrates and also provides a much-needed guiding light in considering the life of man.

A synoptic view of the lives of animals can perhaps be given only in a work produced by a single author, but this necessarily involves the limitations imposed by his ignorance, of which I have been acutely conscious. I must apologize to those scientists who find their work misquoted (or omitted altogether) and also to any students who are consequently misled. We have added many more bibliographical references so that readers can consult original sources.

In mitigating the dangers of inaccuracy I have been fortunate to have the advice of those listed on page vi on their special topics. Several of them found it hard to accept my point of view, but they have all helped to remove many errors; I alone am responsible for those that remain. It may be some consolation that a single author can provide his own point of view on controversial topics.

Above all I am grateful to Dr M. Nixon. Without her help this revision would not have been possible. She has sought out the most recent papers on an immense variety of topics and has undertaken all the laborious work of editing and correcting. We are both of us grateful to Miss P.R. Stephens for help in many ways, and also to R.M. Young who acts as our secretary.

As usual we have to thank all those at the Oxford University Press who have given us so much kindly help.

London   J.Z.Y.
May 1981

Acknowledgements

We offer many thanks to the following experts for their help and advice. Dr J.N. Ball kindly read the whole in manuscript and many useful suggestions, especially on endocrinology.

J.N. Ball	J.J. Hooker
A.d'A. Bellairs	P.D. Jenkins
Q. Bone	J. Jewell
M.R. Clarke	K.A. Kermack
C.B. Cox	D.R. Kershaw
E.J. Denton	R.D. Martin
D.T. Donovan	C. Patterson
A.W. Gentry	D.W. Snow
P.H. Greenwood	H.G. Vevers
A. Hallam	J.E. Webb
J.P. Hearn	K.E. Webster
J.E. Hill	M. Whitear
M. Mills	H.P. Whiting.

We would also like to thank the many authors, publishers, and learned societies who have given permission for the reproduction of illustrations and have provided original material. Full acknowledgement is given in figure underlines and in the bibliography.

From the Preface to the First Edition

THE history of textbooks is often dismissed by the contemptuous assertion that they all copy each other—and especially each other's mistakes. Inspection of this book will quickly confirm that this is true, but there is nevertheless an interest to be obtained from such a study, because textbooks embody an attitude of mind; they show what sort of knowledge the writer thinks can be conveyed about the subject-matter. It may be that they are more important than at first appears in furthering or preventing the change of ideas on any theme.

The results of the studies of scholars on the subject of vertebrates have been summarized in a series of comprehensive textbooks during the past hundred years. Most of these works are planned on the lines laid down by the books of Gegenbaur (1859), Owen (1866), and Wiedersheim (1883), lines that derive from a pre-evolutionary tradition. This partly explains the curiosity that in spite of the great importance of evolutionary doctrine for vertebrate studies, and vice versa, vertebrate textbooks often do not deal directly with evolution. They derive their order from something even more fundamental than the evolutionary principle. The essential of any good textbook is that it should be both accurate and general. As Owen puts it in his Preface: 'In the choice of facts I have been guided by their authenticity and their applicability to general principles.' The chief of the principles he adopted was 'to guide or help in the power of apprehending the unity which underlies the diversity of animal structures, to show in these structures the evidence of a predetermining Will, producing them in reference to a final purpose, and to indicate the direction and degrees in which organisation, in subserving such Will, rises from the general to the particular'. He confessed 'ignorance of the mode of operation of the natural law of their succession on the earth. But that it is an "orderly succession"—and also "progressive"—is evident from actual knowledge of extinct species.'

These principles were essentially sound, and Owen's treatment was to a large extent the basis of the work that appeared after the Darwinian revolution. In English,

following the translation of Wiedersheim's book by W. N. Parker (1886) we have H. J. Parker and Haswell's work, now in its 6th edition. The books of Kingsley and Neal and Rand are in essentially the same tradition, though they incorporate much new work, especially from the neurological studies of Johnston and Herrick. Further exact studies on these same general morphological lines made possible the books of Goodrich (1930) and de Beer (1935), which have provided the morphological background for the present work. Throughout these works on Comparative Anatomy the emphasis is on the evolution of the form of each organ system rather than on the change of the organization of the life of the animal as a whole.

Meanwhile many other treatises appeared dealing with the life and habits of the animals, rather than with morphological principles. Among these we may mention Bronn's *Tierreich* (1859 onwards), the *Cambridge Natural History*, and many works dealing with particular groups of vertebrates. The palaeontologists produced their own series of textbooks, mainly descriptive, such as those of Zittel and Smith Woodward, culminating in Romer's admirably detailed and concise book, to which the present work owes very much. The results of embryological work have been summarized by Graham Kerr (1919), Korscheldt and Heider (1931), Brachet (1935), Huxley and de Beer (1934), and Weiss (1939), among others. Unfortunately there has been little summarizing of what is commonly called the comparative physiology of vertebrates. Winterstein's great *Handbuch der vergleichenden Physiologie* (1912) covers much detailed evidence, but comes no nearer than do the comparative anatomists to giving us a picture of the evolution of the life of the whole organism.

All of these books deal in some way with the evolution of vertebrates, and yet curiously enough they speak of it very little. It is hardly an exaggeration to say that they leave the student to decide for himself what has been demonstrated by their studies. Huxley's *Anatomy of Vertebrated Animals* (1871) is an exception in that it deals with the animals rather than their parts, and at a more popular level. Brehm's *Thierleben* (1876) gives a picture of the life of the animals, though in this case not of their underlying organization. Kükenthal's great *Handbuch der Zoologie* has the aim of synthesizing a variety of knowledge about each animal-group, and some of the volumes dealing with vertebrates make fascinating reading—notably that of Streseman on birds. But the size of the work and the multiplicity of authors make it impossible for any general picture of vertebrate life to appear from the mass of details.

The position is, then, that we have good descriptions of the structure, physiology, and development of vertebrates, of the discoveries of the palaeontologists and accounts of vertebrate natural history, but that there is no work that attempts to define the organization of the whole life and its evolution in all its aspects. Indeed, none of these works defines what is being studied or tries to alter the direction of investigation—all authors seem prepared to agree that biological study is adequately expressed through the familiar disciplines of anatomy, physiology, palaeontology, embryology, or natural history. In passing, we may note the extraordinary fact that there are no detailed works on the comparative histology or biochemistry of vertebrates—surely most fascinating fields for the future, as is, indeed, hinted by the attempts that have been made in older works, such as that of Ranvier (1878), and the newer ones of Baldwin (1937 and 1945).

The present book has gradually grown into an attempt to define what is meant by the life of vertebrates and by the evolution of that life. Put in a more old-fashioned way, this represents an attempt to give a combined account of the embryology, anatomy, phy-

siology, biochemistry, palaeontology, and ecology of all vertebrates. One of the results of the work has been to convince me more than ever that these divisions are not acceptable. All of their separate studies are concerned with the central fact of biology, that life goes on, and I have tried to combine their results into a single work on the way in which this continuity is maintained.

A glance through the book will show that I have not been successful in producing anything very novel—others will certainly be able to go much farther, and in particular to introduce to a greater extent facts about the evolution of the chemical and energy interchanges of vertebrates, here almost omitted! However, I have very much enjoyed the attempt, which has provided the stimulus to try to find out many things that I have always wanted to know.

For any one person to cover such a wide field is bound to lead to inexactness and error in many places. I have tried to verify from nature as often as possible, but a large amount has been copied, no doubt often wrongly. Throughout, the aim has been to provide wherever possible an idea of the actual observations that have been made, as well as the interpretations placed upon them. A proper appraisal of general theories can only be reached if there is first a knowledge of the actual materials, which is the characteristic feature of scientific observation. A book such as the present has value only in so far as it leads the reader to make his own observations and helps him to know the world for himself.

J.Z.Y.

1950

Contents

Chapter 5   Fishes

Chapter 6   Evolution and adaptive radiation of Chondrichthyes

Chapter 7   The mastery of the water. Bony fishes

Chapter 8   The evolution of bony fishes

1 Evolution of life in relation to climatic and geological change

1. The need for generality in zoology

THE aim of any zoological study is to know about the life of the animals concerned. Our object in this book is, therefore, to help the reader to learn as much as possible about all the vertebrate animal life that has ever existed. Thinking of the great numbers of types that have lived since the first fishes swam in the Palaeozoic seas, one might well be appalled by such a task: to describe all these populations in detail would indeed demand a huge treatise. However, in a well-developed science it should be possible to reduce the varied subject-matter to order, to show that all differences can be understood to have arisen by the influence of specified factors operating to modify an original scheme. Animal and plant life is so varied that it has not yet proved possible to systematize our knowledge of it as thoroughly as we should wish. Thinking, again, of the variety of vertebrate lives, it may seem impossible to imagine any general scheme and simple set of factors that would include so many special circumstances. Yet nothing less should be the aim of a true science of zoology. Too often in the past we have been content to accumulate unrelated facts. It is splendid to be aware of many details, but only by the synthesis of these can we obtain either adequate means for handling so many data or knowledge of the natures we are studying. In order to know life – what it is, what it has been, and what it will be – we must look beyond the details of individual lives and try to find rules governing all. Perhaps we may find the task less difficult than expected. Even an elementary anatomical and physiological study shows that all vertebrates are built upon a common plan and have certain similarities of behaviour. Our object will be to come to know the nature of this plan of life, of structure, and action, to show how it is modified in special cases and how each special case is also an example of a general type of modification.

Since the problem arises from the variety of animals that have lived and live today, our central task is obviously to inquire into the reason for the existence of so much difference. If vertebrate life began as one single fish-like type, why has it not continued as such until now? Why, instead of numerous identical fishes, are there countless different kinds, while descendants of most unfish-like form are found living out of the water and even in the air and under the ground?

To put it in a way more familiar, though perhaps less clear: what are the forces that have produced the changes of animal form? Knowing these forces, and the original type, it would be possible to construct a truly general science of zoology, with sure premises and deductions. Even if we cannot reach this end, we should at least try, hoping that after investigation of the biology of vertebrates it will be possible to retain something more than a mass of detailed information. At the end of such a study, if we deal with the subject right, we should surely be better able to answer some of the fundamental biological questions. We should be able to say something about the nature of evolution and of the differences between types, to know whether there have been rhythms of change at work to produce these differences, and also – the acid test of any true science – to forecast how these changes are likely to proceed in the future.

2. What is evolution?

The superficial answer to the question 'Why are there so many different vertebrates?' is that they have been generated by a process of Evolution. Unfortunately this much-used word is ambiguous and even the best biologists seem unwilling to define it. Darwin did not use it in the first edition of the *Origin*. Indeed it was used to refer to ontogeny (literally 'unfolding') until nearly the end of the nineteenth century.

A simple definition sometimes used is that 'Evolution is a change in the genetic make-up of populations'. But every population changes its genes from minute to minute as individuals are born and die. Is all this to be called 'evolution'? We more commonly use the word to talk about sequences of adult forms, that is of phenotypes, especially about the series of animals and plants revealed by palaeontology. What is the connection between these long-term alterations and changes of the genotype that are going on all the time?

Living things are improbable steady-state systems. They exist in environments that change from minute to minute, day to day and over the years and centuries. What enables them to continue on this unlikely course? Briefly, it is the information they inherit which allows them to take actions to prevent death. They can do this on various scales. Each individual is an *agent* selecting from minute to minute what is best to do in the changing circumstances. He can choose 'wisely' because his DNA (deoxyribonucleic acid) provides him with receptors tuned to respond to changes that are likely to occur. With this information he sets in action the enzymes, muscles, glands, and many other organs that are provided by the DNA. How this system came into being is the question of the Origin of Life, which we cannot discuss here (see *Introduction to the Study of Man*). Its result is that every form of life, bacterium, plant, or animal, can continue to meet the demands of its varying environment if it shows adequate variety of actions and sufficient capacity to collect the information needed to act correctly. Each individual is able to do this by virtue of its particular range of sensory and motor capacities.

In this way survival is possible under a limited range of change of circumstances. But the information in the DNA also provides for *reproduction*, producing continually a series of slightly different individuals. This allows for life to continue much longer, by producing new types capable of meeting the situations that result from variations that occur in the climate or other factors (p. 23). This continual change of living organization is the process that we call evolution.

The basic 'cause' of evolution is therefore the tendency of all living things to strive to survive. They succeed in spite of varying conditions because every part inherits information that allows it to *adapt* to the circumstances it is likely to find. A bacterium can switch on production of a new enzyme, a muscle grows stronger with use and a brain learns a new response. The DNA provides every individual with many such ways of 'learning' during its lifetime. But information acquired in this way during life is not passed on to the next generation. The major changes in evolution depend upon differential survival of those genotypes that provide the best information.

The pressure to acquire better sources of information and ways to adapt is thus itself a factor making for change. We shall find evidence that animal types rarely remain stable, there has been a continuous series of extinctions and replacements throughout vertebrate history. These are partly due to the repeated alterations of climate and other conditions (p. 25). But at each stage there are signs that the new types had capacities that made them more efficient than the old. New means of coping with the environment appear, involving greater complexity of organization. In particular vertebrates have developed increasing powers of *adapting* their tissues especially through their senses and nervous systems. This increase of information as to how to survive is the main sense in which there has been progress during evolution (p. 584).

3. Questions about evolution

Nearly all biologists believe that evolution has been the result of some form of natural selection of hereditary variations as postulated by neo-Darwinism (Mayr 1976; Dobzhansky, Ayala, Stebbins, and Valentine 1977). But palaeontologists, who follow both the large and small changes of organisms, have long felt that some questions remain to be answered. Recently molecular biologists, geneticists, and ethologists have raised further problems. Everybody agrees that evolution has occurred, that living forms have changed, but there are still many questions about the agencies that have produced the change (Gould 1977; Stanley 1980). Study of the life of vertebrates should help to answer these questions. We may list them as follows:

(1) We readily understand that evolution involves alteration of the genetic make-up of populations. But is *any* change 'evolution'?

(2) Is all evolutionary change adaptive, or may some of it be due to random chance?

(3) Can small microevolutionary changes explain macroevolution, large alterations of the whole plan of organisms, as when fishes became amphibians or reptiles became birds?

(4) At the opposite extreme can selection explain the numerous small differences that molecular biologists have found between proteins and other macromolecules?

(5) Do the changes follow any clearly defined sequence or direction? Can we detect progress in evolution? What is meant by referring to 'higher' and 'lower' organisms?

(6) As Gould asks, 'What is the tempo of organic change? Does it proceed gradually in a continuous and stately fashion, or is it episodic?'

We shall hope to find answers to some of these and other 'eternal questions' as we study each group of vertebrates in turn and try to understand the processes that have been at work, modifying the basic vertebrate organization.

4. Is variation between demes the basis of evolution?

Every species contains a number of distinct groups of inter-breeding individuals or demes, more or less isolated from each other by mere distance or physical barriers. Thus the western rattlesnake *Crotalus viridis* shows nine 'geographic races' (Savage 1977) (Fig. 1.1). They differ in body stripe, scales, and colour and where races meet there are intergradations. Endless examples of this sort could be given and often it is possible to identify the character of each deme as due to adaptation to local conditions. Sometimes a character changes gradually with distance and this is known as a 'cline'.

If a group formed by selection or in any other way remains isolated for a sufficient length of time its genetic make-up is likely to become incompatible with that of other demes; they become mutually infertile and a new species is formed. This may happen quite rapidly. Lake Nabugabob in Africa has been separated from Lake Victoria for less than five thousand years but contains five endemic species of the cichlid fish *Haplochromis*, each derived from a different parent species in the main

FIG. 1.1 Geographic variation in the Western rattlesnake (*Crotalus viridis*). (After Savage 1977.) (Narrow regions within broken lines indicate areas of intergradation of races.)

lake (Greenwood 1965). This rapid speciation may have occurred because the numbers are small. Conversely many species are the same on both sides of the Isthmus of Panama although the Caribbean and Pacific Oceans have been separated for 5–6 million years. This is especially true of the pelagic species, with numerous individuals. Many of the intertidal and shallow water animals have formed geminate (twin) species, slightly different on the two sides of the isthmus. Evidently isolation and small numbers are among the conditions that promote change. Clearly they are not the only factors determining the *direction* of change, if indeed can it be said to have a direction at all?

5. Genetic drift

The orthodox neo-Darwinian adaptationist position is that natural selection decides which phenotypes and hence which genotypes shall survive. The varieties of organic form according to this view have been produced by the changing demands of the physical and biological environment upon each deme. This is still the basic assumption of the great majority of biologists, even including some who wish nevertheless to emphasize that other factors are also at work in determining organic organization (Gould and Lewontin 1979). These alternative agencies undoubtedly play a part, some in accelerating, others in retarding evolutionary change. The most thoroughly established of them is *genetic drift*. Isolated populations are often small and the first step to speciation may occur in them by pure chance without the action of any selective force at all (Wright 1931, 1968–1978; see Gould and Eldredge 1979). A deme founded by a small number of colonists may continue to maintain the characteristics of the founders and such random differences may persist if the group remains isolated and if there is no strong selection against any of them. Computer modelling has shown that distinctions between groups at least as great as those found in nature can occur purely by a Markov chain of random stochastic changes, that is a sequence where each event is partly dependent on the outcome of previous ones. Figure 1.2 shows how a sort of model trilobite, a 'triloboid', changed by random steps on a computer in five traits, two of its 'head', one of the 'thorax', and two of the 'tail'. If fossils like these were found it would be concluded that taxa A and B had been selected for large size, while absence of tail had evolved separately in C and D and so on. Such simulations do not show that evolution *has* been random, but they warn that the hypothesis of randomness needs to be carefully excluded by appropriate tests of its probability. Palaeontological series can sometimes be tested in this way, but it is not easy (Feller 1968).

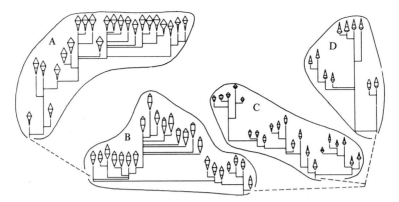

FIG. 1.2. Computer simulation of evolutionary divergence by random stochastic changes. Each of the three parts of the 'triloboid' at the bottom was allowed to accumulate small random changes. The results have been arbitrarily separated by the 'taxonomist' into four 'species'. (From Raup 1977.)

6. Can microevolution alter basic organization (Bauplan)?

Another criticism of the 'adaptationist programme' is that selective influences acting upon some overall feature such as body size may produce 'non-adaptive' changes in particular parts of the body by allometry (p. 447) or the multiple (pleiotropic) action of genes on different parts of the body. This leads us to the important point that organisms should be considered as integrated wholes, fundamentally not decomposable into independent and separately optimisable parts (Gould and Lewontin 1979). This is indeed often forgotten by people when trying to show the adaptive significance of particular features. There are certainly many situations in which features due to individual mutants have been *proved* to be of selective advantage, for instance the industrial melanism of moths. But these are all *superficial* changes. The evolution of altogether new types involves changes of the whole organization. Yet emphasis on the whole, which is offered by such workers as Riedl (1975) as the basis for an alternative to the 'adaptationist programme', is unconvincing. It is not clear what is meant by saying that there are whole patterns of organization existing through 'universal requirements' and finding no direct explanation through adaptation to environmental requirements. Of course it is true that the basic plan of construction of an organism (Bauplan) limits the possibilities of adaptational change. This is particularly obvious during the early stages of development, which are remarkably resistant to evolutionary changes. This fact is indeed the basis of the fundamental embryological law, formulated by von Baer (1828), that embryos are more alike than adults. It was misinterpreted by Haeckel (1866) into his much quoted 'biogenetic law', that ontogeny recapitulates phylogeny.

Adaptational changes mostly come relatively late in ontogeny. Genes controlling external features such as colour can rapidly become incorporated into the population. Mutants affecting early embryological stages survive only in the laboratory. An organism must adapt to its surroundings as best it can with its given Bauplan. Orthodox Darwinian theory holds that fundamentally new organizations can only appear very gradually. But some believe that macroevolution proceeds by sudden jumps in small populations (Stanley 1980). Such quantum evolution might occur by eliminating the later stages of ontogeny by neoteny (p. 71).

7. Is the evolution of molecules adaptive? The importance of polymorphism

Another set of criticisms of the adaptationist programme has developed from the studies of molecular biologists. The composition of the macromolecules of species can be compared in various ways. The gross proportion of nucleotide pairs that are different in the DNA can be estimated by discovering the extent to which hybrid molecules can be formed between single strands of DNA from two organisms (Dobzhansky, Ayala, Stebbins, and Valentine 1977). Table 1.1 shows that the divergence increases with evolutionary distance as would be expected, but the rate of nucleotide change per year is not constant. Nevertheless, phylogenies can be constructed from such data (Fig. 1.3).

Differences between individual proteins can be measured in various ways. Gel electrophoresis and immunoligical methods can be used, with some qualifications, to reveal phylogenetic connections even among closely related species. The most direct method of comparison however depends upon determining the full sequence of amino acids. This is laborious but it shows that the rate of change varies (Table 1.2). Thus haemoglobin and

TABLE 1.1. Rates of nucleotide change in the evolution of primate DNA. (After Kohne, Chiscon, and Hoyer 1972)

DNAs compared	Millions of years since divergence	Per cent nucleotide differences	Per cent changes per million years	Nucleotide changes per year
Man and:				
Man	0	0	—	—
Chimpanzee	15	2.4	0.08	1.6
Gibbon	30	5.3	0.09	1.8
Green monkey	45	9.5	0.10	2.0
Capuchin	65	15.8	0.12	2.4
Galago	80	42.0	0.26	5.2

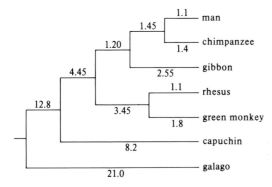

FIG. 1.3. Phylogeny of some primates based on the number of nucleotide-pair substitutions as estimated from the properties of DNA hybrids.

TABLE 1.2. Average rates of evolutionary change of various proteins. Number of amino-acid substitutions per 100 links (PAM units) (McLaughlin and Dayhoff 1969)

Type of protein	Number of PAM units per 100 million years
Fibrinopeptide	90
Growth hormones	37
Immunoglobulin	32
Ribonuclease	33
Haemoglobin	14
Myoglobin	13
Gastrin	8
Encephalitogenic proteins	7
Insulin	4
Cytochrome c	3
Glyceraldehyde 3-PO$_4$ dehydrogenase	2
Histone	0.06

cytochrome c have changed little throughout mammalian history (Table 1.3). In general the compositions of the proteins reflect the affinities that are known from palaeontology, but with striking anomalies, for instance the position of snakes and turtles (Fig. 1.4). Rapidly evolving proteins such as carbonic anhydrases or fibrinopeptides allow determination of affinities between closely related species (Fig. 1.5).

The worrying question that arises is whether the differences between molecules affect survival rates, or are selectively neutral. Clearly molecular differences are not always to be found even between very different animals. Thus man and the chimpanzee are said to be 99 per cent identical in their polypeptides. Conversely many enzymes occur in a variety of slightly different forms, differences in single amino acids being controlled by multiple alleles. It is estimated that more than 40 per cent of the loci are polymorphic for amino acid substitutions, with an average of four genes per locus. Some biochemists have argued that there can be no selective advantage in such a high degree of polymorphism and that these characteristics are 'purely accidental'. Studies of the small differences in the sequence structures of proteins and their rates of evolution have therefore led to the 'neutrality theory', that the differences are due to random fixation, of selectively neutral amino acid replacements, controlled only by mutation rates (Kimura 1968, 1977; Hartley 1979). This may be true of some proteins, perhaps many, but there is now evidence that enzymes differing only slightly have optima at different temperatures. It is unwise to assume that any small difference has no selective advantage. Survival is possible only because there is endless variety.

In natural populations of fishes such as the minnow, *Fundulus*, there are gradations in the frequency of polymorphic genes. These clines are correlated with temperature gradients and presumably have a functional significance (Powers and Place 1978). In some

TABLE 1.3. Number of amino-acid differences between haemoglobin and cyctochrome c chains of man and various other organisms (Dayhoff 1969)

Haemoglobin	Number of differences in:	
Species pair	Alpha chains	Beta chains
Human–chimpanzee	0	0
Human–gorilla	1	1
Human–Rhesus monkey	4	8
Human–spider monkey	—	6
Human–horse	18	25
Human–cattle	17	25
Human–sheep	21	26–32
Human–goat	20–21	28–33
Human–pig	18	24
Human–llama	—	21
Human–mouse	16–19	25
Human–rabbit	25	14

Cytochrome c Species pair	Number of differences
Human–Rhesus monkey	1
Human–horse	12
Human–cattle, sheep	10
Human–dog	11
Human–rabbit	9
Human–chicken, turkey	13
Human–pigeon	12
Human–snapping turtle	15
Human–rattlesnake	14
Human–bullfrog	18
Human–tuna fish	21
Human–dogfish	24
Human–fruit fly	29
Human–screw-worm fly	27
Human–silkworm moth	31
Human–wheat	43
Human–*Neurospora*	48
Fruit-fly–screw-worm fly	2
Fruit fly–silkworm moth	15
Fruit fly–tobacco hornworm moth	14
Fruit fly–dogfish	26
Fruit fly–pigeon	25
Fruit fly–wheat	47

situations polymorphism may provide for immunity, or may increase heterosis (hybrid vigour). Again, ethologists find that many aspects of selection depend on the relative frequency of characteristics. Thus predators concentrate on common varieties of prey and overlook rare ones. Birds that hunt 'intelligent' prey are more polymorphic for colour than those exploiting unintelligent or poor-sighted animals (Paulson 1973).

These are particular examples, but the possible advantages of polymorphism are endless. No species lives in *an* environment, but in a biospace with many subenvironments. The demands on action for survival can only be met by groups of animals that have many slightly different ways of adapting to the prevailing conditions. The information needed to meet such demands is provided by polymorphism. Since we still know little about the relationship between genome and adult characteristics we cannot yet determine which of the characteristics of molecules may have selective advantages.

8. Evolution follows sustained trends

It is a remarkable fact that the fossil record shows that individual features have often continued to change consistently in the same direction for long periods, so that we can attach a meaning to the phrase 'the rate of evolution'. The frequency of changes among demes such as those we have considered, whether random or adaptive such as the colour of snakes, could be measured. But this would be a mere measure of the number of changes, and would be quite a different concept from the rate of change in a consistent direction. In practice palaeontologists do in fact frequently find that changes have occurred in one direction over very long periods. The evolution of the teeth and legs and size of horses is a classic example and there we can relate the evolution to external circumstances (p. 540). But the phenomenon is quite general, 'Almost all fossil sequences long enough to be called "sustained" show prevailing tendencies in some characters and over some part, at least, of the sequence. Trends are thus extremely common in palaeontological data' (Simpson 1953). Clearly therefore we have to look hard for correlations between such changes and external circumstances (or conceivably internal tendencies). Moreover since sequences of evolutionary change are common we have to discuss possible means of measuring their rates.

9. Methods of measuring rate of evolutionary change

Rate of change over a given period can be assessed by estimates of changes in the following: (i) genetic factors; (ii) molecules; (iii) morphological characters; (iv) number of species, genera, or other taxonomic units; (v) rate of expenditure of energy, transfer of information, or other metabolic characteristic. The first and last of these are biologically the most instructive but they are impossible to measure on numerous animals or in extinct populations.

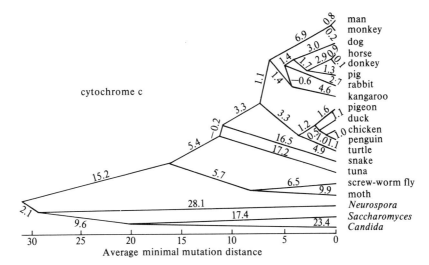

FIG. 1.4. Phylogeny based on the number of nucleotide differences inferred from the amino-acid sequences of cytochrome c. (After Fitch and Margoliash 1967.) The figures give estimated numbers of nucleotide substitutions.

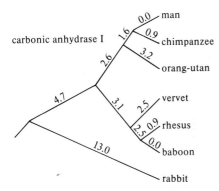

FIG. 1.5. Phylogeny of some primates based on the number of nucleotide substitutions inferred from the differences in the sequences of 115 amino-acids in carbonic anhydrase I. (After Tashian, Goodman, Ferrell and Tanis 1976.)

9.1. Rates of genetic change

As Simpson points out the ideal measure would be 'the amount of genetic change in continuous ... populations per year ...' (1953). This has indeed been measured for some genes in a few populations of *Drosophila* over a few generations (Dobzhansky and Spassky 1947). It is obviously impossible for fossil populations.

9.2. Rates of change of molecules. Do they provide an evolutionary clock?

If the neutrality theory were correct the rate of change of proteins in evolution would depend only on mutation rates. These are approximately constant for a given gene, so the amount of difference between proteins would allow us to make an evolutionary clock, using

some dated phylogenetic event to calibrate the scale (Fig. 1.6). Unfortunately as evidence of protein sequences accumulates it becomes clear that their rates of evolution are not even statistically constant (p. 8). Nevertheless, protein changes averaged over many proteins and organisms do occur at approximately constant rates. Figure 1.6 shows the nucleotide substitution for seven proteins between pairs of species that diverged at known times. The line from the origin to the earliest point fits most of the points except those at the bottom, which lie below it. These are for primates, whose proteins thus seem to have evolved more slowly than those of the other animals. The line shows a rate of only 0.41 nucleotide substitutions per million years for

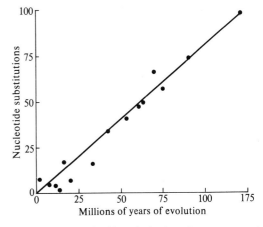

FIG. 1.6. Total nucleotide substitutions in seven proteins calculated for 15 pairs of species whose ancestors diverged at the times shown on the abscissa. (After Fitch 1976.)

the genes controlling these seven proteins. This curiously low figure suggests that molecular evolution is slower than that of the whole organism. The subject is very new and much is still uncertain. Rates have been determined only by using palaeontological data – there is no sign yet of a reliable molecular evolutionary clock. In practice, therefore, we must measure changes in the physical characters, mainly structural, which are in any case the only ones that are available over many species and in extinct populations. Phenotypic characters at least reflect in the main those of the genes.

9.3. Rate of morphological change

Rates of change of linear or other dimensions can be estimated in suitable cases such as the teeth of horses (p. 542). Haldane (1949) suggests that changes should be considered on a percentage rather than an absolute basis, for instance by considering the time needed for a unit increase in the natural logarithm of a variate or one standard deviation (s.d.) or one part in 1000. Change of one s.d. per million years (Ma) might be called a 'darwin'; the change in the teeth of the horse (p. 540) being then at a rate of 40 millidarwins. A serious problem is that such a method involves subjective decisions as to which characters are important. Various methods have been devised for combining the scores of a number of different features, as has been done for instance in the evolution of the lungfishes (Westoll 1949, p. 221) or in the various lines of therapsid reptiles as they approach the mammalian condition (Olson 1944). Multivariate measurements can now be analysed by sophisticated mathematical methods. The degree of relationship that they reveal can then be displayed on suitable matrices (Oxnard 1975) (p. 481). This has been done in a few cases such as the limbs of primates but not yet sufficiently often to allow determination of the rates of evolution.

9.4. Rates of appearance and extinction

Perhaps the most interesting conclusions about rates of evolution come from estimates of the rates of appearance and disappearance of species, or more usually genera, families or other higher taxa (see Hallam 1977a). Taxonomists, in spite of obvious weaknesses, are in fact the best judges of the differences between collections of living or fossil organisms. Their conclusions about classification are the best available and can be used to determine the relative rates of change in lineages.

Such 'taxonomic rates' show that evolutionary change has not proceeded always at the same rate. Many geologists, following Charles Lyell in the last century, have held a 'uniformitarian' doctrine of constancy of change. Lately there has been almost a return

to the 'Catastrophism' of Georges Cuvier, not in any anti-evolutionary sense but as 'the concept of an evolving universe in which erratic changes, conditioned by pre-existing states, take place at greatly fluctuating rates' (Newell 1967). The abrupt replacement of old faunas by new is so common in the fossil record as to suggest general and recurring causes capable of simultaneously affecting different groups of animals. There have undoubtedly been periods of revolution in animal organization. Recent evidence of sudden climatic changes also suggests the conditions under which new types might have a great advantage (p. 29). Over the whole long time of vertebrate evolution there have been quite dramatic variations in frequency of extinction of families in distinct groups of vertebrates (Fig. 1.7). Moreover there seem to have been parallel extinctions even between reptiles and ammonites (Fig. 1.8). Several of these periods of change correspond to the boundaries between major geological periods (p. 9). This is, of course, not surprising since palaeontologists made boundaries at these times because of the drastic changes in the fossils. In looking for 'causes' of these changes it is therefore of first importance to know what happened to the climate (p. 23). Were the massive extinctions due to sudden onset of unfavourable conditions, such as we know occurred in Pleistocene times? (p. 24). This would perhaps leave ecological niches empty for occupation by new types when conditions again became favourable. Data from all animal groups together certainly show increasing diversification after each period of mass extinction (Fig. 1.9).

However, there are various interpretations of the possible relationship between diversity and environmental change (p. 573) and it is not yet possible to make very definite statements. Undoubtedly the relation is complex, but yet there is evidence that both animals and environment change discontinuously, perhaps rhythmically (p. 571). On the other hand the probability of extinction of a genus is nearly constant regardless of its age, at least among rodents (Fig. 1.10). This seems to mean that the survival time of a genus depends upon stochastic cumulative random deterioration of its relation to the environment (van Valen 1973; Gingerich 1977).

10. Quantum evolution

The adjustment of demes to changes in their surroundings goes on continuously. As a result all widespread species show many local races (p. 465). There is evidence, however, that species as a whole are stable entities and they often survive for a very long time. An ingenious study of this has been made by examination of the earliest records of species of mammals living at the

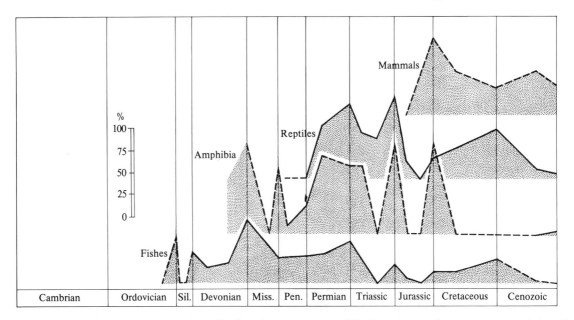

Fig. 1.7. To show the percentage extinction of families of the main groups of fossil vertebrates. There is a strong correlation of extinction peaks near the ends of the Devonian, Permian, Triassic, and Cretaceous periods. (Newell 1967.) (······ 1–10 families; ——— 11–50 families; - - - - > 50 families.)

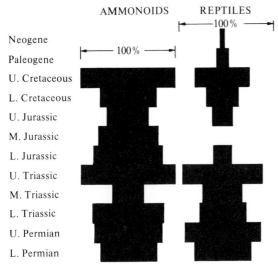

Fig. 1.8. There is parallelism in the extinction of families of ammonoids and reptiles. (Newell 1967.)

present time (or some convenient recent geological time). A graph of the survivorships of 'chronospecies' determined in this way can hardly be very inaccurate since the data for the Pliocene–Pleistocene period are relatively complete. Figure 1.11 shows that species of late Cenozoic mammals survived for 1.2 Ma on average (see also p. 577). There is no reason to think that

extinction was less rapid at this time than at any other but if the species live so long how does evolutionary change occur? Most of the orders of mammals, including flying bats and aquatic whales, evolved from insectivorous ancestors during a Palaeocene–Eocene period of not more than 15 Ma. Yet how could the wholly new types have evolved in this way if each species lived for over a million years? While they were evolving there can only have been 'ten or so successional chronospecies of average duration' (Stanley 1978). The conclusion that some workers reach from such data is that there must have been pulses of rapid evolution, the process called *quantum evolution* (Simpson 1944). Most change, they think, must take place during a period of adaptive radiation to form new species that is brief compared with the total duration of the species. Such periods may be called genetic revolutions (Mayr 1976). We have already seen that numerous species may be evolved very quickly as in Lake Nagubabob (p. 3). Presumably most of them also die out so soon that they are never recorded in the geological record as 'chronospecies'.

Similar considerations have been held to show that the level of difference that constitutes a new genus also arises rapidly, and rarely. 'Despite the detailed study of mammalian lineages of the European Pleistocene, not a single example has been recorded of phyletic transition from one genus to another' (Stanley 1978). The recent

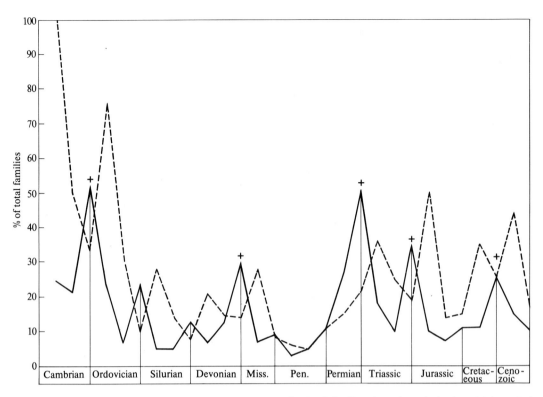

FIG. 1.9. Percentage of first (- - - -) and last (———) appearances of animal families through geologic time. Main periods of extinction(+) were near the end of the Cambrian, Devonian, Permian, Triassic and Cretaceous periods. Miss., Mississippian; Pen., Pennsylvanian. (Newell 1967.)

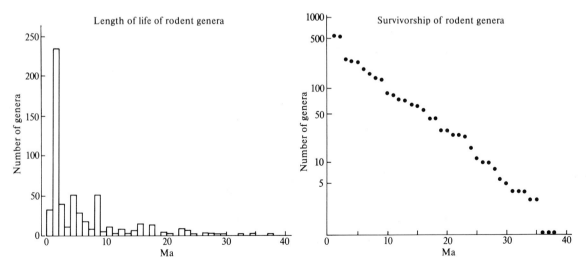

FIG. 1.10. Survivorship in Rodentia. (Left) Duration of fossil record of genera. One genus lived for 38 million years, and the average length of life of a genus is 5.85 million years. Almost half of the genera survived for only 2 million years. (Right) Cumulative plot of the number of surviving genera. The probability of extinction of a rodent genus is nearly constant, regardless of its age. (After Gingerich 1977.)

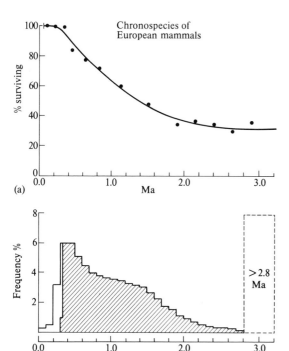

FIG. 1.11. (a). Each point shows the percentage of species occurring in a stratum of known age that survived essentially unchanged until the beginning of the Würm period (40 000 years ago). (b) Histogram of the same data showing the percentage survivors for each 100 000 years. The unshaded area at the left shows the extinctions that have occurred since the Holstein stage. Only 1 per cent of all the species lasted for less than 0.35 Ma. (Stanley 1978.)

evolution of elephants is particularly well known and shows that the three new genera that are recognized as arising from the ancestral *Primelephas* all appeared at the same time (Fig. 29.9). Of course the definition of species and genera is an arbitrary procedure for fossils, but the general conclusion reached by some workers is that much evolution proceeds by a series of quantum jumps, perhaps often in quite small populations whose divergent characteristics have allowed them to occupy a new niche or to re-occupy one left by an environmental change. According to this view, called by Stanley (1975) 'species selection', '. . . macroevolutionary trends are not a result of gradual orthoselection, but arise from a "higher level selection" of certain morphologies from a random pool of speciation events produced by punctuated equilibria' (Gould and Eldredge 1977). Species selection depends upon the validity of 'Wright's rule' – the claim that speciation is essentially random with respect to the direction of a macroevolutionary trend (1967). 'In this higher level process species become

* Indicates fossil genera.

analogous to individuals, and speciation replaces reproduction' (Stanley 1975).

11. Do species originate gradually?

Much evidence is brought forward to support the above quotation. There is, however, a contrary view about the relationship of macro- and micro-evolution. While many palaeontologists believe that there is evidence of evolution by sudden jumps others point to known examples of gradual change of one species and even genus into another (see Dobzhansky, Ayala, Stebbins, and Valentine 1977). Thus the commonest early Eocene mammal *Hyopsodus* became differentiated into several distinct species by gradual change in size of tooth and most probably of the whole animal (Gingerich 1977) (Fig. 1.12). Critics of this example hold that the data are not sufficient to establish the unidirectional trend. In any case they show a length increase of the tooth of 28.8 per cent, only a change rate of 10.1 per cent per million years which is a rate 'invisible in ecological time' (Gould and Eldredge 1977).

However, other examples are claimed to provide sufficiently complete fossil series between species and even genera. Some workers therefore maintain that the apparent quantum jumps and a 'punctuated' picture of phylogeny are artefacts of the incompleteness of the record and/or of the studies made of it. Clearly much depends upon the time scale of the deposition of the strata. Meanwhile the debate continues.

12. Classification

We have now discussed some aspects of the process by which it has come about that vertebrates are so varied. If we are to pursue the aim of finding a meaningful description of all the different sorts of them that are living now and have existed we must decide upon some way of putting them into groups or taxa, the process of taxonomy. At least 42 000 species of vertebrates exist today and there have been far more in the past. We cannot possibly describe each species separately in one book. The problem of classification for the student is how to put the animals into groups so that by giving a relatively small number of descriptions we can gain some understanding of them all. This is not quite how the research worker sees the problem, for he has the task of reducing to order *all* the variety that he meets in nature. What he does may be called microtaxonomy, describing species, sub-species, races, even individual variations. Putting these into larger groups, macrotaxonomy requires a rather different approach.

All biologists agree that in principle the basic unit of classification is the species, which is the group of interbreeding individuals reproductively isolated from

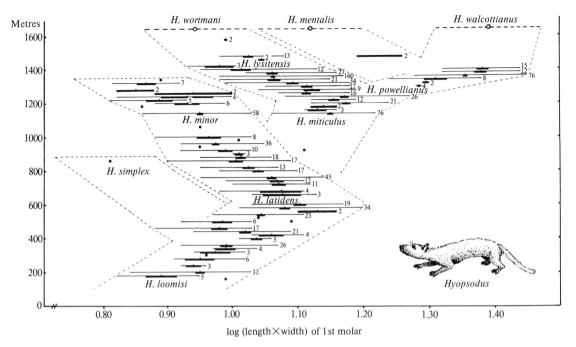

FIG. 1.12. An example of gradual evolution of species as shown by fossils collected in one place. Stratigraphic record of the Early Eocene condylarth *Hyopsodus* [p. 518] in north western Wyoming. Change of tooth size in samples from 6 m intervals in or near a measured stratigraphic section. The small species *Hyopsodus loomisi* became larger gradually through time, until it differed sufficiently to be recognized as a different species *H. latidens*. *H. 'simplex'* was another early species derived from *H. loomisi*. *H. latidens* apparently gave rise to both *H. minor* and *H. miticulus*, which in turn gave rise to *H. lysitensis* and *H. powellianus*. Note the regular pattern of divergence in tooth size (and by inference body size) in pairs of sympatric sister lineages. The vertical slash is the sample mean, solid bar standard error, horizontal line total range, small number is sample size. (From Gingerich 1977.)

other such groups. We may call this the *biospecies* (Cain 1963). In practice however there are obvious difficulties in defining species strictly in this way. The naturalist who observes differences in behaviour in the field, or the morphologist finding differences among specimens in the museum, have no means of deciding whether the groups they recognize are interfertile. So these groups are not strictly defined biospecies but are '*morphospecies*'. The palaeontologist is in an even worse position because he knows that the group of specimens that he finds is part of a 'phyletic lineage', which cannot be broken up into species by any logical criteria. So he divides the lineage into '*palaeospecies*', also called *chronospecies*, using criteria that are admittedly arbitrary. These are some of the problems of microtaxonomy, but since in this book we can seldom deal with individual species we are more concerned with macrotaxonomy by which species are grouped into larger taxa such as genera or classes, by virtue of certain common characteristics.

Again all biologists are agreed that since species have appeared by divergence the system of classification into larger taxa should be hierarchical, following the course of evolution. Indeed many workers, following Hennig (1966; see also Kavanaugh 1972) believe that once we have discovered the branching pattern (or 'cladogram') of evolution of any set of animals, say fishes, then the best classification will be immediately apparent. Groups that have separated recently will be classed together; time and amount of branching will provide the whole basis of classification. Unfortunately things are not so simple. Even apart from the imperfections of the fossil record, description of groups formed by a purely 'cladistic' classification is not necessarily the best way of giving the information we need. The main reason for this is that evolution proceeds at varying rates (p. 7). After a branching the two new groups will each have some characteristics that were there before, usually called 'primitive' but now often labelled 'plesiomorphic' (meaning 'old-featured'). The two will also have characters that are new and hence 'derived' or 'apomorphic' (meaning 'new featured'). The logic of cladistic principles insists that after two or more branch points we should classify together those groups that share derived characters and are therefore called 'synapomorphic' meaning 'sharing new features'. Where this principle may lead is shown in Fig. 1.13 (Mayr 1976). Suppose that after groups B and C have separated from A by some small amounts, D then separates from C and proceeds to change rapidly. Cladists say that we must group C and D together, as distinct from B. But obviously for our purposes it would be much more economical and sensible to describe the characters of B and C together and D separately. In fact we should consider 'symplesiomorphy' when it is greater than synapomorphy (if we must use these long words). To give an example, cladistic classification puts crocodiles with birds together as a group distinct from other reptiles.

The mere fact of branching does not necessarily produce a new type that for our purposes demands complete description. A new name and description become desirable when a group differing *widely* from the old in structure, habits, and habitat has been formed. We need in fact to deal with *grades* of evolution as well as its lines or *clades*. New grades have obviously appeared in the course of vertebrate evolution, for instance when fish acquired jaws, amphibia came on land, birds flew, and so on. Those who advocate pure cladistics object that such definitions of grades are arbitrary and cannot be justified in principle. Unfortunately one must agree with them. There can be no fixed rules by which to decide when to use these higher taxa. Indeed there is much controversy as to whether new 'grades' arise by the same process as microevolution, or by some speeding up of change owing to special circumstances (p. 8). However, pure cladistic classification is equally illogical since it ignores the fact that evolution certainly does not always proceed at the same rate in all branches of a lineage. If it did there would be no amphioxus or lung fish or tree shrews left today (p. 580). Nature produces variety that defeats our attempts at logical analysis. The classification we shall use will be a combination of grades and clades, tailored to follow as nearly as we can the varying speeds of vertebrate evolution.

There is an agreed system of names for the groupings that are used in animal taxonomy and these may be illustrated by the example of man:

Subspecies	*sapiens*
Species	*sapiens*
Genus	*Homo*
Subfamily	Homininae
Family	Hominidae
Superfamily	Hominoidea
Suborder	Anthropoidea
Order	Primates
Infraclass	Eutheria
Subclass	Theria
Class	Mammalia
Superclass	Gnathostomata
Phylum	Chordata
Subkingdom	Metazoa
Kingdom	Animalia

When classifying an animal certain of these taxa are 'obligates'. Every animal must be put into a Phylum, Class, Order, Family, Genus, and Species. If a wholly new creature was found it would have to have all of these, even though it was the only member of each.

Further subdivisions are used wherever they seem to be necessary. Indeed 'cladistic' classifications make use of a great number of them, such as 'brigade', 'cohort', or 'legion', in the attempt to follow a strictly logical system. By convention the names of some groupings are always given the same endings thus:

Superfamilies	- oidea
Families	- idae
Subfamilies	- inae
Tribes	- ini
Subtribes	- ina

It is important to realize that all these groupings are entirely arbitrary, except for species (strictly only for biospecies). There is no principle by which it can be decided, for instance, whether man and gorilla should be put in one single family or in two. There have been attempts to show that the time of branching point should determine the rank. Thus according to Hennig groups present before the Cambrian are given the rank of phyla or subphyla (1966). Those originating between the Cambrian and Devonian are classes, between Carboniferous and Permian are orders, between Triassic and early Cretaceous families, Cretaceous and Oligocene tribes, and those after the Miocene genera. Such a system would involve absurd changes in the terminology that has gradually grown up. In practice we are forced to recognize that the categories are arbitrary. It is inevitable that genera, families, and so on as they are

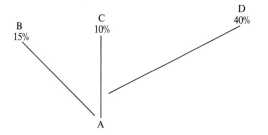

FIG. 1.13. Taxon C is more clearly related to B than to D even though it shows a more recent common connection with D. (Mayr 1976.)

identified by mammologists are not at present exactly comparable to those of ichthyologists or entomologists. Perhaps some day they will be.

The classification we have used is in the main the conventional one, following that of A. S. Romer of Harvard in the third edition of his *Vertebrate paleontolgy* (1966), but we have not hesitated to leave him where new information has indicated possible improvements.

13. What do we mean by 'higher' and 'lower' animals?

Whatever the rate of evolution, we have to enquire whether there is evidence for any overall change, in the sense that vertebrate organization has become 'more efficient'. When we refer to some types as 'higher', we do not usually mean simply 'later'. The sequences of evolution do not involve only a replacement of the animals occupying a given niche by others of slightly different form. At least in some cases the newcomers seem to have been more efficient and for that reason to have displaced the previous inhabitants (p. 581).

Is it then that evolution is continually throwing up new types that are 'better' than the old and therefore displace them from their niches? This is one type of question that we must often have before us while studying the Life of Vertebrates. The detailed evidence of course must come from the fossils, but interpretation of such questions about changes of 'efficiency' demands attention to the possible lives of the extinct creatures in the light of what is known of their living relatives (Chapter 32). Many biologists find discussion of questions of relative efficiency to be distasteful, irrelevant, and even misleading. It is certainly easier to put such questions aside, for indeed each animal is excellent in itself. Such sceptics would want to deny all meaning to the statements that fishes are higher animals than amphioxus or that monkeys are higher than mice. But what about monkeys and men? Probably we all feel that there is *some* sense in calling birds and mammals higher vertebrates. Throughout the book we shall try to examine this problem and perhaps come up with some answers to these difficult questions of values.

14. Colonization of new habitats

To do this we have to look for evidence of the colonization of *new* niches as evolution proceeds. Geological changes may themselves throw up novel habitats. The appearance of new organisms provides fresh opportunities for others. The colonization of the land by plants such as mosses and liverworts some 420 Ma ago was the necessary prelude to the possibility of animal life on land. The evolution of roots and

transport systems allowed for taller plants such as club mosses, horsetails, and ferns, and there followed the evolution of animals that reached up to them. Conifers pushed the canopies still higher and the appearance of the flowering plants 100 Ma ago provided many new niches and new problems. Vertebrates whether herbivorous or carnivorous have co-evolved with the plants and invertebrates on which they depend. The permanent excess of population has driven them to seek out new habitats, and their adaptable behaviour and efficient homeostatic mechanisms have allowed them to undertake such adventures.

Occupation of new niches often requires special structures or physiological processes. To provide these the genotype must produce new information or at least modification of the old. In spite of recent advances in genetics we still cannot provide numerical estimates of the amount of informational change involved say in converting a walking limb into a wing. There is very little knowledge of the relationship between regulator genes and the phenotypic features of vertebrates that ensure survival. Yet it is phenotypes not genes that are 'selfish' and compete for survival. There is much more DNA in all cellular organisms than in bacteria, but amounts of DNA do not help us far with our problem since all higher organisms seem to have more than is needed. There is enough in an insect or a man for 5–6 million genes, but there are only at most 50 000 'enzyme genes' (Mayr 1976). The rest of the DNA may be involved as 'regulator genes', and perhaps change in these has been a major factor in the origin of broadly new types and hence of new higher systematic taxa (macroevolution) as held by those who believe in 'punctuated evolution'. But probably most prefer to consider that all change consists essentially of microevolution.

So we have really no means of measuring the amount of information involved in the genome of any higher organism. Nevertheless, it seems likely that during vertebrate evolution there has been an increase in the complexity of the organization by which they live so that higher organisms are more improbable and therefore have to do more things, to be more 'clever', allowing them to survive in conditions where their ancestors would have perished (Young 1938). One of our main tasks will be to see whether there is evidence for any such increase in information transfer throughout the series of vertebrates.

15. The increasing complexity of life

The acquisition of new matter, and hence growth and reproduction, occurred in the earliest organisms by relatively simple means, as it still does today in the bacteria, lower plants, and some protozoa. It is not easy to provide rigid criteria for the definition of 'simple; certainly some of the chemical changes involved are very complex, but the whole system of a bacterium can, with meaning, be said to be simple. The number of parts that it contains is relatively small and the number of 'adaptive' actions that it can take is limited. A population of bacteria in a suitable medium obtains its raw materials by diffusion; the chief device that it uses to secure these materials is to provide a large number of spores, so that some may come to rest in suitable surroundings. Such a life is in principle measurably more simple than that of a vertebrate, whose system includes many special devices for obtaining access to the raw materials that it needs. We can say that a species of bacteria operates with less information than a vertebrate. Their life system involves fewer selective choices between possible courses of action. Bacteria of any one species are able to alter their enzymes in some ways to suit the substrates available, but their life does not depend upon the differentiation into numerous cell types each with its special functions. The variety of information available in the 'higher' genotypes enables them to take actions that ensure survival under conditions where the 'lower' organisms would die. Of course, each type has its own special 'niche' and the comparison of higher and lower would be easier if we could show exact quantitative differences between habitats.

16. The progression of life from the water to more difficult environments

In general, the new environments colonized have involved ever wider departures from that watery one in which life first arose. This is shown most strikingly if we contrast the simple way in which the means of life are obtained by a marine bacterium with the complicated activities that go to maintain a man alive in a city. Yet all living systems, even those that have changed most markedly since their first origin, are still watery, and must have salts and nitrogenous compounds with which to make proteins and so on. Perhaps, indeed, the basic plan of the living activity differs less in the various types than one might suppose. 'Protoplasm' is certainly not identical in all creatures, but it may be that it differs less than do the outward forms that support it.

In order to provide the conditions necessary for the maintenance of such a watery system, in very different environments, many auxiliary activities have been developed. This has involved increased information in the DNA, allowing new differentiation of parts (p. 584). More complex reactions to the surroundings involve a greater number of decisions and changes of course of action. So we could say that higher organisms have

acquired more information than lower ones and they have the means to acquire more still. This added complexity of the higher animals and plants, enables them to undertake what can be called 'more difficult' ways of life. In order to do this their activity must also be physically greater than is necessary in more lowly types. It may be presumed that more energy is transferred to maintain a given mass of living matter in the less 'easy' environments, and in that sense the higher animals are less efficient than the lower, but obviously this is a faulty criterion of efficiency.

According to this conception, then, evolution has involved a change in the relationship between organism and environment. Life has come to occupy places in which it did not exist before. Perhaps the total mass of living matter has thus been greatly increased. It must not, of course, be supposed that every evolutionary change has produced an increase in complexity in this way; examples of 'degeneration' are too well known to need quoting. We have, however, a clear impression that through the years there has been, in general, some change in animals and plants and that in a sense some of the later organisms are 'higher' than the earlier. It is hardly possible to deny that there is some meaning in the assertion that man is a higher animal than amoeba. Our thesis attempts to specify more clearly what we can know about this evolutionary change, by saying that it consists of a colonization by life of environments more and more different from that in which life arose. This colonization was made possible by the gradual acquisition of a store of instructions enabling adjustments to be made by which life can be maintained in conditions not tolerable before.

It is not easy to enumerate the complexity of any animal or to define quantitatively the nature of its relations with its environment, and for this reason it is difficult to prove our thesis rigorously. This book nevertheless makes an attempt to show how the organization of vertebrate life has become more complex since it first appeared, and that the increasing complexity is related to the adoption of modes of life continually more remote from the simple diffusion of substances from the sea. Of course, even the earliest vertebrates had already departed a long way from the first conditions of life and were quite complex organisms. However, in the history of their life through nearly 500 million years since the Ordovician period we can trace considerable further changes in complexity. During this time vertebrate life has left the sea to live in fresh water, on swampy land, and finally on dry land and in the air. It has produced special types able to support life by such an astonishing variety of devices that we cannot possibly specify them all. We shall only direct attention to a few, and thus attempt to obtain an impression of the scheme of life of the vast hordes of vertebrate animals, which, in one shape or another, have swarmed and still swarm in the waters and over the earth. We shall try to discern whether there is reason to suppose that all this variety is related in some way to changes in the surrounding world, and we may therefore finish this introduction by a brief survey of the evidences for climatic and geographical changes such as may have been responsible for the changes in organic life.

17. The changing surface of the earth

The fact that the geography and climate of the world has not always been the same began to be known in the nineteenth century, but only in the last 20 years has the full extent of the change become appreciated. Yet in spite of all the new information collected by geophysicists and geologists it is only possible to give a preliminary outline of the facts that the biologist wants to know. To understand why animals and plants have changed, and how they have become spread over the earth, we need knowledge of the distribution and composition of the land, sea and atmosphere at various times in the past, as well as information about the temperature and other physical conditions.

Most changes in climate and geography occur with such long periods that they are without appreciable effect on individual organisms, but may greatly affect the history of the race. The idea of geographical change is made familiar by the fact that coastlines and river courses have altered appreciably in historical times. We often hear stories of destruction of some houses or of a village by the sea, though it may come as a shock to learn that the sea level has changed so much that England and France were connected by land 8000 years ago, and that man-made instruments fished up from the Dogger Bank in the North Sea show that it was an inhabited peat bog

6000 years BC. Such changes in sea level are signs of the changes in extent of ice caps and of 'diastrophic movements', which are major features of long-period geological evolution.

Geologists have long known that there have been great changes in the extent of land and sea. The causes of these are still not fully known but are probably related to changes in the flow of heat from within the earth, which produces movements of portions of the earth's crust ('plates'), spreading of the sea floor, collisions between continents, and the resulting uplifting of mountains (p. 18). These periodic changes in the heat flow may be the cause of the repeated elevation and sinking of land masses (Fig. 1.14). Invasion of the edges of the continents by the seas is known as transgression (positive eustasy), the reverse process is regression (negative eustasy). Unfortunately there is no agreement amongst geologists as to whether these processes proceed in any regular cyclical fashion. Changes in sea level may be due to a combination of features (Donovan and Jones 1979). (1) Variations in climate alter the volume of ice that rests upon the land surface (p. 23). (2) Accumulation of sediment due to weathering alters the balance of land and sea, though this is of lesser importance. (3) Movements of the earth's crust produce variations in the volume of the ocean ridges (p. 19). (4) Heat flow changes under the ocean ridges may be the most important factor. High heat flow renders basalt more buoyant, hence displacing more sea water onto continents. Following the changes in level the sea leaves more or less of the continental shelf uncovered. Such upward and downward movements have been common and have profoundly influenced the climate. Oceanic climatic influences tend to produce a damp, equable climate, with large areas of marsh and forest. When the land stands higher, or is more laterally extensive, extremes of climate develop, some parts being cold,

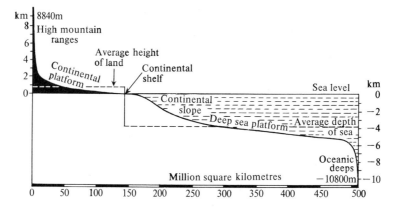

FIG. 1.14. Curve showing the area of the earth's solid surface in relation to the sea level. (From A. Holmes.)

others forming large, dry interior plains, this is known as greater 'continentality' of climate. Undoubtedly there have been great changes in extent of land and sea and also periods of extensive mountain building (orogenesis). Then the action of frost, wind, and rain breaks up and carries away the surface of the land, at a rate of the order of 30 cm per 4000 years, the processes known as weathering and denudation. The material is deposited in the river beds and in the lakes and in the shallow seas of the continental shelf, and even in the deep sea at the mouths of submarine canyons (sedimentation) (Fig. 1.15). The deposition builds the sedimentary rocks, which may be many hundreds of metres in thickness, the whole continental platform continuing to sink for long periods, perhaps with intervals during which it becomes raised above the water. Fossil remains are usually the result of the preservation of the harder parts of animals in sedimentary deposits, and the most complete series of fossils are likely to be those of animals living in the seas.

18. Components of the earth

The earth consists of a central dense core, whose outer part is liquid iron, with a radius of about 3470 km, surrounded by a less dense, but more solid mantle of 2870 km thickness. The outermost layer is a thin crust of materials of lesser rigidity and density, which is an average of 5 km thick below the sea floors and 33 km below the continental surfaces. These depths are known from the distribution of the propagation of earthquake shock waves, which changes velocity at the boundary between mantle and crust, known as the Mohorovičić discontinuity (*Moho* for short).

Above the crust are the still lighter water and atmosphere, whose evolution has greatly influenced life and been affected by it. The earliest atmosphere contained reducing gases such as methane, carbon monoxide and ammonia, and the earliest living things, formed more than 3 000 Ma ago, were probably anaerobic. The oldest known evidences of life are 'stromatolites' rocks formed from blue–green algae 3.4×10^9 years ago. Such organisms produced free oxygen, which then reacted with methane and ammonia to form carbon dioxide, water, and nitrogen. Water and nitrogen also probably appeared by 'outgassing' from the earth during its chemical evolution. The free oxygen and

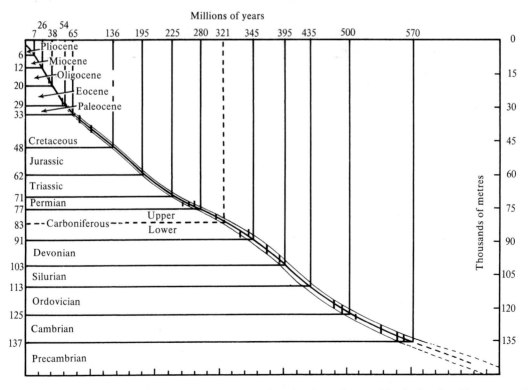

FIG. 1.15. The maximum thickness of sediment in each period is plotted against estimates of the absolute date. The error attached to these determinations is shown by the marginal lines. Apparently the rate of sedimentation has not been constant. (Modified from A. Holmes.)

nitrogen were thus largely the product of organisms, directly or indirectly, during the first 2 billion years of life. By 1.8 billion years ago there was enough oxygen to make red iron oxide deposits and so presumably sufficient for aerobic animal life to begin. However, the earliest animals with skeletons appear only much later, and rather suddenly, in Cambrian time 600 Ma ago.

The composition of the ocean is of special interest to biologists since life first became possible because of conditions in the sea, and then evolved there for perhaps 2000 Ma, that is to say for the greater part of its whole history. Living things still retain in their ionic make-up certain characteristics of the sea, indeed some authors have interpreted the blood plasma of vertebrates as a relic of the Palaeozoic sea (p. 107). The salts in the oceans probably come partly from chemical processes in the mantle, added to by erosion from the land.

19. The earth's magnetic field

The earth's magnetic field is probably the product of differential core motions orientated by the spin. This field shows large changes, which may have important influences on living things. The two poles have irregular variations in position and they reverse their polarity quite frequently (by geological time standards). There have been about 15 reversals in the last 5 Ma. The earth's magnetic field shields the surface from cosmic radiation, which has large effects on life and especially on genetic mutation. The field is weakened during periods of reversal of polarity, which are completed within a few thousand years, and the resulting high mutation rate may possibly be the source of sudden extinction and appearance of new types of life, though others doubt this. There is some evidence of sudden changes in the fossil microfaunas at times of rapid reversal (p. 26).

20. Plate tectonics and continental drift

The presence of core, mantle, crust, sea, and atmosphere was probably established over 3 billion years ago, but the distribution of land over the earth's surface was then very different from now. The changes that have occurred in the positions and levels of the continents have been major factors influencing the course of evolution of all forms of life, including the vertebrates. The conception that the continents have moved relative to each other and to the poles was put forward in its modern form by the German meteorologist Alfred Wegener in 1912. It became widely accepted in the 1950s and 60s as a result of the studies of changes in direction of magnetization of rocks especially those of the ocean floors. Since about 1965 the concept of continental drift has been incorporated into the wider view of the earth's surface known as plate tectonics (= building plates). The whole outer surface is at present divided into about eight major lithospheric plates, and some minor ones. The plates carry both the continents and the seas and include the crust and 20–40 km of upper mantle (Fig. 1.16). The margins of the plates are marked by the zones of earthquakes and volcanic activity. These margins are places where new crustal material is either being formed, as at the oceanic ridges, or forced beneath the continents (Fig. 1.17). The effect of the constant addition of crust from the mantle below is to cause slow rotation of each plate about an axis passing through the earth's centre. The motor for these movements is probably thermal convection in the mantle. This continual outpouring of new material explains why the crust below the ocean floor is thin and much younger than that beneath the continents, and younger near the ocean ridges than further away.

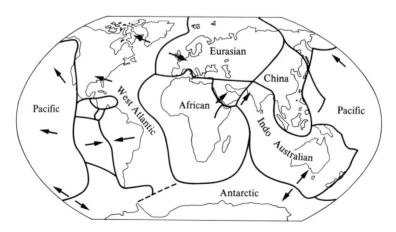

FIG. 1.16. Approximate boundaries of the plates into which the surface of the earth is divided, showing also the directions in which the crust is spreading. (After Tarling and Tarling 1977.)

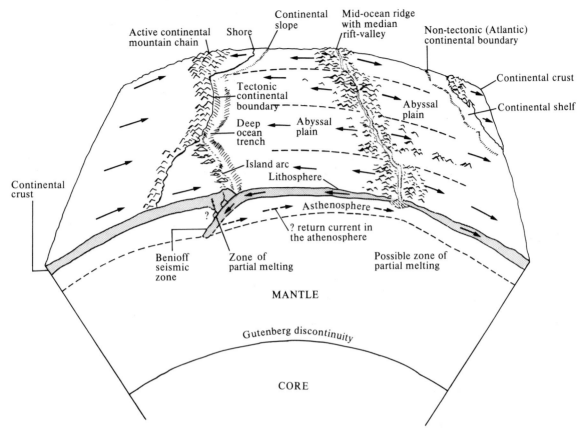

FIG. 1.17. Hypothetical diagram to show the concepts of plate tectonics. The arrows indicate the motion of the lithosphere plates relative to the mid-ocean ridge. (From R. Mason in *The earth and its satellite* (ed. J. Guest). Hart Davies, London, 1971.)

The collision of the plates folds up the surface of the continents, forming the mobile belts (orogenic belts) that make mountain chains such as those down the west of the America's and the Alps and Himalayas. The effects of rain and frost then weather away the mountains, depositing their material on the continental shelfs, which thus contain the majority of fossils. The less dense continents 'float' on the denser material beneath, in so-called isostatic equilibrium, and loss of weight from them by erosion allows them to rise, altering the distribution of land and sea. Conversely addition of sediment will cause crust to be depressed, as on many continental shelves and slopes, but tectonic forces are needed to promote initial subsidence.

21. Evolution of the continents

The early history of the lithosphere plates is not known. Perhaps the earth was at first coated by a thin skin, which became broken into many small fragments by widely distributed volcanic activity. Detailed knowledge of the distribution and climate of the continents goes back

only to the Devonian period, about 375 Ma ago, when there were probably two large land masses (Smith and Briden 1977) (Fig. 1.18). Earth movements had undoubtedly been continuous during the earlier periods, leading to extensive changes (Seyfert and Sirkin 1979). In the early Cambrian Europe and America were widely separated by a proto-Atlantic ocean. They were then gradually pushed together and met at the end of the Silurian. This explains the similarities of the older rocks and fossils of Scotland, Greenland, and Newfoundland, and some of the differences between those of Scotland and England.

The detailed history of later movements of the continental blocks is better known but much is still controversial. We may follow the account given by Irving (1977) based upon a new analysis of palaeomagnetic data. Throughout the Devonian period there were two separate continental masses (Fig. 1.18). That to the north is known as Laurasia, and includes the St. Lawrence area of North America, Europe, and parts of Asia. The southern mass is called Gondwana (after an

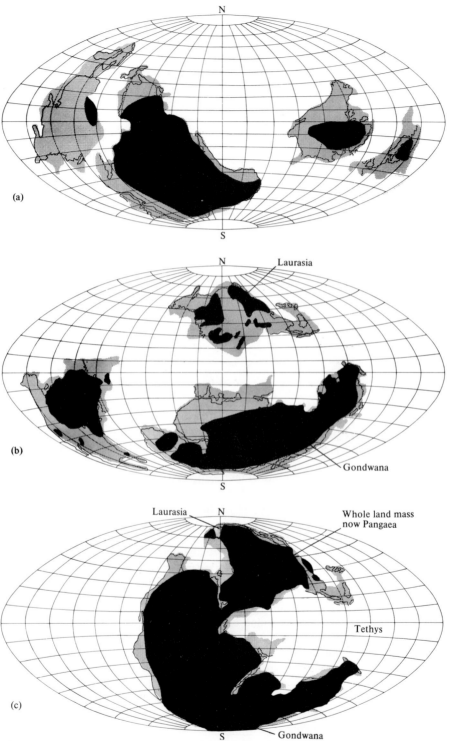

FIG. 1.18a–f. Continental drift since the Cambrian. Maps showing the movements of the major continent plates, relative to one another and to the pole. Darkest shading indicates land. Parts of the plates covered by sea are shown lighter. Areas of glaciation are shown only in the Pleistocene. (a) Cambrian, 510 Ma; (b) Devonian, 380 Ma; (c) Permian, 250 Ma; (d) Cretaceous, 100 Ma; (e)

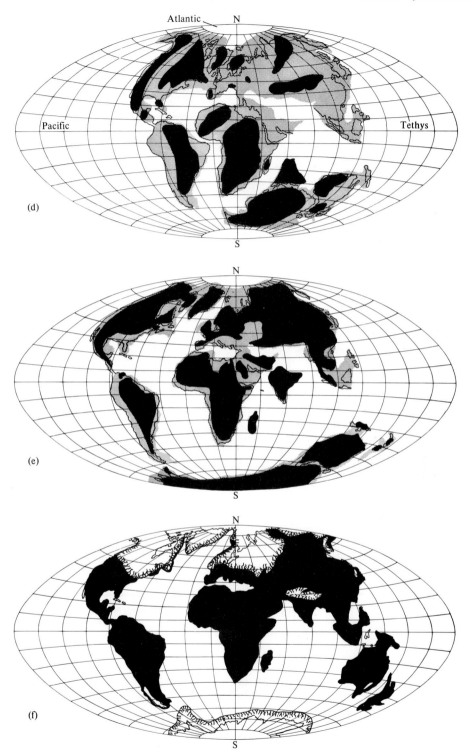

Eocene, 50 Ma; (f) Pleistocene, 40 000 years ago. (With permission from Nicholas Hall from *The illustrated origin of species by Charles Darwin* abridged by R. E. Leakey (1979). Hill and Wang, New York.)

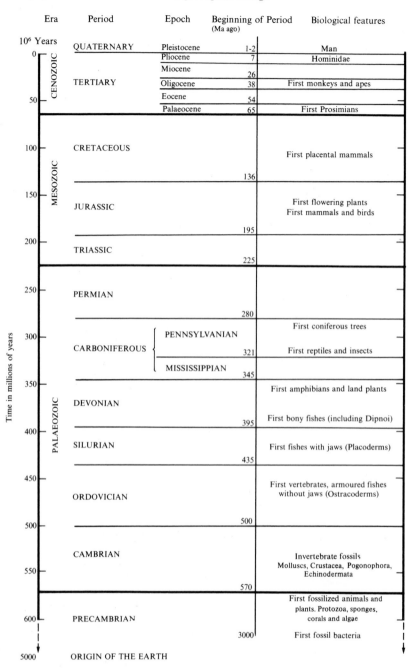

FIG. 1.18G. Geological time. The time-scale follows the convention of the Geological Society in *The Phanerozoic time-scale*. A symposium dedicated to Professor Arthur Holmes **120 S** (1964). (A supplement to the *Quarterly Journal of the Geological Society, London*.)

Indian tribe) and it included what later became South America, Africa, parts of southern Europe and Arabia, India, Australia, and Antarctica. This union presumably explains the fact that 200 Ma ago there was a range of similar large reptiles living in east Africa

and South America. The sea, of uncertain extent, between the two continents is called the Hercynian Ocean.

The two continents then converged and met during the middle Carboniferous to make a single supercontinent, horseshoe-like in shape, known as Pangaea. The Laurasian part at that time lay across the equator and had a tropical climate with lycopsid and sphenopsid ferns (horse-tails), which formed the coal beds of Europe and North America. Much of it was covered by shallow seas, connected with the large ocean to the east named Tethys after the Greek 'mother of seas', the daughter of Oceanus. Although the Gondwana and Laurasian plates were joined where north west Africa met North America and southern Europe, yet shallow seas often covered the link, preventing the migration of land animals but allowing marine ones to move between the Tethys and the Pacific Oceans. Gondwana lay far to the south, parts of it near the pole, and much of it became glaciated over periods as long as 100 Ma. The glacial deposits known as tillites were associated with seed ferns of the *Glossopteris* type, which were thus typical of cold climates, and have been found as fossils in India, East Africa, and Antarctica and other parts of what was Gondwanaland. There were at times large glaciers in the Congo, Brazil, and Australia. There is also evidence of great variation in temperature, such as those we have good records of during the recent Pleistocene Ice Ages. The Carboniferous and early Permian show more evidence of cyclic sedimentation than any other geological period and shallow seas advanced and retreated many times. Some coal and *Glossopteris*, as well as fossil amphibia and reptiles, have been found from Triassic deposits (210 Ma ago) then close to the South Pole. Most of Gondwana was dry land, which was gradually eroded and sediments many kilometres thick accumulated along the eastern margins of South Africa and Australia.

Pangaea last for some 90 Ma (Fig. 1.18c) and for most of this time there were few reversals of the geomagnetic field. Then in the late Permian and Triassic many reversals occurred, probably due to major changes in mantle convection. This Permo-Triassic revolution, the end of the Palaeozoic, was one of the main times of change in living things (p. 26).

The increased plate movements and volcanic activity led gradually to the break up of Pangaea. There is evidence that about 160 Ma ago there were very large laval flows on all the Gondwana continents except South America (Tarling and Tarling 1977). Then Gondwana rotated anti-clockwise, and Europe and north America separated from it and moved northwards so that late in the Jurassic period the Tethys opened between Africa and Eurasia to form the forerunner of the Mediterranean Sea.

The Jurassic period was a time of favourable conditions for land animals and plants. In the Cretaceous a rift appeared dividing Africa from the other parts of Gondwana, South America began to separate from Africa and the Tethys Sea opened to the Atlantic. The seas advanced from the south, reaching south west Africa about 120 Ma ago and Nigeria 105 Ma ago, finally separating Africa from South America 90 Ma ago. The north Atlantic began to open about 55 Ma ago and north east Eurasia moved away from north America throughout the early Tertiary. Northern Alaska rotated anti-clockwise so that the Atlantic rift system was developed and the North pole was no longer covered by continental crust. India rotated away from Madagascar and began to move northwards about 100 Ma ago, finally making contact with Asia 25 Ma ago when the great Oligocene and early Miocene orogenesis of the Himalayan–Alpine fold-belt began. Australia and Antarctica moved away from Africa together for some 25 Ma before beginning to separate 45 Ma ago.

22. Changes of climate

Evidence of marked changes of climate is the finding in England and other regions now temperate of animal and plant remains appropriate to warmer or colder conditions (corals and woolly rhinoceros, for instance). There is thus every reason to think that great changes from hot to cold and wet to dry conditions have occurred, in conjunction with the changes in latitude and in level of the land (Frakes 1979; Laporte 1979; Lockwood 1979).

These fluctuations in geography and climate are obviously of great importance to the biologist. In order to be able to assess the influence of such changes on life we must know more about the rates at which they occur, and careful study shows that some of the climatic changes are rhythmic. Rhythmic changes of climate are, of course, very familiar to us in the cycles of days, months, and years, and the immense importance of these short-period changes for animal and plant life must not be forgotten.

Here we are more concerned with changes of longer periodicity of which the best known are fluctuations of the amount of solar radiation received at any given part of the earth's surface (Fig. 1.19). These are likely to be especially important since plants, and hence ultimately animals, depend for their energy on sunlight. The cycle of fluctuations in the number of sun-spots (11.4 years) involves change in amount of radiation, and this is associated with some biological cycles, for instance the distribution of the growth rings made by trees.

Major alterations in climate have probably been the most important factors determining changes in environment and hence the sequence of vertebrate evolution (Fig. 1.20). Many influences have been suggested as

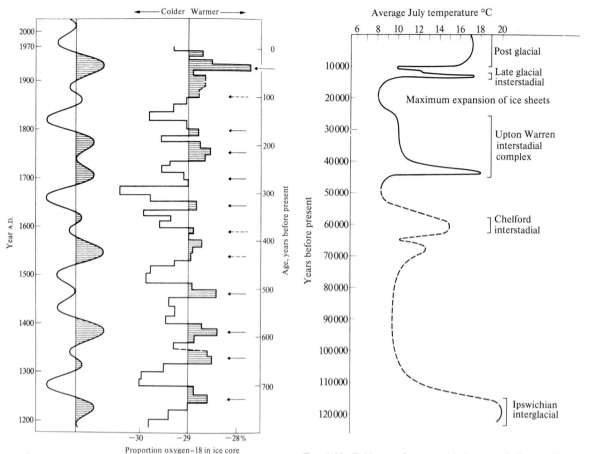

FIG. 1.19. Variations in solar radiation. On the right ^{18}O values of increments of an ice core from north Greenland plotted against time since the deposition of ice. The hatched areas correspond to relatively warm periods. On the left is a synthesis of the two harmonics (78 and 181 years) that dominate the step curve. The curve from 1970 suggests the probable future climatic development. (Dansgaard, Johnsen, Clausen, and Langway 1971.)

FIG. 1.20. Evidence of very rapid changes of climate. Fluctuations in the average July temperature, estimated from fossil assemblages of insects in lowland Britain since the Eemean (Ipswichian) interglacial. (Coope 1975.)

causing these changes and they can be classified as (1) Extra-terrestrial, including happenings in the sun and even elsewhere in the universe. (2) Variations of the earth's orbital parameters, leading to alteration of the radiation received. (3) Terrestrial causes, including continental drift due to the changes within the earth and in its crust and consequent changes in the oceans and atmosphere. These last influences have already been discussed, the effects of the other two are more difficult to assess.

23. Changes in solar output and events in the galaxy

The sun generates energy by the conversion of hydrogen to helium, and the seething motions in its interior produce changes in its output (which is known rather paradoxically as the solar constant). The sunspots are darker areas, whose number waxes and wanes with a period of about 11 years, with a further 80 year cycle of very high and very low peaks. There is still debate over the important question of whether there are variations of the solar constant over much longer periods. Models now developed based upon the unexpectedly small number of neutrinos reaching the earth assume that there are such changes and these would explain global climatic phenomena that must have been very important for vertebrates, such as the world-wide ice ages.

Further agents of change may be events in more distant parts of the galaxy, such as the explosion of supernovae (see Russell 1979). These colossal events radiate in a few weeks as much energy as ten billion suns. They occur about one every 50 years per galaxy, which means that one near enough to affect the earth takes place every 70 Ma. It is suggested that such events as the extinction of the dinosaurs at the end of the Cretaceous may have been the result of such an explosion (p. 27). At that time up to 75 per cent of all species of animals became extinct (Russell 1979), which is an effect greater than would be expected if half of the world's nuclear stockpile was exploded.

24. Changes in the earth's orbit

The variable orbital parameters of the earth are: (1) The change from circle to ellipse in a cycle of $\sim$ 93 000 years. (2) The axis wobbles, so that the season of closest approach to the sun varies, with a cycle of 'precession' of 21 000 years. (3) The tilt of the axis varies in relation to the plane of orbit with a cycle of 4000 years. These three cycles interact to produce variations in the distribution of solar radiation known as the Milankovičić mechanism (Fig. 1.19). These radiation curves agree with the fluctuations of temperature that can be shown to have occurred during the past million years (p. 28). Presumably there have been similar extensive fluctuations throughout geological history, with marked effects on the flora and fauna.

These various influences, some of them cyclic, interact to produce complicated changes in climate both over the whole globe and locally. Unfortunately the only record available of the alterations that have taken place throughout the history of the earth is embodied in the material of the crust, which is itself continually changing its position and composition. It is ironic that the appearance of new fossil types often provides the only reliable source of information about the past climatic variation that may have generated them. Questions about global alterations of physical conditions over long periods can only be provisionally answered by model studies such as those of neutrinos we have mentioned.

25. A summary of climatic history
25.1 Palaeozoic and Mesozoic periods

Putting all the evidence together it is possible to give a very rough and superficial outline of the changes of climate and extent of land and sea in the period since the Cambrian (Fig. 1.21). Throughout the early Palaeozoic periods the fossils are entirely those of aquatic animals, except for some traces of land plants and arthropods after the Ordovician. The oldest remains of vertebrates are fish scales from the Ordovician (p. 190).

During the later part of the Precambrian there were several very extreme glacial periods, affecting several continents, at different times. This is probably evidence of a global cooling of the earth, due to extraterrestrial influences. The subsequent slow warming was accompanied by the appearance of metazoan fossils, marking the beginning of the Phanerozoic period with the Cambrian, at about 570 Ma ago. Subsequent generalized changes in climate are summarized in Figure 1.22. During the Cambrian and Ordovician the temperature was rather higher than at present as shown by the presence of evaporite deposits, which are typical of desert regions. There were many coral reefs in places that were then at relatively high latitudes and these can only form in warm water. There is evidence, however, of some glaciation near to the poles. There was much marine transgression throughout the Cambrian, but this, like the early Ordovician, was a relatively stable period. There was more alternation of extent of the seas during the later Ordovician, with wide regression and glaciation at the Silurian boundary. Then through the Silurian there was melting of the polar ice and wide transgression of the sea over the land, giving warm conditions. These continued with fluctuations through the Devonian, which shows abundant reefs, carbonate deposits, many evaporites, and few coal beds. There were now many land plants and the first forests

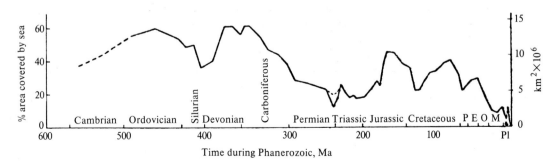

Fig. 1.21. Changes in the area of the Soviet Union that has been covered by sea during the Phanerozoic, in relative and absolute terms. E. Eocene; M. Miocene; O. Oligocene; P. Palaeocene; Pl. Pliocene. (From Hallam 1977*b*.)

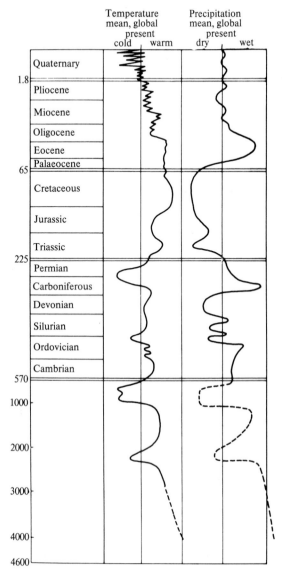

FIG. 1.22. Generalized temperatures and precipitation throughout the history of the earth. Shown as departures from present means. Absolute time scale progressively expanded. (After Frakes 1979.)

appeared at this time. Here also are found the first signs of insects and of vertebrate terrestrial life, in the form of fossil lung fishes and amphibians (p. 220).

The Carboniferous was essentially a period of high rainfall everywhere, especially in low latitudes. These conditions were probably due to the presence of large equatorial oceans, whose low reflectivity (albedo) caused increased heat absorption and evaporation, followed by precipitation. The coal measures are the remains of the forests of spore- and seed-bearing plants

that were then produced, and the land conditions evidently favoured the life of the Amphibia.

These conditions were followed at the beginning of the Permian by a very severe glaciation, similar to that of the recent Pleistocene, with ice-sheets covering several of the continents. The cause of this is uncertain but may have been the air currents produced by the presence of a large land mass at the South pole (Fig. 1.18c). At the end of the Permian the global climate became warmer everywhere, perhaps due to melting of the ice cap increasing the area of the oceans as the supercontinent Pangaea moved into low latitudes.

The Permo-Triassic transition was one of the major changes in geological history, involving both invertebrates and vertebrates, it marks the boundary between Palaeozoic and Mesozoic faunas. There were many extinctions of marine fishes (but not those of fresh water) and also of amphibians and reptiles (Pitrat 1973). Half of the major groups of the Permian are not present in the Triassic (Newell 1967). Perhaps these extinctions and replacements were a result of the many magnetic reversals at that time, with increased radiation and mutation. More probably they followed the large marine regression (Schopf 1974; Simberloff 1974).

The early Triassic climates seem to have been like those of the later Permian, cool and humid. Later it became warmer and drier than at any other time in earth's history. Coral reefs extended at least 10° further from the equator than at present. Mean annual temperatures were perhaps 10 °C warmer than now. The climate was also more uniform and this may explain the fact that animals became less diversified.

The Jurassic was a period of more uniform and equable conditions than any since. There was essentially a single landmass of Pangaea, with the Tethys Ocean to its east and the Pacific to the west, forming one sea (Figs. 1.18c and 1.23). There was little tectonic and orogenic activity and the land was not very mountainous. The climate was milder than today with no polar ice-caps, so that types of ferns and other plants that cannot tolerate frost were found far to the north and south. Evaporites and coral reefs extended to high latitudes (Fig. 1.23). Lung fishes, now tropical, were then world wide. Tropical and subtropical zones were wider than today and the gradients of mean surface air temperature from the equator to the poles are estimated to have been as little as 22 °C against the present 42 °C. The sea level rose throughout the period from occupying less than 5 per cent of the continental area to 25 per cent, but there were several advances and regressions of these shallow seas, leading to considerable changes in the climate (Hallam 1975).

The Cretaceous period, during which the thick chalk

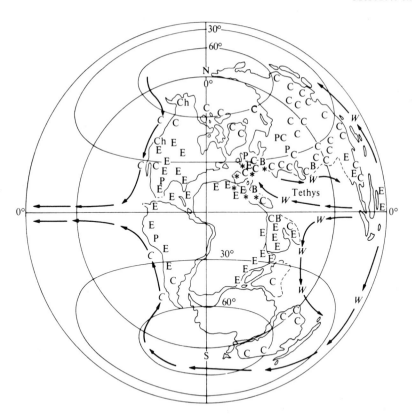

FIG. 1.23. Palaeogeography of the Jurassic period. The positions of the continents and the distribution of reefs (*). evaporites (E), coal (C). bauxite (B). and phosphorite (P). The flow (→) of warm (*W*) and cold (*C*) ocean currents. (After Frakes 1979.)

deposits were laid down, probably lasted for rather more than 60 Ma. More abundant data are available than for earlier periods. Shallow seas were more widespread than at any time since the Palaeozoic. Their sea floors are still largely preserved and have been sampled by the Deep Sea Drilling Project (Fig. 1.24). Palaeotemperatures have been determined from the ratios of the proportion of the isotopes of oxygen in fossils (p. 29). The great equatorial oceans retained the sun's heat and gave a warm, damp climate (Fig. 1.19). Figure 1.25 shows from various determinations that there was an overall rise in temperature to levels 10–15 °C warmer than today and then a marked fall at the end of the period. The figures for Russia show quite rapid lesser fluctuations, which must have had large effects on the flora and fauna. Evaporites and corals were widespread in the Cretaceous and coals were formed at high latitudes. The sea level rose by as much as 250 m, probably due to increased floor spreading as the continents broke up.

25.2 The end of the Mesozoic

The disappearance of the dinosaurs has made the Cretaceous–Tertiary boundary the most famous of all transitions. Recent palaeoclimatic studies have indeed shown that there were rapid changes at that time and they led to many extinctions and replacements. These included the disappearance of the ammonites and belemnites and the differentiation of the birds and placental mammals and radiation of the teleostean fishes and of the cephalopods with reduced shells, the squids and octopuses. However, it is unwise to make simple statements about the relation of climate to these events. We cannot trace in detail the numerous changes of climate that must have taken place, probably in different directions in different parts of the world. Some sauropods and numerous ornithischian dinosaurs persisted to the very close of the Mesozoic. On the other hand, pterodactyls had died out earlier and so had other groups of the dinosaurs. Even the marine faunas were affected; for instance the ichthyosaurs and plesiosaurs died out before the end of the Cretaceous and were followed by the mosasaurs (p. 294), which had a sudden period of success lasting for some millions of years. Then the remaining dinosaurs died out on land, as did the mosasaurs.

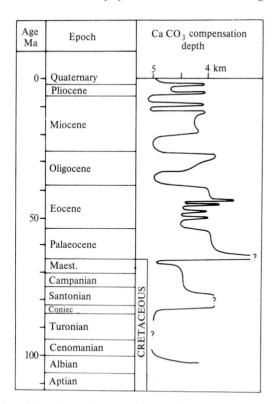

FIG. 1.24. Fluctuations in calcium carbonate compensation depth in the Atlantic and Caribbean. Maest., Maestrichtian; Coniec., Coniacian. (Modified from Frakes 1979.)

A striking feature of the climatic changes was their speed and sharp alternation between extremes. This is well shown by the data derived from the calcareous oozes, which are formed at lesser depths in the ocean when the climate is warmer. These data indicate a very sharp fall of as much as 5 °C at the end of the Cretaceous followed perhaps less than 5 Ma later by a rapid and sharp rise in the Palaeocene to temperatures higher than any since (Fig. 1.24). The fall may have been the cause of the striking changes in the flora and fauna at that time, coming after a period of 180 Ma in which organisms had become adapted to warmer conditions.

25.3 Cenozoic climates

The Tertiary or Cenozoic period is divided as shown in Table 1.4. The names were originally given by Lyell to indicate the percentages of modern species of shells; for curiosity these latter are given (approximately) in the third column.

During the Palaeocene, Eocene, and Oligocene epochs (known together as the Palaeogene) there were further rapid alternations of warm and cooler periods (Fig. 1.26). At some periods during the Eocene the

TABLE 1.4. The names of the Epochs of the Cenozoic, or Tertiary Period were given by Charles Lyell. The times given for the beginning of each epoch are approximate

Epoch	Time from beginning of epoch to present (Ma)	Per cent modern species of shells
Recent		100
Pleistocene ('most recent')	1–2	90
Pliocene ('more recent')	7	50
Miocene ('less recent')	20	20
Oligocene ('few recent')	38	10
Eocene ('dawn recent')	54	5
Palaeocene ('ancient recent')	65	0

Arctic flora showed warm, wet conditions, followed quickly by conditions like those at present. It is important not to be over impressed with any simple account of climatic changes. The periods of time involved are enormously long and it is unsafe to assume that conditions remained constant for any length of time that can be easily imagined, or even that conditions varied at a constant rate. For example, in Yellowstone Park, USA, there are exposures of the remains of Eocene tree trunks and these are arranged in layers, showing that at least twenty forests grew up and were covered by volcanic ash one after the other. Each of these eruptions presumably produced a major revolution for the animals and plants in the area concerned; we do not know how wide that area may have been. Obviously no broad generalization about the presence of 'humid conditions and forests' throughout the Eocene can give us any clear picture of the ecological conditions even in one area. The geological history shows us that conditions were continually changing, sometimes fast (p. 30), but usually only slowly in comparison with the duration of animal lives. We can well imagine that these slow changes were responsible for producing new conditions and hence new types of life, but the data of the rocks are too obscure to show us the detailed circumstances of the emergence of any particular type.

The hot, dry conditions of the Cretaceous were gradually reversed throughout the Palaeogene (Figs. 1.24 and 1.26). There were many oscillations, with

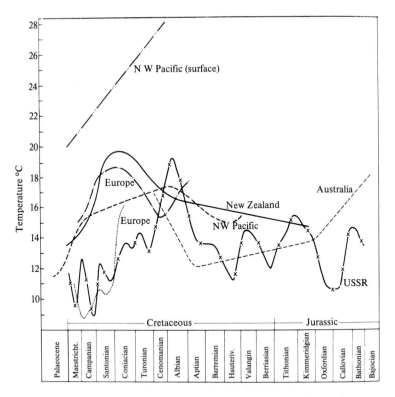

FIG. 1.25. Variations in temperature in the Mesozoic as recorded by various authors in dfferent places. Corrected oxygen isotope data. (After Frakes, 1979.)

sharper falls in the middle and late Eocene leading in the Oligocene to glacial conditions in the Antarctic for the first time since the Palaeozoic.

There was some warming at the end of the Oligocene and early Miocene, giving climates warmer and wetter than today. During the early part of the Miocene the continents lay at least 100 m higher than at present and the great mountain chains of the Western Americas and the Alps–Himalayas had been formed, conditions that probably favour differentiation of land animals. There was a marked cooling in the middle and later Miocene, leading to an Antarctic ice cap 13 Ma ago causing a fall in sea levels of up to 59 m and drier conditions generally. These arid climates were unsuitable to the growth of forests. Several types of animal suitable for life on open prairies appeared. These included the cursorial herbivores and carnivores and the bipedal anthropoids who were the ancestors of man. In the latest Miocene there was a rise of temperature lasting to the end of the Pliocene. Then began the marked cooling with formation of an Arctic ice cap at 3.2 Ma ago initiating the oscillations of the Pleistocene glaciations. The early period from 3.2 to 2.4 Ma was especially cold. One study recognizes 26 alternations of warming and cooling in the

Pacific from 3.2 Ma to the present. Since 1.6 Ma there have been four especially intense glaciations corresponding to the classical ice ages (Fig. 1.27 and Fig. 1.22). These alternations show a major periodicity of 100 000 years, with lesser cycles of 41 000 and 23 000 years probably as a result of the changes of orbital parameters (p. 25).

During this last period there have certainly been rapid and substantial changes in global climate (Fig. 1.27). These have been followed in detail by measuring the ratio of the isotopes of oxygen in cores of Greenland ice. The proportion of ^{18}O present depends upon the amount of ice, because this is relatively poor in the heavy isotope, and so gives an estimate of the climate at any time. Warm and cold conditions alternate with a periodicity that can be attributed to the changes in the orbital parameters of the earth (Fig. 1.19). The fossil Foraminifera show changes with these variations in temperature and salinity. Besides these larger rhythmic cycles there have been very rapid short-term alterations, at least locally (Fig. 1.19). The change from full glacial to the present interglacial climate took only a few thousand years, beginning about 14 000 years ago. Again, the Greenland ice-sheet record shows a change

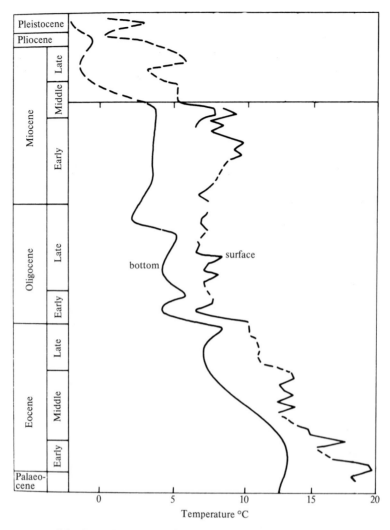

Fig. 1.26. Palaeotemperatures of the Cenozoic. Estimated from oxygen isotope studies; after the middle Miocene they are not estimates. The bottom temperatures are from N. Pacific, the surface sub-Antarctic. (After Frakes 1979.)

about 90 000 years ago from warmer than today to full glacial in 100 years. This was more than a local event for there is evidence of rapid cooling at this time also from the Foraminifera in cores taken from the Gulf of Mexico and in stalagmites in caves in France (see Lockwood 1979).

26. Correlations of climate and evolutionary change

Recent work therefore gives evidence both of long period climatic changes and of rapid fluctuations. The very fact of climatic change is probably a major factor generating evolution of animals and plants. This very brief survey of geological history can hardly do more

than remind us of the depths of our ignorance even today. We see enough to be sure that climatic conditions have varied throughout the millions of years, and also sometimes very rapidly, but we cannot yet see sufficient details to allow us to discover whether there is any rhythm of major cycles. It is easy to talk glibly of 'Carboniferous forests' or 'arid conditions of the Permian', forgetting that these periods lasted for a time that we can now roughly record in numbers but not properly imagine in terms of our experience, although we are among the longest lived of animals. The evidence suggests that conditions did not remain stable for great lengths of time but fluctuated markedly, either irregularly or with complicated rhythms of greater and

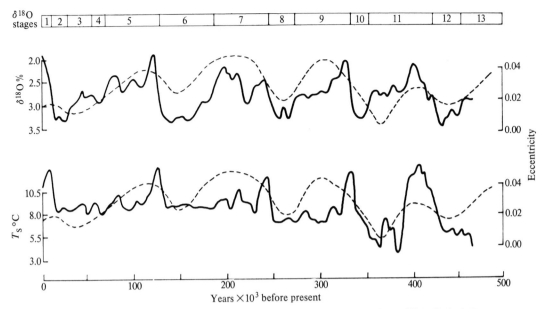

FIG. 1.27. Pleistocene Ice Ages. (Above) Variations in ratios of isotopes of oxygen (solid line) ($\delta^{18}O$). (Below) Temperature at sea surface, curve derived from faunal data. In both graphs the dotted lines show the eccentricity of the Earth's orbit. (After Frakes 1979.)

lesser magnitude. We must not forget that very profound 'climatic' changes occur every day, others every year, and some every eleven years. It is not impossible that these shorter-period changes, necessitating continual readjustment of animal and plant life, have been as important as the slower changes in producing evolution. Even the lengths of these shorter periods may have changed. In the Silurian the day lasted about 21 hours and the year 421 days.

27. Summary

To reduce to order our knowledge of vertebrate life we shall try to discover its general organization and then examine the factors that have produced all the varied types. The pattern of organization we have to study is that of the animal as an active system maintaining itself in its environment. This tendency to maintenance and growth is the central 'force' that produces the variety of life. The opportunity for change is provided by the fact that reproduction seldom produces an exact copy of the parent, and thus a range of types is provided. The tendencies to grow, to produce excess offspring, and to vary lead animals to colonize new environments and produce the variety of life. As evolution has proceeded animals have come to occupy environments differing ever more widely from the sea in which life probably arose. Life in these more difficult environments is made possible by the development of special devices, making the later animals more complex than the earlier and in this sense 'higher'. It remains uncertain what influences have been responsible for producing the changes in organic form. Geological evidence shows that there have been many changes in climate and geography, some of them quite fast but most proceeding at very slow rates in comparison with the rhythms of individual animal lives. It is uncertain whether evolutionary changes follow these slow geological changes, or are a result of competition and the instability imposed on living things by climatic rhythms with shorter periods, such as those of days, years, and the sunspot cycles.

2 The general plan of chordate organization: Amphioxus

1. The variety of chordate life

THE Chordata occupy a greater variety of habitats and show more complicated mechanisms of self-maintenance than any other group in the whole animal kingdom. They and the arthropods and the pulmonate molluscs have fully solved the problem of life on the land – which they now dominate. This domination is achieved by most delicate mechanisms for resisting desiccation, for providing support, and for conducting many operations that are harder in the air than in water. By even more wonderful devices the body temperature is raised and kept uniform and thus all reactions accelerated. Finally, use is made of this high metabolism for the development of the nervous system into a most delicate instrument, allowing the animal not only to change its response to a given stimulus from moment to moment, but also to store up and act upon the fruits of past experience.

Besides these more developed types of chordate that dominate the land and air there are also great numbers of extremely successful aquatic and amphibious types. The frog is often referred to as a somewhat lowly and unsuccessful animal, but frogs and toads are found all over the world. The sharks and bony fishes share with the squids and whales the culminating ecological position in the food chains of the sea, while the bony fishes are the only animals that have achieved considerable size and variety in fresh water. Among the still more lowly chordates the sea-squirts take a very important, though not dominant, position among the animal and plant communities that occupy the sea bottom, but they have never entered fresh water.

One could continue indefinitely with particulars of the amazing types produced by this most adaptable phylum. Yet through all their variety of structure the chordates show a considerable uniformity of general plan, and there can be no doubt that they have all evolved from a common ancestor of what might be called a 'fish-like' habit. In the very earliest stages only the larva was fish-like, and the life-history probably also included a sessile adult stage, such as the tunicates still show today (p. 62). This bottom-living phase was then eliminated by paedomorphosis, the larvae becoming the adults. Therefore the essential organization of a chordate is that of a long-bodied, free-swimming creature. All the other types can be derived from such an ancestor, though in some cases only by what is often called 'degeneration'.

2. Classification of chordates

We may conveniently divide the Phylum Chordata into four subphyla:

  Subphylum 1: Hemichordata
   Balanoglossus; Cephalodiscus; Rhabdopleura
  Subphylum 2: Cephalochordata (= Acrania)
   Branchiostoma (Amphioxus)
  Subphylum 3: Tunicata (= Urochordata)
   Ciona, sea-squirts
  Subphylum 4: Vertebrata (= Craniata)
   Superclass 1: AGNATHA
    Class 1: Cyclostomata. Lampreys and hag fishes
    Class 2: *Cephalaspidomorphi. *Cephalaspis*
    Class 3: *Pteraspidomorphi. *Pteraspis*
    Class 4: *Anaspida. *Birkenia, *Jaymoytius*
   Superclass 2: GNATHOSTOMATA
    Class 1: *Placodermi. *Acanthodes*

Class 2: Elasmobranchii. Dogfishes, skates, and rays
Class 3: Actinopterygii. Bony fishes
Class 4: Crossopterygii. Lung fishes
Class 5: Amphibia
Class 6: Reptilia
Class 7: Aves
Class 8: Mammalia

* Fossil forms.

3. Amphioxus, a generalized chordate

It has long been realized that despite their great variety all these types show certain common features, often referred to as the *typical chordate characters*. It is better to regard these not as a list of isolated 'characters' but as the signs of a certain pattern of organization that is characteristic of the group. There is much reason to suppose that this basic chordate organization was that of a free-swimming marine animal, probably feeding by the collection of minute particles. We are fortunate in having still alive a little animal, amphioxus, the lancelet, which possesses nearly all of these features. Study of amphioxus will go a long way to show the basic plan on which all later chordates are built, and, indeed, gives us a strong indication of what the early chordates may have been like.

Though it can swim freely through the water, amphioxus is essentially a burrowing animal, and many of its special features are connected with this habitat. It lives in the sand, at small depths, and has been found round the edges of all the oceans of the world. Evidently, in spite of its simplicity, it is a successful type. It is found on British coasts and, indeed, the first individual described was sent (preserved) from Cornwall to the German zoologist P. S. Pallas, who supposed it to be a slug and in 1774 called it *Limax lanceolatus*. It was first figured and given the name *Amphioxus* (two-ended) *lanceolatus* by William Yarrell in 1836. However, the name *Branchiostoma* (= gill mouth) had been given in 1834 by O. G. Costa and by the rules of priority this is the official name of the genus. We may keep amphioxus as a common name. Some twenty or more species of

Branchiostoma are recognized, and in addition there is a group of about six species referred to the genus *Asymmetron*. These resemble *Branchiostoma* in general organization, but they have gonads only on the right side.

The adult *Branchiostoma lanceolatum* is usually rather less than 5 cm long and has the typical fish-like organization, whose main external features are related to the methods of locomotion and feeding (Fig. 2.1). The body is elongated, and flattened from side to side. The skin has no pigment, and the muscles can be easily seen as a series of blocks, the myotomes, serving to bend the body into folds. As the name implies, the body is pointed at both ends; there is no recognizable head separated from the body. Indeed, there are no separate eyes, nose, or ears, and no jaws, so that the fundamental plan of chordate organization appears in almost its fullest simplicity from one end of the body to the other. The front end is, however, marked by a series of buccal cirri, which form a sieve around the opening of the oral hood and are provided with receptor cells.

Although the animal has a large number of gill slits these do not appear externally, being covered by lateral folds of the body, which enclose a ventral space, the atrium, opening posteriorly by an atriopore. These metapleural folds give the body a triangular shape in transverse section. The alimentary canal opens posteriorly by an anus, in front of the hind end of the body, thus leaving a definite tail – a region of the body not containing any part of the alimentary canal.

The general arrangement of the organs can best be understood by considering the body as consisting of two tubes, the outer skin (ectoderm) and the inner alimen-

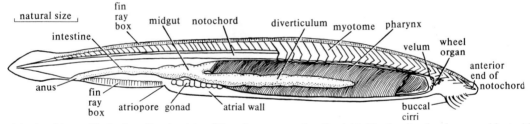

FIG. 2.1. Amphioxus. The body wall and atrial wall have been removed on the right side, showing the pharynx, midgut with its diverticulum, and intestine. The oral hood has been cut away on the right, leaving the buccal cirri, wheel organ, and velum. Scale is one third natural size.

tary canal (endoderm), with a space between (the coelom) lined by a third layer (the mesoderm). This arrangement is actually found during the course of development (Fig. 2.24). The mesoderm at first forms thin layers, the somatopleure applied to the outer body wall and the splanchnopleure to the gut. Very soon the inner layer becomes much thickened where it is applied to the nerve cord and notochord, and here it forms the myotomes, or muscle blocks. In this dorsal part of the mesoderm the coelom, known here as the myocoele, soon becomes obliterated, leaving the ventral splanchnocoele around the gut. Besides the muscle that forms in the myotomes, non-myotomal muscles develop in the somatopleure and splanchnopleure. These are not divided into segments and are innervated by the dorsal nerve roots.

4. Muscle fibres and movement

The adult myotomes are blocks of striated muscle fibres, running along the body, separated by sheets of connective tissue, the myosepta (also known as myocommas). This repetition or segmentation is characteristic of the organization of all chordates. The myosepta do not run straight down the body from dorsal to ventral side but are ≫-shaped (Fig. 2.1). However, each muscle fibre runs straight from before backwards, and the contraction of the whole myotome therefore bends the body. A full discussion of the means by which forward motion is achieved will be given on p. 115.

The muscle fibres are in the form of flat plates or lamellae (Fig. 2.2) (Flood 1968). Each plate stretches between the myosepta and contains longitudinally orientated cross-striated myofibrils. The muscles are connected with the nerve cord in an unusual way, not by motor nerve fibres but by thin processes of the muscle

cells themselves, which were thought by earlier workers to be ventral roots. Each passes medially to join the spinal cord, and there makes a cholinergic synapse with axons of central neurons. There are two types of muscle lamella. Those near the lateral surface are narrow, contain irregular myofibrils, many mitochondria, and glycogen. Their 'ventral root' fibres are thin. The other lamellae are wide, composed mainly of myofilaments and have thick ventral root fibres. The two kinds probably correspond to the slow (sarcoplasm-rich) and fast muscle fibres of higher chordates (Fig. 2.3).

Amphioxus can swim equally well forwards or backwards. Movement is produced by serial contraction of the myotomes resulting in transverse motion of the body inclined at varying angles in such a way as to result in forward propagation. In forward swimming each myotome contracts after that in front of it – the effect being to produce an ∽-bend that moves backwards through the water as the animal moves forward (Fig. 2.4) (Webb 1973, 1976). In backward swimming the wave begins at the tail and moves forward.

For our present purpose the point is that the contraction is serial, that is to say, it depends on the breaking up of the longitudinal muscle into blocks. It was probably the need for division of the musculature that led to the development of the segmentation, and this, affecting primarily the muscles, has come to influence a great part of chordate organization.

Contraction of the longitudinally arranged muscle fibres will only produce a sharp bending of the body if there is no possibility of shortening of the whole. To prevent telescoping, an incompressible and elastic rod, the notochord, runs down the centre of the body. It is often stated that this is a 'supporting structure', but, of course, an animal such as a fish in water needs no

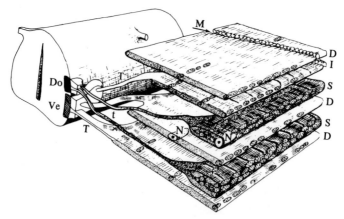

FIG. 2.2. Diagram of the muscle types in amphioxus. The superficial, slow, sarcoplasm-rich (S); the deep, fast, sarcoplasm-poor (D); and intermediate (I) types of muscle. Each sends sarcoplasmic extensions (or 'ventral root fibre' t, T) towards the dorsal (Do) or ventral (Ve) central motor endplate of the spinal cord. (M) myoseptum; (N) cell nucleus. (Flood 1968.)

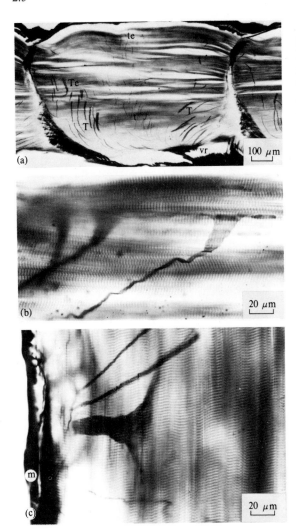

(a)

te

Te

T

T

vr   100 μm

(b)   20 μm

(c)

m

20 μm

FIG. 2.3. Longitudinal section of amphioxus cut at the level of the notochord. (a) There are several thick (T) and thin (t) 'ventral root' fibres inside the muscle. Note intensely stained ventral root fibre bundle (vr) and myosepta. (Te) and (te) 'terminal' expansions of the thick and thin fibres. (b) Each thin ventral root fibre shows a small expansion near the lateral surface of the myomere. (c) Coarse ventral root fibres 'end' in large expansions far from the lateral surface of the muscle. (m) myomere; (arrow) ventral root fibre. (Flood 1968.)

'support'. Nor is the notochord a lever to which muscles are attached, as they are to the bones of many higher forms. No muscles pull on it directly, though the myoseptas are attached to its sheath. Its function is to prevent the shortening of the body that would otherwise be the result of contraction of longitudinal muscles. For this purpose it has developed into a unique 'hydroskeleton' whose stiffness is variable under the control of the nervous system (Fig. 2.5).

The notochord is composed mainly of a series of flattened plates surrounded by a fibrous sheath of Müller's cells. The plates are arranged in a regular manner with their flat surfaces in the transverse plane of the body. Each plate develops as a highly vacuolated cell, the nuclei being later pushed aside alternately to the dorsal or ventral edge (Fig. 2.5). This structure is well suited by the turgidity of its cells, enclosed in the sheath, to resist forces tending to shorten the body. The plates are in fact muscle cells, each crossed by striations showing isotropic and anisotropic bands in polarized light. Electron micrographs show thick and thin filaments, the former being striped with a period of 14.5 nm, similar to that of the paramyosin of the catch muscle of some molluscs such as clams (*Pecten*). The plates contract under electrical stimulation, increasing the internal pressure and stiffness of the notochord. Tails from these notochordal muscles pass to the nerve cord and make cholinergic synapses with the axons of notochordal motoneurons. Probably adjustments of stiffness are made to allow swimming in either direction, the trailing end always shows the greater oscillations (Fig. 2.4). The capacity to move in either direction may be important for burrowing in sand as is the fact that the cord extends from the very tip of the head to the end of the tail, projecting, that is to say, beyond the level of the myotomes.

Amphioxus lives and feeds in the sand and probably does not often swim free in the water since the body is not adapted for fast directional movements though it can maintain 60 cm/s for 50 s but then stops suddenly. This is 13 body lengths (L) per second compared with 10 L/s in the trout (p. 116). There is also a slow mode of swimming (1–3 L/s), probably using the slow muscle. Lancelets may move for distances of 1 km or more, following changes in the sandy bottom, but movements are not well orientated. There are no elaborate fins such as those of later fishes, which ensure static stability like the feathers on an arrow, or are movable, to allow active control of the direction of swimming (p. 120). Indeed fins would be an obstacle to burrowing. There is a low dorsal ridge, which continues behind as a small caudal fin (Fig. 2.6). There are no definite paired fins, but the metapleural folds might perhaps be considered comparable to the lateral fin folds from which all vertebrate limbs are probably derived (p. 153). They are distended with coelomic fluid, and, together with the dorsal ridge, probably serve to protect the body during the rapid dives by means of which the creature enters the sand. The larvae of lampreys swim in a similar way (p. 78).

5. Skeletal structures

Around the notochordal sheath is a further layer of

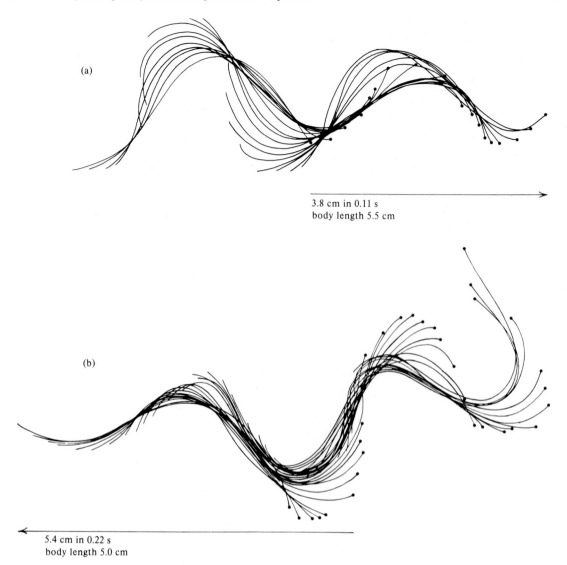

(a)

3.8 cm in 0.11 s
body length 5.5 cm

(b)

5.4 cm in 0.22 s
body length 5.0 cm

FIG. 2.4. Successive positions of amphioxus when swimming. (a) Head first (●) and (b) tail first, seen from above and traced from cinefilm shot at 45 m/s. (Webb 1973.)

gelatinous material containing fibres. There are no cells within this material but it is secreted by cells around the outside, which retain the epithelial arrangement of the mesoderm from which they were derived. This connective tissue continues as a sheath around the nerve cord and above this into a series of structures known as fin-ray boxes, which support the dorsal ridge. These boxes are more numerous than the segments and each contains some 'cartilage'. The relationship of these structures to the fin supports of vertebrates is obscure. Other skeletal rods occur in the cirri around the mouth and in the gill bars.

6. Skin

The epidermis differs from that of vertebrates in being very thin, composed of a single layer of cells, ciliated in the very young larva and with the outer border highly cuticularized in the adult (Fig. 2.7). It is not known whether this cuticle contains a substance similar to the keratin produced by the many-layered skin of later forms. There are receptor cells but no glands or chromatophores in the skin.

Below the epidermis is a fibrous cutis, and below this again a gelatinous material containing fibres, the sub-

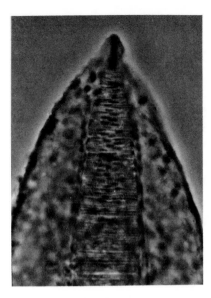

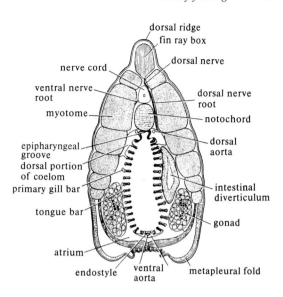

FIG. 2.5. Notochord of amphioxus to show the flattened cells with their nuclei pushed laterally, alternately going to the dorsal or ventral edge. (With permission of J. E. Webb.)

FIG. 2.6. Transverse section through amphioxus in the region of the pharynx.

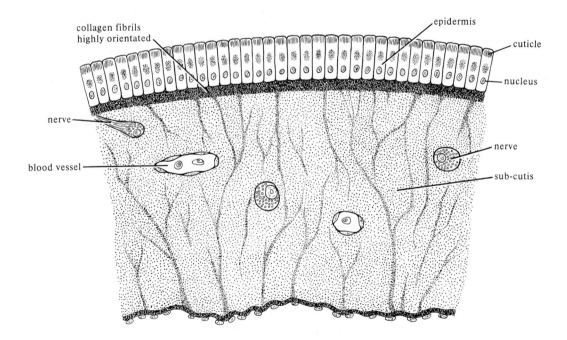

FIG. 2.7. Section through the skin of amphioxus. (After Krause.)

cutis. Both these layers are secreted by scattered cells having some similarity to the fibroblasts of higher forms. They contain a system of cutaneous canals, with endothelial lining.

7. Mouth and pharynx and the control of feeding

Amphioxus obtains its food by extracting small particles

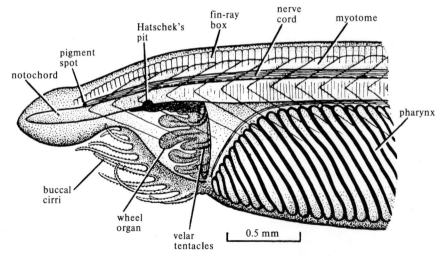

FIG. 2.8. Anterior end of amphioxus, from a stained and cleared whole mount of a young animal.

from a stream of water, which it draws in by means of cilia. In all animals that use cilia for this purpose a very large surface is provided (e.g. lamellibranchs, ascidians), and the pharynx and gill bars of amphioxus occupy more than one-half of the whole surface area of the body. Special arrangements are made for the support and protection of this ciliated surface, the wall of the

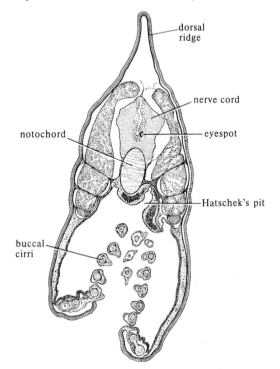

FIG. 2.9. Transverse section through the front end of amphioxus.

pharynx being so greatly subdivided that it needs the protection of an outer layer, the atrium.

The mouth lies covered by an oral hood whose edges are drawn out into buccal cirri, provided with sense cells some of which are mechanoreceptors. When stimulated by touching they elicit a velar reflex by which the tentacles come together and close the inhalent aperture and the velar ring sphincter closes (Fig. 2.8). When feeding the cirri are curved to form a funnel-like sieve preventing the entry of large particles. Around the mouth itself there is a further ring of sensory tentacles, the velum. The oral hood contains a complex set of ciliated tracts, the 'wheel organ' of Müller, and this plays a part in sweeping the food particles into the mouth (Figs. 2.8 and 2.9). Near its centre is a groove, Hatschek's pit, formed as an opening of the left first coelomic sac to the exterior (p. 48).

The main operation of food collection is performed by the pharynx, a large tube, flattened from side to side, whose walls are perforated by nearly 200 oblique vertical slits, the number increasing as the animal gets older. The slits are separated by bars containing skeletal rods and further subdivision is provided by cross-bars (synapticulae). Since the bars slope diagonally many of them are cut in a single transverse section, but it must be remembered that they are essentially the vertical portions of the main walls of body and pharynx, where these have not been perforated by a gill slit. Such a portion of the body wall must contain a coelomic space and this can in fact be seen in the original or primary gill bars. However, an increase of the ciliary surface is produced by downgrowth of secondary or tongue bars from the upper margin, dividing each primary slit; these secondary bars contain no coelom. The coelomic spaces in the primary bars communicate above and below with continuous longitudinal coelomic cavities (Fig. 2.6).

There are cilia on the sides and inner surfaces of the gill bars, the lateral ones being mainly responsible for driving the water outwards through the atrium and thereby drawing the feeding current of water in at the mouth. There is a complicated plexus of nerve cells and fibres in the walls of the pharynx and changes in the rhythm of the cilia occur, but details of the nervous connections are not known. In the floor of the pharynx lies the endostyle, containing columns of ciliated cells, alternating with mucus-secreting cells, which produce sticky threads in which food particles become entangled. Various currents then draw the sticky material along until it reaches the midgut. The frontal cilia of the gill bars produce an upward current, driving the mucus from the endostyle into a median dorsal epipharyngeal groove, in which it is conducted backwards. The cilia of the endostyle also move mucus along the peripharyngeal ciliated tracts, behind the velum, to join the epipharyngeal groove. Radioactive iodine is concentrated by the cells of one of the columns of the endostyle and secreted with the mucus. These may be regarded as the precursors of thyroid cells. They serve to produce iodinated mucoproteins, which are then absorbed farther down the gut (p. 103). Homogenates of amphioxus show that mono- and di-iodotyrosine are present, as well as tri-iodothyronine (T_3) and thyroxine (T_4). Unlike higher vertebrates T_3 is more abundant than T_4. Metamorphosis of axolotls was produced by implants of the pharynx of about 50 amphioxus, but not by the tails. There is no evidence that these iodine compounds have an endocrine activity in the animal itself.

The pharynx narrows at its hind end to open dorsally into a region best known as the midgut, the name stomach being inappropriate. A large midgut diverticulum reaches forward from this region on the right-hand side of the pharynx. From its position this organ is often called the liver, but it is the seat of the production of digestive enzymes. Zymogen cells, similar to those of the midgut, are found in its walls. Some of these appear to be protein secretors, with rough endoplasmic re-

ticulum and secretory granules. Others contain much glycogen and lipid and may be compared to liver cells. Concentration of mitochondria at both apical and basal locations suggest transport through the cells of the caecum. Its strong dorsal and ventral ciliation maintains in it a circulation of food materials and secretion, and its cells are capable of phagocytosis as well as secretory activity. Amphioxus thus combines intracellular with extracellular digestion, doubtless in connection with its microphagous habit. Particles placed in the diverticulum are swept backwards and join the main food cord that passes through the midgut (Fig. 2.10).

The hind end of the midgut is marked by a specially ciliated region, the ileo-colon ring, whose cilia rotate the cord of mucus and food. The movement is transmitted to the portion of the food cord in the midgut and presumably assists in the taking up of the enzymes that emerge from the diverticulum. Extracellular digestion takes place in the midgut and the enzymes responsible have been studied by Barrington (1938). The pH of the contents varies from 6.7 to 7.1. An amylase is present in extracts of the diverticulum, midgut, and hindgut, but not in those of the pharynx. Lipase and protease are present in the same regions, the latter having an optimum action at about pH 8.0, being, that is to say, a tryptic type of enzyme. There is no sign of any protease with an acid optimum, similar to the pepsin of higher forms.

Behind the ileo-colon ring the intestine runs as a straight hindgut to the anus. Absorption of food takes place here, and perhaps also in the midgut, apparently partly by intracellular digestion, since ingested carmine particles are taken into the cells.

The feeding current is regulated by the rate of beat of the cilia and the degree of contraction of the inhalent and exhalent apertures. The walls of the atrium contain an elaborate system of afferent and efferent nerve fibres. The receptor cells are ciliated and have no axon but are in contact with processes of large nerve cell bodies, lying beneath the atrial epithelium and sending axons in by way of the dorsal roots (Fig. 2.11). The motor fibres also pass through the dorsal roots and run without synapse

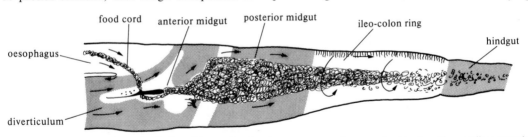

food cord   anterior midgut   posterior midgut   ileo-colon ring

hindgut

oesophagus

diverticulum

FIG. 2.10. Currents in the midgut gland of amphioxus, showing the appearance when an animal is placed in a medium containing carmine particles. Arrows show the chief ciliary currents. (After Barrington 1938.)

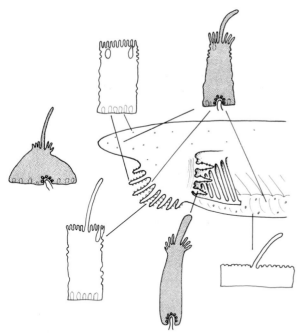

FIG. 2.11. The distribution of sensory cells (shaded) of the anterior region of adult amphioxus. Non-sensory cells unshaded. The sensory cell on the right is from the outer surface of a larva. (Bone, Q. and Best, A. C. G. (1978). *Journal of the marine biological Association of the United Kingdom* **58**, 479–86.

to the cross-striated fibres of the pterygial muscle, which forms the floor of the atrium. The stream flowing into the pharynx is tested by the receptors of the velum and atrium, and if noxious material is present, the water is expelled by closing the atriopore and contracting the pterygial muscle, producing a 'cough'. The system can distinguish between suspensions of food material and inorganic particles. When sufficient food has been taken, collection is suspended until it has been digested (Bone 1960).

The atrial nervous system probably regulates spawning as well as feeding. It has been compared with the sympathetic system of craniates but there are almost no close similarities. The nerve cells in it are receptors and there is no sign of the peripheral synapse on the efferent pathway that is so characteristic of the true autonomic system. The plexus in the wall of the gut also consists of sensory cells (p. 101). The atrial system is developed in relation to filter feeding and has perhaps been completely lost in higher forms that propel the food along the gut by the action of muscles and have developed the autonomic nervous system to control them.

8. Circulation

The blood vessels of amphioxus show the fundamental plan on which the circulation of all chordates is based (Rähr 1979) (Fig. 2.12). Slow waves of contraction occur in various separate parts in such a way as to drive the blood forwards in the ventral vessels, backwards in the dorsal ones. Below the hind end of the pharynx there is a large sac, the sinus venosus, into which blood from all parts of the body is collected. From this there proceeds forwards a large endostylar artery (truncus arteriosus or ventral aorta) from which spring vessels carrying blood up the branchial arches. At the base of each primary bar there is a little bulb, functioning as a branchial heart. From the gill bars blood is collected into paired dorsal aortae, which join behind the pharynx. From the paired and median aortae blood is carried to the system of lacunae that supplies the tissues.

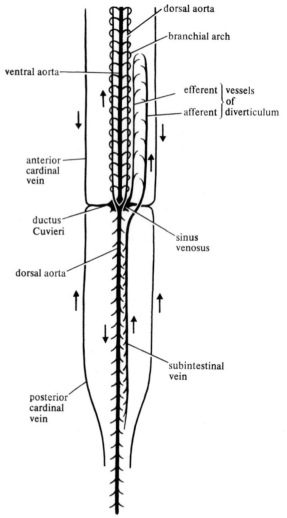

FIG. 2.12. Diagram showing the circulation of amphioxus. (After Grobben and Zarnik.)

There are no true capillaries. From the lacunae blood is collected into veins, the most important of which are the caudals, cardinals, and a plexus on the gut. The cardinals are a pair of vessels in the dorsal wall of the coelom, and they collect blood from the muscles and body wall. They lead to the sinus venosus by a pair of vessels, ductus Cuvieri, which pass ventrally and across the coelom to join the sinus venosus on the floor of the gut. The caudal veins join the plexus on the gut, from which blood is collected by a large subintestinal vein running on to the liver; from here another plexus leads to the sinus venosus.

Contractions arise independently in the sinus venosus, branchial bulbs, subintestinal vein, and elsewhere. The rhythms are very slow (once in two minutes), irregular, and apparently not co-ordinated by any control system.

The blood is colourless and is not known to contain any respiratory pigment. It contains no cells. Presumably the tension of dissolved oxygen acquired by simple solution is sufficient for the small energy needs of the animal, which spends most of its life at rest. It is by no means certain that any oxygenation of the blood takes place in the gills. Orton (1913) has suggested that since these, through their cilia, do much of the work of the body, the blood actually leaves the gills less rich in oxygen that when it enters them. Oxygenation probably takes place chiefly in the lacunae close to the skin, perhaps especially those of the metapleural folds.

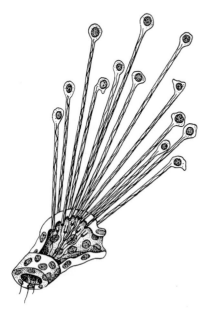

FIG. 2.13. Solenocytes of amphioxus showing the nuclei, long flagella, and the openings into the main excretory canal which leads to the atrium. (After Goodrich 1902.)

9. Excretory system

It used to be thought that one of the most mysterious features about the organization of amphioxus was the presence of flame cells, comparable with those found in platyhelminthes, molluscs, and annelids. These so-called nephridia lie above the pharynx. To each primary gill bar there corresponds a sac, opening by a pore to the atrium and studded with numerous elongated flame cells, which were formerly supposed to be solenocytes (Fig. 2.13). These 'flame cells' do not open internally, but are in close contact with special blood vessels (glomeruli) whose walls separate the 'flame cells' from the coelomic epithelium. Assuming that there are 200 of these nephridia, each with 500 'solenocytes' 50 μm long Goodrich (1902), who provided the most accurate classical information about these organs, showed that the total length available for excretion is no less than 5 m. It is assumed that excretion takes place by diffusion through the flame cell wall, the liquid being driven down the tube by cilia. Coloured particles injected into the bloodstream are not excreted by the nephridia.

Electron microscopy has shown that these 'flame cells' are in fact modified coelomic epithelial cells. Indeed they are quite similar to the podocytes that line the renal capsule of vertebrates. They have therefore been given the cumbrous name of cyrtopodocytes. The feet of these cells covering the glomerular blood vessels are joined by a slit-membrane. There is no basement membrane between blood and coelomic spaces as there is in the solenocytes of polychaetes. So electron microscopy has solved one of the major problems that worried such excellent comparative anatomists as Goodrich (see Welsch 1975).

The brown funnels are blind sacs at the front of the atrium, invaginating into the epibranchial coelom. They are probably receptor organs. Some parts of the atrial wall may perform excretory functions. Masses of cells in the atrial floor, the atrial glands, contain granules that may be excretory but perhaps have been taken up from the food current. In the gonads, especially the testes, there are large yellow masses, containing uric acid, which are extruded with the gametes.

10. Nervous system

Amphioxus possesses a hollow dorsal nerve cord similar to that of vertebrates. This is somewhat modified at the front end, but it is not enlarged into an elaborate brain (Fig. 2.14). The nervous system is connected with the periphery in a unique way. The muscle fibres send their own processes to the nerve cord, forming the structures that all earlier workers called ventral roots (p. 34). The fibres were considered to be nerve fibres and the fact that

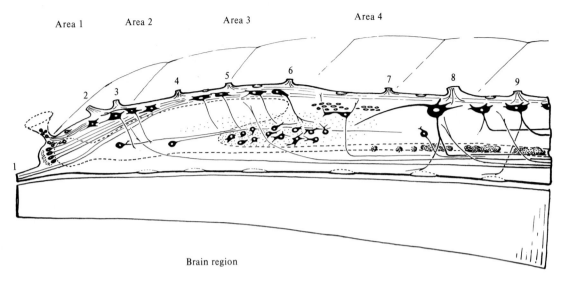

Area 1   Area 2   Area 3   Area 4

Brain region

FIG. 2.14. Diagram of the brain and anterior spinal cord of amphioxus. (Guthrie, D.M. (1975). *Symposium of the zoological Society of London* No. 36, 43–80.)

they are continuous with the muscle fibres led Boeke (1935) and others to the view that all nerves are 'continuous' with their muscles. This supported the view of these 'reticularists' that there are continuous fibrils running throughout the nervous system, contrary to the concept of synapses that was so essential for Ramón y Cajal (1906) and Sherrington (1906). Boeke was an accomplished microscopist and the evidence of continuity that he observed in various tissues was often well founded, as in this case. Only with the coming of the electron microscope was it finally proved that most nerve fibres are separated by a membrane from the sarcoplasm (see *Life of Mammals*, p. 62). The dorsal roots run out between the myotomes and carry all the afferent fibres of the segment and motor fibres for the non-myotomal muscles of the ventral part of the body including the 'trapezius' muscle of the floor of the atrium, also the muscles around the anus and the genital sacs (Bone 1961) (Fig. 2.15).

The fibres of the peripheral nerves differ from those of vertebrates in that they have no myelin sheath. The nerve trunks are surrounded by an epineurium with connective tissue cells but there seem to be no Schwann cells accompanying the nerve fibres (Bone 1958).

The afferent fibres of the dorsal roots are unique among chordates in that the cell bodies (Retzius cells) are not collected into spinal ganglia but mostly lie within the central nervous system. They form two continuous columns along the cord (Figs. 2.16 and 2.17). Their peripheral processes make the subepithelial plexuses of the skin and centrally they end in connection with the motoneurons or with various interneurons. In addition,

on the head and tail there are peripheral receptor cells, sending fibres centrally, also complicated encapsulated organs in the metapleural folds (Bone 1960). There are some large multipolar sensory nerve cells just beneath the atrial epithelium and very many sensory cells in the walls of the diverticulum and midgut (Fig. 2.18). These cells have many branched dendrites and an axon that runs through a dorsal root to the spinal cord.

The spinal cord has only a narrow lumen and its elements are arranged as in vertebrates, namely, ependyma close to the canal, cell layer ('grey matter'), and outer fibrous layer ('white matter'). The glia is mostly in the form of ependyma, with cell bodies near the central canal and end-feet on the outer membrane. There are no blood vessels in the cord. The neurons are not arranged in horns as they are in vertebrates. The somatic motor neurons have a curious form, with a broad process attached to the central canal and a terminal bush in the region of ending of the muscle tails (Fig. 2.19). The 'dendrites' are apparently collaterals of the main trunk. The visceral motor neurons are of more conventional form with dendrites and an axon in the dorsal root (Fig. 2.20).

The most conspicuous cells are the giant Rohde cells, which lie dorsally in the anterior and posterior parts but are absent from about the thirteenth to thirty-ninth segments (Fig. 2.16). Each of these cells has many dendrites, branching in the region of entry of the dorsal root fibres, and a single axon, which runs backwards in the front part of the body, forwards in the hind, passing in each case for the whole length of the cord. These axons probably make connection with the somatic

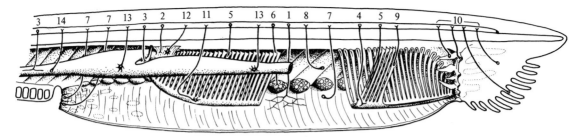

FIG. 2.15. The atrial nervous system of amphioxus. The numbered components are listed below. (Bone 1961.)

	Peripheral cell bodies	Location	Assumed function
Motor systems			
1. pterygial plexus	—	pterygial muscle	motor to cross-striated muscle fibres
2. atriocoelomic plexus	—?	trapezius muscle	motor to cross-striated muscle fibres
3. vasomotor system?	—?	blood vessels?	vasomotor
4. subendostylar plexus	—?	endostyle	secretomotor
5. branchial plexus	—?	gill bars	ciliary motor and motor to smooth muscle fibres
6. gonad plexus	—	gonad sacs	motor to smooth muscle fibres
Sensory systems			
7. Pterygial plexus	unipolar	pterygial muscle	sampling waterflow, or particle concentration of water stream
8. Parietal plexus	unipolar	atrial walls	as pterygial
9. branchial plexus	unipolar and multipolar	gill bars	?
10. buccal and velar system	unipolar	buccal cirrhi and velum	tactile and chemo receptive
11. subendostylar plexus	multipolar	endostyle	?
12. atriocoelomic plexus	multipolar and unipolar	atriocoelomic funnels	water sampling
13. foregut system	multipolar	foregut	?
14. hindgut system	unipolar and multipolar	hindgut	?

motor cells. The most anterior Rohde cell is the largest and sends a median giant fibre ventrally for the length of the cord close to the visceromotor cells, which probably produce the 'coughing' movements of the atrium (p. 39). This axon is the fastest in the animal, conducting at 5 m/s.

Amphioxus responds to all stimuli by movements of 'flight'. There are no isolated or local movements; the effect of any stimulus such as touch on the side of the body is to produce waves of myotomal contraction. These may, however, vary from strong waves going the whole length of the body to single rapid twitches. The Rohde cells participate in the spread of these waves. It seems likely that the arrangement ensures that touch on the anterior part of the body, normally exposed when feeding, produces backward movement (i.e. withdrawal into the sand) but touch on the hind part the reverse movement of emergence and escape.

At the front end the central canal is enlarged to form a cerebral vesicle (Fig. 2.14). The whole neural tube is hardly wider here than in the region of the spinal cord and there is no thickening of the walls, which are indeed mostly formed of a single layer of ciliated epithelial cells (Fig. 2.21). This is a striking indication of the lack of cephalization of the animal. The brain may be divided into four regions (Fig. 2.14). The most anterior receives the first dorsal roots, carrying fibres from the receptors of the oral hood and tentacles, from a depression known as Kölliker's pit and from a pigmented eye spot ('macula'). The second region, between nerves 2 and 4, contains large dorsal cells and many small ones. The third region, between roots 4 and 6, also contains many large dorsal cells with descending axons. The fourth part of the brain, between nerves 6 and 7, is mainly composed of longitudinal tracts. In the ventral part of the third region is the infundibular organ (Fig. 2.21) composed of

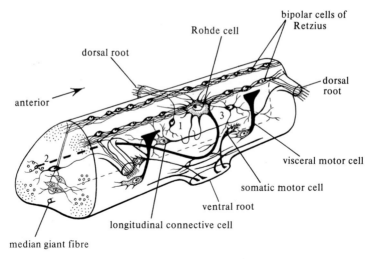

FIG. 2.16. Stereogram illustrating the structure of the spinal cord in an adult amphioxus. The receptor system is made up of a more or less continuous column of bipolar cells of Retzius, together with smaller cells of various types. These receptor cells (1, 2, and 3) can be regarded as equivalent to the dorsal root ganglion cells of vertebrates. The other type of receptor cell is the giant Rohde cell which has a large axon and elaborate dendritic system. Some of these cells probably possess a peripheral axon running in the dorsal root.

The visceral motor cells are arranged segmentally, one per segment. The somatic motor cells lie at a different level in the cord from the ventral roots. Other cells in the cord are internuncials of various types. (After Bone 1958.)

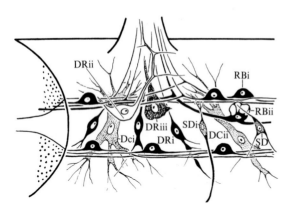

FIG. 2.17. The organization of some of the different types of neutron at a dorsal root of amphioxus. The Rohde cell has been omitted. Dci, DCii, dorsal commissural cells (Dci is multipolar); DRi, dorsal root cells; DRii, cells which occur segmentally, singly in each segment; DRiii, dorsal root 3; RBi, Retzius bipolar cells and RBii a second type; SD; small dorsal cell and SDi is type 1. (Bone 1960.)

tall cells with long cilia, which beat in the opposite direction to those of the rest of the vesicle. From them fibres run backwards down the cord. The organ is also the site of origin of Reissner's fibre (Fig. 2.21). This is a thread of non-cellular material, present in all vertebrates at the centre of the neural canal. It is secreted at the front end and then passed backwards and is often collected and absorbed in a sac at the hind end of the spinal cord.

In vertebrates it arises from secretory ependymal cells of the subcommissural organ, lying dorsally in the diencephalon (Fig. 2.22). The infundibular organ of amphioxus is clearly not exactly similar, yet the Reissner's fibres are clearly comparable; an interesting problem in homology.

A further complication is that the cells of the infundibular organ contain material that stains with the Gomori method, and is similar to the neurosecretory material found in the fibres of the hypophysial tract (Fig. 2.22). The organ thus seems to occupy a central position in the control system as a receptor, originator of nerve fibres, and of two sorts of secretion. There may be much to be learned from this about the origin and significance of the control systems of the diencephalon.

The brain has largely inhibitory influences on the cord. A suspended lancelet shows periods of spontaneous swimming and these last longer after removal of the brain. A decerebrate animal responds more readily than a normal one to mechanical stimulation or to light, and remains active for longer. Stimulation of the brain can suppress the activation produced in Rohde cells by ascending volleys. On the other hand at the beginning of swimming, electrical activity starts in the brain 200 ms before the middle region of the cord.

Adjacent to the central canal there are rows of photoreceptor cells. Each has a partial covering of pigment and an irregular array of photoreceptive membranes. Their connections and functioning are not

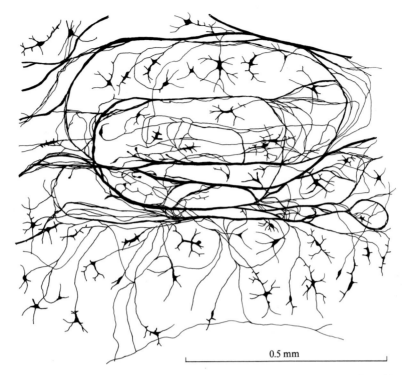

0.5 mm

FIG. 2.18. The course of axon bundles and position of cell bodies on the roof of the midgut of amphioxus. (Bone 1961.)

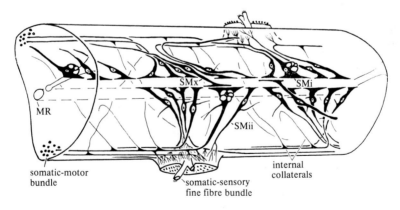

FIG. 2.19. The organization of the somatic motor system of amphioxus. Viewed from an oblique dorsal aspect. MR, median giant fibre; SMi, largest of the somatic motor cells; SMii, smaller somatic motor cell; SMx, 'crossed' somatic motor cell. (Bone 1960.)

known in detail. The 'macula' in the brain consists of a collection of these cells but the lowest threshold for movement responses is to illumination of the cord, not the head.

Amphioxus is therefore provided with receptor and motor systems that serve to keep it in its sedentary

position, able to collect food from the current that it makes by the cilia (p. 39). There are mechanisms that help it to make appropriate movements of escape when it is touched or when the body (but not head) is illuminated. The touch receptors of the buccal cirri produce rejection of large particles and those of the

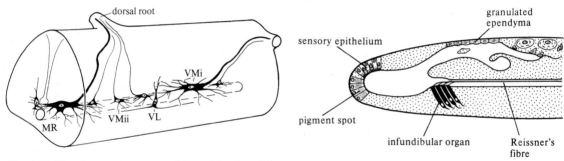

FIG. 2.20. The visceromotor system of the atrial region of the cord of amphioxus. MR, median giant fibre; VL; ventral longitudinal cells; VMi, large visceromotor cell; VMii, smaller visceromotor cells. (Bone 1960.)

FIG. 2.21. Diagram of the anterior end of the nervous system of amphioxus.

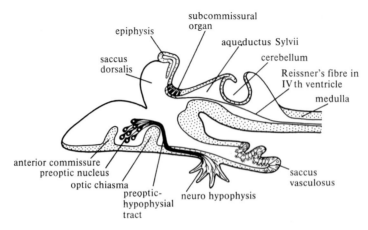

FIG. 2.22. Diagram of the anterior end of the nervous system of a fish, *Polypterus*. (Olson and Wingstrand.)

velum are chemoreceptors. The infundibular organ may be some form of gravity or pressure receptor. By means of these receptor organs and its simple movements of swimming, burrowing, and closing the oral hood, the animal is maintained, probably chiefly by trial and error (phobotactic) behaviour, in an environment suitable for its life. There are none of those elaborate mechanisms that we find in higher chordates for 'seeking' special environments or for so 'handling' or managing them that they may prove habitable by the animal. Amphioxus must take and leave the world very much as it finds it.

11. Gonads and development

The gonads of amphioxus are hollow segmental sacs with no common duct. Each sac develops from mesoderm cells, perhaps originally from a single cell, at the base of the myotomes in the branchial region, the genital cells themselves developing on the walls (Fig. 2.6). The sexes are separate and the genital products are shed by

dehiscence into the atrium, the aperture by which they escape closing and the gonad developing afresh.

Extrusion of the gametes in amphioxus occurs at Naples in spring, on warm evenings following stormy weather. Fertilization is external and development then occurs free in the water. Numerous eggs are produced and they are small but yolky. Complex flowing movements take place in them after fertilization, and cleavage is then rapid and complete, producing a blastula composed of a dome of somewhat smaller and a floor of rather larger cells (Fig. 2.23). These latter then invaginate to make the archenteron, opening by a wide blastopore, which later becomes the anus. At about this stage the gastrula becomes covered with flagella, by which it rotates within the egg case.

The creature now elongates and its dorsal side flattens and eventually sinks in to form the neural tube (Fig. 2.24). At about this time the dorsal side of the inner layer begins to fold near to the front end, in such a way as to make a pair of lateral pouches. The walls of these

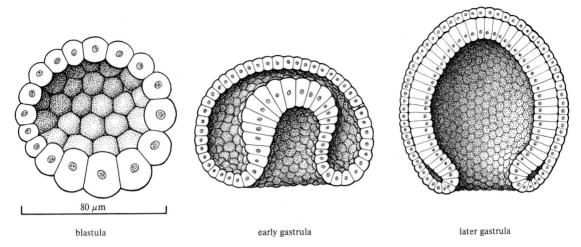

FIG. 2.23. Three stages in the development of amphioxus as seen in stained preparations.

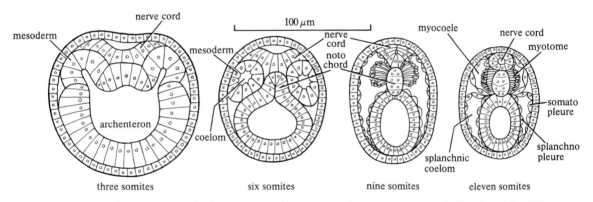

FIG. 2.24. Further stages in the development of amphioxus as seen in transverse sections. (After Hatschek 1893.)

pouches are the future mesoderm and the cavity is the coelom. As in other early chordates, therefore, the coelom is continuous at first with the archenteron. The roof of the archenteron also arches up dorsally and forms the notochord, the gut wall being completed by the approximation of the edges of the remaining portion of the inner layer, which is now the definitive gut wall or endoderm.

The analysis of the processes of development now enables us to say something of the forces by which these formative foldings and cell movements are produced. The formation of the neural tube, mesoderm and notochord and the completion of the gut roof all involve an upward movement of cells towards the mid-dorsal line. This process of 'convergence' is a very marked feature of the development of all chordates.

As the animal elongates, further mesodermal pouches are produced, each separating completely from the endoderm and from its neighbours. The cells of each pouch push down ventrally on either side of the gut, the outer ones applying themselves to the body wall to form the somatopleure, the inner to the gut wall as splanchnopleure (Fig. 2.24). The inner wall of the mesoderm on either side of the nerve cord thickens to form the myotome, and a tongue of cells growing up between this and the nerve cord forms the sheaths of the latter and probably also the fin-ray boxes and other 'mesenchymal' tissues. The upper part of the coelomic cavity, the myocoele, becomes separated from the ventral splanchnocoele. Whereas the former becomes almost completely obliterated, the latter expands to form the adult coelom, the cavities between the adjacent sacs breaking down.

While this differentiation of the mesoderm has been proceeding the animal has elongated into a definitely fish-like form. The neural tube is a small dorsal canal,

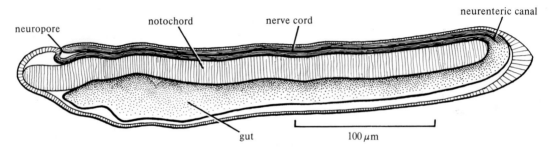

FIG. 2.25. Young amphioxus shortly after hatching.

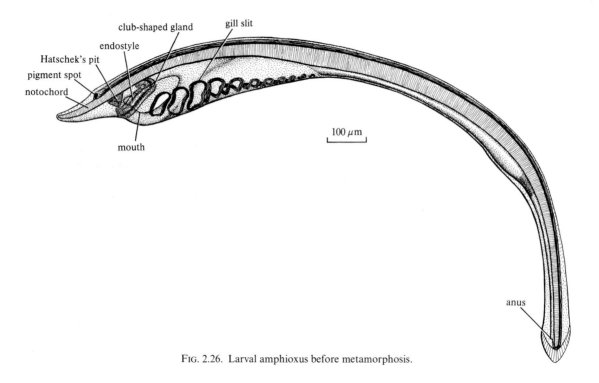

FIG. 2.26. Larval amphioxus before metamorphosis.

opening by an anterior neuropore and continuous behind through a neurenteric canal with the gut (Fig. 2.25). The larva hatches when only two gill slits have been formed and swims at the sea surface by means of its ciliated epidermis, turning on its axis from right to left as it proceeds with the front end forwards.

The mouth now appears as a circular opening and then moves over to the left side and becomes very large. From this time onward the whole development is markedly asymmetrical, presumably in connection with the spiral movement and method of feeding. The first gill slit also forms near the midline but moves up on to the right side (Fig. 2.26). At about the same time the right side of the pharyngeal wall develops into a V-shaped thickening, the endostyle. Behind this there

forms a tube, the club-shaped gland, joining the pharynx to the outside and formed by the closure of a groove in the side of the pharynx. The significance of this organ is still obscure; it is presumably connected with the feeding process, which begins at this stage. It has been thought to represent a gill slit.

The first two coelomic pouches differentiate, asymmetrically, at this time. That on the right becomes the coelomic cavity of the head region, while the left one acquires an opening to the exterior and a heavily ciliated surface. This is perhaps also connected with the feeding systems and becomes developed into Hatschek's pit of the adult. Its interest to the morphologist lies in the fact that the first coelomic cavity opens to the exterior in other early chordates and in some vertebrates (p. 171).

The pit has thus some claim to be considered the equivalent of the hypophysial portion of the pituitary gland.

Further gill slits develop in the mid-ventral line and move over on to the right side until fourteen have been so formed. Meanwhile, a further row of eight slits appears *above* that already formed. These are the definitive slits of the right side and presently the larva proceeds to become symmetrical by movement of eight of the first row of slits over to the left side, the remainder disappearing. At this 'critical stage' with eight pairs of slits the larva pauses for some time before further changes. It is interesting that this is the time at which it most nearly represents what might have been an ancestral craniate, with eight branchial arches (p. 125). Further slits are then gradually added in pairs on both sides. Each slit becomes subdivided, soon after its formation, by the downgrowth of a tongue bar. The atrium is absent from the early larva. Metapleural folds then appear on either side and are united from behind forwards to form a tube below the pharynx. The larvae live for 2–6 months in the plankton and may be widely distributed by ocean currents. Finally the larva sinks and rests on the bottom while undergoing the migration of gill slits that constitutes its metamorphosis.

The development of amphioxus, like its adult organization, shows us many features of the plan that is typical of all chordates and was presumably present in the earliest of them. Thus the cleavage, invagination, and mesoderm formation recall those of echinoderms and other forms similar to the ancestors of the chordates, and also show a pattern from which all later chordate development can be derived. Unfortunately we cannot pursue this study as far as we should like because of the difficulty of investigating the development of amphioxus. Modern embryologists aim at tracing the morphogenetic movements by which the organism is built, and ultimately at discovering the forces responsible for these processes. We still remain ignorant of the details of these movements, and can only guess that the system of cell activities by which an amphioxus is built represents quite closely the original set of morphogenetic processes of vertebrates.

There are, of course, some special features connected with the method of life of the larva, and especially with its asymmetry. The strange sequence of gill formation, the immense left-sided larval mouth, perhaps the club-shaped gland, and Müller's organ, may show considerable modifications of relatively recent date. However, the earliest chordates probably fed by means of cilia and were planktonic, so we must not too hastily assume that even these asymmetrical features are novelties.

The division of the mesoderm of amphioxus into a series of sacs presents an interesting problem. The segmentation of the mesoderm of vertebrates is restricted to the dorsal region. In the lowest chordates (see p. 52), as in their pre-chordate ancestors, there are three coelomic cavities, but it is probable that the many segments of vertebrates arose in order to provide a set of muscles able to contract in a serial manner for the purpose of swimming. Their segmentation would thus be a relatively late development, not related to the segmentation of annelids, which divides the whole body into rings. Accordingly the ventral part of the vertebrate coelom usually remains unsegmented. But in amphioxus (and in the lamprey) it is subdivided from its first appearance and only becomes continuous later. The best interpretation of this condition is to suppose that in order to provide a series of myotomes a rhythmic process subdividing the mesoderm was adopted. In its earliest stages this affected the whole mesoderm, ventral as well as dorsal, but later became restricted to the dorsal region. New morphogenetic processes may often pass through stages of refinement and simplification in such ways.

12. The basic chordate organization

Amphioxus provides us, then, with a valuable example of a chordate that retains the habit of ciliary feeding, which was probably that of the earliest ancestors of our phylum. No doubt in connection with this, and the bottom-living habit, there are many specializations; the enormously developed pharynx with its atrium, the asymmetry, and so on; but the general arrangement of the body is almost diagrammatically simple, and it may well be that amphioxus shows us a stage very like that through which the ancestors of the craniates evolved. Perhaps next the larva remained longer in the plankton and became mature there. The larvae of some acraniates shown signs of such a change (p. 72).

This might give rise to a suspicion that amphioxus is not an ancestral type but a simplified derivative of the vertebrates, perhaps a paedomorphic form. It possesses, however, sufficient peculiar features to make this view unlikely. Neoteny might explain the regular segmentation, separate dorsal and ventral roots, and other features, but can hardly account for the method of obtaining food, or for the condition of the skin. It may be, therefore, that amphioxus shows us approximately the condition of the early fish-like chordates, living in the lower Ordovician some 500 million years ago, and that it has undergone relatively little change in all the time since.

3 The origin of chordates from filter feeding animals

1. Invertebrate relatives of the chordates

WE have seen in the organization of amphioxus the plan of chordate structure as it may have existed in the earliest Palaeozoic times. Before proceeding to discuss the later forms that evolved from animals of this sort we may first look yet farther backwards to discuss the origin of the whole chordate phylum from still earlier ancestors. The great difficulty of such an enquiry is itself a stimulus and a challenge. The fish-like form developed a very long time ago. Scales (denticles) and bone have been found in early Ordovician strata, so the origin of the chordates must be sought far back in the Cambrian, perhaps nearly 600 million years ago (see p. 22 for geological time scale).

The first step in our enquiry, however, before discussing these early forms, should be to find out, if possible, which of the main lines of invertebrate animals shows the closest affinity with the chordates. Almost every phylum in the animal kingdom has been suggested, including the arachnids and nemertines. Many people still suppose that the annelids and arthropods, because of their metameric segmentation, are related to the chordates, but closer examination shows that the similarities are superficial. The segmentation of these annulate animals is an almost complete division of the whole body into rings, and all the organ systems are affected by it to some extent. In chordates only the dorsal myotomal region is segmented; even the mesoderm is not divided in its ventral region in most animals. Moreover, the whole orientation of the body differs in the two groups. The vertebrate nerve cord is dorsal to the gut, in annulates the nerve cord is below and the 'brain' above. The blood circulates in opposite directions, the limbs are based on quite different plans, and so on. Attempts have been made to get over these difficulties by turning the invertebrate upside down! Patten (1912) and Gaskell (1908) carried such theories to extremes and tried to show a relationship of chordates with the eurypterids, heavily armoured arachnids of the Cambrian and Silurian. These animals show a certain superficial resemblance to some early fossil fishes, the cephalaspids

of the Devonian (Fig. 4.58), and these workers, with great ingenuity, claimed to find in them evidence of the presence of many chordate organs.

The safest evidence of affinity is a similarity of developmental processes: animals that develop very differently are unlikely to be closely related. The development of modern annulates is utterly different from that of chordates. The cleavage by which the fertilized egg is divided into blastomeres follows in annulates a 'spiral' plan, in which every blastomere arises in a regular way and the future fate of each can be exactly stated. In later annulates, such as the arthropods, this plan is complicated by the presence of much yolk, but even in these animals the cleavage does not resemble that of chordates, which is radial or 'irregular', the cells not forming any special pattern. This characteristic has been used to divide the whole animal kingdom into two major groups, Spiralia and Irregularia.

The next stage of development, gastrulation, by which the ball of cells is converted into a two-layered creature, also occurs very differently in the two groups. Our knowledge of the mechanics of the processes by which this change is produced is still imperfect, in spite of recent advances, but in lower chordates it occurs by invagination, the folding in of one side of the ball of cells to form an archenteric cavity communicating with the exterior. In annulates this is never seen; the cells that will go to form the gut migrate inwards either at one pole or all round the sphere and only later form themselves into a tube, which comes to open secondarily to the outside. It is probable that when we know more of the forces by which the gastrulation is produced the difference will appear even more marked than it does from this crude and formal statement that gastrulation in chordates is by invagination, in annulates by immigration.

Another basis for comparison has been found by some people in the fate of the blastopore. In chordates, echinoderms, tunicates, and hemichordates it becomes the anus and the mouth is a new opening. So these animals are grouped together as Deuterostomia (secondary mouth), distinct from the other non-vertebrates where the blastopore forms the mouth, hence Prostomia (primary mouth). Actually the mouth and anus form in various ways and it seems a pity to use this distinction alone for classification. Yet it corresponds to the different types of cleavage and to a third feature, namely the method by which the mesoderm and coelom are formed. In echinoderms, hemichordates, tunicates, and chordates this third layer is produced by separation from the endoderm, so that the coelom is at first continuous with the archenteron and is said to be an enterocoele. In annulates cells separate in various ways to form the mesoderm and a coelom then arises within this solid mass as a schizocoele. It is true that in some, indeed many, of the higher chordates the coelom is never continuous with the archenteron, but its method of development shows it to be a modified enterocoele.

In all these points of development the chordates differ from the annulates, but resemble the echinoderms and their allies. Further features support this latter relationship. One of the most important of these is that the echinoderm-like animals, and some of the early chordates, have a larva with longitudinal ciliated bands, very different from the trochophore larva, in which the bands run transversely round the body, which is found in the other line of animals. The nervous system of annulates consists of a set of ganglionated cords, whereas in echinoderm-like animals it is a diffuse sheet of cells and fibres below the epidermis. The nerve cord of the chordates can be derived from the latter but not easily from the former condition. Many further points could be cited, for instance, the presence of a mesodermal skeleton in both chordates and echinoderms, but not in annulates.

The mechanism for providing energy liberation also shows systematic differences. Regeneration of adenosine triphosphate (ATP) is accomplished from phosphoarginine by its kinase in all arthropods and molluscs and this seems to have been the primitive method, occurring in many bacteria, protozoa, coelenterates, and platyhelminths. On the other hand phosphocreatine and its kinase provide the sole phosphagen system in amphioxus and chordates. Both systems are found in annelids, echinoderms, and tunicates (Fig. 3.1). In the sperm of echinoderms and polychaetes only creatine kinase is found. This may have the advantage that a rich energy source is found for locomotion while avoiding competition for the production of arginine-rich histones. Once evolved, creatine was found more efficient as it is easily formed from the abundant glycine and this frees the scarce arginine for protein synthesis. If the use of creatine was characteristic of early larval life it may well have become the exclusive substrate after neoteny (Watts 1975).

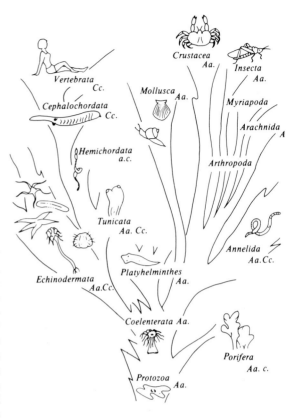

Fig. 3.1. Phylogenetic tree to show the known distribution of arginine (A) and arginase (a) creatine (C) and creatinase (c) in the animal kingdom. (After Watts, D. C. (1968). *Advances in comparative Physiology and Biochemistry* **3**, 1–114.)

In the study of evolution it is not sufficient merely to make formal comparisons, we must try to find out and compare the plan of development and organization common to all members of two groups, a technique often requiring great knowledge and good sense. When this is done in the present case it will be found that the essential plan of development of annulates involves spiral cleavage, gastrulation by immigration, the blastopore forming the mouth, and a coelom formed as a schizocoele, a trochophore-like larva, and full segmentation of the mesoderm. It is exceedingly unlikely that such animals have given rise to chordates with their very different development, which we may crudely define as showing radial cleavage, gastrulation by invagination, a blastopore that becomes the anus, coelom arising as an enterocoele, and larva of echinoderm type.

Extending this method we may divide the whole world of Metazoa by similar criteria into two groups (1) the Spiralia also called Polymera or Prostomia and (2) the Irregularia, Oligomera or Deuterostomia. The first group includes the arthropods, annelids, molluscs, and platyhelminthes. The second group contains, in addition to the chordates, the echinoderms, brachiopods, polyzoans (Ectoprocta; or Bryozoa), graptolites (Graptolithina) and *Phoronis* (Phoronida). The animals in this second group seem at first sight to be very different from the chordates in outward form, but the farther we look into their fundamental organization, the more we become convinced that the ancestors of the fish-like animals are to be found here. By study of the living relics of the early chordates it is possible to trace the history of this strange change with some plausibility although some zoologists still find that 'The idea of bringing such unlike animals together is amazing' (Brien 1974).†

2. Subphylum Hemichordata (= Stomochordata)

Class 1: Enteropneusta, acorn worms
   Balanoglossus; Glossobalanus; Ptychodera; Saccoglossus
Class 2: Pterobranchia (= feather gills)
   Cephalodiscus; Rhabdopleura

In the Hemichordata are placed animals of two types, the worm-like *Balanoglossus* and its allies (Enteropneusta) and two sedentary animals, *Cephalodiscus* and *Rhabdopleura* (Pterobranchia). The Enteropneusta are mostly burrowing animals (Figs. 3.2 and 3.3) varying in different species from 2 cm to over 2 m long. Several genera are recognized (e.g. *Balanoglossus, Saccoglossus, Ptychodera*) and they occur in all seas. *Saccoglossus* occurs around the British coast. The body is soft, without rigid skeletal structures, and divided into proboscis, collar, and trunk. The animals are very fragile and it is difficult to collect specimens in which the hind part of the trunk ('abdomen') is intact. The proboscis, collar, and trunk each contain a coelomic cavity, and the coeloms of the proboscis and collar are distensible by intake of water through a single proboscis pore and paired collar pores. The skin is richly ciliated all over the body. The outer epithelium is thus unlike the squamous, layered skin of higher forms (Fig. 3.4). It contains numerous gland cells, whose secretion is very copious, so that the animals are always covered with slime. A characteristic feature is an unpleasant smell, resembling that of iodoform, which possibly serves, like the mucus, as a protection. In *B. biminiensis* this is due to 2, 6-dibromophenol, which may serve as a disinfectant of the burrow. Iodine is present in the hepatic regions, probably organically bound (Barrington 1974).

Below the skin is a nerve plexus receiving the inner processes of receptor cells and containing ganglion cells (Fig. 3.4). Deep to this are muscles running in various directions. It is said that the animal moves by first pushing the proboscis and collar forward through the sand and then drawing the body after it. Protrusion of the proboscis cannot, however, be very vigorous. It may perhaps be produced by ciliary action distending the coelom as is usually stated – more probably by circular muscles, but these are weak. Numerous longitudinal muscles are present, however, in the proboscis and trunk and are partly attached to a plate of skeletal tissue in the collar. This tissue is attached to the ventral side of a forwardly directed diverticulum of the pharynx. The wall of this is thick, composed of vacuolated cells, and bears a certain resemblance to a notochord (Fig. 3.5). A notochord extending throughout the length of the body would be disadvantageous for an animal whose main movements are lengthening and shortening. It is possible that the diverticulum and plate found in the collar represent the remains of a notochord, serving as a fixed point by which the body is drawn forward on to the proboscis. However, many prefer to call it a 'stomochord' to avoid too close a comparison with the notochord. The external cilia probably play a considerable part in locomotion; possibly they are the chief

†The Pogonophora, which are worm-like animals with no gut, were thought to be related to chordates until it was found that they have a segmented tail piece carrying setae (Southward 1975). Great ingenuity has also been used to show that chordates are descended from nemertine worms (Willmer 1975).

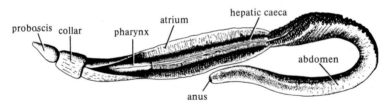

Fig. 3.2. *Balanoglossus*, removed from its tube and seen from the dorsal side. (After van der Horst.)

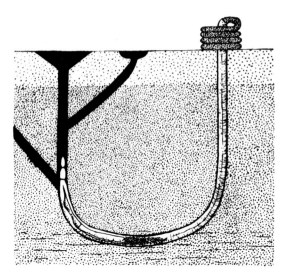

F<small>IG</small>. 3.3. *Balanoglossus* in its tube in sand. (After Stiasny, G. (1910). *Zoologischer Anzeiger* **35**, 561–5.)

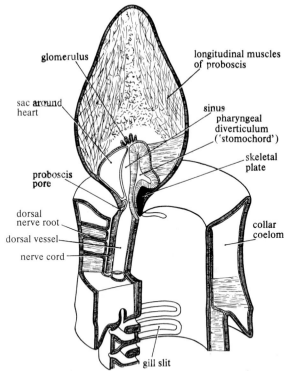

F<small>IG</small>. 3.5. Diagrammatic section of front end of *Balanoglossus*. (Modified after Spengel.)

burrowing organs, the muscles serving mainly to perform escape movements.

The mouth lies in a groove between the proboscis and collar (Fig. 3.6). The proboscis contains many mucus-secreting cells and the food particles are captured on its surface and conveyed by ciliary currents to the mouth. In the anterior part of the trunk there is a wide pharynx, opening by a series of gill slits (Figs. 3.5 and 3.7). These resemble the gills of amphioxus in the presence of a supporting skeleton in the gill bars; there are also tongue bars dividing the slits from above, and horizontal synapticulae strengthening the gill arches. The slits open

in some species into an atrium formed by lateral folds, usually turned upwards to leave a long mid-dorsal opening. In other species each slit opens to a gill pouch. There are no gills and the whole branchial apparatus perhaps assists in the process of feeding, probably by serving to filter off the excess water from the material already collected on the proboscis, which often consists of large amounts of sand or mud. Relative to the size of

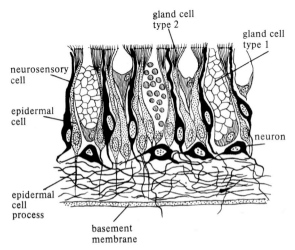

F<small>IG</small>. 3.4. Section of the epidermis of an enteropneust. (After Bullock, van der Horst, and Grassé.)

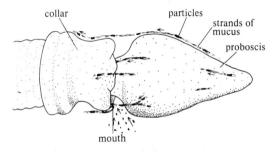

F<small>IG</small>. 3.6. Feeding-currents on proboscis of *Glossobalanus*, shown by placing the animal in water containing carmine particles. The particles are either taken directly into the mouth, or are caught up in strands of mucus and passed backwards. (From Barrington 1940.)

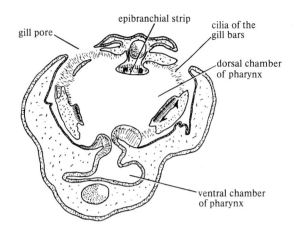

FIG. 3.7. Transverse section of the pharynx of *Glossobalanus*. (From Barrington 1940.)

the animal the pharynx is less extensive than in amphioxus, presumably because ciliary surfaces are provided on the outside and also large masses of sand are forced into the mouth during locomotion.

There is no endostylar apparatus, but the ventral part of the pharynx is often partly separated from the rest (Fig. 3.7). Along this groove the matter ingested is passed to a straight oesophagus and intestine opening by a terminal anus. There is no true tail in the adult but a post-anal region is present in some species during development. Numerous hepatic caeca in the anterior part of the intestine can be seen from the outside as folds of the body wall, often highly coloured.

The blood system consists of a complex set of haemocoelic spaces, communicating with large dorsal and ventral vessels (Fig. 3.8). The former enlarges into a sinus anteriorly and this is partly surrounded by the wall of a pericardial cavity, which contains muscles and may be said to be the heart, though clearly lying in a very different position from that of other chordates. From the sinus, vessels proceed to the proboscis and round the pharynx to the ventral vessel. The blood is said to move

forwards in the dorsal and backwards in the ventral vessels. The front of the sinus forms a series of glomeruli, covered by a region of the proboscis coelom specialized to form excretory cells, the nephrocytes, some of which drop off into the coelom. The blood is red in some species but usually colourless. It contains a few amoebocytes.

The nervous system is one of the most interesting features of Enteropneusta. It resembles that of echinoderms in consisting of a sheet of nerve fibres and cells lying beneath the epidermis all over the body (Fig. 3.4). This sheet is thick in the mid-dorsal and mid-ventral lines, and in the dorsal part of the collar region it is rolled up as a hollow neural tube, open at both ends (Fig. 3.5). These unmistakable resemblances not only to the subepithelial plexus of echinoderms, which is not centralized, but also to the hollow dorsal nerve cord of vertebrates are most instructive, showing the affinity of the groups and the origin of the general plan of the vertebrate nervous system. There are no organs of special sense, unless this is the function of a patch of special ciliated cells on the collar. Receptor cells all over the body send their processes into the nerve plexus (Fig. 3.4), on the primitive plan of neurosensory cells found elsewhere in vertebrates only in the olfactory epithelium and the retina. The plexus is remarkable in receiving fibres from the outer ciliated epithelial cells, which thus represent the ependyma, the earliest form of neuroglia (Fig. 3.4). Nothing is known of the organization of pathways or of the connections with the muscles. The collar nerve cord contains giant nerve cells whose axons proceed backwards to the trunk and forward to the proboscis (Fig. 3.9). They are probably responsible for rapid contractions.

Bullock investigated the behaviour of the animals and found one clear-cut reflex, namely, a contraction of the longitudinal muscles in response to tactile stimulation (1940). Isolated pieces of the body are able to show reflex responses, moving away from light or tactile stimuli. Such local actions are an interesting sign of the un-

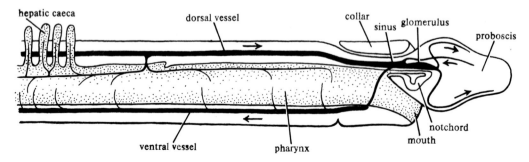

FIG. 3.8. Diagram of the blood system of *Balanoglossus*. (After Bronn.)

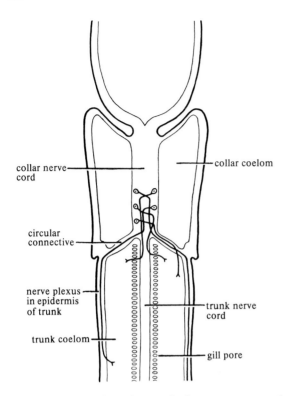

collar nerve cord

collar coelom

circular connective

nerve plexus in epidermis of trunk

trunk nerve cord

trunk coelom

gill pore

FIG. 3.9. Diagram of certain tracts in the nervous system of *Balanoglossus*. (Bullock 1944.)

sacs developing from cells just outside the coelom. These proliferate and bulge into the coelom, covered by the somatopleure. They acquire a cavity and each opens by a narrow duct to the exterior, fertilization being external. The development is remarkably like that of echinoderms. Cleavage is holoblastic and resembles that of amphioxus and ascidians, gastrulation is by invagination, and the coelom is formed as an enterocoele, later becoming subdivided into proboscis, collar, and trunk coeloms. Hatching occurs to produce a pelagic tornaria larva, with a ciliated band that has exactly the relations found in the dipleurula larva of echinoderms. The band passes in front of the mouth, down the sides of the body, and in front of the anus (Fig. 3.10). It then divides into more dorsal and ventral sections, exactly as in the production of the bipinnaria larva of a starfish. This arrangement differs essentially from the rings of cilia that pass round the body in the trochophore larva found in the annelids and other spirally cleaving forms. In later tornaria larvae there is, however, in addition to the longitudinal bands always a posterior ring of stout cilia (telotroch), and in large oceanic forms (which may reach 8 mm in length) the longitudinal band itself is prolonged into prominent tentacle-like loops (Fig. 3.11). The cilia of the posterior ring are purely locomotive, while those of the band set up feeding currents converging to the mouth. As the larva becomes larger the ciliary surface needed for locomotion and feeding has to increase relatively faster than the increasing mass of the body, the latter following the cube but the former only the square of the linear dimensions. Accordingly the cilia of the locomotive ring become broadened and flame-like, while the convolutions of the longitudinal (feeding) band reach fantastic proportions. In some types, however (*Saccoglossus*), the pelagic phase is brief and the telotroch alone is formed. Finally the larva sinks, becomes constricted into three parts, and undergoes

centralized nature of the nervous system, and similar actions are found in echinoderms. A further sign of lack of special conducting pathways is that stimulation of flaps of body wall, partly severed from the rest, produces generalized contraction, proving that conduction can occur in all directions. The dorsal and ventral nerve cords do, however, act as quick conduction pathways, and contraction of the trunk following stimulation of the proboscis is delayed or absent if one, and especially if both, cords have been cut.

Perhaps the most interesting behaviour observed was the activity shown by isolated bits of proboscis, collar, trunk, or portion of trunk. These organs may move around vigorously in an exploratory manner; evidently the main nerve cords are not necessary for the initiation of action, as is the central nervous system of higher chordates.

There are nerve fibres in the walls of the pharynx and oesophagus, where peristaltic movements have been observed. Their relationship to the rest of the nervous system is unknown. They may represent the beginnings of an autonomic nervous system.

The sexes are separate in enteropneusts and the gonads resemble those of amphioxus in being a series of

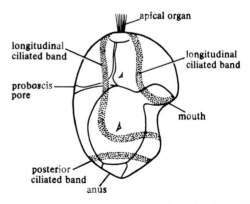

apical organ

longitudinal ciliated band

longitudinal ciliated band

proboscis pore

mouth

posterior ciliated band

anus

FIG. 3.10 Young tornaria larva, seen from the side. (After Stiasny.)

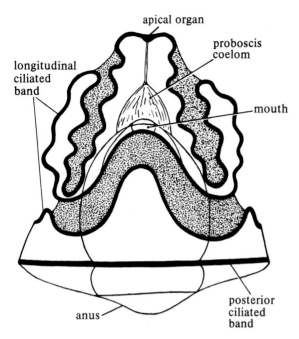

FIG. 3.11. Older tornaria larva seen from ventral surface. (After Stiasny.)

metamorphosis into the worm-like adult. This development is so like that of an echinoderm that it would be necessary to consider the enteropneusts to be related to that group even if no other clues existed. Such close similarity in the fundamentals of development must have a genetic basis.

These animals thus provide a very remarkable and sure demonstration that the chordates are related to the echinoderms and similar groups. The general arrangement of the nervous system as a subepithelial plexus, as well as the whole course of the development, show the affinity with the invertebrate groups, whereas the hollow dorsal nerve cord and the tongue-barred gill slits are by themselves sufficient to show affinity with the chordates, this affinity being also perhaps suggested by other features, such as the 'notochord'. As we have seen already, affinities are not to be determined by single 'characters' but by the general pattern of organization of animals and especially of their development. The organization of the enteropneusts is certainly highly specialized for their burrowing life, but showing through the special features we can clearly see a plan that has similarity with both the echinoderms and the chordates. The special value of study of these animals is that it proves decisively that an affinity between these groups exists. Exactly how they are all related is a more speculative matter, which we shall deal with later (see p. 69) (see Barrington 1965 for their biology).

3. Class Pterobranchia

These are small, colonial, marine, sedentary animals, which show some signs of the general echinoderm-chordate plan of organization we have been discussing. *Cephalodiscus* (Fig. 3.12) has been found on the sea bottom at various depths, mainly in the southern hemisphere: there are several species. The colony consists of a number of zooids held together in a many-chambered gelatinous house. The zooids are formed by a process of budding, but do not maintain continuity with each other. Each zooid has a proboscis, collar, and trunk; there are coeloms in each of these parts, and proboscis and collar pores. The collar is prolonged into a number of ciliated arms, the lophophore, by means of which the animal feeds. There is a large pharynx, opening by a single pair of gill slits, which serve as an outlet for the water drawn in by the cilia of the tentacles for the purpose of bringing food. The intestine is turned upon itself, so that the anus opens near the mouth. A thickening in the roof of the pharynx corresponds exactly in position with the stomochord and contains vacuolated cells. The blood system consists of a series of spaces arranged on a plan similar to that in *Balano-*

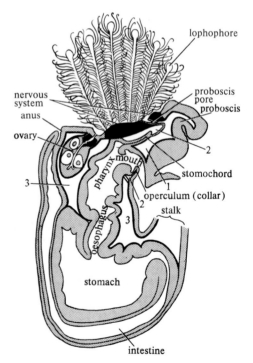

FIG. 3.12. Longitudinal median section of *Cephalodiscus*. 1–3 denote the three coelomic body cavities. (Modified after Harmer, S. F. (1904.) *Cambridge Natural History*, 7. Macmillan, London.)

glossus. There is a dorsal ganglion in the collar, but this is not hollow. The gonads are simple sacs and development takes place in the spaces of the gelatinous house. Gastrulation is by invagination at least in some species and the coelom is formed as an enterocoele. The larva somewhat resembles that of ectoproctous polyzoa, which is not closely similar to the echinoderm larvae, but could be derived from the same plan.

Rhabdopleura occurs in various parts of the world, including the North Atlantic and northern part of the North Sea. The zooids are connected together and have proboscis, collar, and trunk, ciliated arms, coelomic spaces with pores (not 'nephridia' as is sometimes stated) and stomochord, but no gill slit. The nervous system is in a very primitive state with the cell bodies and fibres confined within the epithelial layer. There is a collar ganglion consisting of a peripheral rind of nerve cell bodies and a central neuropil with synapses containing vesicles of several sizes (Dilly 1975). There are motor terminals on muscle fibres and ciliated cells. There is a short larval stage with the cilia not in bands. Gastrulation is by invagination and the blastopore does not form the mouth (Dilly 1973).

The Pterobranchia thus show undoubted signs of the enteropneust–chordate plan of organization and provide also an interesting suggestion of possible affinities with the lophophorate coelomates, the Polyzoa, Bra-chiopoda, and Phoronida. Like the Pterobranchia the Polyzoa Ectoprocta are sessile, with mouth and anus pointing upwards. They feed by means of the cilia borne on a horseshoe-ring of tentacles (the lophophore); but there is no division into proboscis, collar, and trunk, and no tripartite coelom. The nervous system is in the condition of a subepithelial plexus, which is folded, around the base of the lophophore, to form a hollow tube – a remarkable point of similarity to the chordates. Even though it is difficult to compare this tube exactly with the nerve cord of chordates, it is at least evidence of the organization of the nervous system on a plan that allows of such folding. It is probable that the modern pterobranchs are the surviving members of the ancient group of graptolites extending from the Cambrian, but these are known only from their fossil skeletons.

Although it would be unwise to suggest close relationship between the polyzoans and the pterobranchs, the similarities are sufficient to suggest that the chordates arose from sedentary lophophorate creatures, feeding by means of ciliated tentacles. The evidence encourages us to look for the presence somewhere in the line of vertebrate ancestry of an animal with this habit. The difficulties of this view arise when we come to consider how the fish-like organization of a free-swimming animal first appeared, a question better dealt with after consideration of the tunicates.

4. Subphylum Tunicata (= Urochordata)

    Class 1: Ascidiacea (= sac animals)
      Ciona; Botryllus; Clavelina; Amaroucium; Dicarpia; Megalodicopia
    Class 2: Thaliacea
      Salpa; Doliolum; Pyrosoma; Cyclosalpa
    Class 3: Larvacea (= immature forms) (= Appendicularia)
      Oikopleura

The typical tunicates are the sea-squirts, bottom-living filter-feeders in which there is no obvious trace of the fish-like form at all. However, they develop by way of a larva, the ascidian tadpole, which has a notochord and other chordate features. In the Larvacea there has been a process of neoteny by which the tadpole stage becomes sexually mature.

Sea-squirts are sac-like creatures living on the sea floor and obtaining their food by ciliary action. They can also absorb amino acids from sea water. Often the separate individuals are grouped together to form large colonies, but in *Ciona intestinalis*, common in British waters, the individuals occur separately, and this is possibly the primitive condition for the group. The whole of the outside of the body is covered by a tunic or test, in which there are only two openings, a terminal mouth and a more or less dorsal atriopore, both carried upon siphons (Fig. 3.13) (Berrill 1950). The tunic serves for protection and support. It is a living tissue supplied by blood vessels with blind ends. It is composed largely of fibres of a carbohydrate, tunicin, closely related to cellulose, embedded in a ground substance of acid mucopolysaccharide, with some protein. It has therefore several similarities to vertebrate connective tissue. It is secreted by the epidermis but wandering morula cells (vanadocytes) assist in fibre formation (p. 60). In some tunicates, calcareous secretions of various shapes are found in the tunic. The mantle that lines the tunic is covered by a single-layered epidermis. Ascidians are often brightly coloured, the pigment being either in the tunic or the underlying body, which shows through the transparent tunic. The colour can change, at least over a period of some days. Little is known about the origin of the pigment, but it is sometimes derived from that of the blood and may lie in special cells.

The mantle is provided with muscle fibres running in

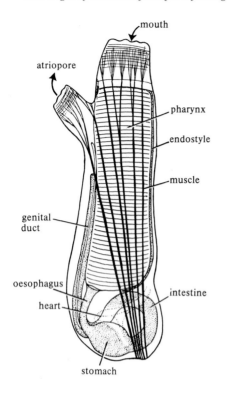

FIG. 3.13. Diagram of structure of *Ciona*. (After Berrill 1950.)

various directions but mainly longitudinally, and serving to draw the animal together, with the production of the jet of water from which the animals derive their common English name.

The greater part of the body is made up of an immense pharynx, beginning below the mouth and forming a sac reaching nearly to the base (Fig. 3.13). The sac is attached to the mantle along one side (ventral) and is surrounded dorsally and laterally by a cavity – the atrium. This pharynx is a food-collecting apparatus; its walls are pierced by rows of stigmata (gill slits) whose cilia set up a food current entering at the mouth and leaving from the atriopore. The entrance to the pharynx is guarded by a ring of tentacles, which may be compared with the velum of amphioxus. The stigmata are very numerous vertical cracks, all formed by subdivision of three original gill slits. Tongue bars grow down to divide each slit and then from each tongue bar grow horizontal synapticulae (Fig. 3.14). This arrangement has clear resemblance to that of amphioxus and results in the production of a pharyngeal wall pierced by numerous holes. Immediately within the stigmata there is a series of papillae, provided with muscles and cilia. There is an endostyle, which has three rows of secretory cells on each side, separated by rows of ciliated cells and with a single median set of cells with very long cilia (Fig. 3.15). The secretion of the inner layers of the endostyle is mainly of protein, which is then combined

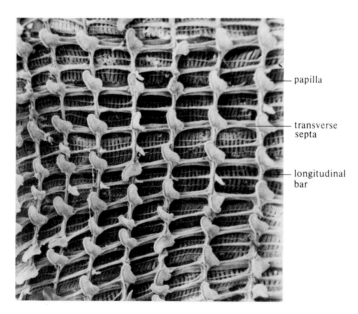

FIG. 3.14. Scanning electron micrograph of the branchial wall of *Ciona*. (Field width 2100 μm.) Fiala-Medioni, A. (1978). *Acta Zoologica* **59**, 1–9.)

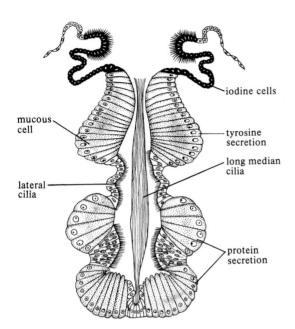

FIG. 3.15. A transverse section of the endostyle of *Ciona*. The blackened cells showing the effect of treatment with [131]I. (As demonstrated by Barrington, E. J. W. and Thorpe, A. (1960). *General and comparative Endocrinology* **5**, 373–85.

with iodinated tyrosine provided by a layer of iodine-uptake cells (Fig. 3.15). There may also be an addition of mucopolysaccharide. The main product of the endostyle is a sheet of iodinated protein which is caught up on the papillae, whose muscles move them rhythmically, spreading a curtain over the inside of the pharynx. Food particles are caught in this layer, which moves upwards and is then passed back to the oesophagus by the cilia of a dorsal lamina or of a series of hook-like 'languets'. The whole curtain is then digested, including the captured micro-organisms and iodinated protein. Autoradiographs made from tunicates that have been provided with isotopes of iodine show that the compounds formed are mainly 3-mono-iodotyrosine (MIT) or 3, 5-di-iodotyrosine (DIT). There is little or none of the tri-iodothyronine (T_3) and thyroxine (T_4) compounds found in amphioxus and craniates (p. 102). The iodo-protein curtain evidently has some special properties as a collector of food. Iodine is even more abundant in the outer layers of the tunic but here it forms scleroproteins, as in the exoskeletal structures of molluscs and insects. The iodine-binding cells of the endostyle show some responses to mammalian thyroid stimulating hormone, but there is no evidence of any hormonal effects of the secretion within the sea-squirt itself.

The extensive ciliated surface of the pharyngeal wall ensures the passage of large volumes of water inwards at the mouth and out at the atriopore. Rapid change of the water is also produced by periodic muscular contractions (p. 60). The pressure of the exhalent current is sufficient to drive the water that has been used well away from the animal.

The oesophagus leads to a large 'stomach' with a folded wall containing gland cells, which produce mucus and digestive enzymes. These include much amylase, invertase, small amounts of lipase, and a weak protease of the tryptic type. The organ is therefore not to be compared with the stomach of vertebrates. There is no cellulase. From the stomach a rather short intestine leads upwards to open by a rectum near the atriopore. The intestine is the absorptive region of the gut and is lined by mucus-secreting and absorptive cells. The curtain of iodinated protein is formed into a food cord by the cilia of the oesophagus and passed on into the stomach containing acid. It is softened either here or at entry to the intestine and mixed with the enzymes. Digestion is entirely extracellular and there is no evidence of uptake of solid particles. Food is moved along the gut almost wholly by the cilia, there is little muscular action. The pyloric gland is a system of tubules ramifying in the wall of the alimentary canal and ending in blind swellings, this system opens into the gut near the junction of the stomach and intestine. The tubes are lined by a simple epithelium whose cells break down to form a holocrine secretion, this does not contain enzymes but may help to break up the food cord.

The heart lies below the pharynx and is a sac, surrounded by a pericardium and communicating with a system of blood spaces derived from the blastocoele. None of these spaces have a true endothelial lining. The biggest is a hypobranchial vessel below the endostyle, from which branches pass to the pharynx. From the branchial sac blood collects to form a dorsal vessel which receives blood from the body and sends it to the alimentary canal. From this blood collects to a visceral vessel joining the hind end of the heart. The circulation thus follows the same general plan as in vertebrates but there are no capillaries and no valves. The blood circulates in a closed system of spaces pumped by peristaltic contractions passing from one end of the heart to the other, first in one direction for 2–3 minutes and then the other. The heart is a tube of simple myoendothelial cells and the contraction is myogenic, controlled by pacemakers at the two ends and electrical conduction occurs at 13 mm/s through low resistance junctions. There is no nervous innervation of the heart (see Goodbody 1974). The periodic reversal, characteristic of the tunicates, may be connected with the need to supply blood to the tunic as well as to the branchial sac and viscera.

The blood plasma is colourless but contains several types of corpuscles, some of which are phagocytes, others lymphocytes, possibly producing antibodies (p. 60). Many of the blood cells contain orange, green, or blue pigment (in different species). The green and other pigments are remarkable in that some contain vanadium, others iron, titanium, tantalum, or niobium (Carlisle 1968). The vanadocytes contain much sulphate and the metal is associated with a chain of pyrrol rings. The concentration of vanadium may be a million times greater in the ascidian than in the sea. It is taken up by the branchial epithelium and later moved from there by the blood cells. The heavy metal chromogens of these morula cells are powerful reducing agents held in the reduced state by sulphuric acid. They break up in the test and function there in the polymerization of simple carbohydrates to form the polysaccharides of the test (p. 57). The haemovanadin is able to reduce cytochrome but it remains uncertain what part if any the pigment plays in respiration. The blood turns blue in air but cannot take up more oxygen than can sea water.

The blood is isotonic with sea water, but has little sulphate. Ascidians appear to have little or no power of regulating their osmotic pressure; none of them is found in fresh water. They are not even able to colonize brackish waters or those of low salinity. For example, they are rare in the Baltic Sea, from which only six species have been reported. Only one species, *Molgula tubifera*, has been reported from the Zuider Zee.

A possible reason for this inability to regulate the internal composition is the need to expose a large surface to the water. Moreover, there are no tubular excretory organs such as could be used to maintain an osmotic gradient. Ninety-five per cent of the nitrogen is excreted as ammonia. Some is stored as purines in nephrocytes in the blood and elsewhere, contained in the concretions within the cytoplasm. They may be stored in the epicardia until the animal dies. The vanadocytes may serve to excrete sulphate.

There has been much debate as to whether the tunicates possess a coelomic cavity. The heart develops from a plate of cells arising early from the mesoderm and lying between ectoderm and endoderm. This becomes grooved and folded to make the heart itself and the pericardium. The irregular system of haemocoelomic spaces around the pharynx and elsewhere is usually said to consist of 'mesenchyme' and to be derived from the blastocoel and therefore not coelomic, but its walls are mesodermal. The situation is complicated by the presence of a pair of outpushings from the pharynx, the epicardia, or perivisceral sacs, which end blindly on either side of the heart (Fig. 3.16). Berrill (1950) and others have suggested that these epicardia

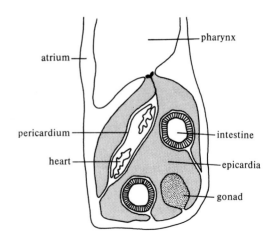

FIG. 3.16. Section through base of *Ciona*.

may be compared with coelomic cavities. Their function in the open condition in which they are found in *Ciona* is perhaps to allow sea water to circulate about the heart and hence to help excretion (and respiration?). In other ascidians the epicardium loses its connection with the pharynx. The closed sac functions in some cases as an excretory organ, containing concretions of uric acid, whereas in other animals it becomes the main source of the cells that make the asexual buds.

The central nervous system consists of a round, solid ganglion (Fig. 3.17), lying above the front end of the pharynx. The ganglion has a layer of cells around the outside and a central mass of neuropil and is therefore quite unlike the nerve cord of a vertebrate. From the ganglion nerves proceed to the siphons, other parts of the mantle, muscles, and viscera. There are few or no nerve cells in the wall of the gut, which contains no muscles except around the anus. The cupular organs are mechanoreceptors abundant around the atrial but not the branchial siphon. Each consists of a group of ciliated cells whose cilia pass into a cupula. Vibrating probes (25–400 Hz) cause siphon closure and body retraction when placed 4 cm from the atrial siphon but only when 2 cm from the branchial one (Bone and Ryan 1978). These responses persist after removal of the brain and must therefore depend on a peripheral nervous connection, but this has not been seen.

Movement consists mainly of contraction and closure of the apertures. A light touch of either siphon causes closure proportional to the strength of the stimulus and this continues after removal of the ganglion. Stimulation just inside either siphon produces closure first of the other one, then of the one stimulated and finally contraction of the body, ensuring that a jet of water sweeps out the aperture that received the stimulus.

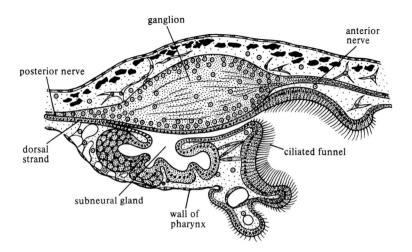

Fig. 3.17. Longitudinal section of the ganglion and subneural gland of an ascidian. (After Bertin, from Grassé.)

These crossed reflexes do not depend upon the integrity of the ganglion, but are much slowed by ablation of it (Florey 1951).

The surface of the body is sensitive to changes in light intensity, and these are followed by local or total contractions, according to their extent. After removal of the ganglion the wider reflexes can no longer be obtained but local responses continue, suggesting the presence of nerve cells in the body wall. Electrical stimulation also provides evidence of this. One shock may produce only a small response but if a second shock follows shortly afterwards there is marked facilitation and a large contraction occurs. These responses are also seen after removal of the ganglion. The various parts of the body are not all equally sensitive to light, the highest sensitivity being in the region of the ganglion. The 'ocelli' are cup-like collections of orange-pigmented cells around the siphons; they may be responsible for the phototropic orientation of the body (Millar 1953).

The neuromuscular system thus appears to function mainly as a reflex apparatus for producing protective movements in response to certain stimuli. This is the role that might be expected of it in an animal that remains fixed in one place. The 'initiative' for food-gathering activities comes from the continuous action of the cilia of the pharynx but these stop beating at intervals as a result of nervous control (Mackie, Paul, Sleigh, Singla, Williams 1974). The nervous system shows little sign of those continuous activities that produce the varied and 'spontaneous' acts of behaviour in higher forms. Nevertheless, it would be unwise to suppose that the nerves are only activated by external stimuli. There are some suggestions that even in these simple animals rhythmical activities are initiated from within. The food-collecting

operations of the pharyngeal wall involve rhythmical movement of the papillae by their muscles. Further, in many species of ascidians there are regular contractions of the siphons and body musculature in rotation, with a frequency of once every 5–8 min, increasing, however, to 1 in 2 min in the absence of food. These may be some form of 'hunger' contraction, directed towards the obtaining of food if more water is moved than by the ciliary current. Their presence is a striking warning of the dangers of assuming that even the simplest nervous system operates only when stimulated from outside.

Colonial ascidians show only slight signs of co-ordinated action. After a strong local stimulus a wave of siphon contraction and ciliary arrest spreads through a colony of *Distaplia* but only at about 1 cm/s. There is no clear evidence of nervous connection and the conduction is probably mechanical (Mackie 1974).

The neural gland is a sac lying beneath the ganglion and opening by a ciliated funnel on the roof of the pharynx (Fig. 3.17). It arises mainly from the ectoderm of the larval nervous system. It has been suggested that the neural gland may be compared with the infundibulum and hypophysis of vertebrates. There is an obvious similarity with Hatschek's pit of amphioxus (p. 48). Both may be receptor organs, testing the water stream and also producing mucus. The subneural gland has also been held to have a similarity to the pituitary in that it controls the release of gametes. When eggs or sperms of the same species are present in the water, signals from the neural gland are said to produce discharge from the gonad. The pathway of the signals is then partly hormonal, partly nervous. Discharge is produced by injection of extract of neural gland, but this acts through the ganglion, since it produces no effect if the nerves leading from this (and the dorsal strand) are cut. Other experiments have not supported this whole hypothesis and the possible function of the gland in reproduction remains to be clarified (see Goodbody 1974).

Further similarities with the pituitary have been claimed, such as the presence of vasopressor and oxytocic substances in the subneural gland. However, the oxytocin-like substance is present elsewhere in the tunicate and in any case differs from that of vertebrates. The best evidence shows that there is no antidiuretic or melanocyte stimulating hormone (Dodd and Dodd 1966). It cannot therefore be claimed that the relationship with the pituitary is clear, but it seems likely that there is some. The position of the gland suggests a connection with the feeding mechanism and there is a daily cycle of secretion, whose product may perhaps be mixed with the food cord. These are signs that, as in the thyroid, a pharyngeal mucus-secreting organ stimulated by the environment has evolved into a glycoprotein-secreting endocrine organ, controlled by substances reaching it in the blood (Barrington 1959).

5. Development of ascidians

Tunicates are hermaphrodite, the ovary and testis being sacs lying close to the intestine and opening by ducts near the atriopore. Fertilization is external in the solitary forms but internal in those that form colonies, the development in the latter taking place within the parent. The details of cleavage and gastrulation show a remarkable general similarity to those of amphioxus. Indeed, the whole development is so strikingly like that of chordates that it establishes the affinities of the tunicates far more clearly than the vague indications of a chordate plan of organization seen in the adult. The result of development is to produce a fish-like creature, the *ascidian tadpole*, which is immediately recognizable as a chordate (Fig. 3.18). The cleavage is total and produces a blastula with few cells, whose future poten-

tialities are already determined. Gastrulation by invagination follows and the creature then proceeds to elongate into the fish-like larva. This possesses an oval 'head' and long tail, the latter supported by a notochord formed by cells derived from the archenteric wall. Forty of these cells make up the entire rod, becoming vacuolated and elongated by swelling.

On either side of this notochord run three rows of muscle cells, eighteen on each side, derived from mesoderm that arises from yellow-pigmented material already visible in the egg and later forming part of the wall of the archenteron. Other cells of this tissue migrate ventrally to make the pericardium, heart, and mesenchyme. The muscle cells contain cross-striated myofibrils. Each muscle cell receives a nerve end plate and transmission is cholinergic. The muscle cells are also linked electrically by gap junctions.

The nervous system is formed by folds essentially similar to those of vertebrates, making a hollow, dorsal tube, extending into the tail and enlarged in front into a cerebral vesicle, within which are lens cells and visual cells and a single statocyst cell carrying an otolith (Figs. 3.18 and 3.19) (Eakin and Kuda 1971). The visual cells project through a pigmented cup and their packed microvilli presumably provide the directional information for the tropisms of the animal (Fig. 3.20). Nerve fibres proceed only to the front end of the rows of muscles and the rest of the cord contains no nerve cells or fibres (Fig. 3.19). Ciliated sensory cells of the epidermis of the tail send axons rostrally to the dorsal nerve cord. They may function as proprioceptors.

The larva takes no food and the gut is not well developed. There is a pharynx with usually a single pair of gill slits opening into an atrium, which develops as an ectodermal inpushing. Below or around the mouth various types of sucker are formed.

The whole process of development occupies only one or two days, and the larva, in the species in which it is set free, is positively phototropic and negatively geotropic

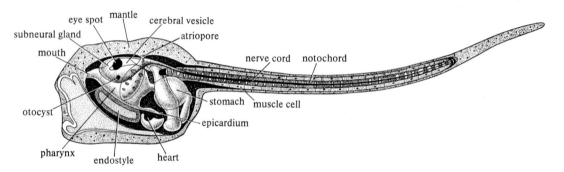

FIG. 3.18. Ascidian tadpole of *Clavelina*.

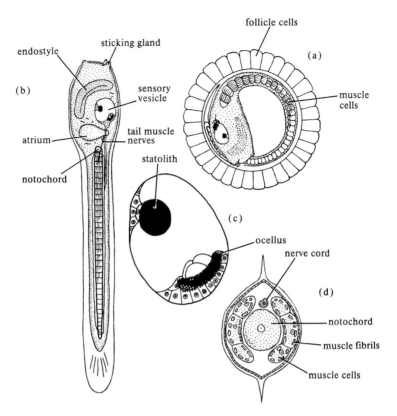

FIG. 3.19. The ascidian tadpole (*Ascidia* or *Ciona* type). *a*, tadpole ready to hatch; *b*, tadpole; *c*, cerebral vesicle; *d*, cross-section of tail. (After Berrill 1955.)

and so proceeds to the sea surface. But its life here is also limited. Within a day or two, depending on the conditions, its tropisms reverse so that it passes to the bottom, turns to any dark place and thus finds a suitable surface. It attaches by the suckers, loses its tail, develops a large pharynx, and grows into an adult ascidian. Presumably its short life in the chordate stage is sufficient to ensure distribution, and the simple nervous system serves to find a place in which to live (Cloney 1978, see Berrill 1975 for reproduction).

In addition to the sexual reproduction, tunicates have great powers of regeneration and also often multiply by budding. The tissue of the bud may be outer epicardial, mesenchymal, pharyngeal, or atrial. The epidermis develops only more tissue like itself and all the other tissues are formed from the inner mass. This occurs by a process of folding to make a central cavity; the nervous system, intestine, and pericardium are then formed by further foldings. The bud thus begins in a condition comparable to a gastrula but develops directly into an adult, without passing through the tadpole stages. The fact that a complete new animal is thus formed from one

or two layers shows that the separation into three layers during development does not involve any fundamental loss of potentialities, as would be required if the 'germ layer' theory held rigorously. The germinal tissue of the bud is not necessarily derived from that of the parent.

6. Various forms of tunicate

Besides some 2000 species of sessile tunicates, about 100 species have become secondarily modified for a pelagic life. These pelagic animals are perhaps all related, but the whole subphylum is conveniently subdivided into three classes, Ascidiacea, Thaliacea, and Larvacea.

7. Class Ascidiacea

The typical sessile ascidians are found in all seas. They may be divided into those that live as single individuals (Ascidiae simplices) and those forming colonies (Ascidiae compositae). Both types include many different forms, however, and the division is not along phylogenetic lines. The colonial forms produced by budding may consist simply of a number of neighbouring individuals (*Clavelina*) or of a common gelatinous test in

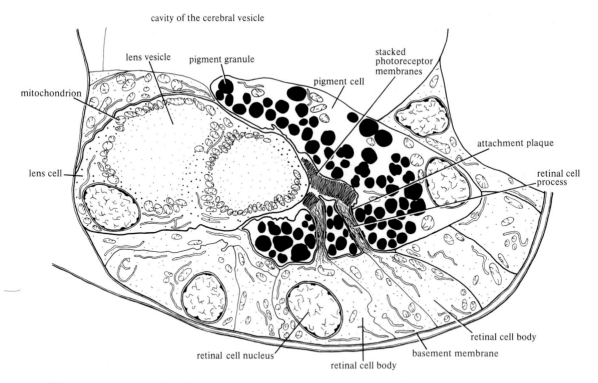

FIG. 3.20. Transverse section of ocellus of the free-swimming tadpole stage of the sea squirt *Ascidia nigra*. (Drawn from an electron micrograph by P. N. Dilly.) The ocellus is situated in the posterior wall of the cerebral vesicle. It consists of three parts, a lens cell, a pigment cell, and a retina. The lens cell usually contains three lens vesicles, which are spheres of cytoplasm bounded by mitochondria. The pigment cell contains granules of melanin, which protect the photoreceptor from stray light. The retinal cells have processes that penetrate the pigment cell. They are similar to vertebrate rods, composed of a pile of membranes closely applied to the inner edge of the lens cell.

which the individuals are embedded (*Botryllus, Amaroucium*). The form of the body is related to the type of bottom upon which they are found; there has thus been an adaptive radiation within the group; a great variety of habitats is available for bottom living creatures, and the animals become adapted accordingly. Many colonial forms are hermaphrodite but there is never self-fertilization within the colony. When two colonies come into contact they first fuse but then rejection occurs. The histocompatability system differs from that of vertebrates in that a second challenge shows no evidence of a memory (Marchalonis 1977).

Most of the species live in the littoral zone, but ascidians are an important element in the benthic abyssal fauna. Indeed they become both more varied and more numerous at greater depths (Monniot and Monniot 1978). In the Bay of Biscay eight species have been found at 2000 m, but at 4000 m there are 31, and the number of individuals increases in the same way. Some abyssal forms are similar to those in littoral zones, others have very long stalks and branchial sacs without

cilia (Fig. 3.21). They may feed by bending over to allow the deep-sea bottom currents to flow through them (Fig. 3.21) a suggestion made by George Bidder to explain the feeding of Hexactinellid sponges (see Young 1974). Some deep-sea forms are active predators. In the Octacnemidae the oral siphon forms two lips which can close to trap crustacean prey (Fig. 3.22). In the Sorberacae the oral siphon forms a lobed proboscis able to catch live prey. There is no branchial sac but a huge stomach (Fig. 3.23).

Many ascidians probably live only for a short time, becoming mature in their first year and dying thereafter. In some species the animals live over a second winter, during which they become reduced in size, growing and budding again in the following spring (*Clavelina*).

8. Class Thaliacea

These are pelagic tunicates living in warm waters. They have circular bands of muscle, enabling the animal to shoot through the water by jet propulsion. In *Doliolum* and its allies (Figs. 3.24 and 3.25) the muscle bands pass

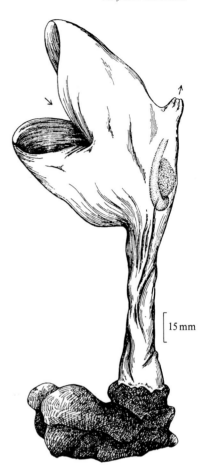

FIG. 3.21. A solitary abyssal ascidian *Dicarpia*. Scale 1 mm.
(Monniot and Monniot 1978.)

FIG. 3.22. An abyssal ascidian *Megalodicopia* in which the oral
siphon is developed into two large lips. The lips can close to
make a trap or form a basket open to the external medium.
(Tokiaka, A. (1953). *Ascidians of Sagami Bay collected by His
Majesty the Emperor of Japan.* Tokyo.)

right round the body (Cyclomyaria), whereas in *Salpa*
the rings are incomplete (Hemimyaria). The mouth and
atriopore are at opposite ends of the body. The tunic is
thin and, like the rest of the body, transparent. The
connections of two types of neuron have been studied in
the brain of salps. The motoneurons produce bursts of
8–15 action potentials correlated with movement. The
pacemaker neurons fire at shorter intervals and these
can be influenced by input from the eye or the impulse
system of the skin (p. 68). Salps can swim either
forwards or backwards (Bone, Anderson, and Pulsford
1980).

The life history of these forms involves a remarkable
alternation of generations. In *Doliolum* the ascidian
tadpole develops into a mother or nurse zooid (oozooid). This by budding gives rise to a string of daughter
zooids, which it propels along by its muscles. The
daughter zooids are of three types: (1) sterile, nutritive,
and respiratory individuals, the trophozooids, permanently sessile on the parent; (2) sterile nurse forms,
which are eventually set free (phorozooids); (3) sexual
forms (gonozooids, Fig. 3.24), nursed and carried by the
phorozooids until sexually mature, when they also
break loose.

In *Salpa* the sexual form (blastozooid), produces only
a single egg, which develops within the mother without
passing through a tadpole stage, nourished by a diffusion placenta, whose cells also migrate into the
developing embryo. This becomes the asexual oozooid
and produces a long chain of blastozooids, which it tows
about until these break away by sections (Fig. 3.25).

The pelagic colonial *Pyrosoma* of warm seas consists
of a number of individuals associated to form an
elongated barrel-shaped colony. The mouths open
outwards and the atria inwards into a single cavity with
a terminal outlet from which a continuous jet emerges.

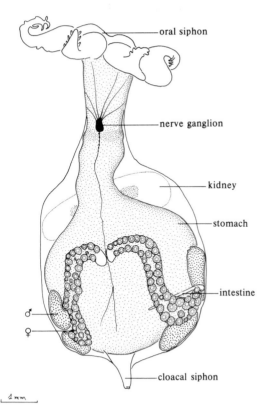

oral siphon

nerve ganglion

kidney

stomach

intestine

♂
♀

cloacal siphon

1 mm

FIG. 3.23. Another abyssal tunicate, *Gasterascidia*. These animals are carnivorous and have a huge stomach; they lack a branchial sac and are hermaphrodite. (Monniot and Monniot 1978.)

The mode of budding from the epicardium and other features suggest an affinity with *Doliolum* and *Salpa*, but *Pyrosoma* also resembles the ascidians in that its zooids are all sexual and capable of budding. The yolky eggs develop within the parent, without forming a larva. The outstanding characteristic of the creatures is the powerful light that they shine. This is produced in photogenic organs on each side of the pharynx. The photogenic cells contain curved inclusions about $2\,\mu m$ in diameter (Fig. 3.26). These are considered by some to be symbiotic luminescent bacteria, but this is doubtful. The light is so powerful that when large masses of *Pyrosoma* occur together the whole sea is illuminated sufficiently to allow reading a book. A remarkable feature of the phenomenon is that the light is not produced continuously but only when the animal is stimulated, as by the waves of a rough sea. If one individual is stimulated others throughout the colony may show their lights, but the mechanism of this effect is not known and the groups of cells that form the luminescent organs receive no nerves. Other types of animal with luminescent bacteria emit light continuously. The sudden flashes of light probably serve as a dymantic reaction (p. 228), giving protection against enemies by producing a flight-reaction in the same way as do sudden manifestations of colour or black spots by other animals. It has been observed in the laboratory that colonies of *Pyrosoma* that are dying and do not light up may be eaten by fishes, whereas any that light up when seized may then be dropped.

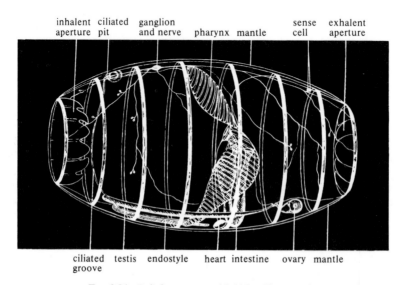

inhalent   ciliated   ganglion   sense   exhalent
aperture   pit   and nerve   pharynx   mantle   cell   aperture

ciliated   testis   endostyle   heart   intestine   ovary   mantle
groove

FIG. 3.24. *Doliolum*, gonozooid. (After Neumann.)

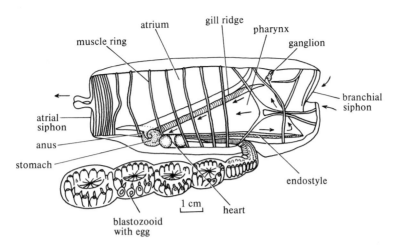

FIG. 3.25. *Cyclosalpa affinis*, oozooid with chain of five wheels of blastozooids. (In Berrill 1950; after Ritter and Johnston.)

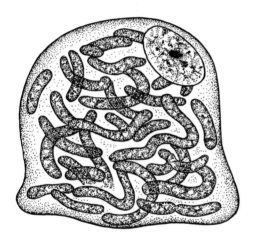

FIG. 3.26. Photogenic cell of *Pyrosoma*. (After Kukenthal.)

9. Class Larvacea

The Larvacea (Figs. 3.27 and 3.28) are small neotenous tunicates that are a very common and important component of the plankton since they feed by filtering minute nanoplanktonic organisms. Instead of a test, each individual builds a 'house', by secretion from a special part of the skin, the 'oikoplastic epithelium'. The tail is a broad structure held at an angle to the rest of the body; its movement produces a current in which the food is carried and caught by a most elaborate filter arrangement in the house (Fig. 3.27). Water enters the house by a pair of posterior 'filtering windows' and is passed through a system of filter pipes and surfaces in the part of the house in front of the mouth. The minute flagellates of the nanonplankton are stopped by these pipes, and at frequent intervals the tail beat stops and

the food particles are sucked back to the mouth. The surface of the feeding filter is a cellular secretion with very regular pores (Fig. 3.29). The filter becomes clogged after a few hours, the house is then abandoned and a new one secreted.

The pharynx has two gill slits, also an endostyle and peripharyngeal bands. The general organization is that of a typical ascidian tadpole, and there can be no doubt that these forms have arisen from other tunicates by the acceleration of the rate of development of the alimentary organs and gonads so that the metamorphosis and normal adult stage are eliminated. This may, of course, have happened long ago, so that the modern Larvacea are not closely related to any living forms, but the fact that they differ in many ways from known ascidian tadpoles does not invalidate the hypothesis; it would be expected that many special features would be developed during evolution after the paedomorphosis. Garstang (1928), however, believed that there is sufficient evidence to show that the Larvacea are related to the Doliolidae and suggested an ingenious hypothesis by which the

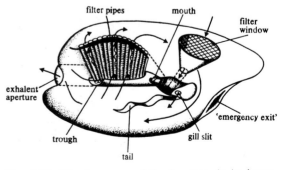

FIG. 3.27. *Oikopleura*, one of the Larvacea, in its house, showing the feeding-currents. (After Garstang 1928.)

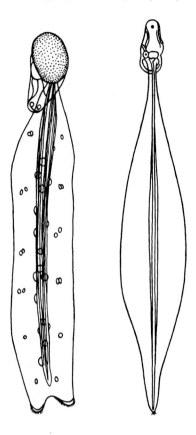

appendicularian home could be derived from the doliolid test, the animal itself remaining attached at the front end by gelatinous threads, which came to make the filter tubes (Fig. 3.30).

The tail is a highly developed organ, serving for locomotion, nutrition, and in the building of the house. It has a wide, continuous fin and is supported by a notochord of 20 cells. Bands of 10 large striped muscle cells extend down each side, giving an appearance that has been compared with metameric segmentation. The small number of the cells makes any such comparison very difficult. Moreover, the muscles are not developed from anything resembling myotomes. The nerve cord is a hollow tube with ganglionic thickenings, each containing one to four nerve cells. From these cells fibres proceed to the muscles and to the skin in a series of roots that usually remain separate, the motor being more dorsal.

The sense organs of *Oikopleura* consist of two remarkable mechanoreceptors on the sides of the trunk. These Langerhans cells bear very long non-motile cilia (70 μm). These are unique among tunicates in being secondary sense cells. They do not have their own axons but receive gap junction synapses from fibres originating in cells of the ganglion, which also make contact with neighbouring epithelial cells. Impulses can be set up in these axons either by mechanical stimulation of the Langerhans cells or by conduction of electrical impulses through the skin of the animal. These impulses produce rapid bursts of swimming movements (Bone and Ryan 1979).

FIG. 3.28. *Appendicularia* seen from the side and from below. (Up to 4 mm in length.) (After Lehmann.)

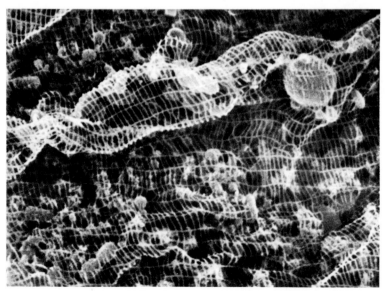

FIG. 3.29. The feeding filter of *Oikopleura* seen in detail with the scanning electronmicroscope. (Field width 6 μm.) (Flood, Per R. (1978). *Experientia* **34**, 173–5.)

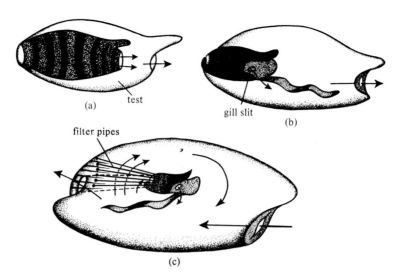

FIG. 3.30. Sequence of stages by which the Larvacea may have been evolved from a doliolid type. (a) Thaliacean type of individual in its test. (b) Paedomorphosis has occurred so that a tailed creature is found in the test. (c) The tadpole has moved away from the inhalent aperture, leaving a series of threads that become the filter pipes, the inhalent aperture becoming exhalant and vice versa. (After Garstang 1928.)

10. The formation of the chordates

We can now recapitulate the points that we have established about the origin of the chordates and attempt to piece together the evidence to show the sequence of events that led to the production of a free-swimming fish-like animal (see Bone 1979). The hemichordates and acraniates are related to the echinoderms. This is established by the similarities in early development (cleavage, gastrulation, mesoderm formation) (p. 46). The members of all three groups have three separate coelomic cavities arising by out-pushing from the endoderm (enterocoely). The most anterior coelom on the left side develops a pore, associated with a pulsating vesicle. These conditions of the coelom surely indicate common ancestry, moreover the premandibular coelom of some craniates also opens by a pore (p. 55). The tunicates have no coelom but their development is strikingly like that of the other three groups. It is interesting that tunicates possess lymphocytes and show immunity reactions (p. 64). No data are available for hemichordates but rejection of allografts and xenografts is reported from echinoderms (Marchalonis 1977). The larva of enteropneusts is very like the dipleurula, and there are many other points of general morphological and biochemical similarity between early chordates and echinoderms, especially the arrangement of the nervous system and presence of a mesodermal skeleton.

The echinoderms we have to consider are not the modern starfishes and sea-urchins, which are relatively active animals, but their sessile Palaeozoic ancestors. These were sedentary, often stalked animals, the carpoids, cystoids, blastoids, and crinoids, feeding by ciliary action. Surviving animals of related phyla, such as Polyzoa Ectoprocta and Phoronida suggest that the ancestor for which we are looking may have possessed a ciliated lophophore for food collecting. For purposes of dispersal its life-history presumably included a larval stage with a longitudinal ciliated band, similar in plan to that of the auricularia.

Granted that chordates are related to echinoderms we have to consider how the transition took place. There has recently been an attempt to show that certain plated echinoderm fossils, known as cornutes (Cornuta), were ancestral to the earliest armoured fishes. The conventional view is that these cornutes were stalked creatures like other carpoids (Fig. 3.31). Jefferies (1975), however, considers that they lay flat upon the muddy bottom and crawled along by means of the 'stalk'. He claims to have found evidence of gill slits, notochord, dorsal nerve cord, and even a fish-like brain and spinal cord. Only paleontologists can evaluate these claims but Figure 3.31 gives a reconstruction of one of these animals and Figures 3.31A, 3.32 and 3.33 provide some indication of the evidence (Jefferies 1973). Probably most workers would feel that too much interpretation is involved. The 'brain' is much more fish-like than one would expect in the Cambrian. The plates of the skeleton of echinoderms are all formed of calcite

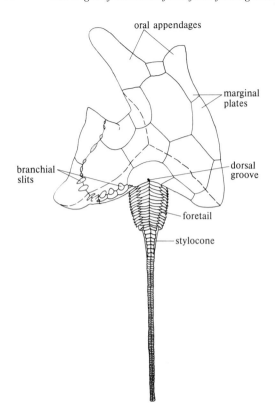

oral appendages

marginal plates

branchial slits

dorsal groove

foretail

stylocone

FIG. 3.31. Reconstruction of the Middle Cambrian fossil *Ceratocystis perneri*. The seven regular holes on the left side of its body are supposed to be gill slits. (Jefferies, R. P. S. (1969). *Palaeontology* **12**, 494–535.)

FIG. 3.31A. *Cothurnocystis*. Latex cast of a fossil from the Upper Ordovician, Scotland. (Published with permission of the Trustees of the British Museum (Natural History) and Dr R. P. S. Jefferies.)

whereas chordate skeletons are phosphatic. However, the view of the series of holes as gill slits seems plausible and would indicate that filter-feeding arose in the adult stage. *Cephalodiscus*, which is in some ways the most primitive of surviving chordates, has gill slits as well as a lophophore. This suggests that the pharyngeal mechanism was substituted for the lophophore as a means of feeding in the adult stage. Graptolites, though known only from their fossil skeleton, were probably somewhat like *Cephalodiscus* in the Cambrian and Permian. There are other possible interpretations. It has been suggested that *Cephalodiscus* was derived from a larval enteropneust (Burdon-Jones 1953). However, it is possible that ciliary mechanisms developed in the pharynx first to deal with food collected outside by tentacles or proboscis. Later the pharynx became developed into a self-contained feeding mechanism, making unnecessary the tentacles, which provide a tempting morsel for predators. The adoral band of cilia of the auricularia probably serves to carry food into the mouth, and for

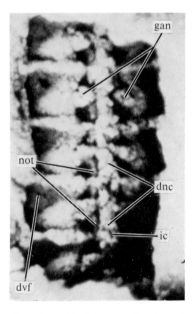

gan

not

dnc

ic

dvf

FIG. 3.32. Natural mould of the ventral surface of *Mytrocystites* showing dorsal ossicles of hind stem, giving a representation of the soft parts. *dnc*, 'dorsal nerve cord'; *dvf*, dorsoventral facet; *gan*, 'ganglion', *ic*, interossicular canal; *not*, 'notochord'. (Jefferies 1973.)

Fig. 3.33. *Lagynocystis*. The branchial apparatus in the front wall of the median atrium showing the anterior aspect of the branchial bars and their hollowness. *br but*, branchial buttress; *calc br b*, branchial bar (calcitized); *h br b*, hollow branchial bar of the median atrium; *imb st*, imbrication step for integument plates; *rect*, rectum. (Jefferies 1973.)

this purpose it is actually turned in to the floor of the pharynx. Garstang (1894) suggested that the endostyle has been derived from this loop of the adoral band.

The pharyngeal method of food collecting thus replaced the tentacles in the adult and the whole apparatus of an endostyle and an atrium to protect the gills became developed. We may notice here the remarkable similarity of this arrangement of the pharynx in tunicates, amphioxus, and cyclostome larvae, and the partial similarity in *Balanoglossus*.

The tunicates show us a stage in which branchial feeding has fully replaced tentacle feeding in a sessile adult. But they have a larva that is beyond all question a fish-like chordate. If the adult tunicate has evolved from a modified lophophore-feeding creature, how has the ascidian tadpole arisen from the auricularia larva? Garstang's auricularia theory, first propounded in 1894, provides a possible answer. As a ciliated larva grows its means of locomotion becomes inadequate because the ciliated surface increases only as the square of the linear dimensions, the weight as the cube. Muscular locomotion is not subject to this difficulty, and some of the starfish larvae actually show flapping of the elongated processes, movements that presumably assist them to remain at the surface. Garstang suggests that the fish-like form arose by development of muscles along the sides of the elongated body, the ciliated bands being

pushed upwards and eventually rolled up with their underlying sheets of nerve plexus to form the neural tube. The adoral ciliated band might then well be the endostyle (Fig. 3.34).

This theory may seem at first sight fantastic. It is necessarily speculative, but it has certain strong marks of inherent probability. It violates no established morphological principles and certainly enables us to see how a ciliated auricularia-like larva could be converted by progressive stages into a fish-like creature with muscular locomotion, while the adults, at first sedentary, substituted gill slits and endostyle for the original lophophore. The alternative is to suppose that the ascidian tadpole arose as a purely tunicate development, providing sufficient receptor and muscular organs to allow for the finding of suitable sites on the bottom (Berrill 1955).

We may plausibly regard the adult tunicate organization as directly derived from that of sessile lophophore-feeding creatures, and the larval organization as descended from an echinoderm-like larva. There is no need, on this view, to regard the sessile adult tunicate as a 'degenerate' chordate. The problem that remains is in fact not 'How have sea-squirts been formed from vertebrates?' but 'How have vertebrates eliminated the sea-squirt stage from their life-history?' It is wholly reasonable to consider that this has been accomplished

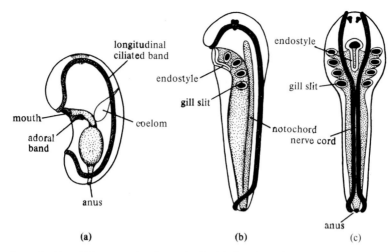

FIG. 3.34. To show the method by which a protochordate animal might have been derived from an echinoderm larva such as the auricularia. *a*, Auricularia in side view; *b*, protochordate in side view; *c*, same, dorsal view. (After Garstang 1928.)

by paedomorphosis. Advance of the time of development of the gonads relative to that of the soma is well known to occur in certain special cases such as the axolotl. The example of the Appendicularia shows that a similar process can happen among tunicates! Various workers have stressed the differences between the ascidian tadpole and the adult appendicularian, in attempts to show that the two are not comparable. But the differences, though considerable, are superficial: the similarity of organization is profound. Any sensible biologist with an understanding of the way in which the characteristic forms of animals arise by change in the rate and degree of development of features can see how the Appendicularia may represent modified ascidian larvae.

The appendicularians do, indeed, carry certain characters of the 'adult' sea-squirt, in particular they have gill slits, though of simple form. Nothing is more likely, however, than that some features of the sessile adult would be adumbrated in its larva and capable of fuller development therein if advantageous. Larva and adult, it must be remembered, possess the same genotype; the remarkable feature in all animals with metamorphosis is the difference between the two stages, not the similarity. Any characteristic may appear at either larval or adult stage or be transferred by evolutionary selection from one to the other. There is no serious objection to the view that the early adult free-swimming chordates arose by paedomorphosis of some tunicate-like metamorphosing form. If the creatures abandoned the habit of fixation it would be possible for characters previously present separately in larva and adult to become combined in a single stage. This is indeed what has happened in the Appendicularia.

Strangely enough, one of the chief difficulties of this theory is to find the position of the enteropneusts. Since the larva is still in the ciliated-band stage there should be no sign of organs characteristic of the muscle-swimming, fish-like pro-chordate. Yet such signs are present in the adult *Balanoglossus*; there is a hollow nerve cord and a sort of notochord. These features seem to suggest that the group has at one time possessed a free-swimming, fish-like stage. An alternative is that these features first arose as adaptations to adult life. The only part of the nerve plexus to be rolled up in *Balanglossus* is in the collar, where there are giant nerve cells concerned in the control of movement. Bone (1979) suggests that the folding is a special development to protect these cells. Perhaps the 'notochord' is also a special development for the attachment of the muscles of the proboscis. So it is possible that gill slits, hollow nerve cord, and notochord all arose first in the *adult* stage and the capacity to produce them were later incorporated into the free-swimming larva. Gill slits are formed in some tornaria larvae before metamorphosis begins.

There is strong reason to suppose that what we may call the Bateson (1886)–Garstang (1928) theory of the origin of chordates is correct. There is little doubt that chordates are related to the sessile lophophore-feeding type of creature rather than to any annulate, and we can reconstruct the course of events by which the lophophore-feeder may have come to have a pharynx with gill slits and its larva to have muscles, and later a notochord and a nerve tube. Then by paedomorphosis the sessile stage disappeared and the free chordates began their course of evolution. There are some reasons for supposing that a type such as amphioxus could have

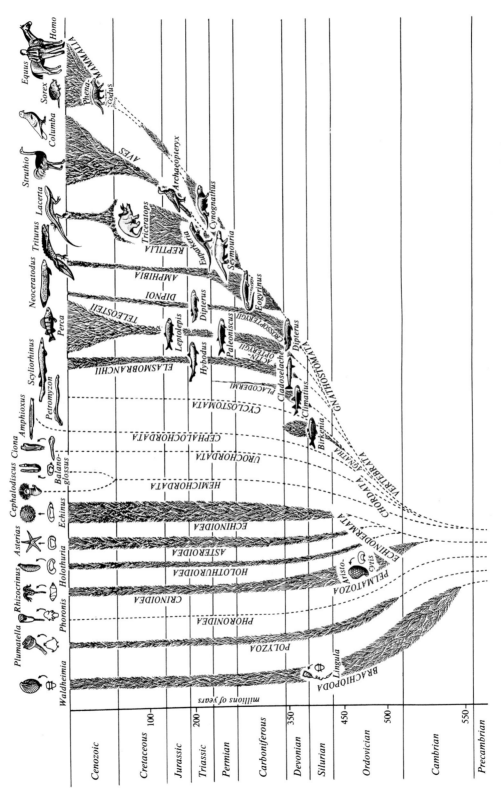

FIG. 3.35. Table to show the probable times of origin and affinities of the chordates and related groups of animal.

been derived from a creature not distantly related to the simpler Appendicularia and this in turn from a neotenous doliolid or some similar ancestral type.

We need not, however, follow the theory into its details, which are speculative. The whole treatment provides a conspicuous example of close morphological reasoning, allied with proper consideration of general biological principles, and establishes with some probability the main outlines of the origin of our great phylum of active creatures from such humble sedentary beginnings.

Can we see in the production of the first fish-like creatures clear signs of an 'advance' in evolution? In acquiring the power of active muscular locomotion the animals became able to live and feed in a variety of habitats, either at the sea surface or on the bottom. Forms with a sedentary adult stage are limited by the necessity for the presence of a sea bottom of suitable character. The larvae were evolved to make sure of reaching such conditions. But whereas suitable situations on the bottom are not common, and are liable to change, the sea surface provides a generalized habitat in which there is often abundant food, though no doubt also strenuous competition for it. Paedomorphosis in this case, as in others, allows the race to eliminate from its life-history the stage passed in a 'special' environment, which is difficult to find. Although the fish-form that was thus produced proved to have great possibilities for further evolution, the change was not at first a strikingly progressive one (Fig. 3.35). The surface of the sea is perhaps the most general of all environments; possibly it was the seat of the origin of life. Races that have devised means of living on the sea bottom may therefore be said to have advanced, because they have invaded a more difficult habitat. To abandon the sedentary life might in this sense be regarded as a retrograde step. The peculiar feature of the early fishes, however, was that they developed powers of active movement in a relatively large organism provided with efficient receptors, and by making use of the feeding mechanism developed at first by the bottom-living adult were able to live successfully at the sea surface. They acquired their dominance at this stage not by invading new habitats but by developing effective means of living in the richly populated plankton zone.

4 The vertebrates without jaws. Lampreys

1. Classification

Phylum Chordata
Subphylum 4: Vertebrata (= Craniata)
    Superclass 1: AGNATHA (without jaws) (= Cyclostomata)
    Class 1: Pteraspidomorphi (fin-shield) (= Diplorhina (double nose))
        Subclass 1: *Heterostraci (different shells). Upper Cambrian – Upper Devonian
          *Pteraspis; *Psammosteus; *Eglonaspis; *Anetolepis
        Subclass 2: *Thelodonti (nipple-teeth) (= Coelolepida). Lower Silurian – Middle Devonian
          *Thelodus; *Lanarkia
    Class 2: Cephalaspidomorphi (head shield form) (= Monorhina (single-nose))
        Order 1: Osteostraci (bone-shell) (= Cephalaspida). Upper Silurian – Lower Devonian
          *Hemicyclaspis; *Kiaeraspis; *Cephalaspis
        Order 2: Anaspida (no shield). Upper Silurian – Upper Devonian
          *Birkenia; *Jaymoytius
        Order 3: Galeaspida (helmet shield). Devonian
          *Galeaspis
        Order 4: Cyclostomata (round-mouth). Upper Carboniferous – Recent
          Suborder 1: Petromyzontidae (stone-sucker)
            Petromyzon; Lampetra; Ichthyomyzon; Geotria; Mordacia; *Mayomyzon
          Suborder 2: Myxinoidea (slimy one)
             Myxine; Bdellostoma; Eptatrietus
    Superclass 2: GNATHOSTOMATA (jaw mouth)

*Fossil forms.

2. General features of vertebrates

All the remaining chordates are alike in possessing some form of cranium and some trace of vertebrae; they make up the great subphylum Vertebrata, also called Craniata. The organization of a vertebrate is similar to that of amphioxus, but with the addition of certain special features. A few of these novelties may now be surveyed, with emphasis on those that provide the basis for the capacity to live in difficult environments that is so characteristic of the vertebrates. Firstly the front end of the nervous system is differentiated into an elaborate brain, associated with special receptors, the nose, eye, and ear. Through these receptors the vertebrates are able to respond to more varied aspects of the environment than are any other animals. Some of them have the ability to discriminate between visual shapes and colours, and in the auditory field between patterns of tones, also between a host of chemical substances. The motor organization allows the performance of delicate movements to suit the situations that the receptors reveal. The swimming process, by the contraction of a series of muscle blocks (myotomes), is itself perfected by improvements in the shape of the fish, allowing rapid movements and turns. Besides the median fins there develop lateral paired ones, serving at first a stabilizing and steering function and then converted, when the land animals arose, into organs of locomotion on the ground or in the air and finally, in the shape of the hands, into a means of altering the environment to suit the individual.

The brain itself, at first mostly devoted to the details

of sensory and motor function, comes increasingly to preside, as it were, over all the bodily functions, and to give to the vertebrates the 'drive' that is one of their most characteristic features. The skull is developed as a skeletal thickening around the brain, probably at first mainly for protection, but later serving for the attachment of elaborate muscle systems. The study of vertebrates is especially identified with study of the skull, because in so many fossils this is the only part preserved.

The food of the earliest vertebrates was collected by ciliary action, but this habit has long been abandoned and only in rare cases today (such as plankton-feeding fishes, flamingoes and some whales) does the food consist of minute organisms. The pharynx of most vertebrates is small, there are relatively few gill slits and these are respiratory. In all except the most ancient forms the more anterior of the arches between the gills became modified to form jaws, serving not only to seize and hold the food but also to 'manipulate' the environment.

The blood system shows two of the most characteristic vertebrate features, namely, the presence (1) of a heart that has at least three chambers and thus provides a rapid circulation, and (2) of haemoglobin within corpuscles, serving to carry large amounts of oxygen to the tissues. The efficiency of this system must have been a major factor in producing the dominance of the vertebrate animals. In the air-breathing forms, and especially the warm-blooded birds and mammals, the respiratory and circulatory systems allow the expenditure of great amounts of energy per unit mass of animal, so that quite extravagant devices can be used, allowing survival under conditions that would otherwise not support life.

The excretory system consists of mesodermal funnels, leading primarily from the coelom to the exterior. It may be that this type of kidney arose in connection with the abandoning of the sea for fresh water (p. 83). Probably all but the earliest vertebrates have passed through a freshwater stage, and it is significant that all except the hag fish, *Myxine*, have less salt in their blood than there is in sea water. Elaborate devices for regulation of osmotic pressure have been developed, and the mesodermal kidneys play a large part in this regulation.

This outline only gives a few suggestive features of vertebrate organization. The details differ bewilderingly in the various types and it is our business now to survey them. In the earliest forms the special mechanisms are absent or at least function only simply. Passing through the vertebrate series we find more and more devices adopted, along with more and more delicate co-ordination between the various parts, culminating in the

extremely highly centralized control of almost every aspect of life that is exercised by the mammalian cerebral cortex.

3. Agnatha

The earliest vertebrates, while showing most of the characteristic features of the group, differ from the rest in the absence of jaws and are therefore grouped together in a superclass Agnatha (without jaws), distinguished from the remaining vertebrates, which are therefore called Gnathostomata (jaw-mouthed). The only living agnathous animals are the Cyclostomata (ring-mouthed) lampreys and hag fishes, but the first vertebrates to appear in the fossil series, mostly heavily armoured and hence known as 'ostracoderms' (bonyskinned), found in Cambrian, Ordovician, Silurian, and Devonian strata, also show the agnathous condition, and have some other features in common with the Cyclostomata (p. 107). This group of agnathous vertebrates shows some interesting experimentation in methods of feeding, before the jaw-method became adopted. The modern cyclostomes are parasites or scavengers, in the adult state, but as larvae the lampreys still feed on microscopic material, using an endostyle resembling that of amphioxus in many ways, but making use of muscular contraction rather than ciliary action to produce a feeding current. The methods of feeding of the ostracoderms probably included shovelling detritus from the bottom.

The Cyclostomata are therefore worth special study as likely to show us some of the characteristics possessed by the earliest vertebrate populations.

4. Lampreys

The most familiar cyclostomes are the lampreys, of which there are various sorts found in the temperate zones of both hemispheres (see Hardisty and Potter (1971, 1972) for the biology of lampreys). All lampreys have a life-history that includes two distinct stages: the ammocoete larva lives in fresh water, buried in the mud, hence its name meaning sand-sleeper. The larva is microphagous but the adult lamprey has a sucking mouth, and if it feeds at all it is parastic on other fishes. *Lampetra fluviatilis*, the lamprey (Fig. 4.1), is a typical example, common in Great Britain. The adult is an eel-like animal about 30 cm long and lives parasitically in the sea, probably for about $2\frac{1}{2}$ years. It then migrates to fresh water where it breeds. The eggs develop into the ammocoete larva, which lacks the sucker and lives buried in the mud, feeding on micro-organisms for about five years. After metamorphosis the young adult migrates downstream to the sea. Species with these double migrations are said to be anadromous. In some

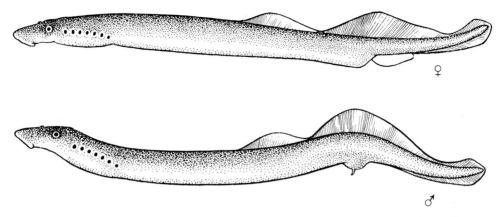

FIG. 4.1. Brook lampreys, *Lampetra planeri*. Ripe female with anal fin fold, and ripe male. Note shape of dorsal fin and presence of copulatory papilla in male. (Curves due to fixation.)

lampreys the adult phase of parasitism in the sea has been lost (p. 103).

The adult *Lampetra fluviatilis* is dark on the back, and white below. The surface is smooth, with no scales. The skin is many-layered (Fig. 4.2). The outermost cells have a striated cuticular border. Mixed with these epithelial cells the lamprey, like most aquatic vertebrates, has many gland cells used for producing slime. It has been claimed that the skin and flesh of lampreys produce toxins and may produce gastro-intestinal poisoning in humans. Could this be the origin of the tradition that King John died of a surfeit of lampreys in 1216? In any case anglers often use lampreys for bait. The skin secretions may also have an antibacterial action, protecting the larva when buried in the mud. Below the epidermis lies the dermis, a layer of bundles of collagen and elastin fibres, running mostly in a circular direction. This tissue is sharply marked off from a layer of subcutaneous tissue containing blood vessels and fat, as well as connective tissue. There are pigment cells in the

dermis and a thick layer of them at the boundary of dermis and subcutaneous tissue. The chromatophores are star-shaped cells whose pigment is able to migrate, making the animal dark or pale. This change is especially marked in the larva and is produced by variation in the amount of a pituitary secretion (p. 94).

The head of the lamprey bears a pair of eyes and a conspicuous round sucker. On the dorsal side is a single nasal opening, and behind this there is a gap in the pigment layers of the skin through which the third or pineal eye can be seen as a yellow spot. There are seven pairs of round gill openings, which, with the true eyes (and some miscounting or perhaps inclusion of the nasal papilla), are responsible for the other name 'nine eyes'. There is no trace of any paired fins, but the tail bears a median fin, which is expanded in front as a dorsal fin. There are sex differences in the shape of the dorsal fins of mature individuals and the female has a considerable anal fin-fold, not supported by fin rays (Fig. 4.1).

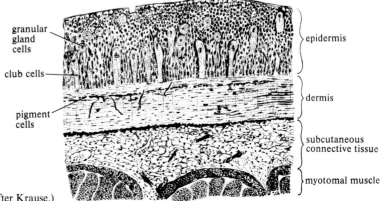

FIG. 4.2. Section of skin of lamprey. (After Krause.)

The lamprey swims with an eel-like motion, using its myotomes in the serial manner that has been mentioned in amphioxus and will be discussed later (p. 115). The waves that pass down the body are of short period relative to the length, so that the swimming is mechanically inefficient; lampreys show great activity, but their progress is not rapid. The animal often comes to rest, attaching itself with the sucker to stones (hence the name, 'suck-stone') or to its prey. In this position water cannot of course pass in through the mouth, but both enters and leaves by the gill openings. When swimming the backward jet of water may assist in locomotion.

The trunk musculature consists of a series of myotomes separated by myosepta (myocommas). Each myotome has a $\gg$-shape, instead of the simple $\gg$ of amphioxus. The muscle fibres run longitudinally and they are striped, but of a somewhat peculiar fenestrated type. In both lampreys and hag fishes there are two types of muscle fibre, as in higher fishes, probably fast and slow in action, innervated by large and small nerve fibres, respectively (p. 34). There are also sensory fibres, probably proprioceptive, at the outer edges of the myosepta.

5. Skeleton of lampreys

The skeleton of lampreys consists of the notochord and various collections of cartilage. This latter is partly of the typical vertebrate type, that is to say, consists of large cells in groups, separated by a matrix of the protein chondroitin sulphate, which they secrete. In other regions a tissue containing more cells and less matrix is found, the so-called fibrocartilage, and this more nearly resembles fibrous connective tissue and serves to emphasize that no sharp line can be drawn between these tissues. There is also, in the larva, a tissue known as mucocartilage, which is an elastic material serving more as an antagonist to the muscles than for their attachment.

The notochord remains well developed throughout life as a rod below the nerve cord. It consists of a mass of large vacuolated cells, enclosed in a thick fibrous sheath (Fig. 4.3). The rigidity of the whole rod depends on the turgor of the cells and it often collapses completely in fixed and dehydrated material (Fig. 4.15). No doubt in life it serves, like the notochord of amphioxus, to prevent shortening of the body when the myotomes contract.

The notochordal sheath is continuous with a layer of connective tissue, which also surrounds the spinal cord and joins the myosepta and thus eventually the subcutaneous connective tissue. Within this connective tissue there develop certain irregular cartilaginous thickenings that are of special interest because they may be

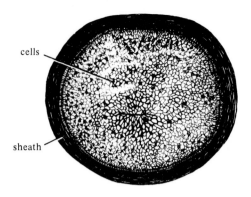

cells

sheath

Fig. 4.3. Transverse section through notochord of lamprey. (After Krause.)

compared with vertebrae, perhaps with the basidorsal element (p. 115). They lie on either side of the spinal cord (Fig. 4.4), that is to say, above the notochord, and consist either of one nodule on each side of the segment, through the middle of which the ventral nerve root emerges, or of two separate nodules, with the nerve between them. These protovertebrae contain haemopoietic tissue (p. 83). Rods of cartilage extend dorsally and ventrally into the fins, but are not attached to the 'vertebrae'.

The lamprey skull shows, even in the adult, the basic arrangement found only in the embryo of higher vertebrates. The floor is formed of paired parachordals on either side of the notochord and in front of this paired trabeculae. Attached to this base is a series of incomplete cartilaginous boxes surrounding the brain and organs of special sense (Fig. 4.5). To this skull is attached the skeleton that supports the sucker and gills. The arrangement of the skull differs considerably from that of later vertebrates. The cranium has a floor around the end of the notochord, and in front of this there is a hole containing the pituitary gland. The side walls are strong but the roof is composed only of a tough membranous fibrocartilage. The auditory capsules are compact boxes surrounding the auditory organs at the sides. The olfactory capsule, imperfectly paired, is also almost detached from the cranium. Other ridges of cartilage lie below the eyes and there is a complex support for the sucker.

The skeleton of the branchial region consists of a system of vertical rods between the gill slits, joined by horizontal bars above and below them. This cartilage lies outside the muscles and nerves and is therefore difficult to compare with the branchial skeleton of higher fishes, which lies in the wall of the pharynx. The elastic action of the cartilages produces the movement of inspiration. A backward extension of the branchial basket forms a box surrounding the heart.

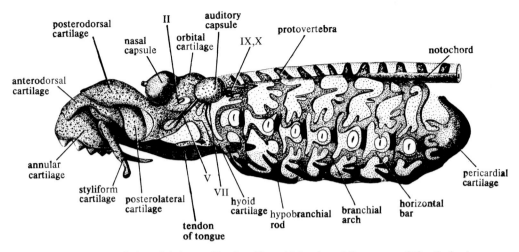

FIG. 4.4. Lateral view of skeleton of head and branchial arches of *Petromyzon*. (After Parker.)

6. Alimentary canal of lampreys

The sucker is bounded at the edges by a series of lips, which besides being sensory serve also to make a tight attachment when the lamprey sucks (Fig. 4.6). In the sucker are numerous teeth, whose arrangement varies in the different types of lamprey. These teeth are horny epidermal thickenings, supported by cartilaginous pads, and are therefore not comparable with the teeth of

vertebrates, which are derived mainly from mesodermal tissues (Fig. 4.7). The caps of keratin are replaced at frequent intervals, keeping the teeth sharp during the parasitic phase. They become blunter before spawning. Sharper and larger teeth are borne on a movable tongue, which is used as a rasp (Fig. 4.8).

An annular muscle runs round just above the lips of the sucker and presumably serves to narrow the margin and hence to release the fish. The remaining muscles are

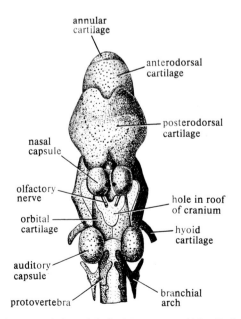

FIG. 4.5. Dorsal view of skull of *Petromyzon*. (After Parker.)

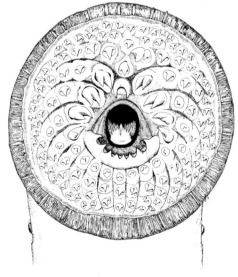

FIG. 4.6. The sucker of *Petromyzon marinus* showing the dentition in the feeding stage: the outer circular lip, the teeth, and at the centre the tongue, also with teeth. (Hubbs and Potter 1971.)

mostly attached to the tongue and base of the sucker. The largest of these muscles, the cardioapicalis muscle, is attached posteriorly to the cartilage surrounding the heart and in front is prolonged into a conspicuous lingual tendon, which is attached to the tongue and serves to pull it backwards (Fig. 4.8). Presumably the action of this muscle deepens the oral cavity and is thus the main agent securing attachment of the sucker. There is a collar of circular fibres around the front end of the cardioapicalis muscle, serving to lock the tendon and maintain the suction. Dorsal and ventral to the main tendon are groups of muscles that rock the tongue up and down to produce a rasping action. The muscles of the sucker are all derived from the lateral plate and are innervated from the trigeminal nerve; their fibres are striated.

The mouth is a small opening above the tongue and leads into a large buccal cavity. At the hind end this divides into a dorsal passage, the oesophagus, for the

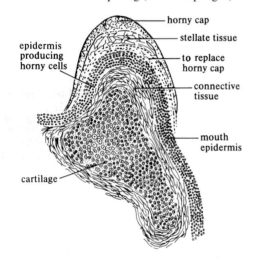

FIG. 4.7. Section through tooth of lamprey. (After Hansen, from Kükenthal.)

food, and a ventral respiratory tube, which leads to the gill pouches but is closed behind. At the mouth of the respiratory tube is a series of velar tentacles, corresponding exactly in position to those of amphioxus, and serving to separate the mouth and oesophagus from the respiratory tube while the lamprey is feeding. The seven branchial sacs are lined by a folded respiratory epithelium and surrounded by muscles, and these, together with the elastic cartilages and appropriate valves, ensure the pumping of the water tidally, in and out of the external openings. In front of the first sac is the remains of an eighth pouch, whose surface is not respiratory. The active movement is exhalation, followed by passive elastic recoil of the branchial basket. The uptake of oxygen is very low in ammocoetes, higher in the adult.

The 'salivary' or buccal glands are curious organs of which little is known. They are a pair of pigmented sacs, embedded in the hypobranchial muscles. Each has a folded wall, from which a duct proceeds forward to open below the tongue. The salivary glands produce a secretion that prevents coagulation of the blood of the fishes on which the lamprey feeds (Fig. 4.9). The nature of this secretion is not known, but it rapidly turns black on exposure to the air and the glands for this reason appear to be pigmented. It has been observed that in lampreys taken from fishes the intestine is filled with red corpuscles, and there is therefore no doubt that they feed mainly on the blood of their prey. However, there are reports of muscles, scales, and other materials in the gut and probably a lamprey will eat various parts of its prey and also scavenge dead material. Both vision and olfaction are used in finding the prey or other food. Once attached a lamprey remains so for several days and the fish often dies.

Following the opening of the Welland Ship Canal between Lake Ontario and Lake Erie in 1929 sea lampreys (*Petromyzon marinus*) have penetrated all the Great Lakes and become a very serious pest, reaching Lake Superior by 1946. Plankton and bottom-feeders

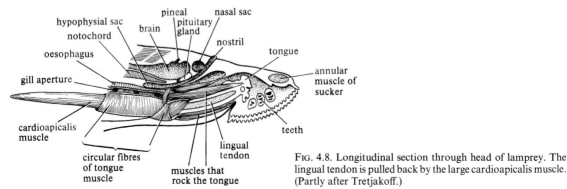

FIG. 4.8. Longitudinal section through head of lamprey. The lingual tendon is pulled back by the large cardioapicalis muscle. (Partly after Tretjakoff.)

FIG. 4.9. Brook trout (*Salvelinus fontinalis*) being attacked by recently transformed sea lampreys (*Petromyzon marinus*) in an aquarium. (By courtesy of R. E. Lennon in Hardisty and Potter 1971.)

were reduced and the two local carnivores Lake Trout (*Salvelinus*) and Burbot (*Lota*) almost eliminated. The primary food fishes, such as Chub (*Coregonus*) then showed a temporary increase when their predators disappeared but they in turn were depleted by the lampreys, first the larger then the smaller. A very small fish, the Alewife (*Alosa*) first appeared in 1949 and came to dominate the lakes. A Sea Lamprey Control Committee was established in 1946 and has reduced the predation by use of electric barriers to migration and chemical destruction of ammocoetes. Fish stocks have partly recovered by re-stocking and the operation is a successful experiment in fish population control.

The oesophagus (foregut) leads directly into a straight intestine (midgut); there is no true stomach in lampreys (Fig. 4.10). The surface of the intestine is increased by a fold, the typhlosole, running a somewhat spiral course. There is a liver, gall-bladder, and bile-duct of typical vertebrate plan, but no separate pancreas. However, in the wall of the anterior part of the intestine of the larva there are large patches of cells that resemble those of the acini of the pancreas of higher forms and contain secretory granules (Fig. 4.11). Barrington (1936, 1942) has shown that extracts of this region have a high proteolytic power, the enzyme being of the tryptic type, with its optimum between pH 7.5 and 7.8. Some of this tissue is collected in the walls of short diverticula, reaching forwards. The situation is therefore essentially similar to that found in amphioxus, and we may regard

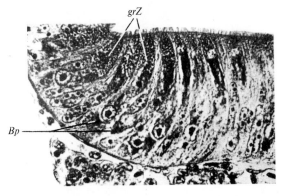

FIG. 4.11. A transverse section through the epithelium of the anterior part of the intestine from a larval lamprey to show the zymogen cells with large nucleoli and eosinophil granules (*grZ*). (*Bp*) basophil granules. (× 400.) (From Luppa, H. (1964). *Zeitschrift für mikroskopischanatomische Forschung* **71**, 85–113.)

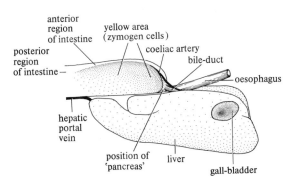

FIG. 4.10. Mid-gut of larval lamprey. (From Barrington 1942.)

these patches of zymogen cells, or the diverticula, as the forerunners of the exocrine portions of the pancreas. In the lampreys the endocrine portion, not yet identified in amphioxus, also appears. Around the junction of the foregut and intestine are groups of follicles that do not communicate with the lumen of the intestine. These 'follicles of Langerhans' (Fig. 4.12) were, appropriately enough, first seen by the discoverer of the islets in higher forms, and Barrington has shown that following destruction of this tissue by cautery there is a rise in blood-sugar. Moreover, after injection of glucose, vacuolation of the cells occurs. We may safely conclude that these cells are involved in carbohydrate metabolism, but only one type of cell is present, similar to the B cells known in mammals to produce insulin. No A cells, which produce glucagon, are present.

7. Blood system of lampreys

The blood vascular system is arranged on the same general plan as in amphioxus but there is a well-developed heart. This lies behind the gills and can be considered as a portion of the subintestinal vessel, folded into an S-shape and divided into three chambers. The heart is suspended in a special portion of the coelom, the pericardium, whose walls are supported by cartilage allowing the heart to act as a suction pump. In the larva the heart first appears as a straight tube and owing to an abnormality of development it sometimes fails to develop its S-shape. Contractions can nevertheless be seen in these abnormal hearts, passing from behind forwards along the straight tube. Similarly in the normal heart contraction proceeds in the chambers

from behind forwards. The most posterior chamber is a thin-walled sinus venosus, into which the veins pour blood. This leads to an auricle (atrium), also thin walled, lying above the sinus. The atrium passes blood into the ventricle below it, a thick-walled chamber, providing the main force for sending the blood round the body.

The heart receives nerve fibres from the vagus nerve and contains nerve cells, some of which give a chromaffin reaction (p. 139). The heart rate accelerates with stimulation of the vagus nerve or with acetylcholine, adrenaline, or noradrenaline. Reserpine causes the heart to stop, but acetylcholine then re-starts it. These properties show several differences from those of the hearts of other vertebrates (see *Life of Mammals*). In *Myxine* there are no nerves to the heart or nerve cells in it and acetylcholine has no effect.

Blood leaves the ventricle by a large ventral aorta, running forwards between the gill pouches, to which it sends a series of eight afferent branchial arteries. These break up into capillaries in the gills, and efferent branchial arteries collect to a pair of dorsal aortae, running backwards, which join and form the main dorsal aorta. This passes down the trunk and carries blood to all the parts of the body by means of series of segmental arteries and special vessels to the gut, gonads, and excretory organs. A curious feature is that many of these arteries are provided with valve-like cushions at the point at which they leave the main trunks (Fig. 4.13). It may be significant that such valves are not found where the efferent branchials join the dorsal aorta, nor at the points of exit of the renal arteries, so that perhaps the valves serve to reduce the pressure in the majority of the arteries, while leaving it high in those to the kidneys. The removal of large quantities of water is an important

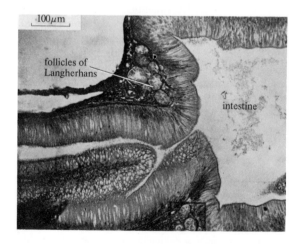

FIG. 4.12. A longitudinal section through the junction of the oesophagus and the anterior intestine showing the follicles of Langerhans embedded in the submucosa in the angle between. (From Morris, R. and Islam, D. S. (1969). *General and comparative Endocrinology* **12**, 72–80.)

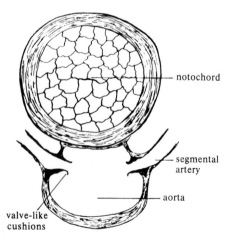

FIG. 4.13. Valve-like cushions at the origin of segmental arteries of a lamprey. (From Kükenthal, after Keibal.)

problem in all freshwater animals and is facilitated by a high pressure in the kidneys. This must be difficult to maintain in an animal with a branchial circulation and hence a double set of capillaries.

The venous system consists of a network of sinuses, with contractile venous hearts in various places. There is a large caudal vein, dividing where it enters the abdomen into two posterior cardinals. These run forward in the dorsal wall of the coelom, collecting blood from the kidneys, gonads, etc., and opening into the heart by a single ductus Cuvieri on the right-hand side, this being the remains of a pair found in the larva. Anterior cardinals collect blood from the front part of the body, and there is also a conspicuous ventral jugular vein draining venous blood from the muscles of the sucker and gill pouches. Besides the veins proper there is a large system of venous sinuses, especially in the head. Blood from the gut passes by a hepatic portal vein through a contractile portal heart to the liver, from which hepatic veins proceed to the heart.

The blood of lampreys, like that of all vertebrates except a few deep-sea fishes, contains the respiratory pigment haemoglobin, enclosed in corpuscles, here nucleated, large and spherical. The sequences of two haemoglobins have been identified and up to 35 per cent of the residues are similar to those of man. This arrangement immensely increases the oxygen-carrying power of the blood. Haemopoietic tissue occurs in the intestinal wall of the larva and this has been regarded by some as representing a 'proto-spleen'. In the adult the blood-forming tissue lies in the protovertebral arches and represents the bone marrow. White corpuscles resembling lymphocytes and polymorphonuclear cells occur, produced by lymphoid tissues in the kidneys and elsewhere. There is no distinct system of lymphatic channels. There are no compact thymus, spleen, or lymph nodes in the adult, but the larva has foci of lymphoid cells in the branchial region. This 'protothymus' undergoes involution at metamorphosis. In lampreys and hag fishes the lymphoid tissue responds to antigenic stimulation. They develop delayed allergic reactions and show allograft immunity and clear cellular and humoral immunological memory. Lampreys have an unusually simple immunoglobulin, not containing the usual light and heavy chains but probably forming a tetramere of four monomeric units without covalent bonds, and a rigid α-helix configuration.

8. Excretory system and osmoregulation in lampreys

The excretory and genital systems of vertebrates consist of a series of tubes opening from the coelom to the exterior and serving to carry away both excretory and genital products. This plan of organization is quite different from that found in amphioxus and represents a new acquisition by the vertebrates. It is not clear whether the excretory or genital component of the complex is the primary one, nor indeed why they are associated. The gonads develop from the walls of the coelom in all animals possessing that cavity; and some people hold that the coelom represents an enlargement of a sac that at first served purely as a gonad. Genital ducts leading from the coelom to the exterior are common in invertebrates, and we may guess that at their first appearance the urinogenital tubules of vertebrates served only for genital products.

The conversion of these tubules to excretory purposes may have been a result of the adoption of the freshwater habit. The blood of lampreys contains a higher concentration of salts than the surrounding water when they are in fresh water, and a lower concentration than that outside when in the sea. When the animals are in the river they must deal with the tendency for water to flow in. This water must be removed without losing salt; accordingly in most freshwater animals, including vertebrates, we find some system by which the separation can be achieved.

The region that gives rise to the kidney during development lies between the dorsal scleromyotome and the more ventral lateral plate mesoderm; it is known as the nephrotome. This tissue differentiates during development from in front backwards, making a series of segmental funnels, opening into a common archinephric duct (Fig. 4.14). The most anterior funnels open into the pericardium; usually there are four of these in a freshly hatched larva, opening into a single duct, which reaches back to an aperture near the anus. Close to each funnel there develops a tangle of blood vessels, the glomerulus (Figs. 4.15 and 4.16). Presumably the osmotic flow of water into the body is relieved by the pressure of the heart-beat forcing water out from the glomeruli into the coelomic fluid, whence it is removed by the funnels, with the aid of their cilia. The tubules become longer and twisted after hatching and may perhaps serve for salt reabsorption.

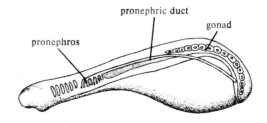

FIG. 4.14. Diagram to show arrangement of the pronephros in a freshly hatched lamprey. (After Wheeler.)

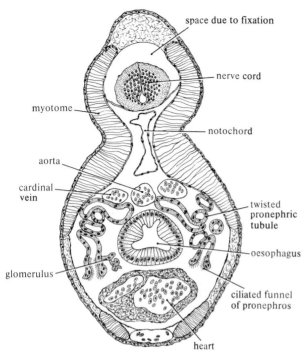

FIG. 4.15. Section through newly hatched larva of *Lampetra* behind the pharynx.

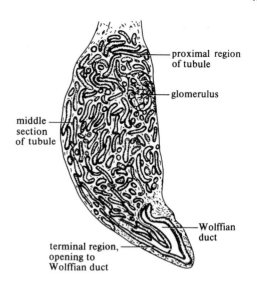

FIG. 4.17. Transverse section of kidney of *Lampetra*. (After Krause.)

These anterior funnels constitute the pronephros. As the animal grows they are replaced by a more posterior set, the mesonephros. There is, however, a gap of several segments in which no tubules appear (Fig. 4.17), a strange and unexplained discontinuity, common to all vertebrates. The pronephric tubules gradually disappear and finally in the adult all that remains of the organ is a mass of lymphoid tissue. Meanwhile the mesonephros develops as a much larger fold, hanging into the coelom and containing very extensive winding tubules. These do not open to the coelom (at least in the adult) but each to a small sac, the Malpighian corpuscle, which contains a portion of the coelom and the glomerulus. This is

obviously more efficient than a few funnels for allowing the heart to pump excess water out of the blood and down the tubules. The latter themselves have become greatly elongated and make up the main bulk of the organ (Fig. 4.17). The segmental arrangemnt is therefore much obscured and as extra glomeruli are added it disappears completely. The mesonephros extends at its hind end as the animal grows, until it forms the adult kidney, a continuous ridge of tissue reaching back to the hind end of the coelom. Besides the excretory apparatus the kidney also contains much lymphoid tissue and fat, and it probably plays a part in the formation and destruction of red and white corpuscles.

The mechanism of osmoregulation is remarkably similar to that of teleostean fishes, considering some 500 million years of independent evolution. In freshwater the excess of water that moves in osmotically is removed by passing a filtrate of the blood from the pronephric funnel or glomerulus, down a long kidney tubule in

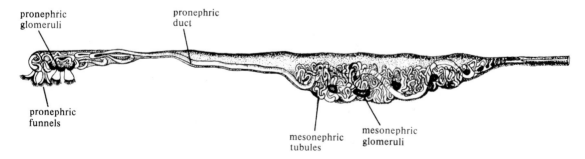

FIG. 4.16. Kidney system of a 22-millimetre larva of *Lampetra*. (After Wheeler.)

which the salts are re-absorbed by tall columnar cells with prominent microvilli. The urine is thus hypotonic to the blood, which maintains a concentration of NaCl of about 120 mM/l. Lampreys can also take up sodium and chloride, even from solutions much more dilute than the blood, by special ion-uptake cells (ionocytes) on the gills (Fig. 4.18). In the sea these cells (or others in a similar position) excrete ions, maintaining the blood hypo-osmotic to sea water. This is done by the same mechanism as in teleosts; water is swallowed and the monovalent ions are absorbed and then the excess of them removed by the ion excretory cells. Divalent ions are excreted by the gut or kidney. With this mechanism a specimen of *Petromyzon* was found to have 360 mM/kg of total ions, about a third of the value of the sea water. During the anadromous spawning migration the gut degenerates and ion-uptake cells appear. The animals can then no longer osmoregulate if returned to sea water.

9. Reproductive system of lampreys

The gonads are unpaired ridges medial to the mesonephros. Primordial germ cells, set aside very early in development, migrate into these ridges and develop into eggs or sperms. The differentiation of the gonad occurs relatively late in lampreys, so that in young ammocoetes the organ is 'hermaphrodite', containing developing oocytes and spermatocytes together. The ripe ovary consists of ova each surrounded by single-layered follicular epithelium, which finally ruptures and liberates the egg into the coelom, whence it escapes by pores to be described presently. The testis consists of a number of follicles containing sperms; it is unique among vertebrates in that the follicles have no ducts; when ripe they rupture into the coelom, which becomes filled with spermatozoa and these escape, like the ova, by pores.

These apertures by which the gametes escape are similar in the two sexes and consist of short channels, one on each side, leading from the coelom to the lower end of the kidney duct (Fig. 4.19). The channels normally open only a few weeks before spawning, but injections of oestrone or anterior pituitary extract will cause perforations of the ducts in young lampreys, indeed even in the ammocoete larvae.

Fertilization is external, but there are modifications of the cloaca in the two sexes to assist in ensuring fertilization and proper placing of the eggs in the 'nest' (p. 100). The lips of the cloaca of the ripe male are united to form a narrow penis-like tube. The cloacal lips of the female are enlarged and often red; in addition she has an anal fin-fold. These sex differences, which develop shortly before spawning, can also be initiated by injection of anterior pituitary extracts (p. 95).

10. Nervous system of lampreys

The nervous system of the cyclostomes is very much better developed than that of amphioxus and shows the

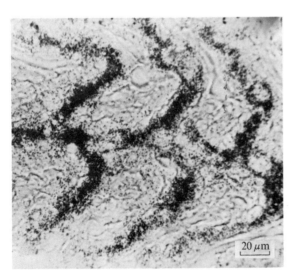

FIG. 4.18. Localization of ^{36}Cl in ammocoete gill demonstrated by autoradiography. Silver grains are located in blood vessels and cells at the base of the inner platelet area. (From Patinawin, S. (1967). Ph.D. thesis, University of Nottingham.)

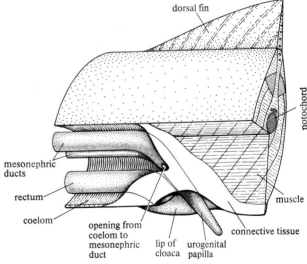

FIG. 4.19. Cloacal region of fully adult *Lampetra*. (After Knowles.)

characteristic plan that is present in all vertebrates. The essence of the vertebrate nervous organization may be said to be that it consists of large numbers of neurons and is highly centralized. The brains of vertebrates contain much larger aggregates of nervous tissue than are to be found in any other animals, except perhaps cephalopods, and this tissue produces by its actions the most characteristic features of vertebrate life. Vertebrates are active, exploratory creatures, and their behaviour is much influenced by past experience. The plan found in the lamprey provides an introduction to that of other vertebrates. As compared with amphioxus there has been a very high degree of cephalization. The front end of the spinal cord is enlarged into a complicated brain, and the nerves connected with a number of the more anterior segments have become modified to form special cranial nerves (for review see Rovainen 1979).

The spinal nerves differ from those of all other vertebrates in that the dorsal and ventral roots do not join. The ventral roots contain large motor fibres passing to the myotomes and also small fibres going to the myosepta and perhaps serving as proprioceptors. The dorsal roots consist largely of sensory fibres with bipolar cell bodies collected into dorsal root ganglia, including proprioceptor fibres from the myotomes. It is not known whether the dorsal roots also contain any efferent fibres. The autonomic nervous system is diffuse and shows some generalized and some special features. The gut is mainly innervated by the vagus, which extends far back along the intestine. There is little contribution of fibres from the spinal nerves to the alimentary canal, since this has no mesentery, being attached only at its cranial and caudal ends. There are, however, numerous fibres from the spinal nerves to the rectum, ureters, and cloacal region, and numerous postganglionic neurons are found here. Nerve cells are also found in the intestinal plexuses.

Isolated fibres running in both dorsal and ventral roots may represent a sympathetic system. Many of these run directly to their endings, for instance in the arteries, without interpolation of neurons. Cells presumed to be autonomic are found in the buccal and branchial mucosae, the endostyle and walls of the blood vessels and heart. Some of them are shown by fluorescence microscopy to be monoaminergic fibres. The gut musculature is said to be contracted by acetylcholine and relaxed by catecholamines. The autonomic system is therefore even more scattered than in elasmobranchs (p. 147). There is also a remarkably abundant network of nerve cells and fibres in the subcutaneous plexus of the skin of cyclostomes, which may be part of the autonomic nervous system concerned with regulation of the large venous sinuses there, and perhaps with the extrusion of mucus, especially in myxinoids.

None of the nerve fibres in the nervous system of cyclostomes are provided with myelin sheaths and in this they resemble the nerves of amphioxus. Conduction is slow in such non-medullated fibres. The large Müller's fibre of the spinal cord conduct only at 5 m/s (20 °C), although their diameter is more than 50 μm. The much smaller but medullated fibres of a fish or frog conduct at over 50 m/s and the largest ones of mammals at over 120 m/s.

The spinal cord is of a uniform transparent grey colour and is flattened dorsoventrally, apparently to allow access of oxygen, and metabolites, no blood vessels being present within the cord. However, vessels are present in *Myxine* in which the cord is also flat. The nerve cell bodies lie, as in higher vertebrates, towards the centre, but the synaptic contacts are not made in this 'grey' matter but at the periphery, in what would correspond to the white matter of higher forms. The outer part of the cord is thus made up of a neuropil – formed of the terminations of the incoming sensory fibres and the dendrites of the motor cells. These cells (Fig. 4.20) lie in the ventral part of the cord, their axons running out to make the large fibres of the ventral roots and their dendrites passing to all parts of the peripheral regions of both the same and the opposite sides of the cord. They are thus presumably able to be stimulated directly by impulses in the processes of the afferent fibres that end in these regions.

Motoneurons for the myotomes can be readily studied by intracellular electrodes. Other cells that have been identified include motoneurons for the fins, sensory dorsal cells and interneurons with axons proceeding either forwards or backwards. Some of these cells probably provide reciprocal inhibition during swimming movements (Rovainen 1974).

In the young larva connections are simple (Fig. 4.21). Sensory cells lying within the cord and known as Rohon – Beard cells are found, as in many young vertebrates. In lampreys they persist in the adult even though there are then also dorsal root ganglia. Mauthner's fibres of the larva arise from a pair of cells with dendrites in the neighbourhood of the eighth cranial nerve. Their axons cross and descend throughout the cord. They probably serve to produce quick escape reactions and are present in the early stages of many fishes and amphibians. Later, direct control of the spinal cord from the brain is obtained through the large Müller's fibres, originating from giant cells in the reticular formation of the brain, whose large dendrites (Fig. 4.28) receive fibres from several higher centres. These fibres thus provide an uncrossed final common pathway to the spinal cord. The dendrites of the motor cells branch around them

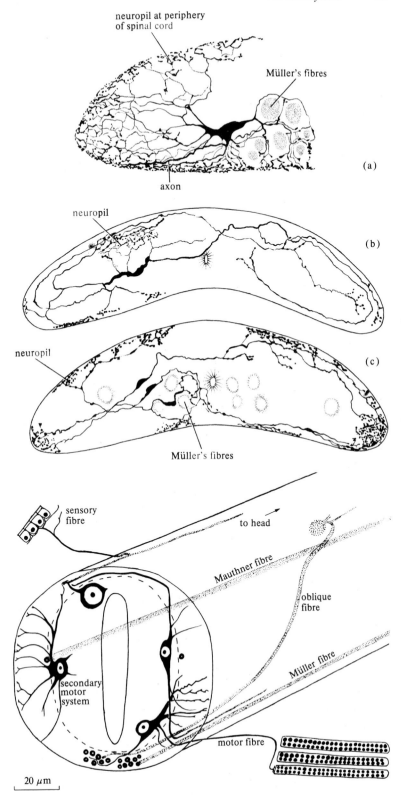

Fig. 4.20. Cells of the spinal cord of the larva of *Lampetra*. (a) and (b), large motor-cells, with dendrites spreading to the opposite side; (c), small cells with widespread dendrites, but no axon. (After Tretjakoff.)

neuropil at periphery of spinal cord

Müller's fibres

(a)

axon

neuropil

(b)

neuropil

(c)

Müller's fibres

sensory fibre

to head

Mauthner fibre

oblique fibre

secondary motor system

Müller fibre

motor fibre

Fig. 4.21. The spinal cord of pro-ammocoete stage of lamprey. (From Whiting, H.P. (1955). In *Biochemistry of the developing nervous system* (ed. H. Waelsch). Academic Press, London.)

20 μm

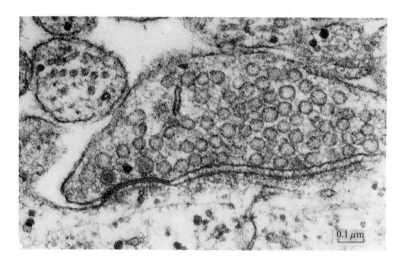

Fɪɢ. 4.22. Micrograph showing a morphologically mixed synapse in the midbrain of a larval sea lamprey. (Rovainen 1974.)

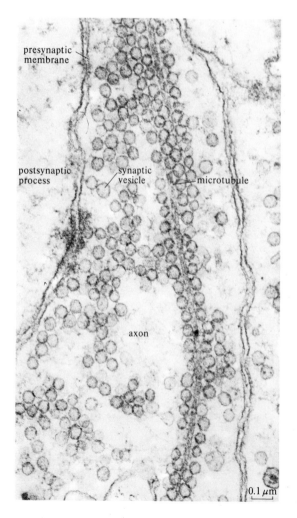

(Fig. 4.20) receiving synapses that are unusual in utilizing dual electrical and chemical transmission (Fig. 4.22). Another interesting feature is that at these synapses there are aggregations of vesicles attached to neurotubules (Fig. 4.23).

In lampreys the brain is very much smaller relative to body weight than in any other vertebrate and they also have the smallest forebrain (Figs. 4.24 and 4.25) (Ebbesson and Northcutt 1976). The brain itself (Fig. 4.26) is built on the typical vertebrate plan, as an enlargement of the front end of the spinal cord, with thickenings and evaginations corresponding to the various organs of special sense. Although we know little of its internal functional organization in lampreys, it is probably not far wrong to regard it as chiefly consisting of a series of hypertrophied special sensory centres; thus the forebrain is connected with smell, midbrain with sight, hindbrain with acousticolateral and taste bud systems. The forebrain and olfactory sense are moderately well developed in adult lampreys, as is the visual sense, with its chief centre in the midbrain. The auditory and acousticolateral systems are not very well marked, and the cerebellum is small. Taste is also much less developed than in the higher fishes (p. 184).

Fɪɢ. 4.23. Longitudinal section of an axon of the spinal cord of *Petromyzon marinus* showing a point of synapse with a postsynaptic nerve process. Synaptic vesicles converge on the presynaptic membrane but are randomly distributed in the axoplasm. A single microtubule is flanked by synaptic vesicles. (From Smith, D. S., Järlfors, U., and Beranek, R. (1970). *Journal of Cell Biology* **46**, 199–219.)

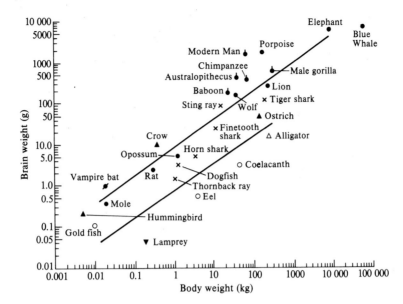

FIG. 4.24. The brain: body weight relationship of some vertebrates. The elasmobranchs have a higher ratio than do teleosts. Mammals and birds cluster round the upper line and lower vertebrates the lower one, but the elasmobranchs lie closer to the upper line. (Data from Jerison 1973 and Ebbesson and Northeutt 1976.)

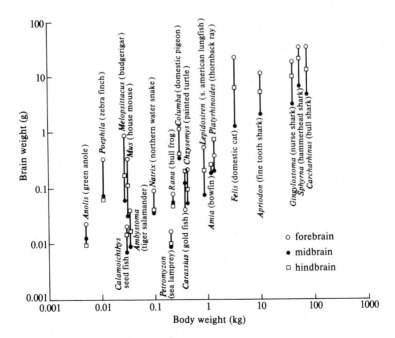

FIG. 4.25. The weight of the three major divisions of the brain from 26 vertebrates. (From Ebbesson and Northcutt 1976.)

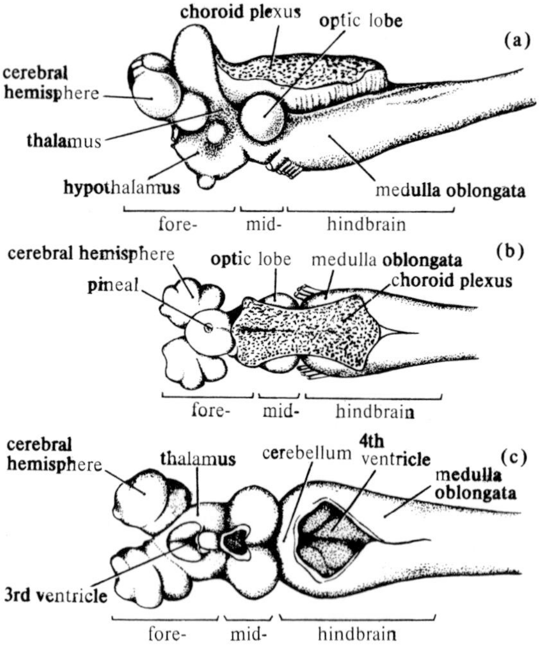

FIG. 4.26. Brain of the lamprey. (a) side view; (b) dorsal view with choroid plexus intact; (c) after removal of choroid. (After Sterzi.)

Parts of the brain

Forebrain	Cerebral hemispheres
(prosencephalon)	(telencephalon)
	Between-brain
	(diencephalon)
Midbrain	Optic lobes
(mesencephalon)	
Hindbrain	Cerebellum
(rhombencephalon)	(metencephalon)
	Medulla oblongata
	(myelencephalon)

The upper surface of the brain is covered by an extensive vascular pad, the choroid plexus or tela choroidea (Fig. 4.26). This extends into the ventricles of the brain at three points – into the third ventricle of the diencephalon, into the iter (duct) leading through the midbrain from third to fourth ventricles, and into the fourth ventricle itself. The roof of the brain is thus non-nervous in these regions. In later vertebrates the choroid extends only into the third and fourth ventricles. These vascular membranes of the brain are highly developed in lampreys because of the absence of blood vessels in the brain. The meningeal tissue contains glucose-6-phosphatase and large amounts of glycogen, from which it can produce glucose on incubation and pre-

sumably does so during periods of metabolic stress.

From the lower part of the mid- and hindbrain arise all the cranial nerves except the olfactory and optic. These nerves follow the same plan as those of gnathostomes but they are difficult to make out by dissection in the lamprey and will be left for consideration in connection with the dogfish, in which they can easily be dissected. The cranial nerves represent nerves similar to the dorsal and ventral nerve-roots of the trunk, much modified as a result of the special development of the head (p. 126). They carry afferent fibres from the skin of the head and gills and motor fibres for moving the eyes, sucker, and branchial apparatus.

From the relative sizes of the parts of the brain it can be seen that the various special sensory centres are still small. The largest part of the brain is the medulla oblongata, which is well developed because of the extensive sucking apparatus, innervated from the trigeminal nerve.

The forebrain consists of a pair of cerebral hemispheres all parts of which receive olfactory fibres. They show the basic structure found in all vertebrates, namely a dorsal pallium, mediodorsal hippocampus and medioventral corpus striatum (Fig. 4.27). Neurons are arranged in groups towards the centre, there is thus no cerebral cortex. The lateral ventricles open by the foramina of Munro into a median third ventricle, whose walls constitute the diencephalon or between-brain (Fig. 4.28). This diencephalon connects the forebrain with the midbrain and includes the thalamus which receives a few fibres from the optic nerves and may send information from vision and other senses to the cerebral hemispheres. The main visual centre is the roof (tectum) of the midbrain. The optic tracts of adult lampreys mostly end there and it is a highly differentiated, stratified region. Besides the optic fibres it receives also impulses from fibres ascending from the spinal cord and others from the auditory and lateral line centres. The midbrain is therefore undoubtedly one of the most important parts of the brain in lampreys, though nothing is known in detail of its functions. Its cells control movements of the animal, by means of fibres that run to make connection with the dendrites of the Müller's cells. Other fibres from the optic tectum reach to various parts of the brain, and it is probable that its activities are closely correlated with those of many other regions.

The ventral region of the diencephalon, the hypothalamus, is well developed in all vertebrates as a central organ controlling visceral activities and the internal life of the organism. Nerve fibres from the supraoptic nucleus of the hypothalamus proceed to the neural lobe of the pituitary and, as in other vertebrates,

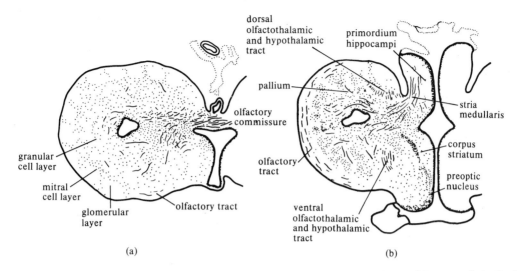

FIG. 4.27. Transverse sections through (a) the olfactory bulb and (b) the cerebral hemispheres of *Lampetra fluviatilis*. (From Nieuwenhuys, R. (1967). *Progress in Brain Research* **23**, 1–64.)

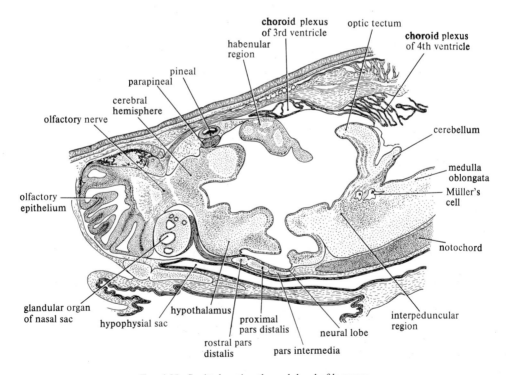

FIG. 4.28. Sagittal section through head of lamprey.

these fibres are filled with granules of neurosecretory material.

The simple cerebellum is a ridge across the front end of the medulla. It receives fibres from the vestibular and lateral line nerves and also from the spinal cord,

midbrain, and hypothalamus. It sends fibres to the motor centres at the base of the midbrain and medulla (Figs. 4.29 and 4.30). These connections are typical of all vertebrates.

Reissner's fibre is a continuous strand of material

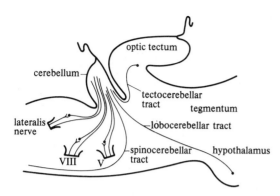

FIG. 4.29. Diagram of the afferent connections of the lamprey cerebellum, projected on a sagittal plane. (From Nieuwenhuys, R. (1967). *Progress in Brain Research* **25**, 1–93.)

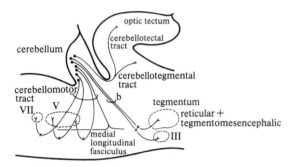

FIG. 4.30. Diagram of the efferent connections of the lamprey cerebellum, projected on a sagittal plane. (From Nieuwenhuys, R. (1967). *Progress in Brain Research* **25**, 1–93.)

produced by a group of secretory ependymal cells in the diencephalon known as the sub-commissural organ. It passes down to the end of the spinal cord. It is apparently continually produced and then broken up by ependymal cells, the products passing outwards perhaps serving to nourish the cord, which has no capillaries (Fig. 4.31).

11. The pineal eyes

In lampreys, besides the usual pair of eyes, there is also the pineal organ or third eye (also called the epiphysial, or median eye) attached to the roof of the diencephalon. It is better developed in these animals than in any other living vertebrate except perhaps certain reptiles. The organ is actually not median but consists of an un-equally developed pair of sacs, that on the right, the pineal, being larger and placed dorsal to the mor-phologically left sac, the parapineal (Fig. 4.32). The sacs develop by evagination from the brain and remain connected with the dorsal epithalamic or habenular region of the between-brain by two stalks. The two

organs are similar in structure, consisting of irregular flattened sacs with a narrow lumen. The inner walls contain receptor cells; their bases make contact with ganglion cells whose axons run to unequal right and left habenular ganglia. In addition there are pigment cells which make migratory movements, being arranged differently under conditions of illumination and dark-ness. The significance of these photomechanical changes is unknown but they demonstrate that the pineal cells are sensitive to light.

The structure of these pineal organs shows that they consist of portions of the diencephalic wall where the ciliated cells of the ependyma are specialized as photore-ceptors. They show the same general plan as the paired eyes, but with no differentiated lens or other dioptric apparatus. It has been possible to find out something of the part that these organs play in the life of the lamprey. When a bright spot of light is directed upon the pineal region of a stationary ammocoete larva movement is usually initiated, but only after illumination for many seconds. Moreover, these movements can be elicited even after the pineal organs have been removed! In the larval lamprey the paired eyes are deeply buried below pigmented skin, so the movement is not likely to be due to them; indeed it continues when they too have been taken out! Evidently there must be still other receptors, able to respond to changes of light intensity in the wall of the diencephalon. This recalls the fact that photore-ceptors are found within the substance of the nervous system of amphioxus. This power of response to changes of illumination has been retained in the verte-brates, and persists in some as yet unknown cells in the brain, even after the paired and pineal eyes have become specialized for light reception. The whole study is of special interest as showing the stages by which the eyes may have been evolved as evaginated portions of the diencephalon. Higher fishes also show the power of responding to changes of illumination after the paired eyes and epiphysis have been removed (p. 175).

If the pineal eyes are not essential for the initiation of movement, what is then their function? In the am-mocoete larva there is a daily rhythm of change in colour, the animals becoming dark in the daytime and pale at night. The diurnal rhythm of colour change may be connected with the nocturnal activity of ammocoete larvae. They have been shown to move only at night. After removal of the pineal eyes this change no longer occurs: the animals remain continually dark (Fig. 4.33). This effect on the colour is produced by the action of influences from the pineal, perhaps passing to the pituitary gland. However, the pineal itself produces a substance that contracts melanophores. Pineals taken from paling ammocoetes were placed under the skin of

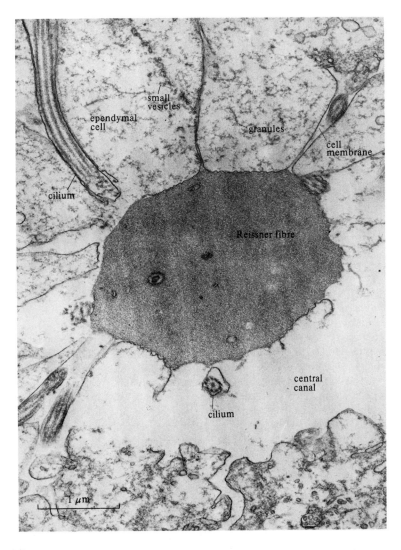

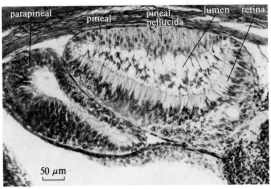

FIG. 4.32. Sagittal section of the pineal complex of *Geotria australis* undergoing metamorphosis. (Eddy 1972.)

others kept in the dark and produced local pallor. Homogenized pineals from pale lampreys caused pallor in larval *Xenopus* closely comparable to that produced by the hormone melatonin which is produced by the pineal organ in mammals (p. 479).

The pineal also influences metamorphosis of lampreys which does not occur after its removal. But this effect cannot be wholly dependent on light since ammocoetes kept in the dark proceed normally to metamorphosis. The pineal also has an influence on reproduction.

12. Pituitary body and hypophysial sac

The lower portion of the diencephalon, the hypothalamus, forms a prominent pair of sacs, the lobi inferiores, which contain a partly separated diverti-

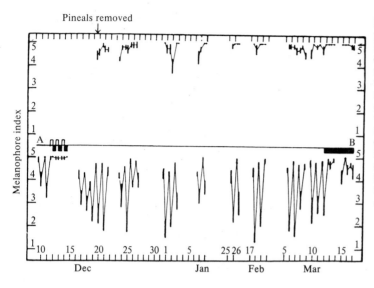

Pineals removed

Melanophore index

FIG. 4.33. Colour changes of larval lampreys, measured by the melanophore index (see Fig. 11.2). Animals kept out of doors except as shown along the line *AB*, where rectangles above the line show illumination with electric light and below the line total darkness. Normal animals show a regular daily rhythm, becoming pale at night. Reversal of normal day and night illumination stops the change. On 19 December the pineal eyes were removed from five out of the ten individuals and these thereafter remained dark (upper chart); the other five continued to show the normal rhythm, until placed in total darkness. (After Young, J. Z. (1935). *Journal of experimental Biology* **12**, 254–70.)

culum of the third ventricle and end below in the infundibulum (Fig. 4.28). The pituitary gland (hypophysis) is pressed against the underside of the hypothalamus (Fig. 4.34). There is no differentiated median eminence, but a posterior thickening of the infundibular floor forms a neural lobe (pars nervosa) with a structure comparable with the neural lobe of higher vertebrates. The glandular portion of the pituitary (adenohypophysis) is a mass of secretory cells in which three parts can be recognized. Anteriorly are the rostral and proximal regions of the pars distalis, separated by connective tissue from the posterior pars intermedia, meta-adenohypophysis, in association with the neural lobe (Fig. 4.34). After experimental removal of the pars intermedia lampreys become permanently pale in colour, showing that, as in other vertebrates (p. 138), a melanophore-stimulating hormone (MSH) is liberated into the blood by this gland, the secretion being under the control of the pineal eye by way of the hypothalamus. The lamprey pituitary has been shown

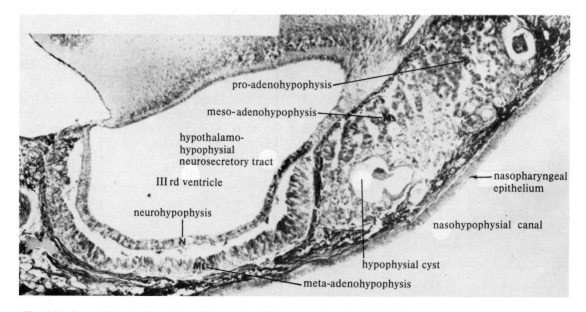

pro-adenohypophysis

meso-adenohypophysis

hypothalamo-hypophysial neurosecretory tract

IIIrd ventricle

neurohypophysis

N

nasopharyngeal epithelium

nasohypophysial canal

hypophysial cyst

meta-adenohypophysis

FIG. 4.34. Longitudinal section of the pituitary gland of *Lampetra fluviatilis*. (× 87) (From Larsen 1972.)

to contain a 'water balance' hormone, arginine vasotocin, as well as MSH. Arginine vasotocin is probably
produced in neurosecretory cells of the pre-optic nucleus of the brain and carried down the axons that end in
the neural lobe (Fig. 4.35).

The proximal pars distalis contains basophil cells
producing a gonadotropic hormone; gametes and secondary sex characters do not develop after hypophysectomy. The opening of the coelom and swelling of the
cloaca can be induced by injection of mammalian
gonadotropin, even in ammocoetes. There are acidophil
cells in the proximal pars distalis possibly producing a
corticotropic hormone. Basophil cells of the rostral pars
distalis increase at metamorphosis. They have been held
to be thyrotropes but there is no evidence of thyrotropin
and metamorphosis of lampreys is not dependent on the
thyroid.

There is no clear evidence of a hypophysial portal
system making a neurovascular link between hypothalamus and adenohypophysis. The pituitary can
function independently. Lampreys in which it has been
transplanted away from the brain develop secondary
sexual characters and mature gametes. However, some
fibres of the neurohypophyseal tract terminate close to
blood vessels in the floor of the hypothalamus just above
the pars distalis. This is the region where the median
eminence develops in higher vertebrates, and it is
possible that releasing factors may pass, either directly
through the few blood vessels which here run into the
pars distalis, or through the connective tissue between
the hypothalamic floor and the pars distalis (Fig. 4.35).

The lamprey pituitary, therefore, shows many of the
main features found in other vertebrates but probably
lacks the full complement of hormones and has a rather
simple control system, perhaps mainly by feedback from
other tissues rather than regulation by hypothalamic
releasing factors. There is at least one gonadotropin, but
no clear evidence of adrenocorticotropin, somatotropin, lactotropin, or thyrotropin. However, there are
several types of cell in the pars distalis and it may be that
further research will show evidence of secretion of these
hormones, perhaps in less differentiated forms.

The pituitary of lampreys is also unique because of
the development of a nasohypophysial sac (Fig. 4.36).
Characteristically in vertebrates the pituitary body
develops by the formation of a pocket of buccal
ectoderm, whose walls then become folded, so that the
part in front of the lumen becomes the pars anterior,
that behind the pars intermedia. In nearly all vertebrates
the lumen then loses its connection with the exterior. In
lampreys the hypophysial rudiment is continuous with
that of the olfactory epithelium. The latter then moves
dorsally and the two remain connected throughout
larval life by a strand of cells. At metamorphosis this
acquires a lumen and forms a tube extending from the
nostril, below the pituitary and brain. Because of its
development this is sometimes called the nasohypophysial tube but some workers doubt that it represents
the cavity of the hypophysis and prefer the name
nasopalatine or nasopharyngeal canal.

The nasal sac, which is very simple in the larva, shows
much internal folding in the adult, increasing the
sensory surface, which has been shown to serve for
detection of prey or quality of water (Fig. 4.28). Inside
the single nostril, guarded by a valve, are openings into
the nasal sacs, which are cavities with folded walls. The
olfactory receptor cells give axons that make up the
olfactory nerves, entering the olfactory bulbs on the
anterior end of the hemisphere. Behind the nasal sacs lie
numerous glandular follicles opening into the sac in the
larva, but completely closed in the adult. They may be
comparable to Jacobson's organ but their function is
unknown (p. 263).

The nasohypophysial tube proceeds back behind the
pituitary to a closed sac lying between the first pair of gill
pouches (Fig. 4.36). During the movements of respiration this sac is squeezed and water is expelled with
some force through the nostril. When the gills relax
water flows in at the nostril, and in this way the olfactory
organ is provided with samples. If the nasohypophysial
opening is closed with a plug of plasticine the lamprey
no longer reacts to solutions, for instance of alcohol, to
which it normally responds by freeing its sucker and
swimming away.

By using a special tank to measure activity it has been
shown that *Petromyzon marinus* does indeed respond to
the odour of other fish (Fig. 4.37). The lampreys showed
a circadian rhythm of activity. During the period of

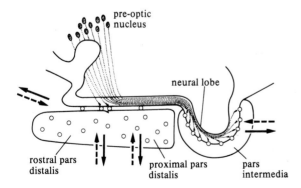

Fig. 4.35. Pituitary of lamprey. (From Jasinski, A. (1969).
General and comparative Endocrinology Suppl. 2, 510–21.)

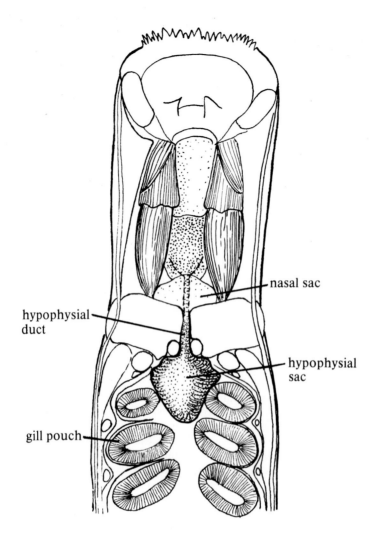

F<small>IG</small>. 4.36. Dissection of lamprey from the ventral surface after injection of coloured gelatine to show the outline of the nasohypophysial sac and its duct, which is shown dotted where it runs upwards between the nasal sacs. Contraction of the branchial apparatus squeezes the sac, so that water is drawn in at each relaxation.

inactivity in the circadian cycle water from an aquarium containing trout was introduced and the animals became active and continued to swim for some time while locating the source of the odour. Animals rendered anosmic (without smell) moved randomly under the same conditions. One of several amines, isoleucine methyl ester, isolated from water containing the prey, was particularly effective (and acted also upon shark and teleostean predators).

13. Adrenal tissue in lampreys

There are no compact adrenal bodies but cells representing the two parts of higher forms are found widely scattered. The interrenal (cortical) cells form groups beneath the cardinal veins in the region of the pronephros. They contain large vacuoles with much phospholipid but have not been shown to produce steroid hormones, nor have these been proved to occur in the blood. Mammalian ACTH is said to cause hyperplasia of the cells and cortisol and aldosterone decrease the Na^+ losses of lampreys migrating upstream.

The suprarenal (medullary) tissue is abundant in the form of chromaffin cells in the walls of the heart and great vessels. They show a green fluorescence after treatment with formaldehyde (indicating presence of catecholamines) and contain dense-cored vesicles. Ad-

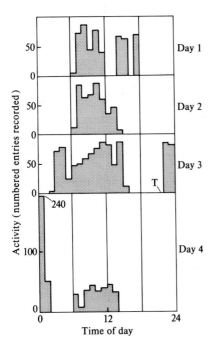

FIG. 4.37. *Petromyzon marinus* responds with vigorous loco-motor activity to the introduction of the odour of trout at point T during the period of inactivity in the circadian cycle. (From Kleerekoper, H. and Mogensen, J. (1963). *Physiological Zoology* **36**, 347–60.)

renaline and noradrenaline occur in the tissue of the heart.

14. Lateral line organs of lampreys

The lateral line receptors, peculiar to fish-like verte-brates, are little patches of sensory cells found along certain lines on the head and trunk. They are all innervated by cranial nerves, those on the body and tail being served by a special backward branch of the vagus nerve. The receptor cells carry long hairs and are thus able to detect either movement of the water relative to the fish or of the fish itself. Objects moving nearby set up disturbances that may also be detected (p. 183). In the lamprey the lateral-line organs are very simple (Fig. 4.38), being open to the exterior and not sunk in a canal as in higher forms. The rows are somewhat irregular, especially those on the body.

15. Vestibular organs of lampreys

The labyrinth may be considered as a specialized portion of the lateral-line system, concerned with re-cording the position of the head and angular accele-rations. There is no evidence to decide whether lampreys can respond to sound, but responses to vibration have been recorded from the eighth nerve perhaps arising

from special hair cells in the macula neglecta (Fig. 4.40). The labyrinth develops by an in-pushing of the wall of the head, and this then becomes closed off from the exterior. Internal foldings divide up the sac into a number of chambers, which differ somewhat from those of gnathostomes (Fig. 5.34). There is a large central vestibule, into which open below several partially se-parate sacs, provided with patches of sensory hairs (Lowenstein, Osborne, and Thornhill 1968) (Fig. 4.39). These correspond, from in front backwards, to the maculae of the utricle, saccule, and lagena of higher forms (Fig. 4.40). The hairs of the maculae are loaded with otoliths. There are only two broad semicircular canals, corresponding to the anterior and posterior vertical canals of other vertebrates. Each contains an ampulla with a complex ridge, the crista with three components set approximately at right angles. The wide canals provide a large volume of endolymph suitable for allowing detection of the slow angular accelerations that are likely to occur in a lamprey. Fishes that move faster have much narrower (and longer) canals (p. 145). Elec-trical recording shows that the nerve from *each* ampulla of the lamprey can be excited by rotation around all the three axes of space and these signals allow the perfor-mance of simple vestibular-ocular reflexes by which movements of the eyes stabilize images on the retina when the fish turns (Rovainen 1976).

The macula and cristae carry complexes of sensory hairs, including a kinocilium and several stereocilia (Fig. 4.41). Each group is functionally polarized, giving a response to bending in the direction away from the stereocilia. A curious cross-striated structure extends from the basal plate and may serve to carry excitation to the synaptic endings of the nerve fibres, each of which is provided with a synaptic bar surrounded by vesicles. The bending of the stereocilia probably depresses the

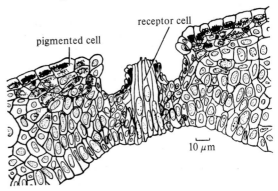

FIG. 4.38. Section of lateral line organ of tail of adult *Lampetra*. Pigmented cells surround the pit and receptor cells do not show long hairs. (After Young, J. Z. (1935). *Journal of experimental Biology* **12**, 229–38.)

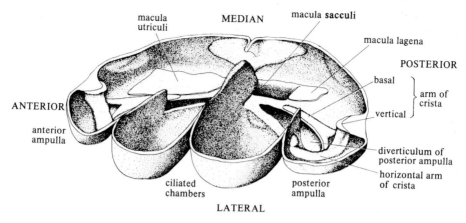

FIG. 4.39. Diagrammatic view of the structure of the lower half of the left labyrinth of *Lampetra fluviatilis*. (From Lowenstein *et al.* 1968.)

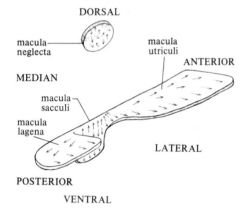

FIG. 4.40. Diagram showing the orientation of the sensory cells of the right maculae of *L. fluviatilis*. The arrowheads show the position of the kinocilium. (Lowenstein *et al.* 1968.)

basal foot of the kinocilium, setting up an excitatory generator potential. Movement in the opposite direction inhibits spontaneous discharge.

Opening into the vestibule are two large sacs covered with cilia (Fig. 4.39), whose beat produces complicated counter currents in the dorsoventral plane. It has been suggested that these may act as a gyroscope, compensating for the absence of a horizontal canal but their function remains unknown. The twisted plate-like maculae contain hair cells orientated in different directions (Fig. 4.40). These are therefore potential gravity receptors for detecting linear accelerations.

In *Myxine* the condition is even simpler, there being only a single vertical semicircular canal (Fig. 5.34). However, this has cristae at both ends. The macular system also does not show the characteristic subdivisions but is a single *macula communis*.

16. Paired eyes of lampreys

The structure of the paired eyes is similar to that in other vertebrates but shows a number of unusual and perhaps primitive features. They are formed, like the pineal eyes, by evaginations of the wall of the diencephalon; the so-called optic nerve is therefore not really a peripheral nerve but a portion of the brain; it should strictly be called the optic tract. The eyes are moved by extrinsic muscles arranged in a somewhat unusual manner. Accommodation is effected by a process found in no other vertebrates. The cornea consists of two distinct layers, separated by a gelatinous substance. Attached to the outer cornea or 'spectacle' is a cornealis muscle, apparently of myotomal origin, which flattens the cornea and pushes the lens closer to the retina (Fig. 4.42). There is an iris, outlining a round pupil, which does not change in diameter under different illuminations. Most species of lampreys are diurnal animals. They are said to move towards white objects and probably use both the eyes and the nose to find their prey. In the ammocoete larva the paired eyes are buried below the pigmented skin and the animal makes no movements when light is shone on to this region.

17. Skin photoreceptors

Like many lower vertebrates the lamprey has light-sensitive cells in the skin, as well as those in the eyes. These receptors are abundant in the tail and if a light is shone on to this region the animal rapidly moves away (Fig. 4.43). If the spinal cord is cut just behind the head and a light then shone on to the tail, the *head* will be seen to move. This suggests that the impulses are carried forwards by means of the lateral line nerves, which is confirmed by the fact that if these latter are sectioned,

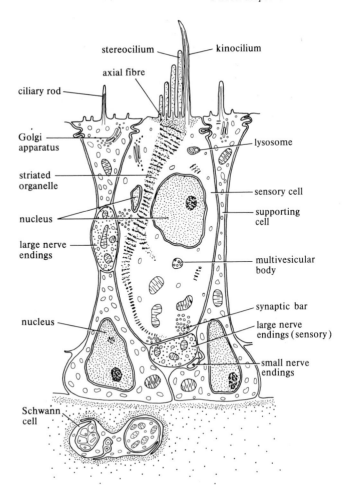

FIG. 4.41. Diagram of the structure of a sensory hair cell from the vestibular organ of a lamprey. (After Lowenstein, O. and Osborne, M. P. (1964). *Nature, London* **204**, 197–8.)

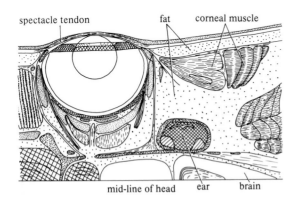

FIG. 4.42. Horizontal section of the eye of a lamprey. (From Walls 1942.)

leaving the spinal cord intact, then no movements follow when the tail is illuminated. This sensitivity of the lateral line organs to light is not found in other fish-like

vertebrates. Indeed the receptors are not strictly lateral line organs but pigmented epidermal cells. The sensitivity curve shows a sharp peak at 530 μm, this being the region of the spectrum at which light penetrates farthest into sea water. The pigment is probably a porphyropsin. The sensitivity increases up to 30-fold by dark adaptation (Steven 1950, 1963).

In hag fishes (*Myxine*) the head and cloacal regions are more sensitive to light than is the rest of the body. The impulses from the skin are conducted through the spinal nerves in these animals, not the lateral line nerves.

It has been claimed that *Petromyzon marinus* produces an electric field in the water around its head (Kleerekoper and Sibakin 1956). It is said to arise near the eyes and may be used for orientation.

18. Habits and life history of lampreys

We have very little information about the life of lampreys during the time that they are in the sea. They

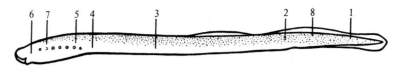

Fig. 4.43. Ammocoete larva of *Lampetra planeri,* showing the effect of shining a narrow beam of light on to various parts of the side of the body. Illumination at 1, 2, or 8 is followed by movement after a few seconds, but no movement follows illumination at points 3–7. (From Young, J. Z. (1935). *Journal of experimental Biology* **12**, 229–38.)

are often caught attached to other fishes. The up-river migration of *Lampetra fluviatilis* occurs in the autumn, when they are caught in traps on the way, for use as food. Large numbers used to be taken (over a million from the Thames in 1898) but have been much reduced by pollution and damming. However, annual catches of up to half-a-million are still reported for some European rivers. The spawning migrations of lampreys may take them for thousands of kilometres. They are said to perform remarkable feats of climbing, leaping from stone to stone and hanging on by their suckers. During this period of migration some lampreys assume brilliant orange and black colour patterns. On the other hand, lampreys land-locked in the lakes of New York (*Petromyzon marinus unicolor*) feed in fresh water and ascend only a few kilometres up streams to breed.

Once in the river the anadromous lampreys do not feed again but live over the winter on the reserves accumulated in the form of fat, especially under the skin and in the muscles. During the winter the gonads ripen progressively and the secondary sexual characters begin to become apparent only in February. The females then develop the large anal fin, while in the male a penis-like organ appears (Fig. 4.1) and the base of the dorsal fin becomes thickened.

Spawning occurs in the spring and is preceded by a form of nest-building. The males probably arrive first and serve to attract the females by olfaction. French fishermen catch *Petromyzon marinus* by placing a mature male in a trap and by the next day it is crammed with females. Spawning is usually at a place below a weir where the water is shallow and rather swift, and the bottom both stony and sandy. Stones are then dragged by the mouth in such a way as to make a small depression. Fertilization is secured by a process of copulation in which the male fixes by the sucker on to the fore-part of the female and the two then become intertwined and undergo rapid contortions, the eggs being squeezed into the water, while sperms are ejected through the 'penis' (Fig. 4.44). Fertilization is therefore external, but the sperms must be placed very close to the eggs, for they remain active only for about one minute after entering the fresh water, which provides the stimulus that activates them. The eggs and sperms are not all laid at once; mating is repeated several times until

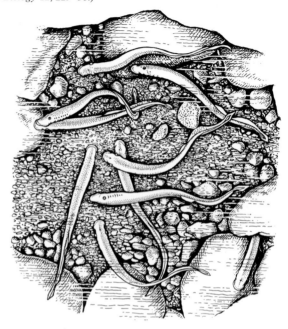

Fig. 4.44. Spawning lampreys seen in their nest. (After Gage.)

all the products have been shed, after which the animals are exhausted and soon die. The movements of the animals stir up the sand in the nest (this is probably the function of the anal fin of the female) ensuring that the eggs are covered up as they are carried away by the current.

19. The ammocoete larva

The eggs contain a considerable quantity of yolk, but their cleavage is total and gastrulation and neurulation proceed in a manner not unlike that of the frog. After about three weeks the young hatches as the ammocoete larva, about 7 mm long. At first this is a tiny transparent creature, but its larval life lasts for a long time, during which it grows into an opaque eel-like fish, up to 170 mm long (Fig. 4.43).

This portion of life is spent buried in the mud, the animals emerging only occasionally at night to change their feeding-ground, presumably if the mud is not sufficiently nutritious. There is no sucker, the mouth being surrounded by an oral hood rather like that of amphioxus (Fig. 4.45). The paired eyes are covered by muscles and skin. The head at this stage is little sensitive to light, but the animal quickly begins to swim if the tail is illuminated. We have seen already (p. 100) that in lampreys there are photoreceptors in the tail, connected with the lateral line nerves. In the larva these are the main photoreceptors, and they ensure that the animal lies completely buried.

If a number of larvae are left in a vessel with a layer of mud on the bottom they rapidly disappear and remain hidden indefinitely, the heads perhaps just visible in small depressions made by the rhythmic respiratory movements. When disturbed they always swim with the head downwards and in contact when possible with the ground. This habit leads them to burrow rapidly. It is not known whether they have other receptors to guide them to mud rich in possible food organisms. The nasal and hypophysial sacs are poorly developed in the larva, and the sense of smell can hardly serve this purpose.

Feeding takes place by the intake of water through the mouth and the separation of small food particles from it in the pharynx (Fig. 4.45). For this purpose there is used a great quantity of mucus, secreted by the surface of the velum as threads. These condense to form a cord in the centre of the pharynx, supplemented, at least in the young larva, from the endostyle (Fig. 4.45). This endostyle is a most remarkable organ, forming early in development as a sac below the pharynx (Fig. 4.46). It consists of a pair of tubes, on the floor of which there are four rows of secretory cells (Fig. 4.47). There is a single opening to the pharynx, by a slit at about the middle of the length. As development proceeds the inner rows of cells at the hind part of the organ become coiled upwards, and at the end of larval life the endostyle therefore forms a very large mass below the pharynx, composed of tubes lined partly by secretory and partly by ciliated cells. Probably no enzymes are secreted by the endostyle, its function being to produce mucus. Although it resembles the endostyle of amphioxus in the arrangement of the secretory columns, there is a differ-

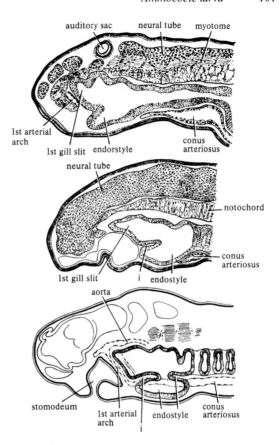

FIG. 4.46. Development of the endostyle of the lamprey. Sagittal sections through the head at three stages. *i*, inpushings which cut off the endostyle from the pharynx. (After Dohrn, from Kükenthal.)

ence in that the organ in the ammocoete larva is not an open groove. There is, however, a ciliated groove in the floor of the pharynx, that is to say, on the roof of the endostyle (Fig. 4.47).

The details of the feeding-currents of the ammocoete larva are not understood. An important difference from the arrangement in amphioxus is that the current is produced by muscular rather than ciliary action. The velum, a pair of muscular flaps, provides the main current when the animal is at rest. The branchial basket can also be expanded and contracted by an elaborate system of muscles. It is not easy to observe how the food particles are taken up from the current, but apparently a strand of mucus occupies the whole of the centre of the pharynx and rotates as it passes backwards into the oesophagus catching particles. Mucus is passed out through the gill openings to form a bladder-like structure around the head and it may be a major function of the endostyle to produce this.

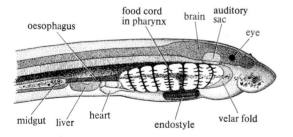

FIG. 4.45. Young ammocoete larva of lamprey fixed while feeding on green flagellates and detritus and then stained and cleared.

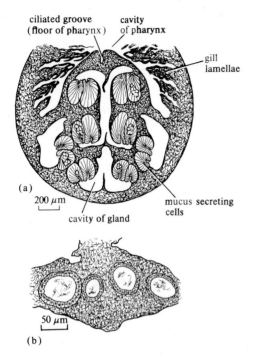

ciliated groove (floor of pharynx)   cavity of pharynx

gill lamellae

(a)

200 μm

cavity of gland   mucus secreting cells

50 μm

(b)

FIG. 4.47. (a) Transverse section of endostyle of ammocoete larva. (b) Transverse section of thyroid follicles of adult. (After Young, J. Z. and Bellerby, C. W. (1935). *Journal of experimental Biology* **12**, 246–53.)

Evidently the system enables the animals to feed efficiently on the small unicellular algae and bacteria of the mud. In amphioxus the ciliated pharynx, occupying a considerable proportion of the whole surface, is only able to support a tiny creature, but the muscular feeding system of the ammocoete allows a relatively small pharynx to feed an animal 170 mm long and weighing up to 10 grams. This use of muscles for moving the gills was evidently an important step in chordate evolution. It allowed the animals to escape from the limitation of size imposed by the ciliary method of feeding. After the development of jaws to form a still more efficient feeding mechanism the rhythmic movement of the branchial apparatus persisted for the purpose of respiration. We cannot be certain about changes which occurred so long ago, but it seems likely that the respiratory movements of a fish were first introduced to provide food rather than oxygen.

The endostyle therefore shows the survival of the primitive feeding methods of chordates, but it also undergoes at metamorphosis an astonishing change into a thyroid gland. The mucus-secreting columns shrink and the whole organ becomes reduced to a row of closed sacs, lying below the pharynx (Fig. 4.47(b)). Each of these sacs is lined by an epithelium, containing a

structureless 'colloid' substance, and is therefore closely similar to a thyroid vesicle. Moreover, experiments have shown that extracts of this organ contain iodine and exert an accelerating effect on the metamorphosis of frog tadpoles. Although nothing is known of the part played by the secretion of this gland in the life of the adult lamprey, we may safely conclude that we have here the conversion of an externally secreting feeding organ into a gland of internal secretion. The actual mucus-secreting cells are not transformed into those of the thyroid follicles, these latter are derived from epithelial cells in the wall of the larval organ. One cannot avoid speculating on this extraordinary change of function. It may perhaps be significant that the endocrine gland that regulates basal metabolism (the thyroid) is derived from the part of the feeding system that in the earliest chordates was responsible for providing the raw materials of metabolism. Experiments with radioactive iodine show that this element is concentrated in certain cells of the larval endostyle (Fig. 4.48). Moreover, after addition of the anti-thyroid substance thiourea to the water there is hypersecretion by the endostyle. Thyroxine has been extracted from the gland and it probably has an endocrine function as well as secreting mucus, though no one has ever produced any changes in larval lampreys by administering thyroid hormones. Iodinated secretion can be detected by autoradiography in the lumen of the endostyle and is presumably passed to the intestine and digested. Iodine is thus bound and synthesized into thyroid hormones by a part of the gut in the larva, showing a stage between digestion and endocrine secretion, which is achieved in the adult. It is doubtful whether the pituitary produces a thyroid-stimulating hormone at either stage. Lampreys thus show, as larvae, a stage in which the accumulation of iodoproteins, previously widespread, becomes concentrated in the pharynx. Perhaps at this site there were already cells specialized for halide transport (cf. the ionocytes of teleosts, used for osmoregulation, p. 168). In adult lampreys and all higher chordates the iodoprotein is secreted into the blood under the control of blood-borne signals (Fig. 4.49). The change may well be related to developments in the regulation of metabolism, which, in the animals with a fully endocrine thyroid becomes more nearly independent of variations in the external supply of iodine.

The great change in the endostyle is only part of the complete metamorphosis by which the ammocoete larva changes into an adult lamprey. It is not known what triggers the change. The mouth becomes rounded and its teeth, tongue, and complex musculature develop. The paired eyes (previously buried) appear; the olfactory organ becomes internally folded, and the olfac-

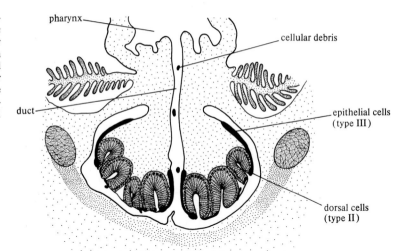

Fig. 4.48. Autoradiograph of endostyle of ammocoete larva of *Petromyzon marinus* at level where it is connected by a duct to the pharynx. The dense stain shows distribution of protein-bound [131]I. The radioactive clumps of cellular debris in the glandular lumen and in the duct suggest that the material represents a holocrine secretion, which will probably be absorbed in the intestine.

nerve and tracts much enlarged. The nasohypophysial sac grows backwards to the gills. In the pharynx the gills develop into sacs opening to the branchial chamber. Changes also take place in the intestine. The yellow-brown colour of the larva gives place to the black with silver underside of the adult. The animal more and more frequently leaves the mud and finally migrates to the sea to begin its parasitic life.

20. Paired species of lampreys, a problem in systematics

Besides the river lampreys, such as *Lampetra fluviatilis* (Linn.), which show this characteristic migratory life-

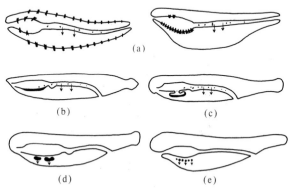

Fig. 4.49. Diagram to show distribution of iodoproteins, at first in exoskeletal structures, as in many invertebrates and in tunicates (a). Some of this material is concentrated in the pharynx. This tendency is exaggerated in amphioxus and the ammocoete larva. In the adult lamprey and later animals this pharyngeal material forms the thyroid. (a) Many invertebrates and tunicates; (b) Amphioxus; (e) Ammocoetes; (d) Metamorphosis of ammocoetes; (e) General vertebrate type. (After Gorbmann, A. (ed.) (1959). *Comparative endocrinology*. Wiley, New York.)

history, there are also in both hemispheres small, non-parasitic brook lampreys ('prides'), such as *L. planeri* Bloch, which remain throughout their life in fresh water and never feed in the adult state. These prides are very abundant in many English rivers and streams, but since the greater part of their life is passed in the ammocoete stage they are not often seen. The larvae remain in the mud probably for five years or more and undergo metamorphosis in late summer and autumn. The characteristic of this type of lamprey is that the adults never migrate and never feed. The gonads are already well developed at metamorphosis and ripen during the winter. Spawning takes place in March or April and the animals then die.

There has been much dispute about the status of these freshwater races. In structure the adult *Lampetra planeri* is nearly identical with an adult *L. fluviatilis*, except that the latter is much the larger and has sharper teeth. A marked difference is that the late ammocoetes of river lampreys contain far more oocytes than those of brook lampreys (Fig. 4.50). This is because of extensive degeneration (atresia) and disappearance of oocytes in the latter, the remaining oocytes being more advanced. The brook lamprey is thus prepared at metamorphosis for a rapid maturation thereafter. Its condition is therefore more that of curtailed adult life, perhaps not strictly to be called neoteny. Crossing of the two sorts could presumably never take place in nature, on account of the size difference, but by artificial stripping of the adults cross-fertilization in both directions can easily be achieved. Unfortunately the hybrid larvae have never been reared to maturity; we cannot therefore say whether the small size and failure to migrate of the *planeri* forms are inherited characters or are produced by the influence of the environment. The effect of the

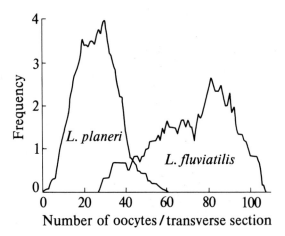

FIG. 4.50. Frequency curves for the numbers of oocytes in transverse sections (in a standardized region) of ammocoetes of *Lampetra fluviatilis* and *L. planeri*. (Hardisty and Potter 1971.)

non-migratory condition is to enable the prides to colonize very fully rivers that, because of effluents and other obstacles, they would be unable to occupy if a migration to the sea was necessary. By this process of acceleration of the development of the gonads a dangerous stage in the life-history has been avoided.

Similar pairs of migratory and non-migratory forms of lamprey are found in Japan, North America, and Australia. Indeed, the condition appears to be developing independently in several river systems in the United States (Morman 1979). Since it may be difficult for the brook lampreys to spread from one river system to another it is possible that many of the *planeri* forms have evolved separately, perhaps quite recently. If so, this is a remarkable example of a similar response produced in different parts of a population by a similar environmental stimulus, in this case probably the effluents. This process of alteration in the relative times of metamorphosis and sexual maturity (paedomorphosis) has occurred also in certain amphibians (the axolotl) and in tunicates (Larvacea). Similar changes in rates of development may have been essential factors in the development of the whole chordate phylum (p. 69).

In one race, found in Italy, ammocoetes with mature gonads have been reported. However, in most of these lampreys the paedomorphosis is only partial: metamorphosis does take place, but is immediately followed by maturity. Since in mammals, injections of anterior pituitary extracts accelerate development of the gonads, it was thought possible that complete neoteny might be produced by making such injections into larvae of *Lampetra planeri*. No completely sexually mature ammocoetes have yet been produced by this method, but following the injections the larvae assumed the secon-

dary sexual characters, which are normally shown only at maturity, namely, swelling of the cloaca, opening of the pore from coelom to exterior, and the changes in body form. No signs of metamorphosis were produced by these injections and we are left without information as to the cause of that change in the lamprey. In Amphibia even very young larvae undergo metamorphosis when treated with thyroid extracts, but similar treatment of ammocoete larvae has failed to produce any change. Further investigation of the problem should be very interesting, since it seems likely that the differences between the *fluviatilis* and *planeri* forms are the result of an endocrine factor accelerating the onset of sexual maturity in the latter. The fact that the change is occurring in various parts of the world adds further interest to this example of evolution in progress.

Besides all these relatively small lampreys, there is a much larger form, the sea lamprey, *Petromyzon marinus*, Linn., reaching to over a metre in length. This animal differs from *Lampetra* in body form, structure of sucker, and other features, as well as in size. Like most other groups of animals lampreys therefore present several problems of nomenclature. Linnaeus included the three types that occur in Europe in the one genus *Petromyzon*; since they are all rather alike in shape this is in some ways a reasonable procedure. But are we then also to include in the same genus forms that differ more widely, such as those occurring in America and in the southern hemisphere? As so often happens, systematists have chosen the course of splitting up the Linnaean genus, even though several of the resulting genera have only one species. Thus J. E. Gray suggested the genus *Lampetra* for the brook and river lampreys, keeping *Petromyzon* for the larger species of sea lamprey (1851). Other genera have been added, such as *Ichthyomyzon* Girard from N. America, held to be the most primitive of living lampreys, and *Mordacia* Gray and *Geotria* Gray for the forms from the southern hemisphere (Chile, Australia, and New Zealand). Such generic distinctions are an advantage when they call attention to large differences. For instance, it is a striking fact that lampreys are found in temperate waters of both hemispheres, but not in the tropics, and it is interesting to learn that the forms from New Zealand, Australia, and South America (there are none in South Africa) show distinct peculiarities. Thus *Geotria* possesses a large sac behind the sucker.

A special problem of nomenclature arises from the fact that the river and brook lampreys are almost identical in structure and differ mainly in size, time of sexual maturity, and habits. A further complication is that the germ-cells of the two races allow cross-fertilization, although this probably never occurs in nature! We may take Dobzhansky's (1951) definition of species as 'groups of populations which are reproductively isolated to the extent that the exchange of genes between them is absent or so slow that the genetic differences are not diminished or swamped', and in this sense we may retain the specific names *L. fluviatilis* and *L. planeri* for the two populations.

Well preserved fossil lampreys have recently been found in the Pennsylvanian coal measures near Chicago (280 million years old). *Mayomyzon* was remarkably similar to modern lampreys with seven gill pouches (Fig. 4.51).

21. Hag fishes, order Myxinoidea

The hag fishes, *Myxine*, *Bdellostoma*, *Polistotrema*, and *Eptatretus* (Fig. 4.52), are animals highly modified for sucking. They live buried in mud or sand and eat polychaetes and other invertebrates, as well as scavenging dead fishes. The eyes are functionless rudiments, though the animals are sensitive to changes of illumination through skin receptors. There are sensory tentacles around the mouth, and the teeth and sucking apparatus are well developed. They have efficient jaws which bite laterally to tear and fragment the prey. They burrow into the bodies of dead or dying fishes. As many as 123 *Myxine* have been taken from a single fish. Since the introduction of trawling they have become less common in the North Sea, where they used to be a serious source of loss to fishermen by their attacks on fishes caught in drift nets or on lines. They seem to find fish when they are dying or just dead. Entering by the mouth of their prey they eat out the whole contents of the body, leaving a sac of skin and bones. When they are themselves caught on lines (for instance, with a salted herring bait) the hook is swallowed so deeply that it may be found near the anus!

The gills are modified into pouches (6–14 in *Bdellostoma*, 6 in *Myxine*), opening by tubes into the pharynx, and to the exterior (Fig. 4.53). In *Myxine* all the tubes are joined and open by a single posterior aperture on each side. Water enters at the nostril and is pumped backwards by a muscular velum through the gill chambers and out behind. There is also a single posterior oesophagocutaneous duct on the left side, which is probably closed during normal respiration but is opened to allow expulsion of large particles. If the nostril is closed experimentally with a plug no water enters by the mouth or posterior apertures but the fish survives well, presumably respiring through the skin.

The circulation is largely an open system of large sinuses and there are accessory hearts in the portal system, cardinal veins and tail. The gut resembles that of lampreys in the absence of a stomach. There are

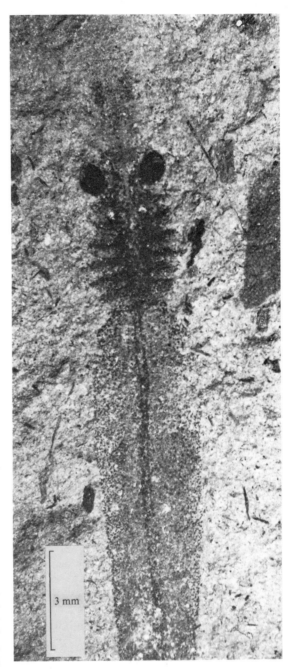

FIG. 4.51. *Mayomyzon pieckoensis*. Dorsoventral aspect. (From Bardack and Zangerl 1971.)

zymogen cells throughout the intestine but no concentration of them at the anterior end, nor any intestinal caeca. Cells similar to the B cells of the endocrine pancreas are collected around the bile-duct.

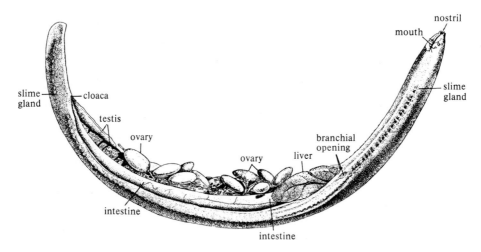

FIG. 4.52. *Bdellostoma*, partly dissected, showing ovary with eggs. (After Retzius, from Kükenthal.)

The thyroid gland consists of a long series of sacs formed by evagination from the floor of the pharynx.

The pituitary shows some curious features, and perhaps represents an even earlier stage than that of lampreys (Holmes and Ball 1974). The adenohypophysis consists of clusters of cells embedded in connective tissue, seems poorly vascularized and with no distinction between pars distalis and intermedia. It is completely separated from the neurohypophysis by thick connective tissue (Fig. 4.54). The adenohypophysis contains basophil, acidophil, and chromophobe cells, which may produce undifferentiated versions of the two main protein-like pituitary hormones, the ACTH – MSH type and the STH – LTH type. There is no evidence for any thyrotropic or gonadotropic hormones (glycoproteins) in *Myxine*. There is no system of hypophysial portal vessels in *Polistotrema* (living off Chile) or *Myxine* and these animals, living in a uniform environment, show no seasonal cycles in the gonads. The Pacific form *Eptatretus* does, however, have some

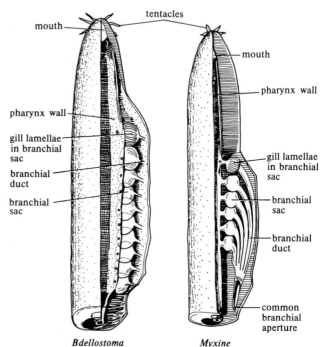

FIG. 4.53. Arrangement of gills. (After Dean 1895.)

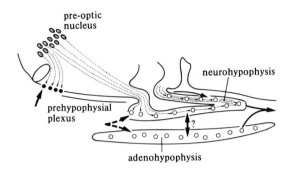

pre-optic
nucleus

neurohypophysis

prehypophysial
plexus

?

adenohypophysis

FIG. 4.54. Myxinoid pituitary. A diagram of a sagittal section to show the hypothalamo-hypophysial vascular and neuro-secretory links. The ? indicates that the direction of blood flow in the vertical vessels between the neurohypophysis and adenohypophysis is unsettled. Broken arrows, arteries; solid arrows, veins; thin arrows, possible portal veins. (From Jasinski, A. (1969). *General and comparative Endocrinology* Suppl. 2, 510–21.)

portal vessels and is said to enter shallow water and has a breeding season, possibly involving control of the pituitary by the brain.

The secretion of the neurosecretory cells of the hypothalamus also shows unusual conditions. Some of the axons of the neurohypophysial tract enter the neurohypophysis but others end in a unique *prehypophysial plexus* behind the optic chiasma (Fig. 4.54). The axons contain neurosecretory material and it is possible that both releasing factors and neurohypophysial hormone(s) are here passed directly into the systemic circulation. Since the blood is isotonic with sea water there is presumably less need for regulation of its composition by pituitary hormones (p. 84).

Myxinoids have a unique system of large subcutaneous venous sinuses, provided with smooth muscles, and an elaborate plexus of nerve fibres. The function of this system is not known.

Down the sides of the body are pairs of slime glands, able to secrete large amounts of very sticky mucus, which may be protective and is said also to be produced under the operculum to hasten the end of a dying fish that the hag has attacked.

A curious difference from the nervous system of lampreys is that the dorsal and ventral roots join, though the details suggest that the union is not similar to that found in gnathostome vertebrates. The brain shows several features of reduction and simplification and no pineal eyes are present. There is only one semicircular canal in the ear (p. 98). The kidneys show a more generalized condition than in any other vertebrate in that the pronephros persists in the adult and is hardly marked off from the mesonephros, so that an almost continuous series of funnels and glomeruli can be

recognized. Moreover, there is a regular series of mesonephric glomeruli, a pair in each segment.

The development is known only in *Bdellostoma*, where the egg is yolky and cleavage partial, leading to the formation of an embryo perched on a mass of yolk. It is often stated that these animals are protandric hermaphrodites, because individuals are found in which the front end of the gonad contains eggs, whereas the hind part is testis-like (Fig. 4.52). No ripe sperms have ever been found in this region, however, and, moreover, individuals with fully testicular gonads do occur. Since it is known that in other vertebrates (including the lampreys) the gonads go through a hermaphrodite stage during development it seems likely that *Bdellostoma* is not a functional hermaphrodite but that the double-sexed gonad shows a rather late persistence of the indeterminate stage.

The hag fishes all live in the sea and their blood differs from that of other chordates in that it is isosmotic with sea water. However, the individual ions are regulated; sodium and phosphate exceed their values in sea-water, and the other ions are present in lower concentration. It is often assumed that fishes, with their glomerular kidneys, evolved in fresh water. However, the very earliest fragments of armoured agnathans are from marine Cambrian and Ordovician deposits that may be littoral or marine and it might be that the condition of the blood and kidney of *Myxine* is that of the earliest agnathans and that the glomerulus was not evolved as an adaptation to freshwater life, as is often supposed.

22. Fossil Agnatha, Heterostraci, the earliest known vertebrates

The organization of the lampreys and hag fishes shows that they preserve many characteristics from a very early stage of chordate evolution, probably that of about the Silurian period. Their special interest for us is in giving an insight into the organization possessed by the vertebrates before jaws were evolved. However, no doubt many changes have gone on during cyclostome evolution and we must not suppose that all Silurian vertebrates were like lampreys. Indeed, we may now complete our picture of this stage of evolution by examining the fossil fishes known to have existed at that period. We shall find them superficially so different from modern cyclostomes that only careful morphological comparison reveals the similarities. The inquiry will show us once again how a common plan of organization can be found in animals of very different superficial form and habits.

The earliest known craniate chordates are the agnathous ostracoderms known as Heterostraci or Pteraspida (*heteros* – different, *ostraco* – shell; *pteron* –

wing, *aspis* – shield). They first appeared in the Upper Cambrian 540 million years ago and survived until the Upper Devonian, 150 million years later. The earliest of them were marine bottom-dwellers but later some forms became nektonic and invaded various marine and freshwater habitats. The group was thus widespread and successful, but was ultimately gradually replaced by gnathostome fishes during the Devonian (see Moy-Thomas and Miles 1971; Halstead 1973).

The Heterostraci were mostly only a few centimetres long, but a few species reached 1.5 m. They had a wide anterior region covered by a carapace (Fig. 4.55). Behind this was a scaly trunk and tail. A heavy exoskeleton is present in many early craniates and has considerable similarity in the agnathans and gnathostomes (p. 152). The plates and scales are formed by a combination of more superficial denticles (placoid scales) and underlying bone. The denticles are like teeth, as the name implies, indeed the teeth of all later chordates are the only remnants of this tissue that was once found all over the body. Denticles are composed of the hard tissue dentine, produced by odontoblasts lying around a pulp cavity and with processes radiating outwards (Fig. 5.13). Over the surface is the hard shiny enamel, produced by ectodermal ameloblasts. Denticles may lie free and separate in the dermis, as in the dogfish (p. 124). In the ostracoderms, however, they were often joined together to form plates, and attached to the underlying bone (Fig. 4.56).

The lateral line sense organs (neuromasts p. 183) lay

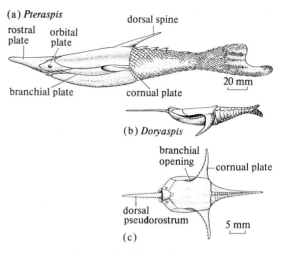

FIG. 4.55. Fossil pteraspids. (After Halstead 1973.)

in pits opening by pores between the ridges of dentine. They may lie in canals within the bone, where there are also vascular channels, but the inner layer of bone is compact (Fig. 4.56). The bone of Heterostraci is called isopedin, or aspidin, and is held by some to be an acellular, primitive type (Ørvig 1967; Tarlo 1969).

The heavy armour of the early forms must have made swimming difficult and probably ungainly. Besides the cranial plate there were others ventrally and covering the gills (Fig. 4.55). There were probably seven gill pouches but with a posterior branchial aperture. The

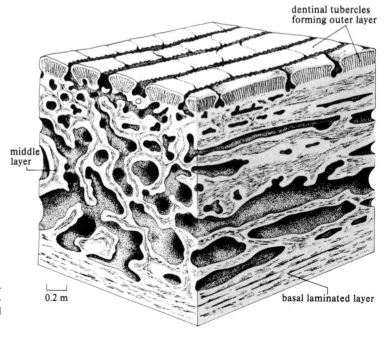

FIG. 4.56. Diagram to show the exoskeletal structure of **Psammosteus meandrinus*. (After Kiaer in Moy-Thomas and Miles 1971.)

FIG. 4.57. Restoration of shield of (a) *Hibernaspis macrolepis*, of (b) *Eglonaspis rostrata* and of (c) *Olbiaspis coalescens*. (a) and (b) dorsal view; (c) lateral view. (After Obruchev in Moy-Thomas and Miles 1971.)

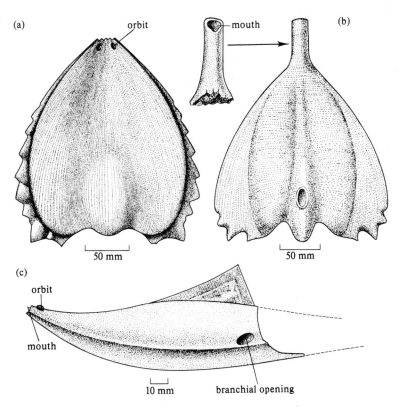

form of the gill pouches is uncertain. The tail was hypocercal (p. 122) with the lower lobe longer. There were no movable paired fins but in later forms there were rigid cornual plates, which may have served as gliding planes (Fig. 4.55).

The mouth was partly surrounded by long plates, suggesting that it formed a protrusible apparatus, that could be pushed out as a kind of scoop or shovel whereby mud and decaying refuse could be taken off the bottom, for it seems likely that such were the food and habits.

There may also have been a velar pump (as in lampreys, p. 80) and there is some evidence of a sac-like endostyle. In some of the later heterostracans the mouth was a tube at the very front end (Fig. 4.57).

These animals mostly had paired eyes and also a median pineal eye. They had paired nasal sacs, but the position of their opening does not appear from the fossils, at least there is no sign of a single hole. There were only two semicircular canals. The brain, if correctly interpreted, is rather like that of lampreys, with a choroidal roof to the midbrain.

The thelodonts (coelolepids) are little-known ostracoderms perhaps related to the Heterostraci but covered with small placoid-like scales or spines, which are often found isolated in Silurian and early Devonian rocks.

23. Osteostraci. Cephalaspids and anaspids

These are fossil agnathans that show even more similarity to the modern cyclostomes than the pteraspids. Besides the median pineal foramen their heads show another hole, which is presumed to be a single median nostril. The Osteostraci are the best known of all the ostracoderms and many details of their internal anatomy have left impressions on the inside of the flattened head shield. The rest of the body was fish-like, with an upturned tail (heterocercal, see p. 115) covered with heavy bony scales (Fig. 4.58). Pectoral fins were present in later forms and seem to have evolved several times independently within the group from lateral fin-folds (p. 155). They would no doubt increase the control of swimming. They were not yet present in *Tremataspis* from the Silurian.

On the dorsal surface of the shield are two median holes, one behind the other, which served a nasohypophysial opening and a pineal eye. The whole outline of the cranial cavity is preserved and shows a brain remarkably like that of a lamprey, with a nasohypophysial canal below it (Fig. 4.59). There were paired eyes and only two semicircular canals. Long tubes leading through the shield contained the cranial and spinal nerves, which can be reconstructed in detail (Fig. 4.60).

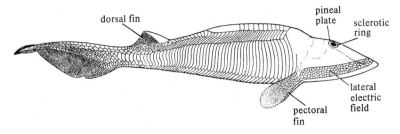

FIG. 4.58. A cephalaspid restored, *Hemicyclaspis*. (From Stensiö.)

They are said to resemble lampreys in the separation of the dorsal and ventral nerve roots and in other features. On the underside of the shield is a series of ridges, which outline a set of ten pairs of branchial pouches. The first of these lies far forward at the sides of the mouth and the ridge in front of it is probably the premandibular arch; it carries the profundus nerve (p. 128), which was large. The ventral surface of the head was flat and covered with small scales. Probably the gills were pouches, as in lampreys. Canals for the aorta, epibranchial arteries, and some features of the veins and heart have been preserved.

The mouth was a slit at the extreme front end with which the animals may have scooped decaying matter from the lake floor. On the dorsal surface there are sunken areas, covered by small scales, known as the median and lateral fields, and supposed by some to have contained electric organs. They were apparently served by a very rich blood supply and a system of wide canals leads to the vestibular region. These canals might have contained nerves, but Watson (1954) makes the far more likely suggestion that they housed tubular extensions of the labyrinth and served to carry pressure waves to the ear, perhaps providing a substitute reinforcement for the defective lateral line system.

We therefore know in some respects as much about these fossils as of many living fishes. They show in the complete segmentation of the head the most primitive condition known among craniates. Many of their features are very like those of modern lampreys and there can be little doubt that, as Stensiö (1958) suggests, the latter represent their surviving descendants, which have lost the bony shield.

The Anaspida (mostly Silurian) are placed near the Cephalaspids but they are less well known. They were small fishes (up to 15 cm in length) covered with rows of

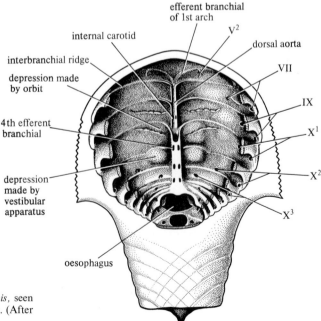

FIG. 4.59. Head shield of the cephalaspid *Kiaeraspis*, seen from below. Roman numerals indicate cranial nerves. (After Stensiö.)

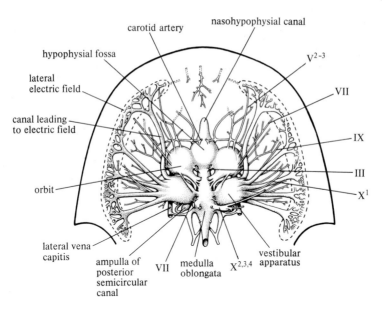

FIG. 4.60. Cast of endocranium and system of canals in the head shield of *Kiaeraspis*. Roman numerals indicate cranial nerves. (After Stensiö.)

bony scales (Fig. 4.61). The tail shows a lower lobe larger than the upper ('hypocercal'). This would presumably serve to drive the head end upwards perhaps to compensate for the weight of its armour. The anaspids possessed a curious ventral or ventrolateral fin-fold (Fig. 4.60) or perhaps a series of them. There were large paired eyes, median holes presumed to be nasal and pineal and a series of up to fifteen small round gill openings.

Jaymoytius, from the Silurian of Scotland is a small, little ossified creature that is probably an anaspid (Fig. 4.62). There was a continuous unconstricted notochord and a hypocercal tail. The body shows a series of transverse structures that have been interpreted as myotomes, but are probably thin anaspid scales. There were lateral fin-folds, presumably representing the continuous pectoral and pelvic fins (p. 153). There was an annular cartilage round to the mouth and a series of up to fifteen branchial pouches, surrounded by a branchial basket very like that of lampreys (p. 78). This form is of great interest and has been considered to be the most primitive of the 'vertebrate' series of which we have knowledge. It is suggested that it might be the ammocoete larva of an ostracoderm.

The agnathan fossils of China formed a separate radiation in the Devonian, and are placed here as a distinct order, the Galeaspida (Halstead, Liu, and P'an 1979). The bony covering of the flattened, front part of the body was drawn out into lateral spines. It is

penetrated by paired openings for the eyes and a large median one which is probably for a nasal or nasohypophyseal canal. The structure of the brain and other parts can also be observed in these fossils (Fig. 4.60A). There were two semicircular canals. There are up to 20 structures on either side which have been called gills but may be somites (Halstead 1979).

The affinities of these ostracoderm fossils with each other and with the cyclostomes have been much disputed. Lankester claimed that pteraspids were related to cephalaspids 'because they are found in the same beds, because they have a large head shield and because there is nothing else with which to associate them'. At the other extreme Stensiö holds that we have sufficient evidence to assert that the pteraspids have given rise to the myxinoids, and the cephalaspids to the lampreys. Except for the absence of jaws there is indeed little in common among the fossil forms.

The Agnatha were the first animals of the chordate type to become large, and they apparently all did so by feeding on the detritus on the bottom. They evolved into various types, mostly rather heavily armoured and perhaps slow-moving forms. The lampreys and hag fishes have been derived from early Agnatha by the evolution of a sucking mouth, perhaps with loss of the bony skeleton and paired limbs. However, it was the unknown forms that evolved a biting mouth that made the next great advance in vertebrate evolution.

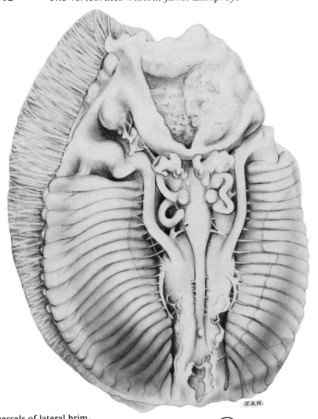

FIG. 4.60A. *Duyunaspis paoyangensis* from Peking. Drawing of dorsal view showing mineral infilling of canals and spaces in endocranial connective tissue which have become exposed subsequent to the removal of the original calcified tissues. Identification of structures indicated in the accompanying diagram. (Halstead 1979.)

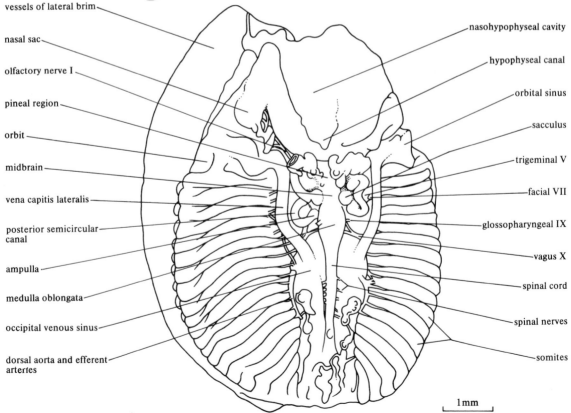

vessels of lateral brim

nasal sac

olfactory nerve I

pineal region

orbit

midbrain

vena capitis lateralis

posterior semicircular canal

ampulla

medulla oblongata

occipital venous sinus

dorsal aorta and efferent arteries

nasohypophyseal cavity

hypophyseal canal

orbital sinus

sacculus

trigeminal V

facial VII

glossopharyngeal IX

vagus X

spinal cord

spinal nerves

somites

1mm

FIG. 4.61. Lateral view of restoration of *Pharyngolepsis oblongus*. (After Ritchie in Moy-Thomas and Miles 1971.)

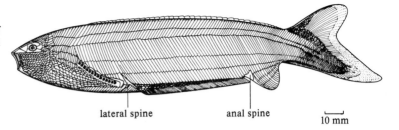

lateral spine   anal spine

10 mm

Jaymoytius kerwoodi

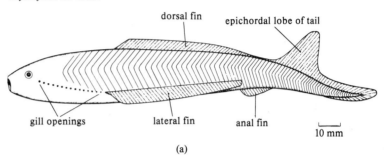

dorsal fin   epichordal lobe of tail

gill openings   lateral fin   anal fin

10 mm

(a)

sclerotic cartilage   notochord

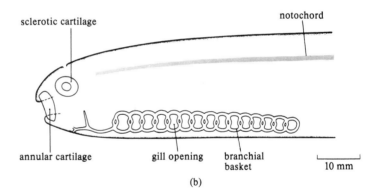

annular cartilage   gill opening   branchial basket

10 mm

FIG. 4.62. **Jamoytius kerwoodi* (a) lateral view of restoration and (b) of the head region. (After Ritchie in Moy-Thomas and Miles 1971.)

(b)

5 Fishes

1. Gnathostomata

Up to the end of the Silurian period, 395 million years ago, all the known vertebrates were agnathous, with mouths suitable only for filtering or feeding on soft invertebrates (or as parasites). The appearance of biting jaws in the new race of Gnathostomata was a revolution that enabled the vertebrates to become more active carnivores, feeding on each other and perhaps on their previous enemies the scorpion-like Eurypterids. There are few clues as to the ancestry of the gnathostomes and by the early Devonian various types were already present, some that were probably ancestral to those alive today and others very different.

Living fishes with jaws mostly fall into two well-marked classes, the cartilaginous fishes (Chondrichthyes) such as the sharks and rays, and the bony fishes (Osteichthyes) including the familiar ray-finned fishes and the lung fishes (Fig. 5.1). These two groups arose in the late Devonian. Before that time various other types of fish dominated the waters. We shall recognize two major groups (1) the acanthodians provided with spines, perhaps related to ancestors of the bony fishes and placed as a subclass of them (p. 189) and (2) the placoderms, considered to be closer to the cartilage fishes although well provided with bone, which we shall treat as a distinct class (p. 150).

The arrangement of major groups that we shall follow is therefore:
Superclass 2. GNATHOSTOMATA
  Class 1. *Placodermi
  Class 2. Chondrichthyes
    Subclass 1. Elasmobranchii
    Subclass 2. Holocephali
  Class 3. Osteichthyes
    Subclass 1. *Acanthodii
    Subclass 2. Actinopterygii
    Subclass 3. Sarcopterygii
Authors differ greatly in the way they arrange the main divisions of the animal kingdom. Some of the suggested alternative classifications will be discussed later.

To become familiar with the significant features of fish life we shall first describe the cartilage fishes, including the familiar dogfish. The earliest gnathostomes (placoderms and acanthodians) will be considered later.

2. The elasmobranchs: introduction

In all parts of the sea there are to be found members of the class Elasmobranchii (literally 'plate-gilled' fishes), including sharks ranging from monsters of 17 m long to the common dogfish *Scyliorhinus caniculus* of 30–60 cm, and the smallest sharks, *Squaliolus* and *Euprotomicrus* of less than 30 cm. Nearly all the fishes in the group are carnivores or scavengers: the skates and rays are bottom-living relatives, feeding mostly on invertebrates. Although elasmobranchs are not quite so fully masters of the water as are the bony fishes, they are yet well

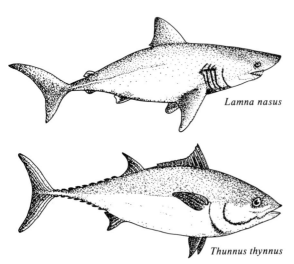

Lamna nasus

Thunnus thynnus

Fig. 5.1. Above is a cartilage fish, a mackerel shark (*Lamna nassus*) and below a bony fish, the blue-finned tunny (*Thunnus thynnus*). The two are only very distantly related but show convergence in shape and form. (After Marshall 1965.)

enough suited to that element to survive in great numbers in all oceans. Perhaps the skill and cunning of a shark is exaggerated by the frightened boatman or bather, who is apt to mistake a keen nose and the persistence of hunger for intelligence, especially when he is faced at intervals with a well-armed mouth; but the sharks have a large brain and their active, predacious habits enable many of them to live by eating the perhaps more elaborately organized bony fishes.

Evidently such active creatures have changed considerably if they have been evolved from the heavily armoured and probably slow-moving agnathous vertebrates that shovelled up food from the bottom of Palaeozoic seas. It used to be supposed that these elasmobranch or cartilage fishes represent a very primitive stock, but we now realize that there have been great changes since the biting mouth was first evolved; we cannot be sure that any features we find in the elasmobranchs were possessed by the earliest gnathostomes.

In elasmobranchs, which are mostly heavier than water, the tail is heterocercal with the dorsal lobe larger than the ventral, giving an upward lift. This type of tail allows for steering in the vertical plane and, together with the flattened pectoral fins, compensates for the lack of buoyancy (p. 120). There are two dorsal fins, which secure stability against rolling, and also assist in making possible the vertical turning movements.

The muscles for the production of the swimming movements are a serial metameric set, with longitudinal fibres, essentially like those of the lamprey or amphioxus. The central incompressible axis is no longer a simple rod. The notochord becomes replaced during development by a series of vertebrae. Each of these consists of a centrum carrying dorsally a neural arch around the spinal cord and ventrally either a pair of ribs or a haemal arch (in the tail) (Fig. 5.2). The centrum is formed from layers of cartilage laid down around the notochord or within it. The vertebrae are slightly hollow at both ends (amphicoelous), the cavities between them contain pads, the intervertebral discs, derived from the notochord. The vertebrae are held together by ligaments but not articulated together by complex facets as they are in land animals (p. 230). As waves of contraction pass down the body each centrum is first compressed and then stretched. To meet these stresses the centra are reinforced by longitudinal bridges of calcification, placed radially as seen in cross-section and each thickens peripherally, where the stressing is greatest (Fig. 5.2).

Between the main neural and haemal arches there are accessory cartilages called interdorsals and interventrals (Fig. 5.3). Some authorities suppose that each vertebra is formed from four fundamental parts, basidorsals, basiventrals, interdorsals, and interventrals. It is hard to apply this theory to the vertebrae of other fishes (p. 166) and probably the interdorsals and interventrals in modern chondrichthyes are secondary, they are not present in early sharks or placoderms.

The muscle fibres exert pull upon the skin of a shark so that the internal pressure increases more than tenfold from slow to fast swimming. The skin thus acts as an 'external tendon', transmitting muscular force and displacement to the tail. The great length of this exotendon gives it a mechanical advantage greater than that of the endoskeleton (Wainwright, Vosburgh, and Hebrank 1978).

The median and paired fins are supported by cartilaginous rods, the radials, and their edges are strengthened by horny fin-rays, the ceratotrichia. The radials of the paired fins form a series attached to larger rods at the base. These more basal rods are attached to a 'girdle' of cartilage embedded in the body wall. The pectoral girdle is a hoop extending some way round the body, but the pelvic girdle is simply a transverse rod in the abdominal wall. The origin of these girdles and of the fins will be discussed later (pp. 153, 165).

3. The swimming of fishes

The propulsive forces that move a fish through the water are usually produced by the longitudinal muscle fibres of the myotomes (myomeres), which are blocks of longi-

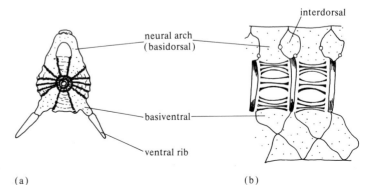

FIG. 5.2. Trunk vertebrae of the shark, *Lamna*, (a) seen from in front and (b) from the side. (After Goodrich 1930.)   (a)

(b)

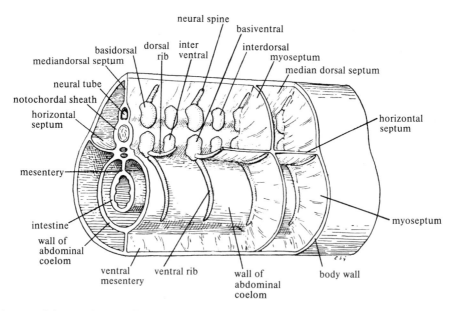

FIG. 5.3. Diagram of the organization of a vertebrate showing the relationship of the axial skeleton to the transverse and longitudinal septa. (After Goodrich 1909.)

tudinal muscle fibres, placed on either side of an incompressible central axis, the notochord or vertebral column, and attached to connective tissue partitions, myosepta (myocommas). The myomeres form closely packed segments with ≪ or ≷ shapes and a horizontal septum divides the dorsal and ventral halves. The effect of these shapes is that contraction of the muscle fibres bends the body laterally but not dorsoventrally. There is a thin outer layer of red muscle fibres, but the central mass is of white muscle fibres. With these two types of muscle fibre the fish can meet the conflicting demands of economy in low-speed cruising and short bursts of maximum speed (Bone 1978). With the support possible in water fishes can afford to carry a large mass of fast white fibres used only occasionally for attack or escape. The red muscles contain many mitochondria and much myoglobin, they respire aerobically using fat, have a rich blood supply and are not readily fatigued. The white fibres are specialized to contract fast with maximum power, they respire anaerobically using glycogen, and can build up a lactic acid debt for a short time. They have a less rich blood supply than red muscles. Electromyography of dogfish has shown that the deep fast fibres become active only for up to two minutes. In higher teleosts the fast fibres may be active in continuous swimming even at low speeds and this may be connected with their multiple innervation (see below). There is evidence that in some fishes the lactate produced by fast fibres can be oxidized by 'co-operative metabolism' in the red fibres. This lactate can also be used to drive the

ion pumps of the gills. In the more specialized teleosts (such as scombrids) the white fibres are arranged as interlocking cones so that they lie obliquely to the backbone and can all shorten at the same speed, allowing an optimum power output. The cones are attached to tendons which oscillate the tail foil. The power source is thus isolated from the caudal propeller, which is a hydrodynamically efficient plan.

As a fish gets larger the power that the muscles can produce increases roughly in proportion to volume, that is to length cubed, whereas the power needed to overcome drag varies with length squared. The speed produced actually increases with length from a 0.35 to a 0.60 power depending upon the Reynold's number and other factors.

Muscle forms 40–60 per cent of the body mass in fishes. They can afford to carry more than terrestrial animals and they need more to move in the denser medium. The muscles differ in many ways from those of the other vertebrates (Bone 1978). Each twitch fibre of most elasmobranch muscles receives two motor nerve fibres in a single motor end plate near its end. In teleosts the twitch fibres receive many nerve fibres along their length instead of single motor end plates. The significance of these interesting differences is not known.

Neuromuscular spindles, present in all tetrapods, are absent from fishes. It is doubtful how far animals with no 'posture' have need for them. In rays there are beaded proprioceptor endings both in parallel with muscle fibres and in series at their ends and these have some of the properties of spindles (Fig. 5.4). In dogfishes there are coiled receptors, the Wunderer corpuscles, in the outer ends of the myosepta and elsewhere, which are sensitive to frequency and amplitude of flexure.

4. The hydromechanics of fish propulsion

The contractions of the myotomes produce forward locomotion by setting up a sinusoidal oscillation of parts of the body about the axis of progression. There is still controversy as to the way in which the parts of the body and the fins act upon the water (see Blight 1976). In long-bodied forms such as lampreys, eels, and dogfishes the oscillation of each segment shows a slight phase lag behind the segment in front of it. This has often been interpreted as producing forward movement by the passing of waves of contraction down the body, making a series of inclined planes directed backwards and oscillating from side to side (Gray 1957, 1968). This is the pure undulatory mode of propagation, called anguilliform after the common eel *Anguilla*. The whole body participates, though the amplitude increases to-

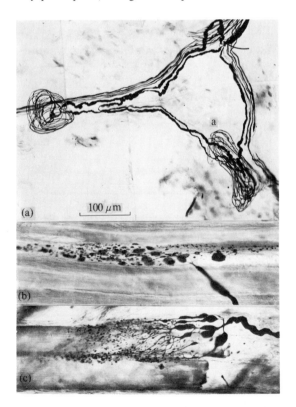

(a)

(b)

(c)

100 μm

Fig. 5.4. The three proprioceptive endings known from elasmobranchs. (a) Wunderer corpuscles found in the flank of the body in *Scyliorhinus*. (b) Stretch receptor endings from among the slow fibres of the pectoral fin of a ray. (c) Similar receptor from caudal myotome surface in ray. (Bone 1978.)

wards the tail. In this form of propulsion much energy is wasted in the production of a vortex at the tail. In most fast moving fishes this loss is reduced, and the vortex turned to advantage, by limiting the undulation to the posterior third of the body. This mode of propulsion is called carangiform, after the teleostean family Carangidae (see Lighthill 1975).

In a study of the problem Blight shows that there are rarely more than two or three flexures in the body at one time and forward progression can take place simply by bending from side to side (1976, 1977) (Fig. 5.5). The principle involved can be well seen in the earliest movements of amphibian tadpoles. The body passes from one S form to the opposite one by contraction on one side while retaining the bend of the tail (Fig. 5.6). 'In effect, the animal bends itself into a flexure to one side at the fore end and out of a flexure to the other side at the rear' (Blight 1977). Electrodes implanted at different levels in fact show simultaneous contraction all down the body in the earliest movements of a tadpole (Fig. 5.7). A little later there is some delay more posteriorly when swimming slowly, but again simultaneity in rapid movement. 'The requirements for continuous development of a propulsive thrust is that the animal be in a continuous state of transition between opposite flexures and always be in possession of a pair of contralateral bends. It must therefore begin a third bend before it has lost the first' (Blight 1977).

The chief difference between the swimming of a newt embryo and a typical adult fish is that the amplitude of the oscillation of the head is reduced by introduction of a phase lag (Fig. 5.5). The flexure is reversed at the head before it is completed at the tail so that the animal always has at least two bends in the body at the same time, sometimes three, instead of the simpler condition of alternating one with two. This has some hydrodynamic significance but its chief advantage may be that it allows stabilization of the head for vision. In very rapid escape movements the head is in fact often allowed to oscillate.

The magnitude of the forward thrust generated by a fish depends, among other things, upon (i) the area of the surface of the body and fins, (ii) the speed and extent of their lateral movements, and (iii) their angle of attack. These factors vary with the shape of the body and the action of its muscles. The faster moving (carangiform) types of fishes have (i) large caudal fins, (ii) a much smaller length of body relative to its depth, (iii) less flexibility (as their many small bones show). In the fastest, such as the tunnies or mackerel sharks the body muscles transmit most of their pull to the tail stalk so that the caudal fin moves from side to side. Nevertheless 'the propagated character of the undulation is retained.

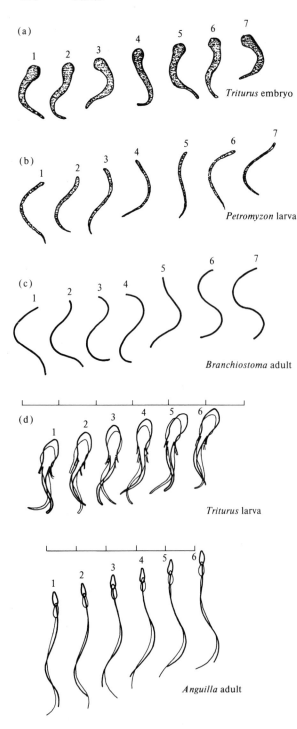

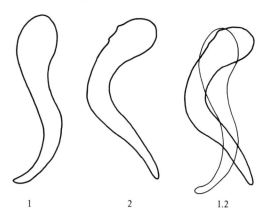

FIG. 5.6. Two succeeding positions from the swimming cycle of a *Triturus* embryo, separately and superposed, showing how contraction on the right bends the fore part of the body and straightens the tail. (Blight 1977.)

The caudal fin changes its angle of inclination as it moves from side to side in such a way that its movement always has a backward-facing component. (One may think of the resistances to this backward-facing component as providing the thrust, but this approach is almost impossible to make quantitative or indicative of

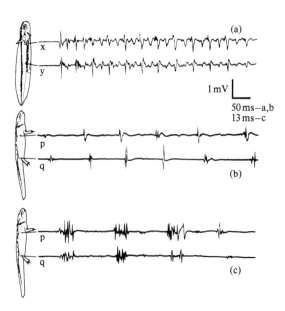

FIG. 5.7. Electromyograms made during swimming activity of *Triton*. (a) Embryo, using monopolar wire electrodes, showing ipsilateral responses as large spikes and contralateral ones smaller. There is no longitudinal delay recorded over the length of the trunk between x and y. (b) Bipolar electrodes, showing only ipsilateral activity with longitudinal delay along the trunk between p and q. (c) Same larval stage as in b, swimming more vigorously with return to simultaneity. (Blight 1976.)

FIG. 5.5. Swimming movements traced from ciné film. Time intervals between succeeding positions is 16 ms in a, b, and d; 13 ms in c; and 125 ms in e. In d and e pairs of subsequent positions are superposed, the preceding positions with finer lines. (After Blight 1977.)

F<small>IG</small>. 5.8. The vortices cast off by the caudal fin trailing edge as the fish moves to the left. A jet-like, streamline pattern is induced. This was traced from streamlines experimentally obtained for such a configuration of vortices. (Goldstein 1938.)

how much propulsive energy is wasted to make a vortex wake)' (Lighthill 1975).

In this mode of propulsion no one complete wave length is present at any time and therefore side forces are not cancelled out as they are in anguilliform loco-motion. This would produce wake vortices and large recoil yawing movements of the head. These undesir-ables are avoided by the very narrow base of the tail because this is the region where wave amplitude in-creases fast, which would produce large local side forces on the water. Secondly, side-slip or yaw is reduced by increasing the depth of the body and fins.

The typical fish tail acts as an oscillating hydrofoil with swept back shape, designed to produce maximum thrust with minimum drag, having a high aspect ratio (fin span^2/fin area) and a fork or sickle shape. This lunate type of tail evidently produces the maximum speed and efficiency, though the precise reason for this are still not understood. It seems that 'the most hydromechanically advantageous configuration has the leading edge bowed forward but the trailing edge straight' (Lighthill 1975).

The precise nature of this advantage is still a matter for investigation but Lighthill makes some very interest-ing suggestions about its relation to the vortex wakes. These may serve to produce thrust instead of drag, inducing a jet-like motion where momentum is related to the thrust on the fish (Fig. 5.8). The effect as shown in this figure is two-dimensional but with high aspect ratio lunate tails of certain outline it would become three dimensional. Fishes swimming in schools may improve further on this and share the thrusts, either by swimming directly above each other or in a diamond pattern where the vortices may help to drag them forward (Fig. 5.9) (Lighthill 1975).

The shape of the body also has an important influence on the effect of the fish on the water and hence on the turbulence in the flow of water and the resistance that must be overcome. Gray showed the strange paradox that in a dolphin the resistance cannot be that of a rigid model towed at the speed at which the animal moves, since this would require that the muscles generate energy at a rate at least seven times greater than is known in the muscles of other mammals (1936). By watching the flow of particles past the body of fish-like

models he found that movements such as those pro-duced in swimming accelerate the water in the direction of the posterior end, and this would greatly reduce the turbulence. Various other suggestions have been made as to how drag may be reduced by boundary layer control. The water expelled from the gills perhaps keeps the layer charged with sufficient energy to avoid sepa-ration. In turning, a fish closes the gills on the inner side and concentrates all the efflux on the outside bend, where the tendency to boundary layer separation would be greatest. There have been many suggestions that mucus secretion also has this function (as well as protecting against infection). It is possible that small quantities of high polymers damp turbulence in the surface layer.

The shape of the body of a fish, like that of a ship, can be analysed into a shorter forebody, or entrance, and a larger afterbody, or run, the division being at the position of greatest cross-sectional area. The greater backward taper is the essence of streamlining. It reduces the drag due to turbulence for the reasons shown in Fig. 5.10. The drag due to skin friction is probably less important than that due to turbulence but many swift fishes have smooth skins with few or no scales (e.g. sword fishes). Actual speeds have been measured by making fishes swim in the water-filled rim of a moving wheel. Larger fishes swim according to the relation:

$$\text{Velocity } (V) = \tfrac{1}{4} (L (3f - 4))$$

for dace. Thus with increasing length (L) the fish

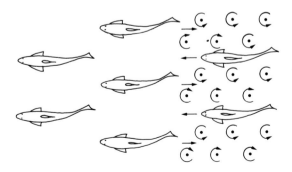

F<small>IG</small>. 5.9. Diamond-shaped fish schooling pattern. (Lighthill 1975.)

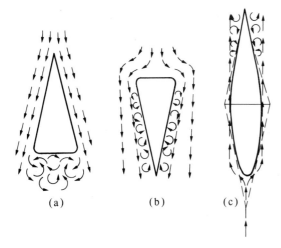

FIG. 5.10. A cone produces a greater drag when pulled with the apex forwards, because of the turbulence in its wake. Minimum turbulence is produced by a blunter forebody and a longer pointed afterbody. In this way the flow remains aturbulent along the greatest possible surface. (From Marshall 1965.)

maintains a given speed with fewer beats (f). Maximum recorded speeds vary from 0.8 km/h for gobies (*Gobius*) to 70.8 km/h for tunny (*Thunnus*). High speeds are maintained only for a few seconds, probably due to accumulation of lactic acid.

Something is known of the nervous mechanism responsible for the production of the swimming waves. An eel can swim if its whole skin has been removed. If a region of the body is immobilized by a clamp, swimming waves can pass along past the immobile section. Therefore the rhythm is determined by some intrinsic activity of the spinal cord and not by any mechanism such as proprioceptor impulses arising in active muscles and causing others to contract.

Experiments in which the spinal cord was cut across show that in the eel the rhythm is only initiated when suitable impulses reach the cord either from spinal afferents or from the brain. Thus the spinal eel can be made to swim either by fixing a clip on to its caudal fin or by electrical stimulation of the cut end of the spinal cord. Though the cord requires such afferent stimuli for its functioning, they do not determine the frequency of the rhythm, which bears no relationship to that of the applied stimuli.

In the dogfish the isolated spinal cord is able to initiate rhythmic swimming. After transection behind the brain the posterior portion of the fish exhibits continuous swimming movements for many days. Light touch on the sides of the body inhibits these movements, but some sensory impulses are necessary for their initiation; after complete deafferentation, by section of all the dorsal roots, the movements cease.

The information available does not yet enable us to understand fully how the swimming rhythm is initiated and maintained, nor how it is influenced by the brain. It would be very interesting to have further knowledge on these topics, especially because the locomotor rhythms of land animals are probably based on the serial contractions of their fish ancestors.

5. Equilibrium of fishes in water; the functions of the fins

Making use of the methods of investigation of aeronautical engineers, studies have been made of the forces that operate to keep a fish stable as it moves through the water, or allow it to become temporarily unstable and hence to change direction. Instead of attempting to study a living or dead fish moving in water, Harris (1936) made models and supported them in a wind-tunnel in an apparatus suitable for measuring the forces at work in the various directions. Such a method, in which no compensating movements of the fins are allowed, makes it possible to investigate the so-called 'static stability' of the fish, that is to say, to see whether the body and fins are so shaped as to provide forces that tend to bring the fish back into its previous line of movement after it has deviated in any direction. Any body such as a fish or aeroplane is said to be in stable motion if when it veers slightly from its line of progress the new forces produced upon its planes tend to restore the original direction of motion.

The forces acting on the fish are measured along three primary axes, longitudinal, horizontal, and vertical. Deviation from the line of motion about the longitudinal axis is known as rolling, about the transverse axis as pitching, and about the vertical axis as yawing (Fig. 5.11). The forces along these three axes are known as drag, lateral force, and lift.

In order to discover the effect of the median fins and tail on the stability, these fins were removed, the heterocercal tail being replaced by a cone having the same taper as the actual caudal fin. The model was then placed in the wind-tunnel with a wind at 64 km/h, which corresponds to a motion of 4.8 km/h in water. The lateral force was measured when the body was made to yaw at various angles. The results showed that the equilibrium in this plane is quite unstable; a slight turn off the direct course would produce a turning moment tending to increase still further the deflection (Fig. 5.12). This is a well-known property of all airship hulls, and is known as the 'unstable moment' of the hull. It is corrected in the airship by the addition of suitable horizontal and vertical fin surfaces at the rear end, so

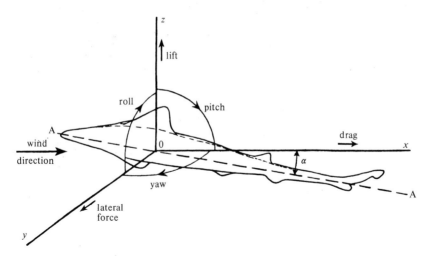

FIG. 5.11. Diagram of model of the dogfish *Mustelus*, showing the conventional terms for describing deviations of motion. The longitudinal axis *x* is that of the wind tunnel and *y* (horizontal) and *z* (vertical) are at right angles to it. The arrows show the direction known as positive rolling, pitching, and yawing, which occur about the *x*, *y*, and *z* axes respectively. α is the angle of attack between the axis of the model and the *x* axis. (From Harris 1936.)

that the airship becomes in effect a feathered arrow. The forces operating on the fins tend to bring the body back into the original line of motion.

The fins of the fish operate in a similar manner. If the experiment is performed with a model to which all the fins behind the centre of gravity have been added, namely, the caudal, anal, and second dorsal fins, it is found that the curve for the yawing moment now has a steep negative slope (Fig. 5.12 (b)), that is to say, every deviation produces forces that tend to give directional stability. With the first dorsal fin also in position the model possesses a remarkable neutral equilibrium (Fig. 5.12 (c)). Deviations by as much as 10° produce no resultant yawing moment about the centre of gravity. The form of the dorsal fins is therefore definitely such as to maintain stable swimming and prevent yawing.

Turning of a fish is produced either by the propagation of a wave down one side only of the body or by asymmetrical braking with the pectoral fins (see below). The former type of turn has been investigated by Gray in the whiting, where there is a large caudal fin. This gives great lateral resistance, so that the first part of the turn is executed by bending the front part of the fish on the tail as a fulcrum. This enables the animal to turn through 180° within a circle of the diameter of its own length. After removal of the caudal fin the turns are much less effective.

In both elasmobranchs and teleosts the dorsal fins are well developed in the active swimmers. In most elasmobranchs they are fixed, but in many teleosts the dorsal fin can be raised up or folded down, and it is observed that the fin is raised during turning. This would have the

effect of increasing the yawing moment produced by asymmetrical action of the body muscles or by unilateral braking with the pectoral fins.

Since the body is so markedly flexible in the lateral plane and there are powerful muscles available for turning it in this direction, the part played by the fins in determining the stability is important mainly when the body is held straight. The fish thus has the double advantage of great stability (by keeping the body straight) and great controllability (by bending it). In a body unable to change its shape in this way, stability and controllability would be inversely related. This is the case for the stability of the fish in the vertical plane, in which the body is little flexible. Fig. 5.12 (d) shows the positive slope of the curve for the pitching moment and clearly the equilibrium in this plane is quite unstable. The pectoral fins contribute more than any others to movement in this plane, and since they lie in front of the centre of gravity they greatly increase the stability. The fish must be able to alter direction in the vertical plane, and it has apparently sacrificed static stability for controllability. The equilibrium in this plane is a dynamic one, controlled by the movable pectoral fins, and it is so unstable that only a small movement of these fins is necessary to produce a deflecting force that restores the original direction of motion.

The pectoral fins, lying in front of the centre of gravity, tend to produce a movement of positive pitch, that is to say, they force the head upwards. This effect is normally compensated by a component produced by the heterocercal tail. The upper lobe of this is rigid and the lower is flexible and bends so that it is not vertical and

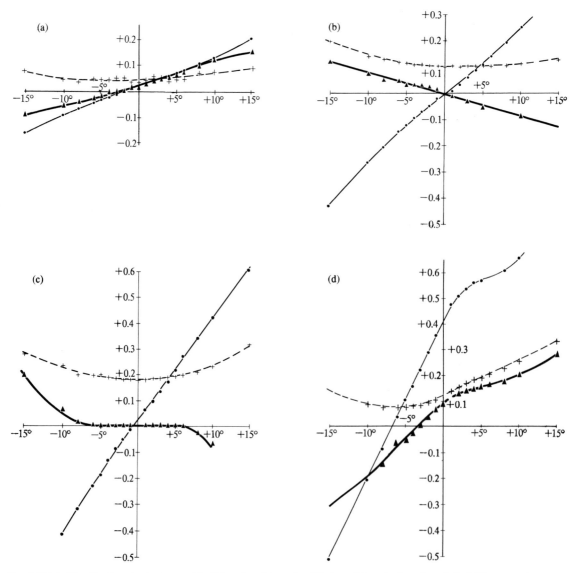

FIG. 5.12 (a). Results of yawing test on model of *Mustelus* without fins. The lateral force is plotted as a light full line, drag force as a light broken line; yawing moment about centre of gravity as full heavy line. Abscissae show the angle of attack in degrees, ordinates the lateral force and drag in pounds weight, yawing moment in in-lb × 1/10. (b) Yawing test similar to (a) but with the fins behind the centre of gravity in place. (c) Yawing test with all median fins in place. (d) Pitching test on model of *Mustelus* with all fins intact and pectoral fins set at an angle of incidence of 8°. Lift force is shown as a light full line, drag force as a light broken line, pitching moment about the centre of gravity as a heavy full line. (From Harris 1936.)

produces a lift force on the tail, giving, of course, negative pitch. After amputation of the hypocaudal lobe and anal fin a dogfish swims continually along the bottom of the tank: in order to compensate for the absence of negative pitch the pectoral fins are held horizontally and hence there is no moment to counteract the weight of the fish. If the pectoral fins are then also removed the anterior end of the body is pointed upwards, often so much so as to cause the fish to swim with its head out of the water. This is the result of an over-strenuous attempt to compensate, by raising the head, for the negative pitch produced by the tail. The system is no longer suitable for making the continuous adjustments necessary to ensure stability. The heterocercal tail allows for versatility in the orientation of the forward thrust from the tail (Thomson 1976). Its action can be modified to produce either a horizontal swimming or a powerful epibotic (head up) or a hypobotic (head down) movement. The shark can make steep dives or climbs in the water to attack the prey and shear off bits of it. The dorsal angle of thrust is large in the big, swift epipelagic sharks, least in bottom living and littoral forms like the dogfish.

This analysis makes it clear why a heterocercal tail is found in almost all the primitive swimming chordates; it is almost a necessity for an animal with a specific gravity in excess of the medium and little flexibility in the vertical plane. The component of positive pitch could be provided by the flattened head or by continuous lateral fin folds, such as may have been present in early fishes, and adjusted by the limited flexibility possible in the fin. The development of movable pectoral fins confers much greater control. Since the useful portions of a fin fold for this purpose would be those well in front of and behind the centre of gravity, we can perhaps see the reason why the intervening portion has become lost. In the modern sharks the pelvic fins have little influence on the stability and are perhaps retained only for their modification as claspers.

Elasmobranchs adopt various devices to reduce their specific gravity. The liver contains unsaturated oils such as squalene ($C_{30}H_{70}$) and may constitute one-fifth of the weight of a shark but is less in skates and rays. The blood contains less salt than the sea and the osmotic deficiency is made up by urea and the even lighter trimethylamine oxide (p. 136). The loss of calcified tissue may originally have been determined by similar factors.

6. Skin of elasmobranchs

Being swift and predatory animals, more attackers than attacked, the sharks do not possess a very heavy external armament. The epidermis is often covered with close-set scales (*Scyliorhinus*). It also contains mucus-secreting cells, giving protection against infection and regulating permeability. Beneath this is a thick dermis of connective tissue with fibres arranged in spirals some right-handed, others left-handed, making a meshwork like a hammock, of great strength and flexibility, able to maintain the shape of the body and allow bending. Scattered over the skin are the characteristic denticles or placoid scales (Fig. 5.13). Each of these consists of a pulp cavity, around the edge of which lies a layer of odontoblasts secreting the calcareous matter of the scale, known as dentine. This has a characteristic structure resulting from the fact that the odontoblasts send fine processes throughout its substance. The outside of the dentine is covered by a layer of enamel, secreted by the ameloblasts of the overlying ectoderm. Part of the outer layer differs from true enamel,

however, and is of mesodermal origin and is known as chirodentine. There is also some acellular bone at the base of the teeth. The outer (enamel) layer has the hardness of steel on Mohs scale (Moss 1977). Usually the denticles pierce through the ectoderm, after which no further enamel can be added to their surface. Obviously the scales are similar to teeth, which are indeed to be considered as specialized denticles developed on the skin of the jaws. It has often been supposed that the denticle is the primitive type of fish scale, from which others have been derived, but it now seems more likely that the earliest covering was a continuous layer, later broken into large scales, from which the denticle was ultimately derived (p. 190).

The skin also gives protection to the fish by its colour, produced by a layer of chromatophores beneath the epidermis. Many sharks have a spotted or wavy pattern, which breaks up their visible outline as they move in the water, especially near the surface. They are able to change their colour, though only slowly, becoming darker on a dark background (see p. 138).

7. The skull and branchial arches

In general organization a dogfish follows closely the fish plan, which we have already considered. Most of its special new features are in the head, and we may now turn to a consideration of the organization of the head and jaws of a gnathostome vertebrate. The jawless vertebrates of the Silurian and Devonian included animals of various types, but the vertebrates began to flourish and increase more abundantly with the appearance of creatures with jaws in the late Silurian. From this stage onwards we have to follow the parallel history of numerous orders and families, as the vertebrate plan of structure became adapted for various habitats. It seems likely that the development of a biting mouth greatly increased the range of possibilities of vertebrate life. The most obvious use of a mouth is for attacking other animals, but it may also have been used to collect plant food from all sorts of situations where it would not be available to the microphagous or shovelling Agnatha. Probably the mouth was also early used for defence, and in this way influenced the whole bodily organization, making unnecessary the heavy armature that is so characteristic of many early vertebrates. Modern research has shown that the armour has become progressively reduced along various lines of fish evolution. Older ideas of comparative anatomy regarded the cartilage fishes as showing a primitive stage, preceding the appearance of bone. We now realize that this is the opposite of the truth and that the dogfish and its relatives represent a higher type, able to defend themselves by mobility, by biting, and by efficient

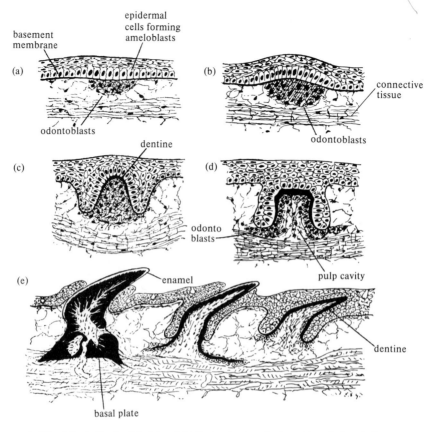

FIG. 5.13. Development of denticles in the dogfish. (a) and (b) first gathering of odontoblasts below the basement membrane and the epidermal cells that will become modified. (c), first deposition of dentine. In (d) there is more dentine and a pulp cavity. In (e) are the stages in the formation of enamel and the basal plate, while the denticle cuts the epidermis. (From Goodrich 1909.)

sensory and nervous organization. Heavy defensive armour is a primitive form of protection for animals, as for man.

Besides its use in feeding and defence, the mouth can also be used as a means of 'handling' the environment, for instance in the nest-building activities of many fishes. Indeed, it is difficult for us to realize the utility of the jaws for an animal not provided with any other means of seizing hold of objects.

The development of the mouth to a point at which it could be used in these varied ways was, therefore, a very important stage in evolution. Recognition of the Gnathostomata as a separate group of animals is far more than a matter of classificatory convenience, it marks the achievement of the possibility of life in a greatly increased range of environments.

Morphological analysis enables us to see how this biting mouth was produced, by modification of one or more of the gill slits. The main differences that separate the gnathostome from cyclostome vertebrates are there-

fore in the head and its skeleton. Although the modern elasmobranchs show the skull and jaws in a modified and reduced condition, they provide by their simplicity a good starting-point for discussion. The 'skull' of a dogfish consists of a series of cartilaginous boxes surrounding the brain and receptor organs (Fig. 5.14). The nasal capsules, orbital ridges, and auditory capsules are largely fused with the main cranium, producing a single continuous structure, the chondrocranium. It is interesting to consider how this structure has arisen during the process of cephalization. Presumably parts of it represent the modified sclerotomes of trunk regions. We shall see presently that there is strong evidence that the head has arisen by modification of a segmental arrangement such as is seen in the trunk; the morphogenetic processes that build the skull must therefore be related in some way to those of the vertebrae. The first rudiment of the skull in the embryo consists of two pairs of cartilaginous rods, the parachordals and trabeculae (Fig. 5.15). The former lie on either side of the noto-

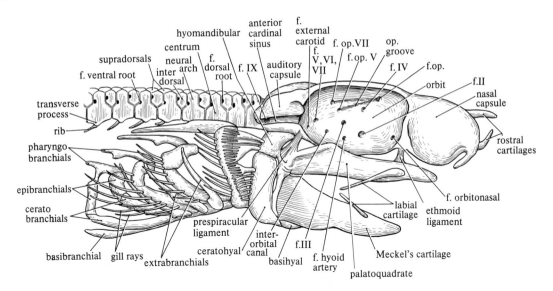

Fig. 5.14. Skull and branchial arches of the dogfish (*Scyliorhinus*). *f.* foramen; *op.* ophthalmic; *op. groove* for *op.V.*; *op.V.*, *op.VII*, opthalmic branches of *V* and *VII*; *II–IX*, foramina for cranial nerves. (After Borradaile.)

chord, the trabeculae in front of the notochord. These first rods fuse up to make a continuous plate; from this grow sides and roof, completing the cartilaginous neurocranium around the brain. Meanwhile cartilaginous capsules form around the nose, eyes, and ear, and become joined to the neurocranium. Posteriorly, behind the auditory capsules, the cranium is completed by the addition of a number of segmented elements, evidently modified vertebrae.

The problem is, therefore, to determine the nature of the pro-otic part of the skull. Before we can settle this we must consider the visceral or branchial arches. These are pairs of rods of cartilage developed in the walls of the mouth and pharynx, between the gill slits. In the dogfish each typical branchial arch (Fig. 5.14) consists of a series of four pieces, the pharyngo-, epi-, cerato-, and hypo-

branchials. Ventrally some of the arches join a median basibranchial plate. These rods lie in the pharynx wall and on their outer sides carry a series of projecting rods, the branchial rays and extrabranchial cartilages, whose function is to support the lamellae of the gills (p. 132).

There are five such branchial arches, differing only slightly from each other. In front of these lie two arches, the hyoid and mandibular, which, though modified, are obviously of the same series. The hyoid more nearly resembles a typical branchial arch. Its most dorsal element, the hyomandibular cartilage, is a thick rod attached dorsally to the skull by ligaments and at its lower end forming the support for the hind end of the jaw. It apparently corresponds to the epibranchials. The more ventral elements, cerato- and basihyal, resemble the corresponding members of more posterior arches.

Fig. 5.15. Diagram of skull of selachian embryo seen from left side before fusion of the main cartilages; cranial nerves black, numbered; arteries cross-hatched, *ic.* internal carotid artery; *o.* ophthalmic artery; *or.* orbital artery; *I–X* cranial nerves; *VIIh.* and *VIIp.* hyomandibular and palatine branches ·respectively of facial nerve. (From Goodrich 1930.)

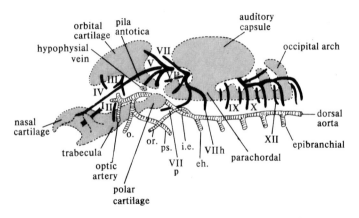

The jaws themselves (mandibular arches) depart more widely from the form of a typical branchial arch, but the two thick rods of which each is composed, the upper palatopterygoquadrate bar and the lower Meckel's cartilage, are recognizably members of the branchial series. Looking at the whole apparatus with a thought to the embryological processes that have produced it, with as it were a manufacturer's eye, we can see at once that the jaws and hyoid arch have been produced by a modification of the processes that make the branchial arches.

8. The jaws

Study of the serial relationship of the jaws and branchial arches gives us an understanding of the course of evolution of the mouth. We may suppose that the ancestors of the gnathostomes possessed a nearly terminal mouth, either on the front end of the body or on the ventral surface. The pharynx was pierced by a series of gill pouches, beginning shortly behind the mouth and separated by arches, each containing a set of cartilaginous bars (Fig. 5.16). There is some evidence that this condition persisted in the cephalaspids (p. 110), where there is found to be a series of ten pairs of gill slits, beginning far forward on either side of the mouth. The muscles moving the more anterior parts of the pharynx wall and the anterior arches could be called into play to help in the collection of food. In this way the mouth came to be used for prehension, and the grasping jaws of the gnathostomes appeared as the more anterior arches became modified to allow more efficient seizing, and the skin over them was modified to form the teeth. The mouth probably shifted backwards during this process and its lateral edges joined the first gill slit. The rods supporting the posterior wall of that slit thus became bent over into the characteristic position of the vertebrate jaws. The ventral mouth allows the nostrils to sample the food in front of it. Feeding on the bottom becomes possible with special methods for detecting the prey, such as electroreception (p. 207).

There is some uncertainty as to the means of support of the jaws in the earlier stages of their evolution. The front end of the palatopterygoquadrate bar is attached to the cranium in the dogfish by the ethmopalatine 'ligament'. In most elasmobranchs the hind end of the upper jaw is not fixed to the cranium but is slung from the latter by the hyomandibula and by a prespiracular ligament. This means of support, known as *hyostylic*, was for long supposed to have been the original one. But the earliest gnathostomes (the acanthodians) do not have this arrangement (p. 190), indeed, their hyoid arch is an almost typical branchial arch, not modified to support the jaw. In the primitive condition one would

not expect the hyoid arch to have any connection with the mandibular. In the acanthodians the jaw is supported by direct attachments to the cranium at its hind as well as front end, a condition known as *autodiastylic*.

The early elasmobranchs themselves do not have a hyostylic jaw support, but an arrangement in which the upper jaw is both attached to the cranium and also supported by the hyomandibula. This *amphistylic* condition persists to-day in the primitive shark *Hexanchus*. Apparently the jaws, which at first swung from the skull, later became fixed at the hind end to the hyoid, and this finally became the only means of support posteriorly. The advantage of this last arrangement is presumably that it allows a wide gape for swallowing the prey whole. As the sharks sought to eat larger and larger fishes, those in which the hind end of the upper jaw was less firmly fixed to the skull were the more successful and so the hyostylic condition was achieved. The upper jaw of a shark is protruded during feeding.

If this theory of the origin of the jaws is correct we may expect to find some trace of a cartilaginous support for the side wall of the pharynx in front of the original first gill slit; a premandibular visceral arch. Many sharks have two pairs of labial cartilages in this position, which have been held to represent arches. However, there are strong grounds for believing that the trabeculae cranii represent the dorsal parts of these original branchial arches. The trabeculae are the rods lying on each side in front of the parachordals and contributing to the floor of the skull (Fig. 5.16). Many points indicate that these rods are not part of the axial skeleton. The main axis of the body presumably ends at the front end of the notochord, that is to say, at the level of the front ends of the parachordals. Indeed there is much confirmatory evidence to show that this level represents the end of the segmented part of the body, everything in front of this level being as it were pushed forward from above or below. The trabeculae have exactly the relations to the most anterior nerves and blood vessels that would be expected of visceral arches. Confirmation of the theory comes from the discovery that the cartilage of the front part of the trabeculae, like that of the visceral arches, is formed by material streaming down during development from the neural crest, that is to say, from ectoderm.

The branchial arches, hyoid, jaws, and trabeculae thus all constitute a single series, the result of the working of a repetitive or rhythmic process, appropriately modified at each level.

9. Segmentation of the vertebrate head

The rhythmicity or metamerism seen in the cartilages can be traced throughout the structure of the head.

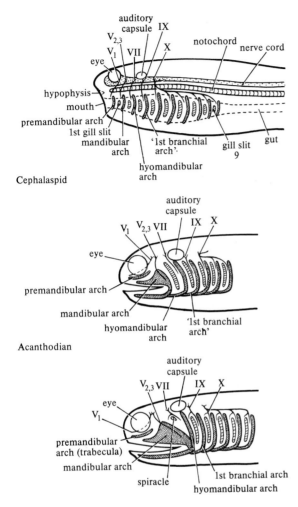

FIG. 5.16. Diagrams to show the condition of the visceral arches and jaws in early vertebrates. The Roman numerals indicate the cranial nerves. V_1, profundus nerve; $V_{2,3}$, trigeminal nerve. (Modified after Westoll.)

Although in higher vertebrates the head appears as a distinct structure, separated from the body by a neck, yet there is every reason to think that it has arrived at that stage by gradual modification of the anterior members of an originally complete metameric series. The jaws, the receptor organs, and the brain have become developed at the front end of the body, producing what zoologists conveniently if pretentiously call cephalization.

The fundamental segmentation of the head is not very easily apparent to superficial observation; the working out of its details is an excellent exercise in morphological

understanding. Recognition of the segmental value of the various structures also makes them the more easily remembered. For instance, the nerves found in the head have been named and numbered for centuries by anatomists in an arbitrary series:

I.	Olfactorius	olfactory
II.	Opticus	optic
III.	Oculomotorius	oculomotor
IV.	Trochlearis (patheticus)	trochlear
V.	Trigeminus	trigeminal
VI.	Abducens	abducent
VII.	Facialis	facial
VIII.	Acousticus	auditory
IX.	Glossopharyngeus	glossopharyngeal
X.	Vagus	vagus
XI.	Accessorius	accessory
XII.	Hypoglossus	hypoglossal

Morphological study has shown that these nerves are not isolated structures, each developed independently, but that they represent a regular series of segmental dorsal and ventral roots of the head somites. The satisfaction and simplification given by this generalization is one of the clearest advantages of morphological insight. More important still, such understanding of the morphology of a structure shows us how to look for the morphogenetic processes that produce it; such knowledge of how organs are made is an essential step in mending or remaking them (Jollie 1977).

The idea of the essential similarity of structure of the head and trunk was early developed by Goethe, who tried to show that the mammalian skull is a series of modified vertebrae. Unfortunately this view cannot be maintained in detail and the theory was brought to ridicule by T. H. Huxley and others. The segmental value of the skull floor and sides is not at all easy to determine; the parachordals arise as a pair of unsegmented rods on either side of the notochord.

10. The pro-otic somites and eye muscles

Ideas about the segmentation of the head were first correctly formulated by F. Balfour (1878). In his studies of the development of elasmobranchs he showed that three myotomes, the pro-otic somites, can be recognized during development in front of the auditory capsule (Fig. 5.17). The auditory sac, pushing inwards and becoming surrounded by cartilage, then breaks the series of myotomes, so that several are missing in the adult, though the series is complete in the embryo.

If this analysis is correct we should be able to recognize that the nerves of the head belong to a series of dorsal and ventral roots, similar to that in the trunk, the ventral roots being those for the myotomes and the

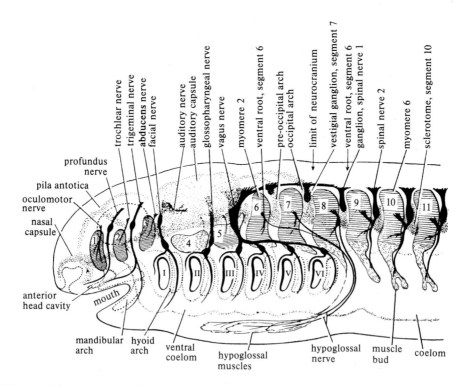

FIG. 5.17. Diagram of the segmentation of the head of a dogfish. *I–VI* gill slits; *1–3* pro-otic segments; *4–11* post-otic segments (4 disappears). (Ventral root segment 8 wrongly labelled 6.) (From Goodrich 1930.)

dorsal roots, running between the myotomes, carrying sensory fibres for the segment and motor fibres for any non-myotomal musculature present (p. 34). In the spinal region the dorsal and ventral roots join, but this is not the primitive condition (witness amphioxus and the lampreys), and in the head region the earlier state of affairs is retained, the dorsal and ventral roots remain separate. Presumably the arrangement we find in the head today was laid down in very early times, in the Silurian period or earlier, when the dorsal and ventral roots were still separate. The head, in spite of its specializations, preserves for us a relic of that ancient condition.

The branchial nerves, such as the glossopharyngeal, show clear signs of this condition. Each has the small pre-trematic branch in front of the slit, a larger post-trematic branch behind it, and a pharyngeal branch to the wall of the pharynx. The pre-trematic branch usually contains mostly sensory fibres from the skin, the pharyngeal branch visceral sensory fibres, including those from taste buds. The post-trematic branch contains both motor and sensory fibres. In addition to these more ventral branches the branchial nerves also usually provide dorsal rami to the skin of the back.

The three pro-otic somites become completely taken up in the formation of the six extrinsic muscles of the eye, arranged similarly in all gnathostome vertebrates. The four recti roll the eye straight upwards, downwards, forwards, or backwards, and the two obliques, lying farther forward, turn it, as their name suggests, upward or downward and forward (Fig. 5.18). Of these muscles the superior, anterior, and inferior rectus and inferior oblique are all derived from the first myotome and are innervated by the oculomotor (third cranial) nerve. The superior oblique, innervated by the trochlear nerve (fourth cranial), is the derivative of the second and the posterior rectus (external rectus of man), innervated by the abducens (sixth cranial), of the third somite. These three nerves are evidently the ventral roots of the three pro-otic somites. At some early stage of vertebrate evolution all the myotomal musculature of the front part of the head became devoted to the movement of the eyes. The muscles originally forming part of the swimming series became attached to a cup-like outgrowth from the brain.

Most of the rest of the musculature of the head, including that of jaws and branchial arches, is derived from the somatopleure wall of the coelom and is therefore lateral plate or visceral musculature. This lateral plate muscle is indeed better developed in the

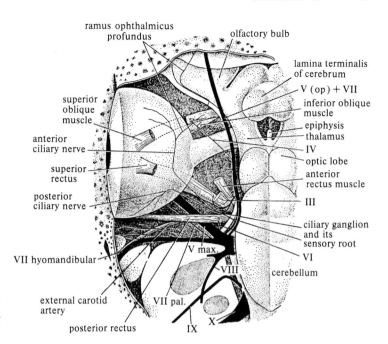

ramus ophthalmicus
profundus

olfactory bulb

lamina terminalis
of cerebrum

V (op) + VII

inferior oblique
muscle

epiphysis

thalamus

IV

optic lobe

anterior
rectus muscle

III

ciliary ganglion
and its
sensory root

VI

cerebellum

VIII

V max.

VII hyomandibular

superior
oblique
muscle

anterior
ciliary nerve

superior
rectus

posterior
ciliary nerve

external carotid
artery

VII pal.

posterior rectus

IX

X

Fig. 5.18. The brain and orbit of the dogfish. *V.* (op) and *VII*, superficial ophthalmic branch of trigeminal and facial; *II* to *X*, cranial nerves. (After Young 1933.)

head than in the trunk, where all the muscles, even of the more ventral parts of the body, are formed by downward tongues from the myotomes. The lateral plate origin of the jaw muscles at once gives us the clue to the nature of some more of the cranial nerves, the fifth, seventh, ninth, and tenth. These nerves all carry ganglia containing the cell bodies of sensory fibres and these are comparable to the spinal dorsal root ganglia. But the nerves also transmit motor fibres to the muscles of the jaws and branchial arches. They are in fact mixed roots, just as we have seen that the primitive dorsal roots should be, carrying the sensory fibres for the segment and motor fibres for the non-myotomal muscles (p. 41).

11. The cranial nerves of elasmobranchs

These nerves are more easily studied in elasmobranchs than in any other vertebrates, because of the relatively soft and transparent cartilage through which they run. We may therefore take this opportunity to examine the whole series of cranial nerves in some detail in the dogfish, beginning with the oculomotor nerve, the first ventral root. Examination after removal of the brain will show clearly in any vertebrate, including man, that this nerve arises from the ventral surface, at the level of the hind end of the midbrain (optic lobes). This is not true of the trochlear or pathetic nerve, which emerges from the dorsolateral surface of the brain but nevertheless is the ventral root of the second segment. Its cells of

origin lie close behind those of the oculomotor nerve, in the ventral part of the brain. The reason for the dorsal emergence is that the muscle lies dorsally and the nerve has been modified so as to reach its muscle by running partly within the tissues of the brain. The third ventral root (abducens), which is very short, is clearly ventral.

In looking for the dorsal roots that correspond to these three segments we have to examine the trigeminal, facial, and auditory nerves. The trigeminal of the dogfish, like that of man, has ophthalmic, maxillary, and mandibular branches (Fig. 5.18), but can be shown to represent the dorsal roots of the two first segments. The ophthalmic branch is a sensory nerve carrying fibres for skin sensation from the snout. The maxillary branch supplies sensory fibres to the upper jaw, whereas the mandibular is a mixed nerve to the skin and muscles of the lower jaw. Besides these main branches there is also a small but important sensory branch from the trigeminal to the eyeball (Fig. 5.18). This joins a motor root from the oculomotor nerve where the latter swells slightly to form a ciliary ganglion. Two ciliary nerves then carry motor and sensory fibres to the eyeball. In some specimens a branch of the more anterior ciliary nerve leaves the eyeball anteriorly, runs between the oblique muscles, and out of the orbit again to end in the skin of the snout (Fig. 5.18). Though this branch is small and inconstant in the dogfish, its course corresponds exactly with that of a much larger nerve in the related shark *Mustelus* and in skates and rays. In these animals there are two ophthalmic branches of the trigeminal nerve; one, having a course similar to that of the main nerve in the dogfish, is the ramus ophthalmicus superficialis; the second, the ramus ophthalmicus profundus, runs across within the orbit, gives off the long ciliary nerve to the eyeball, passes between the oblique muscles, and leaves the orbit for the skin of the snout. In higher vertebrates the nasociliary nerve and the long ciliary, innervating the eyeball, represent the profundus, while the rest of the ophthalmic nerve of mammals corresponds to the superfical ophthalmic of elasmobranchs.

The relations of these nerves to other structures shows that the so-called trigeminal nerve really includes the dorsal roots of two segments combined. The profundus can be traced back in development to a nerve that is obviously the dorsal root of the first somite, of which the oculomotor nerve is the ventral root. Indeed it may be noticed that the profundus and oculomotor partly join, at the ciliary ganglion. The ramus ophthalmicus profundus and the oculomotor nerve thus constitute the dorsal and ventral roots of the first or premandibular somite, whose corresponding branchial arch is presumably the trabecula cranii (p. 126). The dorsal root does not show the full structure of a branchial nerve, presumably because there is no gill slit. The profundus represents only the dorsal branch of a typical branchial nerve, innervating the skin.

The ramus ophthalmicus superficialis, and the maxillary and mandibular branches together constitute the dorsal root of the second pro-otic somite, whose ventral root is the trochlear nerve. The corresponding gill arch is the mandibular (palatopterygoquadrate bar and Meckel's cartilage), whose gill slit we have suggested has been incorporated with the edge of the mouth. The trigeminal nerve shows considerable similarity to a branchial nerve, its maxillary branch represents the pre-trematic and the mandibular the post-trematic ramus, while the ophthalmicus superficialis is the dorsal branch to the skin. There is no pharyngeal branch. An anomalous feature of the trigeminal is that it contains sensory fibres whose cells of origin lie within the brain (mesencephalic root). These fibres are perhaps from the proprioceptors from the masticatory muscles and eye muscles. The latter mostly run with the eye muscle nerves (another anomaly!).

The dorsal root of the third segment, whose ventral root is the abducens, includes the whole of the facial and also the auditory nerve. The facial is a large mixed nerve in the dogfish. Its ophthalmic branch runs to the snout, carrying mainly fibres for the organs of the lateral line system that lie there. A large buccal branch supplies sensory fibres to the mouth and a palatine branch joins the trigeminal. A small prespiracular branch carries sensory fibres from the skin in front of the spiracle, and the main portion of the nerve continues behind the spiracle as the hyomandibular nerve, dividing up into motor branches for muscles of the hyoid arch and sensory ones for the skin of that region.

This nerve is obviously the branchial nerve to the spiracle; we can safely say that the facial and abducens are the dorsal and ventral roots of the third or hyoid segment. The auditory (vestibular) nerve is included as part of the dorsal root of the third somite because the otic sac is formed by sinking in of a portion of the ectoderm within the territory of the facial nerve. The labyrinth still communicates with the surface of the head in the adult dogfish by a canal, the aquaeductus vestibuli. The nerve that innervates the otic sac, whatever complexities it may acquire, is to be regarded morphologically as a portion of the dorsal root of the hyoid segment.

The segmental nature of the structures in the pro-otic region can therefore be made out without serious difficulty. The disturbance introduced by the otic capsule makes the segmental arrangement of the more posterior region of the head somewhat confused. The series of dorsal roots is uninterrupted; the ninth (glossopharyngeal) nerve is the dorsal root of the fourth segment of the series and runs out through the cartilage of the auditory capsule. The dorsal roots of the succeeding segments are then fused to form that very puzzling nerve the vagus. The branches it sends to the gills are clearly typical branchial nerves, but why should they all come off together from the medulla oblongata, and if there is any advantage in this union, why is the ninth nerve not also so incorporated? Above all, why does the vagus send two branches far outside the segments of its origin, the lateral line branch carrying fibres to the organs right to the tip of the tail and the visceral branch fibres to the heart, stomach, and sometimes the intestine?

Evidently these 'wanderings', from which the vagus gets its name, began very long ago. The nerve reaches as far back in cyclostomes as in any other vertebrates. It is easy to understand that if visceral functions are to be directed from the medulla oblongata there is an advantage in having sensory impulses sent direct to that region of the brain and motor impulses sent out thence to the viscera. It may be that these advantages allowed the centralization of these visceral functions, while the need for serial contraction of the swimming muscles led to the retention of the segmental arrangement of the spinal cord. It is an interesting thought that but for the swimming habits of our ancestors our nervous system might by now consist of a central ganglion with nerves passing from it direct to all the organs. Indeed we are tending in that direction, as the spinal cord shortens and becomes more and more nearly a simple pathway between the brain and the periphery.

However this may be, the vagus is certainly a nerve compounded of the dorsal roots of several segments and it is a mixed nerve, containing both receptor and motor fibres. Some of the more posterior rootlets of this series are separated off in higher animals (not the dogfish) to form the eleventh cranial nerve, the accessorius or spinal accessory, which in mammals sends motor fibres to certain muscles of the neck, the sternomastoid, and part of the trapezius. Its motor nature has led some to suppose that this nerve is a ventral root, but these muscles are derived from lateral plate musculature and the accessorius represents the motor portion of the hinder dorsal roots of the vagus series.

The ventral roots of this post-otic region have become much reduced. Several myotomes are always missing completely, so that there are no ventral roots corresponding to the glossopharyngeal and first three or four vagal segments. The more anterior of the surviving post-otic somites are to be found not in the dorsal region but ventrally, as the hypoglossal musculature of the tongue. The muscle buds have grown round into this position behind the gill slits, and the nerve (hypoglossal) that innervates them represents the ventral roots of the more posterior segments of the vagus-accessorius series (Fig. 5.17). The origin of this nerve from the floor of the medulla is a clear sign that it is a ventral root.

Thus the entire series of cranial nerves is:

Segment	Arch	Dorsal root	Ventral root
Premandibular	Trabecula	R. op. profundus V	Oculomotorius III
Mandibular	Palatopterygoquadrate bar and Meckel's cartilage	Rr. op. superficialis, maxillaris, and mandibularis V	Trochlearis IV
Hyoid	Hyoid	Facialis VII	Abducens VI
		Acousticus VIII	
1st Branchial	1st Branchial	Glossopharyngeus IX	(absent)
2nd Branchial	2nd Branchial	Vagus X	
3rd Branchial	3rd Branchial	+	Hypoglossus XII
4th Branchial	4th Branchial	Accessorius XI	
5th Branchial	5th Branchial		

Two cranial nerves have not yet been considered, the first, olfactory, and second, optic. Our thesis is that all connections between centre and periphery are made by means of a segmental series of dorsal and ventral roots and therefore these nerves, too, should be fitted into the series. No embryological or other studies have enabled this to be done and the reason in the case of the optic nerve is quite clear. It is not morphologically a peripheral nerve at all. The eye is formed as a vesicle attached to the brain; the optic 'nerve' therefore develops as a bundle of fibres joining two portions of the central nervous system; in fact it is now usually called the optic tract, not the optic nerve.

This reasoning will not apply to that very peculiar and interesting structure the olfactory nerve. This is unique among all craniate nerves in consisting of bundles of fibres whose cell bodies lie *at the periphery*. The cells of the olfactory epithelium, like the sensory cells in invertebrates and some of those of amphioxus, are neurosensory cells, that is to say, their inner ends are prolonged to make the actual nerve fibres that pass into the brain. This fact does not by itself solve the problem of fitting the nerve into the series of dorsal and ventral roots, but it reminds us that the nerve is very ancient, and suggests that it does not fall into the rhythm of the rest of the series because it precedes the other cranial nerves either in time or space, or perhaps even both. The olfactory nerve may have existed before any segmental structure appeared, possibly as the nerve of sense organs on the front end of the ciliated larva which we suppose gave rise to our stock (p. 71). Alternatively we can say that the olfactory nerve is as it is because it lies in front of the region over which the segmenting process operates; it is, as it were, 'prostomial'. If we wish we can hold both these views together.

There are one or two other exceptions to the rhythmic arrangement of nerves, perhaps more difficult to account for than the first and second cranial nerves. If all connections between centre and periphery are made by dorsal and ventral roots what is the status of the fibres that run down the infundibular stalk to reach the cells of the pituitary body? This glandular tissue, derived from the epithelium of the hypophysial folding of the roof of the mouth, is undoubtedly a peripheral organ. It receives its nerve fibres direct from the brain so presumably we must say that the pituitary, like the nose, is prostomial, lying in front of the segmental region, and this is reasonable enough from its position. There is good reason to believe that it is an extremely ancient organ, already present in the earliest chordates.

A still more puzzling exception is the nervus terminalis. This is a small bundle leaving the brain ventrally behind and below the olfactory nerves and running to the olfactory mucosa or to the accessory olfactory organ of Jacobson, where this is present (p. 263). In some vertebrates it carries a small ganglion. The fibres are probably afferents and they run backwards through the brain tissue to the pre-optic nucleus of the hypothalamus. A possible clue to its origin is that this is the region of the brain where the morphologically ventral region of the neuraxis ends (p. 126). The nervus terminalis may represent the ventral olfactory nerve, the much larger main nerve being morphologically dorsal.

A further puzzle of some importance which may be mentioned here is the course of the proprioceptor fibres for those muscles that are supplied purely by ventral roots. The eye muscles contain proprioceptor organs and C. S. Sherrington and others have shown that the afferent fibres connected with these may run to the brain through the third, fourth, and sixth nerves, that is to say, through ventral roots. Similarly, it has been shown that there are afferent fibres in the hypoglossal nerves in mammals. Conversely it is now known that there are efferent fibres running from the brain to many receptor organs. For example, such fibres run in the auditory nerve. To pursue these questions farther would lead us into discussion of the factors that control the making of connections within the nervous system. Here we are concerned only with analysis of the plan that produces the main outlines of the structures in the head, a plan which, with all its modifications, is essentially segmental.

12. Respiration

The function of the branchial arches is not merely to support the gills but to allow the movements of the pharynx wall by which the respiratory current of water is produced. It is for this reason that the jointed system of rods is present. Ventilation of the gills is brought about by two pumps, a buccal pressure pump acting by contraction of the cavities of the mouth ahead of the gill resistance, and suction pumps behind them. The suction is produced by expanding the outer part of each gill chamber and closing the slit by a flap. Closure of the mouth is by the adductores mandibulae, and of the buccal and branchial cavities by the many small muscles. Opening and expansion are largely produced elastically. The buccal pressure pump and the branchial suction pump act in a sequence and maintain a flow of water over the lamellae of the gills throughout the respiratory cycle. During swimming the current of water is maintained simply by keeping the mouth open. Each gill arch carries one series of filaments facing forwards and

another backwards and the latter meet those of the arch behind (Fig. 5.19). Attached to the filaments are numerous thin plates, the lamellae, through which blood flows inwards, that is in the opposite direction to the water. This counter-flow arrangement increases the proportion of oxygen that can be taken up by the blood.

13. The gut of elasmobranchs

The digestive system of sharks shows several changes from the plan found in lampreys, especially the presence

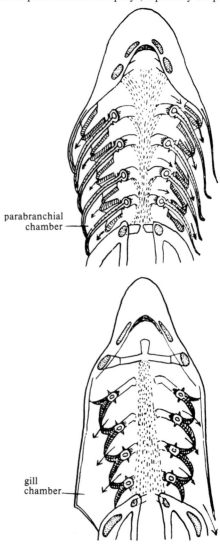

FIG. 5.19. The gill system of a shark (above) and a teleost (below). The gill filaments are cross-hatched. On the left of each figure, the parabranchial chambers of the shark, or the gill chamber of the teleost, are shown in the expanded phase, water being sucked in from the mouth. The expulsion of water from the gill system is shown on the right. (After Marshall 1965.)

of a true stomach, characteristic of gnathostomes. Apparently little or no digestion goes on in the mouth and pharynx. The teeth consist of rows of backwardly directed denticles (Fig. 5.20). They are carried on special folds of skin lining the jaws and are continually replaced as they are worn away on the lips. The replacement of milk teeth by permanent teeth in mammals is a relic of the serial replacement in a fish. The teeth are often pointed and their edges sharp. Small prey is swallowed whole but bites are taken from larger fishes by shaking the head so that the teeth cut lumps of flesh from prey too big to swallow (Fig. 6.8). The teeth are specialized for many different modes of feeding. The more primitive forms are long-jawed for grasping (*Chlamydoselachus*). From this there have evolved cutting, gouging, crushing, and sucking types with jaws and muscles modified accordingly (p. 154). In many sharks the jaws make short lever arms to give a powerful bite from the quadratomandibular muscles. They may damage their own teeth but they are quickly replaced (Moss 1977). The 'gill rakers' are rods attached to the branchial cartilages serving to prevent the escape of prey. The basihyal supports a short non-protrusible tongue.

The wall of the pharynx is lined by a stratified epithelium on to which open numerous mucous glands, sometimes complex. The mucus serves to assist the passage of the food, but probably has no strictly digestive function, though the salivary glands of higher

vertebrates no doubt originate from a modification of these mucous glands.

The pharynx narrows to an oesophagus with thick striped muscle walls, leading to the stomach. We have seen that in cyclostomes the oesophagus opens directly into the region of gut that receives the bile and pancreatic secretion. The stomach, which we now meet for the first time, has probably been formed as a special portion of the oesophagus. It probably evolved with the jaws, serving originally as a receptacle for the large pieces of food, or even whole fishes, which could now be swallowed. The mucous glands became modified to produce acid, since this prevents bacterial decay. Finally, an enzyme, pepsin, was evolved able to digest proteins in acid solution. In the dogfish this condition has been fully established and the stomach has essentially the structure and functions found in all higher vertebrates. However, in the gastric glands of all non-mammalian vertebrates only one type of cell is recognizable, there are no separate pepsin-secreting and acid-producing cells. Nevertheless, there is a pepsin-like enzyme present and the contents are acid. The stomach is divided into two parts, a descending cardiac and ascending pyloric limb (Fig. 5.21). The former is a large sac giving rhythmic contractions. In it the whole prey is reduced to liquid form and then passed forwards along the narrow pyloric division. Only liquid material is allowed to pass the pylorus, presumably because the spiral valve of the intestine restricts its lumen. The forward movement of chyme is regulated by the sympathetic ganglia (p. 147).

There is a powerful pyloric sphincter, and immediately beyond this the bile and pancreatic duct open. The liver is a large two-lobed organ, receiving the hepatic portal blood from the gut. It serves as a storage organ containing much glycogen and fat and sometimes the hydrocarbon squalene giving buoyancy (p. 123). The liver probably also plays a part in the destruction of red blood corpuscles. Bile is carried away to a gall-bladder, from which a bile-duct leads to open at the front end of the spiral intestine.

The pancreas, hardly recognizable as a distinct organ in the lamprey, forms in the dogfish an elongated body between the stomach and intestine. It contains both exocrine and endocrine cells and its duct enters the intestine shortly below the pylorus. The 'small' intestine of elasmobranchs is of a peculiar form, being short but with its surface greatly increased by the presence of a tightly wound flap or ridge, the 'spiral valve'. The intestinal contents are alkaline and contain trypsin, amylase, and lipase. There is no constant fauna of commensal bacteria. Absorption presumably takes place wholly in this organ, for the remaining length of

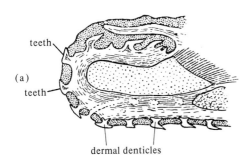

teeth

(a)

teeth

dermal denticles

(b)

FIG. 5.20. Sections through the lower jaws of A, embryo of dogfish (*Scyliorhinus*), and B, sand-shark (*Odontaspis*), showing the transitions between dermal denticles and teeth. (From Norman 1931.)

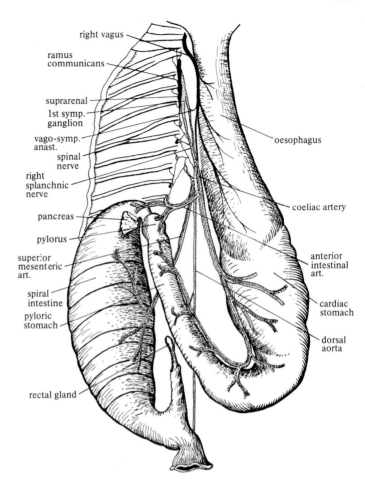

FIG. 5.21. Alimentary canal and sympathetic system of the ray (*Raja*). The nerves and suprarenal tissue have been blackened with osmium tetroxide. The spleen lies between the two parts of the stomach and has been removed. The pancreas has been cut short. (Young 1980.)

gut consists only of a short rectum. Into this opens the rectal gland, which contains branched glands secreting sodium chloride, and much lymphoid tissue (p. 83).

14. The circulatory system

The heart develops as a specialization of the subintestinal vessel between the place where it receives the veins from the liver and the body wall and the gills, which are to be supplied under high pressure. It consists

of a single series of three main chambers, sinus venosus, atrium, and ventricle, all of which are muscular, and there is also a muscular base to the ventral aorta, the conus arteriosus, provided with valves (Fig. 5.22).

The five afferent branchial arteries carry blood to the gill lamellae, whence it is collected by a system of four efferents and connecting vessels into a median dorsal aorta, carrying blood to all parts of the body. Oxygenated blood is supplied to the head from three

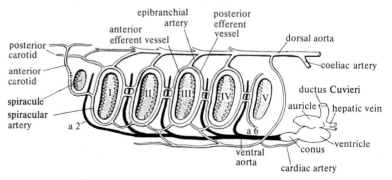

FIG. 5.22. Diagram of the branchial circulation of an elasmobranch fish. a_{2-6}, five afferent vessels from ventral aorta; *I–V*, branchial slits. (After Goodrich 1930.)

sources. (1) From the top of the first gill a carotid artery leaves the efferent branchial and runs forwards and towards the midline: it then divides into an external carotid to the upper jaw and internal carotid to the brain. (2) The dorsal aorta divides at its front end into branches, which join the carotids before their division. (3) From the vessel that collects blood from the first gill arises a hyoidean artery, carrying oxygenated blood to the spiracle. From here the hyoidean artery runs on as the anterior carotid (Fig. 5.22) across the floor of the orbit to join the internal carotid within the brain-case.

The heart is supplied by a cardiac artery arising from the dorsal aorta behind the gills. The blood pressure in the ventral aorta is 300–400 mm water, and there is a drop of 100–200 mm water across the gills. The circulation is slow, with a mean circulation time as low as 2 min. In fish the venous return is largely a consequence of movement forcing blood into a system of very large sinuses. The pericardium is almost completely enclosed in a cartilaginous framework by the basibranchial plate above and pectoral girdle below it. This produces a negative pressure of -20 mm to -50 mm of water each time the ventricle empties and this serves to draw blood from the veins. The pericardioperitoneal canal, leading from the pericardium to the abdominal coelom is a passage narrow behind and so acting as a valve allowing fluid to pass from the pericardium but not into it.

A caudal sinus from the tail opens into a renal portal system above the kidneys. From the latter, and from the muscles of the back, blood is collected into the pair of very large posterior cardinal sinuses, lying on the dorsal wall of the coelom. Above the heart these receive the openings of other large sinuses, such as the anterior cardinal sinus, running above the gills and collecting blood from the head by way of an orbital sinus, and the jugular, lateral cardinal, subclavian, and other sinuses from the body wall. Blood then passes round the oesophagus in the two ductus Cuvieri into the sinus venosus, where hepatic sinuses also open.

The resistance offered by a vessel to flow within it decreases with approximately the fourth power of the diameter, therefore the large size of these vessels substantially assists in allowing return to the heart. The heart muscles, like any others, require antagonists; they can contract in one direction only, and each chamber therefore needs to be actively dilated. This feature of the fish heart was already known to William Harvey in 1649. The fish heart consists of a series of three muscular chambers, presumably because the low venous pressure is able to dilate only a chamber with very thin walls, such as the sinus venosus. Contraction of the sinus is weak but helps to inflate the auricle (atrium), which is also under negative pressure. The auricle inflates the ven-

tricle, which thus constitutes the third step in this serial pressure-raising system. Finally there is a fourth step from the ventricle to the muscular conus arteriosus, where a series of three valves open in sequence maintaining pressure in the ventral aorta after it has fallen in the ventricle (Fig. 5.23). In the land animals, where most of the blood only passes through a single set of capillaries, a two-step system (auricle and ventricle) is sufficient for each part of the circulation. There is a gradient of elastic tissue in the walls of the arteries along the ventral aorta, through the afferent and efferent branchioles to the dorsal aorta. This ensures a nearly continuous flow of blood to match that of the water over the gills (p. 132). Valves at the exit of each segmental artery close when the vessels are compressed by the muscular swimming wave.

Little is known of the control of the circulation but it is probably less elaborate than in higher animals. The heart beat is myogenic and the cardiac muscle resembles that of higher vertebrates. The sinus venosus will beat when isolated and is probably the pacemaker. There is a cardiac branch from the vagus ending in an elaborate plexus in the sinus venosus (Fig. 5.26). Stimulation of this nerve slows the heart. There is no anatomical or

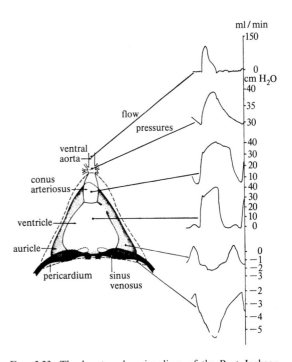

FIG. 5.23. The heart and pericardium of the Port Jackson shark *Heterodontus*. On the right is the flow profile from the ventral aorta and the pressure profiles from the pericardium and cardiac chambers (the pressure profiles were not recorded simultaneously). (From Satchell 1971.)

physiological evidence of a sympathetic nerve to the heart, but abundant sympathetic fibres run to the arteries. Small doses of adrenaline cause prolonged rise of blood pressure.

There are receptors in the efferent branchial vessels and in the postbranchial plexus above the cardinal veins (p. 139). Nerve impulses from these receptors can be recorded in the vagus at each systole and are increased by raising the blood pressure. Their reflex effects are to slow the heart and respiration and decrease the blood pressure, perhaps for protection of the gill capillaries. These reflexes are presumably the ancestors of the carotid sinus and similar reflexes of land vertebrates.

15. Immunology in fishes

Elasmobranchs and teleosts produce antibodies composed of light and heavy peptide chains like those of mammals. Sharks can produce antibodies to viruses or to foreign cells. However, the response is slow and the immunological memory less perfect than in mammals. Many studies of transplantation have been made, using scales. Autotransplants are readily accepted, but homografts are rejected and are followed by immunological memory and enhanced secondary reactions. Fishes also have blood group systems with iso-antibody properties similar to the human ABO system.

Curious 'natural' transplantations can occur in fishes, however. Males of ceratioid angler fishes (p. 215) make tissue and blood connections with their females. Parasitic copepods (*Lernea*) are accepted by the tissues of their hosts and may even burrow into the heart. These tolerances are unexplained.

16. Osmoregulation and excretory system

One of the major features of elasmobranch life is that they produce very few young, which develop for a long time in a protected environment, either within the mother or in an egg case. This long developmental period involves the problem of disposing of nitrogenous waste. Ammonia would be toxic and the device adopted is to convert it to the soluble but innocuous urea. The habituation of the embryonic tissues to the urea is carried over to the adult, and indeed proves to have certain advantages for the regulation of osmolarity and buoyancy in an environment rich in unwanted inorganic ions (Bentley 1971; Pang, Griffith, and Atz 1977).

The blood of the elasmobranchs differs from that of all other vertebrates except *Latimeria*, Dipnoi, and some estuarine crab-eating frogs in its very high content of urea. As measured by the depression of the freezing-point the blood is hypertonic to the surrounding sea water (say, 3.5 per cent NaCl). But there is far less salt in the blood than in the sea, in fact only about 1.7 per cent

NaCl. The excretion of salt to maintain this low level is performed not by cells of the gills as in teleosts but by the rectal gland. The cells of this sac are actively secretory, with many mitochondria and produce an average of 0.47 ml per kg/h of a fluid in which each ml is equivalent to 1.1 ml of sea water in respect of NaCl. Curiously, if the rectal gland is removed a dogfish still maintains its normal plasma level, at least for three weeks. Yet there is no increased chloride secretion by the kidney. The mechanism concerned is not known but may involve a reduced NaCl permeability of the gills or excretion by chloride-secreting cells which occur there.

In normal elasmobranchs, therefore, the blood is nearly isotonic with the sea but its composition is regulated (homeosmotic). This arrangement is apparently a legacy of the fact that the ancestors of the elasmobranchs were originally freshwater animals (p. 152). The return passage to the sea has been accomplished through the device of retaining urea and trimethylamine oxide (TMAO). The gill surfaces, in which alone the blood comes into close contact with sea water, are relatively impermeable to urea, but this substance penetrates freely into the tissues, as it does in other animals. Elasmobranch tissues if placed in sea water are therefore in contact with a strongly hypertonic medium. They are so habituated to the presence of urea that they are unable to function unless it is present in a concentration that would be toxic to most animals. Urea is synthesized in the liver mainly via the ornithine cycle, involving at least five specific enzymes. The serum proteins of elasmobranchs are all globulins. The absence of albumens, which in mammals have important osmotic influences, may be connected with the high concentration of urea.

This arrangement has presumably been responsible for the fact that few of the elasmobranchs have returned to fresh water. They include the Bull Shark (*Carcharinus*) found in Lake Nicaragua, the Zambesi, and the Ganges, also the Sawfish (*Pristis*) which is found in the Perak River (Malaysia) and in Lake Nicaragua and hundreds of miles up the Mississippi. The blood of these forms has much less urea and rather less salt than their marine relatives. The kidney tubules resorb urea and salts and the urine is hyposmotic to the blood. The rectal glands are reduced.

In the ordinary marine elasmobranchs the high urea concentration is maintained by the presence of a special section of the kidney tubules that absorbs urea and TMAO, leaving a hyposmotic urine. The urinary apparatus is a mesonephros and these fishes show a considerable specialization in that the urinary functions of this organ are separated from its generative ones in the male. The hinder part of the kidney is sometimes called opisthonephros (the term metanephros should be used only for the definitive kidney of amniotes) which has a different method of development. It consists of a mass of tubules ending in very large glomeruli, and a section of very long tubules. It has the function of urea resorption though no special segment is devoted to this. All the tubules join to form a series of five urinary ducts and these enter a urinary sinus, opening to the cloaca. The sinus can be compared functionally with a bladder but it is a mesodermal structure, derived from the main kidney duct, and is not strictly comparable to the endodermal bladder of tetrapods. The urinary sinus is a small organ but the volume of urine secreted is much larger than in most marine teleosts. It contains about the same concentration of sodium and chloride as the blood. The secretion of the rectal gland has about the same volume as the urine but with concentrations of sodium and chloride twice that of the blood and slightly higher than sea water. On account of the high osmolality of the blood, water enters by osmosis through the gills without the expenditure of free energy and elasmobranchs probably do not need to drink (as marine teleosts must do, p. 169). Sodium chloride and other salts enter by diffusion too, and need to be excreted. The kidney tubules remove the divalent ions, sulphate and phosphate from the blood.

17. Genital system

The genital system is highly specialized to allow internal fertilization and the production of a few very yolky eggs. The majority of species are viviparous, with varying degrees of placentation. Eggs are laid only by chimaeras, *Raja* and a few families of sharks, including *Scyliorhinus*. There is a single large ovary, from which the eggs are carried by the cilia of the peritoneum to a pair of funnels lying on either side of the liver behind the heart (Fig. 5.24). These are apparently formed from pronephric funnels and the Müllerian duct (oviduct) separates from the original nephric duct. In the adult it becomes a thick-walled muscular tube, bearing a swelling in oviparous species, the nidamental gland, the upper part of which produces albumen, the lower the horny egg case.

The testes are paired and sperms are collected at their front ends by vasa efferentia leading into the anterior or reproductive portion of the mesonephros. This consists of a much coiled, thick-walled, vas deferens, whose glands produce material that aggregates the sperms into spermatophores. The vas expands into a broader ampulla (seminal vesicle), which at its lower end gives off a forwardly directed blind diverticulum, the sperm sac, developmentally the lower end of the Müllerian duct,

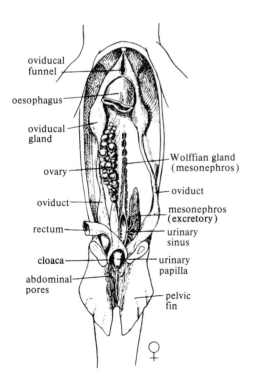

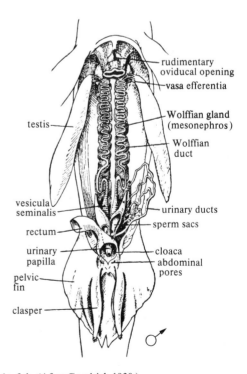

Fig. 5.24. Urinogenital systems of the dogfish. (After Goodrich 1930.)

reduced of course in the male though small funnels are still visible at the upper end.

Transmission of the sperms is produced by a large and complicated pair of claspers. These are modified parts of the pelvic fins of the male, forming grooved, scroll-like organs. They contain cartilages that can be erected by muscles and inserted into the female oviduct and anchored there by a cartilage at the tip. Sperm, extruded from the cloaca, is then flushed down the groove by sea water from a pair of siphon-like sacs. The walls of the sacs secrete large amounts of 5-hydroxytryptamine, which initiates powerful contractions of the oviduct (in *Squalus*). The mechanism of erection is operated by nerves and may involve the liberation of adrenaline; experimental injection of that substance will produce erection, and it is perhaps significant that the male possesses a reserve of adrenaline-producing tissue (see opposite p.).

Development of elasmobranchs is by partial cleavage, producing a blastoderm, perched on top of a large mass of yolk. In *Scyliorhinus* and *Raja* the egg is protected by an elaborate egg case, the 'mermaid's purse', made of keratin, within which development proceeds until the yolk has been used up.

In viviparous elasmobranchs the oviduct forms a 'uterus' but many species are aplacental, the young developing either at the expense of the yolk or by intra-uterine cannibalism as in the porbeagle shark (*Lamna*) whose embryos are active predators, snapping at what-ever they touch. In some species the uterine wall secretes embryotrophe or 'uterine milk'. There may be secretory processes (trophonemata) that enter through the spir-acles and secrete into the gut (*Trygon*). The yolk sac forms placentae in *Mustelus* and other sharks, with interdigitation of foetal and maternal tissues and blood circulations separated by as few as three layers. Studies with [³H]-glucose show that diffusion is also possible from the mother to the intestine of the foetus via the yolk sac. In general the egg-laying (oviparous) species are small animals, living inshore near the bottom, whereas the larger, pelagic sharks are viviparous.

18. Endocrine glands

The pituitary contains the usual parts, the pars distalis (anterior), pars intermedia, and neurohypophysis with a large pituitary cleft. In addition there is a partly separate ventral lobe, possibly corresponding to the tetrapod pars tuberalis but more probably part of the distalis and containing a gonadotropin and a thyrotropin (Fig. 5.25). The gonads of the dogfish retrogress after removal of the ventral lobe. Lactotropic hormone and adrenocorticotropic hormone (ACTH) are present (Holmes and Ball 1974).

The neurointermedia receives abundant fibres of the neurohypophysial tract. It contains melanophore-stimulating hormone (MSH). Some arginine vasotocin and various (at least three) neutral octapeptides are present but their functions are not known. Hypotha-lamic control of parts of the pituitary is by a portal system passing through the median eminence, but this does not directly reach to the ventral lobe. Yet extracts of the hypothalamus influence the cells of the ventral lobe. Possibly the control is by way of the systemic bloodstream by a relay in the pars distalis. Unlike teleosts there is no direct neurosecretory innervation of the pars distalis.

The thyroid is formed by a downgrowth from the floor of the pharynx, to which it often remains attached by a narrow stalk containing a small ciliated pit, a reminder that the organ was once a ciliated mucus-secreting gland. Little is known of its function but it shows cyclic changes related to seasonal migration and reproduction. After its removal from dogfishes no yolk is laid down in the eggs.

The adrenal tissue is especially interesting because the two parts, so closely associated in mammals, are here

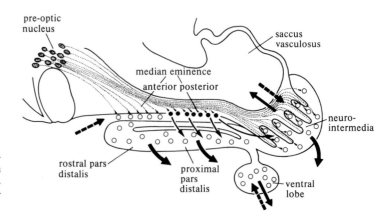

FIG. 5.25. Selachian pituitary. Solid ar-rows, veins; broken arrows, arteries; thin arrows, portal veins. (After Jasinski, A. (1969) *General and comparative. Endocri-nology* Suppl. 2, 510–21.)

found widely separated. A segmental series of glands, the suprarenals, are rich in noradrenaline. They project into the dorsal wall of the posterior cardinal sinus and can be seen when it is opened (Fig. 5.26). The more anterior ones are fused to form an elongated structure on either side of the oesophagus. The sympathetic ganglia are closely associated with these suprarenal bodies, as would be expected from their common origin from cells of the neural crest. The segmental series continues along the whole length of the abdomen, the more posterior members being embedded in the kidney tissue (Fig. 5.27). These posterior suprarenal bodies are larger in the male than in the female, but only the central part of the male glands shows the chromaffin reaction with chrome salts that indicates the presence of adrenaline. The peripheral portion of each gland appears to consist of non-functioning cells, possibly a reserve used only during reproduction (see p. 138).

The part of the adrenal corresponding to the cortex of mammals is represented in elasmobranchs by the interrenal bodies, lying medially in some species, paired in others, in the kidney region (Figs. 5.27 and 5.28). The cells of these organs resemble cortical adrenal cells. Since they are not in contact with the suprarenals at any point, it would seem that the association of the two parts is not necessary for their functioning, at least in these animals. Removal of the interrenal is always fatal. The gland is stimulated by 'stress' or by mammalian ACTH. Extracts of it prolong the life of adrenalectomized rats. There is evidence that it influences carbohydrate metabolism and activity of the gonads but not electrolyte balance. Its steroid is mainly the unusual 1-α-hydroxycorticosterone, and there is probably no aldosterone.

The endocrine tissue of the pancreas is not in the form of islets but as an outer layer along some of the ducts. This is the position that the tissue occupies early in development in mammals. There is evidence for the

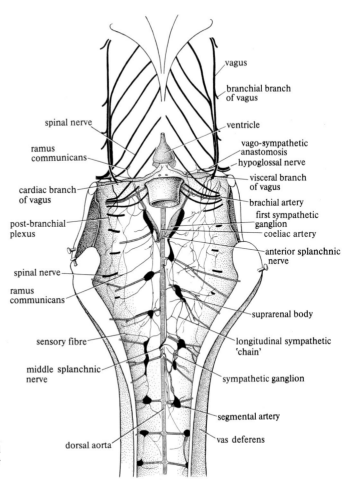

FIG. 5.26. Dissection of suprarenal bodies and sympathetic nervous system of the dogfish. (After Young 1933.)

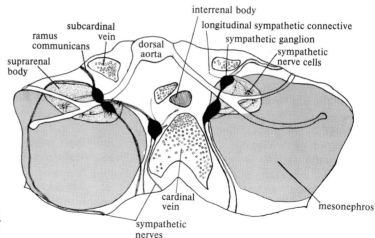

FIG. 5.27. Diagram of transverse section through hind region of mesonephros of dogfish. (From Young 1933.)

presence of insulin and glucagon. After removal of the pancreas of a dogfish the blood sugar increases and hypophysectomy reverses this.

The ultimobranchial gland below the pharynx has a follicular structure and produces calcitonin (p. 172).

The gonads contain endocrine tissues producing steroid hormones. There are interstitial cells in the testis which produce testosterone, androstenedione, and progesterone, as in mammals. The ovarian follicle produces hormones in two different ways. The postovulatory follicle becomes luteinized in *Scyliorhinus* and other oviparous species and produces progesterone. In the ovoviviparous *Torpedo*, however, the corpora lutea are formed from atretic follicles (i.e. those where the egg has been absorbed). The progesterone they produce probably stimulates growth of the uterine folds and inhibits further oogenesis during gestation.

19. Nervous system

The brain is larger in elasmobranchs in relation to body weight than in any other anamniote vertebrates and indeed approaches that of birds and mammals (Fig. 4.24) (Ebbesson and Northcutt 1976; Northcutt 1977). This is not fully explained by the absence of bone, since that tissue is only 21 per cent more dense than cartilage. The forebrain is also relatively large (Fig. 4.25). The organization of the brain is characteristically different from that of both the cyclostomes and bony fishes (Fig. 5.18). The forebrain is large and has cerebral hemispheres thickened both in floor and roof, whereas in teleosts the roof is thin. The hemispheres are wide relative to their length and the end of the unpaired portion of the forebrain between the hemispheres, the lamina terminalis, is also much thickened. Attached to

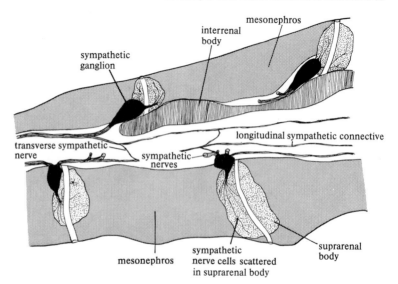

FIG. 5.28. Diagram of arrangement in hinder mesonephric region of dogfish. (From Young 1933.)

the ends of the cerebral hemispheres are large olfactory bulbs and there are also large nasal sacs. Evidently the olfactory sense is well developed in these animals and they depend greatly on it for hunting. Degeneration studies show that the olfactory tracts do not project to all parts of the telencephalon as often stated, but only to a restricted ventrolateral region. Tertiary olfactory fibres also reach only to limited areas and much of the hemisphere receives other sensory information through projections from the thalamus. Efferent pathways from the hemisphere proceed to the thalamus, hypothalamus, optic tectum, and various medullary centres. The connections of the shark forebrain are thus much more complex than had previously been supposed and are basically like those of mammals. A curious difference is that both the ascending and descending pathways are crossed.

There is of course no cortical arrangement of tissue in the hemispheres. The cells form thick masses around the ventricle (Fig. 5.29). The roof is quite thick and contains decussating fibres in the midline. The sides and floor make up the main bulk of the organ, the lateral wall being known as the striatum, its upper part the epistriatum. The medial wall is known as the septum and its upper portion is often referred to as the primordium hippocampi, having a position similar to that of the hippocampus of mammals. The dorsal portion of the roof or dorsal pallium is especially well developed in galeoid sharks. There is extensive migration of its cells away from the ventricle to form inner and outer laminae (Fig. 5.30). The inner, central nucleus, is especially well developed. This region receives projections from thalamic nuclei of the opposite side, bringing signals from visual, lateralis, and trigeminal sources. This is an interesting higher cerebral centre but neither its structure nor its functions are well known. Centres that are similar in position, connections and perhaps functions are found in teleosts, reptiles, birds, and mammals. Has

this important centre developed in parallel from the basic gnathostome genome? After removal of the forebrain the sense of smell is lost but the fish shows no obvious disturbance of posture, locomotion, or behaviour.

The thalamus is now known to receive inputs not only directly from the retina but also from the optic tectum, spinal cord, cerebrum, and cerebellum (Fig. 5.30) (Ebbesson 1972 *a*, *b*). The various projections partly overlap and there are no clear thalamic nuclei. The spinal and cerebellar inputs are diffuse and limited to the medial regions suggesting the mammalian intralaminar region.

The lower part of the between-brain, the hypothalamus, is as well developed (relatively) in these animals as in mammals. Its hind part (inferior lobes) receives impulses from the forebrain (the 'fornix' of higher vertebrates) and gustatory pathways from the medulla. Its efferent fibres run to reticular centres. The more anterior part of the hypothalamus lies above the pituitary and contains the supraoptic nucleus, whose axons form the hypophysial tract, ending in the neuro-intermediate lobe. The supraoptic cells of all vertebrates are large and contain granules of neurosecretory material that is passed down the axons and liberated in the pituitary neural lobe. The anterior hypothalamus is a higher centre for visceral control, regulating, for example, circulation, respiration, and many metabolic activities.

Attached to the hind end of the hypothalamus of fishes is a peculiar organ, the saccus vasculosus, with folded, pigmented walls, forming a sieve-like maze of cavities. Various functions have been suggested for it including secretion and depth perception. It may serve for resorption of the cerebral fluid. In mammals this occurs in granulations all over the surface of the cerebral hemispheres but the meningeal surface is simple in sharks and without a subarachnoid space. The large intracranial space between the brain and the cranial

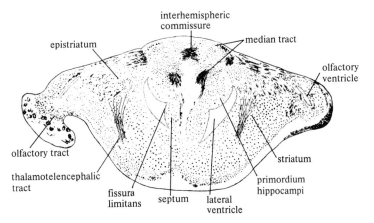

FIG. 5.29. A cross-section through the forebrain of a shark. (After Kappers, Huber and Crosby 1936 and modified following Ebbesson 1972*a*.)

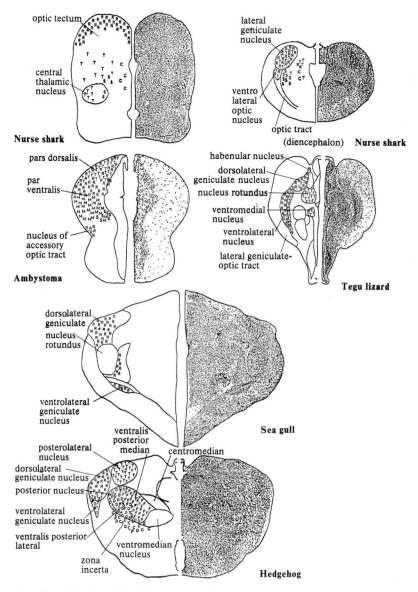

FIG. 5.30. Cross-section of the midbrain and the thalamus of the nurse shark and the thalamus of various vertebrates. The areas of input and projection are labelled as follows: A. auditory; C, cerebellar; D, dorsal columnus; tactile etc.; F, trigeminal; M, medullary; R, retinal; T, tectal. (After Ebbesson 1972*b*.)

cartilage is filled with a fluid that is similar to plasma, unlike the ventricular fluid, which contains no protein. If the dye Evans blue is injected into the cerebral fluid of a shark it does not appear over the surface but rapidly enters the saccus. The epithelial cells of the cavity are separated from the flattened capillary endothelium only by a basement membrane, as in other organs engaged in active transport, including the choroid plexus, kidney tubules, and ciliary body of the eye. However, cerebral fluid is probably also removed by bulk flow around the

cranial roots. The saccus is one of the few characteristic features that sharks share with bony fishes.

The midbrain, as in cyclostomes and teleosteans, is very large. The optic tracts end in its roof (tectum opticum) after complete decussation below the brain. The cells of the tectum are arranged in a complicated pattern of layers (Fig. 5.31). Other sensory centres that send tracts to the optic lobes are the olfactory (cerebral hemispheres), acousticolateral, cerebellar, gustatory, and probably also the general cutaneous centres of the

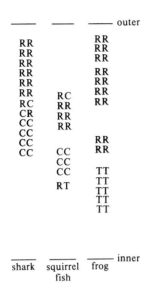

FIG. 5.31. Lamination of afferent fibre systems in the tectum. C, telencephalic input; R, retinal input; T, tectal input. (Modified after Ebbesson and Northcutt 1976.)

spinal cord. Efferent tracts from the midbrain roof go to the base of the midbrain (tegmentum) and extend backwards into the medulla, perhaps into the spinal cord. The efferent midbrain fibres have direct influence on the spinal cord, and electrical stimulation of points on the tectum opticum produces various movements of the fins. Forced movements follow injury to the midbrain. Sharks are able to learn visual discriminations between light and dark or horizontal and vertical stripes and this capacity persists after extensive lesions to the tectum, so presumably the forebrain is involved.

The cerebellum is a large organ in elasmobranchs, as in all animals that move freely in space. It is large and convoluted in most galeoid sharks such as *Lamna* (but not in *Scyliorhinus*) and also in skates and rays, but is smooth in squaloids. It consists of an unpaired corpus receiving spinal pathways, paired auricles with vestibular inputs and lateral line lobes, also paired. The internal organization of the cerebellum is like that of other vertebrates except that there are no basket cells and the dendrites of the Purkinje cells of the lateral line lobes are not flattened in one plane. The connections are unusual. There is probably no inferior olive and no climbing fibres. The granule cells are driven by mossy fibres, which in the vestibulocerebellum are primary vestibular fibres. The Purkinje cells are therefore probably driven only through their dendritic spines. Their axons project to cerebellar nuclei from which fibres pass down to the reticular and motor nuclei and upwards to the thalamus. This suggests that the arrangement of

parallel fibres is the essential feature of this cerebellum.

Injury to the cerebellum produces no obvious deficiencies unless the underlying cerebellar nuclei are damaged. The cerebellum therefore does not initiate or program movements but regulates the balance between excitatory and inhibitory control of motor programs determined elsewhere (Young 1978). Changes in the spontaneous discharge of Purkinje cells (18 per s) follow cutaneous stimulation only if the fish moves (Eccles, Táboříková, and Tsukahara 1970).

The medulla oblongata is the region from which most of the cranial nerves spring and especially those that regulate the respiration and visceral functions. In mammals this control is indirect, but in fishes the nerves that spring from the medulla directly innervate the respiratory muscles of the gills and floor of the mouth. It is no doubt for this reason that the centre for the initiation of the respiratory rhythm developed in the medulla.

Electrodes placed in contact with the surface of the brain record electroencephalograms (EEG) essentially similar to those of mammals. In the telencephalon there is an alpha rhythm at 4–9 Hz, modified by scents (see below). In the mesencephalon larger potentials at 5–11 Hz were recorded in the dark, replaced by smaller, irregular ones in the light. The EEG is thus essentially similar in elasmobranchs, teleosts, and amphibia (pp. 177, 258) (Gilbert, Mathewson, and Rall 1967).

20. Chemical senses of elasmobranchs

The sensory apparatus of elasmobranchs is largely dominated by the chemical senses. Sharks and rays find their prey mainly by smell. The paired nasal sacs have much folded walls. Water enters by a single opening but this may be partly divided by a fold, making a groove opening to the mouth so that the respiratory movements draw water through the nose. There are taste buds scattered over the wall of the mouth and pharynx. It has been shown experimentally that, as in higher animals, these are receptors for sampling the food after it has been brought close to the animal, whereas the nose acts as a distance receptor. Smell and taste are therefore different senses for a dogfish, as for us. By training fishes to discriminate between various substances it can be shown that those that we should smell are detected by the nose in the dogfish, but its organs of taste, like ours, can discriminate only between a few qualities, including salt, sour, and bitter. Sharks can no longer find food if their nostrils are blocked by cotton wool. Recordings from indwelling electrodes show that the EEG rhythms from the forebrain of a free-swimming shark are modified a few seconds after introducing 0.001 M betaine to the tank. At the same time the fish raised its head and began to swim. Amino acids and other breakdown products of tissues produced similar responses.

21. Eyes

The eyes are well developed in sharks and no doubt serve as an important means of finding the prey and avoiding enemies (Gruber 1977). The retina of most species consists largely of rods but cone-like receptors have been reported. The lemon shark (*Negaprion*) shows two stages during adaptation (Purkinje shift) and probably has good day and night vision. There are cones and a fovea in *Mustelus* and *Myliobatis*. The system in many species probably sacrifices acuity for sensitivity, allowing detection of a moving object in dim light. Receptive fields of ganglion cells are large (up to 3 mm) and sensitivity increases 100 000-fold during dark adaptation reaching levels similar to those found in cats, 10 times lower than in man. Behind the retina there is usually a reflecting layer, the tapetum lucidum, composed of orientated plates of guanine. This reflects up to 90 per cent of the light, more than in any other animals. The tapetum may be provided with pigment cells, which expand in the light but contract in darkness, increasing sensitivity in the dark but avoiding eye shine in the day. The retinae of mesopelagic sharks, like those of teleosts, contain a golden pigment, chrysopsin, tuned to the narrow wavelength of light there.

The lens is elliptical and very dense as in all fishes, since it must perform the whole work of refraction. It is provided with a protractor-lentis muscle, presumed to produce active accommodation for near vision by swinging the lens forward. Elasmobranch eyes, unlike those of teleosts have upper and lower lids, often mobile and sometimes with an opaque nictitating membrane (Carcharinidae). The iris is peculiar in those elasmobranchs that hunt by day; when it narrows it divides the pupil into two or more slits by means of a flap, the operculum. The muscles of the iris are better developed in elasmobranchs than in most bony fishes and the pupil makes wide excursions except in deep-sea species. The sphincter iridis muscle, which narrows the pupil, works as an independent effector. It is stimulated to contract by light, but its movements are not dependent on any extrinsic nervous mechanism. The radial dilator fibres, which open the pupil, receive motor fibres from the oculomotor nerve (Kuchnow 1971). The closure of the iris when illuminated is relatively slow. After the whole eye has been taken from the head, in the dark, the sphincter, being an independent effector, still closes when illuminated. The muscle, being without nerves, is not affected by any of the usual drugs that mimic action of the autonomic nervous system, though some of these affect the innervated dilatator muscle. We have therefore the curious situation that no 'autonomic' drugs applied to the isolated dark adapted eye cause closure of the pupil; this can only be produced by illumination (Fig. 5.32).

The pineal of elasmobranchs contains photoreceptors. Although there is no obvious window above it light penetrates here more readily than elsewhere. This median eye has a sensitivity well below the intensity of moonlight. Its function is uncertain.

22. Ears

The ear of elasmobranchs contains receptors concerned with detection of linear and angular acceleration, position with reference to gravity, and perhaps with hearing (Popper and Fay 1977). There are three pairs of semicircular canals, each with an ampulla containing a ridge, the crista, carrying receptor cells, whose hairs are embedded in a gelatinous cupula that closes the ampulla. The crista behaves as a highly damped torsion pendulum, bending with movement of the fluid. The receptors discharge impulses continuously and during angular rotations the frequency in the appropriate ampulla is increased by rotation in one direction and inhibited in the other, initiating compensatory movements of the eyes and fins (Fig. 5.33). The semicircular canals of many fishes have a somewhat greater diameter and radius of curvature than those of birds and mammals (Fig. 5.35). The larger inertial fluid masses

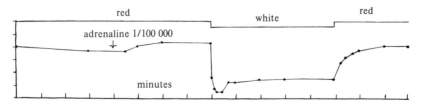

FIG. 5.32. Movements of margin of pupil of an isolated iris of the shark *Mustelus*, followed by plotting with a camera lucida and here shown mm magnified 53 ×. Addition of adrenaline causes slight dilation of the already dilated pupil and illumination then causes closure Acetylcholine even in concentrations of 1 in 10 000 has a similar *dilatory* effect. (From Young, J. Z. (1933). *Proceedings of the Royal Society, London* **B112**, 228–49.)

FIG. 5.33. Compensatory posture of fins and vertical deviation of eye in fish tilted on to its right side. (From Lowenstein, O. (1936). *Biological Reviews* **11**, 113–45.)

provide the sensitivity appropriate to the slower turning due to the water and the absence of a distinct neck. Skates and rays have especially large semicircular canals, these are also large in teleost fishes that turn slowly such as eels, sea horses, and flatfishes (Fig. 5.34). Conversely they are narrow in fast swimmers and the internal radius is often less in the horizontal canal than in the others, presumably to provide sensitivity for rapid turns in the yawing plane. The efficiency of the canals for

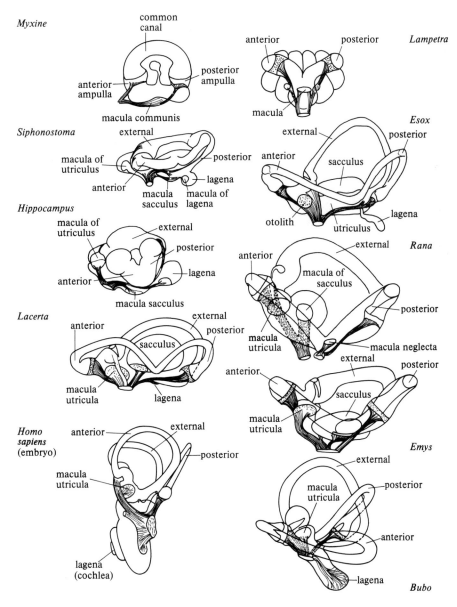

FIG. 5.34. The labyrinths of animals from the major groups of vertebrates. (After Retzius 1881.)

recording angular accelerations depends upon their small diameter (r) and radius of curvature (R). These increase only very little with the size of the animal (Fig. 5.35) and always remain in the same ratio to each other (Jones 1974).

The otolith organs include three patches of receptor cells in partially distinct sacs, the utricle, saccule, and lagena. The endolymphatic duct is an open canal and in some species serves to admit sand grains, which are attached to the maculae as gravity receptors. The utricle seems to be the main receptor producing appropriate orientation in relation to gravity. All three maculae have units that respond to tilt and linear acceleration. The lagena shows a maximum discharge rate near the normal position of the head and thus serves as an 'into level' receptor. The others show maximum activity when the head is out of true. The lagena and utriculus do not respond to vibration but there are units in the sacculus of rays giving nerve impulses up to 120 Hz, and vestibular microphonics up to 750 Hz occur. At high intensity there is much synchronization of units. The macula neglecta is another candidate for hearing. It is a small area very sensitive to vibration and lying in a special canal that opens through a hole in the cranium and is covered by a membrane. Thus the ear may function as a vibration receptor. Sharks are attracted to pulsed low frequency noises such as those made by a struggling fish (or man!). Their threshold for such frequencies is much lower than ours. They can be trained to respond to low thresholds over a range of up to 1000 Hz. It is not clear how hearing is achieved in the absence of a swim bladder nor whether the ear or lateral line is responsible.

There is a well-developed system of lateral line organs, whose function is considered later (p. 183).

Some of the organs of this system on the head are highly modified in elasmobranchs to form the ampullae of Lorenzini, long canals filled with jelly and ending blindly in sacs lined with receptors (Boord and Campbell 1977). These are insulated by tight junctions from the extracellular space and function as electroreceptors. They are sensitive to weak d.c. and low frequency a.c. fields in the water so that they can detect a live flatfish or other prey hidden in the sand, by the electrical changes during its respiratory movements or heart beat. They are also sensitive to light tactile stimulation or local changes of pressure and they increase their discharge of nerve impulses with very slight falls in temperature, but it is uncertain whether in life they act as mechano- or thermoreceptors as well as electroreceptors.

The lateral line system consists of isolated neurons, the pit organs, and neuromasts in fluid-filled canals opening to the surroundings by pores. Both sorts respond to movements nearby due to combined water displacements and pressure waves (the near field). This could provide information about the fishes own movements as well as that of animals and inanimate objects. However, a dogfish swims normally after the lateral line has been cut.

No doubt elasmobranchs, like other animals, have many senses referred to the skin, such as we call touch, pain, and the like, but few studies of these exist. There has been no satisfactory demonstration of any intramuscular endings in the myotomes of fishes. In fact the only specialized sensory endings in the trunk of a dogfish are corpuscles under the skin. These discharge impulses during every swimming cycle near to the time of maximum velocity and thus provide information about the frequency and angle of bending of the body (Roberts 1969). This information is essential for swim-

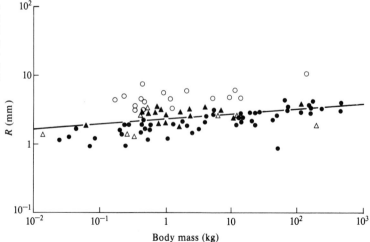

Fig. 5.35. The slight change in the radius of curvature (R) of the semicircular canals with the body weight of the animal. Closed circles – mammals; open circles – fishes; closed triangles – birds; open triangles – reptiles. The points are from 93 specimens obtained from 87 species. (From Jones and Spells 1963.)

ming, which continues after severing the spinal cord only if some dorsal roots are intact.

23. Behaviour of sharks

Precise observations of sharks in the sea have been limited for obvious reasons. When they see divers, sharks either move away or come unpleasantly close. When cornered against a reef-face *Carcharinus* and other sharks give a specific 'agonistic' display, which has been called 'hunching' (Gruber and Myrberg 1977) (Fig. 5.36). Various forms of social and dominance behaviour have been seen. Little is known about courtship and there are only four reports of copulation! Experiments with conditional responses and operant conditioning show that young nurse sharks can learn a brightness discrimination as fast as other vertebrates (Fig. 5.37) (Aronson, Aronson, and Clark 1967).

24. Autonomic nervous system

The sympathetic system of elasmobranchs consists of an irregular series of ganglia, approximately segmental, lying dorsal to the posterior cardinal sinus and extending back above the kidneys. These ganglia contain motor nerve cells (postganglionic cells) whose axons end in the smooth muscles either of the arterial walls or of the viscera. The cells themselves are controlled by preganglionic nerve fibres whose cell bodies lie in the spinal cord and whose processes run out in the ventral spinal roots and rami communicans (Fig. 5.38). In teleosts and other vertebrates some of these preganglionic fibres pass into the splanchnic nerves and run on to end in the coeliac and anterior mesenteric ganglia. No such 'solar plexus' is present in elasmobranchs. Again in most vertebrates the sympathetic ganglia send post-

FIG. 5.36. Agonistic display of the grey reef shark on the left seen from the side, front, and dorsally. On the right are the same views of the shark when not in display. (After Gruber and Myrberg 1977.)

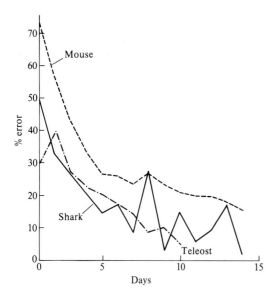

FIG. 5.37. Three vertebrates on a brightness discrimination problem show similarity in rate of learning. (From Aronson *et al.* 1967.)

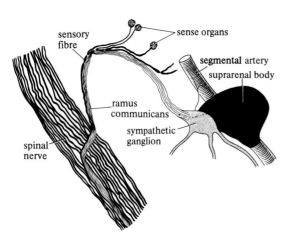

FIG. 5.38. Drawing of a sympathetic ganglion and related structures in a dogfish. (After Young 1933.)

ganglionic fibres back to the spinal nerves for distribution to the skin ('grey rami communicans') but these are absent in elasmobranchs and correspondingly there is no evidence of sympathetic control of skin functions (e.g. chromatophores); a very different condition is found in bony fishes (p. 185). Another peculiarity of the sympathetic system of elasmobranchs is that it does not extend into the head. This condition is unique among vertebrates, but it is not clear whether it is primary or the result of a secondary loss.

In mammals it is usual to recognize a parasympathetic system acting in antagonism to the sym-

pathetic, but this is not easy to define in the elasmobranchs. The vagus, it is true, is well developed, with branches to the heart and gut, but little is known of autonomic fibres in the other cranial nerves, or of a special 'sacral' parasympathetic system. Movements of the gut are controlled by the sympathetic ganglia, the vagus playing little part. Stimulation of the sympathetic at relatively high frequency (10 Hz) produces contraction first of the pylorus, then pyloric stomach and finally, a minute later, of the cardiac stomach (Fig. 5.39) (p. 134). Lower frequencies produce inhibition, followed by strong rebound contraction when the stimulus ceases (Fig. 5.40). Experiments show that these actions are mimicked by either 5-hydroxytryptamine or adrenaline but not by acetylcholine. They do not therefore fall easily into the usual category of either the sympathetic or parasympathetic divisions. Moreover there is no evidence that they are 'antagonized' by any action of the vagus. Again, the heart is controlled by inhibitory fibres from the vagus but has no sympathetic innervation, though there are sympathetic fibres to the arteries. A ciliary ganglion connected with the oculomotor nerve is present as in other animals, but there is no sense in which it can be called antagonistic to the sympathetic system, since the latter does not extend into the head.

The system of internal regulation by autonomic nerves is evidently very different from that of mammals. There are some reasons for thinking it may involve less elaborate mechanisms for regulation of homeostasis by

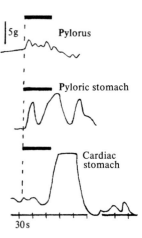

FIG. 5.39. Responses of the pylorus and the two parts of the stomach of a dogfish (*Scyliorhinus*) to stimulation, at the bars, of the right 1st sympathetic ganglion with 10-ms square waves at 10 Hz. (Young 1980.)

FIG. 5.40. Responses of the cardiac stomach of the dogfish to stimulation of the sympathetic with 10-ms square waves. At low frequencies there is inhibition followed by rebound contractions. At higher frequencies there is a delayed contraction after the end of stimulation (Young 1980).

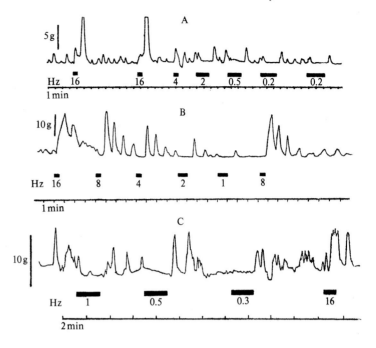

antagonistic control systems. Such comparisons are always difficult to make and we need further investigation of the facts. The life of elasmobranchs is well enough controlled to allow them to be among the dominant predators of the sea, but they have not been a success in fresh water or on land.

6 Evolution and adaptive radiation of Chondrichthyes

1. The early gnathostomes

THE first fishes with jaws appeared in the Upper Silurian, about 400 million years ago presumably descended from agnathans. They were heavily armoured, with a covering of thick rhomboidal scales. The earliest of all were the acanthodians (*Climatius*, Fig. 6.1), most or all of which lived in fresh water. Shortly after appear the placoderms, some freshwater and others marine. There is debate about the affinities of these two groups with later fishes. One common view now is that the acanthodians gave rise to the bony fishes and the placoderms to the cartilage fishes, by loss of the bone. We shall therefore deal with the placoderms here and the acanthodians in connection with the bony fishes (p. 189).

2. Classification

Superclass 2: GNATHOSTOMATA
  Class 1: Placodermi (plate skin). Lower–Upper Devonian
    *Order 1: Arthrodira (jointed pieces). Silurian–Devonian
      Coccostus; *Rhynchodus*
    *Order 2: Ptyctodontida (folded teeth). Devonian
    *Order 3: Phyllolepida (leaf scales). Middle–Upper Devonian
      Phyllolepis
    *Order 4: Petalichthyida (flattened fishes) (=Macropetalichthyida). Devonian
      Lunaspis
    *Order 5: Rhenanida (=Stegoselachii) (Rhine fishes)
      Gemuendina
    *Order 6: Antiarchi (ancient ones)
      Bothryolepis
  Class 2: Chondrichthyes (cartilage fishes). Middle Devonian–Recent
    Subclass 1: Elasmobranchii (plate gills)
    *Order 1: Cladoselachii (branched tooth). Devonian–Permian
      Cladoselache; *Goodrichthys;* *Diademodus;*
      Ctenacanthus
    *Order 2: Pleuracanthodii (side-spines) (= Xenacanthida). Devonian–Triassic
      Xenacanthus (= *Pleuracanthus*)
    Order 3: Selachii. Devonian–Recent
      Suborder 1: Hybodontoidea (humped teeth)
      Hybodus
      Suborder 2: Heterodontoidea (different teeth). Jurassic–Recent
      Heterodontus
      Suborder 3: Hexanchoidea (six gills) (= Notidanoidea). Jurassic–Recent
      Hexanchus; Chlamydoselachus; Heptranchus
      Suborder 4: Galeoidea (sharks). Jurassic–Recent
      Scyliorhinus; Mustelus; Cetorhinus;
      Carcharodon; Rhinocodon

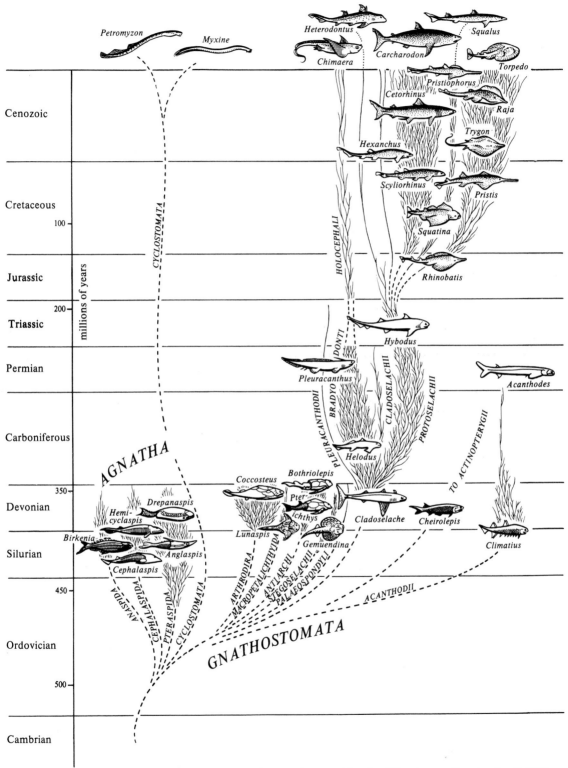

Fig. 6.1. Chart of possible evolutionary relationships. Some authors would prefer different connections (see Jarvik 1980).

Suborder 5: Squaloidea (dirty shark). Jurassic–Recent
   Squalus; Squatina; Pristiophorus; Alopias
Suborder 6: Batoidea (rays and skates)
   Raja; Rhinobatis; Pristis; Torpedo;
   Trygon Myliobatis; Mobula
Subclass 2: Bradyodonti (slow tooth) (=Holocephali). Devonian–Recent
   *Helodus; *Iniopteryx; Chimaera;
   Hydrolagus; Callorhynchus

3. Placodermi

This class includes the earliest gnathostomes except for the acanthodians (p. 189). They were rather strange fishes, mostly heavily armoured, as the name implies, and many with sharp spines. They were mostly bottom-living animals, dorsoventrally flattened and somewhat like modern rays (see ** Gemuendina*, Fig. 6.1). Many lived on invertebrates or by shovelling mud, like their agnathan ancestors. Most lived in the sea, but some in fresh water.

The armour consisted of a bony head-shield, movably jointed with a trunk-shield. The shields are composed of many bones, but their pattern is not like that on the head of later fishes. The body behind the shields was mobile and with a heterocercal tail (p. 115). It was usually naked but sometimes covered with thick scales. The plates were made of bone arranged in three layers; sometimes with dermal denticles covering the outer layer. Lateral line canals run in grooves in the bone.

The jaws are quite unlike those of later fishes, consisting of bony plates with sharp edges (but usually no teeth) and were often jointed to the head-shield. The upper jaw bars (palatoquadrate and mandible) were mostly not ossified, and therefore the mode of jaw support is not known. The spiracle was probably a typical gill slit. There were paired pectoral and pelvic fins and often large pectoral spines attached to the trunk plate. In the ptyctodontids the males had pelvic claspers and reproduction was presumably like that of sharks. This feature is one of the reasons for considering that the placoderms were related to the ancestors of the chondrichthyes. They also show some similarities in the snout and skull to the Holocephali (p. 158). However, there are so many differences that we cannot learn much from the placoderms about the origin of later groups.

4. Characteristics of elasmobranchs

The organization of a shark used to be considered to show the earlier stages of fish evolution, but we have seen evidence that this is a mistake (p. 115). The sharks and skates and rays are highly developed creatures; in particular, the absence of bone is a secondary feature. They have been able to give up their defensive armour because of the development of other means of pro-

tection, swift swimming, good sense organs and brain, and powerful jaws. We can now examine the history of these changes and study the varied creatures that can be classified as elasmobranchs. As usual in examining such histories we must try to discover evidence about the forces that have operated to produce the changes of type, and look for signs of any consistent trends, persisting for long periods of years.

The elasmobranchs form a very compact group, nearly always marine and of predaceous habit, hunting by smell. They have a high concentration of urea in the blood, no bone in the skeleton (though the cartilage may be calcified), no operculum over the gills, and no lung or swim bladder. The hyomandibula plays a part in the support of the jaw. The tail is usually heterocercal. The pectoral fin is anterior to the pelvic and the latter is usually provided with claspers, fertilization being internal. The body is more or less completely covered with placoid scales (denticles) and these are specialized in the mouth to form rows of teeth, serially replaced. The intestine is short but with its absorptive surface increased by a spiral valve. The typical cartilage fishes with these characters may be placed in the subclass Elasmobranchii to distinguish them from an early aberrant offshoot the Bradyodonti, represented today by the peculiar rat fishes, *Chimaera*, and rabbit fishes, *Hydrolagus* (p. 158).

5. Palaeozoic elasmobranchs. Cladoselachii

The selachians are among the most numerous of the various predatory animals in the sea. There have been many branches of this line and we may now survey the history of the group from its first appearance. The characters we have used in our definition mark the elasmobranchs off from the earliest-known gnathostomes (Fig. 6.1). The elasmobranchs were possibly derived from some early placoderm, but the earliest evidence of the existence of true sharks is in the form of isolated teeth and scales from Middle Devonian deposits ('**Cladodus*'). These early teeth are of a characteristic type known as 'cladodont', with a flattened disc-like base carrying conical cusps of which the central is the largest (Schaeffer 1967) (Fig. 6.2). The body form was much like modern sharks (Fig. 6.3). The jaws were

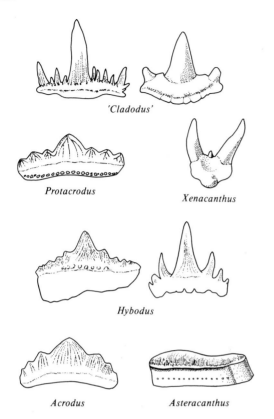

Fig. 6.2. Fossilized teeth of early cladodont and hybodont elasmobranchs. (From Schaeffer 1967.)

body wall (Fig. 6.3). It has been suggested that this was the earliest condition of the pectoral fin, perhaps showing its derivation from a continuous or extended fin-fold (Fig. 6.5). This theory has the advantage that it agrees with the embryological development of the fin by concentration of a series of segments (Fig. 6.6). It also seems likely that anterior and posterior fins expanded in the horizontal plane would be necessary for stabilization (p. 120). Moreover, this theory of the origin of paired fins has the great advantage that it compares them with the median fins, which are also continuous folds. It has been argued, however, that the cladoselachians are very far from the earliest known fishes and that in both ostracoderms (p. 76) and placoderms fins are known that have a narrowly constricted base. We cannot yet say for certain what has been the course of evolution of the paired fins, but the fin-fold theory has much plausibility.

The tail of cladoselachians was heterocercal but with nearly equal upper and lower lobes. It was probably stiff, since the radials were unsegmented, allowing little adjustment. Many of these early sharks had strong spines in front of the dorsal fins. They had no claspers.

The cladoselachians represent the ancestral Devonian sharks, from some of which all later forms have been derived. Animals of similar type were fairly common in late Devonian and Carboniferous seas. *Goodrichthys* reached a length of 2.5 m (Fig. 6.7).

The pleuracanthodians (*Xenacanthus* = *Pleuracanthus*) were a specialized group of freshwater carnivores (Fig. 6.3). The tail was straight (diphycercal) and the paired fins had become modified accordingly (see p. 120). The axis was completely freed from the body wall to give a paddle-like fin, with pre- and postaxial rays, a type known as archipterygial (Fig. 6.5), because it was once supposed to be ancestral to all

relatively long (Fig. 6.4) extending forward to the snout and attached to the postorbital process of the cranium and also by the hyomandibula (amphistylic). There are no vertebral centra, so that the notochord remained unconstricted. The pectoral fins in some of these early sharks had a broad base not sharply marked off from the

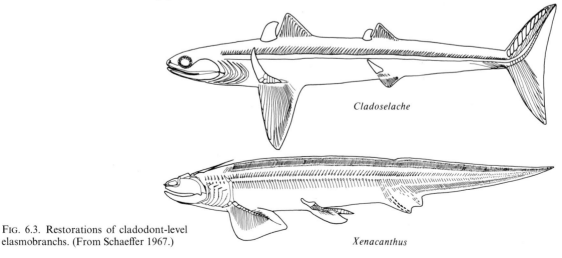

Fig. 6.3. Restorations of cladodont-level elasmobranchs. (From Schaeffer 1967.)

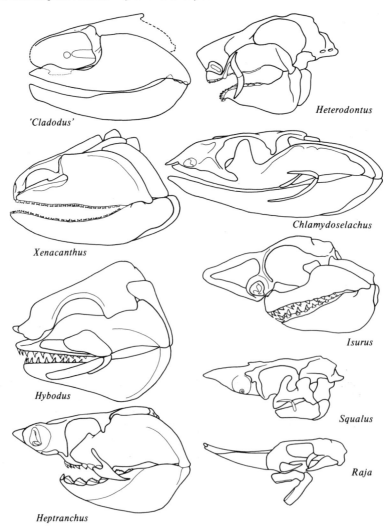

FIG. 6.4. Fossil and modern elasmobranch skulls seen from the side. (From Schaeffer 1967).

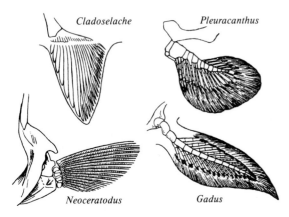

FIG. 6.5. Pectoral fins of various fishes. (From Norman 1931.)

others. A large spine on the head gives the group its name. Claspers were present. These animals were common in the Carboniferous and Lower Permian, but in subsequent times they disappeared without leaving descendants.

6. Mesozoic sharks. Selachii

The elasmobranchs of the type that survive today, the order Selachi or neoselachians, had already appeared in the Upper Devonian probably descended from a clado-selachian ancestor. The Mesozoic neoselachians still retained some of the earlier characteristics and these may be called protoselachians and are placed in a separate suborder Hybodontoidea.

After flourishing in Palaeozoic seas sharks seem to

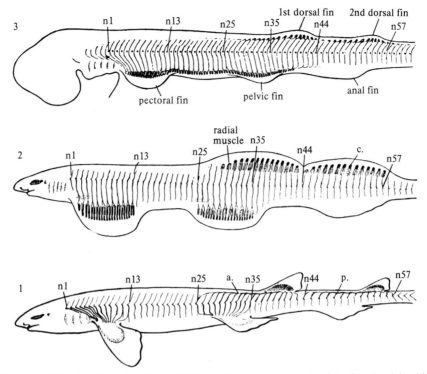

FIG. 6.6. Development of the fins of the dogfish. 1, Adult showing the nerve-supply of the fins; 2, adult with the fins shown expanded and their nerves and muscles shown as if concentration had not taken place; 3, a 19-mm embryo, showing the actual condition. *a*. anterior collector nerve of first dorsal fin; *c*. (black) cartilaginous radial partly hidden by the radial muscle; *n*. 1–57, spinal nerves and ganglia; *p*. collector nerve of second dorsal fin. (From Goodrich, E. S. (1906). *Quarterly Journal of the Microscopical Society* **501**, 333–76.)

have become less numerous during the Permian and Triassic. During this period there was probably little fish life in the sea and the stock seems only to have survived by adopting a varied diet, including invertebrate food. The protoselachian or hybodont sharks of this period showed some changes in the direction of modern forms (Fig. 6.7). The lower jaw was rather short. In early forms it articulated with an otic process of the palatoquadrate but was free in some later ones. In some types there were only five branchial arches (*Hybodus*). The notochord was still unconstricted but there were long narrow neural arches and sometimes haemal arches. The pectoral fin was free from the body wall posteriorly and had three distinct basal elements. It was therefore probably more freely adjustable than in cladoselachians. The radials in the hypochordal (lower) lobe of the tail were segmented.

The hybodont capacity to eat molluscs was further developed in some related Carboniferous sharks, which acquired a dentition of flattened crushing plates – hence 'pavement-toothed sharks', such as *Helodus*. The upper jaw became fused with the brain-case, allowing more powerful crushing action.

Heterodontus, the Port Jackson shark of the Pacific is often regarded as a surviving member of the hybodont group though other workers place it with the galeoids. We shall keep it in a distinct suborder. Its characteristic teeth, are found back to the Middle Jurassic. As the name implies they are of two types, pointed ones in front and flattened ones, for crushing molluscs, behind. There is total cleavage of the yolk of the egg. The meroblastic (partial) cleavage, typical of modern elasmobranchs and teleosts, was therefore a relatively late development and other survivors of the Mesozoic period besides *Heterodontus* also show holoblastic cleavage (cf. 138). The palatoquadrate of *Heterodontus* extends forwards to the snout. As in hybodonts the palatoquadrate is attached to the brain-case in front and by the hyomandibular behind, this is usually considered to be a form of amphistyly. However, this and other primitive features of *Heterodontus* do not necessarily separate it from the galeoid sharks (Compagno 1977).

The transition to the modern level of shark organization in Jurassic times or earlier involved shortening of the jaws and adoption of a hyostylic suspension. The axial skeleton acquired calcified centra with neural and

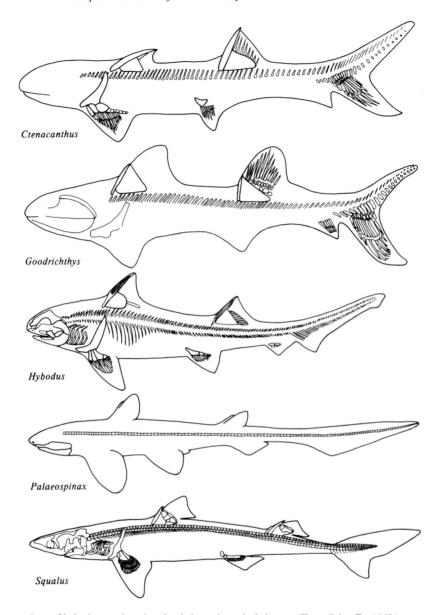

Ctenacanthus

Goodrichthys

Hybodus

Palaeospinax

Squalus

Fig. 6.7. Restorations of hybodont and modern-level elasmobranch skeletons. (From Schaeffer 1967.)

haemal elements. Changes in the teeth and body form occurred mainly in two directions.

In the suborders Hexanchoidea, Galeoidea, and Squaloidea, the true sharks, the teeth all became sharp and the animals swift swimmers. In the suborder Batoidea, on the other hand, the teeth remained flattened and sometimes became highly specialized for a mollusc-eating diet (Fig. 6.8), producing the flattened bottom-living creatures, the skates and rays. The stages of this transition can be followed, and some of the intermediate

types still exist. Thus in *Rhinobatis*, the banjo-ray (Fig. 6.1), the pectoral fins are enlarged but still distinct from the body. Almost identical creatures have been found in Jurassic rocks. It is probable that several separate lines showed this flattening of the body, and that the Batoidea are not really a homogeneous group.

7. Modern sharks

The sharks are of three types all dating from the Jurassic. The Hexanchoidea (Notidanoidea) have re-

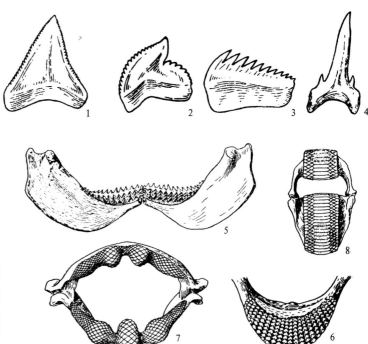

FIG. 6.8. Teeth of various elasmobranch fishes. 1, Man-eater (*Carcharodon*); 2, tiger shark (*Galaeocerdo*); 3, comb-toothed shark (*Hexanchus*); 4, sand-shark (*Odontaspis*); 5, blue shark (*Carcharinus*); 6, nurse shark (*Ginglymostoma*); 7, guitar fish (*Rhina*), 8, eagle-ray (*Myliobatis*). (After Norman 1931.)

mained nearly at the hybodont level for 150 million years and show many primitive features, such as an amphistylic jaw and an unconstricted notochord. There are six or seven gill slits, which may be a secondary development, since there are five in *Hybodus*. *Hexanchus* and *Heptranchus* are long-bodied, slow-moving sharks from warm waters. They are viviparous but without placentae. *Chlamydoselachus*, the frilled shark, lives in deep water and feeds on cephalopods.

The suborder Galeoidea includes 73 per cent of all living species of sharks, mainly inhabiting shallow tropical and warm temperate seas. They have two dorsal fins, not supported by spines. Here belong the dogfishes *Scyliorhinus* and *Mustelus*, both mainly bottom-living animals feeding on a mixed diet, including crustaceans and molluscs. *Carcharinus*, the grey sharks up to 4 m long, are among the most abundant and economically important of all. They occur in all seas and some fresh waters and are widely used for food, oil, and leather. They attack tuna and other food fishes and also humans. In *Mustelus* the teeth are square and flat and are used for crushing. In *Cetorhinus*, the basking shark, the predaceous habit of the group has been abandoned in favour of straining small food directly from the plankton by means of special combs on the gills (gill rakers), an arrangement recalling that of the whalebone whales. The great effectiveness of this method of feeding may be seen in the length of 10.5 m or more attained by some of these sharks. They maintain a neutral buoyancy

by means of large quantities of oil and the hydrocarbon squalene, stored in the liver. Basking sharks produce very numerous small eggs, which develop within a 'uterus', but without placentae. *Rhincodon*, the whale shark, is also a plankton feeder and becomes very large. It is not closely related to the basking sharks. It moves up and down vertically, the mouth open, sucking in plankton. In this group there are also many of the fiercest man-eating sharks, such as *Carcharodon*, often 9 m long, found in many seas. Some fossil forms of this genus are estimated to have reached a much greater length, possibly of 27 m.

The smaller suborder Squaloidea includes the sharks of cold and deep waters. There is a spine in front of each dorsal fin. They are not, however, otherwise different in habits from the other sharks. The spiny dogfish (*Squalus*) is a well-known type and here belong also the sawsharks (*Pristiophorus*) and a group of bottom-living forms, the angelfish or monkfish (*Squatina*), which acquire a superficial similarity to the skates and rays. *Alopias*, the thresher, is said to differ from most sharks in that instead of seizing the prey as it is presented, it hunts systematically, several sharks working together and using their whip-like tails to drive smaller fishes such as mackerel into shoals, where they are then seized.

8. Skates and rays

The skates and rays, have become specialized for life on the bottom of the ocean in shallow waters, feeding mainly on invertebrates, and usually having blunt teeth (Fig. 6.4). Locomotion is no longer by transverse movements of the body but by waves that pass backwards along the fins. In the earlier stages, such as *Rhinobatis*, the banjo-ray, which has existed from the Jurassic period to the present, the edges of the fins are still free and the tail is well developed. In *Pristis*, another sawfish type, outwardly similar to *Pristiophorus* and known since the Cretaceous, the head is drawn out into a long rostrum armed with denticles. Its use is uncertain but the head strikes from side to side among shoals of fishes. There are species of sawfishes in India, China, and the Gulf of Mexico that live in fresh water. In *Raja*, first found in the Cretaceous, the pectoral fins are attached to the sides of the body and the median fins are very small. In the more recent *Trygon* and other sting-rays the tail is reduced to a defensive lash, the dorsal fin persisting as a poison spine. In the eagle-rays (*Myliobatis*) the teeth are flattened to form a mill able to grind mollusc shells (Fig. 6.8). The sea-devils (*Mobula*) have expansions of the fins at the front of the head, which they use to direct plankton to the mouth, where it is sucked in. The fins in *Torpedo*, the electric ray, extend so far forward that the front of the animal presents a rounded outline. The animal is protected by a powerful electric organ, formed by modified lateral plate muscle, innervated by cranial nerves. Several species of *Raja* have weak electric organs perhaps used for guidance (p. 205).

Life on the bottom has produced many further modifications in the skates and rays. In those that live in shallow and hence well-illuminated waters the colour of the upper surface is often elaborate, the underside being white. In certain species of *Raja*, for example, there is a pattern of black and white marks, which probably serves to break up the outline of the fish.

The eyes of the skates and rays have moved on to the upper surface of the head and are protected by well-developed lids. In most forms the pupil is able to vary widely in diameter and often has an operculum by which the aperture can be reduced to two small slits.

There is a special modification of the respiratory system so that water is drawn in not through the mouth but by the spiracle, which is provided with a valve that shuts at expiration, as the water is forced out over the gills. The Batoidea have therefore developed many special features for their bottom-living habits and have diverged among themselves into many varied lines. They have been very successful and are among the commonest fishes in the sea.

The method of ensuring stability in the pitching plane adopted by elasmobranchs (p. 120) necessitates a certain flattening of the front end of the animal. It is not therefore surprising that this tendency is often exaggerated and has several times produced flattened bottom-living creatures, such as the skates and rays. The Actinopterygii show the opposite tendency, to lateral flattening (p. 201).

9. Bradyodonti (= Holocephali)

Finally we must consider the Holocephali, including the rat-fishes (*Chimaera*, Fig. 6.9), rabbit-fishes (*Hydrolagus*) and elephant-fish (*Callorhynchus*). They preserve today some features of elasmobranch life in Palaeozoic times though in other respects they are aberrant. They live close to the bottom and feed on molluscs and other invertebrates. The tail is long and thin and they move by sweeping movements of the large pectoral fins. There is an erectile spine in front of the dorsal fin, sometimes poisonous. There is no stomach and the mouth is a small aperture surrounded by lips, giving the head a parrot-like appearance. The teeth are large plates firmly attached to jaws, and the upper jaw is remarkable in being fused to the skull (holostylic), giving the group its name, the hyoid arch being free. There are no denticles. The gut is a simple straight tube between pharynx and anus. A short oesophagus leads directly to a broad spiral intestine and this to a short rectum. It is possible that this is a primitive condition before the evolution of a stomach. The Holocephali also differ from the elasmobranchs in the presence of an opercular flap attached to the hyoid arch. There are extra claspers in front of the usual pelvic ones and a further copulatory organ on the head of the male, the cephalic clasper. The notochord is unconstricted and the vertebrae reduced to separate nodules. The lateral line canals are open (a primitive feature) and specially developed on the underside of the snout, presumably to detect food. The cleavage is holoblastic, another primitive character.

Many of the internal features resemble those of selachians, for instance the conus arteriosus, and urino-genitals in which there are separate urinary and spermatic ducts. The blood contains much urea and there is a rectal gland. The brain has a peculiar shape on account of the large size of the eyes, which almost meet above the brain, so that the diencephalon is long and thin.

Chimaera extends back to the Cretaceous but forms that were probably related are found in the Carboniferous (*Helodus*, *Iniopteryx*). The various types are grouped together as Bradyodonti. They had flat, grinding teeth, and some gut contents have been found to include remains of the shells of brachiopods, crustacea and crinoid echinoderms. The 'holostylic' attachment of the mandible may be connected with this grinding function. It is difficult to believe that the complete hyoid arch can be anything but a primitive feature. The

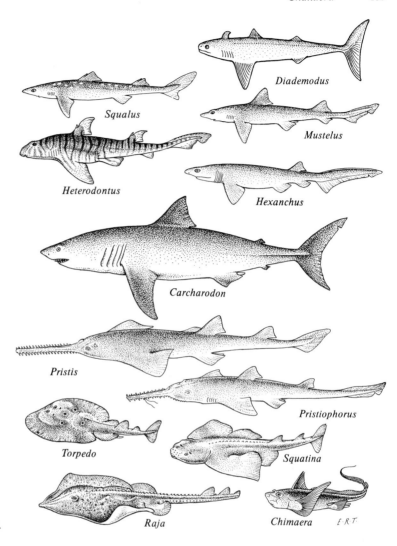

FIG. 6.9. Various elasmobranch fishes.

bradyodont dentition recalls that of the hybodont protoselachians, from which some people consider the Holocephali may have been derived (p. 154).

The affinities of the Holocephali and their relatives are still debated. Some workers have placed them with the placoderms and even considered that they are 'living arthrodires' (p. 152).

10. Tendencies in elasmobranch evolution

The elasmobranchs have been in existence ever since the Devonian, and for much of this long period of nearly 400 million years we can follow their changes with some accuracy. This type of fish was first formed by loss of the heavy bony armour of the earliest gnathostomes, associated with the adoption of a rapidly moving and carnivorous habit. The resulting shark-like form has remained with relatively little change through the whole history of the group; cladoselachians from the Devonian are in many ways like modern sharks but we can see some signs of an 'improvement' in the capacity for this carnivorous way of life. Modern elasmobranchs show an impressive variety of methods of feeding (p. 156).

If our interpretation of the evidence is right, however, the modern shark type has been evolved from the Devonian type through a heterodont stage. During the late Permian and Triassic there was little fish food for the sharks and they appear to have taken to living on invertebrates. Eating this diet was presumably easier for animals possessing the two types of teeth described on page 157, and the animals also became rather flattened with their life on the bottom. On the reappearance of numerous fishes in the sea, in the Jurassic, some of these heterodonts resumed the shark-like habit, lost the crushing teeth, and developed into the varied fish-eating types alive today. Others of the heterodonts, however, became still more specialized for bottom life, as the modern skates and rays (Zangerl 1973). We can therefore contrast the 'morphotypes' of Palaeozoic and modern elasmobranchs (Fig. 6.10). The former have much dermal armour and fin spines, no centra, and fins supported to their edges by cartilaginous fin rays. The mouth is terminal and the palatoquadrates are attached to the neurocranium. Teeth are all alike. There is much prismatic calcification and some bone. The modern forms have no fin spines, calcified centra, but no bone, reduced cartilaginous fin rays, ceratotrichia as fin supports, mouth subterminal with variable jaw support often involving the hyomandibula. There are various tooth shapes.

The persistent tendency in all this is to eat other animals of some sort. When fishes are available sharks will eat them, and the bodily organization for doing so seems to have been evolved at least twice. Similarly other members of the same stock ate molluscs and crustacea and became modified for this. At a time when circumstances forced the animals to strive in one direction those with a particular bodily type, say, broad, 'heterodont' teeth were selected. When fish food again became available those animals born with quicker habits and sharper teeth were able to eat the fish and the shark type returned. However, we can recognize distinct advances in the later types. The shortening of the jaws and transition from amphistyly to hyostyly allowed for greater force and for protrusion to seize the prey. Teeth were developed suitable for sawing and slicing rather than for holding and tearing (though this still continues). The calcified centra gave greater stability than the unconstricted notochord. The increased mobility of the pectoral and caudal fins provided increased manoeuvreability. It is easy to say that these are assertions that cannot be proved rigorously, but it is surely foolish to deny that they are plausible. Together they all add up to a greater variety of detailed control (for instance in the fins), presumably needed to match the mobility of teleostean prey.

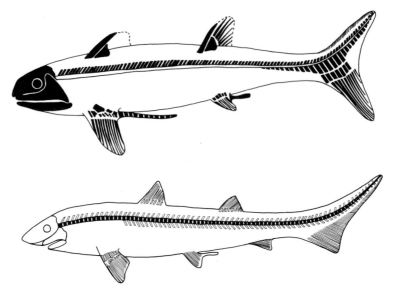

FIG. 6.10. Morphotypic design of a Palaeozoic elasmobranch (above; black shows prismatic calcification of skeleton), and a modern elasmobranch (below; black is non-prismatic calcification of vertebral centra) whose fins are mainly supported by ceratotrichia. (Zangerl 1973.)

7 The mastery of the water. Bony fishes

1. Introduction: the success of the bony fishes

THE acanthodians from the late Ordovician, Silurian, and Devonian were gnathostome fishes with bony skeletons. The presumed descendants of the acanthodians can be divided into two groups: the crossopterygians, the lobed-fin or lung fishes, including Devonian forms that led to the amphibia, and actinopterygian or rayed-fin fishes, culminating in the modern bony fishes. In Devonian times the Crossopterygii and Actinopterygii were very alike and both contained bone. The term bony fishes or Osteichthyes is often applied to these two groups together, since they have some features in common and distinct from the elasmobranchs.

The great group of Actinopterygii, which, for all the importance of the elasmobranchs, must be reckoned as the dominant fish type at the present time, includes most of our familiar fishes, perch, pike, trout, herring, and many other types of 'modern' fish. In addition there are placed here some surviving relics of the stages that have been passed before reaching this condition, such as the bichir, sturgeons, bow-fin, as well as related fossil forms.

Many groups of animals have been successful in the water; crustacea, for instance, are very numerous and so are cephalopod molluscs and echinoderms, but the success of the bony fishes surpasses that of all others. From a roach or perch in a stream, to a huge tunny or a vast shoal of herrings in the sea, they all have the marks of mastery of the water. They can stay almost still, as if suspended, dart suddenly at their prey or away from danger. They can avoid their enemies by quick and subtle changes of colour. Elaborate eyes, ears, and chemical receptors give news of the surrounding world and complicated behaviour has been evolved to meet many emergencies. Reproductive mechanisms may be very complex, involving elaborate nest-building and care of the young; social behaviour is shown in shoaling movements, which may be accompanied by interchange of sounds.

Bony fishes abound not only in the sea but also in fresh water, which has never been effectively colonized by cephalopods or elasmobranchs. They can exist under all sorts of unfavourable or foul-water conditions and a considerable number of them breathe air and live for a time on land. Perhaps the majority are carnivorous, but others feed on every type of food, from plankton to seaweeds.

To whatever feature of fish life we turn we find that the bony fishes excel in it in a number of ways in different species. It is small wonder that with all these advantages they are excessively numerous. There are some 575 species of living elasmobranchs, but more than 20 000 species of bony fish have been described.

The number of individuals of some of the species must be really astronomical. For instance more than a million tons of small gadoid fishes, such as sprats, are caught every year. This approaches a million million in number and the whole population can hardly be less than 10^{16}. Again, it is estimated that a thousand million blue-fish collect every summer off the Atlantic coast of the United States and, being very voracious carnivores, they consume at least a thousand million million of other fishes during the season of four months. This gives some idea of the tremendous productivity of the sea, and of the way the bony fishes have made use of it. Needless to say, man has also made considerable use of the bony fishes, which indeed provide, with the elasmobranchs, a not inconsiderable portion of the total of human food.

2. The trout

Salmo trutta, the brown trout, may be taken as an example of a bony fish; we shall also refer at intervals to conditions in other common freshwater fish such as the dace (*Leuciscus*) and perch (*Perca fluviatilis*). There is considerable confusion about the various types of trout and their close relatives the salmon. The brown trout is abundant in rivers and streams throughout Europe and is commonly about 20 cm long at maturity, though it may grow larger. It is grey above and yellowish below, with a number of dark spots scattered along the sides of the body (Fig. 7.1).

The body form is typical of that of teleostean fishes in being short, narrow in the lateral plane but deep

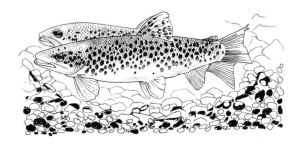

FIG. 7.1. Male and female brown trout (*Salmo trutta*) spawning. The male is quivering – a short sequence of rapid shudders of the whole body which excites the female. (After Jones, J. W. (1959). *The salmon*. Collins, London.)

dorsoventrally, in fact more obviously streamlined than the shape of elasmobranchs. The movements of a trout do not at first sight obviously involve the bending of the body into an **S** ; nevertheless, the method of swimming is essentially by the propagation of waves along the body by contraction of the longitudinally directed fibres of the myotomes (p. 34).

The tail differs from that of elasmobranchs in being outwardly symmetrical, though internally there are still traces of the upturned tip of the vertebral column (Fig. 7.2). Besides the typical caudal 'fish-tail', supported by bony rays, there are two dorsal fins and a ventral fin, but the hinder dorsal fin differs from the others in having no rays to support it and is called an adipose fin, because of its flabby structure. The paired fins are rather small and it is from their structure that the whole group derives the name Actinopterygii or rayed-fin fishes. There is no lobe projecting from the body and containing basal fin supports, as there is in the fin of lung

fishes. All the basal apparatus of the fin is contained within the body wall and only the fin rays project outwards, as a fan. The pelvic fin of bony fishes often lies relatively far forward; in the trout, however, it is unusually far back, just in front of the anus; in other types it may be level with the pectoral fin, or even anterior to it (Fig. 7.2). The significance of the shape of the body and fins in swimming will be discussed later (p. 200).

The skin consists of a thin epidermis and thicker dermis, the former has stratified squamous layers and some contain keratin (Mittal and Whitear 1979). It contains mucous glands. The mucus of some eels and other fishes has remarkable powers of precipitating mud from turbid water. The mesodermal dermis provides an elaborate web of connective tissue fibres. It also contains nerves, chromatophores, and scales. The latter are thin overlapping bony plates, covered by skin, that is to say, they do not 'cut the gum' as do placoid denticles. The exposed part of each scale bears the pigment cells, which control the colour of the animal, in a manner presently to be described. The bone of the scales is absorbed at intervals by scleroclasts, making a series of rings, which, like the growth-rings on a tree, are due to the fact that growth is not constant but occurs fast in the spring and summer and hardly at all in the winter. The age of the fish can therefore be determined from these rings (Fig. 7.3), or from the similar markings on the ear stones (p. 180). While an adult salmon is in fresh water no growth occurs, leaving a spawning mark on the scale.

The head of the trout shows some of the most specialized and typical teleostean features (Fig. 7.4). There are two nostrils on each side, but no external sign

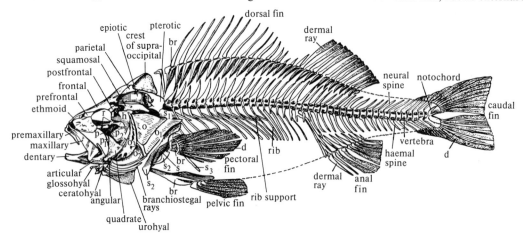

FIG. 7.2. Skeleton of the perch. *br*, basal and radial cartilages; *d*, dermal fin rays; *h*, hyomandibula; *o*, operculum; o_1, suboperculum; o_2, interoperculum; p_1, ectopterygoid; p_2, metapterygoid; *r*, suborbital ring; *s*, shoulder girdle; s_1, dorsal process of shoulder girdle; s_2, outer rim of shoulder girdle; s_3, posterior process of shoulder girdle. The bones labelled frontal and parietal are now given other names (p. 164). (After Zittel, from Dean 1895.)

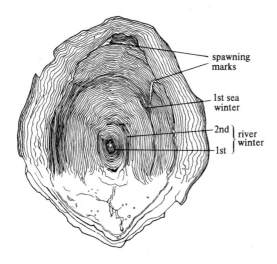

FIG. 7.3. Spawning marks are the result of erosion or absorption of the scale margin due to a calcium deficiency during the fasting period. (After Jones, J. W. (1959). *The salmon*. Collins, London.) 1. 1st river winter. 2. 2nd river winter. 3. 1st sea winter.

of ears. The mouth is very large and its edges are supported by movable bones, to be described below. The maxillary and mandibular valves are folds of the buccal mucosa, serving to prevent the exit of water during respiration. The tongue, as in selachians, has no muscles, but may carry teeth and taste-buds. Behind the edge of the jaw is the operculum, a flap covering the gills

and also supported by bony plates. In connection with these special developments of jaws and gills the skull has become much modified and has developed complex and characteristic features (Fig. 7.2).

3. The skull of bony fishes

The main basis for the skull is a chondrocranium and set of branchial arches, exactly comparable to those of the elasmobranchs. In the early stages of development there is a set of cartilaginous boxes around the nasal and auditory capsules, brain and eyes, and a series of cartilaginous rods in the gill arches. Bones are then added in two ways: either (1) as cartilage bones (endochondral bones) by the replacement of some parts of the original chondrocranium, or (2) as membrane or dermal bones, laid down as more superficial coverings and considered to be derived from a layer of scales in the skin. This outer position of the bones can be clearly seen in many cases by the readiness with which the membrane bones can be pulled away from the rest of the skull.

The skull bones are arranged in a regular pattern, whose broad outlines can be seen in all fishes and in their tetrapod descendants. However, there are many confusing variations and the naming of bones has given much controversy. No generally acceptable theory has yet appeared, perhaps because we know little of the factors that cause separate bony elements to develop. There is some evidence of relations between bones and teeth and bones and lateral line canals and the pattern of the latter

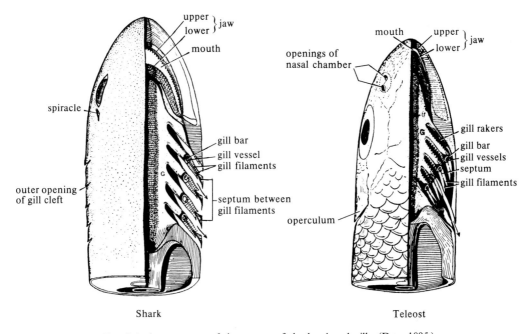

Shark   Teleost

FIG. 7.4. Arrangement of the organs of the head and gills. (Dean 1895.)

may play a large part in determining the plan of the skull (p. 242). Provisionally we may recognize four classes of dermal bones (1) canal bones, (2) tooth bones, (3) 'ordinary' bones, whose determination is unknown, (4) extra bones, filling special areas (Wormian bones).

The arrangement of the numerous bones is made less difficult to understand and remember if they are considered in the following order. First the endochondral ossifications in the original neurocranium, then the dermal bones that cover this above and below; next the endochondral bones formed within the original cartilaginous jaws, then the dermal bones that cover the edges of the jaws, and finally the ossifications in the branchial arches and pectoral girdle, which latter is in bony fishes attached to the skull.

The endochondral ossifications may be considered by beginning at the hind end of the skull: here the floor ossifies as the basioccipital, the sides as the exoccipitals, and the roof, over the spinal cord, as the supra-occipital bone; these posterior bones are not well marked off from each other in the adult skull. In the auditory capsule are five separate otic bones, of these the epiotic and pterotic can be seen externally (Fig. 7.2).

The floor in front of the basioccipital is occupied by a basisphenoid bone and the walls above this by alisphenoids. The eyes nearly meet in the midline and the orbits are here separated only by a thin orbitosphenoid. The only more anterior part of the chondrocranium to ossify is the region between the nasal capsule and the orbit, forming the ectethmoid.

The dermal bones that cover this partly ossified neurocranium include a pair of frontals and a median supraethmoid. Behind these are large paired parietals and small paired postparietals. These names have been inferred from study of crossopterygians and early amphibians (p. 242), which showed that the homologies earlier accepted were wrong. Figure 7.2 carries the old nomenclature in which the large paired bones were called frontals. Around the eyes is a ring of circumorbitals, and on the floor of the skull two median bones, the parasphenoid and vomer.

The jaw bones are numerous, including both endochondral and dermal elements, and the relation of the method of support to that found in other animals is not clear. In the embryo palatopterygoquadrate bars and Meckel's cartilages are seen. The upper jaw bears inward projections, which extend towards the chondrocranium and probably represent the traces of an autodiastylic means of support (see p. 126). But the effective support in the adult is achieved by the ossified hyomandibular cartilage. The palatopterygoquadrate bar ossifies in several parts and palatine, pterygoid, mesopterygoid, metapterygoid, and quadrate bones

appear, some partly formed in membrane (Fig. 7.8B). The only part of Meckel's cartilage to ossify is the articular bone, at the hind end. The actual edges of the jaws are supported by membrane bones, the premaxilla, maxilla, and jugal, covering the upper jaw. The dentary covers most of the lower jaw, except for a small bone, the angular, which lies on the inner side at the posterior end.

The hyomandibular bone runs from an articulation with the otic capsule to the upper end of the quadrate. The symplectic is a small separate ossification at the lower end of the hyomandibula. The rest of the hyoid arch is present as epi-, cerato-, and hypo-hyals, which support a large tongue often carrying teeth. Bony fishes only rarely possess an open spiracle and immediately behind the hyoid arch are attached the bones supporting the operculum that covers the gills. The branchial arches are formed of several pieces, as in elasmobranchs but with a different arrangement (see p. 162), each being ossified separately.

The effect of this complicated set of bones is to provide an efficient apparatus for the protection of the brain and sense organs, support of the jaws and teeth and of the respiratory apparatus. Teeth are found on the vomers, palatines, premaxillae, maxillae, dentary, and on the tongue. Covering the typical dentine (orthodentine) is a layer of harder vitrodentine, poor in organic matter and perhaps derived partly from ectodermal ameloblasts. The teeth are usually spikes pointing in a backwards direction, used to prevent the escape of the food and not usually for biting or crushing. They may, however, form plates or be firmly attached to the bones. Folds of the mucous membrane, supported by cartilage and carrying gill rakers, are found in species that feed on small prey.

4. Respiration

Limitations are imposed on the respiration of fishes by the facts that water is 800 times more dense than air and the dissolved oxygen is 30 times more dilute. The cost of respiration is therefore high and may be up to 30 per cent of the total metabolism.

Respiration is produced by a current passing in a single direction, as in elasmobranchs, namely, in at the mouth and out over the gill lamellae, but the mechanism by which the current is produced is somewhat different. The pumping action is produced by a buccal pressure pump and opercular suction pumps, resulting from sideways movements of the operculum, enlarging the branchial cavity. The branchiostegal folds, below the operculum, prevent inflow of water from behind. When the operculum moves inward dorsal and ventral flaps in the throat prevent the exit of water forwards.

The gill lamellae differ from those of elasmobranchs

in the great reduction of the septum between the respiratory surfaces (Fig. 7.4). This has the effect of leaving the lamellae as free flaps, increasing the surface available for respiration.

The filaments and their very thin lamellae are arranged to form a system of pores through which the water flows (Fig. 7.5). There is a counter-current arrangement so efficient that water leaving the gills contains only one-fifth of the original oxygen content (in a trout) (Fig. 7.6). If water is made to pass in the opposite direction only one-fifth of that normally extracted is removed.

The area of the gills varies greatly, being relatively larger in a more active species. The rate of respiration, controlled by a medullary centre, will increase with exercise or if the oxygen content of the water is low. But the most active rate of respiratory exchange is only some four times the standard rate (as against 20 times in man and 100 times in insects). The area of the respiratory surface is thus an important limiting factor in the movement and growth of fishes. During activity of a fish lactic acid accumulates in the blood and the pH falls. The fish is thus able to display a considerable burst of activity and then to repay the oxygen debt over a long subsequent period. Receptors assisting in the control of respiration are found in the pseudobranch (Laurent 1974). This is the modified first gill whose lamellae contain many receptors and ionocytes (chloride-secreting cells). The receptors vary their discharge with changes in the blood, including P_{CO_2}, P_{O_2}, pH, and

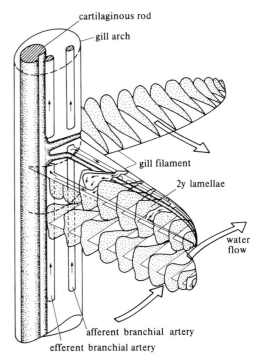

FIG. 7.6. Diagram showing the pattern of blood flow over the gills of a teleost. (After a diagram by F. Conte in Hoar and Randall 1970.)

osmotic pressure, and is thus partly comparable to the carotid body of mammals. The pseudobranch may also have other functions (p. 178).

Many fishes breathe air, especially those living in shallow, stagnant water in hot climates (see p. 212). Some that feed near the surface may also take gulps of air (*Gambusia*, the mosquito fish), but others can come onto the land when pools dry up (*Anabas*, the climbing perch). Some can even aestivate by burrowing in mud such as the catfish, *Saccobranchus*. *Clarias* leaves the water at night to hunt on land.

5. Vertebral column and fins of bony fishes

The vertebral column of bony fishes performs the same functions as in other fishes, namely, to prevent shortening of the body when the longitudinal muscles contract. It has, however, become very complicated and with the ribs and neural and haemal arches forms an elaborate system serving to maintain the body form under the stresses of fast swimming (Fig. 7.2). Like other parts of the skeleton it is extensively ossified, and the necessary lateral flexion is obtained by division of the column into a series of sections joined together. Typically there is one such section (vertebra) corresponding to each segment, but in the tail region of *Amia* there are twice as many vertebrae as segments.

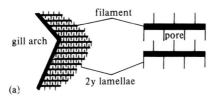

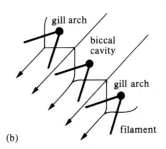

FIG. 7.5(a). Diagram of the general structure of the gill of a teleost in transverse section. The secondary lamellae form pores allowing the water flow. (b) The flow over the gills. (After Hughes, G. M. (1966). *Journal of experimental Biology* **45**, 177–95.)

Each vertebra consists of a centrum, neural arch and neural spine, and in the tail region, in addition, haemal arch and haemal spine. These parts are formed partly by ossification of cartilaginous masses, like those of elasmobranchs, and partly by extra ossification in the sclerogenous tissue around the notochord and nerve cord and between the muscles. The vertebrae are intersegmental, the middle of each lying opposite the myocomma that separates two muscle segments.

The centra are concave both in front and behind (amphicoelous), and in the hollows between them are pads made of the remains of the notochord, an arrangement that allows the column to resist longitudinal compression and yet remain flexible. Similar flat or concave articulations of the centra are found in other aquatic vertebrates from the elasmobranchs to the whales. Extra processes on the front and back of the vertebrae ensure the articulation and are comparable to the zygapophyses found in tetrapods. The ribs, which are so prominent in the backbone of many fishes, are of two sorts; pleural ribs between the muscles and the lining of the abdominal cavity, and more dorsal intramuscular ribs. Both sorts are attached to the centrum. The bony rods attached above the neural and below the haemal arches are often called neural and haemal spines, though it is doubtful whether they correspond to the neural spines of land vertebrates. They form the supporting rods or radials of the median fins and are usually divided into two or three separate bones in each segment. In addition to these radials the fins are also supported by a more superficial set of bony rods, the dermal fin rays (dermotrichia or lepidotrichia), which may be considered as modified scales and accordingly lie superficial to the radials (Fig. 7.2). These dermal fin rays are usually forked at their tips. They make an extra support for the fin margin and to them are attached the muscles that serve to throw the fin into folds.

In the tail region the internal skeleton is not quite symmetrical and shows signs of origin from an animal with a heterocercal tail. The notochord turns up sharply at the tip, so that the neural spines are very much shorter than the haemal spines, known here as hippural bones. The final portion of the notochord is often surrounded by a single ossification, the urostyle, and the whole makes a rigid support for the dermotrichia of the tail. Such a tail with internal asymmetry but external symmetry is said to be homocercal.

The myotomes are arranged in a complicated pattern having the effect that contraction of each affects a considerable section of the body (Fig. 7.7); in fast swimmers such as the tunny each myotome may overlap as many as nineteen vertebrae. Between the lateral and ventral muscle masses there is in many fishes a layer of red muscle (p. 116) and this is especially well developed in the tunnies and bonitos.

The paired fins are similarly supported by ossified radials, covered by dermal fin rays. At the base the radials are connected with 'girdles' lying in the body wall. The pectoral girdle (Fig. 7.2) consists of a cartilaginous endoskeletal portion in which ossify the scapula, coracoid, and sometimes mesocoracoid, while dermal bones, large cleithrum, and one or more small clavicles, become attached superficially. Above these a further series of dermal bones, the supraclavicle and posttemporal, attach the pectoral girdle to the otic region of the skull.

The pelvic girdle is very simple, consisting only of a single bone, the basipterygium.

6. Alimentary canal

The food of the trout consists mainly of small invertebrates such as *Gammarus, Cyclops*, and other crustaceans, and aquatic insects and their larvae, together with the fry of other fishes, and perhaps sometimes larger pieces of 'meat'. The food is sucked in by expansion of the mouth cavity assisted in some teleosts by protrusion of the premaxillae, these are

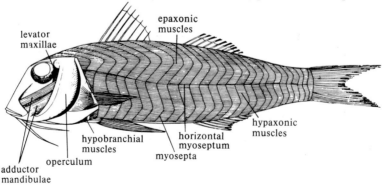

Fig. 7.7. Muscles of a teleostean fish, mainly based on *Mullus*. (From Ihle.)

connected only by joints to the maxillae which rotate forwards as the mouth opens (Fig. 7.8A). In the preparatory phase the buccal and opercular cavities are decreased. Then in Phase 1 of suction they are expanded

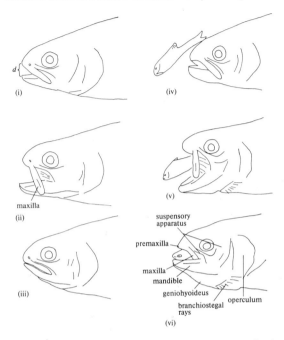

FIG. 7.8A. Jaw movements of the trout (*Salmo gairdneri*). Drawings made from frames of a ciné film to show a 'coughing sequence' from a normal fish (i, ii, and iii); these movements are identical with those during feeding. The other frames show a feeding attempt by a fish from which the maximandibular ligaments were removed bilaterally (iv, v, and vi). (From Lauder, Jr 1979.)

with the mouth closed. In Phase 2 it is opened and the prey sucked in. During Phase 3 it is closed again, generating a positive pressure forcing water over the gills. The actions may be repeated as the prey is partly ejected and then crushed between the jaws. The wide gape is achieved by the special joints and timed action of the muscles that allow elevation of the cranium as well as depression of the mandible and hyoid. The strike is an invariant response governed by a preprogrammed output from the brain without feedback. Negative pressures up to $-600 \, cm \, H_2O$ may be achieved (Lauder 1980).

A suction mechanism was already present in palaeoniscids and is found in *Polypterus* and *Amia*, but the negative pressures were at first small. In modern Teleostei the maxilla is not attached to the dermal cheek bones and its swing forwards prevents the inflow of water at the sides. Food is then mostly swallowed whole, being helped down the pharynx by the mucous secretions, but these, as in elasmobranchs, contain no enzymes. The entrance to the stomach is guarded by a powerful oesophageal sphincter, no doubt serving to prevent the entry of the water of the respiratory stream. The stomach is not sharply divided into cardiac and pyloric portions as it is in elasmobranchs, and the intestine is long and coiled and has no spiral valve. The duodenum is beset by a number of wide-mouthed pyloric caeca, serving to increase the intestinal surface (Fig. 7.8B). The intestine and caeca are lined throughout by a simple columnar epithelium and there are no specialized multicellular glands such as the Brunner's glands or crypts of Lieberkühn of mammals. The exocrine pancreas consists of numerous diffuse glands in

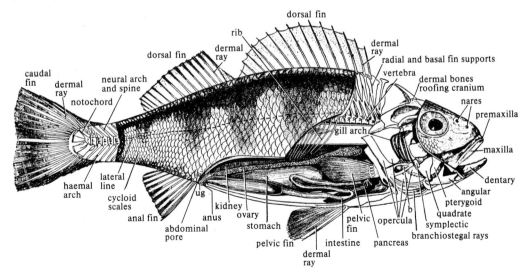

FIG. 7.8B. The general anatomy of a teleostean fish. *b*., bulbus arteriosus; *ug*., urinogenital opening. (From Dean 1895.)

the mesentery. The endocrine portion, however, forms one or more separate masses of tissue very rich in insulin. After their removal a fish shows hyperglycemia and glycosuria. There is no gland attached to the rectum.

7. Swim bladder

Dorsal to the gut is a very large sac with shiny, whitish walls, the swim bladder, filled with gas. A narrow pneumatic duct connects this with the pharynx in the more primitive forms. The origin and functions of the swim bladder will be discussed below (p. 213); it serves as a hydrostatic organ giving the fish neutral buoyancy, so that it can remain suspended in water without the continual swimming needed by most sharks. Small adjustments by the fins remain necessary to prevent turning upside down.

8. Circulatory system

The general plan of the circulation is similar to that of an elasmobranch (Fig. 7.9), that is to say, there is a single circuit and all the blood passes through at least two sets of capillaries. The heart contains a series of three chambers, sinus, auricle, and ventricle but the muscular conus arteriosus is absent, there being only a thin-walled bulbus arteriosus at the base of the ventral aorta. The walls of the bulbus are elastic but not muscular, and study of its action by means of X-rays shows that it is dilated by the ventricular beat and then contracts, thus maintaining the pressure against the capillaries of the gills. The ventral aorta is short, but the arrangement of the afferent and efferent branchial vessels is essentially as in elasmobranchs.

The blood pressure in the ventral aorta is less than 40 mm Hg in most fishes at rest, and in the dorsal aorta between two-thirds and three-quarters of this. The venous pressures are around zero, the pericardium being fibrous but not rigid as it is in elasmobranchs (p. 135). There is no communication between the pericardial and peritoneal chambers. There is a vagal cardiac depressor nerve and some evidence of adrenergic (?sympathetic)

fibres also in the vagus. Nerves to the peripheral blood vessels pass through the gray rami communicantes (p. 147). There is therefore probably a more detailed regulation of the circulation than in elasmobranchs.

There is a well developed lymphatic system beneath the skin and in the muscles and viscera. Lymphoid tissue is abundant in various organs but there are no lymph nodes along the vessels. There is a large spleen concerned with haemopoiesis, which also proceeds in the kidneys. The red cells are smaller in bony fishes (8–10μm) than in elasmobranchs (up to 20 μm). Some very active scombroid fishes have blood counts higher than man (6.48×10^6 cells/mm^3 in *Acanthurus*). On the other hand in some Antarctic fish, such as *Trematomus*, there are as few as 0.66 cells/mm^3, while in the translucent ice-fish, *Chaenocephalus*, there is no haemoglobin at all.

The three main categories of leucocytes, granulocytes, monocytes and lymphocytes, all occur in fishes. A fourth category, thrombocytes, replaces the platelets of mammals as the source of prothrombin for clotting.

9. Kidneys and osmoregulation

The kidneys are mesonephric in the adult and consist of an elongated brown mass above the swim bladder. The ducts of the two kidneys join posteriorly and are swollen to form a bladder which, being mesodermal, must be distinguished from the endodermal cloacal bladder of tetrapods. The urinary duct opens separately behind the anus, there being no common cloaca.

Nitrogenous elimination is a function mainly of the gills, which excrete as ammonia and urea more than six times as much nitrogen as the kidneys. The latter excrete creatine, uric acid, and the weak base trimethylamine oxide, which is present in large amounts in the blood of marine teleosts.

One of the most striking features of the life of bony fishes is that they occur both in fresh water and the sea, and many, such as the trout itself, can move from one to the other. It is supposed by some that the earliest gnathostomes were freshwater animals (p. 190), and the

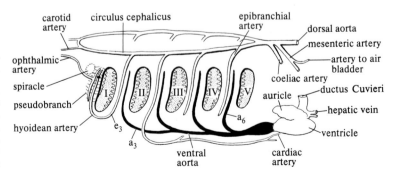

Fig. 7.9. Diagram of the branchial circulation of a teleostean fish. a_{3-6}, four afferent vessels from ventral aorta; e_3, efferent vessel of first branchial arch; s, position of spiracle (closed); $I-V$, five branchial slits. (The hyoidean artery is the afferent vessel of the pseudobranch and the ophthalmic artery is its efferent vessel: the pseudobranch represents the hyoidean gill.) (From Goodrich 1909.)

bony fishes might be said to show evidence of this in that the concentration of salt in their blood is always less than in the sea, in the neighbourhood of 1.4 per cent NaCl against 3.5 per cent outside. In fishes in fresh water the blood is more dilute, about 0.6 per cent NaCl, but is, of course, more concentrated than the surrounding medium, which contains only traces of inorganic ions. Freshwater fishes are able to take up salts from the water through the gill surfaces.

The need to conserve salts led to the development of the filtration system in the kidney of glomeruli and tubules with no less than six distinct regions. Some of these are for resorbtion and others excrete divalent ions. The result is a urine almost free of monovalent ions. Various special devices are adopted in fresh water for minimizing the tendency to gain water and lose salt. The skin is little vascularized and probably makes an almost waterproof layer. The production of mucus assists in this waterproofing, and abundant mucus is secreted when an eel is transferred from salt to fresh water: the full change cannot be made suddenly without killing the fish.

In marine teleosts the problem is the opposite one of keeping water in, or keeping out salt. The usual kidney mechanism is clearly ill suited for this and it is found that the glomeruli are few, or often completely absent from the kidneys. The tubules may have only two regions, serving to remove divalent ions especially magnesium sulphate. This no doubt reduces the loss of water, but is not enough by itself to solve the problem, which is met by drinking water and salts and excreting the salts. For this purpose special chloride-secreting cells (ionocytes) are present in the gills and it has been shown that the amount of oxygen they use, and hence the work they do in diluting the blood, is proportional to the difference of concentration between the inside and the outside. In fresh water the cells actively absorb sodium and chloride. In salt water sodium is extruded against the electrochemical gradient. This involves much energy. For example in *Fundulus* in sea water as much as 35 per cent of the body's exchangeable sodium must be removed per hour. Correspondingly in several fish an increase in the activity of gill Na^+–K^+-activated adenosine triphosphatase occurs in passing from fresh to sea water. This adaptive change is much reduced by hypophysectomy and appears to depend on ACTH and adrenal steroids.

A marine fish is able to drink and absorb sea water in spite of the fact that this is more concentrated than its blood. The ioncytes dispose of the excess salts. It remains to explain the means by which a solution passes against the osmotic gradient from the cavity of the gut to the blood; the membranes here must have some

special properties of which we are ignorant. Sodium and chloride enter with the water but magnesium is excluded and may be precipitated in the intestine.

Osmoregulation is effected by the combined control action of the pituitary, adrenal, corpuscles of Stannius, and urohypophysis (urophysis). After hypophysectomy freshwater teleosts lose salt and die unless treated with both ACTH and prolactin, the former increases sodium uptake and the latter prevents its loss (p. 171). After removal of the corpuscles of Stannius the calcium concentration in the blood increases due to lack of excretion of it.

10. Reproduction

The genital system is nearly completely separated from the excretory in both sexes. The testes (soft roes) are a large pair of sacs opening into the base of the urinary ducts. The ovaries (hard roes) are also elongated in the trout and the eggs are shed free into the coelom (Fig. 7.10) and passed to the exterior by abdominal pores. This condition is unusual among teleosts, in most the ovaries are closed muscular sacs, continuous with the oviducts.

The gonads produce steroid hormones similar to those of elasmobranchs and tetrapods, 11-ketotestosterone and other androgens from groups of interstitial cells, and progesterone and oestrogens from the theca interna or granulosa cells of the ovarian follicles. The yolk itself contains much oestrogen in some species. In some fishes the gonadal hormones are responsible for producing the secondary sexual characteristics, but the hormonal mechanisms are very varied. Some fishes continue reproductive behaviour after gonadectomy.

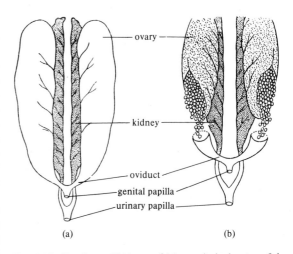

(a)   (b)

FIG. 7.10. Ovaries and kidneys of (a) a typical teleostean fish and (b) a trout and some other fishes where the eggs are shed into the coelom. (From Norman 1931.)

Fertilization is external and the eggs of the trout are shed in small pits or depressions in the sand; being sticky, they become attached to small stones. The eggs are very yolky and cleavage is therefore only partial, forming a cap of cells, the blastoderm, which eventually differentiates into the embryo. After hatching, the young fish may still carry the yolk sac and obtain food from it for some time, while beginning to eat the small crustaceans and other animals that are its first food.

11. Races of trout and salmon and their breeding habits

There is considerable confusion about the various races of trout and their allies the salmon. In both trout and salmon the adult originally spent the greater part of its life in the sea but returned to the rivers to breed. Trout and salmon that do this are still abundant on the West Atlantic coasts and ascend all suitable rivers to breed, the process being known as the 'run'. But the trout has produced many races of purely fluviatile animals, living either in lakes (where they often become very large) or rivers, and never returning to the sea to breed. These freshwater races differ in small points from each other and are given various common names (phinock, Severn, Loch Leven trout, brook trout, etc.). There can be no doubt that interesting genetic differences between these forms exist, but they have not yet been fully studied. The salmon are much less prone to form purely freshwater races, though such are known.

During the breeding season characteristic changes take place in the fishes and differences between the sexes appear. In the salmon the jaws become long, thin, and hooked, especially in the male. The animals make pairs and the male fights with others that approach the female. As the gonads ripen, the other parts of the fish, which were well supplied with fat at the beginning of the run, become progressively more watery. Finally, spawning takes place, the female laying the eggs in a shallow trough (redd), which she has 'cut' in the gravel by movements of her tail, while the male sheds sperms over them. She then covers the eggs with gravel by further cutting movements. The young male salmon (parr), which have not yet been to the sea, may become sexually mature. They accompany the fully grown fish, hanging around the cloacal region and shedding their sperms at the same time as the large male. It is possible that this development of a kind of third sex serves to increase the variability of the population. The spent parr eat some of the eggs and they then proceed to grow, migrate to the sea, and return later.

Male trout will follow a spawning salmon and fertilize her eggs if her own male is not looking. Hybrids formed in this way can develop, but are said to be less fertile than the normal types; indeed the males are wholly sterile.

After fertilization the salmon are very exhausted (known as kelts); the males seldom return to the sea. The females, however, may recover and after a period in the sea return to breed again, and this process may be repeated several times.

Very young trout or salmon are known as alevins or fry and remain mostly among the stones (Fig. 7.11). When they emerge they are called parr and have a number of characteristic parr-marks along their sides. After two to four years spent as parr in fresh water salmon acquire a silver colour and pass to the sea as smolts. Once at sea they may swim to Greenland and tagged fish have been captured some 3000 km from their natal rivers. Young salmon returning for the first time to breed are called maidens. If they have spent only one and a half years in the sea they are called grilse and may then return to the sea as kelts. Others ascend for the first time after three years or more at sea.

It is well established that salmon nearly always return to breed in the river in which they were born, and it is certain that they may journey for considerable distances in the sea. The Chinook salmon of the northwestern United States migrate downstream as young smolts and may move up to 6000 km away in the Pacific before returning to the same small stream five years later. The mechanisms by which these migrations are initiated and guided are only partly known. They probably involve endocrine changes, for example the thyroid is very active in the smolt as they begin to migrate. The return to the home river may be a result of olfactory condition-

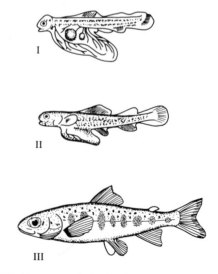

Fig. 7.11. Three stages in the development of the salmon. I and II are alevins; III, parr with marks along its side. (From Norman 1931.)

ing (see p. 184) but navigation in the sea may be by a sun compass mechanism (Hasler 1971).

12. Pituitary gland

The gland differs from that of other vertebrates in the method of control by the hypothalamus. The median eminence is often said to be absent, but is probably actually present within the pituitary in a modified form as the anterior neurohypophysis (Fig. 7.13). The blood supply of the pars distalis passes through this region, but there is no portal system external to the pituitary. The other unusual feature is that cells of the pars distalis are directly innervated by hypothalamic neurosecretory fibres (Holmes and Ball 1974).

In more primitive ('isospondylous') fishes there is a distinct communication between the mouth and a pituitary cleft and this duct remains open in a few adult fishes such as *Elops* (Fig. 7.12) and *Hilsa*. The remains of it can be seen as the follicles of lactotropes in the rostral

pars distalis (Fig. 7.13). In higher teleosts the pituitary arises as a solid mass and there are no follicles.

The adenohypophysis surrounds the neurohypophysis and interdigitates with it. It is divided into three regions, rostral pars distalis, proximal pars distalis, and pars intermedia, which surrounds the pars nervosa (Fig. 7.13). The first part contains acidophil lactotropes and corticotropes, and in some species basophil thyrotropes. The proximal pars distalis contains acidophil somatotropes and basophil gonadotropes and sometimes basophil thyrotropes. The arterial supply from the internal carotid arteries enters the anterior neurohypophysis and there makes a primary capillary plexus. This then continues into the pars distalis as a system of sinusoids among the secretory cells, the secondary plexus. There is therefore a portal system of short capillaries within the pituitary rather than the long external portal vessels of other vertebrates.

Two sorts of nerve fibre run from the hypothalamus to the pituitary. Type A are peptidergic and end mainly around the capillaries or pituicytes of the neurohypophysis, whereas type B are at least mainly aminergic and many of them pass on to innervate the cells of the pars distalis. It is still uncertain how these fibres control secretion but there is evidence that extracts of carp hypothalamus stimulate secretion of gonadotropin not only in carp pituitary but also in that of sheep. Conversely sheep gonadotropin-releasing factor will activate release of carp gonadotropins. There is probably only a single gonadotropin, which resembles luteinizing hormone (LH), but when injected into mammals it produces some effects of FSH as well as LH. But only mammalian LH (and not FSH) will produce gametogenesis and steroid secretion in fishes.

The prolactin of fishes differs somewhat from that of tetrapods and is concerned in osmoregulation, reducing the exchange of sodium across the surface of the body

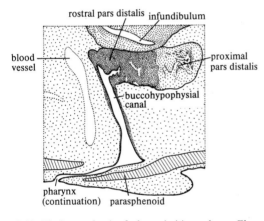

FIG. 7.12. Pituitary gland of the primitive teleost *Elops*, showing the persistent Rathke's pouch in the form of a hollow buccohypophysial canal, piercing the parasphenoid bone. (After Olsson.)

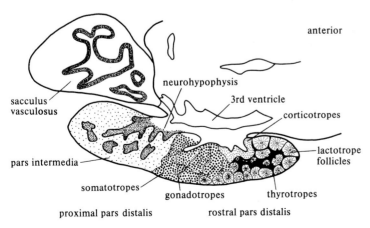

FIG. 7.13. Pituitary of eel (*Anguilla anguilla*). Diagram of midsagittal section. (After Olivereau, M. (1967). *Zeitschrift für Zellforschung Mikroskopische Anatomie* **80**, 286–306.)

and gills. In some species it also increases the water permeability of the gills and increases renal water excretion. The lactotropes become more active when migrant fishes enter fresh water. This response is not controlled by the hypothalamus and the prolactin cells (lactotropes) of a pituitary transplanted ectopically, away from the brain, are activated by passage from dilute sea water to fresh water.

The pars intermedia produces melanophore stimulating hormone (MSH also called intermedin) which collaborates with the sympathetic system in the control of colour change. Two types of cell are present and the lobe probably influences other activities, including osmoregulation, control of calcium levels and perhaps reproduction.

Two octapeptides occur in the neurohypophysis, arginine vasotocin (AVT) and a homologue of oxytocin known as isotocin or ichthyotocin (IT). These are neurosecretory products of the A fibres of the pituitary. The secretory material and pituitary content of AVT vary with passage from salt to fresh water but details of its function are not known.

13. Other endocrine glands

The thyroid tissue is not aggregated into a compact gland but forms scattered masses along the ventral aorta. Its hormones appear to be identical with those of mammals, including tri-iodothyronine and thyroxin. Thyroid follicles are often found in the kidneys, heart, eye, and elsewhere in the body of fishes, especially those deprived of iodine. The thyroid is necessary for the maturation of the gonads and shows cyclic activity. There is no clear evidence that it stimulates respiratory metabolism or growth or is involved in osmoregulation, its function in a teleost, thus remains uncertain.

The suprarenal and interrenal tissues are partly associated in masses around the thickened walls of the posterior cardinal veins. Because of the difficulty of isolating these tissues there is little information as to their function. They produce various steroid hormones but probably not aldosterone, which may be a tetrapod speciality for regulation of electrolytes. Adrenalectomized eels (a drastic operation) fail to survive, in fresh water they lose sodium and in sea water they gain it. In fresh water they fail to produce the usual copious urine, administration of cortisol corrects this (but not aldosterone). The corpuscles of Stannius are groups of gland cells dorsal to the kidneys, they have been held to be related to the adrenals, but their nature is still uncertain. They do not produce steroids. They are derived from the nephric duct, contain two types of cell, and may be concerned with ionic regulation, especially of calcium.

The ultimobranchial gland is a mass of cells, developed from the floor of the last branchial pouch. It lies beneath the oesophagus and produces the polypeptide calcitonin. The amino acid content of this is different from that of mammals and it is 25 times more active. As yet its function in teleosts is not known. It probably facilitates calcium transport. There are no parathyroids in fishes.

The endocrine pancreas is unusual in that much of it occurs separated from exocrine tissue as isolated masses (Brockmann bodies). Moreover these may occur widely scattered in the abdomen, for instance in the spleen and even in the ovaries. Owing to the ease of separation this islet tissue was one of the earliest sources of insulin (especially from *Lophius*). Several types of cell are present and the glands produce glucagon as well as insulin.

At the hind end of the spinal cord of fishes is a small lump consisting of masses of secretion produced by neurosecretory cells of the spinal cord and hence called the urophysis (Figs. 7.14 and 7.15). In function it appears to be connected with salt regulation. Injection of hypertonic NaCl produces hypersecretion, the products accumulating at the cut surface if the cord has been severed. Extracts contain at least two hormones (urotensins I and II), which may control aspects of

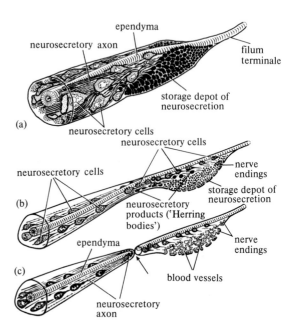

FIG. 7.14. Urohypophysis of A, eel; B, loach (*Misgurnus*); C, the same in a loach after sectioning the spinal cord and injecting hypertonic saline. (After Enami, N. (1959). *Symposium on Comparative Endocrinology* (ed. A. Gorbmann). Wiley, New York.)

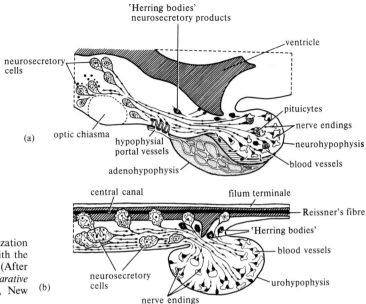

FIG. 7.15. Comparison of pattern of organization of the caudal neurosecretory system (a) with the hypothalamo-hypophysial system (b). (After Enami, N. (1950). *Symposium on Comparative Endocrinology* (ed. A. Gorbmann). Wiley, New York.)

osmoregulation, cardiovascular function, and reproduction. The neurosecretory fibres carry action potentials as well as transporting the hormone.

14. The brain of bony fishes

The parts of the brain concerned with the various special senses are differently developed according to the habits of the fish. The telencephalon is mainly concerned with olfaction but is a relatively small part of the brain, as shown in all the four cyprinids in Figure 7.16. The optic lobes are large, especially in the visual feeders. Where

palatal mouth tasting is important there are large lobes at the entry of the facial nerves. The gurnards (Triglidae) have chemical receptors in the pectoral fins and special lobes at the front of the spinal cord (Fig. 7.17).

The telencephalon differs from that of all other vertebrates. There are paired hemispheres but no lateral ventricles and the whole roof is a thin sheet of ependyma (Fig. 7.19). This condition is known as 'eversion' and is the opposite of the 'inverted' or thick-roofed forebrain that is found in those bony fishes close to the line of tetrapod descent (Fig. 7.19). Many authors have tried to

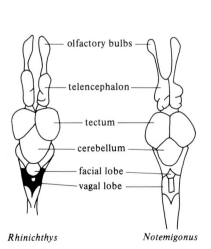

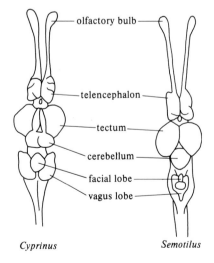

FIG. 7.16. Dorsal view of the brain of four cyprinids. (After Evans, H. E. (1952). *Journal of comparative Neurology* **97**, 133–42.)

compare regions of the hemisphere with those of higher vertebrates but the cell groups are poorly marked and recent workers tend to use a nomenclature restricted to teleosts (Fig. 7.18) (Nieuwenhuys 1967). Two main regions are recognized, a ventral area dominated by fibres from the olfactory tracts and a dorsal one receiving projections of optic and perhaps other senses through the diencephalon. The dorsal region becomes progressively more developed in the sequence through Chondrostei and Holostei to Teleostei (see Ebbesson and Northcutt 1976; Ebbesson 1980).

The action of the telencephalon is exerted through tracts to the hypothalamus, there is no direct pathway to motor centres. Electrical stimulation of it produces feeding behaviour in goldfish. After removal of the telencephalon there are disturbances and weakening in courtship and reproductive behaviour in various teleosts. The effect of the hemisphere is therefore to organize, facilitate, and integrate innate mechanisms elsewhere in the brain. Different parts of it influence particular functions such as aggression, sexual activity or parental care (Aronson and Kaplan 1968). After lesions to the telencephalon there were no obvious sensory or motor deficits but learning of some maze and other tasks was impaired though never altogether prevented. Indeed in some brightness discriminations goldfishes without forebrains performed better than controls. The forebrain thus functions as a non-specific arousal mechanism that modulates the operations of the

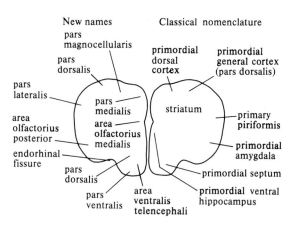

FIG. 7.18. An outline of the telencephalon of a cyprinid showing the nomenclature. (Nieuwenhuys 1967.)

tectum and other centres, perhaps acting to intensify and focus the effects of reinforcement during learning. After removal there is regeneration of the olfactory tract or telencephalon of *Lebistes* and *Gasterosteus* (but not of goldfish). Normal courtship patterns and previously learned conditioned responses then return.

The diencephalon is not large and most of the optic tract fibres end not here, but in the midbrain. However, there is a well-marked lateral geniculate nucleus in some species. It receives fibres from the optic tract and sends axons to the telencephalon. The roof is everted to form a pineal body, which contains photoreceptor cells whose axons proceed along the stalk. There is some doubtful evidence that its function is related to reproduction as it is in mammals. The photoreceptors of the pineal are especially numerous in some deeper-dwelling species (McNulty 1979). Other parts of the diencephalon may also contain receptors sensitive to light. The minnow *Phoxinus* has a transparent patch on the head in this

FIG. 7.17. The innervation of the taste system of searobin (*Triglidae*). The photograph above of the searobin shows the three anterior, movable pectoral fin rays on which the taste sensors are located. Below is the nervous system of the searobin showing the innervation of the three fin rays. (After Bardach and Villars 1974.)

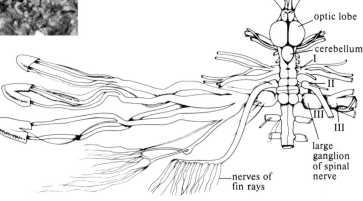

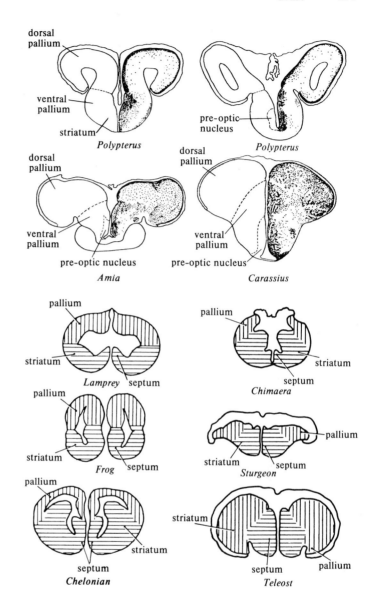

FIG. 7.19. The everted forebrains of actinopterygian fishes. In *Polypterus* the eversion leads to complete union posteriorly (top right) so that a space outside the brain is enclosed to form a false ventricle. There is little migration of cells away from the ventricle in *Polypterus*, more in *Amia*, and much more in teleosts (*Carassius*). The area that probably corresponds to the corpus striatum is labelled in *Polypterus*. (After Nieuwenhuys, R. (1966). In *Evolution of the forebrain: phylogenesis and ontogenesis of the forebrain* (ed. R. Hassler and H. Stephan). Thieme, Stuttgart.)

Below are transverse sections of the forebrain of various vertebrates to show the condition of inversion (thick roof) in lamprey, frog and chelonian, and eversion (thin roof) in *Chimaera*, sturgeon, and a teleost. (After Kappers, Huber, and Crosby 1936.)

region, and it has been found possible to train the fish to give appropriate responses to changes of illumination even after removal of the paired eyes and the pineal body. Evidently there are light-sensitive cells in other parts of the walls of the diencephalon, besides those that become evaginated to form the eyes. Experiments on lampreys also showed the presence of such cells (p. 92).

The hypothalamus is an important centre in fishes, receiving a major part of the output from the telencephalon through the medial and lateral forebrain bundles. It also receives fibres from the gustatory and acoustico-lateral line systems. It sends fibres forwards and also backwards to the medulla. There is a conspicuous neurohypophysial tract leading to the pituitary. In *Lophius* it is 2 cm long and its axons have been shown to conduct action potentials as well as neurosecretion. Many fishes have a large saccus vasculosus (p. 138).

The midbrain is often the largest part of the brain. The cells, spread out over its roof (tectum opticum), are not all collected round the ventricle but have migrated away to make an elaborately layered system. Into this midbrain cortex there pass not only the great optic tracts but also ascending tracts from the sensory regions of the spinal cord, lateral line system, gustatory systems, and cerebellum. Large motor tracts pass back towards the spinal cord; the details of their endings have not been traced, but they certainly exercise control over motor functions. The signals reaching the tectum in the optic nerve have already been coded by processing in the retina. They contain information about colour, form, and movement and there are cells in the tectum that are units coded for these features. It is not known how the system stores the abstract descriptions of the shapes that a goldfish can learn to recognize (Sutherland 1968). Electrical stimulation of the optic lobes produces well co-ordinated movements of local groups of muscles, for instance those of the eyes or fins. It can hardly be doubted that this well-developed midbrain apparatus thus controls much of the behaviour of the fish and is able to mediate quite elaborate acts of learning and other forms of more complex behaviour. After removal of the tectum of one side a minnow is blind in the opposite eye. There is a point-to-point mapping of the retina on the tectum and if the optic tract is cut and allowed to regenerate this projection is exactly replaced. When a goldfish is trained to respond to some visual stimulus the learning process occurs in the midbrain and continues unaffected after removal of the forebrain. Conversely, olfactory learning takes place in the latter and is undisturbed by injury to the tectum opticum.

The base of the midbrain (tegmentum) contains motor centres. Electrical stimulation here produces abrupt and massive responses of the locomotor apparatus, very different from the sequences of co-ordinated movements that appear after stimulation of the roof of the tectum.

The Mauthner cells are a large pair near the roots of the vestibular nerves, with axons passing down the spinal cord and making synapse with motoneurons. They are activated by electrotonic synapses and probably serve to produce the quick dashes, which are characteristic escape movements of many fishes. They are reduced in eel-shaped and benthic forms and also, strangely, in the large, fast-moving scombroids. It may be that the high internal temperature of these fishes makes nervous conduction so rapid that giant fibres are not necessary, in any case they move for most of their lives. It has also been suggested that a sudden burst of speed might make the tail of a tuna fish 'stall'.

The cerebellum is very large in teleosts, especially in the more active swimmers, and a forwardly directed lobe of it, the valvula cerebelli, extends under the midbrain. Various disorders of movement have been reported after removal of the cerebellum, such as swaying when moving quickly. Presumably it plays an important part, as in other vertebrates, in producing precise and correctly timed movements. It is enormous in the Mormyridae covering all the rest of the brain. Here it may assist in direction-finding by electrical pulses, perhaps acting as a timing device.

Surface recordings of the electrical activity of the brain of the goldfish show an alpha rhythm of 4–8 Hz in the telencephalon (Fig. 7.20). There is a 7–14 Hz rhythm in the mesencephalon. With repetitive flashing this was inhibited and a high frequency, low amplitude activity appeared. The cerebellum shows still higher frequencies (up to 180 Hz), also altered by light flashes. The medulla shows a rhythm at 1 Hz corresponding to the respiration. This was one of the earliest signs of electrical activity ever recorded in the brain, by Adrian and Buytendyk in 1931.

15. Receptors for life in the water

The features of the environment that are relevant for life are very different in air and water. Man is so well used to the air that it is not easy to appreciate fully the conditions underwater, where changes of illumination, though obviously important, provide less detailed evidence of the sequence of distant events than they do in air. We can say that light carries less information for a fish; to put it in another way, fewer distinct choices between alternative behaviour pathways are made on the basis of visual clues by a fish than by a man.

On the other hand, the water around the fish provides mechanical stimuli both at low and high frequency that are more closely related to distant events than is generally true in air. Sound travels with much less decrement through water. Both hearing and touch are of great importance in the water and the lateral line system provides a system of touch that is perhaps wholly outside our experience. Localization of nearby objects by such a sense, perhaps assisted by echo-location by water movements, provides the fish with many relevant clues. It is interesting that these receptors are connected with a very large cerebellar system, perhaps concerned with measuring time differences. Electroreception is a source of valuable clues for aquatic animals and is widely adopted by fishes, often including a system for echo-location.

Chemical changes in the water also provide much information and both taste and smell are well developed. That smell is analysed by a distinct system in the forebrain, not directly related to the gustatory system, is one of the fundamental principles of control of vertebrate behaviour. Distant chemical changes provide the first clue to the presence of food, a mate, or an enemy, whereas the detailed finding of these involves eyes, ears, touch, and an accurate timing system. There thus arises the distinction between the systems for initiation of action in the forebrain (emotive) and for its fulfilment (executive) by centres farther back.

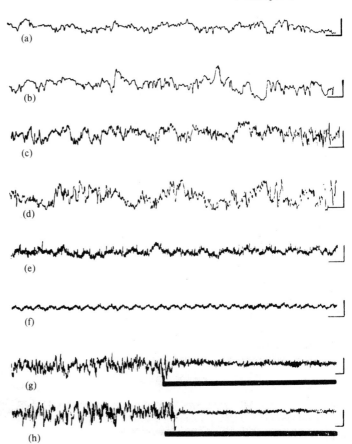

Fig. 7.20. Recordings from various parts of the brain of the goldfish (*Carassius*). Recordings are: (a) from two points of the right telencephalon; (b) from the left and right telencephalon; (c) from two points of the right mesencephalon; (d) between the left and right mesencephalon; (e) from the cerebellum; (f) from the medulla; (g) from the right mesencephalon; (h) from the left mesencephalon. (In (g) and (h) the horizontal black line indicates rapid repetitive light flash (75 cycles/s).) Calibration: vertical lines indicate $150\,\mu V$ and horizontal 1 s. (Schadé and Weiler 1959.)

16. Eyes

An animal provided with suitable receptors can obtain various sorts of information about the environment from the changes in illumination. Control of the whole physiology to follow the rhythm of day and night may have been the original reason for the development of photosensitivity in the diencephalon (see p. 92). At the stage of evolution reached by teleosts information is gained from the fact that light varies in frequency (*colour*) and intensity (*brightness*) and that it is reflected from many substances, revealing their *movement* and *shape*. The greatest sensitivity of the fish eye is often in the yellow-green, which is the wavelength that penetrates farthest into the water (p. 204). In order to extract the maximum of information at high as well as low intensities it is necessary to adjust the sensitivity, and hence the signal/noise ratio. For this purpose teleosts have developed retinas with distinct rods, cones, and twin cones and in many there is a temporal area composed of numerous thin cones (e.g. in *Blennius*). The pupil usually varies little in diameter, and adjustment of sensitivity is by migration of pigment between the

receptors and contraction of a 'myoid' segment of the latter. In bright light the pigment expands, the cones contract forward, towards the light and the rods contract back, beneath the pigment. These photomechanical changes thus serve the same end as changes of pupil diameter in other vertebrates.

The photochemical change in the rods of marine fishes is the same as that of land vertebrates, namely the breakdown of the rose-coloured 'visual purple' (rhodopsin) first to the yellow retinene and then to colourless vitamin A_1. In freshwater fishes there is a different pigment porphyropsin, or visual violet, which breaks down to vitamin A_2. Intermediates between these are found in fishes that migrate between the sea and fresh water. In some fishes there is evidence of capacity for colour vision (Cyprinidae). In these pigments maximally sensitive in the red, green, and blue have been identified, each in a single cone. Neurons with maximum responsiveness in these ranges are found in the midbrain roof. Fishes living in shallow waters often have a regular pattern of alternating twin cones and single cones, apparently suitable for detection of rapid movement.

Those that live deeper have simpler patterns, with only twin cones.

In all fishes there is a very large, dense, spherical lens, to which is attached a retractor lentis muscle (campanula Halleri). It is innervated through the oculomotor nerve and ciliary ganglion and at its origin is a richly vascular pigmented structure, the falciform process, presumed to be nutritive (Fig. 7.21). The choroid gland is a rete mirabile with a counter-current system similar to that found in the gas gland (p. 214). This serves to produce a high oxygen tension for the retina (which is avascular). The blood for it passes first through the pseudobranch (p. 165) which contains carbonic anhydrase and serves to prevent the access of carbon dioxide, which would otherwise accumulate, as it does in the swim bladder. The lens has a graded refractive index, much higher at the centre, thus avoiding spherical aberration, without any means of stopping down (Fig. 7.22). As the fish grows the refraction of parts of the lens must be changed continuously, but the mechanism for this is unknown. The ratio between the radius of the lens and its focal length is constant at 1:2.55 (Matthiessen's ratio). This ensures that the cone of light focussed on the vitreal surface remains within each rod by internal reflection. There is, therefore, little need for accommodation and it is uncertain whether this occurs. The eye is usually said to be myopic at rest and to be accommodated for distant vision by pulling the lens nearer to the retina, but this would focus it only on to one area, perhaps that of highest cone density (Locket 1977). This may be true for forward vision in the trout, but laterally even distant objects are in focus when the eye is at rest. The muscle perhaps pulls the lens posteriorly giving distant vision forwards. The system of accommodation probably varies with the habitat.

It is unwise to generalize about teleostean eyes, for they are very varied. Whereas the trout, like most, has a round pupil, which varies little if at all in size, other fishes, whose eyes are more exposed to light from above, have a more mobile iris. In flat-fishes and the angler-fishes, such as *Lophius*, and the Mediterranean *Uranoscopus*, the star gazer, the iris has an 'operculum' and is very muscular; its movements are controlled by nerves and not, as in selachians, by the direct effect of light. The sympathetic system sends branches into the head in these animals (Fig. 7.32) and its fibres cause contraction of the sphincter of the iris, whereas fibres in the oculomotor nerve cause contraction of the dilator, the opposite arrangement to that in mammals. In the eel the pupil is also capable of wide changes of diameter, but here the control is mainly by the direct response of the circular sphincter iridis muscle to light incident upon it. The pupil of the isolated eye of an eel closes when illuminated and reopens again in darkness (Fig. 7.23). Presumably because of its lack of nervous control this iris is not affected by many of the usual 'autonomic' drugs. For instance, closure will occur in the presence of atropine and the dark-adapted pupil remains un-

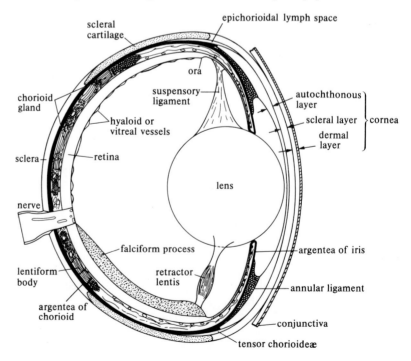

FIG. 7.21. Diagrammatic vertical section of a typical teleostean eye. Not all the structures here shown are found in all species. (From Walls 1942.)

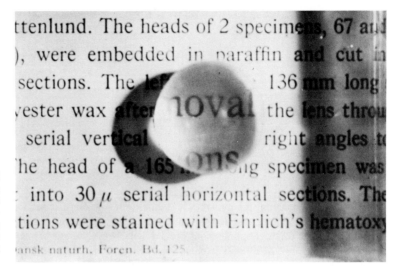

Fig. 7.22. Lens from a deep-sea gadoid (*Mora moro*) immersed in sea water, in a transparent container, resting on a printed page. The undistorted magnified image extends to the edges of the spherical lens, showing it to be free from spherical aberration. (From Locket 1977.)

changed when placed in a solution as strong as 1 per cent pilocarpine, but then closes immediately on illumination (Fig. 7.24). The isolated pupil of *Uranoscopus*, however, closes when pilocarpine is applied (Fig. 7.25), and in this case the sphincter muscle is innervated by sympathetic nerve fibres. Adrenaline also causes the sphincter to contract and acetylcholine in moderate concentrations causes dilatation.

The eyes may be small or absent in fishes living in caves, muddy waters, or the very deep sea, the bathypelagic zone (below 1000 m) where no light penetrates. Mesopelagic fishes (200–1000 m), however, often have exceptionally large eyes, with special devices for increasing sensitivity. There may be aphakic apertures, spaces around the lens, allowing light to enter from all directions. In the mesopelagic myctophid *Tarletonbeania* the supraorbital photophore shines into a ventral aphakic aperture, probably allowing adjustment of the luminescence of the fish to match the down-welling light (p. 209). In the deep sea form *Ipnops* there is no lens and the eye is reduced to a cornea and retina spread out flat on the head!

Binocular vision is especially important for predators in dim light where monocular clues do not allow judgement of distance. Mesopelagic fishes have adaptations that allow wide monocular vision, for the search and approach phases of predation, and then binocular vision for the snap. Some, such as *Argyropelecus*, have tubular eyes with the retina differentiated into two or more parts with different functions (Fig. 7.26). The lens is disproportionately large (collecting much light) and only the main part of the retina lies at the focal distance and conforms to Matthiessen's ratio. This allows good vision in a small binocular field. The accessory retina on the medial wall does not conform to the ratio (except at the base) and has a simple structure but is presumably adequate for detection of presence of prey or enemies, light reaches it partly through the transparent cornea, which extends down the lateral side of the eye cup.

The retinae of mesopelagic fishes are suited to the spectral band in the blue-green (475–80 nm) mainly transmitted through oceanic waters. They are of golden colour with pigments absorbing at around 485 nm. The density of the pigment of teleosts is high enough to absorb 95 per cent of the light likely to fall on the retina. In deep-sea elasmobranchs it is only of half that density,

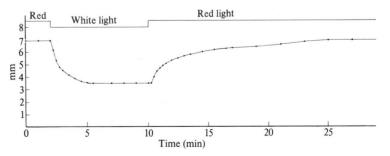

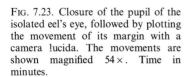

Fig. 7.23. Closure of the pupil of the isolated eel's eye, followed by plotting the movement of its margin with a camera lucida. The movements are shown magnified 54×. Time in minutes.

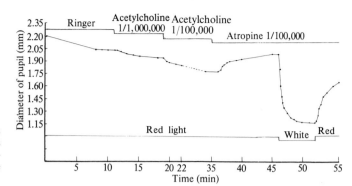

FIG. 7.24. Changes of diameter of isolated eel's iris in Ringer solution with the addition of various drugs. Acetylcholine produces little closure, atropine some opening. Light still produces closure after application of atropine.

but as they have a reflecting tapetum they may still catch the full quota of quanta and indeed gain from the reduced noise due to lesser spontaneous breakdown of pigment (Denton and Nicol 1964). In many mesopelagic fishes a high density of pigment is secured by the presence of very long rods. There may also be several banks of rods, as many as six in *Bathylagus*. It is interesting that the rods of the inner banks do not reach the pigment epithelium (Fig. 7.27), and it is a question how they are formed and absorbed and their pigment regenerated (see *Life of Mammals* p. 378). The banking does not produce a greater thickness than is achieved in other species by lengthening the rods, but it may increase sensitivity in another way. If the myoids and outer segments act as light guides then the effect of the bank will be to allow a greater number of rods to discharge into each ganglion cell.

The tropical fish *Anableps* lives with the head half out of water and the eyes are adapted for use in both media. The upper part of the cornea is thickened, the iris provides two pupils, the lens is pearshaped, and there are two retinas in each eye.

17. Ear and hearing of fishes

The ear provides receptors that ensure the maintenance of a correct position of the fish in relation to gravity and to angular accelerations. In addition, in many species it serves for hearing. The inner ear is completely enclosed in the otic bones. There is a perilymphatic space only in those species that hear well.

Each ear sac consists of a pars superior, the utriculus and semicircular canals, and pars inferior, the sacculus and lagena (Fig. 7.28). The maculae carry otoliths with characteristic names and shapes. Their seasonal growth rings indicate the age of the fish.

The sensitive macula of the utricle lies horizontally, with the lapillus resting upon it, whereas the maculae of the saccule and lagena are vertical. These receptors with otoliths have double or triple functions. At rest they act as static receptors, signalling the position of the fish in relation to gravity and setting the fins and eyes in appropriate positions. In movement, together with the semicircular canals, they signal angular accelerations, initiating compensatory movements (p. 144). Thirdly, some of the otolith organs respond to sonic vibrations.

In the fishes that hear well there is a connection between the swim bladder and the ear. This may be either direct, by means of fine canals extending forwards (in Clupeidae and others) or indirectly by a chain of modified vertebrae, the Weberian ossicles (Fig. 7.29). This latter arrangement is found in the freshwater Ostariophysi, which hear particularly well. The re-

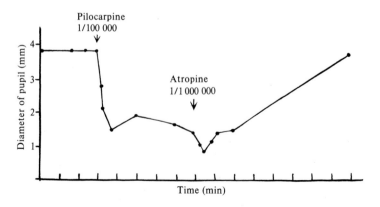

FIG. 7.25. Movements of margin of pupil in isolated iris of *Uranoscopus* in an isotonic solution. Pilocarpine produces closure and atropine opening of this pupil whose sphincter muscle is innervated by *sympathetic* nerve-fibres. (From Young, J. Z. (1933). *Proceedings of the Royal Society, London* **B112**, 228–49.

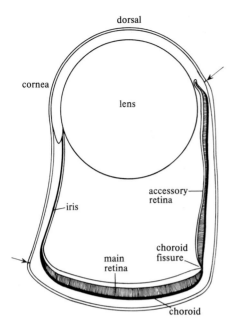

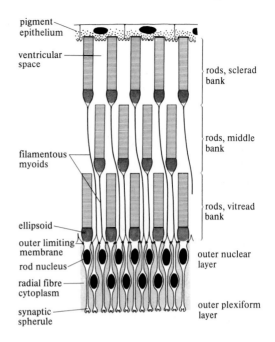

FIG. 7.26. Diagram of the tubular eye of a deepsea fish *Argyropelecus*. The lens almost fills the dorsal part of the eye. The cornea is thin and its limits are shown by the two arrows. The main retina conforms to Matthiessen's ratio but the accessory one does not. (After Locket 1977.)

FIG. 7.27. Diagram of a multibank retina from a deepsea fish. Specializations are seen in those parts of the rods which lie between the outer limiting membrane and the pigment epithelium. (After Locket 1977.)

ceptors are in the inferior part of the ear (saccule and lagena; Fig. 7.28). The sagitta carries a special wing projecting into the cavity and is so suspended as to serve to amplify vibrations. Near to it is a thin portion of the wall of the sac, which would favour the passage of variations of pressure transmitted to the endolymph by the ossicles.

These Ostariophysi respond to sounds between about 60–5000 Hz but thresholds are lowest over the range 200–600 Hz. After removal of the pars inferior responses continue only up to 120 Hz. If the swim bladder is punctured a minnow can still respond, but only up to 3000 Hz and with a sensitivity diminished by more than fifty times.

Minnows can be trained to discriminate between warbled notes separated by $\frac{1}{4}$ tone. Non-ostariophysid fishes have mostly a much lower upper limit of hearing and lower capacity for discrimination. The Mormyridae however, approach the minnows in this respect and here there is a special isolated portion of the air bladder within the otic bone.

In the best cases the sense of hearing of fishes thus approaches that of man, in spite of the absence of a coiled cochlea and basilar membrane with fibres of different lengths. Clearly the discrimination of tones cannot here depend upon differential resonance as the theory of Helmholtz requires. Discrimination is based on the presence of units able to follow the sound wave

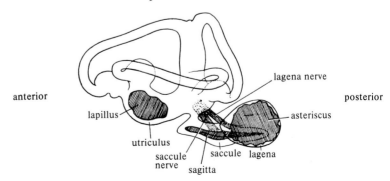

FIG. 7.28. Diagram of ear of the minnow *Phoxinus*. (From Frisch, K. V. (1938). *Zeitschrift für vergleichende Physiologie* **25**, 703–47.)

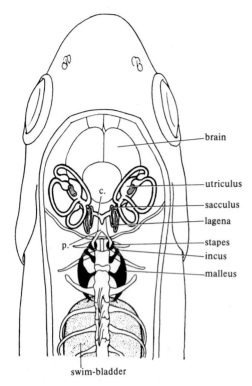

brain

utriculus

sacculus

lagena

stapes
incus

malleus

swim-bladder

FIG. 7.29. Position of ear in Ostariophysi and its relation to the three Weberian ossicles, which are shown in black. *c*, transverse canal between the two sacculi; *p,* sinus impar (perilymphatic space). (From Frisch, K. V. (1938). *Zeitschrift für vergleichende Physiologie* **25**, 703–47.)

with impulses in phase with it. The hearing of fishes is probably often used to detect the hydrodynamic sounds produced by the swimming movements of others. Sharks can certainly find their prey in this way (they will attack a hydrophone emitting pulses of broad-band noise). Fishermen have long practised the art of listening for fishes and attracting them by sounds and refined methods for doings this are now being increasingly developed.

The field around a vibrating object in water, such as the tail of a fish, can be characterized by frequency and by the magnitude of, and phase relationships between, pressure and displacements. By these features it is possible to distinguish between a 'far-field' component or propagated acoustic wave and a 'near-field' component or non-propagated wave, involving larger displacements than in the far field but no pressure changes. The extent of the near field declines as the reciprocal of the frequency, therefore fishes that respond to high frequencies, such as goldfish or herring, have effective pressure-displacement transducers. Conversely low frequency sources in the immediate neighbourhood are

best detected by the displacements that they give. In clupeids there is therefore also a system that allows determination of pressure, displacement and frequency and hence of the amplitude and distance of a source. This is ensured in the herring by a gas-filled bulla in the ear, with a good low-frequency response that transforms pressures into displacements. These can stimulate the sense organs of the utriculus and lateral line canals which are also connected to the bulla. The relative magnitudes of the pressure-produced displacements and direct displacements of liquid in the canals can be compared by the fish to give the distance and direction of an object producing the disturbances. This should allow them to distinguish between their conspecifics and a predator, such as cod, or a food object (Denton, Gray, and Blaxter 1979).

The ear bulla is kept filled by a very narrow (7 μm) duct leading from the swim bladder. The latter has itself properties appropriate to serve as a reservoir but not for buoyancy, including a valved stomach duct but no gas secretion organ. Its wall has a compliant central section that is able to collapse under pressure as the fish moves deeper, leaving gas always available in the end sections. There is also a very thick layer of guanine making the walls unusually impermeable to gases. Thus, once swallowed, enough gas is retained to maintain auditory function for months in spite of diurnal vertical migrations of 100 m or more, by gas passing in both directions between the swim bladder and the auditory bulla. The auditory bulla also allows extraction of information about changes in depth as the fish swims, though not of the maintained static pressure level. Herring larvae without a swim bladder do not adapt and may be able to determine absolute depth. They become more sensitive when the bullae fill with gas.

18. Sound production in fishes

A surprisingly large number of fishes can produce sounds audible to ourselves, and these noises are used for shoaling, to bring the sexes together, or to warn off or startle enemies (Tavolga 1971). Only a small number (one or two) of distinct signals are produced, usually by differences in time intervals. There are no finely graded series of signals in fishes as there are in birds (but not in amphibia or reptiles). Among the loudest of the sounds is that produced by the drum-fish (*Pogonias*) of the Eastern Atlantic. The 'whistling' and other noises of the 'maigre' (*Sciaena*) are supposed to be the origin of the song of the Sirens, since they can easily be heard above the water. In both these fishes the sounds are made mostly if not wholly in the breeding season. In others such as siluroids and *Diodon*, the noise is associated with the presence of spines and may be a warning. In

Congiopodus the nerves that innervate the muscles of sound production also supply muscles that raise the spines (Packard 1960).

The mechanism for sound production is very varied, involving either stridulation by the vertebrae (some siluroids), operculum (*Cottus*, the bull-head), pectoral girdle (trigger-fishes), teeth (some mackerel and sun-fish), or phonation by the air bladder. The latter may be involved either by its use for 'breathing' sounds in physostomatous forms (p. 212) or as a resonator when beaten by the ribs by special muscles (toad-fishes, *Opsanus*). These have very thin red fibres capable of maintaining frequencies up to 200/s. This makes the whole swim bladder vibrate and function like the surface of a high-fidelity loud speaker. Organs for producing sound are common in benthopelagic fishes living on the continental slopes but are absent from mesopelagic and bathypelagic relatives. The sound is often produced only by males but a large sacculus of the inner ear is found in both sexes. Sound is also produced by some freshwater bottom-living forms and this seems to be most useful where there is an interface between water and solid structures.

19. The lateral line organs of fishes

The lateral line organs (neuromasts) consist of sensory cells bearing 'hairs' embedded in a gelatinous cup, the cupula, which is displaced by water movements. They occur either seperately on the surface or sunk into lateral line canals, which open at intervals by pores through the scales. The organs are part of the same system as the inner ear and are innervated by nerve fibres of the vestibular nerve (VIII). These fibres then run out with the branches of the seventh, ninth, and tenth cranial nerves. Besides the main canal running down the body and served by the lateral line branch of the tenth cranial nerve, there are also lines following a definite pattern on the head, namely, supra- and suborbital lines, a line on the lower jaw, and a temporal line across the back of the skull. The canals on the head are innervated mainly from the seventh, partly from the ninth cranial nerve. The nerve fibres enter the very large acoustico-lateral centres of the medulla and valvula cerebelli.

Fishes possess the capacity to react to an object moving some distance away in the water ('distant touch sense') and this is reduced or absent after section of the lateral line nerve. The organs allow detection of the direction and velocity of a moving object and perhaps of its size. Presumably the moving object sets up currents in the water, which move the fluid (or mucus) in the canals. It has also been suggested that the canals serve to record displacements produced by the swimming movements of the fish itself, but this has not been proved. Fishes deprived of the lateral line show no muscular inco-ordination, although if blind they collide frequently with solid objects. The organs on the head detect the flow pattern of the water in front and the 'shadow' produced by solid objects there. It has often been suggested that these organs serve for hearing, perhaps at low frequencies. The receptors of the lateral line organs and the inner ear are similar in structure and both are acoustic displacement receptors. But the neuromasts are 'near-field' detectors whereas the inner ear in ass-ociation with the swim bladder is a 'far-field' detector (p. 180).

Study of the electrical activities of these organs in rays has shown that they discharge impulses all the time, even when not under the influence of any external stimulation (Fig. 7.30). By passing currents of water along the tubes Sand (1937) showed that a tailward flow checks and a headward flow accelerates this 'spon-taneous' discharge of impulses. Such changes in the streams of impulses arriving at the brain could, no doubt, form the basis for initiation of movements of the

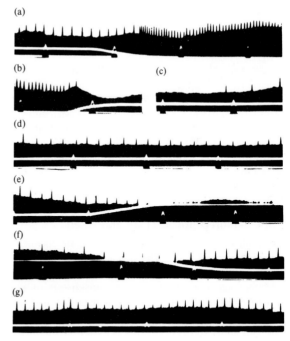

FIG. 7.30. Responses of a single end organ in a lateral canal of a ray, shown with an oscillograph after amplification. Time signal 1.0 s intervals. The movements of the continuous white line show (a) the beginning of a headward flow, increasing the frequency of discharge; (b) the end of this flow; (c) return of spontaneous discharge after an interval of 28 s; (d) spon-taneous discharge 60 s later; (e) beginning of a tailward perfusion, inhibiting the discharge; (f) the end of this perfusion; (g) the spontaneous discharge 10 s later (From Sand, 1937.)

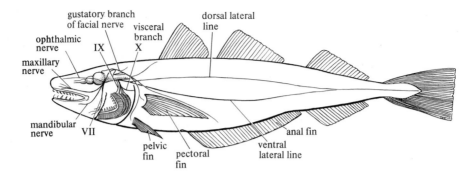

FIG. 7.31. Dissection of whiting to show the cranial nerves, and especially the nerves for the taste-buds. *VII*, hyomandibular branch of facial; *IX* glossopharyngeal; *X*, visceral branch of vagus.

fish. The lateral line nerves contain efferent fibres, which are not spontaneously active but discharge when the fish moves. They are not a feedback since they are not excited by natural stimulation of the lateral line organs. They inhibit the discharge of the sense organs and thus protect them from over stimulation during movement (Russell and Roberts 1972). In fast moving fishes all the neuromasts are in canals but in bathypelagic forms they are all on the surface, on stalks, giving greater sensitivity. Canals are absent in slow swimmers and bottom dwellers like the angler-fishes. The lateral line system must certainly be of great importance in aquatic life, for it is found in all types of fishes and also in the early Amphibia and in the aquatic larvae of modern members of that group. The distant touch receptors can obviously be used in many ways not only to locate moving objects and water currents but perhaps also to serve for echolocation, by computing the time relation of reflected waves set up by the fish itself.

20. Chemoreceptors. Taste and smell

As in all vertebrates, there are two separate chemical senses, taste and smell. The former serves mainly to produce appropriate reactions to food near the body, such as snapping, swallowing, or movements of rejection. Smell, on the other hand, is a 'distance sense', by which the whole animal is steered. The distinction between the two types of receptor is somewhat obscured in bony fishes by the fact that taste-buds are not restricted as they are in mammals to the tongue and pharynx but may occur on the barbels and all over the body. They are innervated by branches of the seventh, ninth, and tenth cranial nerves, which may reach far backwards (Fig. 7.31). In some species it has been shown that the fish is able to turn and snap at a piece of food placed near the tail. This power is lost if the branches

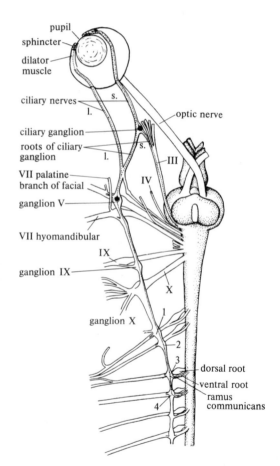

FIG. 7.32. Diagram of ventral view of the sympathetic system of the front part of the body of *Uranoscopus,* showing the fibres in the sympathetic and oculomotor that are responsible for the pupillary light reflex. (After Young, J. Z. (1931). *Quarterly Journal of microscopical Science* **74**, 491–535.) *l,* long ciliary nerve; *s,* short ciliary nerve; *III–X,* cranial nerves with their sympathetic ganglia *1–4.*

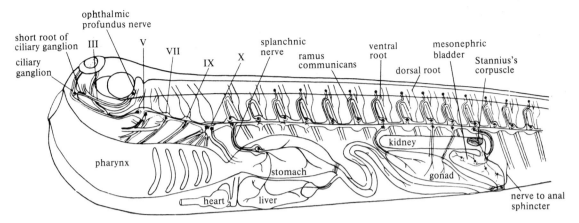

FIG. 7.33. Diagram of the autonomic nervous system of *Uranoscopus* seen from the side. (From Young, J. Z. (1931). *Quarterly Journal of microscopical Science* **74**, 491–535.)

from the cranial nerves are cut. In mammals taste-buds serve to discriminate only four qualities (salt, sour, bitter, and sweet), most of our so-called 'tasting' being in reality the smelling of the food in the mouth. In fishes, besides the four taste qualities, others are discriminated by the taste-bud system, and it has been shown that the minnow (*Phoxinus*) continues to make such discriminations after the forebrain has been removed. Other chemical discriminations are made by the nose, however, and can only be performed with an intact forebrain. Thus *Phoxinus* tastes and smells the same classes of substances as man does. The taste-buds are exceedingly sensitive, the threshold for sweet substances being 500 times and for salt, 200 times lower than in man. On the other hand, some substances that are very bitter for us produce little reaction in *Phoxinus*. In addition to smell and taste fishes also have a common chemical sense all over the skin mediated by free nerve endings and small receptor cells supplied by spinal nerves. It's thresholds are high, but it may serve to produce avoidance of polluted water.

In many fishes the nose is one of the chief receptors (macrosmatic). There are two nostrils on each side, allowing for the sampling of a stream of water (Fig. 7.4). The nose does not communicate with the mouth, except in a few fishes that live buried in the sand (*Astroscopus*).

The sense of smell is used to find food and for recognition of the sex of members of the same species. Minnows can recognize the odours of other individuals of the same species and they can be trained to give distinct reactions to extracts made from the skin of other species of fish living in fresh water. In the presence of 'alarm substances' produced by damaged skin of a member of the same species, minnows (and other Ostariophysi) show a 'fright reaction', scattering and

refusing food. The alarm substance is produced by special cells and released only if the skin is damaged.

The state of development of the nose is very varied. It is large in macrosmatic solitary predators such as *Anguilla* and in many schooling species that also have well-developed eyes (*Phoxinus, Gobio*). Daylight predators, on the other hand, are microsmatic (*Esox, Gasterosteus*). Other evidence shows that fishes can discriminate between the smells of water plants and between the waters of different streams. The olfactory bulb produces spontaneous rhythmical electrical changes (as in Fig. 7.20) and these vary when salts or other substances are introduced into the nasal cavity. Neurons in the olfactory bulb of salmon responded differently to water from the homepool they had left years before and that of distant places. It is suggested that this is a result of their imprinting on the water where they were hatched.

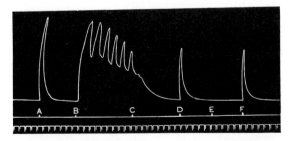

FIG. 7.34. Tracing of contractions of the muscle of the urinary bladder of *Lophius*, attached to a lever. At A, D and F faradic stimulation of vesicular nerve. Drugs added at B to make dilution of acetylcholine of 1/2 000 000; at C adrenaline at 1/500 000; and at E ergotoxine at 1/50 000. Time. minutes. (From Young, J. Z. (1936). *Proceedings of the Royal Society, London* **B120**, 303–18.)

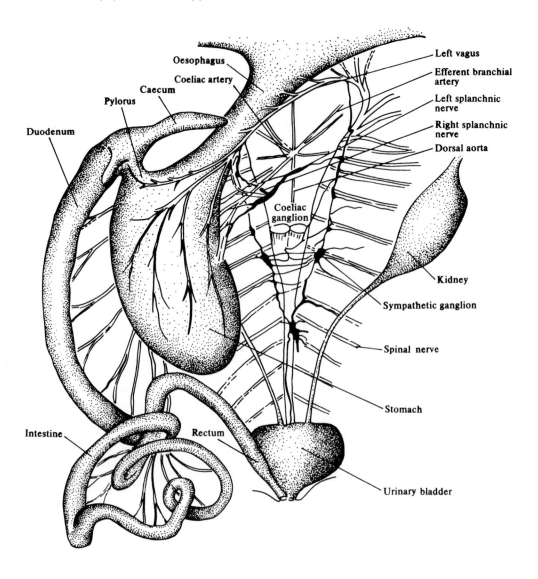

FIG. 7.35. The alimentary canal and autonomic nervous system of the angler fish (*Lophius*).

21. Touch and temperature senses

Touch is, of course, well developed in fishes, and in many species there are special sensory filaments, which presumably serve this sense. They are usually developed around the mouth, as in the catfish; in other fishes, they are modifications of the fins, the pectoral fins of gurnards, which also contain chemoreceptors (Fig. 7.17). Fishes can be trained to swim up a stream when the temperature is increased. The threshold may be as low as 0.03 °C. The receptors are innervated by spinal nerves and are probably not the lateral line organs as some have believed.

22. Autonomic nervous system

The autonomic nervous system of bony fishes is organized on a plan rather different from that both of elasmobranchs and of land animals. There is a chain of sympathetic ganglia, extending from the level of the trigeminal nerve backwards, a ganglion being found in connection with each of the cranial dorsal roots (Figs. 7.32 and 7.33). These ganglia do not receive preganglionic fibres from the segments in which they lie, but by fibres that run out in the ventral roots of the trunk region and thence forwards in the sympathetic chain. This emergence of the preganglionic fibres for the head

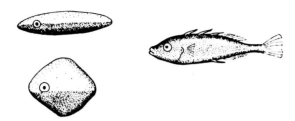

F<small>IG</small>. 7.36. The red belly of the stickleback releases attacking behaviour in other males and following by females. Of the above models only the two on the left acted as releasers. (From Tinbergen, N. (1948). *Wilson Bulletin* **60**, 6–52.)

in the trunk region recalls the arrangement in land animals.

Each trunk sympathetic ganglion, besides receiving a white ramus communicans of preganglionic fibres from its spinal nerve, also sends a grey ramus back to that nerve, this ramus carrying postganglionic fibres to the skin. Some of these fibres control the melanophores, causing them to contract (p. 211). In elasmobranchs there are no grey rami communicans and no sympathetic system in the head (p. 148); the differences between the two groups are therefore very striking.

Little is known about the parasympathetic system of bony fishes. The oculomotor nerve carries fibres to the iris, which work in the opposite direction to fibres from the sympathetic (p. 180). There is also a well-developed vagal system, but so far as is known no parasympathetic fibres in other cranial nerves and probably no sacral parasympathetic system. Electrical stimulation of the vagus nerve produces movements of the stomach of acanthopterygian fishes. In more primitive forms such as salmonids its action may be inhibitory. The splanchnic nerves (sympathetic) are usually inhibitory to the stomach and intestine (Young 1980).

In most of the viscera acetylcholine causes initiation of rhythmic contractions and these are inhibited by adrenaline (Fig. 7.34). In *Lophius* this is true of the stomach, intestine, and bladder, which contracts on stimulation of the hinder sympathetic ganglia (Fig. 7.35). In these fishes it is not possible to divide up the autonomic nervous system into sympathetic and parasympathetic divisions by either anatomical, physiological, or pharmacological criteria. Presumably the two 'antagonistic' systems found in mammals are a late development, allowing for a delicate balancing of activities for the maintenance of homeostasis.

23. Behaviour patterns of fishes

The well-developed receptors and brain of the teleostean fishes constitute perhaps the most important of all factors in giving them their great success. Varied habits and quick actions enable the fish to make full use of the possibilities provided by the special features of their structure – the swim bladder, mouth, and so on. The receptors and brain make it possible for the fish to learn to react appropriately to many features of its surroundings. Thus the eyes besides orientating the fish to movements in the visual field allow the discrimination of wavelengths and distinct reactions to differing shapes (see Bull 1957).

Fishes differ from higher vertebrates in that they grow continuously throughout life and also add new cells to their brains and eyes. The brain of a young larval herring contains 3.5×10^5 neurons and of the largest adult 11×10^6. The first number is about that found in an insect and the larval behaviour of the herring is 'more insect-like than vertebrate-like, being characterized by fixed-action patterns ... limited to the switching on of stereotyped patterns of feeding, shoaling and light-dependent vertical movements' (Packard and Wainwright 1973).

The social behaviour of many species includes the development of special 'releasers', shapes, colours, or postures that are displayed by one individual and elicit specific reactions in another (Fig. 7.36). Shoals of herring may be 1.5 km long and contain some 150 000 000 individuals. The animals are presumably kept together in most species by visual stimuli, though sounds may play a part. Shoaling gives protection to small fishes, and some come together in shoals to breed (herring). There may also be some advantage for the finding of suitable feeding conditions.

There is no doubt that fishes possess great powers of learning. They can form conditioned reflexes involving discrimination of tones, also second-order conditioned reflexes, in which after the animal has learnt to give a certain behaviour in response to a visual stimulus it is then taught to associate the latter with an olfactory stimulus. There are many other examples of such powers, but unfortunately we have as yet little information as to the way in which they are brought about by the brain. Nor have the naturalists provided us with very clear examples of the use of these powers by fishes in nature. There are many tales of carp coming to be fed at the ringing of a bell, and similar powers of association must play a part in the life of fishes in more natural situations. Bull has shown that fishes can be trained to discriminate between very small differences of water flow, temperature, salinity, or pH, and no doubt it is by means of such powers that they normally find a suitable habitat.

The migrations of fishes have attracted much attention, but are still imperfectly understood. They vary from the 'catadromous' downward migration of young animals to the sea and the reverse 'anadromous' movement to breed, to the astounding journeys of the eels, 5600 km westwards from Europe or eastwards from America to their breeding-place in the Sargasso Sea (off Bermuda) and the return of the elvers to the homes of their parents (Fig. 7.37). No one has yet discovered the factors that direct these movements, currents may play a part, but can hardly be the only influence. In preparation for the journey to breed the eel undergoes progressive changes as it approaches the sea. These are seen in the osmoregulatory system with an increase in gill tissue cells secreting chloride, in the thickening of the layers of the skin, in the enlargement of the eyes and the development of a retinal pigment, chrysopsin, found in deep-sea fish. There is also enlargement (increase in capillary length) of the rete mirabile, the gas-secreting organ of the swim bladder. The pigmentation of the body is now dark dorsally and silver ventrally, also an indication of a mesopelagic habitat during this migration.

The planktonic leptocephalus larval stages, long thought to be a separate genus, may last for several years. The body is laterally compressed, leaf-like, and transparent. The eyes, nose, and brain are large but there is almost no ossification in the skeleton and only the mandibular hemibranch bears gill filaments providing blood directly to the eyes and brain. There are no red corpuscles and no spleen. There are teeth but the gut is a straight tube, partly occluded. No food remains have ever been seen in it and nutrition may be from dissolved organic material or minute particulate matter. The kidneys are aglomerular and the high osmolality of fluids expressed from larvae (700 mosmol/kg) suggests that 'Pelagic eel larvae are more in ionic equilibrium with sea water than other marine fishes with the exception of *Myxine*' (Hulet, Fischer and Reitberg 1972).

Leptocephalus larvae are found in some other primitive teleosts, grouped together in the superorder Elopomorpha (p. 196). These are known to have existed in the Cretaceous, before the break-up of the supercontinents (p. 21). This may well be the explanation for their extraordinary migrations from both sides of the Atlantic, involving continually greater journeys and lengthening of the near embryonic existence as a leptocephalus.

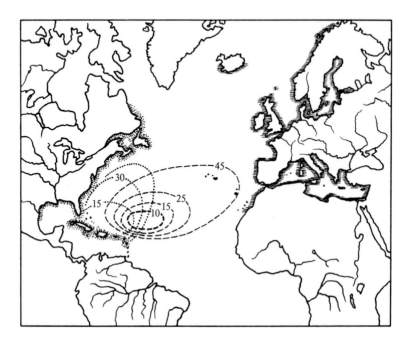

FIG. 7.37. Migrations of the eels. The European species (*A. anguilla*) occurs along the coasts outlined with lines, the American species (*A. rostrata*) where the coast is dotted. The curved lines show where larvae of the lengths indicated (in millimetres) are taken. (After Norman 1931.)

8 The evolution of bony fishes

1. Classification

Class OSTEICHTHYES
  Subclass 1: Acanthodii (spiny). Ordovician–Permian
    *Climatius; *Acanthodes
  Subclass 2: Actinopterygii (ray fins)
    Infraclass 1: Chondrostei (cartilage and bone)
      Order 1: Palaeoniscoidei (ancient fishes). Devonian–Recent
      *Cheirolepis; *Palaeoniscus; *Amphicentrum;
      *Platysomus; *Dorypterus; *Cleithrolepis;
      *Tarrasius; Polypterus, bichir
      Order 2: Acipenseroidei (swift fin). Jurassic–Recent
      *Chondrosteus; Acipenser, sturgeon;
      Polyodon, paddle-fish
    Infraclass 2: Holostei (only bone). Triassic–Recent
      *Acentrophorus; *Lepidotes; *Dapedius;
      *Microdon; *Pholidophorus; Amia, bowfin;
      Lepisosteus, gar-pike
    Infraclass 3: Teleostei (bony fishes). Upper Triassic–Recent
      Superorder 1: *Leptolepimorpha (slender, graceful form). Middle   Triassic–Cretaceous
        *Leptolepis
      Superorder 2: Elopomorpha. Cretaceous–Recent
        Elops, ten-pounders; Megalops, tarpons; Anguilla, eel; Conger, conger eel
      Superorder 3: Osteoglossomorpha (bony tongue). Jurassic–Recent
        *Lycoptera; Mormyrus, elephant snout fish; Pantodon, butterfly fish
      Superorder 4: Clupeomorpha. Cretaceous–Recent
        Clupea, herring
      Superorder 5: Ostariophysi (ossicle fishes). Eocene–Recent
        Cyprinus, carp: Tinca, tench; Silurus, catfish;
        Gobio, gudgeon; Phoxinus, minnow
      Superorder 6: Protacanthopterygii (early spine fin fishes). Upper Cretaceous–Recent
        Salmo, trout; Esox, pike; Astronesthes, deep-sea snaggletooth; Myctophus, lanternfish
      Superorder 7: Paracanthopterygii (nearly spiny fin fish). Eocene–Recent
        Lophius, anglerfish; Photocorynus, deep-sea anglerfish; Lepadogaster, suckerfish; Gadus, whiting
      Superorder 8: Atherinomorpha (smelt fishes). Upper Cretaceous–Recent
        Exocoetus, flying fish; Belone, garfish
      Superorder 9: Acanthopterygii (spiny fin fishes). Upper Cretaceous–Recent
        *Hoplopteryx; Gasterosteus, stickleback; Syngnathus, pipefish; Hippocampus, seahorse; Zeus, John Dory; Perca, perch; Labrus, wrasse; Uranoscopus, star gazer; Blennius, blenny; Pleuronectes, plaice; Solea, sole; Exocoetus, flying fish; Belone, garfish

* Fossil forms.

2. Subclass 1: Acanthodii

The acanthodians, found in freshwater deposits extending from the Ordovician to the Permian but chiefly in the Devonian, are the oldest known gnathostomes. They have been considered as relations of various groups and indeed in earlier editions of this book were placed with the elasmobranchs. These were small fishes with a fusiform body, with heterocercal tail and two, or later one, dorsal fins. The lateral fins consisted of a series of pairs, often as many as seven in all, down the sides of the body. The effect of these in stabilizing the fish would presumably be different from that of a continuous fold, and the problem of the form and function of the earliest paired fins remains obscure (p. 153). The fins were all supported by the large spines from which the group derives its name.

The whole surface of the body was covered with a layer of small rhomboidal scales, composed of layers of material resembling bone, covered with a shiny material similar to the ganoin of early Actinopterygii. On the head these scales were enlarged to make a definite pattern of dermal bones, numerous at first but fewer in the later forms. The pattern of the bones has no close similarity to that of later fishes. The reduced bones of the later acanthodians are related to the lateral line canals, which have an arrangement similar to that in other fishes, but run between and not through the scales and bones of the head. The teeth are formed as a series of modified scales. The skull is partly ossified – important evidence that the boneless condition of elasmobranchs was not typical of all early gnathostomes.

The jaws of acanthodians were attached by their own processes to the skull (autodiastyly) and are remarkable in that four separate ossifications take place in them (two in the upper and two in the lower jaw), making a series of elements similar to that found in the typical branchial arches. At first the mandibular, hyoid, and each of the branchial arches were provided with small flap-like opercula, but in later forms the hyoidean operculum became especially developed and covered all the gills. The nostrils were near the front of the head, like those of Actinopterygii.

The acanthodians have been considered to be related to either or both the Chondrichthyes and Osteichthyes (see Miles 1973). Jarvik (1980) places them close to the Selachii. Possibly they were somewhere near to the ancestry of the stock that gave rise to the Actinopterygii, Crossopterygii, and Dipnoi.

The Actinopterygii include many lines of descent that can be followed with some completeness to their extinction or modern descendants. Various classifications have been suggested and the one used here is

simple but for that very reason obscures the multiplicity of parallel lines. More cladistic classifications have been devised but the older arrangement based on grades is more conveneint. The animals at the earliest level are grouped in the infraclass Chondrostei. Those of an intermediate grade are Holostei, some of which gave rise to the great modern infraclass Teleostei. Various alternative classifications, proposed largely by ichthyologists at the British Museum of Natural History, on a basis of bony structure are discussed in Greenwood, Miles and Patterson (1973).

3. Subclass 2: Actinopterygii; Infraclass, Chondrostei

3.1. Order 1: Palaeoniscoidei

The Devonian and Carboniferous forms are grouped together in the order Palaeoniscoidei, and animals of similar type survive today as *Polypterus*, the bichir of African rivers, which though showing some specializations remains in its general organization near the palaeoniscid level.

A typical Palaeozoic palaeoniscid such as *Cheirolepis* was a long-bodied creature (Figs. 8.2 and 8.3) with a heterocercal tail, single dorsal fin, and pelvic fins placed far back on the body. The pectoral and pelvic fins had broad bases and the radials fanned out from a small muscular lobe, present in all early actinopterygians but lost in later forms. The body was covered with thick rhomboidal scales very similar to those of acanthodians. They articulated by peg and socket joints and have a structure known as palaeoniscoid (Fig. 8.1). The scale is deeply embedded and grows by addition both to the bony or isopedin portion and to the shiny surface-layer, the ganoin, which thus becomes very thick. There is a middle layer of 'pulp' corresponding to the cosmine layer of the cosmoid scale of Crossopterygii and the two types have obvious similarities, though it is not clear how they are related.

The skull was built on a distinctly different plan from that of Crossopterygii, in that there was no joint such as was present in those fishes to allow the front part to flex on the hind. The jaw support was amphistylic, in the sense that the palatoquadrate was attached to the neurocranium by a basal process, but the otic process did not reach the skull and the hind end of the jaw was supported by the hyomandibula. There were even more dermal bones than are found in modern Actinopterygii, arranged so as to form a complete covering for the chondrocranium and jaws. These bones were derived from the original scaly covering of the head and naming and comparing them with the bones of other forms is a matter of some difficulty. Some of the main bones resemble in appearance and shape those found in

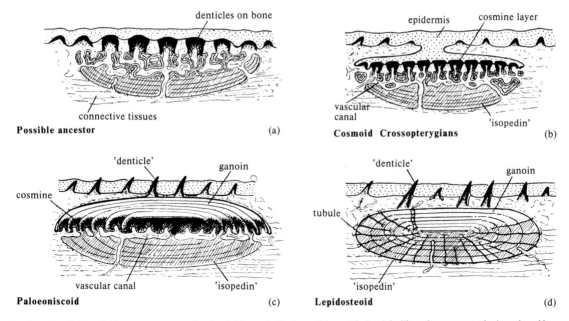

FIG. 8.1. Possible evolutionary sequence of scales. (a) hypothetical condition with denticle-like substance attached to a basal bony plate lying in the connective tissue; (b) 'cosmoid' scale of early crossopterygians; (c) palaeoniscoid scale with layers of 'ganoin'; (d) lepidosteoid scale of the garfish, with tubules. (From Goodrich 1909.)

tetrapods, but there are others for which no such homologues can be found, and sometimes there is considerable difficulty in recognizing even the main outlines of the pattern. The problem is that we have no rigid criterion by which to set about giving names to the skull bones. No system yet discovered is wholly satisfactory, and we must admit to insufficient knowledge of the factors that determine that bone shall be laid down in certain areas and that sutures shall separate these from each other. However, some of the dermal bones lie in relation to the lateral line canals (or rows of neuromasts), which may provide the stimulus to bone formation. The lines are remarkably constant, perhaps because of their function in detecting water movements in relation to swimming, and this is the factor that determines the position of many of the bones. Others fill in the spaces between (anamnesic bones). Yet others may be differentiated in relation to the teeth. However, the number of bones along any one line may vary greatly even in one species (e.g. in *Amia*). The whole pattern is more variable in fishes than in higher vertebrates, but it is usual to consider that the bones of early Actinopterygii resemble those of Crossopterygii and of the early amphibians (Fig. 11.22).

The roof of the skull usually shows a large pair of frontals between the eyes, and parietals behind these. Between the frontals and the nostrils there is a postrostral bone and the front of the head usually also carries a

number of rostral bones, not found in higher forms. Behind the parietals in the midline is a series of extrascapular bones.

The side of the skull of palaeoniscids is covered by numerous bones, including infra-, sub-, and supra-orbitals around the eyes. The outer margin of the upper jaw is covered by premaxillae and maxillae, which are the main tooth-bearing bones. Behind the orbital series of bones the cheek is very variable. Sometimes there is a large bone identifiable as a pre-opercular, with a series of opercular bones behind it. The lower portion of the throat was covered by a series of gular plates. The spiracle in these early forms opened above the opercular bones. The pectoral girdle was attached to the back of the skull by a supracleithrum, below which a cleithrum and clavicle made a series of dermal bones behind the gills, covering the cartilaginous girdle. The roof of the mouth contained a median parasphenoid, with paired vomers in front of it, and a series of pterygoid bones occupied the space between it and the edge of the jaws, the palatine, ectopterygoid, endopterygoid, and sometimes others. Finally the lower jaw, besides the main dentary carrying the teeth, shows many small bones such as the pre-articular and coronoids on the inner surface; splenial, angular, and surangular on the outside.

It will be clear that this skull of *Cheirolepis* may be closely compared with the skull of a crossopterygian or a modern teleostean. The general plan is related to that of the lateral line organs arranged along occipital, supratemporal, and infra-orbital lines. The numerous small bones are evidently similar in the different groups, though it is not easy to assign a suitable name to every one of the more numerous bones of the earlier forms.

These palaeoniscids from the Middle Devonian were rather rare freshwater fishes; they had sharp teeth and probably lived on invertebrates. We have no information about their internal anatomy, but it seems not unlikely that the swim bladder possessed a wide opening to the pharynx still in *Polypterus*, perhaps descended from this stock, and that they breathed air, as did other Devonian fishes. However, they did not have internal nostrils, which are found in the crossopterygians.

During the Carboniferous and Permian the palaeoniscids were numerous, mostly as small, sharp-toothed fishes. Several distinct lines became laterally flattened and acquired an outwardly symmetrical tail and blunt crushing teeth (Fig. 8.3). These characteristics probably indicate a habit of feeding in calm waters, perhaps mainly on corals, and they have appeared several times in the actinopterygian stock (p. 201). Palaeoniscids of this type were formerly placed together in a family Platysomidae, but it is now considered probable that the type arose independently several times; thus *Amphicentrum* is found in the Carboniferous, *Platysomus* and *Dorypterus* in the Permian, and *Cleithrolepis* in the Triassic. Similar forms arose again later among the holosteans and teleosteans and we have, therefore, evidence that this type of animal organization tends to evolve into deep-bodied creatures. *Dorypterus* further resembles modern teleosteans in a great reduction of its scales and in the forward movement of the pelvic fins.

In the Triassic various lines of palaeoniscids approached the holostean form and are loosely grouped as sub-holosteans. Towards the end of the Triassic animals of typical palaeoniscid type became rare. Instead we find more active and speedy, mainly marine descendants, the Holostei (p. 193). The modern bichirs and the sturgeons are survivors that branched off in the Palaeozoic and in spite of subsequent specializations give us some idea of the characteristics of these early Actinopterygii. *Polypterus* (the bichir) and the related *Calamoichthys*, are both carnivorous fishes inhabiting rivers in Africa. The swim bladder shows some similarity to a lung. It forms a pair of sacs lying ventrally below the intestine and opening to the pharynx by a median ventral 'glottis' (Fig. 9.16). This is the arrangement found in lung fishes (except *Neoceratodus*) and in tetrapods, and it seems reasonable to suppose that it has survived in *Polypterus* from Palaeozoic times. However, it is not certain to

what extent the swim bladder is still used as a lung. *Polypterus* may come to the surface to breathe but cannot survive out of the water.

This fish shows many other ancient characteristics. The covering of thick rhomboidal scales, hardly overlapping, gives the animal an archaic appearance; the structure of the scales is 'palaeoniscoid'. In the skin there is a layer of denticles outside the scales. The presence of a spiracle, the arrangement of the skull bones, and many other features suggest that *Polypterus* is essentially a palaeoniscid surviving to the present day. In the intestine there is a spiral valve, which was present in the early Crossopterygii and Actinopteryggii (as judged from fossilized faeces, 'coprolites'). It occurs today not only in the Elasmobranchii and Dipnoi but also in sturgeons and, though much reduced, in *Lepisosteus* and *Amia*. There is a single pyloric caecum in *Polypterus* (the caeca are well developed in sturgeons, *Lepisosteus*, and *Amia*).

In the pituitary the opening of Rathke's pocket persists in the adult, leading to a ramifying cavity in the rostral pars distalis lined by mucous cells. In this and other features (persistent pronephros) there are signs of neoteny. These fishes possess a typical median eminence and portal system, which are absent in teleosts (p. 138). Indeed these structures are very like those of tetrapods. The body is rather eel-like. The tail of *Polypterus* is no longer markedly heterocercal, but shows distinct signs of that condition. We can even find a parallel among Carboniferous palaeoniscids for some of the special features of *Polypterus*. The long body and dorsal fin are found in the fossil *Tarrasius*, which may have been close to the ancestry of *Polypterus*, though it lacks the covering of scales. The pectoral fin in *Tarrasius*, as in *Polypterus*, has a peculiar lobed form, which has been compared with the 'archipterygial' pattern (p. 218) and hence held to show that these animals are related to the Crossopterygii. The resemblance is, however, only superficial and the plan of the fin is essentially actinopterygian. The forebrain eversion makes a false ventricle (p. 178). There is a valvula cerebelli as in teleosts.*

3.2. Order 2: Acipenseroidei

The sturgeons are a second isolated line descending from the palaeoniscids and characterized by reduction of bone. This was already apparent in the Jurassic *Chondrosteus. Acipenser* and other modern sturgeons live in the sea but migrate up river to breed. They may reach a very large size (1000 kg) and since a tenth of this is caviar they are exceedingly valuable. They feed on invertebrates, which they collect from mud stirred up from the bottom by a long snout. The mouth of all sturgeons is small and the jaws weak and without teeth.

* It has recently been strongly argued by Jarvik (1980) that *Polypterus* should not be classified with actinopterygians and has several affinities with coelacanths and dipnoans. He also argues that *Amia* has not evolved from palaeoniscids (p. 190).

The jaws hang free from the hyomandibular and symplectic, and can be swung downward and forward to suck up the prey from the bottom (snails, worms, crustaceans, insects, or fishes). The skull and skeleton is almost wholly cartilaginous, the notochord is unconstricted, and the dermal skeleton much reduced. There is a large swim bladder, but the fish is not neutrally buoyant and the tail is heterocercal. The fish is covered with rhomboidal scales, but on the front of the body there are five lines of bony plates bearing spines, with the skin in between carrying structures similar to denticles. There is an open spiracle. The internal anatomy of the sturgeons also shows various features that have been held to show affinity with the elasmobranchs; for instance, besides the spiral valve there is a conus arteriosus in the heart and a single pericardio-peritoneal canal. The pituitary has a median eminence and portal system with no direct innervation of the cells of the pars distalis. However, there can be no doubt that sturgeons are descended from an early offshoot from the actinopterygian line. They retain some features lost by most members of the line, but resemble the teleosts in other characters, for instance a thin roof to the cerebral hemispheres. The paddle-fishes, *Polyodon*, of the Mississippi and *Psephurus* from China are related to the sturgeon but perhaps not very closely. The snout is flattened and they feed by swimming with the huge mouth open and filtering the water with the gill rakers of the pharynx.

4. Infraclass 2: Holostei

During the later Permian period some palaeoniscids gave rise to fishes of a different type, which replaced their ancestors almost completely during the Triassic and flourished greatly in the Jurassic. We may group together the fishes of this type as Holostei but the term is used variously by different authors and includes several lines, whose relationships are not clear. The earliest holostean, *Acentrophorus from the upper Permian, is much like a palaeoniscid but with a small mouth, shorter, deeper body and slightly upturned tail. This 'abbreviated heterocercal' tail was presumably made possible by the changed swimming habits resulting from the use of the swim bladder as a hydrostatic organ. If the fish floats passively there is no need for a heterocercal tail (p. 115). Similarly in the arrangements that ensure stability in the vertical plane, the head does not need to be flattened to produce an upward lift. The development of the swim bladder has thus made possible the lateral flattening and shortening of the body so characteristic of later Actinopterygii. The body of holosteans was at first covered with thick ganoid scales, but these became thinner in later types. The jaw suspension is character-

istic, the maxilla being freed from the pre-opercular and able to swing forwards (p. 166). As a result the lower jaw could now be protruded forwards in front of the upper and a 'sucking' action, characteristic of teleosts, was evolved, the prey being drawn into the mouth from a distance (Lauder 1979). By a change in the insertion of the adductor mandibulae muscle a more powerful jaw action then became possible. Some of the holosteans achieved crushing teeth and replaced the dipnoans in the early Mesozoic. There are various smaller distinctive holostean features, such as the loss of the clavicle.

We do not know whether fishes of this type arose from a single palaeoniscid stock; it is very likely that the change occurred several times, and that throughout the Triassic and Jurassic there were several lines with these holostean characteristics, evolving separately. During the Cretaceous they became fewer, being replaced by their teleostean descendants, but two fishes of holostean type survive today, *Lepisosteus* the gar-pike (often written *Lepidosteus*) and *Amia* the bow-fin. We shall classify them with fossil forms as the infraclass Holostei. It has been emphasized that the two are not very alike and that *Amia* is the more closely related to the Teleostei (Patterson 1973). These are freshwater fishes, living in the American Great Lakes and other parts of eastern North America, but the group was mainly a marine one, having taken to the sea in the Triassic at a time when other groups were doing the same (palaeoniscids, coelacanths, elasmobranchs). The basic cause of this movement is not known, but perhaps there was an increase of planktonic and invertebrate life on which the fish depended.

Lepisosteus shows a rather primitive structure and must have remained at approximately the Triassic stage. With its complete armour of thick scales (Fig. 8.2) it presents the appearance of a primitive fish. The swim bladder opens to the pharynx and the gar-pikes come to the surface to gulp air. They maintain nearly neutral buoyancy, and can hover with little movement. The tail is nearly symmetrical. There is no spiracle. There are long jaws with which it catches other fishes: the intestine contains a simple spiral valve. Fossils similar to the modern gar-pike are found in the Eocene.

Some of the later holosteans became deep- and short-bodied and developed a small mouth with flat crushing teeth or a beak, for instance *Lepidotes (Triassic – Cretaceous), and *Dapedius (Jurassic). They probably browsed on corals, like the modern parrot fishes (Scaridae). *Microdon and other 'pycnodonts', numerous in the Jurassic and Cretaceous, became laterally flattened, like some palaeoniscids and the modern sea butterflies (Chaetodontidae) (Fig. 9.12) and other coral fishes.

Another line of holostean evolution, developing from

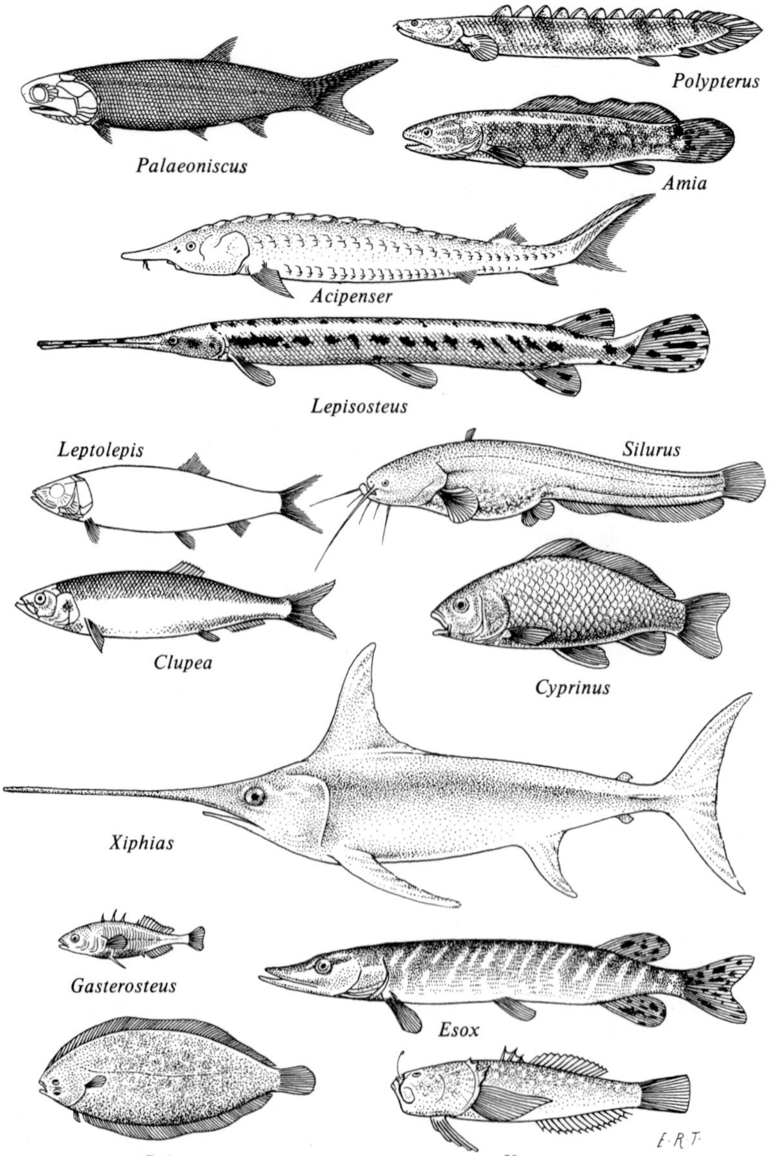

FIG. 8.2. Various actinopterygians.

the original stock, retained the streamlined body and from something like these both the modern teleosteans and the amioids were evolved (Fig. 8.3). *Caturus* (Triassic – Cretaceous) was covered with thick scales, but in *Pachycormus* (Cretaceous) they are thinner; these were active pelagic predators. In *Amia* the scales became reduced to single bony cycloid scales, probably acquired in parallel to those of the Teleostei. Meanwhile other changes took place, the tail fin becoming externally completely symmetrical and the maxilla and other cheek bones reduced. *Amia* has nearly reached the teleostean stage but retains certain primitive features in

the skeleton and the small eggs with holoblastic cleavage. In the pituitary there is still a well developed median eminence but some neurosecretory fibres innervate at least some cells of the pars distalis, as in modern teleosts.

5. Infraclass 3: Teleostei

The groups so far considered have been nearly completely replaced by the Teleostei, fishes derived from earlier holostean stocks, which have carried still farther the tendencies to shortening and symmetry of the tail, reduction of the scales, and various changes in the skull,

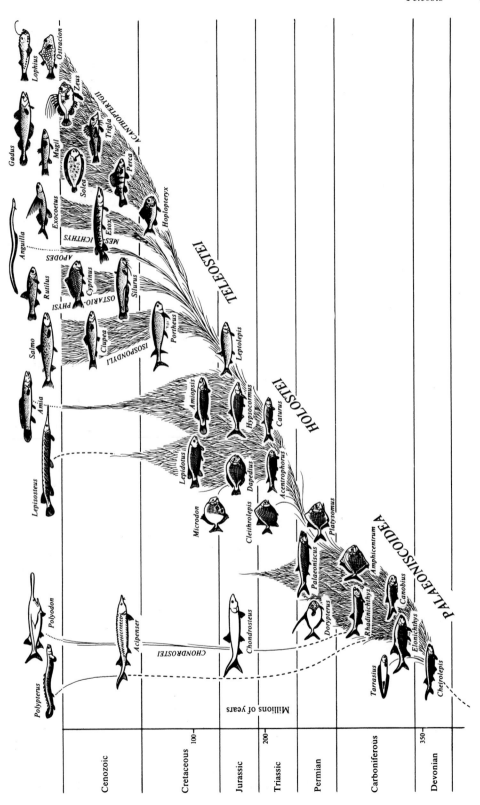

Fig. 8.3. Evolution of the Actinopterygii. A new classification of teleostean fishes is given on p. 196.

such as reduction of the maxilla. The type apparently arose in the sea in late Triassic times, but remained rare until the Cretaceous, by which time several different lines of evolution had already begun. *Pholidophorus* from the Triassic still carried an armour of thick scales and is usually classed as a holostean (Patterson 1973). It may well have given rise to fish such as *Leptolepis* in the upper Triassic which are generally considered to be close to the ancestry of all Teleostei and are placed here as a separate superorder Leptolepimorpha. *Leptolepis* was a long-bodied fish with the pelvic fins placed far back, a skull with a full complement of bones, and a large maxilla. The scales still show traces of the ganoin layer.

From some fish like these leptolepids have been derived more than 20 000 species of bony fish found today. This evolution occurred in a number of parallel lines, several of which were well established by the end of the Mesozoic. In view of this parallel evolution it is

useful to subdivide the Teleostei cladistically, along 'vertical' lines rather than 'horizontal' or 'typological' divisions showing levels of structure. Such a classification has been set out by Greenwood, Rosen, Weitzman, and Myers (1966) and is in the main adopted here and includes nine superorders (Fig. 8.4). The Leptolepimorpha are the early forms, often included among Holostei. The next four superorders used to be grouped together as the Isospondyli, fish with soft rays, as distinct from the spiny finned fishes. They all show primitive features in the large maxilla, which forms the posterior margin of the upper jaw, often the persistence of an open duct to the swim bladder, and the posterior position of the pelvic fins. The scales are cycloid, that is without projections on the free border.

The Elopomorpha are among the most primitive of living teleosts and include the ten-pounders (*Elops*), tarpons (*Megalops*) and the eels. The superorder Osteoglossomorpha extend back to the Upper Jurassic

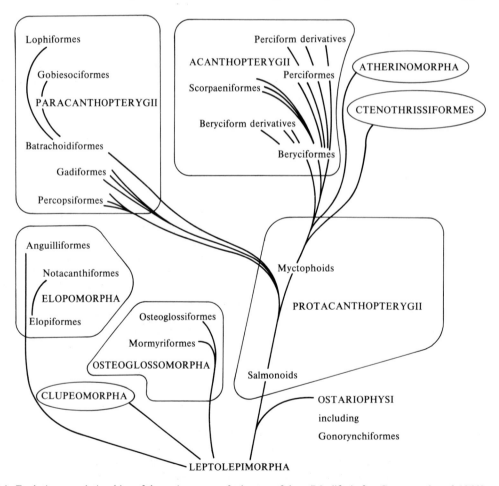

FIG. 8.4. Evolutionary relationships of the main groups of teleostean fishes. (Modified after Greenwood *et al.* 1966.)

(*Lycoptera*) and include the freshwater Mormyrids, the elephant snout fishes. The Clupeomorpha include many familiar and successful forms such as the herring (*Clupea*). The superorder Ostariophysi is a large group of some 5000 species of mostly freshwater fishes. The anterior vertebrae are modified to form a chain of bones, the Weberian ossicles, joining the swim bladder to the ear. Here belong the carp and goldfish (*Cyprinus*), roach (*Rutilus*), and catfishes (*Silurus*). The sixth superorder, the Protacanthopterygii includes fishes such as the trout (*Salmo*) and pike (*Esox*). They are of structure intermediate between that of the more primitive forms and the latest spiny-finned teleosts and hence sometimes called Mesichthyes (Fig. 8.2). Several groups of deep-sea fishes such as the lantern fishes (myctophoids) are included here, but may belong nearer to the Actinopterygii.

The superorder Paracanthopterygii includes the angler fish (*Lophius*), and the related deep-sea anglers, *Photocorynus*, which save energy as floating fish-traps (p. 202). The whiting (*Gadus*) is related to these.

The members of the superorder Acanthopterygii, the spiny-finned fishes, are the most numerous and highly developed fishes, characterized by the stiff spines at the front of the dorsal and anal fins. The premaxilla forms the tooth-bearing margin of the jaw, the toothless maxilla acting as a lever to push the jaw forward to suck in food (p. 166). The duct of the swim bladder is closed, the body shortened, and the pelvic fins far forward. Fishes of this type appeared in the late Cretaceous (*Hoplopteryx*) and the condition may have been evolved along several different and parallel lines. The group includes a vast array of modern types. Here belong the perches (*Perca*), mullet (*Mugil*), wrasse (*Labrus*), John Dories (*Zeus*), blennies (*Blennius*), and gurnards (*Trigla*). The flat-fishes, plaice (*Pleuronectes*) and sole (*Solea*) are further members of this very large order, as is the elongated gar-fish (*Belone*) and the flying fish (*Exocoetus*) (Fig. 8.3).

6. Analysis of evolution of the Actinopterygii

Our knowledge of the history of the Actinopterygii is sufficiently complete to allow more detailed discussion about the process of evolution than has been possible from consideration of the more ancient and less perfectly known groups of fishes. First of all we may emphasize the persistence of change. No actinopterygian fish living in the Devonian is to be found today. *Polypterus* may perhaps be regarded as a palaeoniscid, showing features that were common in the Carboniferous, but it has undergone many changes since that time. Similarly the living sturgeons show the stage of organization present in the Jurassic Chondrostei, but

they also have changed much. *Lepisosteus* is in general structure similar to a Triassic holostean such as *Lepidotes* and *Amia* to a Jurassic one such as *Caturus*, but both have their own more recent specializations. It is not until we come to the Cretaceous and Tertiary history of fishes that types belonging to recognizable modern genera can be found.

The Actinopterygii have therefore been changing slowly and throughout the period of their existence (p. 195). Was this change dictated in some way by a change in their surroundings? Unfortunately we cannot answer this question very clearly. The sea is a relatively constant medium, though not as 'unchanging' as is sometimes supposed. In particular the relative extent of sea and freshwater changes frequently and perhaps because of such a change the early freshwater actinopterygians took to the sea. Probably the life in the sea is not constant over long periods. Almost certainly the available nitrogen, phosphorus, and other essential elements change in amount. We have reason to suspect that there was an increase in the extent and productivity of the sea during the Triassic period. It may be that such gradual changes of the life in the water, depending ultimately on climate or inorganic changes, have been responsible for the continual change of the fish population. However, it cannot be said that we can detect evidence of any such relationship; there is no clear proof that the changes in the fish populations follow changes in the environment. For the present we can only note the fact that change occurs, even in animals living in the relatively constant sea.

Very striking is the fact that as evolution proceeds not merely does each genus change but whole types disappear and are replaced by others. Thus the palaeoniscid type of organization had disappeared almost completely by the Triassic and become replaced by the holostean. A few members retained the old organization and still survive today as *Polypterus* and the sturgeons. Similarly the Holostei, with their abbreviated heterocercal tails, hardly survived into the Tertiary, but were replaced by the Teleostei, only *Lepisosteus* and *Amia* remaining to show the earlier organization.

This replacement of one type by another appears much more remarkable when we reflect that by a 'type', say the palaeoniscid, we do not mean a homogeneous set of similar organisms all interbreeding. Quite the contrary; the palaeoniscids included many separate lines, each with its own peculiarities. When one 'type' therefore is thus replaced by another it must mean either that some one of these many stocks gives rise to a specially successful new population, which ousts the old ones or that *all* the members of the stock are changing their type together. It is not easy with a record such as that of the fish, which is far from continuous, to say which of these is true in any particular case, but we have sufficient evidence to be sure that either of them is possible.

There is no certain example in the Actinopterygii of a single new type replacing all the former ones, but the origin of the Teleostei may perhaps show a case of this sort. It is possible that the Teleostei is a monophyletic group, arising from a single type such as *Leptolepis* and proving so successful that nearly all creatures of previous types soon disappeared.

The Acinopterygii provide also examples of perhaps an even more interesting process, the parallel evolution of a number of different lines. There can be no doubt that from Palaeozoic times onwards several independent lines of fishes have shown similar changes. The tail has become shorter and more nearly symmetrical, the body has become flattened and deepened dorsoventrally, while the pelvic fins have moved forward and the scaly armour has been reduced, all of these being signs of a more effective swimming and steering system (p. 200). As we come closer to modern times and the geological record becomes more complete we obtain more and more critical evidence that such changes have occurred in separate populations. Thus in the descendants of the holosteans we can recognize at least three such lines, that leading to the round-shaped *Microdon*, another leading through *Caturus* to *Amia*, and a third through *Leptolepis* to the Teleostei; probably there were many others. The important point is that although each line possessed peculiar specializations of its own, they all showed some shortening, development of symmetry of the tail, and thinning of the scales.

It is a considerable advance to be able to recognize such tendencies within a group. We begin to see the possibilities of a general statement on the matter. Instead of examining a heterogeneous mass of creatures called Actinopterygii we can recognize an initial palaeoniscid type and state that in subsequent ages this has become changed in certain specified ways and even at a specified rate. Imperfect though our knowledge still is, it enables us to approach towards the aim of our study, to 'have in mind' all the fishes of actinopterygian type.

This is to make the most of our knowledge: there remains a vast ignorance. We cannot certainly correlate this tendency of the fishes to change with any other natural phenomena. Put in another way, we do not know why these changes have occurred. The sea certainly did not stay the same, but it does not seem likely that its variations have been responsible for those in the fishes. It would be very valuable to be able to make a more certain pronouncement on this point, for the case is one of crucial importance. On land the conditions are constantly altering, and therefore we often find reason to suspect that those in the animals are following the environment. But can this be so in the water?

The evolutionary changes in the Actinopterygii certainly involve a definite difference in the whole life. By development of the swim bladder as a hydrostatic organ the animals have become able to remain at rest at any level of the water, and thus, by suitable modification of the shape of the body and fins, to dash about with remarkable agility in pursuit of prey or avoidance of enemies. This has enabled them to dispense with the heavy armour and thus further to increase their mobility. But what made it *necessary* to adopt these changes? Not surely any actual alteration in the sea itself. We must look then for some factor imposed on the situation by the fishes themselves or the neighbouring animals that constituted their biotic environment. Is it the pressure of competition that has been responsible for the change in fish form? It may well be that the presence of an excess of fishes has led them continually to search for food more and more actively, and in new places, with the result that those types showing the greatest ability have survived. Given the initial genetic make-up of the palaeoniscids, further agility is most easily acquired by those fishes in which competition tended to produce shorter tails, thinner scales, and the other characteristics towards which the animals of this group tended.

The fact that the same set of changes can be produced independently from several different populations of approximately similar type (and presumably genetic composition) is strikingly shown by the specialized creatures evolved for life in coral reefs. Animals with rounded bodies and small mouths, sometimes with grinding teeth, have appeared independently several times; in the Carboniferous, *Amphicentrum*; Permian, *Platysomus* and *Dorypterus*; Triassic *Cleithrolepis*; Jurassic *Microdon* and *Dapedius*, and in some modern teleosteans such as parrot fishes (Scaridae) and butterfly fishes (Chaetodontidae) (Figs. 8.3 and 9.12). This is very valuable evidence of the way in which a common factor can work on genetical constitutions that are similar but not identical. In this example the stimulus is a particular set of environmental conditions; in other cases a similar effect may be produced by competition between animals, which was probably the 'cause' of the common changes that affected so many descendants of the palaeoniscids.

The history of these fishes therefore gives plausible ground for the belief that the driving 'forces' that have produced evolutionary change are the tendencies of living things to do three things: (1) to survive and maintain themselves, (2) to grow and reproduce, (3) to vary from their ancestors, all of these operating under the further stress of any slow change in the environment.

Finally we must consider whether this change in the fishes can in any way be considered to be an advance. Several times we have found ourselves implying that this is so, that the later teleosts are 'higher' than their Devonian ancestors. We shall be wise to suspect this judgement as a glorification of the present of which we are part. However, perhaps this danger is less marked when we are dealing with fishes not ancestral to ourselves, whose 'advance' does not therefore bring them nearer to man. The judgement can be put into quite specific terms: the later Actinopterygii are 'higher' than the earlier ones because they are more mobile, quicker, and can live free in the water with lesser expenditure of energy than their ancestors. Unfortunately we have no means of estimating the total amount of biomass of fish matter that is supported by the teleostean organization, but it seems possible that it is absolutely greater than that of any previous type, say the holostean or palaeoniscid. If this is true, the change in plan of structure has perhaps led to an increase not only in fish biomass but in the total biomass of all life in the sea.

The teleostean plan has certainly allowed for the development of a great range of specialization depending on fresh genetic information and fitting the animals to all sorts of situations in the sea and fresh water. We must therefore not forget this adaptability in judging the status of the group: it seems likely that modern teleosts are more varied than any of their ancestors (p. 201). This power to enter a wide range of habitats not previously occupied is perhaps the clearest sign of all that a group has 'advanced'. It is in this sense of gaining information that suits animals to new modes of life that there has been a progress in evolution. It is true that the sea and fresh water have been in existence relatively unchanged throughout the period that we are considering: in a sense the fishes have not found a 'new' environment. But they have found endless new ways of living in the water.

9 The adaptive radiation of bony fishes

THE variety of Actinopterygii is so great that it would be impossible to give a complete idea of it and the best that we can do is to consider various functions in more detail and specify some of the ways in which the animals have become specially modified.

1. Swimming and locomotion

The teleosts have perfected in various ways the process of swimming by the propagation of waves of contraction along the body. The situation is different from that of elasmobranchs on account of the presence of the swim bladder, serving to maintain the fish steadily at a given level in the water. The stabilization of the animal during locomotion has therefore become a wholly different problem, and the fins are correspondingly changed. In the sharks the pectoral fins serve to correct a continual tendency to forward pitching and by adjustment of their position they are used to steer the animal upwards or downwards in the water.

With a swim bladder the fishes have become freed from the tendency to remain at the bottom, which was prevalent in the more primitive forms and is still so common in sharks that it has several times produced wholly bottom-living ray-like types. A fish with a swim bladder needs only very little fin movement to maintain it at a constant depth or to change its depth. The elaborate mechanism of pectoral aerofoils and a lifting heterocercal tail is no longer needed for the maintenance of a constant horizontal cruising plane Harris (1936). Concomitant with the loss of the heterocercal tail in evolution occurs a rapid and tremendous adaptive radiation of the pectoral fin in form and function. A stage in this process seems to have been the use of the paired fins to produce oscillating movements during hovering, and this is still found in *Amia* and *Lepisosteus*, fishes that remain relatively slow and clumsy (Fig. 8.2).

Many of the lower teleosteans are relatively poor swimmers and some of them, like so many elasmobranchs, have become bottom-living. Thus in the catfishes there is a large anal fin, acting, like a heterocercal tail, to give lift and negative pitch. The pectoral fins are used to balance this tendency, very much as in sharks.

In the more specialized teleosteans, however, the pectorals are placed high up on the body and are used as brakes (Fig. 9.1). The plane of the fins' expansion is vertical and they thus produce a large drag force and a small lift force. This lift, of course, tends to make the fish rise in the water when stopping, and there is also a pitching moment, depending on the position of the fin in relation to the centre of gravity, usually positive. That the fish does not rise in the water, or pitch, when it stops is apparently due to the anterior position of the pelvic fins, so characteristic of higher Actinopterygii, which has puzzled many morphologists. Experiments on the sun fish (*Lepomis*) have shown that after amputation of the pelvic fins the fish rises in the water when stopping and raises its head (positive pitch). In fact the pelvic fins are able to tilt the nose downwards. By alterations in their position they can be used to control the rising or diving movements and turning one of them outwards produces rolling (Fig. 9.1). It has been suggested that the pelvic fins function as keels to prevent rolling, but their amputation in *Lepomis* does not produce excessive rolling. Stability in the transverse plane is presumably assured by the dorsal and anal fins. The use of the fins for stopping was also developed in some Mesozoic fishes, for instance in the Triassic coelacanth, **Laugia*, which possessed high pectorals, and pelvic fins in the anterior position.

In the fishes with high pectoral fins, therefore, the pelvics are usually found far forward. In the flying fish (*Exocoetus*), however (Fig. 9.4), high pectoral fins are found with posterior pelvics. In this position the pelvics would tend to help rather than hinder any tendency by the pectorals to produce a rise.

With these increased opportunities for delicate control of movement without the devices of flattening of the front part of the body and a heterocercal tail, the bony fishes have also been able to make many other improvements in the efficiency of their swimming. 'Necking of body shape anterior to the caudal fin, together with extension of depth in other regions through dorsal and

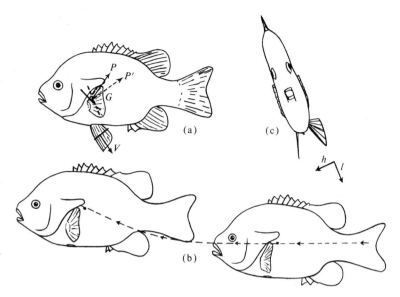

FIG. 9.1. Use of the paired fins for braking. (a) Forces produced by the fins of *Lepomis* during deceleration. The pectoral and pelvic fin planes are represented by the heavy lines, *P* and *V*, the resultant forces on the pectoral and pelvic fin respectively. Dotted line and force *P′*, condition during action of pectoral fins only, pelvic fins being held in 'neutral' position. *G*, position of centre of gravity. (b) Sun-fish, pelvic fins amputated, stopping by extending pectorals. Although the body remains horizontal, the fish rises during the stop. (c) Front view of sun-fish producing a rolling movement by the action of one pelvic fin. *h* and *l*, horizontal, and lateral forces. (From Harris 1938.)

anal fins, have helped to minimise recoil, while the body resistance that thrust must balance is minimised by their good streamlined shape' (Lighthill 1975). The shape of the body is a compromise between the optimum circular outline to give a compact form and reduce turbulences and the need to present a large lateral surface for locomotion. The body is nearly round in the fastest swimmers where the tail produces the main thrust. In tunny-fishes the tail is double-jointed. The tendons of the myotomes are so arranged that the angle of attack increases as the fin moves sideways and is maximal at the line of motion of the fish, reaching an average of 32°. Among other special features of the tail is the capacity to twist the upper and lower lobes in opposite directions.

The speed that can be reached increases with the length of the fish. Cruising speeds, that are maintained for hours are of the order of three to six times the body length per second, the relationship varying with the species. During sudden bursts the speed may be much greater. Thus Bainbridge found that 10 L/s could be maintained only for one second, 5 L/s for 10 s, and 4 L/s for 20 s (in dace, goldfish, and trout) (1961).

The locomotion of each type of fish is adapted to its habits. Most freshwater fishes are 'sprinters' but there are varying degrees of staying power. thus we may distinguish (1) typical sprinters (pike and perch), (2) sneakers (eel) with some staying power, (3) crawlers (rudd, bream), with considerable staying powers for escape, (4) stayers, either for migration (salmon) or for feeding (carp). In the fish with staying powers there is a broad strip of narrow red muscle fibres. The tunnies are the fastest swimmers (up to 70 km/h) and these speeds are possible because the heat produced by the red muscle

fibres keeps their temperature 10–20 °C above that of the water. To prevent loss of heat at the gills the veins and arteries of the branchial muscles form a counter-current heat exchange system (see Sharp and Dizon 1978).

Bathypelagic fishes, living in deep waters, below the thermocline need to overcome special problems. Here currents and turbulence are low, but since the water is cold it is very viscous, making swimming difficult but sinking slow. Many deep-sea fishes have elaborate lures, often luminescent. They may be described as 'floating fish traps'. They often have no swim bladder and achieve an almost neutral buoyancy by great reductions of the skeleton and muscles (e.g. *Ceratias*). The only parts to be well ossified are the jaws. On the other hand, deep-sea fishes that retain the swim bladder have a well-developed skeleton and powerful muscles (Marshall 1960, 1979).

2. Various body forms and swimming habits in teleosts

Departures from the streamlined body form typical of pelagic fishes have been very numerous; in nearly every case they are associated with a reduction in the efficiency of swimming as such and the development of some compensating protective mechanism (Fig. 9.2). Lateral flattening, which is already a feature of all teleostean organization, is carried to extremes in many types. Thus the angel-fish, *Pterophyllum*, often seen in aquaria, is provided with long filaments and a brilliant coloration, which, in its natural habitat (rivers of South America), give it a protective resemblance to plants, among which it slowly moves. The flat-fishes (plaice, solie, halibut, etc.)

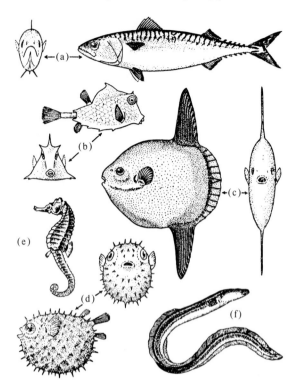

FIG. 9.2. Differences in form of fishes. (a) mackerel (*Scomber*);
(b) trunk-fish (*Ostracion*); (c) sun-fish (*Mola*); (d) globe-fish
(*Chilomycterus*); (e) sea-horse (*Hippocampus*); (f) eel
(*Anguilla*). (From Norman 1931.)

have carried this flattening to extreme lengths. They feed
on molluscs and other invertebrates on the sea bottom
and lie always on one side. The upper side becomes
darker and protectively coloured, the lower side white.
In order to have the use of both eyes the whole head is
twisted during the post-larval period. These forms are
mostly poor swimmers, but their coloration gives them a
remarkable protective resemblance to the background
(p. 210).

The John Dory (*Zeus faber*) has made a different use
of lateral flattening. The fish is so thin that its swimming
is very slow, but being inconspicuous when seen from in
front it can approach close to its prey, which it then
catches by shooting out the jaws.

Flattening in the dorsoventral plane is less common
among teleosts than selachians. The flattened forms are
mostly angler-fishes, of which there are several different
sorts; *Lophius piscatorius* (Fig. 9.3) is common in British
waters. It is much flattened, with a huge head and mouth
and short tail. It 'angles' by means of a dorsal fin,
modified to form a long filament with a lump at the end,
which hangs over the mouth. Swimming, though vigorous, is slow, and protection (both for attack and

defence) is obtained by sharp spines, protective coloration, and flaps of skin down the sides of the body,
which break up the outline. There is even a special fold
of pigmented skin over the lower jaw, serving to cover
the white inside of the mouth.

Other anglers are the star-gazers, *Uranoscopus* of the
Mediterranean and *Astroscopus* from the Western Atlantic seaboard. Their lure is a red process attached to
the floor of the mouth and they lie in wait buried in the
sand, with the mouth opening upwards and only the
eyes showing. The colour is protective, there are poison
spines, and in *Astroscopus* there are electrical organs
located near the most vulnerable spot, the eyes, and
formed from modified eye muscles (p. 205).

Other fishes abandon the swift-moving habit for the
protection afforded by the development of heavy armour, such as that of the trunk-fish (*Ostracion*) and the
globe fish (*Chilomycterus*) (Fig. 9.2). Special spinous
dorsal rays, such as those of the sword fish (*Xiphias*)
may be developed, without loss of the swift-moving
habit; indeed these fish are among the fastest swimmers.
There are many groups in which an elongated body
form like that of the eel has been developed. In *Anguilla*
itself this is associated with the habit of moving over
land. The Syngnathidae, sea-horses and pipe fishes, no
longer swim with the typical fish motion but by passing
waves along the dorsal fins. The long and often grotes-
quely cut-up body form gives a strong protective
resemblance to the weeds among which they live and
feed on mysids and amphipods. The tail of the sea-
horses has lost its caudal fin and is used as a prehensile
organ, being wrapped around the stems of sea-weeds for
attachment.

Evidently the mastery that the Actinopterygii have
acquired in the water has depended to a large extent on
the freedom given by the swim bladder as a hydrostatic
organ. This gives special interest to the question of how

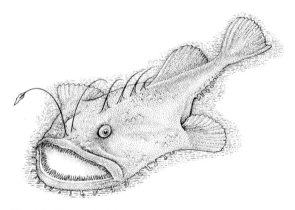

FIG. 9.3. The angler-fish, *Lophius*.

this use first began. If we are right in suggesting that the air or swim bladder was at first a respiratory diverticulum of the pharynx we can suppose that its value as a hydrostatic organ gave special opportunities to its possessors. The fishes born with the organ better developed found themselves with an advantage. This is the orthodox Darwinian interpretation. Those that were active swimmers, continually venturing into new waters, would be able to make full use of the new organ.

Many different fishes are able to jump out of the water, presumably to escape enemies. Salmon and tarpon can jump to 3 m above the water. The flying fishes have special structures to assist in such jumps. In *Exocoetus* (Fig. 9.4 (j)) the enlarged pectoral fins serve for gliding for distances up to 400 m, but in the flying gurnards (*Dactylopterus*) they are actually fluttered up and down, though the flight is feeble.

Several types of fish have the pectoral fin modified to allow 'walking'. The mud-skipper (*Periophthalmus*) uses

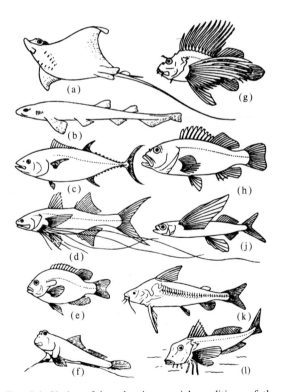

FIG. 9.4. Various fishes showing special conditions of the pectoral fins. (a) eagle-ray (*Myliobatis*); (b) dog-fish (*Scyliorhinus*); (c) tunny (*Thunnus*); (d) thread fin (*Polynemus*); (e) sun-fish (*Lepomis*); (f) mud-skipper (*Periopthalmus*); (g) scorpion fish (*Pterois*); (h) cirrhitid fish (*Paracirrhites*); (j) flying-fish (*Exocoetus*); (k) catfish (*Doras*); (l) gurnard (*Trigla*). (From Norman 1931.)

the pectoral fins, provided with special muscles acting as levers, to chase after crustaceans and insects. The prey is then caught with its teeth. The cornea is curved for vision in air (Sponder and Lauder 1981).

Fishes that live in situations from which they are likely to be carried away develop suckers. Thus in the gobies, found between tide-marks, the pelvic fins form a sucker. The cling fishes (*Lepadogaster*) are another group with the same habit. The remoras have developed a sucking plate from the first dorsal fin and by means of this they attach themselves to sharks and other large fish. In order to catch their food they leave the transporting host, though they also feed on its ectoparasites.

3. Adaptations to various depths

The effects of depth have caused the fishes to vary more in bodily form and physiology than any other vertebrates (Fig. 9.5). The fishes of the sunlit waters of the epipelagic zone are the most familiar to us. Many are predators, terminal members of the food chains leading from the phytoplankton. The mesopelagic zone includes a great number and variety of fishes many migrating up and down hundreds of metres every night to the euphotic zone to feed. These animals must therefore be able to withstand enormous changes of pressure, which increases by one atmosphere every 10 m. They show adaptations to vision in dim light. Many have large eyes, sometimes tubular and with long rods and layered retinas giving special sensitivity (p. 179). They have special means of concealment during the day, often by transparency (when young) or silvery sides and luminous organs facing downwards to conceal their dark outline from below (Denton, Gilpin-Brown, and Wright 1972). A special part of the retina may be devoted to matching the animal's own luminosity with that from above.

In the bathypelagic zone below 1000 m there is constant darkness except for the flashes of bioluminescent animals. Here fishes are fewer. The eyes are small and even degenerate (p. 179). The main receptors are those for smell and taste and the lateral line organs are often on the surface and not in canals. Since the fish hangs motionless there is no noise due to its own movement. There is no longer any need for camouflage by photophores but these may have other uses, as lures by angler fishes or to find a mate, which must be especially difficult in darkness. Some species are hermaphrodite and in ceratioid anglers the males are parasitic, with precocious development of olfactory and visual organs with which to find a mate. The male then becomes parasitically attached by its pincer-like teeth, followed by fusion of tissue and circulatory systems.

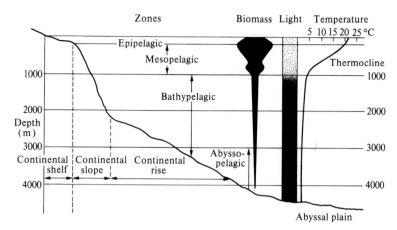

FIG. 9.5. Diagram showing temperature, light penetration, and biomass in the seas. (After Marshall 1971.)

The individuals of both sexes remain immature until they meet, which may occur soon after metamorphosis or much later (Pietsch 1975).

The colour of benthic fishes is often dark brown or black and many crustaceans are red. The malacosteoid fishes have red photophores and red-sensitive retinas, allowing them to see the crustaceans! (Denton, Gilpin-Brown, Wright 1972).

Because of the density of the water, movement requires great energy expenditure, and, many bathypelagic fishes probably move little and function as floating food traps rather than as hunters. There is no water disturbance so the body is flimsy, with little mineral in the bones and weak muscles. The jaws are enormous with folding teeth and a stomach so distensible that the fish can consume prey larger than itself.

The benthic fishes, living entirely on the bottom are usually without swim bladders. Others swim just above the bottom ('benthopelagic') and these maintain neutral buoyancy by a bladder.

4. Structure of mouth and feeding habits of bony fishes

Although perhaps the majority of fishes are carnivorous, there are species with all sorts of other methods of feeding. The more active predators have strong jaws and sharp teeth, such as those of the pike (*Esox*), cod (*Gadus*), and very many others. The teeth on the edge of the jaw serve to bite and catch the prey, those on the walls of the pharynx to prevent its escape if, as is often the case, it is swallowed whole. The teeth can often be first lowered to allow entrance of the prey and then raised to prevent its exit (e.g. in *Lophius*). In connection with this habit the walls of the oesophagus and even stomach are often composed of striped muscle, capable of quick and powerful contraction. In the lophioid anglers the first dorsal spine is modified to make a movable lure. In one species this is a good model of a

little fish and in deep sea it is luminous. This fishing with a bait is a very economical, energy-saving tactic.

Many carnivorous fishes are very fierce. For instance, the blue fish (*Pomatomus*) of the Atlantic move in shoals, cutting up every fish they meet, making a trail of blood in the sea. The barracuda (*Sphyraena*) of tropical waters may attack man. They are said to chase shoals of fish into shallow waters and to keep them there to serve for food as required.

Other fishes feed on invertebrates and are then usually bottom-feeders. Thus the plaice (*Pleuronectes*) has developed chisel-like teeth on the jaws and flattened crushing teeth in the pharynx; it feeds largely on bivalve molluscs. The Labridae (wrasses) also have blunt teeth and eat molluscs and crabs. The sole (*Solea*) has a weaker dentition and eats mostly small crustacea and worms. Fish such as the herring (*Clupea*) that live on the minute organisms of the plankton have small teeth and weak mouths, but are provided with a filtering system of branched gill rakers, making a gauze-like net, comparable with the filtering system found in basking sharks (p. 157), paddle fish (p. 193), and whale-bone whales (p. 499).

Herbivorous and coral-eating fishes have crushing teeth similar to those of the mollusc-eaters; indeed, many forms with such dentition will take either form of food. The parrot-fishes (Scaridae) have a beak and a grinding mill of flattened plates in the pharynx. With this they break up the corals, rejecting the inorganic part from the anus as a calcareous cloud. The Cyprinidae, including many of our commonest freshwater fishes (goldfish, carp, perch, and minnow), have no teeth on the edge of the jaw, hence the name 'leather-mouths'. There are, however, teeth on the pharyngeal floor, biting against a horny pad on the floor of the skull. These fishes are mainly vegetarians, but many take mouthfuls of mud and extract nourishment from the plants and invertebrates it contains.

5. Protective mechanisms of bony fishes

In general teleosts depend for protection against their enemies on swift swimming, powerful jaws, good receptors, and brain. The majority of them have thus been able to abandon the heavy armour of their Palaeozoic ancestors. In many cases, however, subsidiary protective mechanisms have been developed, and are especially prominent in fishes that have given up the fast-swimming habit and taken either to moving slowly among weeds or to life on the bottom. These developments are a striking example of the way in which, following adoption of a particular mode of life, appropriate subsidiary modifications take place, presumably by selection of those varieties of structure that are suited to the actions of the animal.

These protective devices may be classified as follows:
1. Protective armour of the surface of the body.
2. Sharp spines and poison glands.
3. Electric organs.
4. Luminous organs.
5. Coloration.

6. Scales and other surface armour

The typical cycloid teleostean scales have already been described (p. 163). They form a covering of thin overlapping bony plates, providing some measure of protection, but not interfering with movement. The hinder edges of the scales are sometimes provided with rows of spines, and are then said to be ctenoid. In some fishes the scales bear upstanding spines and possess a pulp cavity, which recalls that of denticles. In the tropical globe fishes (or puffers) and porcupine-fishes (_Diodon_) there are very long and sharp spines and the puffers are able to inflate themselves and cause the spines to project outwards, a very effective protective device. In a few fishes there are bony plates making an armour even more complete than that of the Palaeozoic fishes. In the trunk- or coffer-fishes (_Ostracion_) the thick plates form a rigid box, from which only the pectoral fins and tail emerge as movable structures, the former apparently assisting the respiration, the latter the swimming. These fishes live on the bottom of coral pools and have a narrow beak with which they browse on the polyps.

7. Spines and poison glands

More than 200 species of fishes have poison glands associated with spines. Most of them are slow swimmers often living among rocks or corals or in the sand. The poisons are therefore used for protection rather than predation (as in snakes). Thus the European weever (_Trachinus_) lives buried in the sand and has poison spines on the operculum and the dorsal fins. It is suggested that the dark colour of the fins serves as a warning. Some catfishes, scorpion fishes, and toad fishes also have poison spines. The stargazer, _Uranoscopus_, of the Mediterranean and tropical waters has powerful spines with poison glands on the operculum, which inflict a most unpleasant wound if the animal is disturbed by hand or foot while it lies in the sand angling for its prey (p. 202). Many fishes have spines but are not venomous. _Lophius_, the monkfish angler, is also armed with dangerous spines. Several species of catfish have large spines, sometimes serrated. In the trigger fishes (Balistidae and related families) of the tropics one or more of the fins is modified to make a spine that can be raised and locked in that position. These fishes have very brilliant coloration, but since some of them live in the highly coloured surroundings of coral reefs it is not certain that the colours serve as a warning.

More than 700 further species of fishes are poisonous, again mostly reef dwellers. Often the toxicity depends on the diet of the fish, as in the so-called ciguatera toxin of various tropical coralfishes. Tetrodotoxin (TTX) of certain puffers, trunk fishes and others is one of the most powerful poisons known. It kills cats at an oral lethal dose (LD_{50}) of 200 μg/kg, by a specific nerve blocking action. It causes the death of many humans in Japan and elsewhere.

8. Electric organs

Specializations for the production of electric fields outside the body have arisen independently in six families of teleosts and in the torpedoes and rays (Bennett 1971). Strong fields are used for attack or defence, weak fields for electroreception or communication, especially in muddy rivers. The currents are mostly produced by modified muscle end-plates but in the teleostean Stenarchidae by nerve fibres themselves. Strong shocks are produced when the fish is attacked or when prey is detected; weak discharges are usually continuous with variation in frequency for communication.

The only marine electric fish are the elasmobranchs and _Astroscopus_, also a stargazer (Fig. 9.6). In _Torpedo_ the organs are formed from branchial muscles and innervated by cranial nerves. They produce up to 50 V (in water) and are used to stun larger fishes, which are then eaten. The electric organs of rays are in the tail and are weak, producing only a few tens of millivolts. _Astroscopus_ is an angler fish living in the sand on the east coast of America. The organs lie behind the eyes, formed from eye muscles. They produce up to 5 V and are active when small fish pass over, perhaps stunning them.

The Gymnotidae of South American rivers include species with both strong and weak organs. The electric eel, _Electrophorus_ produces a shock of up to 500 V by the

STRONGLY ELECTRIC

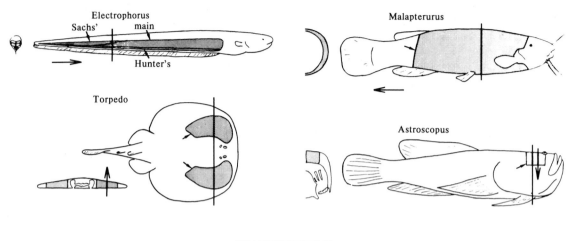

WEAKLY ELECTRIC

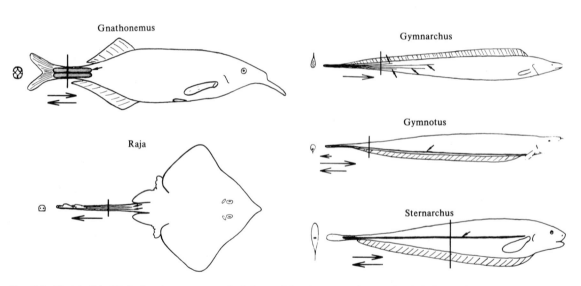

FIG. 9.6. Electric fish. Their electric organs are stippled or solid and cross-sections, at levels indicated by solid lines. (Bennett 1968.) Arrows show the direction of current flow.

electric organs along the sides of the tail, innervated by spinal nerves. These fishes swim mostly by movements of the anal fin, the body musculature being mainly converted to an electric organ. *Gymnotus* produces only a fraction of a volt as impulses of about 50/s. This may accelerate up to 200 or stop altogether ('listening') and these changes are involved in communication.

The stenarchids are gymnotids in which the electric currents are produced not by muscles, but by modified axons. These end blindly in tissue that was probably muscular in their ancestors. The axons run first forward

and then back (Fig. 9.7). They enlarge to about 100 μm and in two regions the nodes of Ranvier are enlarged, with folded axonal extensions (Fig. 9.8). These act as a series capacitor for the current that is set up by the normal, short nodes (Waxman 1975). These organs discharge at up to 1700 Hz, a higher frequency than in any other fish.

In the electric catfish *Malapterurus* the electric organ lies under the skin all along the body. It was formerly supposed to derive from glands, but is actually muscular. The whole organ is innervated by branches of one

(a)

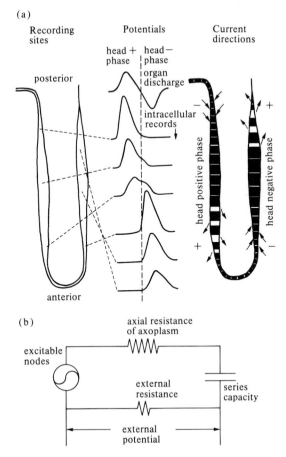

(b)

FIG. 9.7. Axonal electric organ in *Sternarchus*. (a) The top record is of the organ discharge recorded externally. Below are intracellular records at sites along the axon. The narrow nodes in the more proximal part pass inward current in the head positive phase, those in the more distal part in the head negative phase. The large nodes are inexcitable. (b) Equivalent circuit of an electrocyte segment. (After Waxman 1975.)

large neuron on each side and produces shocks of up to 150 V when prey is detected. The mormyrids of tropical African rivers have weak electric organs in the tail; used for electroreception of novel situations and for communication.

The muscles that form electric organs are modified to varying degrees. Some show irregular striations. They are arranged as piles of plates (electroplaques) innervated on one surface only, thus allowing series summation in the pile, while separate piles are in parallel, the whole giving currents of up to 600 W. To be effective the discharge must be synchronous and to ensure this the command neurons are electrically coupled. To allow for differences in conduction time the axons sometimes run exaggerated courses, or differ in

diameter or in end plate delay (Fig. 9.9). It is still uncertain how the fishes avoid shocking themselves. There are sometimes special sheaths of fat, so thick in *Malapterurus* that the nerve fibres are up to 1 mm in diameter.

9. Electroreceptors

All fishes with weak electric organs are very sensitive to applied fields and practise electrolocation (Scheich and Bullock 1974). In addition some catfish and many elasmobranchs have electroreceptors but not electric organs. They may respond to the fields of their own muscles or those of other fishes ('passive electroreception'). Some fishes have two or more sets of receptors, some sensitive to the high frequency used for electrolocation and the others insensitive to this and used passively. The more specialized receptors reach thresholds of 0.2 V/cm or even less.

The electroreceptors are all modified lateral line organs, innervated from the eighth nerve (p. 131). Electric sensitivity is drastically reduced if these nerves are cut. The receptor cells contain synaptic vesicles, whose action maintains a tonic discharge across their synapse with the nerve endings. The applied field depolarizes the inner face of the receptor cell and increases the rate of transmitter release. The receptors are of two types *ampullary*, which open by a pore and *tuberous*, which are closed by a plug. The ampullary receptors have very high resistance walls, the cells being connected by zonular junctions. They are rhythmically active and sensitive to low frequency (0.1–10 Hz) or d.c. stimuli. The tuberous receptors are phasic, often discharging in synchrony with the electric organ. They are sensitive only to high frequencies (300–1000 Hz), giving brief responses to step changes in voltage. They are surrounded by many layers of cells sealed by tight junctions, thus presumably reducing the capacity of the wall.

The pattern arrangement of the receptors and their lengths and other characters allow the fish to obtain directional information. Thus in the ampullae of Lorenzini (Fig. 9.10) the high resistance of the cores ensures that the stimulus to the receptor is greatest if the canal is orientated in the direction of the field and zero if perpendicular to it.

The pattern of impulses produced in response to a fish's own electric discharges varies with the presence of insulators and conductors near by, but the details of electrolocation are still not known. The mormyrids have an exceptionally large cerebellum and the command signal for the electric organ is sent to this and may serve to determine the latency of the receptor response (see p. 176).

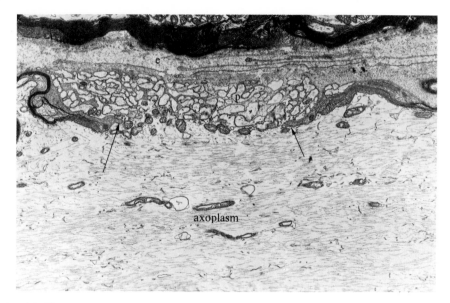

FIG. 9.8. *Sternarchus.* The gap at a large node, showing the surface of the axon extended by many irregular processes (p): *Ax* is the axoplasm, *e*, the extracellular space. (×7500) (After Waxman 1975.)

Various behavioural responses have been seen in electric fish. Aggressive species such as *Electrophorus* will attack metal rods but not insulators. Most weak electric fish swim with the body rigid, to simplify the process of electrolocation. The fish often moves backwards when investigating. Fishes may sense the potentials induced as they swim through the earth's magnetic is reputed to show particular behaviour patterns when an earthquake is impending (see Adey and Bawin 1977).

10. Luminous organs

Fishes of many different families live at great depths and 95 per cent of individuals caught below 50 m are luminescent. The development of luminous organs is therefore a further example of parallel evolution. The

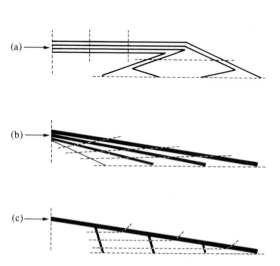

FIG. 9.9. Methods of compensatory delay to ensure simultaneous discharge. (a) Equalization of arrival time by devious pathways. (b) Conduction is faster in thicker axons. (c) Local compensatory delays in thin terminal branches. (Bennett 1968.)

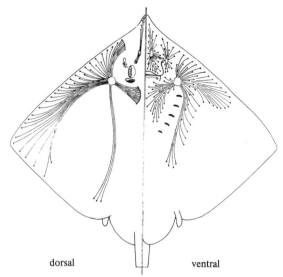

dorsal   ventral

FIG. 9.10. Ampulla of Lorenzini of the skate (*Raja*). The openings of the canals are indicated by dots. (Murray, R. W. (1960). *Journal of experimental Biology* **37**, 417–24.)

organs usually show as rows of shining beads of various colours (mostly blue-green) on the sides and ventral surface of the fish (Nicol 1969). The light intensity of lantern fishes was found to be $1 \times 10^{-6} \mu\text{W/cm}^3$ on a receptor placed at one metre. This should be visible to other fishes at up to 16 m in the very clear mesopelagic water.

In many species the light is due to organs containing luminous bacteria, whose appearance may be controlled by the movement of a fold of skin, or of the whole organ, or of chromatophores. Some teleosts, however, have self-luminous photophores and these are also found in *Spinax* and a few other Squalidae. They are formed from modified mucous glands, and may be provided with reflectors of guanine crystals and even lenses. They can be flashed on and off, probably by sympathetic nerve stimulation. Injection of adrenaline produces flashes in some species.

The luminous organs probably often serve for recognition of the sexes and then show distinctive patterns. They may serve to startle attackers and in a few cases to illuminate the prey. *Pachystomias* produces red light for the purpose, to which most fishes are insensitive. In the deep-sea anglers (*Ceratias*) the luminous tip of the fin is used as a lure. In many mesopelagic fishes the luminous organs are on the underside and serve to hide the silhouette of the fish as seen from below. There may be light organs around the oesophagus or rectum, serving, with reflectors to lighten the whole under surface of the fish. In others the light provides disruptive camouflage (see Herring 1978; Marshall 1979).

11. Colours of fishes

The bony fishes show perhaps the most brilliant and varied coloration of any animals, rivalling even the Lepidoptera and Cephalopoda in this respect. The enormous range of colour and pattern provides an excellent example of the detailed adjustment of the structure and powers of animals to enable them to survive. A great difficulty is introduced into the study of animal coloration by the fact that we are usually ignorant of the capacity for visual discrimination possessed by the animals likely to act as predators. Moreover, it is very difficult for us to obtain this information. When we examine any two objects we are able to say not merely that they are different but that one is red and the other green. A person or animal that is colour-blind may also be able to detect a difference, but yet remain unaware of any distinction of colour; the objects appear to him only as differing in brightness. In order to decide whether animals are able to distinguish between light of two wavelengths we must present them

with objects of different colour but the same brightness.

We are therefore faced with the possibility that some of the colours that appear to us so brilliant are to other animals merely differences of tone, and animals to us conspicuous because coloured, when seen in monochrome, may be protected. Some of the colours of fishes may be only a means of producing a pattern of protective greys, as seen through the eyes of an attacker. However, there is no doubt that some fishes are able to discriminate between illuminated bodies which though of different wavelength reflect light of equal brightness. In the subsequent description of fish coloration we shall not be able to consider predators further, but shall describe the colours as they appear to the eye of a normal man.

The colour of fishes is produced by cells in the dermis (i) the chromatophores and (ii) the reflecting cells or iridoctyes (Fig. 9.11). The chromatophores are branched cells containing pigment granules (melanosomes) which may be either black (melanin) or white, red, orange, or yellow (carotenoids or flavines). The iridocytes contain stacks of platelets of guanine each with an optical thickness close to a quarter wavelength of light so as to produce, where no chromatophores are present, either a white or a silvery appearance. This material is used in the manufacture of artificial pearls, the scales of the cyprinoid *Alburnus lucidus* (the bleak) being used for the purpose. The iridocytes may be either outside the scales, when they produce an iridescent appearance, or inside them, giving a layer, the argenteum, that produces a dead white or silvery colour. By a combination of the chromatophores, and of these with

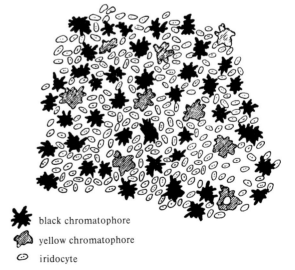

✳ black chromatophore

▨ yellow chromatophore

⌀ iridocyte

Fig. 9.11. Coloration elements in the skin of the upper side of a flounder. (*Platichthys*). (After Norman 1931.)

the iridocytes to produce interference effects, a wide range of colour is achieved. Thus by mixing yellow and black either brown or green is produced. Blue is usually an interference colour. Pelagic silvery fish use mirror camouflage by reflection from their iridocytes (Denton 1970).

The use of colour by the fish may be classified as cryptic or concealing, sematic or warning patterns, and epigamic or sex coloration. Cryptic coloration may be achieved in various ways and may be subdivided into two main types: (1) assimilation with the background, (2) breaking up the outline of the fish. Assimilation is common, but is often associated with some degree of disruption of outline. The absence of all pigmentation in pelagic fishes, for instance the leptocephalus larvae of eels, is an example of assimilation. Fishes living among weeds, such as the sea-horses and pipe fishes, or *Lophius* the angler (Fig. 9.3), often resemble the weeds in colour, and in addition develop 'leaf-like' processes. The colour of many familiar fishes, such as the green of the tench,

may be said to resemble that of the surroundings by assimilation. When we consider the much more numerous examples of patterns involving several colours the distinction between assimilation and disruption is more difficult to draw. Many free-swimming pelagic fishes have the upper side dark and striped with green or blue, whereas the underside is white, the beautiful pattern that is seen in the mackerel (*Scomber*). This gives them protection from above and below, the striping probably making the animal less conspicuous in disturbed water than it would be if of uniform colour. The white underside also serves to lessen any shadows, an important factor for animals that live in shallow water; similar shading is used by land animals.

Devices of spots and stripes are found on fishes that live against a variegated background (Fig. 9.12). The beautiful red and brown markings of a trout are a good example. Flat fishes, living on sandy or gravelly bottoms, adopt a spotted pattern, which gives them a high degree of protection, and we shall see later that they are

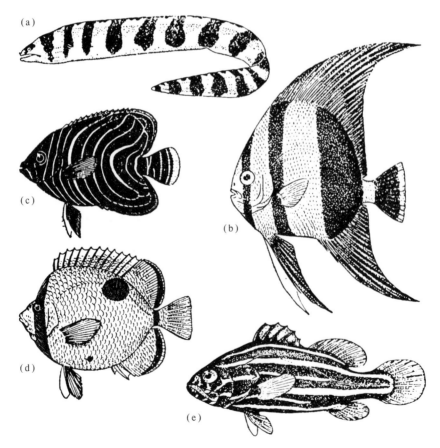

FIG. 9.12. Colour patterns of various tropical fishes. (a) Muraena, *Gymnothorax*; (b) bat-fish, *Platax*; (c) butterfly fish, *Holacanthus*; (d) butterfly fish, *Chaetodon*; (e) perch, *Grammistes*. (From Norman 1931.)

able to change colour to suit the ground on which they rest. The brilliant colours of many tropical fishes probably serve mainly to break up the outline, though no doubt the surroundings in which they live are also brilliant. Great variety of colours may be found on a single fish, especially in the trunk fishes (*Ostracion*), one species of which is described as having a green body, yellow belly, and orange tail, while across the body are bands of brilliant blue, edged with chocolate-brown. Moreover, the female has another colour scheme and was for long considered as a different species!

Colour differences between the sexes are frequent in fishes, the male being usually the brighter. Thus in the little millions fish, *Lebistes*, there are numerous 'races' of males with distinctive colours, but the females are all of a single drab coloration. The genetic factors that produce the various types of male are carried in the Y chromosome. Presumably the colour of the males acts as an aphrodisiac as a part of the mating display, but the significance of the different races is not known.

Sematic or warning coloration involves the adoption of some striking pattern that does not conceal but *reveals* the animal. This type of colouring is found in animals that have some special defence or unpleasant taste (such as the sting of the wasp), and its use implies that animals likely to attack are able to remember the pattern and the unpleasant effects previously associated with it. It is not easy to be certain when colours are used in this way, but it is possible that the conspicuous spots on the electric *Torpedo ocellata* have this function. Among teleosts there is the black fin of the weevers (*Trachinus*), possibly a warning of their poison spines, and the spiny trigger fishes and globe fishes (p. 205) also have conspicuous colours.

12. Colour change in teleosts

In spite of the reputation of the chameleon the teleosts are the vertebrates that change their colour most quickly and completely. The melanophores are provided with nerve fibres (Fig. 9.13), and these cause paling of the skin by a movement of the pigment granules (melanosomes) perhaps guided by microtubules and drawn by microfilaments or along an electrophoretic gradient that exists within the cells (Collis 1979).

The nerve fibres in question are postganglionic sympathetic fibres, leaving the ganglia in the grey rami communicans (Fig. 7.33) to all the cranial and spinal nerves. The preganglionic fibres that operate them, however, emerge only in a few segments in the middle of the body (Fig. 14.18), so that severance of a few spinal roots will affect the colour of the whole body. The transmitter involved is noradrenaline, which produces contraction of the melanophores of isolated scales.

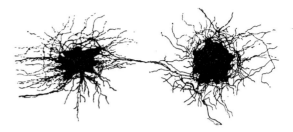

FIG. 9.13. Nerves of the melanophores of a perch. (From Ballowitz.)

Some workers suppose that there are also melanophore-expanding nerve fibres, but the evidence is not convincing (Pye 1964).

There is, however, another agent that causes expansion of melanophores in a wide variety of vertebrates, namely the melanophore stimulating hormone (MSH) of the pars intermedia of the pituitary (see p. 94), and there is evidence that this works also in teleosts. Hypophysectomized specimens of the Atlantic minnow *Fundulus* are nearly always lighter than normal individuals, especially when on a dark background. Injection of posterior pituitary extracts, or placing of isolated scales in the extract, causes expansion of the chromatophores of many teleosts.

It is probable, therefore, that colour change is mainly produced by the nerve fibres tending to make the animals pale and secretion of the posterior pituitary to make them dark, but other influences may be involved. Injection of the pineal hormone melatonin causes aggregation of pigment in some larval fishes, though not in adults. The leucophores (white chromatophores) expand when the melanophores contract. The iridocytes do not change. The yellow and red pigments (in xanthophores and erythrophores) probably change under MSH control.

It is more difficult to decide how external influences are linked with this internal mechanism. Fishes mostly become pale in colour on a light background and vice versa, and the effect is produced predominantly through the eyes. The change in colour begins rapidly but its completion may take many days. We can therefore distinguish between colour changes that are transient ('physiological') and those that are long-lasting (also called 'quantitative' or 'morphological'). The latter depend upon the formation or loss of pigment or pigment cells.

The value of the colour change in bringing the animals to the same tint as their surroundings is considerable. Fishes kept on a light background are very conspicuous for the first few minutes when transferred to a dark one. The fisherman acknowledges this by painting the inside of his minnow-can white, to make the bait conspicuous. In the flat fishes, living on the sand, the protection assured by the colour change is of special importance. It has been suggested that it is possible for a flat fish to assume a pattern similar to that of the ground on which it lies, but it is probable that the degree of expansion of the chromatophores is adjusted to suit the amount of light reflected from the ground; by increasing or decreasing the areas of dark skin, effects approximately appropriate to various backgrounds are produced.

13. Aerial respiration and the swim bladder

Many fishes are able to live outside the water. The excursions on to the land vary from the wriggling of the eel through damp grass to the life of the Indian climbing perch (*Anabas*) spent almost entirely on land. In the eel there is no special apparatus for breathing air (though oxygen may be taken in through the skin). The climbing perch is provided with special air chambers above the gills (Fig. 9.14) and even when in water it comes to the

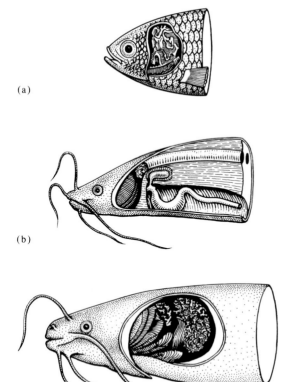

(a)

(b)

(c)

FIG. 9.14. Special respiratory apparatus. (a) in climbing perch (*Anabas*); (b) Indian catfish (*Saccobranchus*); (c) African catfish (*Clarias*). (From Norman 1931.)

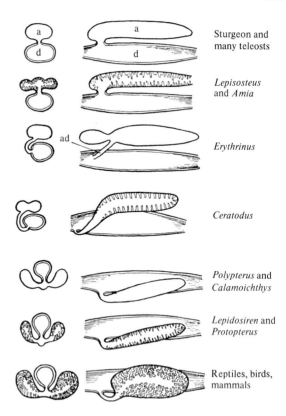

Sturgeon and many teleosts

Lepisosteus and *Amia*

Erythrinus

Ceratodus

Polypterus and *Calamoichthys*

Lepidosiren and *Protopterus*

Reptiles, birds, mammals

FIG. 9.15. Air-bladder of various fishes, seen from in front and from the left side. *a*, air- or swim-bladder; *ad*, air-duct; *d*, digestive tract. (After Dean 1895.)

surface to gulp air and will 'drown' if prevented from doing so, even though it is placed in well-oxygenated water.

Many other fishes gulp air, especially those living in shallow tropical waters, which readily become deoxygenated. There may be other special mechanisms for gaseous interchange. In the Indian catfish *Saccobranchus* there are large air sacs growing a long way down the body from the gill chambers (Fig. 9.14).

The swim bladder, which has contributed so largely to the success of the later teleosts, may have arisen as an accessory respiratory organ, used in the same way as those described above. In all the more primitive teleosts (Isospondyli) the swim bladder preserves in the adult its opening to the pharynx ('physostomatous'), whereas in higher forms it becomes completely separated ('physoclistous'). Survivals of still earlier Actinopterygii have the opening especially well developed, though it varies from group to group (Fig. 9.15). Thus in the sturgeons there is a wide opening into the dorsal side of the pharynx. In *Amia* and *Lepisosteus* the opening is also dorsal and the walls of the sac are much folded and used for respiration. In *Polypterus* the opening is ventral and the bladder has the form of a pair of lobes below the gut. This arrangement recalls that of the tetrapod lungs and is also found in the modern lung-fishes and presumably in their Devonian ancestors, from which we may suppose that the tetrapods arose (p. 221). This ventral position of the swim bladder is one of the features that still leads many zoologists to suppose that *Polypterus* was a member of the crossopterygian line of fishes. It is probable, however, that the affinity is only that which persists between all primitive members of both Actinopterygii and Crossopterygii and is to be taken as an indication that the swim bladder was originally a widely open respiratory sac, or perhaps pair of sacs. Once the power to produce a pharyngeal diverticulum had been developed it is easy to imagine that the actual position of the opening might shift either dorsally, as in the later Actinopterygii, or ventrally, as in the tetrapods. The blood supply of the swim bladder should provide some indications both of its origin and function. In *Polypterus*

and Dipnoi there are pulmonary arteries springing from the last (sixth) branchial arch and presumably containing venous blood (Fig. 9.16). Blood returns to the heart by pulmonary veins. Essentially the same arrangement is found in *Amia*, but in all other Actinopterygii oxygenated blood is supplied to the bladder from the dorsal aorta (or sometimes from the coeliac artery).

14. Buoyancy

Probably, then, the original function of the swim bladder was respiratory, and this may still be its main function not only in *Amia* and *Lepisosteus* but also in some of the physostomatous Teleostei. However, in the majority of teleosts its dorsal position, closed duct, and arterial blood supply show that it has some other function and it has long been supposed that this is concerned in some way with flotation. The swim bladder is absent from bottom-living forms, such as flat-fishes, *Lophius* and *Uranoscopus*, though it may be present in their pelagic larvae. Negative buoyancy can be an

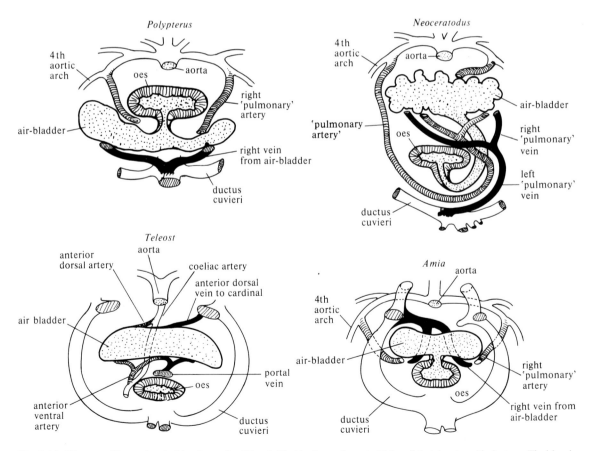

FIG. 9.16. Diagrams illustrating the blood-supply of the air-bladder in a palaeoniscid, lungfish, teleost, and holostean. The blood-vessels are seen from behind, and cut short in transverse section. oes., oesophagus. (From Goodrich 1930.)

advantage in keeping the fish against the bottom. A sucker assists in fishes living in streams, or in coastal pools, for example gobies (*Gobius*).

The bladder is usually filled with gas from glands but in some physostomatous forms by gulping air. The swim bladder acts as a hydrostatic organ by replacing about 5 per cent in marine and 7 per cent in fresh water of the volume of the fish body with gas, making it neutrally buoyant at a particular depth. The gas pressure can be varied but only slowly. Accordingly the bladder is best developed in epipelagic fish living in the upper 200 m. It is absent however from some rapid swimmers such as mackerel. It is present in some mesopelagic ones, living down to 1000 m, but is absent in bathypelagic forms presumably because even small changes in depth would be expensive to compensate. Such fishes move little and achieve nearly neutral buoyancy by lightening the skeleton. The bladder is found again in benthopelagic fishes (Macrouridae, rat tails, grenadiers), which are more robust and move to catch animals living on the relatively rich food supply. The bladder may be present even at the greatest depths (*Bassogigus profundissimus* at 7160 m; Nielsen and Munk 1964).

In fishes near the surface the contents resemble air, but at greater depths there is relatively more oxygen, up to 95 per cent. However nitrogen is the main gas in Salmonidae, even at great depths and in the freshwater whitefish (*Coregonus*) living at 100 m it is pure nitrogen. The secretion of an inert gas at ten times atmospheric pressure must be a physical process. The gases are usually produced by a special red gas gland and the mechanism involves secretion of lactic acid into the blood, causing liberation of oxygen. The blood vessels form a rete mirabile in which the blood flows in opposite directions in the arteries and veins (Fig. 9.17) (Schmidt-Nielsen 1979). The gas gland secretes lactic acid which has a large salting-out effect on teleost haemoglobin, reducing its oxygen-carrying capacity (the Root effect). The blood entering the arterial capillaries has at most 10 ml of O_2 per 100 ml, but the lactic acid causes the tension to be higher in the venous capillaries and so oxygen diffuses back into the arterial capillaries. Very high concentrations can thus be built up if the capillaries are long enough, but it is not known exactly how the gas is discharged into the bladder. The salting-out effect on gaseous solubility produces a similar, but less marked, concentration of nitrogen and even argon. This counter-current multiplication allows secretion into the bladder even against a pressure of 600 atm (at 3000 m). The capillaries are very long in fish and especially so in those living at great depths, reaching 25 mm in *Bassozetus*.

Reduction of the gas pressure can be done rapidly by letting out gas through the duct, if present, and is readily achieved by absorption into the blood. Indeed special mechanisms are needed to prevent loss. Any increase of oxygen tension in the veins causes greater diffusion to the arteries and the rete thus acts as a gas trap. Moreover, the wall is avascular and often has a layer of crystals of guanine, in which oxygen is little soluble, hence the characteristic shiny appearance. Removal of gas is achieved by a special vascular area, the oval, which is often enclosed in a special chamber guarded by a sphincter. During gas secretion the blood vessels of the oval are closed off. Secretion and absorption are controlled by nerve fibres from the vagus and sympathetic nerves.

Adjustments of volume in either direction are slow. Even near the surface a fish could maintain neutral

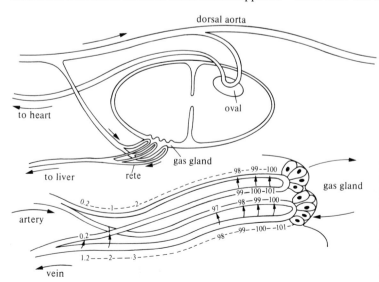

Fig. 9.17. Diagram of the circulation of the swim-bladder and the countercurrent multiplier system. Blood may reach the swim-bladder through the rete mirabile (supplying the gas gland) or by a vessel to the oval (where fine blood vessels can absorb gases). As the gas gland produces lactic acid, the oxygen tension increases in the venous capillary and gas diffuses to the arterial capillary. (After Schmidt-Nielsen 1979.)

buoyancy only if it descended at about 2–5 m/h. After rising the bladder expands and the fish must swim to keep down. Some fishes make large daily migrations, thus lantern fishes (*Myctophidae*) come to the surface at night but spend the day at 300 m or deeper. They are found to be neutrally buoyant when at the surface and probably the bladder is compressed at night. In some lantern fishes the bladder is present only in juveniles; in adults it atrophies and is replaced by waxy esters which give the fish about the same density as sea water.

The various diverticula connecting the bladder with the ear (and the Weberian ossicles, p. 182) are associated with pressure receptors that assist in the control of the bladder. In clupeoids the bladder is especially adapted to assist with hearing (p. 180). Loaches are famous as fish barometers, whose behaviour can be used to predict weather changes.

15. Special reproductive mechanisms in teleosts

The teleosts show great variation in breeding habits (see Breder and Rosen 1966). The eggs are sometimes left to develop entirely by themselves, but may be looked after by one or both parents. Some species of Sparidae and Serranidae are invariably monoecious and self-fertilizing. Some teleosts are hermaphroditic, either synchronously, with sperm and eggs ripe at once, or consecutively, being either male or female first.

The method of association of the sexes is varied and there are numerous devices for bringing the sexes together, such as colour differences, sound production, and the liberation of stimulating substances into the water (pheromones). In some deep-sea fishes the female produces scent, which is detected by the specially large nose of the male who then becomes permanently attached to the much larger female (Fig. 9.18).

Breeding is often preceded by a migration of the fishes to suitable situations and the association into large shoals. The eggs may be classified as either *pelagic,* if they float, or *demersal,* if they sink to the bottom. In the former case they are sometimes provided with an oil globule and are exceedingly numerous. Thus a single female ling (*Molva*) has been calculated to lay up to 30 million eggs and a cod (*Gadus*) 1 million, whereas the herring (*Clupea*), whose eggs sink to the bottom, probably does not lay more than 50 000 eggs and the viviparous guppy (*Poecilia*) carries less than 50 young. The large numbers laid by the pelagic species are presumably an insurance against failure of fertilization and especially against random elimination of the eggs and young.

Demersal eggs, especially of freshwater animals, are usually laid with some special sticky covering, by means of which they are attached to each other and to the

FIG. 9.18. Deep-sea fish *Photocorynus* with parasitic male attached. (From Norman 1931.)

bottom or to stones, weeds, etc. Thus the eggs of many cyprinids (carp, etc.) are attached to weeds. The eggs of salmon and trout, however, though demersal, are not sticky. From depositing eggs on weeds it is only a short step to the building of a nest and guarding of the eggs by one or both parents. Thus the sand goby (*Pomatoschistos minutus*) lays its eggs in some protected spot, where they are guarded by the male, who aerates them by his movements. Quite elaborate nests may be built, as by the sticklebacks (*Gasterosteus*), where the male again stays to care for the eggs. A still further development is the retention of the young within the body. In some catfishes they develop within the mouth of either parent. Other fishes brood the eggs in the mouth (*Tilapia*) or intestine (*Tachysurus*). In pipe fishes and sea-horses the males are provided with special pouches for the young, containing a nourishing secretion which is dependent on the presence of prolactin (p. 365).

Although external fertilization is usual, in two orders of teleosts, cyprinodonts and perciformes some species show internal fertilization and the young then develop within the ovary (*Zoarces, Gambusia, Poecilia*). The mechanisms by which mating and the nutrition of the embryos are assured show some interesting parallels with the conditions in mammals, including the formation of placentae or nutritive material. In *Poecilia* the female adopts a special position of readiness for copulation, and this has been shown to depend partly on an internal factor in the female and partly on a pheromone secreted into the water by the male. The embryos are not attached to the wall of the ovary but develop free in the sac, feeding upon an 'embryotrophic' material, apparently produced by the discharged ovarian follicles, which become highly vascular and remain throughout the several months of 'pregnancy'.

Rhodeus amarus, the bitterling (Cyprinidae), shows

similar association of the sexes at mating, here made
necessary by the fact that the eggs are laid within the
siphon of a swan mussel. For this purpose the cloaca of
the female develops into a tubular ovipositor. This
development takes place under the influence of a
hormone produced by the ovary. Addition of progesterone and related substances to the water containing
the fish causes growth of the ovipositor (Fig. 9.19).

The full growth of the ovipositor and preparation of
the female for spawning depends on the presence in the
water of the male and also of the swan mussel. Water in
which males have been kept stimulates growth of the
ovipositor. When the female is ready to deposit the eggs
she adopts a vertical position in the water and the
spawning male, in full nuptial coloration, swims around
her. An egg passes into the oviduct and erection of the
ovipositor is produced by pressure of the urine, produced by contraction of the walls of the urinary bladder,
the exit being blocked by the egg. The extended ovipositor is thus able to place the egg within the siphon of
the mussel (Fig. 9.19) and the male then immediately
thereafter sheds his sperms over the opening and they
are presumably carried in by the current. The whole
process shows the elaborate interplay of internal devices, pheromones, and external stimuli necessary for
the perfection of this remarkable method of caring for
the young. Yet the various features are all developments
of systems found in other vertebrates.

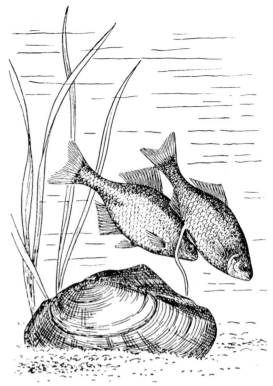

FIG. 9.19. Male and female bitterling (*Rhodeus*) with swan mussel
in which eggs are about to be deposited. (From Norman 1931.)

10   Lung fishes

1. Classification

Class OSTEICHTHYES
  Subclass: Sarcopterygii (fleshy fins)
    Order 1: Crossopterygii (lobe fins)
      Suborder 1: *Rhipidistia (fan-webs). Devonian–Carboniferous
        *Osteolepis; *Sauripterus; *Diplopterax; *Eusthenopteron
      Suborder 2: Coelacanthini (hollow spines) (= Actinistia). Devonian–Recent
        *Coelacanthus *Undina; Latimeria
    Order 2: Dipnoi (double breathing). Devonian–Recent
      *Dipterus; *Ceratodus; Neoceratodus; Protopterus; Lepidosiren

2. Sarcopterygii

The true bony fishes, the Osteichthyes, first appeared at the end of the Silurian and from the beginning three types can be recognized. Besides the ray-finned Actinopterygii that have already been considered there were two types of fleshy-finned fishes, the Sarcopterygii. These were the lung fishes or Dipnoi, and lobe-fins or Crossopterygii, all of which have become extinct except the coelacanth, Latimeria.

Although the lung fishes and their allies are here considered last of all the groups of fishes, because they lead on to the amphibia, it is important to realize that in many features they stand close to the ancestral stock of gnathostomes. It is a mistake to consider them as 'higher' animals than, say, the elasmobranchs or actinopterygians. Only four genera belonging to this group are found at the present time, Neoceratodus, Lepidosiren, Protopterus, the lung fishes of Australia, South America, and Africa respectively, and Latimeria, discovered in 1938 off the east coast of South Africa and near the Comoro Islands off Madagascar (Thomson 1969). These are relics of a group that can be traced back with relatively little change to the Devonian, and at about that period the first amphibia arose from some similar line. The characters of the modern crossopterygians are therefore of extraordinary interest, because they show an approach to the condition of the ancestors of all tetrapods.*

* However, Jarvik (1980) considers that the Dipnoi may be related to elasmobranchs.

3. Rhipidistia

This group of fossils is especially important because it probably contains the ancestors of the tetrapods. It is curious to think that we ourselves are descended from a fish similar to *Osteolepis from the middle Devonian, one of the earliest and most primitive member of the group. In appearance it shows an obvious similarity both to palaeoniscids and to early Dipnoi and it was possibly close to the line of descent from some acanthodian ancestor to both of these groups and to the amphibia. These were freshwater carnivorous fishes. The body was long and the tail heterocercal. A feature distinguishing all early Sarcopterygii from early Actinopterygii was the presence of two dorsal fins in the former. The paired fins have a characteristic scaly lobed form, from which the group derives its name, and the skeleton of the pectoral fin contained a basal element attached to the girdle and a branching arrangement at the tip (Fig. 11.8). This plan is distinctively different from that of the rayed fin of the Actinopterygii, and could also easily have led to the evolution of a tetrapod limb (p. 230).

The body was covered with thick, pitted, rhomboidal scales, with an appearance very similar to that of the palaeoniscid scale. These scales have, however, a characteristic structure known as cosmoid (p. 191). Cosmine is a complex association of hard and soft tissues that covers the bones and scales of some fossil Agnatha, Dipnoi, and Rhipidistia (Thomson 1975). It consists of a layer of dentine enclosing a mosaic of canals and cavities containing sensory neuromasts, opening by pores to the surface (Fig. 4.56). The dentine is covered by 'enameloid' and rests upon the vascular, 'spongy', bone and layered bone or 'isopedin' of the dermal skeleton. The sheets of shiny cosmine do not correspond to the individual bones but are often continuous, thus obscuring the sutures. To allow for growth the cosmine was apparently resorbed and re-deposited. Periodic absorption continued also in adults and may have been associated with use of the mineral store for reproduction or migration. The structure appears to have some relation to that of placoid scales and no doubt the morphogenetic processes that give rise to the isolated pulp cavities of placoid scales are similar to those that produce the cosmoid plates. It is sometimes suggested that the latter are formed of 'fused denticles', but this is of course only a manner of speaking. Denticles do not fuse, but morphogenetic processes may occur in such a way as to produce flat plates of dentine. Indeed, it is possible that the evolution occurred in the other direction, that is to say that the placoid scale is a special case of the cosmoid. The condition in which a substance is formed nearly uniformly all over the surface of the body is a more general one than that in which such

formation occurs only in isolated areas. The relation-ship of the cosmoid to the ganoid scale of early Actinopterygii is not quite clear, but the ganoid type seems to show a reduction of the pulp cavities and development of the shiny surface-layer (p. 191). The early Dipnoi of the Devonian possessed cosmoid scales. In later osteolepids and Dipnoi there has been a thinning of the scales, as among the Actinopterygii, so that the later Dipnoi are covered with thin, overlapping, 'cyc-loid' scales.

The skull of rhipidistians (Fig. 10.1) was well ossified; there was a series of bony plates arranged according to a pattern with a general similarity to that of palaeoniscids (p. 190) and which might well have been ancestral to that of amphibia. The brain-case showed a very striking division into two parts, anterior ethmosphenoid and posterior otico-occipital, with a joint between, allowing some movement, perhaps in taking prey (Thomson 1966). The notochord runs in a canal below the posterior part of the skull and the joint comes where it ends, marking the morphological front end of the body (p. 78). The trabeculae thus remain as separate el-ements in front. There was also a joint across the top of

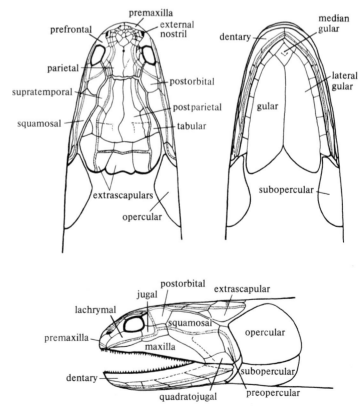

Fig. 10.1. Skull of *Osteolepsis*. (After Säve-Söderbergh, G., in Westoll, T. S. (1943). *Biological Reviews* **18**, 78–98.)

the skull between the parietal and postparietal bones. A movable joint at this level persists in the living coelacanth, and was present also in the earliest amphibians (the ichthyostegids, p. 267).

The palatoquadrate was attached to the brain-case in front and by a basipterygoid process behind, an autostylic arrangement similar to that in amphibia but different from the hyostyly of modern elasmobranchs and actinopterygians. The teeth of osteolepids were simple cones, not flattened plates such as are characteristic of Dipnoi, but the teeth on the palate show a somewhat broad folded surface and each tooth is replaced by another growing up near it, both of these being features found in the earliest amphibia. Sections of the teeth show a peculiar infolding of the enamel to make a labyrinthine structure, which is not found in other fishes but again is characteristic of the teeth of the first amphibians (p. 268).

There was only one pair of nostrils on the surface of the head and an internal nostril or choana, at the front end of the palate, bordered by the premaxillae, maxillae, palatines, and prevomers. These fishes may have breathed air; they certainly also possessed gills, covered with an operculum.

Animals of this sort were abundant in Devonian waters and by the end of that period had diverged into several different lines. It is interesting that the tendencies shown by these lines are similar to those that we discovered in the evolution of the Actinopterygii. Some of the later osteolepids became shorter in body, the tails tended to become symmetrical (diphycercal) and the scales to become thinner and overlapping. *Diplopterax and *Eusthenopteron represent separate lines from the late Devonian, both showing these characters. Probably the development of these features depends on the use of the swim bladder as a hydrostatic organ and the associated changes in the method of swimming.

4. Coelacanths

The osteolepids became rare in the Carboniferous and disappeared after the early Permian, but a line descended directly from them remained common through the Mesozoic and still survives today. These coelacanths (Fig. 10.2) show certain very characteristic features, which enabled the strange fish brought in 1938 to the museum at East London, South Africa, to be recognized immediately as belonging to the group. They are rather deep-bodied animals, with a characteristic three-lobed diphycercal tail. The type first appeared in the late Devonian and was obviously derived from osteolepid ancestry having two dorsal fins, diphycercal tail, lobed fins, and a rhipidistian pattern of skull bones, including in most forms a parietal joint. There was a calcified swim

Fig. 10.2. Coelacanth, *Latimeria chalumnae*. Photograph taken whilst still alive. (89 cm long.) (Kindly provided by A. Locket.)

bladder. *Coelacanthus and other Carboniferous forms lived in fresh water, but *Undina and other Jurassic and Cretaceous types lived in the sea.

The first living specimen of the group was fished off the east coast of South Africa and others have since been caught around the Comoro Islands near Madagascar (Fig. 10.2). All are referred to the genus *Latimeria*. They have been caught near the bottom at moderate depths (150–400 m). Unlike most of their fossil ancestors they are large fishes, weighing up to 80 kg; they are dull blue in colour (Attenborough 1979). The whole body is covered with heavy cosmoid scales. (See Locket (1980) for biology of coelacanth.)

The notochord is a massive unconstricted rod. The skull possesses a well marked joint between a condyle on the hind end of the basisphenoid and a glenoid cavity on the front of the base of the oto-occipital region. This joint, together with fibrous unions between other bones allows of movement of the front part of the head on the hind. A large pair of muscles runs from the parasphenoid up and back to the pro-otic and serves to raise the front part of the head on the hind. Coracomandibular muscles attached to the palatoquadrate have the reverse action and the movement is presumably concerned with catching the prey. There are small teeth on the premaxilla and tip of the dentary, and on a palate but not on the margin of the jaws. *Latimeria* lives on other fishes, apparently swallowed whole by the powerful oesophagus. There is a well-developed spiral intestine.

The swim bladder arises by a ventral opening from the oesophagus and proceeds backwards and dorsally for the whole length of the abdominal cavity. The lumen is very small and the organ is 95 per cent fat. It may serve to reduce the specific gravity. Respiration is by the gills. The heart shows a linear 'embryonic' condition, with the sinus venosus and auricle behind the ventricle. There are four rows of valves in the conus. The serum osmolality is lower than the surrounding sea water and contains much urea as in elasmobranchs and Dipnoi. However, the sodium concentration is lower, similar to that of marine teleosts. The concentration of urea is as high in the urine as in the blood, so the kidney is unable to resorb urea (unlike elasmobranchs) (Griffith Umminger, Grant, Pang, and Pickford 1974). The kidney does not produce concentrated sodium chloride and salt excretion is probably performed by the rectal gland, which resembles that of elasmobranchs. The red cells are large, as also in elasmobranchs, Dipnoi and Amphibia. The eggs are larger than in any other fishes (9 cm diameter and 330 g) and they develop within the oviduct. Five well developed embryos, found in a female of 65 kg, were about 30 cm long and already most of the adult features were present (Smith, Rand, Schaffer, and Atz 1975).

The brain lies far back in the cranium, of which it occupies less than one-hundredth part, the rest being filled with fat. Its structure is somewhat like that of a teleostean, with a thin forebrain roof, and large striatum, but without eversion. There is no valvula to the cerebellum. The pituitary is unique, with an enormously long (up to 10 cm) anterior extension of the hypophysial cavity. This tube has largely fibrous walls containing islets of endocrine cells, and anteriorly it terminates in the floor of the cranium as an isolated endocrine lobe, the buccal pars distalis. This may represent the ventral lobe of the elasmobranch pituitary (p. 138).

There are anterior and posterior nares but both open on the surface of the head and they have nothing to do with respiration (i.e. there is no choana). The rostral organ is a large median sac opening to the surface by three pairs of canals and richly innervated by the superficial ophthalmic nerve. A similar sac occurs in fossil coelacanths back to the Devonian but its function is quite unknown. The eye, inner ear, and lateral line system are well developed.

It is hard to see what features have enabled *Latimeria* to survive with little change since the Jurassic or earlier (see p. 219). It clearly cannot be by special development of the brain or receptors. Its habitat is isolated, but not especially protected and its population seems to be small since even by exceptional efforts so few specimens have been found. Perhaps they are more numerous in deeper waters. In some of its features it shows developments parallel to those of the Teleostei rather than to the Dipnoi, whose remote ancestry it shares. Several of its characteristics are paedomorphic. These can hardly be alone responsible for such a long survival, but some of them also appear in the other survivors from the Palaeozoic, *Polypterus*, sturgeons, and Dipnoi.

5. Fossil Dipnoi

The Devonian Dipnoi were more like their osteolepid relatives than are the surviving modern forms (Fig. 10.3). The early members of this group, such as *Dipterus (Fig. 12.2), showed the typical elongated body, thick cosmoid scales, heterocercal tail, lobed fins, and well ossified skull. The pattern of the bones was obscured by a seasonal deposit of cosmine, this being periodically absorbed to allow of growth.

The individual bones have a certain similarity to those of osteolepids, but there are extra bones that are difficult to name. There was no premaxilla or maxilla, nor any teeth along the edge of the jaw; instead broad, ridged tooth plates were developed on the palate and inside of the lower jaw, presumably as an adaptation for eating molluscs and other invertebrates. These crushing-plates are characteristic of the Dipnoi and preclude even the earliest of them from being the actual ancestors of the amphibia. By the end of the Devonian the Dipnoi were showing changes similar to those of the oesteolepids and palaeoniscids. The body became shorter, the first dorsal fin disappeared, the tail became diphycercal, and the scales lost their shiny surface layer and became thin. The teeth of *Ceratodus appear in the Triassic and were known to geologists long before the related living animal was discovered. There has been very little change in this animal in more than 150 million years, though the

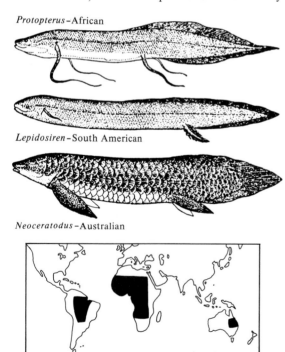

Protopterus – African

Lepidosiren – South American

Neoceratodus – Australian

Fig. 10.3. The three living lung fishes and their distribution. (From Norman 1931.)

recent members are placed in a distinct genus *Neoceratodus*.

The evolution of Dipnoi is especially interesting because the rate of change has actually been measured (Westoll 1949). Twenty-six different characteristics such as proportion of skull, nature of dermal bones or dentition were divided into 3–8 grades. Each fossil genus was thus given a score, the minimum being for characters appearing in the oldest lung fish, the hypothetical ancestor being 0. Actual scores ranged from 4 for the earliest to 100 for two of the living genera. Plotting against time we see that there was an early acceleration of evolution followed by very slow or zero change in the last 150 million years. (Fig. 10.4). It has also been shown that a similar method applied to the evolution of coelacanths gives a curve of similar shape (Schaeffer 1952).

6. Modern lung fishes

The three surviving genera of lung fishes (Fig. 10.3) are mainly inhabitants of rivers (though *Protopterus* lives in large lakes) and they all breathe air. *Neoceratodus* lives only in the Burnett and Mary rivers in Queensland, the pools of which become very low and stagnant in summer. *Lepidosiren* from the rivers of tropical South America, and *Protopterus* from tropical Africa, can survive when the rivers dry up completely. They dig into the mud, leaving a small opening for breathing, and can remain in this state for at least six months. Remains of cylindrical burrows found associated with dipnoan bones show that this habit of aestivation has been adopted by the group at least since Permian times. The three survivors all show similar deviations from the

conditions found in *Dipterus, but *Neoceratodus* has diverged less than the other two. The tail fin is symmetrical (diphycercal) in all three, with no trace of separate dorsal fins. The paired fins are of 'archipterygial' type in *Neoceratodus*, with an axis and two rows of radials (Figs. 10.5 and 11.20). The scapula is covered by clavicles, cleithra, and post-temporals, the latter articulating with the skull. The scales are reduced to bony plates.

The vertebrae are cartilaginous arches, the notochord remaining as an unconstricted rod. In the skull there is also a great reduction of ossification, the dorsal bones consisting of a few bony plates, forming a pattern not obviously comparable with that of other forms (Fig. 10.6). The jaw suspension is autostylic. The food consists of small invertebrates and decaying vegetable matter, which is eaten in large amounts. In the gut there is no stomach and the intestine is ciliated. There are no hepatic caeca, but a well-developed spiral valve is present.

The external nostrils lie just at the edge of the mouth and the internal nostrils open into its roof. The nostrils are used only to 'sniff' water by passing it through the olfactory organs and out through the mouth and opercular slits. The swim bladder is developed into a definitely lung-like structure (there is one in *Neoceratodus*, a pair in each of the others), divided into many chambers. *Neoceratodus* has been observed to come to the surface to breathe air but only when the oxygen tension falls below 83 mm Hg (Johansen, Lenfant, and Grigg 1967) and is said to be able to survive in foul water that kills other fishes, but it cannot live out of the water. *Lepidosiren* and *Protopterus* have been shown to obtain 98 per cent of their oxygen from the air. The wall of the swim bladder of all forms contains muscle and elastic fibres and the cavity is subdivided into a number of pouches or alveoli. In *Protopterus* and *Lepidosiren* the

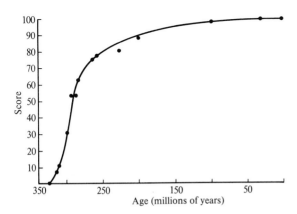

FIG. 10.4. Rate of evolution in lung fishes. Each point represents the index of a single genus obtained by taking 26 characters and rating them with grades of structure. The lowest value was given for the most primitive condition, the highest for the most modern one. (After Westoll and Simpson.)

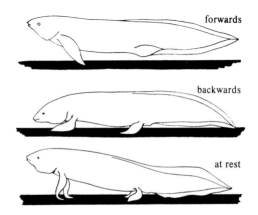

FIG. 10.5. *Neoceratodus,* showing the method of walking on the bottom. (From Ihle, after Dean.)

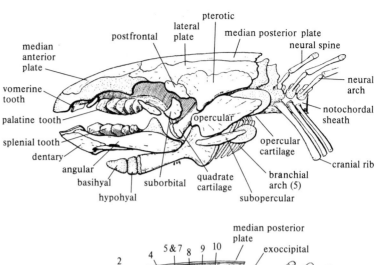

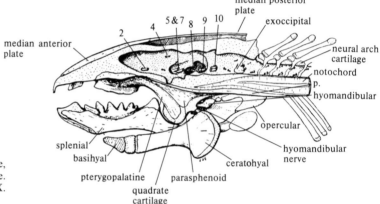

FIG. 10.6. Skull of *Neoceratodus*. Above, lateral view; below, view of medial surface. 2–10 foramen for cranial nerves II–X. (From Goodrich 1930.)

edges of the slit-like glottis are controlled by muscles and there is an epiglottis. Although the lungs lie dorsally their opening is to the oesophagus, ventrally (Fig. 9.15). Air is drawn in through the mouth by lowering its floor and then passed back by a buccal force-pump, the mouth being closed and the tongue pressed against the roof as a seal. The hyoid apparatus and pectoral girdle assist in the movement. Expiration is by the elasticity of the lung tissue, the ribs play no part. The lung is supplied with blood from the last branchial arch in *Neoceratodus* (Fig. 10.7), but in the other Dipnoi there is a more elaborate arrangement. The second and third gill arches bear no lamellae and their afferent and efferent branchial vessels are directly continuous, so that blood flows from the ventral to the dorsal aorta and carotids. The pulmonary artery springs from the dorsal aorta. Blood returns in a special pulmonary vein to the partly separated left side of the sinus venosus. The auricle is partly divided into two and the ventricle is almost completely divided by a ridge and a series of muscular trabeculae in *Lepidosiren* and *Protopterus*, but in *Neoceratodus* it is only partly divided by an auriculoventricular plug (Klitgaard 1978). The ventral aorta is

shortened into a spirally twisted muscular bulbus cordis provided with a system of valves such that the blood from the left side of the auricle is directed mostly into the first two branchial arches, that from the right side into the last two. In this way some separation of pulmonary and systemic circulations is achieved. Indeed, although there is every reason to believe that the mechanism has been in existence for nearly 300 million years, it shows us most clearly a possible intermediate stage between aquatic and pulmonary respiration. It seems likely that the earliest amphibia employed a similar system. There is a coronary artery arising from the anterior efferent branchial arches (Fig. 10.7). A further amphibian feature of the vascular system is the presence of an inferior vena cava, a vessel collecting blood from the kidneys and reaching to the heart by passing round to the right of the gut in the mesentery. The more dorsal cardinal veins, joining the ductus Cuvieri remain present however, and there is a renal portal system.

The adrenals of Dipnoi are represented by two separate masses of tissue. The perirenal tissue of *Protopterus* is a considerable mass of material around the kidney, containing lipid, steroid, and round-cell

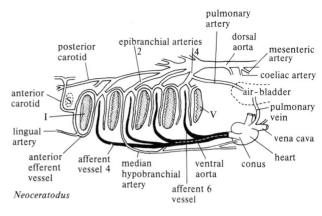

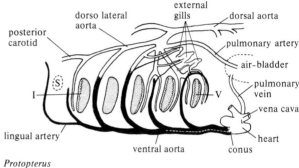

FIG. 10.7. Branchial circulation of A *Neoceratodus* and B *Protopterus*. *S.* position of closed spiracle; *I–V* five branchial slits. The gills are present on the hyoid and next four branchial arches. (From Goodrich 1930.)

(lymphoid) tissues, as well as endothelial and pigment cells. The steroid tissue shows histochemical properties similar to those of mammalian adrenal cortex and undergoes changes after injection of mammalian ACTH. This tissue thus shows a collection of functions, haematopoietic, phagocytic, storage, endocrine, and pigmentary, which may show the starting point of the evolution of the tetrapod adrenal cortex. Cells that give the chrome reaction, the adrenal medullary tissue, lie in the walls of the intercostal branches of the dorsal aorta. This condition could have given rise to that of amphibia.

The pituitary shows a curious mixture of teleostean and amphibian characters. There is no distinct regional separation of cells in the pars distalis as there is in teleosts. There is a well-marked median eminence but also some direct innervation of the pars distalis by hypothalamic fibres. There is a well-marked cleft separating the pars intermedia and distalis and diverticuli of it penetrate the latter, as in sturgeons. The neural lobe secretes arginine vasotocin and mesotocin (in *Protopterus*) and these hormones cause renal diuresis and loss of sodium but it is not clear how this is related to the aestivation of the fish. The amount of neurosecretory material in the lobe is less in animals taken from cocoons than from water.

The arrangement of the urogenital system is similar to that of amphibia and is probably closer to that of the ancestral gnathostome than in any living elasmobranch or actinopterygian. In the male there are vasa efferentia by which sperms are passed through the excretory portion of the mesonephros. In the female eggs are shed into the coelom and carried out by a Müllerian duct, whose opening lies far forward. An interesting feature is that the Müllerian duct is very well developed in the male. This is one of several details (lack of ossification, unconstricted notochord) which raise the suspicion that the living Dipnoi have acquired their special characters by a process of paedomorphosis or partial neoteny, that is to say, becoming sexually mature in an early stage of morphogenesis.

The embryology, again, shows similarity to that of amphibia and dissimiliarity from the other groups of fishes in that cleavage is total and gastrulation takes place to form a yolk plug. There is therefore no blastoderm or extra-embryonic yolk sac. However, the cells of the vegetative pole contribute little to the shape of the embryo and indeed may form a partially separate yolk sac. The larvae show distinct similarity to those of amphibia, especially the larvae of *Lepidosiren* and *Protopterus*, in which there is a sucker and external gills.

The nervous system shows the same affinities as the rest of the organization (Figs. 10.8 and 10.9). The

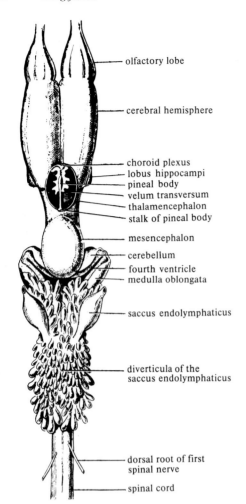

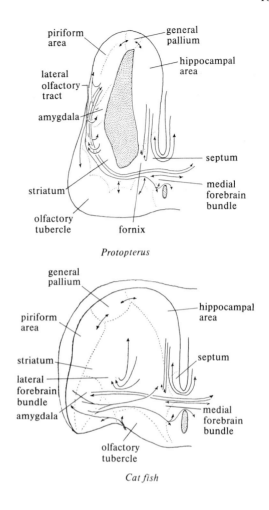

Protopterus

Cat fish

FIG. 10.8. Dorsal view of the brain of *Protopterus*. (After Burckhardt, from Sedgwick, A. (1905). *A student's textbook of zoology*. Sonnenschein, London.)

FIG. 10.9. Diagrams of the telencephalon of a representative teleost and dipnoan. Arrows indicate course of fibre bundles. (Schnitzlein, H. N. (1968). In *The central nervous system and fish behavior* (ed. D. Ingle). University of Chicago Press.)

forebrain is evaginated into a well-marked pair of cerebral hemispheres. The roof of these, though not very thick, is nervous and therefore is definitely of the inverted type, not everted as in Actinopterygii. The optic lobes are little developed, the mesencephalon being hardly wider than the diencephalon. The cerebellum is small. A peculiar feature is the development of the inner ear to form a special lobed saccus endolymphaticus, lying above the medulla oblongata. The significance of this is not known, but it is interesting that similar backward extensions of the ear are found in amphibia. The Dipnoi also resemble urodele am-

phibians in several histological features such as large size of epidermal cells and ciliation of the young larva (Whiting and Bone 1980).

In many features, then, the Dipnoi differ from modern fishes and resemble the amphibia, which evolved from the same stock. The early crossopterygians probably competed somewhat precariously with other fishes, as do the Dipnoi, and indeed many urodeles, today. It was only after tens of millions of years of evolution during the Carboniferous and Permian times that numerous land animals arose, in the form of the later amphibian and reptilian types (Fig. 12.2).

11 First terrestrial vertebrates: Amphibia

1. Amphibia

DURING the later part of the Devonian period a population of osteolepid fishes such as *Eusthenopteron* lived in the freshwater systems and there is every reason to suppose that some of these animals, first crawling from pool to pool and then spending more time on the land, gave rise to the terrestrial populations that we distinguish as Amphibia. They may well first have frequented the shallow water as juveniles to escape the attacks of larger, predatory fishes. Since they already possessed lungs and stoutly-constructed fins, it would not have been difficult for them to venture further, onto the muddy banks. There they would have found a supply of invertebrate food, which provided the opportunity for their evolution into forms better adapted for prolonged stay on land. No doubt the early efforts at life on land were crude. The whole locomotory and skeletal system comes under a completely new set of forces when the support of the water is withdrawn and the effects of gravity become insistent. It is not surprising that these new conditions produced greater changes in vertebrate organization than had occurred in tens of millions of years previously. Nevertheless, so slow is the pace of evolution that the only known Devonian Amphibia, and many of the Carboniferous ones too, still looked and presumably behaved very like fishes. Nearly all of these early amphibians were wholly or largely aquatic, feeding on fish or on aquatic invertebrates. It was not until the Permian that crocodile-like forms such as *Eryops* (Figs. 11.1 and 12.2), and small terrestrial amphibians such as *Cacops* evolved.

Of all the features that arose at this time in connection with the new life on land the presence of pentadactyl limbs is perhaps the most conspicuous. It is appropriate that this should be marked in zoological nomenclature: the Amphibia are the first of the great group of land vertebrates, the Tetrapoda.

The modern Amphibia are quite different from their remote Palaeozoic ancestors, for they have lost their scales in order to use their moist skin for the excretion of carbon dioxide. The resulting permeability of the skin to water vapour has restricted them to wet or moist environments, or microclimates. But these modern forms are by no means a precariously existing remnant. They are quite numerous and successful in the ecological niches that they occupy and make an important element in many food-chains. There are some 2000 species at present recognized, placed in 250 genera. However, contrasting this with the numerous species of teleosts, of birds, and of mammals we shall see that the amphibians, though well adapted for certain situations, do not succeed in maintaining themselves in many different types of habitat. There are desert toads, such as *Chiroleptes* of Australia, but these survive by burrowing and by special abilities, such as the power to hold large amounts of water, associated with loss of the glomeruli of the kidneys.

Modern Amphibia are placed in a separate subclass Lissamphibia with three orders: Urodela (newts and salamanders) retain the original long-bodied, partly fish-like form; the Anura (frogs and toads) have lost the tail and become specialized for jumping; and the Apoda are limbless, burrowing animals found in the tropics. The urodeles and anurans are found as fossils back to the Cretaceous and Triassic respectively, but we have only scanty information about their connection with the earlier amphibians found in rocks from the late Devonian to the Triassic (Chapter 12).

2. The frogs

Perhaps the most successful of all amphibia are those belonging to the genus *Rana*, abundant in every part of

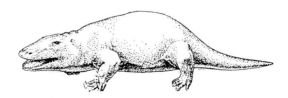

FIG. 11.1. *Eryops*. An early Permian form, with well developed limbs, probably living in swampy areas.

the world except in the south of South America, on oceanic islands, and New Zealand. Ranid frogs are typical of the highly specialized subclass Anura, whose members usually inhabit damp places such as marshes or ditches, living for most of their life in the grass or undergrowth and feeding by catching flies and other insects with their tongue. They are preyed upon by birds, fishes, and especially snakes, and escape from these by their hind legs, used either for jumping or swimming. The young develop as tadpoles in the water, where they are omnivorous. The various species differ in size and small points of colour, though there are also some that depart widely from the usual habits, e.g. *R. fossor* which burrows. *R. temporaria* is the species found in Great Britain, *R. esculenta* is a slightly larger form found on the continent of Europe and occasionally in east England, *R. pipiens* is the common small North American frog; *R. catesbiana* the giant bull-frog, whose body is up to 23 cm long, also lives in North America. *R. goliath* of the Cameroons is over 30 cm long, but is mainly aquatic.

3. Skin

The earliest amphibia possessed the scales of their fish ancestors, but these were lost in later lines, though retained in some Apoda. Some frogs carry dermal plates on the back, however, fused to the neural spines (*Brachycephalus* of Brazil). Some tree frogs, Hylidae, have bony 'casques' on the head, used to plug the entrance to burrows and prevent dessication.

Amphibia differ from reptiles in that the skin is relatively moist and used for respiration; on the other hand, the skin also shows a character typical of land animals in having heavily cornified outer layers. The epidermis therefore consists of several layers in the adult frog and is renewed by a process of moulting at intervals of a few days or weeks. The moult is under the control of the pituitary and thyroid glands and does not occur if either of these be removed, the keratinized cells merely accumulating in those circumstances as a thick skin. However, in the toad *Bufo* moulting seems to be initiated primarily by adrenal steroid hormones. Local

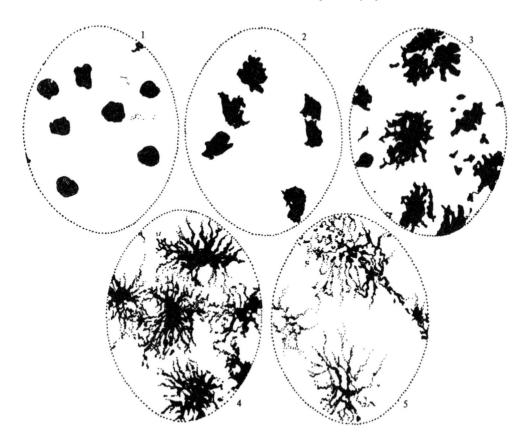

FIG. 11.2. Stages of dispersal of pigment in the melanophores in the web of the frog *Xenopus* as used to assess the melanophore index (stages 1–5). (After Hogben, L. and Slome, D. (1931). *Proceedings of the Royal Society, London* **B108**, 10–53.

thickenings of the epidermis often occur in amphibia, for instance to form the horny teeth by which the larva feeds. Such thickenings are also a conspicuous feature of the warty skin of the toads which mostly have a drier skin and are more fully terrestrial than are the frogs. Various forms of claws and pads occur on the feet. In *Xenopus*, the clawed aquatic frog, they are used to dig for insects in the mud. In *Leptodactylus* they serve for burrowing. Many amphibian larvae have horny teeth. The fact that the epidermis of Amphibia can produce such local thickenings is of interest in considering the origin of feathers and hairs. In larval amphibians the skin is ciliated. The dermis of terrestrial anurans has a special layer of ground substance. The mucopolysaccharides of this probably act like a sponge, retaining water but allowing the passage of solutes, especially oxygen. In *Rana* there is a substantial network of respiratory capillaries (Jasiński and Miodoński 1978).

The glands of the skin are more highly developed than in fishes, and are of two types, mucous and poison glands. Both of these consist of little sacs of gland-cells, derived from the epidermis. The mucus serves to keep the skin moist, this being essential as it is a respiratory surface for gaseous exchange; the secretion may perhaps also serve for temperature regulation. The problem of regulation of temperature is important for all terrestial animals, since air conducts heat much less well than water and therefore violent changes of temperature are frequent. Evaporation produces large influences on temperature and no doubt it was the adjustment of these effects that led to the development of temperature regulating mechanisms in birds and mammals. Frogs in dry air are always found to be colder than their environment, the difference being sometimes as much as 5 °C. It is probable that in some circumstances use is made of this cooling, since tree frogs (*Hyla*) may be found fully exposed to tropical sunlight, which would be expected to raise their temperatures to a lethal level. On the other hand, the loss of water involved by evaporation in this way would presumably soon become serious.

The poison glands or granular glands are less developed in *Rana* than in *Bufo* ('the envenom'd toad') where they are collected into masses, the parotoid glands (not 'parotid') (Habermehl 1974). The poisons are alkaloids with actions like digitalis. The effect on man is to produce an irritation of the eyes and nose; only rarely does it affect the skin of the hands. When swallowed it produces nausea and has an action like atropine on the heart. The poison of *Dendrobates* of Colombia is used on arrows; it acts on the nervous system.

Some Amphibia have characteristic smells, produced by secretions, and these hedonic glands are used to attract the sexes to each other. In some male newts (Plethodontidae) the secretions of these gland cells below the chin serve to calm the female for copulation.

Another use of glandular secretions is to keep the eyes and nostrils free from obstruction. The demands of terrestrial life require the production of numerous such special devices and lead to the complexity that we recognize as an attribute of these 'higher' animals.

4. Colours

The use of colour is also highly developed in Amphibia. The animals are often greenish and the colour is produced by three layers of pigment cells in the dermis (Bagnara 1976). Melanophores lie deepest, and above them are iridocytes, full of stacks of plates of purines such as guanine, which by diffraction produce a blue–green colour. Yellow or red carotenoid xanthophores overlie these and filter out the blue. Change of colour is produced by expansion of the pigment in the melanophores under the action of the melanophore stimulating hormone (MSH) of the posterior pituitary gland (Fig. 11.2). Movements in the other chromatophores can also affect the colour, yellow being produced by disarrangement of the iridocytes or amoeboid movement of the lipophores until they lie between or underneath the iridocytes. Other colours may contribute to the patterns, blue (though rarely) by the absence of the lipophores, red by pigment in the lipophores.

Changes in the melanophores may be of two sorts, primary or direct and secondary or visual. The primary response depends on the direct effect of light on the skin, causing expansion. The secondary effect consists in contraction of the pigment if the animal is illuminated on a light-scattering surface (light background) but expansion (and hence darkening of the animal) when it is illuminated from above on a light-absorbing (dark) background. There are, presumably, distinct responses from different parts of the retina, illumination of the dorsal part producing contraction and of the ventral part expansion of the melanophores. When frogs are kept for a long time on dark backgrounds their melanophores multiply and come to contain extra amounts of melanin. This *morphological colour change* is also the result of the increase of pigment in *epidermal* melanophores, a type of cell that occurs in all vertebrates, and are smaller than the dermal melanophores (which can also multiply). The epidermal melanophores also donate granules of pigment to neighbouring epidermal cells by a mechanism known as cytocrine activity.

The control of the colour change of the dermal cells of amphibians is mediated by variation in the secretion of the pituitary gland. The melanophore stimulating hormone causes expansion of the melanophores and xanthophores but contraction of the light-reflecting iridophores (Bagnara 1976). Paling is thus produced by inhibition of secretion of MSH, probably by the neurosecretory fibres of the hypothalamo-hypophysial tract. There is evidence from fluorescence studies that the endings of these fibres in the pars intermedia contain catecholamines. This also partly explains how injections of adrenaline produce paling of frogs, but this hormone probably also acts directly on the chromatophores.

A further influence on the colour is produced by the pineal hormone, melatonin. Tadpoles become pale when kept in the dark but the response does not occur if the pineal has been removed. Adult frogs do not pale in the dark, nor respond to melatonin.

In Amphibia there is no direct control of the pigment cells by nerve fibres such as are present in bony fishes (p. 209). The colour change is therefore rather slow. After removal of the pituitary the melanophores still show changes correlated with change of incident illumination. The melanophores of isolated tails of *Xenopus* tadpoles expand within 30 min in the dark but contract again after 10 min illumination. This is therefore a direct photochemical response.

The colour patterns adopted are usually cryptic or concealing in their effect, but the colour also has an important influence on the temperature and varies with it and with the humidity, as well as with the incident illumination. The uniform brilliant green of tree frogs makes them very difficult to see among the leaves. On the other hand, *R. temporaria* and other species living among grass show a pattern of dark marks, which breaks up their outline. In other amphibians, however, the colour makes the animal conspicuous, for instance the black and yellow markings of *Salamandra maculosa*. Dendrobatids are very poisonous South American frogs with vivid patterns of green, yellow, blue, and black. Conspicuous colour is often associated with great development of the poison glands and is therefore presumably aposematic or warning coloration, allowing recognition by possible attackers. This correlation is not always found, however; the toad *Ceratophrys americana* is dull coloured but poisonous, whereas *C. dorsata* has a bright pattern but is harmless. *Pseudotriton* is a red salamander that is innocuous but mimics the poisonous forms (Batesian or pseudaposematic mimicry).

Many frogs make a sudden exposure of brightly coloured patches on the thighs when they jump. This flash colour presumably serves to startle and distract the attacker and such colours may be called dymantic or startling. A similar use of colour is made by the cuttlefish (*Sepia*), which may suddenly produce two black spots when alarmed, and also by some Lepidoptera. It is

interesting that the colour used in this way so often takes the form of black spots ('eye-spots'), which have an especially striking quality. In some anurans these colours are irregular dark marks, but in *Mantipus ocellatus* they take the form of definite eye-spots.

The presence of pigment serves to protect the organs from the effects of radiant energy. Dark colour may also assist in thermoregulation both in the adults and in the eggs. Many anurans become lighter in warm, dry

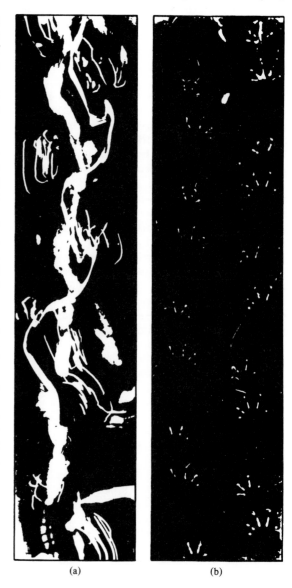

(a)   (b)

FIG. 11.3. Record of the movements of *Ambystoma* walking on a smoked drum. (a) in rapid locomotion (with the body on the ground); (b) in slow locomotion (raised up on the legs). (From Evans 1946.)

conditions than in cold and damp, but amphibians make less use of such regulation than reptiles (p. 276). In some frogs and toads there are sexual differences in colour.

5. Locomotion

The general build of the body is essentially fish-like in stegocephalian and urodele amphibians. Such forms have two means of locomotion. When they are frightened and move fast they wriggle along with the belly on the ground, the effective agent being serial contraction of the segmentally arranged myotomal musculature, by means of which the animal as it were 'swims on land', with the legs hardly touching the ground (Fig. 11.3). When moving deliberately, on the other hand, a newt raises up its body on the legs, which then propel it along as movable levers, the main part of the action being produced by drawing back the humerus or femur, the more distal muscles of the limbs serving to maintain the digits pressed against the ground (Fig. 11.4).

The carrying of the weight on four legs places an entirely new set of stresses on the vertebral column. Instead of being mainly a compression member as it is in fishes it comes to act as a girder, carrying the weight of the body and transmitting it to the legs. This function produces a column whose parts are largely bony and articulated together, flexibility becoming less important than strength. The types of strain involve new muscle attachments and the development of special processes and parts of the vertebrae (p. 277). These changes, however, have not proceeded very far in the amphibians; many urodeles spend much time in the water and their vertebrae often show a lack of ossification,

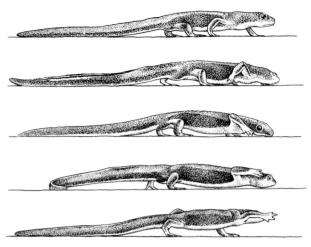

FIG. 11.4. Drawings made from photographs of a newt, *Triturus*, during slow locomotion. (After Evans 1946.)

parts of the notochord persist and provide the main compression member required for swimming.

In the anurans the entire skeletal and muscular system has become specialized for the peculiar swimming and jumping methods of locomotion, by means of extensor thrusts of both hind limbs, acting together. Frogs, and especially toads, also walk on land, bringing into play a set of myotactic (proprioceptor) reflexes that depend on the contraction of the muscles against an external resistance (Fig. 11.5).

FIG. 11.5. Reflexes associated with the transition from swimming to walking in toads. The shaded outlines show successive positions as the animal emerges on to solid ground. The first effective contact is by the left fore-limb whose retraction and extension elicits a crossed protraction reflex in the right fore-limb (L_1), a diagonal extensor response in the right hind-limb (L_2), and a placing response in the left hind-limb (L_3). The right fore-limb then touches the ground and produces corresponding responses R_{1-3}. The left fore-limb in response to stretch of its protractor muscles swings forward and this produces retraction of the left hind-limb (L_4) and protraction of the right hind-limb (L_5). Fixation of the right hind-foot then produces a crossed flexor response (RH_1). (From Gray and Lissmann 1947.)

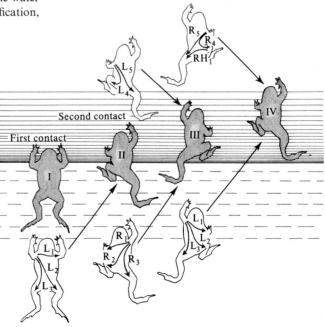

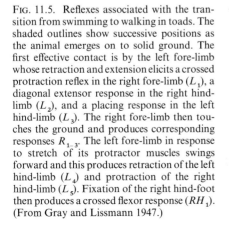

The actions of jumping and walking are possible because of profound changes in the arrangement of the skeleton and muscles. The myotomal muscles no longer perform their primitive function of producing metachronal waves of contraction, and accordingly the vertebral column (Fig. 11.6) has lost its original flexibility. Instead, it acts as a support by which the movement of the hind limbs is transmitted to the rest of the body. There is no longer any sinuous motion and the number of vertebrae is very low (nine in the adult *Rana*), and behind them is an unsegmented rod of bone, the urostyle, ventral to the notochord, which is made up of the fused, elongated bodies of several caudal vertebrae. Shortening of the body is a characteristic feature of the change from aquatic to terrestrial life, and is seen in many lines of amphibian and reptilian evolution. It has proceeded farther in the frogs than in any other tetrapods.

The second to eighth vertebrae of *Rana* are concave in front, convex behind (procoelous), and have large transverse processes but no ribs. In other amphibians they may be amphicoelous or opisthocoelous. They fit together by complex facets, the zygapophyses. The first vertebra has two concave facets for articulation with the two condyles of the skull; its centrum and transverse processes are much reduced. The ninth (*sacral*) vertebra has large transverse processes, which articulate with the ilia of the pelvic girdle. There are free ribs in the primitive frogs *Ascaphus* and *Leiopelma*.

6. Evolution and plan of the limbs

The girdles of the paired limbs have become much changed from their fish-like condition (Figs. 11.7 and 11.8). Their basic pattern is similar in the two limbs and has been retained throughout the whole tetrapod series. Whereas in fishes the girdles are rather small cartilages and bones, the pelvic girdle being restricted to the ventral region of the body, in amphibians they become enlarged in connection with the weight-bearing function of the limbs.

The details of the sequence of stages by which a tetrapod limb arose from a fish fin are still somewhat disputed. It is probable that the ancestral crossopterygian possessed a lobed fin, rather like that seen in *Eusthenopteron* (Fig. 11.7). As the fishes came on land the fin would be used as a lever, giving greater effect to the wave-like motions by which the creature 'swam on land'. The muscles of the limb, contracting in a serial

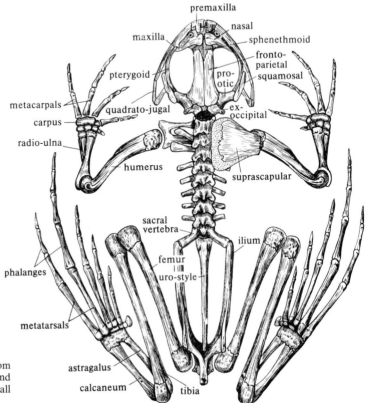

FIG. 11.6. The skeleton of the frog, seen from the dorsal surface; the left suprascapular and scapular have been removed. (After Marshall 1920.)

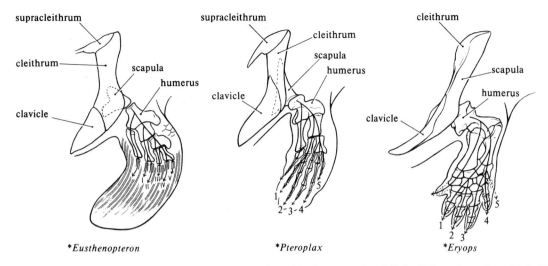

FIG. 11.7. Transformation of crossopterygian pectoral girdle and fin into pentadactyl limb. Oblique front view of left side. (Modified from Gregory and Raven 1942.)

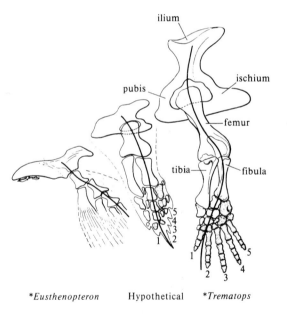

Eusthenopteron   Hypothetical   *Trematops*

FIG. 11.8. Transformation of crossopterygian pelvic fin into tetrapod limb. (Modified from Gregory and Raven 1942.)

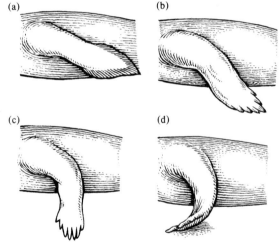

FIG. 11.9. Diagrams illustrating the probable changes in position during the evolution of a pelvic fin into a tetrapod limb. (a) as in *Ceratodus*; (b) double flexure to give knee and ankle joints, leaving foot direct backwards; (c) and (d) rotation of tarsus and digits turning foot forward. (After Gregory and Raven 1942.)

manner, would tend to move it backwards and forwards relative to the body, thus assisting in locomotion. At first the limb perhaps carried only little weight, but as tetrapod evolution proceeded the limbs became elongated and turned under the body, raising it off the ground. To work effectively in this way the limbs came to be held bent down at elbow and knee (Fig. 11.9) and a firm application to the ground was produced by bending outwards at wrist and ankle. Finally the limbs were

brought in to the side of the body by rotation, such that the elbow pointed backward and the knee forward.

These are the changes that must have occurred at some time to produce the full tetrapod condition, but we cannot follow exactly the order in which they took place. Their effect is to convert a paddle-like fin, whose main movements were up and down, and were used for stabilization in the horizontal swimming plane, into an elongated jointed strut, on which the animal can balance, and which can be moved as a lever to produce locomotion.

The limbs and girdles and their muscles show a remarkable constancy of pattern throughout the tetrapods. The muscles of the fins of fishes are concerned mainly with lowering and raising (Fig. 11.20), and they run from a girdle in the body wall to the basal radials in the fin, and between the radials. After the animals came on land the muscles served not only to raise and lower the limbs but also to draw them forwards and backwards; indeed, many fishes already make much movements, including *Protopterus* and the living coelacanth, *Latimeria*. The muscles therefore become arranged around the shoulder, and hip joints into groups serving as adjustable braces, by which the body is balanced on its legs and by whose contraction the latter are moved. Those muscles that draw the limb towards and away from the mid-ventral line can be called medial and lateral braces (adductors and abductors) and the muscles drawing the leg backwards and forwards are posterior and anterior (retractor and protractor) braces. For the attachment of these muscles proximally the pectoral and pelvic girdles, small in fishes, become expanded into plates (Figs. 11.8 to 11.10), and these are divided into a number of characteristic pieces, though the mechanical reason for the division is not clear.

7. Shoulder girdle

The earliest Amphibia such as *Palaeogyrinus* are known from their teeth as labyrinthodonts. They inherited a shoulder girdle almost exactly like that of their osteolepid ancestors except that a new dermal element, the interclavicle, was added to the ventral surface. As in gnathostomes generally (except elasmobranchs) the shoulder girdle was a dual structure consisting of (a) a primary or endochondral component evolved from the basal fin elements of the ancestral fish form. This serves to provide an articulatory surface for the limb as well as points of attachment for the limb musculature. (b) A dermal ring of superficial bony elements. These had sunk inwards and applied themselves to the ventro-anterior surfaces of the endochondral girdle which, consequently, they braced and supported.

The endochondral girdle, consisted of two half rings,

which overlapped in the ventral midline. Each half was a single unit but, by topographical comparison with girdles of later tetrapods, it is often arbitrarily divided into two regions, a dorsal scapula and a ventral coracoid. Between these two regions a screw-shaped glenoid cavity received the humerus. The one endochondral ossification is usually homologized with the scapula of amniotes (Watson 1925). Later forms (e.g. *Seymouria*, *Diadectes*) possessed a second bony element which is generally interpreted as a precoracoid. The endochondral girdle was small in the earliest Amphibia (e.g. *Pteroplax*). In later genera its size progressively increased, presumably to withstand the greater thrust transmitted by the larger limbs of these forms and to provide attachment for the increased mass of brachial musculature.

The dermal girdle consisted, typically, of paired cleithra, clavicles, and interclavicle. The latter, a new element, lay between and often beneath the clavicles and probably formed a locking mechanism preventing the complete separation of the epicoracoid cartilages. The connection between the dermal girdle and the skull was lost presumably to permit greater mobility of the head. This foreshadowed the reduction and loss that was the subsequent fate of the dermal, shoulder girdle elements in tetrapod evolution. Of the modern Amphibia, the Anura alone of recent tetrapods, have retained a cleithrum. Each half of the endochondral girdle consists of a dorsal, bony scapula with a cartilaginous suprascapula, and a ventral coracoid bone connected to an anterior precoracoid cartilage by a mesial epicoracoid cartilage. The precoracoids are invested by the clavicles and, as in all modern amphibians, the interclavicle is absent.

Anuran shoulder girdles may be divided into two broad categories according to whether the two epicoracoid cartilages are fused mesially (a) along their entire lengths (firmisternal condition) or (b) along their anterior edges only (arciferal girdles, Fig. 11.11) (but see Griffiths (1963) for further complication). The latter occurs typically in 'walking', toad-like Anura (e.g. Bufonidae, Pelobatidae) and in the aquatic xenopids. The clavicles are the main struts for keeping the glenoids apart and, consequently, they are well developed and

Fig. 11.10. Suggested protetrapod stage, between crossopterygian and labyrinthodont. (From Gregory and Raven 1942.)

never lost. The coracoids, on the other hand, may only be moderately well developed. Immediately behind their point of fusion the epicoracoid cartilages diverge and overlap and their posterior margins are continued as epicoracoid horns, which run in lateral grooves on each side of the sternum. The posterior tip of each horn has a muscle attachment connecting with abdominal recti.

This type of sternum/epicoracoid system permits a certain degree of independent movement of the girdle halves whilst, at the same time, preventing the epicoracoid cartilages from being forced too far apart. The mechanism clearly facilitates the independent arm movements characteristic of locomotion in the arciferal frogs.

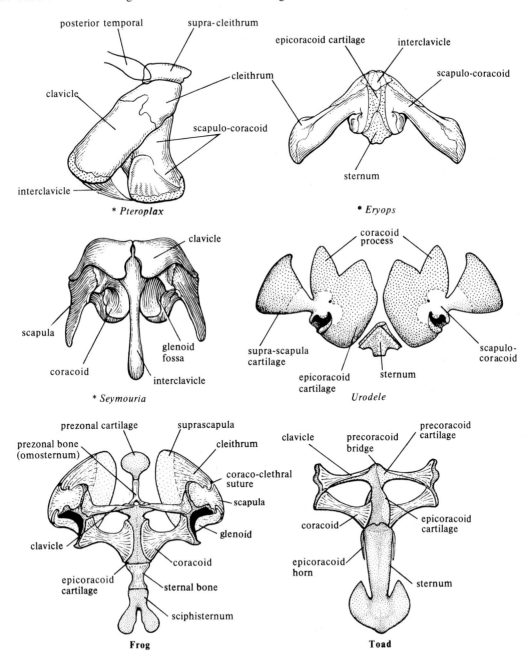

Fig. 11.11. Amphibian pectoral girdles.

The firmisternal girdle is a rigid structure allowing no independent movement of the two halves (Fig. 11.11). It occurs typically in frogs with a jumping habit (e.g. Ranidae, Microhylidae) and provides an excellent landing mechanism. The glenoids are braced apart by the large coracoids. The clavicles and precoracoids are thus deprived of their strutting function and frequently become reduced or even completely lost. No epicoracoid horns are present and the sternum, no longer involved in locking the girdle halves, serves principally for the attachment of pectoral muscles. This function is also performed, in some frogs, by a prezonal (omosternal) element, which is really an extension of the precoracoid cartilages. This structure, although present in some arciferal girdles (e.g. Leptodactylidae, Fig. 11.12), is more usually associated with the firmisternal pattern and a jumping habit.

The shoulder girdles of modern urodeles are greatly simplified, the only ossification being a scapulocoracoid encircling the glenoid. The two epicoracoids overlap broadly, and anteriorly are quite free of each other; posteriorly they are usually rather weakly locked by a cartilaginous sternum. The Apoda, of course, retain no vestiges of either limbs or limb girdles.

8. Pelvic girdle

The pelvic girdle is much larger in land animals than the small ventral cartilages found in fishes. It is formed of three main cartilage bones in all tetrapods (Fig. 11.13), but it is not clear how these originated, nor whether the division has mechanical significance. The dorsal ilium becomes attached to specially modified transverse processes of one or more sacral vertebrae. This ilium can be regarded mechanically as the ossification along a line of compression stress due to the weight-bearing.

The ventral portion of the girdle consists of an

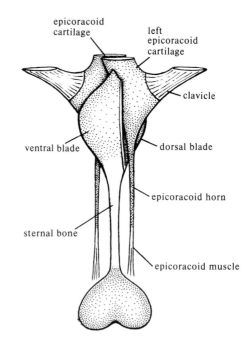

FIG. 11.12. *Leptodactylus prognathus*. Ventral view showing sternal articulation with girdle. The left half of the ventral sternal blade has been removed.

anterior pubis and posterior ischium, the three bones meeting at the acetabulum, where the femur articulates. The girdle thus provides a plate to which the muscles that brace the limb can be attached in such a way as to balance the body on the leg.

In urodeles the pelvic, like the pectoral, girdle becomes reduced and mainly cartilaginous. The pelvic girdle of anurans is highly specialized and unlike that of any other vertebrate. The ilia are very long and directed

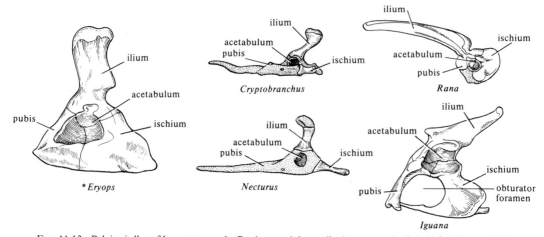

FIG. 11.13. Pelvic girdles of lower tetrapods. Regions mainly cartilaginous are stippled. (After Evans 1944.)

forward to articulate with the transverse processes of the single pair of sacral vertebrae. The base of the ilium is expanded to make the dorsal portion of a disc, of which the pubis is the anterior, the ischium the posterior part, with the acetabulum at the centre. The girdle is thus developed into a long lever for transferring force from the limb to the vertebral column during jumping. The two acetabula lie close together, reducing the effect of differences in the thrust of the two legs (Cox 1966).

Considerable movement is possible at the iliosacral joints, at least in Salientia (Whiting 1961). In *Rana* the ilia may rotate through an angle of over 90° on the sacral ribs in the vertical plane. This movement is used during a strong leap. In *Discoglossus* the sacrum can be turned laterally on the pelvis through 20°. The movement is used both in turning to take food and in locomotion. In *Xenopus* the sacrum can slide backwards and forwards on the pelvis, producing a considerable shortening and lengthening of the whole animal. This movement is probably used in driving into the mud for food.

9. The limbs

The pattern of bones and muscles in fore and hind limbs of tetrapods is surprisingly constant in spite of the various uses to which the limbs are put. Evidently similar morphogenetic processes are at work in both limbs. There are nearly always three main joints in each limb, at shoulder (hip), elbow (knee), and wrist (ankle). The hand and foot provide basically similar five-rayed levers, with several joints in the digits (Fig. 11.14).

The bones of the limbs can be plausibly derived from those of a crossopterygian fin, and indeed the condition

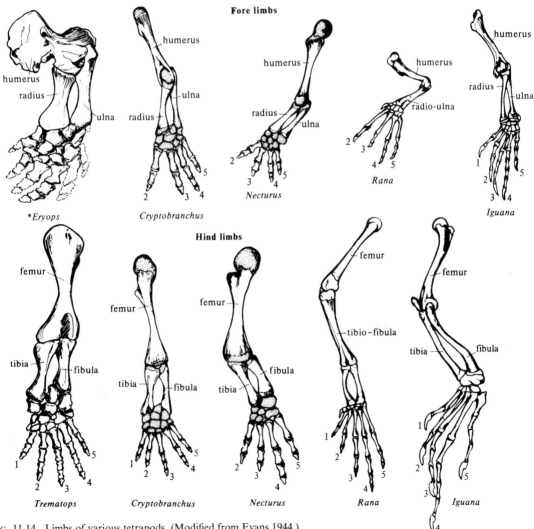

Fig. 11.14. Limbs of various tetrapods. (Modified from Evans 1944.)

in *Eusthenopteron* already distinctly suggests that of the limb of an early amphibian (Figs. 11.7 and 11.8). We know less about the origin of the hind than of the front leg, but the two are so similar that they may be treated together for elementary analysis. There is a basal humerus (femur), articulating distally with two bones in each case, a more anterior (pre-axial) radius (tibia) and a posterior (post-axial) ulna (fibula). These bones articulate at the wrist or ankle with a carpus or tarsus, consisting, in the fully developed condition, of three rows of little bones, namely 3 in the proximal row, about 3 centrals, and 5 distals. Each of the latter carries a digit, composed of numerous jointed phalanges. In naming these bones of the carpus and tarsus it is convenient to call the proximal carpals by their position radiale, intermedium, and ulnare and the tarsals tibiale, intermedium, and fibulare. The centrals and distal carpals may then be numbered beginning with 1 at the pre-axial border in each case. Unfortunately other less explicit systems of naming are in use, as shown in the following table 11.1 which includes the names used in mammals and by human anatomists.

The plan of the carpals and tarsals can well be imagined to have been derived from that of a fin such as is seen in the fish *Eusthenopteron* (Fig. 11.7), which might be said to have humerus, radius, and ulna, carpals, and 7 or 8 digits. In the amphibian *Eryops* (Fig. 11.1) most of the digits radiate from the radius, in later forms mostly from the ulna (Fig. 11.14). Moreover, in the hand of *Eryops* there seem to have been six digits and it is usually stated that the first of these is a prepollex 'not comparable with the pollex of higher forms'.

The effect of this system is to provide a lever that can be held firmly against the ground while it is moved by the muscles running from the girdles to the humerus or femur. In addition the lever is itself extensible by means of its own muscles. Whatever may have been their origin in fishes these muscles in tetrapods work in such a way as to bend each segment up and down. The shoulder and thigh joints usually allow movement in several planes, both towards and away from the midline (adduction and abduction), and forwards and backwards (protraction and retraction). As we have seen, the animal balances at these joints by muscles arranged round them. Movements of rotation are also possible at these, and sometimes at other joints, the distal bone turning about its own axis on the proximal one. Such movements may be very important for the proper placing of the limbs in walking. Pronation is the rotation of the radius about the ulnar bone, so that the manus is directed caudally, supination being the opposite movement. The terms flexion and extension are convenient at certain joints (e.g. the elbow), but have no consistent meaning with reference to the main axes of the body.

The limbs of the earlier amphibians were ponderous affairs, with large bones and widely expanded hands and feet (Figs. 11.7 and 11.10). It is not certain exactly how they were used; probably they were held out sideways, giving a wide base on which the somewhat precarious balance was maintained, the body being often slumped on to the ground. In modern urodeles the limbs retain the full pattern of parts, but with imperfect ossification, as would be expected since they carry little weight.

In frogs, specialized for jumping, the radius and ulna are united and the carpals are reduced in number. There are only four true digits, the first digit (thumb or pollex)

TABLE 11.1. *Plan of the tetrapod carpus and tarsus. (The names used for the bones in man are shown in brackets.)*

	CARPUS		
	Pre-axial		*Post-axial*
Proximal	radiale (scaphoid)	intermedium (lunate)	ulnare (triquetral)
Central		centrale (tubercle of scaphoid)	
Distal	carpal 1 (trapezium)	carpal 2 (trapezoid)	carpal 3 (capitate)	carpals 4 and 5 (hamate)
	TARSUS		
	Pre-axial		*Post-axial*
Proximal	tibiale (talus or astragalus)	intermedium (os trigonum)	fibulare (calcaneum)
Central		centrale (navicular)	
Distal	tarsal 1 (medial cuneiform)	tarsal 2 (intermediate cuneiform)	tarsal 3 (lateral cuneiform)	tarsals 4 and 5 (cuboid)

being reduced. There is, however, a small extra ossification, the prepollex, which becomes well developed as a copulatory organ in the male and may be compared with a similar digit found in some stegocephalians. It is to be expected that in a system of repeated parts, such as a tetrapod limb, multiplications and reductions will be common. It can be imagined that they can be produced by changes in the rhythm of morphogenetic processes, and it is surprising that there is such constancy in number of digits.

The hind legs of frogs are long, giving a good leverage in jumping. The tibia and fibula are united and the proximal row of tarsals is reduced to two, greatly elongated and known as the tibiale (astragalus or talus) and fibulare (calcaneum). The distal tarsals are reduced to a total of three, bearing five 'true' digits and an extra one, the calcar or prehallux.

10. The back and belly muscles

With the change in the method of locomotion the muscular system has become greatly modified from that found in fishes. In urodeles, which still use the old method and hence may be said to swim on land, the dorsal musculature is well developed (Fig. 11.15), but in anurans the dorsal portions of the myotomes, the epaxial musculature, no longer have to produce the locomotory effect by lateral flexion. They remain in frogs only as muscles that bend the body dorsally, serving to brace the vertebral column on the sacrum

(Figs. 11.16 and 11.17). Short muscles run between the vertebrae, and dorsal to these is a continuous sheet of longitudinally arranged fibres, the longissimus dorsi muscle, running from head to sacral vertebra and urostyle. This muscle, though forming a continuous band, is crossed by a tendinous intersection, showing its segmental origin. At the hind end the coccygeo-sacralis and coccygeo-iliacus muscles brace the urostyle on the pelvic girdle.

The pectoral girdle is attached to the axial skeleton by a series of muscles. Rhomboid and levator scapulae muscles run from the suprascapula to the vertebrae and skull. The cucullaris muscle corresponds to the mammalian sternomastoid, running from the skull to the suprascapula; it is derived from lateral plate musculature and innervated by the vagus. The naming of these muscles of the scapula, and indeed all amphibian muscles, meets the difficulty that many of the bundles of fibres are similar in their general course to muscles found in mammals and yet differ sufficiently to raise serious doubts about the wisdom of using the mammalian names. The similarity of arrangement of the limb muscles is so striking throughout the tetrapods that there is probably no harm in keeping to the well-established system of names, but we know so little of the hereditary or mechanical factors that control the arrangement of muscle fibres into 'muscles' that discussion of homologies is difficult.

The hypaxial musculature, formed from the more

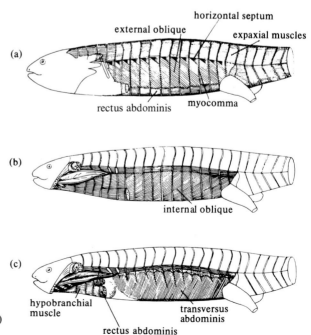

FIG. 11.15. Muscles of larval *Ambystoma*. (a), (b) and (c) show successive layers.

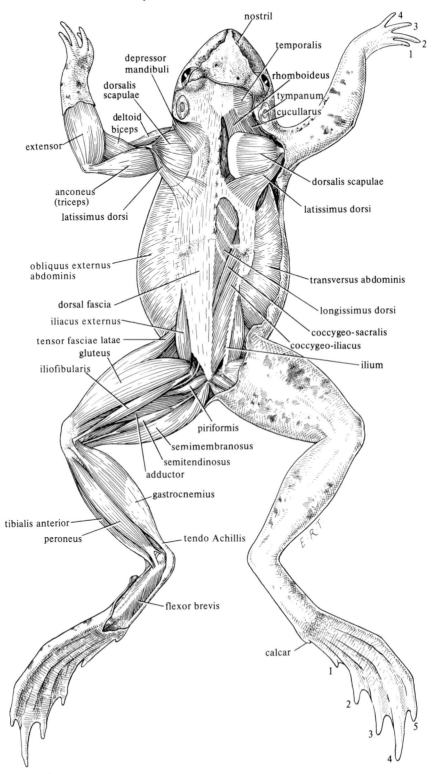

FIG. 11.16. *Rana temporaria* dissected from the back to show the muscles. (Partly after Gaupp 1896.)

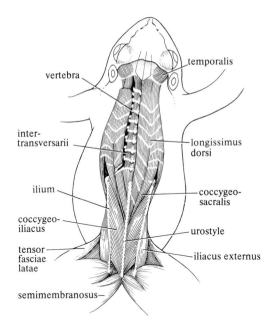

vertebra

temporalis

inter-
transversarii

longissimus
dorsi

ilium

coccygeo-
sacralis

coccygeo-
iliacus

urostyle

tensor
fasciae
latae

iliacus externus

semimembranosus—

FIG. 11.17. Deeper dissection of muscles of back of frog. (Partly after Gaupp 1896.)

ventral portions of the myotomes, is more developed than in fishes and differentiated into several parts, for the purpose of slinging the viscera, which of course need support in air in a way that is unnecessary in water (Fig. 11.18).

These muscles are differentiated into layers whose fibres run in different directions. The plan found, with modifications, in all tetrapods is seen in amphibian larvae and includes four sets of fibres. The external obliques run caudally and ventrally; inside this layer is the internal oblique, running in the opposite direction, and within this again the transversus abdominis running approximately dorsoventrally (Fig. 11.15). The rectus abdominis consists of fibres in the midline running anteroposteriorly.

In the adult frog three of these sets of fibres can be recognized. In the mid-ventral region (Fig. 11.19) are the longitudinally arranged fibres of the rectus abdominis, making a sling between the sternum and the pubis. These fibres are interrupted at intervals by transverse fibrous tendinous divisions, giving an appearance of segmentation. In the mid-ventral line is the tendinous linea alba. The sling formed by the rectus abdominis is supported laterally by thin sheets of muscle fibres running up to the vertebral column, the obliquus externus and transversus abdominis (Fig. 11.19).

In the anterior region the hypaxial muscles have become restricted to the throat, where they form the hyoid musculature, which by raising and lowering the

floor of the mouth is the main agent of breathing. The intermandibular muscle runs transversely between the rami of the lower jaw. Deep to this lie other muscles, including the sternohyoid, close to the midline, which is a forward continuation of the rectus abdominis.

11. The limb muscles

The muscles of the limbs were presumably derived from the radial muscles that moved the fins of fishes. These are formed from the myotomes and they are mainly arranged so as to raise and lower the fin (Fig. 11.20). In modern Amphibia the limb musculature is still partly formed from myotomes (Griffiths 1963). The segmental origin of the limbs is also shown by the fact that they are innervated by branches of the spinal nerves of several segments (2 for the fore limb, 4 for the hind limb in the frog). Presumably the original arrangement was such as to move the limbs in association with the waves of contraction passing down the body. In modern urodeles the limb is brought forward and its joints flexed as the epaxial muscles at the level of its front end contract, and then passes back and extends as the wave of contraction moves past. This may have been the primitive movement, making the limb more useful as a lever during the early attempts to 'swim on land' (Fig. 11.4).

The muscles of the limbs of tetrapods are presumably derived from those that raise and lower the fins of fishes, modified, as we have seen, to brace the limbs and move them, allowing standing and walking. The muscles that run from the girdles to the humerus and femur are therefore able to draw the leg forward and backward, as well as to raise and lower it in the transverse plane. The actions of the various bundles are of course not confined to a single plane: all the muscles running from the back to the humerus can raise (abduct) the upper limb, but the more anterior members also protract, the more posterior retract it. Similarly there is a ventral series whose anterior members work with the anterior dorsal muscles as protractors, although they antagonize the action of raising the whole limb. Moreover, many of the muscles have a rotating action on the humerus and femur. It is, however, possible to consider the muscles of the arm and leg in two great groups; first a more anterior and ventral ('ventrolateral') set serving to draw the limb mainly forward and towards the midline (protraction and adduction) and to flex its more distal joints, second a more posterior, dorsal ('dorsomedial') mass serving mainly to draw the limb backwards and away from the body (retraction and abduction) and to extend its joints.

In the fore limb the proximal members of the ventral group make a sheet of fibres running transversely to the main body axis, attached to the sternum and hypaxial muscles at one end and to the humerus at the other

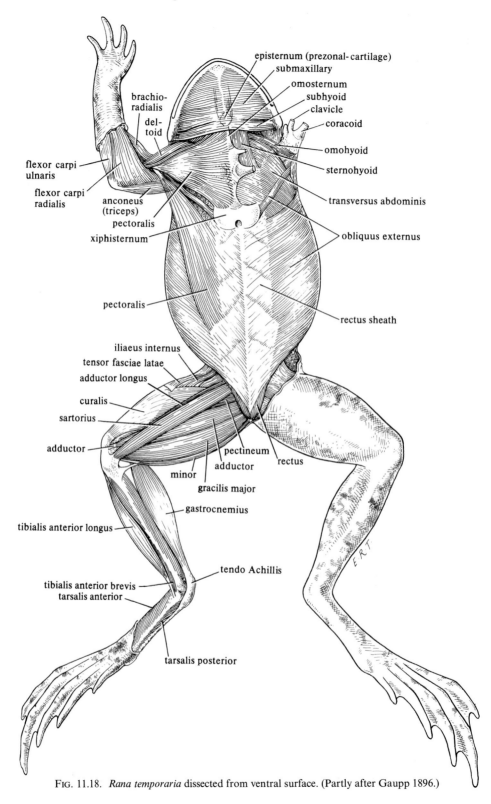

Fig. 11.18. *Rana temporaria* dissected from ventral surface. (Partly after Gaupp 1896.)

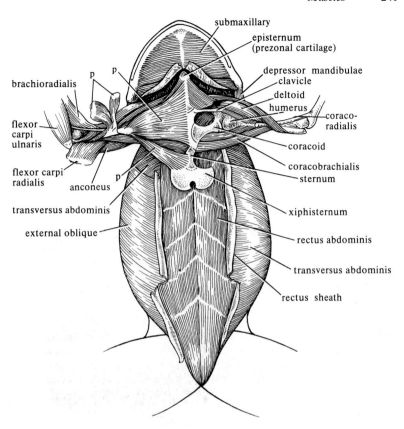

FIG. 11.19. Dissection of muscles of frog from ventral surface; *p*, pectoralis. (Partly after Gaupp 1896.)

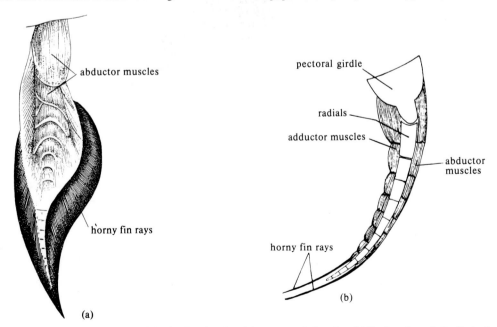

FIG. 11.20. (a) *Neoceratodus*, lateral surface of the fin showing the abductor muscle bundles. (b) Section through the fin in the transverse plane, showing the arrangement of the muscle bundles as abductors and adductors. (From Ihle, after Brauz.)

(Fig. 11.19). Within this sheet can be recognized the deltoideus, pectoralis, coracoradialis, and coracobrachialis muscles. In the limb itself this group is continued, there being, roughly speaking, a set of muscles in each segment that serves to flex it on the next. Thus the brachioradialis flexes the elbow joint and in the forearm the flexor carpi radialis and flexor carpi ulnaris flex the wrist. The flexor digitorum longus muscle arises from the medial epicondyle of the humerus and is inserted by tendons to the carpus and terminal phalanges. Flexor digitorum brevis muscles arise from this tendon for insertion on the digits.

Of the dorsal muscle mass (Fig. 11.16) the latissimus dorsi and dorsalis scapulae are the most proximal, running from the middle of the back to the humerus and serving to abduct and draw back the whole limb. The triceps (anconeus) serves to extend the elbow; in the forearm are extensor carpi ulnaris and radialis and extensors for the fingers.

According to this plan protractor (flexor) muscles lie mainly anterior to retractors (extensors), corresponding to the ancient movement by which the limb was drawn first forward then back as a swimming wave passed down the body. In all tetrapods flexor muscles are in general innervated by spinal roots anterior to those for the extensors. The locomotory movements of the limbs therefore still show the passage of an excitation wave backwards along the spinal cord, a relic of the swimming rhythm of fishes. However, the changes that have taken place in the relative positions of the parts of the limbs make it difficult to follow out this simple pattern in detail. It must also continually be remembered that many muscles produce rotation as well as movement in the main planes of the body.

In the hind limb, muscles of the same two general types can be recognized, namely anterior muscles, which draw the limb forward and flex and adduct its joints, and posterior ones, which draw it back and extend and abduct. The specialization of the main muscle masses has gone much farther, however, so that more individual muscles are found, especially round the hip joint, each serving to move the limb in a special way.

In the thigh (Figs. 11.16 and 11.18) the muscles of the anterior group, lying on the ventral surface, are the pectineus and the adductors, running from the pelvic girdle to the femur and thus serving to move the whole limb inwards (adduction). The sartorius, biceps, semimembranosus, and semitendinosus are two-joint muscles mainly producing flexion at the knee as well as at the hip.

The more posterior and dorsal group of muscles includes the gluteus and tensor fascia lata from girdle to femur, extending the thigh joint, and the very large cruralis (including the rectus femoris and triceps femoris) running from girdle and femur to tibia. This is the main extensor of the knee, being helped by gracilis and semimembranosus. This extension is obviously an important part of the jumping movement of the frog.

In the shank the arrangement of the flexors and extensors into the anterior and posterior groups is much modified. The more conspicuous muscles are the tibialis anterior and peroneus running from the femur to the tarsus so as to flex the ankle joint. Long and short flexors move the toes, as in the fore limb. At the back of the tibio-fibula the gastrocnemius (plantaris longus) runs from the femur to be attached by the tendo Achillis to the tarsus. Its main action is to extend the ankle in the movements of jumping and swimming. Tibialis posterior runs from the tibia to the tarsus. Within the foot there is an elaborate system of small muscles for bending and stretching the toes and abducting them away from each other, so as to expand the web for swimming.

The whole system is designed to produce the characteristic sudden simultaneous extension movement of all the joints of both hind limbs, by which the frog swims in water and jumps on land. The hind limbs can also be used for alternate walking movements, especially in toads (Fig. 11.5).

12. The skull of Labyrinthodontia

The skull of the Devonian and Carboniferous Amphibia was essentially like that of the osteolepid fishes in the arrangement of the bones, but the proportions had been altered so that the pre-optic region was relatively large and the more posterior 'table' of the skull short (Fig. 11.21).

The nasals and frontals, which were small in crossopterygians, were quite long in labyrinthodonts, whereas the parietals were shorter and the post-parietals absent altogether in the later forms. The difference is so marked that for a long time people were deceived in identification of the bones and it was said that the pineal opening lay between the frontal bones in fishes but between the parietal bones in tetrapods. The bones identified as 'frontal' in the fish types were, of course, parietals, whereas the 'parietals' were the post-parietals, which have gone completely from most amphibians, though still present in the earliest Devonian forms (Fig. 11.22). This is an excellent example of how study of changes of proportion can clear up morphological difficulties.

The opercular apparatus covering the gills was lost early in amphibian evolution; perhaps the reduction of the whole posterior part of the head was effected by a single morphogenetic change. In modern Amphibia the skull is much flattened and its ossification reduced, so

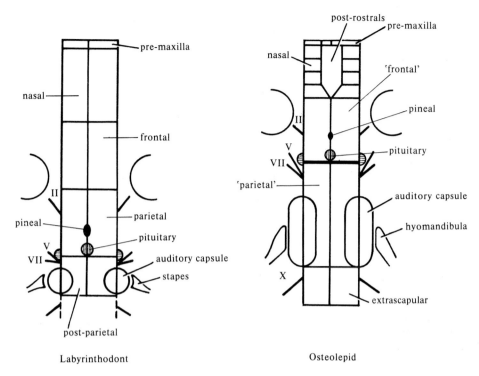

Labyrinthodont   Osteolepid

Fig. 11.21. Diagram of skull bones and other structures. *II, V, VII, X* show emergence of cranial nerves. (After Westoll 1943.)

that large spaces are left; in the earlier forms, however, the skull was of the more usual domed shape and the roof and jaws were covered by a complete set of dermal bones. Presumably the loss of bone was another development producing a reduction of weight advantageous to a terrestrial animal.

Lateral-line organs are present in aquatic amphibians and their position is marked on the bones of the fossil skulls by rows of pits. By using these lines as reference marks it is possible to compare the pattern of the bones on osteolepid and early amphibian skulls and to confirm the remarkable similarity. The main new development found in the skull of early amphibians was correlated with the modification of the Eustachian tube in connection with the sense of hearing, and the need for a sensitive resonator to pick up the air vibrations. Already in the earliest amphibians the opercular coverings of the gills were lost (there was a small pre-opercular bone in *Ichthyostega*) and the spiracular opening thus uncovered acquired a tympanic membrane. The hyomandibular cartilage, no longer concerned (if it ever had been) with supporting the jaw, was modified to form the columella auris, serving to carry vibrations across to the inner ear. In modern urodeles the whole ear apparatus is much modified, there being no tympanum. Instead the columella is fused to the squamosal and the ear thus

receives its vibrations from the ground.

Other small changes in the skull in passing from the fish to the amphibian stage include the increase in size of the lachrymal bone, which also came to have a hole to carry the tear duct, draining the orbit. A series of small bones surrounds the orbit in early amphibians, as in fishes. Large squamosals and quadratojugals support the quadrate. At the back of the skull these early Amphibia possessed various of the small bones that are found in fishes but not in modern amphibians, the supratemporal and intertemporal, post-parietal (much smaller than in crossopterygians) and tabulars. In fact there are numerous small bones, arranged in a pattern clearly recalling that of the fish ancestor, but showing some reductions and less variation than in those very variable fish skulls. This simplification (which was later carried farther), together with some changes in the shape, are the chief transformations that have converted the fish skull into the amphibian skull.

The palate of the early amphibians also resembled that of crossopterygians, showing a complete plate made of vomer, palatines, pterygoids, and ectopterygoids. These bones, as well as the premaxillae and maxillae, often carried folded teeth (hence 'labyrinthodonts'), with a pit for a replacing tooth beside each one, an arrangement similar to that of their fish

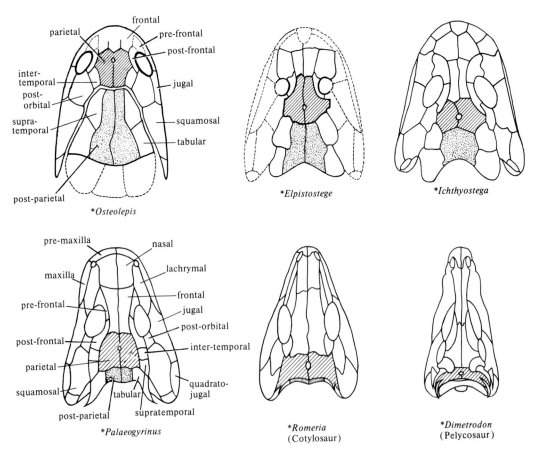

FIG. 11.22. Skulls of a crossopterygian and various early tetrapods to show the shortening of the posterior region. (After Westoll 1943.)

ancestors (p. 218). The internal nostril opened far forward, through the palate. The lower jaw was covered by a number of dermal bones (Fig. 12.1), but the actual jaw articulation was made between cartilage bones, the upper quadrate, and the lower articular.

13. The skull of modern Amphibia

Modern amphibia share several cranial features that distinguish them from typical labyrinthodonts. The number, extent, and thickness of the dermal elements are greatly reduced so that the otic capsules are generally exposed. The orbits and interpterygoid vacuities are large, the mandibular ramus is short and the skull as a whole much flattened. The occiput is shortened so that the hypoglossal nerve emerges behind the skull and (with the few exceptions noted below) the parietal foramen has been lost.

The skull of the frog (Fig. 11.23) shows great reduction and specialization from the early amphibian type. It may be considered as consisting of a series of

cartilaginous boxes or capsules, in whose walls some ossifications occur, partly covered by dermal bones. The cartilaginous boxes, well seen in a tadpole's skull, are the central neurocranium around the brain, and the olfactory and auditory capsules. Ossifications occur especially at the points of compression stress, namely, around the foramen magnum (the exoccipitals), where the auditory capsule joins the cranium (the pro-otic), and at the base of the nasal capsules (the mesethmoid). The paired occipital condyles are found only in modern Amphibia, and are formed by the failure of the basioccipital to become ossified (Fig. 11.24). Paired occipital condyles have also arisen, independently, in the mammal-like reptiles.

The dermal bones covering the roof of the skull are the nasals and frontoparietals, while on the floor is the large dagger-like bone, the parasphenoid, and a small tooth-bearing vomer. The remains of the cartilaginous palatopterygoquadrate bar can be recognized as a rod, covered in front by premaxillae and maxillae, and

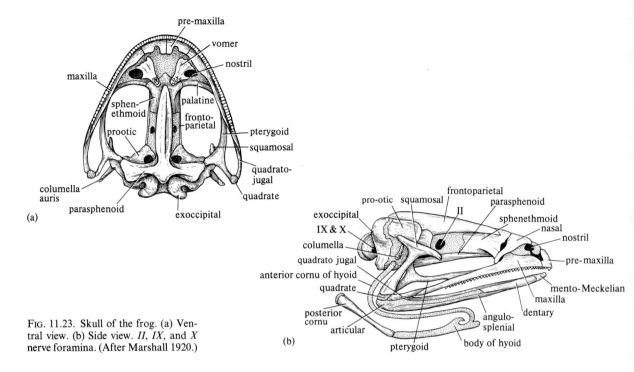

Fig. 11.23. Skull of the frog. (a) Ventral view. (b) Side view. *II, IX,* and *X* nerve foramina. (After Marshall 1920.)

dividing behind into an otic process fixing it to the skull (autostylic) and a cartilaginous quadrate region articulating with the lower jaw. This region is covered by the pterygoid ventrally, the quadratojugal laterally, and the squamosal dorsally. The palatines are membrane bones forming the anterior wall of the orbit. The upper jaw is thus supported by struts formed from the nasals and palatines in front and the squamosal and pterygoid behind, an arrangement that gives a large mouth for respiration and eating insects, combined with the advantages of strength, mobility of the lower jaw, and lightness in weight.

In the lower jaw Meckel's cartilage remains complete but ossified in front as the mento-Meckelian bones. Its hinder part forms the articular cartilage. There are two dermal bones, a large dentary laterally and more medially an angulosplenial (Fig. 11.23).

The visceral arches are well formed in the tadpole but are much modified in the adult frog. In the tadpole the skeleton of the hyoid arch consists of a large pair of ceratohyals attached to a basal hypohyal. As a result of subsequent metamorphosis the ceratohyals later form the long anterior cornu of the hyoid, attached to the pro-otic bone. The body of the hyoid is a plate lying in the floor of the mouth and formed from the hypohyal and from the hypobranchial plate at the base of the remaining arches. The posterior cornua support the floor of the mouth and the whole apparatus assists in

respiration. The sixth and seventh of the series of branchial arches give rise respectively to the arytenoid and cricoid cartilages of the larynx.

The lateral plate muscles of the branchial arches are well developed only as the muscles of the jaws. Certain muscles of the scapula (the cucullaris and interscapularis) are innervated from the vagus and recall the sternomastoid and other muscles innervated by the spinal accessory nerve in mammals.

The muscles of the hyoid arch, innervated by the facial nerves, remain mainly as the depressor mandibulae (Fig. 11.25) running from the back to the angle of the jaw and serving to lower the floor of the mouth. The jaw-closing muscles, the adductor mandibulae muscles, belong to the mandibular segment and are innervated by the trigeminus. They run from the hind end of the jaw to the surface of the skull and squamosal.

The skull and jaws of the frog thus constitute a protection for the brain and special sense organs, a feeding apparatus, and a means of respiration. The heavy protection afforded by the dermal bones of fishes and early amphibians has been largely dispensed with, probably for lightness. The front part of the skull, concerned with the nose, eyes, and brain, has become increased in size and the hind part, originally concerned with the gills and pharynx, greatly reduced. These changes, carried to extremes in frogs, have been in progress throughout the evolution of Amphibia. It is

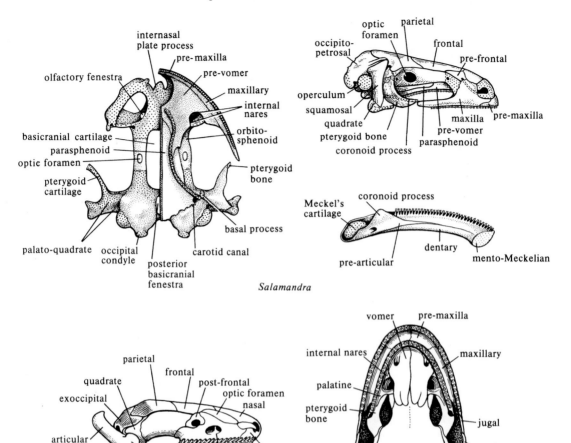

Fig. 11.24. Skulls of amphibians.

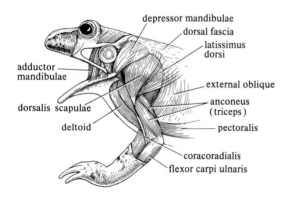

Fig. 11.25. Muscles of head and neck of frog dissected from the side. (Partly after Gaupp 1896.)

not difficult to imagine that they have been the result of rather simple genetic changes, affecting the relative growth of various parts of the skull. We are still far from the knowledge necessary to say exactly what developmental changes have occurred. However, we know enough to imagine how selection through millions of years has changed the quantities of certain substances so as to produce less bony and shorter heads. These enabled their possessors to maintain sufficient mobility to hold a place in a world peopled by the reptiles and other still more active descendants of the early amphibians.

The preceding account, particularly with regard to the osteology, should not be regarded as diagnostic of all anurans. Bufonid skulls are completely devoid of teeth. They possess a supratemporal bone, which fuses with the squamosal and roofs the otic capsule. Hylids frequently develop secondary dermal ossifications to form expanded helmets; this trend also occurs in leptodactylids (e.g. *Calyptocephalus*), where the skull may be completely roofed and sculptured. Pseudoteeth (serrations of the jaw elements) frequently occur on the dentary and pre-articular (e.g. *Amphodus*) but the only modern form to possess true teeth on the lower jaw is *Amphignathodon*. No recent frog retains the large parietal foramen so typical of the fossil amphibia but some leptodactylids and the aquatic xenopids have a small canal perforating the frontoparietal, through which runs a fibro-nervous tract from the pineal organ to the habenular ganglion (Griffiths 1954). The anuran skull is always easily distinguished from those of all other Amphibia by the fact that the frontals are fused with the parietals.

Urodele skulls are, in some respects, less specialized than those of Anura. The frontals and the parietals remain discrete and in certain species both lachrymals and prefrontals are present. In other respects they are clearly more degenerate (or paedomorphic?). No urodele has either a jugal or quadratojugal (except *Tylotriton*) and in perennibranchs even the maxillaries and nasals are lost. Urodeles are further distinguished from frogs (but not from caecilians) by the great size of the prevomers (each consisting really of a prevomer + palatine) and by the possession of a tooth-bearing coronoid, as well as a dentary and a prearticular.

The apodan skull is a much more rigid structure than that of either of the above subclasses and, at first sight, approaches more closely to the ancestral pattern. The number of bones present, however, is no greater than in any of the other modern groups. The overall compactness is effected particularly by the expansion of the nasals and of the marginal elements of the upper jaw and is probably correlated with the burrowing habits of the group. Lower as well as upper jaws carry teeth and a toothed coronoid is present in the mandible.

14. Respiration in Amphibia

The new problems presented by life on land have led to the production of very varied means of respiration among amphibia. In a terrestrial habitat oxygen is available in plenty; the difficulty is evidently to arrange for a regular interchange of air in contact with adequately moistened surfaces. The interchange is provided for in most cases by modifications of the apparatus used in fishes, but pumping air presents new problems and it seems that these are not easily solved, since in many amphibians the skin is used as an accessory respiratory mechanism. The retention of moisture becomes more difficult as the ventilation becomes efficient; probably

for this reason air is often only transferred to the lungs after it has remained for some time in the mouth. We see again that the new way of life, in a medium remote from water, makes it necessary to possess more complicated methods of self-maintenance.

15. Respiration in the frog

The lungs of the frog are paired sacs, opening to a short laryngeal chamber, which communicates with the pharynx by a median aperture, the glottis. The glottis and laryngeal chamber are supported by the arytenoid and cricoid cartilages. The arytenoids guard the opening of the glottis and are moved by special muscles. During breathing the mouth is kept tightly closed, the lips being so arranged as to make an air-tight junction. Air is sucked in through the nostrils by lowering the floor of the mouth by means of the hypoglossal musculature, and can then either be breathed out again or forced into the lungs by raising the floor. The external nares are closed by a special pad on the anterior angle of the lower jaw, supported by the mento-Meckelian bones. This pad is thrust upwards and pushes the premaxillaries apart, so altering the position of the nasal cartilage that the nostrils are closed. This is a special mechanism, found only among anurans. In urodeles the nostrils are closed by valves provided with smooth muscles. Such valves are present in the frog but are said to be functionless.

The movements of the floor of the pharynx are not continuously of the same amplitude. After a period of relatively slight movements the nostrils are kept closed while the throat is lowered. Air is thus drawn from the lungs and then again returned to them once or twice before the nostrils are reopened. The whole procedure presumably ensures the maximum gaseous interchange for the minimum water loss.

This force-pump system, using the buccal cavity as a bellows is clearly derived from the movements of the floor of the mouth of fishes, by which water is passed over the gills. In amphibian larvae water is pumped in this way and there is direct continuity between the mechanism of larva and adult. The basic rhythmic mechanism, centred on the nerve cells of the medulla oblongata, is no doubt the same throughout, but the anurans have improved upon it by the addition of special features, requiring intricate co-ordination of the muscles of the larynx and the apparatus for closing the nostrils.

The skin is very vascular, and especially so in the buccal cavity. It plays a large part in respiration, actually serving to remove more carbon dioxide than do the lungs. There is, however, little power to vary the amount of exchange through the skin, which is therefore constant throughout the year. There is considerable regulation of the exchange in the lungs. The rate of breathing depends, as in mammals, on the effect of the carbon dioxide tension of the blood on a respiratory centre in the medulla. There is also a vasomotor control of the blood supply to the lungs and, through the vagus nerve, of the state of contraction of the latter. By such means the rate of respiratory exchange is greatly increased during the breeding season, and made to vary with the activity of the animal.

16. Respiratory adaptations in various amphibians

The skin and the lungs show many variations according to the habitat of the species, special devices being adopted to enable the animals to live in particular environments. The lungs vary from the well vascularized sacs with a highly folded surface found in the frogs, and especially in the drier-skinned toads, to small simple sacs in some stream-living amphibia. The lung will serve to lift the animal in the water; for this reason it is reduced in the frog *Ascaphus*, which lives in mountain streams in the eastern USA. In newts this hydrostatic function of the lungs is predominant and the inner surface is often quite simple. The lung is entirely lost in stream-living salamanders, such as the European alpine *Salamandra atra* and in the North American plethodontid salamanders. The coldness of the water reduces activity and lowers the need for respiratory exchange to a level at which it can be fully met by the skin. The skin shows increased vascularity in these forms with reduced lungs, capillaries reaching nearly to the outermost layers of the epidermis. In the African frog *Astylosternus*, in which the lungs are vestigal, the male develops vascular papillae on the waist and thighs during the breeding season.

Gills are present in amphibian larvae, and also in neotenous adult urodeles that may be considered as larvae that have failed to undergo metamorphosis (p. 254). The gills are extensions of the branchial arches, and carry branched villi, richly supplied with blood. The spiracular pouch forms the Eustachian tube and the next two or three pouches open as the gill slits. In anurans a fold of tissue, the operculum, grows back from the hyoid region and fuses with the skin behind the gills to make an atrial chamber, opening by a 'spiracle' on one side only. Where the main trunk is long the gill projects and is 'external', whereas in other cases, as in the later frog tadpole, the filaments are directly attached to the arches and are called 'internal'. There are no profound differences between the two types.

17. Vocal apparatus

The production of air-borne sounds by a larynx is another characteristic tetrapod feature. In anurans sound is used as a protective (fear) response and by the male frog as a call to attract the female. Both sexes have vocal organs, but those of the female are much smaller. The noise is produced by the vibration of the elastic edges of a pair of folds of epithelium of the laryngeal chamber, the vocal cords. Air is passed backwards and forwards between the lungs and a large pair of sacs (or a single median sac), the vocal pouches, formed below the mouth. These also serve as resonators, and are developed only in the male. Sound production is controlled from the tegmentum, presumably an area related to respiration. Electrical stimulation here produces some sounds but the full mating calls are produced only from the pre-optic region of the hypothalamus, an area concerned with other aspects of reproduction. Calling is abolished by castration.

Various urodeles make noises, some by means of a larynx others by the lips during inspiration.

18. Circulatory system

The venous and arterial systems are less fully separated in Amphiba than in lung fishes. In anurans and some urodeles the auricles are completely divided by an interauricular septum, blood returning from the tissues and skin to the right, and arterial blood to the left auricle. There is only a single ventricle, but this is provided with spongy projections of its wall, which may prevent the mixing of the blood. The ventral aorta (conus arteriosus), springs from the right side of the ventricle and may thus receive first the less oxygenated blood. The conus arteriosus has transverse valves and a longitudinal 'spiral valve' which directs blood ventrally to the carotid and systemic arches, and dorsally to the pulmonary arch.

The ventral aorta is very short and the arches much modified in the adult (Fig. 11.26). Of the original six that can be recognized the first two disappear, the third on each side gives rise to the carotid artery, the fourth remains complete and forms the systemic arch. The fifth remains also in some urodeles, but disappears in anurans. The sixth arch becomes the pulmonary artery and loses its connection with the dorsal aorta: special 'cutaneous arteries' carry deoxygenated blood from this arch to the skin (Fig. 11.27). It was previously held that these pulmonary arches received the first blood leaving the ventricle because they offer a lower resistance than do the systemic and carotid ones. The pressure in the latter was said to be increased by a special network, the 'carotid gland'. It has also been held that this organ is a receptor, connected with regulation of the blood pressure. The current evidence is that it has no innervation but secretes adrenaline (Gouder and Desai 1966). By use

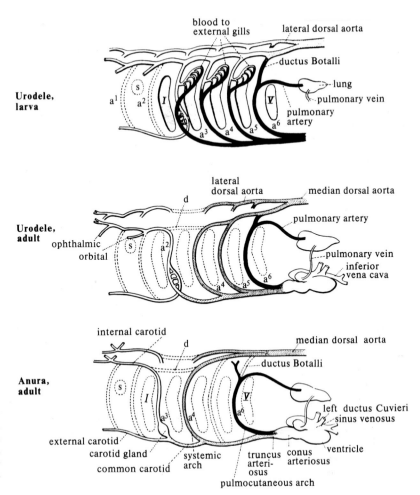

blood to external gills
lateral dorsal aorta
ductus Botalli
lung
pulmonary vein
pulmonary artery

Urodele, larva

a^1 a^2 I V a^3 a^4 a^5 a^6 s

lateral dorsal aorta
median dorsal aorta
pulmonary artery
pulmonary vein
inferior vena cava

Urodele, adult

ophthalmic
orbital

s a^2 a^4 a^5 a^6

internal carotid
median dorsal aorta
ductus Botalli
left ductus Cuvieri
sinus venosus

Anura, adult

s I V a^3 a^4 a^6

external carotid
carotid gland
common carotid
systemic arch
truncus arteriosus
conus arteriosus
ventricle
pulmocutaneous arch

Fig. 11.26. Diagrams illustrating development and fate of aortic arches in Amphibia, left-side view completed. Vessels carrying most arterial blood white, most venous blood black, and mixed blood stippled. a^{1-6}, Primary arterial arches; I, V, gill slits; s. closed spiracle. (From Goodrich 1930.)

of X-ray opaque material and dyes it has been shown that there is no difference between the carotid and other arches in the time of arrival of blood or its pressure (Foxon 1964). The position of the opening of the carotid and right systemic arches determines that they receive blood mainly from the left side of the ventricle. The less well oxygenated blood flows to the pulmonary and left systemic arches and the latter mainly supplies the gut through the coeliaco-mesenteric artery. The heart is provided only with very small coronary arteries, but the spongy walls of the ventricle may allow metabolic exchange.

The venous system (Fig. 11.28) is based on the same plan as that of Dipnoi. The posterior cardinal veins are replaced early in life by a vena cava inferior or postcaval vein, formed from parts of the right posterior cardinal and the hepatic vein, draining into the sinus venosus. Most of the blood from the hind limbs passes through the renal portal system to the inferior vena cava, but

there is an alternative path through pelvic veins and a median anterior abdominal vein, which breaks into capillaries in the liver.

The blood pressure is regulated by the extrinsic nerves of the heart, fibres from the vagus tending to slow and from the sympathetic nervus accelerans tending to speed the beat. The latter nerve is a new development, there being no sympathetic innervation of the heart in fishes (the condition in Dipnoi is unknown). The diameter of the arteries throughout the body is also under control from sympathetic vasoconstrictor and perhaps also vasodilatator nerves. The arterioles in the web of the foot can be seen to constrict when the medulla oblongata is stimulated. Substances extracted from the posterior lobe of the pituitary and from the adrenal medulla also serve to cause constriction of the arteries and perhaps also of the capillaries.

There is therefore a complex mechanism for ensuring that the pressure of the blood is maintained and the flow

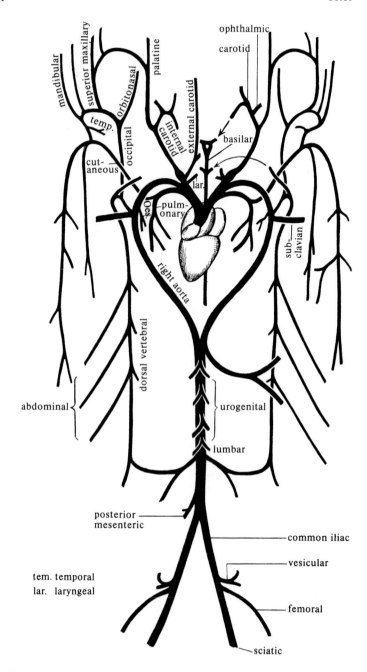

Fig. 11.27. Diagram to show the chief arteries and their anastomoses in the frog. *lar*. laryngeal; *oes*. oesophageal; *temp*. temporal. (After Gaupp 1896.)

directed into the part of the body that requires it for the time being.

19. Lymph vessels

The transfer of substances between the cells and the bloodstream is effected in any vertebrate by a transudation through the walls of the capillaries into the tissue fluids. Under the pressure of the heart-beat water and solutes leave the capillaries, passing through their walls, while proteins remain behind. The blood passing into the venous ends of the capillaries therefore has a high colloid osmotic pressure and this serves to suck back fluid from the tissues. In this way a circulation from the capillaries into the spaces around the cells is produced. Clearly, however, it is essential for this mechanism that the pressure of the ventricular heart-beat shall exceed

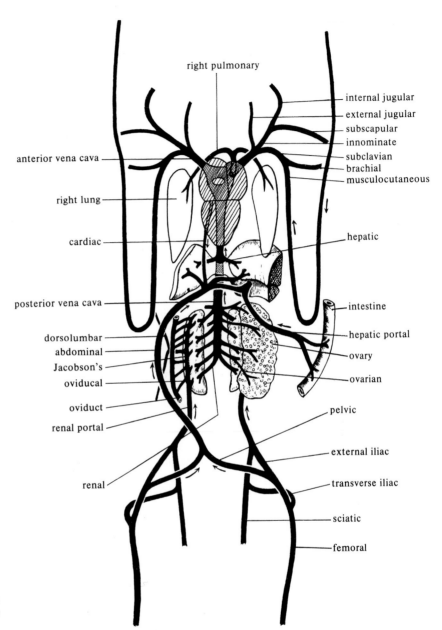

right pulmonary

internal jugular

external jugular

subscapular

innominate

subclavian

brachial

musculocutaneous

anterior vena cava

right lung

cardiac

hepatic

posterior vena cava

intestine

hepatic portal

dorsolumbar

abdominal

Jacobson's

oviducal

ovary

ovarian

oviduct

pelvic

renal portal

external iliac

transverse iliac

renal

sciatic

femoral

FIG. 11.28. Diagram to show the chief veins of the frog. (After Gaupp 1896.)

the colloid osmotic pressure of the blood. This it does by about three times in the frog.

The lymphatic system consists of a set of spaces which, in the frog at least, communicate with the tissue spaces around the capillaries. Injection of gum into the lymphatic system, by increasing the colloid osmotic pressure in the tissue spaces, prevents the back suck of fluid into the venules and hence leads to swelling of the part injected. The lymph spaces in the tissues join to

form larger channels and great sinuses, such as that below the loose skin of the back of the frog. The lymph is kept circulating by the action of lymph hearts. These are sacs of endothelium covered by striated muscle fibres, which contract rhythmically. In the frog there are anterior and posterior pairs of these, opening into veins. The more posterior pair lies on either side of the coccyx and can be seen if the skin is removed. The lymphatic vessels also assist in the process of repair. If, after injury, red cells come to lie in the tissues the lymphatics send out sprouts to pick them up and return them to the bloodstream.

20. Blood

The red corpuscles of amphibia are much larger than those of mammals, reaching in the urodele *Amphiuma* the immense size of 70 μm, the largest in any vertebrate; they nearly always exceed 20 μm. The red cells are formed mainly in the spleen and kidney, and are destroyed, after a life of about 100 days, by the spleen and liver. The bone marrow is a source of red cell formation in *Rana temporaria* but not, except during the breeding-season, in *R. pipiens*. A process of breaking up of the red cells occurs after they have entered the bloodstream, giving a number of enucleated fragments, and this, when the part remaining with the nucleus is small, produces a result like the extrusion of the nucleus during the development of the red cell of mammals. In *Rana* only small portions of the cytoplasm are broken off in this way, but in *Batrachoseps* a large proportion of enucleated corpuscles is produced.

The haemoglobin of the frog has a lower affinity for oxygen than that of mammals, even when at the same temperature, and in this respect is notably less efficient. Also, although the power of the blood to combine with carbon dioxide is great, there is a less delicate regulation of the reaction of the blood than in mammals.

The white cells of Amphibia are of three types, lymphocytes, with a large nucleus and little cytoplasm, monocytes, which are larger phagocytic macrophages, and polymorphonuclear granulocytes. These last may be neutro-, eosino-, or basi-phil and are migratory and phagocytic. Thus the white cell picture with which we are familiar in mammals was evidently established a very long time ago.

There is a globular spleen near the tail of the pancreas. Antibodies are synthesized by cells of the spleen, lymph glands or thymus. Amphibians have well developed systems of cellular and humoral immunity. Allografts of skin are rejected after invasion of lymphocytes and a specific memory develops. After early removal of the thymus rejection is slower but may still occur. Accumulations of lymphoid cells occur in the tongue, gut, and elsewhere.

The blood of frogs also contains numerous small platelets (thrombocytes), which probably break down when in contact with foreign surfaces to produce the thrombin that combines with the fibrinogen of the blood plasma to produce clotting.

21. Urinogenital system

A pronephros develops in the larva, consisting of 2–4 funnels, the nephrostomes opening from the coelom to the pronephric duct. This degenerates at or before metamorphosis and is replaced by the tubules of the mesonephros. In *Rana*, where there is a general shortening of the body, these extend over only a small number of segments and the kidneys are compact. In urodeles and the primitive frog *Ascaphus* the kidneys are elongated and in *Amphiuma* there is one tubule in each segment but in most amphibians many secondary tubules are added. Originally each primary tubule consists of a funnel opening to a tubule from which a diverticulum leads to a sac, Bowman's capsule, whose wall is invaginated by a capillary plexus, the glomerulus, the whole constituting a Malpighian body. Secondary tubules mostly lack the nephrostome and in the frog the funnels do not open into the tubules but directly into veins, serving as part of the circulatory system. In the adult there are some 5000 glomeruli, from each of which a short ciliated tube leads to the proximal convoluted tubule. There follows a second short ciliated region, corresponding in position to Henle's loop of mammals, and leading to a distal convoluted tubule, which joins the Wolffian duct.

The blood supply of the kidney differs from that of mammals in that blood arrives from two distinct sources; the branches of the renal artery run mainly to the glomeruli, those of the renal portal vein to the tubules. This corresponds to the functions now well established for those two parts, namely that the glomerulus filters off water and crystalloids, some of which are then reabsorbed by the tubule. Many details of this process are not clear, however, for instance how the urea concentration in the urine is raised many times above that of the blood.

The frog, having a moist skin, is presumably in constant danger of osmotic flooding with water when it is submerged, and of dessication when on land. The water content of the body is regulated by a complex control system centred on the hypothalamus. Exchanges through the skin, kidneys and bladder are influenced by octapeptides from the neurohypophysis and corticosteroids from the adrenal and perhaps by prolactin (lactotropin) from the adenohypophysis. Flooding is prevented by the efficient functioning of the glomeruli; they allow the frog to excrete as much as one third of its weight of water per day (man 1/50th). The mechanisms for resistance to dessication are less perfect. A frog (*Rana*) was found to lose 13 per cent of its body weight in 2–4 hours whereas in the same conditions it took a week for a snake (*Thamnophis*) to lose the same amount. There is no long water reabsorbing segment in the kidney, the part of the tubule corresponding to Henle's loop being short. A large cloacal (endodermal allantoic) bladder is present (to be distinguished from the mesodermal bladder of fishes) from which water can be reabsorbed. Certain desert amphibia (*Chiroleptes*) conserve water by losing the glomeruli altogether. *Rana cancrivora* is euryhaline and has up to 2.9 per cent of urea in the blood (Gordon *et al.* 1961).

The embryonic gonads have the potential to develop into either sex, the outer cortex forms the ovary, the medulla the testis. Genetic females may produce sperms in old age, which will produce only female offspring. High temperature upsets the balance of sex-determining factors and may turn genetic females into males. In male toads a rudiment of the cortex, escaping the inhibitory action of the medulla, persists as a rudimentary ovary, Bidder's organ. If the testis is removed this will develop and produce eggs. Two species of newts (*Ambystoma*) in Indiana consist entirely of parthenogenetic females. They are triploids derived by hybridization of two bisexual species.

The nuclei of the ovarian eggs of urodeles have especially large chromosomes with high DNA content in which individual genetic regions can be identified by their 'lampbrush' structure. This has allowed various types of developmental problem to be studied and the genetic organization of related species to be compared.

The Müllerian duct, by which eggs are carried to the exterior, develops separately from the Wolffian system in the frog, but arises from the latter during development in urodeles. In this, as in many other features, the frog shows a greater degree of specialization of its developmental processes. The ovaries are sacs formed from folds of the peritoneum, having no solid stroma such as is found in mammals. There are, however, follicle cells around each egg; these presumably produce the ovarian hormones. Sections of an ovary show eggs in various stages of development, but not all those that begin complete their maturation; many degenerating, atretic eggs are found. Ripening of the eggs proceeds under the influence of a hormone produced by the anterior lobe of the pituitary. This in turn is controlled by external environmental factors to ensure breeding in the spring. Suitable injections of mammalian anterior pituitary extracts will ensure ripening of the ovaries and ovulation at any time of year. The 'prolans' excreted in the urine of pregnant women have a similar effect, and the production of ovulation in *Xenopus* was used as a test for the diagnosis of human pregnancy.

Having left the ovary the eggs find their way to the mouths of the oviducts mainly by ciliary action of the latter. The walls of the oviduct are muscular and glandular and secrete the albumen; they are dilated at the lower end to form uterine sacs, in which the eggs are stored until laid.

The testes discharge directly through the mesonephros by special ducts, the vasa efferentia, formed by outgrowths from the mesonephros into the gonad. This is presumably a secondary development from the original vertebrate condition in which the sperms were carried away by the nephrostomes. The fact that the sperms pass through the kidney emphasizes that the Amphibia have diverged at a very early stage of the evolution of the vertebrate stock, and remain still in many respects at a lower level of evolution than the modern fishes, all of which have acquired separate urinary and genital ducts. In *Alytes*, in many ways primitive, the sperms do not, however, pass through the kidney!

In some frogs (*Rana temporaria*) there is a special diverticulum, the vesicula seminalis, leading by several small channels to the lower end of the Wolffian duct. It contains spermatozoa during the breeding season and its appearance suggests a secretory activity.

Most of the Amphibia have failed to effect the complete transfer to land life: they return each year to the water to breed. They may migrate for several kilometres from their home ground to the breeding site, returning to the same pond each year. Toads (*Bufo*) have been known to return for five years to the site of a pond that had been filled in. Many tropical Amphibia have evolved devices by which they avoid the necessity to return to the water to breed (p. 273).

Secondary sexual differences are marked in many species. In frogs the males precede the females to the water and then attract the latter by their vocal apparatus. The calls of the males alternate with each other in a 'chorus' and probably the louder and longer are more attractive to females and more inhibitory to other males. The frog *Eleutherodactylus coqui* from Puerto Rico gives the call indicated by its specific name. Tests with synthesized 'speech' show that the trill 'qui' attracts the females and 'co' repels other males (Wells 1977). In toads the males show alternate periods of sexual activity and 'calling'. External influences such as rainfall initiate the calling, often in many animals at the same time.

The male clings to the back of the female by means of a 'nuptial pad', developed as an extra digit, prepollex, on the hand (p. 237). Injection of male hormones or implantation of testis will cause this organ to develop in young female frogs.

In newts fertilization is ensured by an elaborate courtship. Sperms are made into spermatophores by special pelvic and cloacal glands and there are also abdominal glands, which produce a secretion attractive to the female.

The male newt *Triturus* develops in the breeding season a large crest; black, red, and blue spots; and an orange belly. First he performs a dance until the female approaches. He then retreats and if she follows and touches his tail with her snout he deposits a spermatophore, which she picks up with her cloaca. He next renews his dance and two or three spermatophores are transferred before he has to go to the surface to breathe. The female thus 'tests' the vigour of the male. His chance of fertilizing her depends on how long he can stay submerged and the number of spermatophores he can produce (Halliday 1977). Fertilization then takes place in the oviduct and she lays each egg separately wrapped in a leaf. Sperm may be stored for many months in a spermatheca.

22. Eggs

The eggs are deposited either as compact masses or in strings. The outer egg capsules may be beaten up by the legs of the parents to make frothy 'nests', especially if laid on land or in trees. The mucoid capsules contain substances that protect against macro- and micro-predators. In some newts and toads the eggs contain a neurotoxin resembling the tetrodotoxin of puffer-fishes (p. 205). The pigmented animal poles allow absorption of enough solar radiation to raise the temperature by several degrees.

Cleavage is always holoblastic but the amount of yolk varies, being more abundant in eggs developing to an advanced stage, either in nests or viviparously. The young tadpoles may be very immature at hatching, with external gills, and remain at first attached to leaves, etc. by cement organs. In other species they may have internal gills or even lungs and in a few they hatch fully metamorphosed. In those with a long development the tail is flat and vascular and serves for respiration. Limitation of the tadpole stage has occurred in some species of all families of frogs, indicating the pressure to avoid the need to return to water.

23. Larvae

Larval amphibia usually occupy adaptive zones quite different from the adult. The degree of emphasis on the larva varies. In some species this stage is suppressed and there is direct development into an adult. At the other extreme the aquatic larva becomes sexually mature and the terrestrial phase is eliminated (neoteny). The larval stage is often considered as a survival of the ancestral fish type. It is better to think of it as a special adaptive stage of genotypes that have not yet evolved fully effective terrestrial characteristics. The numerous larvae often provide the possibility of exploiting large food resources, e.g. of plankton or plants. This allows for the great losses involved in metamorphosis and terrestrial life, with advantages also for dispersal. The amphibian

organization thus provides for the exploitation of transitory small ponds of fresh water. The details of tadpole structure and habits differ greatly. Some are nektonic and live on plankton (microphagous), swimming continuously with a large tail. Others are bottom living (benthic) herbivorous polliwogs of shallow waters and a few are predaceous carnivores with large jaws and beaks, some feeding on their smaller sibs (group cannibalism).

24. Metamorphosis

The substitution of adult for larval adaptations obviously involves dangers and the process is achieved rapidly by hormonal control. The main effector hormone is thyroxine T_4. Removal of the thyroid prevents metamorphosis and addition of iodine or T_4 to the water accelerates it. The change is a result of increasing sensitivity to T_4 and the increased amounts of it secreted. The change involves (1) suppression of purely larval organs (e.g. tail), (2) conversion of some larval features to adult state (jaws), (3) development of purely adult characters (legs). The tissues that first become sensitive to T_4 are in the first class (tail fins), the jaws become sensitive next and the hind-limb bud latest.

The increase in thyroid secretion is triggered by thyrotropic hormone (TSH) from the pituitary, whose removal will prevent metamorphosis. The polypeptide hormone prolactin also plays a part in the control, by promotion of larval growth and by inhibition of changes. The whole sequence of metamorphosis is probably regulated by the hypothalamus. The median eminence becomes developed only at pre-metamorphosis. A graft of pituitary away from the hypothalamus produces some TSH, but not enough to allow complete metamorphosis.

The processes of change involve almost every part of the body. Genes that have been operating in the larva are switched off, for instance the haemoglobins are coded by different loci in larval and adult *Rana* (Salthe and Mecham 1974). Larval structures such as the gills, tail, and beak are resorbed, while limbs, eyelids, and lungs are developed. The intestine becomes shorter for life as an insectivore and there are many changes in the liver and kidneys and in metabolism. The Mauthner cells disappear from the brain and the lateral-line organs from the skin.

25. Paedogenesis (neoteny)

In many species of urodeles there are races which become sexually mature in the larval stage (neoteny). In *Necturus* and *Amphiuma* this happens in all individuals (obligate neoteny) and in these metamorphosis cannot be induced by T_4. In other genera such as species of the classic example the axolotl, *Ambystoma*, some or all of the individuals of a given population metamorphose. In these facultative neotenes T_4 generally induces metamorphosis. The thyroids are sensitive to thyroid-stimulating hormone (TSH), but the evidence suggests that the neoteny is due to a low level of thyrotropin-releasing hormone (TRH) probably due to a combination of genetic predisposition with environmental factors.

Lack of iodine has been shown to produce epigenetic suppression of metamorphosis as in an English population of *Triturus*. In other cases paedogenetic populations have been produced artificially by discarding individuals that metamorphose. Selection can therefore obviously produce the condition of return to a fully aquatic cycle, which may be advantageous, especially if water is abundant and terrestrial conditions are unfavourable.

26. Digestive system

Nearly all adult Amphibia feed on invertebrates, mainly insects, but also worms, slugs and snails, spiders, centipedes, and millipedes. The larval stages are usually omnivorous, but they may be cannibalistic, feeding on the tadpoles of the same or other species – an interesting form of provision for the next generation by excess productivity of the mother. There are only minor modifications of particular species in relation to their diet; as regards their food Amphibia occupy a generalized or 'easy' habitat. The fact that they are not particular in choice of diet has no doubt been part of the secret of their success.

The lips are little developed but the tongue is the characteristic organ for catching the food and is one of the special features required for terrestrial life. In *Rana* it is attached to the floor of the mouth anteriorly and flicked outwards by its muscles. To keep it moist and sticky a special intermaxillary gland is present. From the shape of the premaxillae it can be deduced that this gland was present in labyrinthodonts. The saliva contains a weak amylase, which may serve to release sufficient substances for tasting. Special tracts of cilia carry the secretion from the intermaxillary glands to the vomeronasal organ and palatal taste buds (Francis 1961). Another feature made necessary by terrestrial life is the presence of cilia to keep the fluids moving over the oral surfaces. These cilia are absent in aquatic Amphibia. In these the tongue is reduced, food being caught by the 'gape and suck' method.

There are teeth only on the upper jaw of frogs, on the premaxillae, maxillae, and vomers. They are used only to prevent the escape of the prey; few Amphibia bite. Biting teeth are present, however, in the adult *Ceratoph-* *rys ornata*, whose larvae also have powerful jaws and are cannibalistic. The South American tree frog *Amphignathodon* has teeth in the lower as well as the upper jaw and presumably has redeveloped them, a remarkable case of the reversal of evolution. Teeth are also present on the lower jaw of most urodeles.

The oesophagus is not sharply marked off from either mouth or stomach and the latter is a simple tube. Its lining epithelium of mucus-secreting cells is folded and simple tubular glands open at the base of the folds. These glands, unlike those of mammals, are composed of only a single type of cell, which secretes both the acid and the pepsin found in the stomach.

The intestine is marked off from the stomach by a pyloric sphincter. It is relatively short and dilates into a large intestine which is a special development for absorption of water by terrestrial animals. The liver and pancreas have the structure common to all vertebrates and produce juices of the usual type. The intestine of the omnivorous tadpole is more coiled than that of the adult frog. Its length is controlled by the type of food taken, being longer in larvae kept on a vegetable diet. Urodele tadpoles are mostly carnivorous.

The type of food taken by Amphibia depends on what is available. Most species are not particular feeders. However, they can learn with only one or two trials to avoid distasteful insects. Frogs and toads devour large numbers of insects. If the common insects available are pests the amphibian's part in controlling their number works to the advantage of man.

27. Nervous system

The organization of the nervous system of amphibia might be said to be essentially similar to that of fishes. In both groups there are highly developed special centres in the brain, each centre related to a special receptor system. In neither group is there a dominant part, integrating the activity of the whole, as does the cerebral cortex in mammals. The receptor systems include feature detectors for restricted classes of events, eliciting stereotyped response patterns. Correspondingly the nervous system is capable only of limited recombinations compared to those of birds or mammals. The whole central nervous system of *Rana* contains only 16 million neurons, about 6 million in the telencephalon, 1 million in the midbrain tectum (Oksche and Ueck 1976).

The plan of the spinal cord is like that of fishes, but well-marked dorsal and ventral horns are present. The large motor cells of the cord have dendrites that spread for up to 2 mm in the white matter, where their synaptic connections are made in a complicated 'neuropil' (Fig. 11.29). This is a simpler arrangement than is found

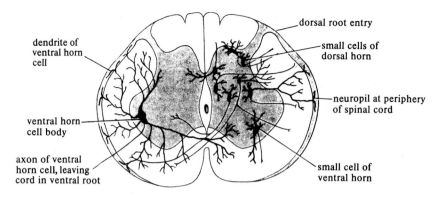

FIG. 11.29. Transverse section of the spinal cord of a frog, showing cells in the grey matter with their axons and dendrites spreading into the 'white' matter. (After Gaupp 1896.)

within the grey matter of the mammalian cord. In larval urodeles there are single sensory neurons in the cord that receive impulses both from the skin and from muscles (Rohon–Beard cells).

The arrangement of the spinal nerves is much modified by the development of the limbs. Ten spinal nerves are found but since an embryonic first one is missing they are usually numbered 2–11 (Fig. 11.30). Two spinal segments contribute to the brachial and four to the sciatic plexuses in the frog. From these plexuses fibres are distributed to the muscles and skin of the limbs (Fig. 11.30).

The clasp reflex of anurans is a spinal reflex. A decapitated male continues to clasp the female and fertilize the eggs! The inhibitory centre for the reflex is in the midbrain tectum, stimulation of which produces immediate release from the female. Castrated frogs or toads do not clasp, but will do so after transection of the brain at the level of the cerebellum.

The brain (Figs. 11.31 and 11.32) resembles that of Dipnoi very strikingly. The prosencephalon is based on an inverted plan (p. 173); the large evaginated cerebral hemispheres, therefore, have a thick roof as well as floor. In the frog there is only a short unpaired region of the forebrain (diencephalon) but this is longer in urodeles. The walls of each hemisphere may be divided into a pallium, medial ventral septum, and lateroventral striatum (Fig. 11.33). The pallium is vaguely divided into a general pallium dorsally, medial (hippocampal) and lateral (pyriform) parts (Fig. 11.33). The cell bodies mostly lie around the ventricle, but in all parts of the hemisphere some are found throughout its thickness. Only in this sense can there be said to be any cortex. (Kemali and Braitenberg 1969). The cells are pyramidal in shape and the connections are made in the outer 'white' matter. Many of them are microneurons whose axons end among the dendrites of the cell.

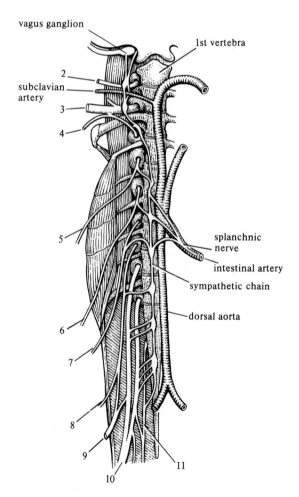

FIG. 11.30. Ventral branches of the spinal nerves (2–11) of the frog. The sympathetic chain is also shown. (After Gaupp 1896.)

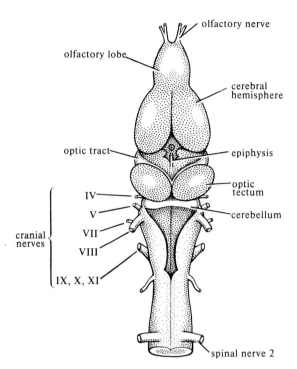

FIG. 11.31. The brain of *Rana*, dorsal view. (Modified from Gaupp 1896.)

Most but not all parts of the hemisphere are reached by olfactory tract fibres, which are the axons of the mitral cells of the olfactory bulb (Fig. 11.33). In the frog there are regions at the hind end of the hemispheres that receive forwardly directed fibres, some probably connected with tactile and others certainly with optic impulses. There is therefore some opportunity for the hemispheres to act as correlating centres, but we have little information as to the functions performed in them. Their connections are made by means of two large tracts, the lateral and medial forebrain bundles, but these reach only to the thalamus, hypothalamus, and midbrain, not back to the cord (Fig. 11.35). Electrical stimulation of the forebrain does not produce movements of the animal; presumably such a crude method, though it may excite a few neurons, cannot imitate the more subtle patterns in which they are normally active, perhaps mainly in connection with olfactory stimuli.

After removal of the cerebral hemispheres frogs became more 'sluggish' and lacked 'spontaneity'. They did not avoid large objects or seek shadow or dig holes or hunt for prey. The symptoms were less severe when the basal parts of the hemispheres were left intact.

Some indication of the function of the cerebral hemispheres is given by the fact that by placing electrodes, connected with a suitable amplifier, upon them

rhythmical changes of potential can be recorded (Figs. 11.33 and 11.34). These are most marked in the olfactory bulb and probably propagate backwards along the hemisphere. The rhythms continue even in a brain that has been removed from the head. They are therefore a sign of some intrinsic activity of the brain, rather than of response to peripheral stimulation.

The diencephalon includes three parts, the epithalamus, thalamus, and hypothalamus. The epithalamus is the dorsal region and receives nerve fibres from the pineal organ (p. 264). It includes asymmetrical habenular ganglia and its roof forms folds of choroid plexus, a sac-like paraphysis, and certain secretory tissues. The thalamus receives collaterals of optic nerve fibres and also fibres from other receptor systems (Fig. 11.35). The amphibian thalamus is unique in that the cell bodies of the neurons nearly all lie around the ventricle, with dendrites projecting outwards. There is no clear separation into nuclei, but whereas afferents from the retina and tectum and probably the cerebral

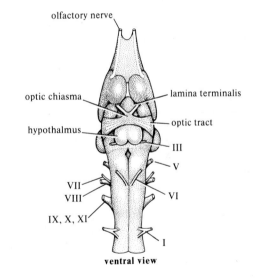

ventral view

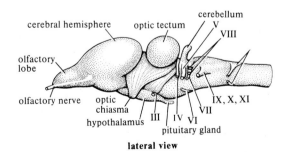

lateral view

FIG. 11.32. Two further views of the brain of *Rana*. (Modified after Gaupp 1896.)

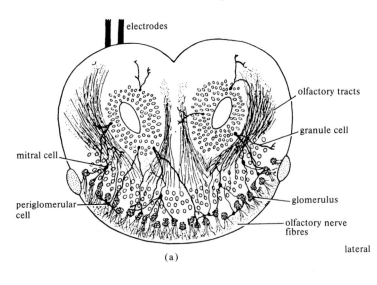

(a)

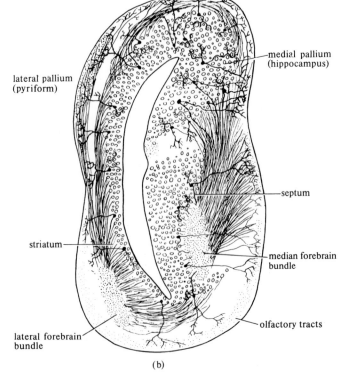

(b)

FIG. 11.33. Diagrams of the structure and probable cell connections in, (a) the olfactory bulb, and (b) the cerebral hemisphere of the frog. In the glomerulus fibres of the olfactory nerve make contact with dendrites of mitral cells. The electrodes are shown as they would be placed for recording the potentials shown in Fig. 11.34. (From Gerard, R.W. and Young, J.Z. (1937). *Proceedings of the Royal Society, London* **B122**, 343–52.)

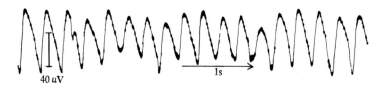

FIG. 11.34. Rhythmical changes of potential between electrodes placed on the surface of the olfactory bulb of the frog as in Fig. 11.33. (After Gerard, R.W. and Young, J.Z. (1937). *Proceedings of the Royal Society, London* **B122**, 343–52.)

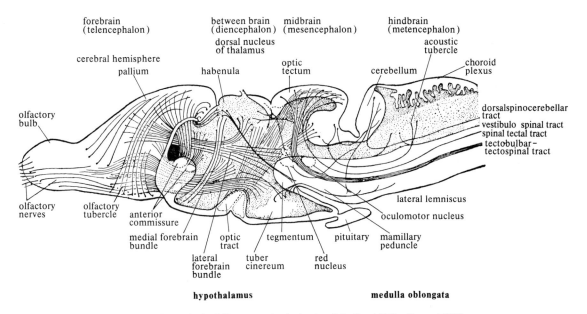

forebrain
(telencephalon)

cerebral hemisphere
pallium

between brain
(diencephalon)
dorsal nucleus
of thalamus

midbrain
(mesencephalon)

hindbrain
(metencephalon)
acoustic
tubercle

habenula

optic
tectum

cerebellum

choroid
plexus

olfactory
bulb

dorsalspinocerebellar
tract
vestibulo spinal tract
spinal tectal tract
tectobulbar –
tectospinal tract

olfactory
nerves

olfactory
tubercle

anterior
commissure

lateral lemniscus

oculomotor nucleus

medial forebrain
bundle

optic
tract

tegmentum

pituitary

mamillary
peduncle

lateral
forebrain
bundle

tuber
cinereum

red
nucleus

hypothalamus

medulla oblongata

FIG. 11.35. Principal fibre tracts in the brain of the frog. (After Papez 1929.)

hemisphere are mixed throughout the more dorsal part, spinal and medullary inputs reach only to the more ventral and caudal region. Efferents from the thalamus project to the midbrain roof and base and to the cerebral hemispheres. The visual input includes 'on' fibres responding to blue light (p. 263).

The hypothalamus is an important centre regulating the whole endocrine system as well as other parts of the brain. It receives large projections from the olfactory system and cerebral hemispheres and sends fibres in the mamillothalamic tract and by the mamillopeduncular tract to the midbrain and medulla. A large part of its output consists of neurosecretory fibres. Axons of the pre-optic nucleus end in the neurohypophysis. Other nuclei send fibres to the median eminence where they regulate the liberation of releasing factors which pass through portal vessels to control the production of tropic hormones by the adenohypophysis.

The midbrain is very well developed and shows many similarities to that of fishes. Its cells do not lie round the ventricle, many have moved out to make an elaborate system of cortical layers. Electrical stimulation of various parts of the optic tectum produces movements of the limbs and other muscles; there can be no doubt that this region plays an important part in behaviour. Most of the fibres of the optic tract end here, and there are also other pathways from the olfactory, auditory, medullary (gustatory?), and spinal regions (Fig. 11.36). Efferent fibres leaving the tectum pass to the midbrain base, medulla, and perhaps back into the cord. This

region therefore has wider connections than any other part of the nervous system.

Recordings from the optic tectum show that some cells respond only to small, moving objects with large contrast and these are probably prey-detectors (Fig. 11.36, see p. 263). Others, lying deeper, were activated by large objects, especially if moving towards the frog. Electrical stimulation of the optic tectum can elicit any or all of the phases of prey-catching, turning, fixation, snapping, gulping, and clearing of the mouth. Stimulation of the pre-tectal region or dorsal thalamus produced turning or jumping away (Ewart and Borchers 1971). The thalamus also inhibits the prey-catching reaction. After lesions to it a frog or toad snaps even at resting objects that it passes as it moves or even at its own legs.

The cerebellum of Amphibia is small and simple, perhaps because these are mostly animals that do not have to adjust themselves freely in space during locomotion, they move mainly in a single plane. There is little need for control of speed or distance of movement, except of the eyes, head, and tongue. The cerebellum of the frog is a single sheet with the molecular layer of granule cells on its anterior face and a line of Purkinje cells behind. Vestibular and spinal cord afferents end on both the Purkinje cells and granule cells. The parallel fibres run right across the whole cerebellum. Efferents pass to the midbrain tectum and affect movements of the eyes and limbs. The circuitry of the cerebellum differs from that of mammals in that there is little inhibition by

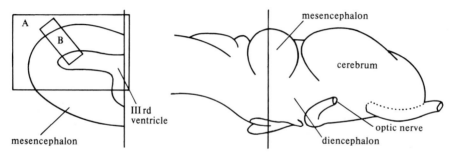

Toad brain - lateral view

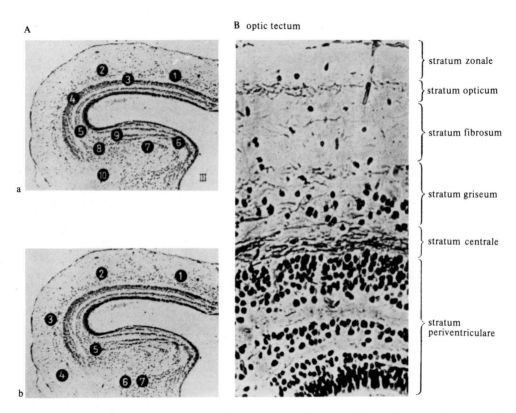

FIG. 11.36. Optic tectum and subtectal areas of toad brain. Top left, section through the right side of mesencephalon. A(a), recording experiments to localize various types of neurons; 1 and 2, neurons with small visual excitatory receptive fields; 3 and 4, large visual fields; 5 and 6, tactile; 7 and 8, acoustic; 9, optic–tactile; 10, optic–vibratory sensitive neurons. A(b), stimulating experiments to localize neurons that elicit turning reactions; 1 and 2, upward contralateral turning; 3, horizontal; 4, downward; 5–7, cleaning of mouth. B, Structure of optic tectum. (From Ewert and Borchers 1971.)

either Golgi or stellate cells which are few (Llinás 1976). The basic functions of the cerebellum are thus mediated by the climbing fibre inputs direct to the Purkinje cells and mossy fibre inputs through the fine parallel fibres. The Purkinje cells have an inhibitory action on the vestibular nucleus and directly on the vestibular hair-cells. The function of the frog's cerebellum is thus probably largely to regulate the vestibulo-oculomotor system, controlling movements of the eyes.

The ventricles of the brain are covered by a lining of ciliated ependymal cells. Some of these are tanycytes (ependymal gliocytes) with processes stretching out into the brain tissue. In various regions the linings form patches of specialized tissue, the circumventricular organs (Fig. 11.37). These have a rich blood supply and systems of neuroendocrine and receptor neurons (Fig. 11.38). They are partly concerned with secretion of the cerebrospinal fluid (CSF) but probably have further functions. The subcommissural organ in the roof of the third ventricle secretes the Reissner's fibre, a strand passing down the spinal cord (see p. 43).

An interesting feature of amphibian brains is their capacity for regeneration after injury, more complete in urodeles than anurans. A whole new hemisphere can be formed after removal, by proliferation of neuroblasts partly from the olfactory epithelium and partly from the sectioned surface.

Regeneration after section of the optic nerve leads to re-establishment of the normal precise topological relations between the retina and optic tectum. This system has been extensively used for studies of the mechanisms that guide the formation of connections in the brain (see Gaze 1974).

28. Endocrine organs

The three main parts of the pituitary are clearly recognizable and the pars distalis is like that of tetrapods, lacking the divisions seen in fishes (see Hanke 1974). There is no persistent hypophysial cleft. Secretion in the pars distalis is controlled by a portal system and

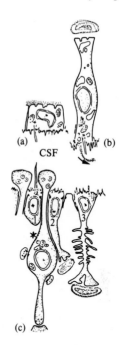

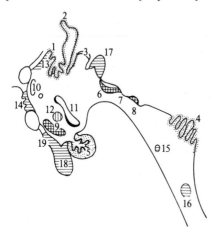

FIG. 11.37. Diagram of the circumventricular system of the vertebrate brain (1) Choroid plexus of the third ventricle; (2) paraphysis; (3) dorsal sac; (4) choroid plexus of the fourth ventricle; (5) saccus vasculosus; (6) subcommissural organ; (7) ependyma of the recessus mesocoelicus; (8) ependyma of the recessus colliculi posterioris; (9) ependymal organ of the infundibular recess; (10) recessus praeopticus organ; (11) paraventricular organ; (12) tuberal organ; (13) subfornical organ; (14) organon vasculosum laminae terminalis; (15) area recessus lateralis; (16) area postrema; (17) pineal organ = epiphysis cerebri; (18) neural lobe of the hypophysis; (19) median eminence. Not all of these circumventricular organs exist in amphibia. (After Oksche and Ueck 1976.)

FIG. 11.38. Cell types of the circumventricular organs. (a) Choroidal epithelium involved in the formation of cerebrospinal fluid (active transport of Na^+). (b) Secretory ependymal cell of the subcommissural organ; the secretory material is formed within dilated cisternae of the endoplasmic reticulum and released (arrow) into the cerebrospinal fluid (CSF). (c) Circumventricular cell complex formed by CSF-contacting neurons and ependymal cells. The CSF-contacting neurons (*) are aminergic, cholinergic, or peptidergic. At the dendritic tip of these bipolar neurons a release of active substances into the CSF may occur. On the other hand, they may function as sensory cells (chemoreceptors?). (1) Epithelial ependymal cell; (2) ependymal cell with a vascular contact; (3) branched ependymal cell named ependymal tanycyte. Note the vascular contacts of this cell. (After Oksche and Ueck 1976.)

releasing factors through a well developed median eminence. It is not directly innervated as it is in teleosts (Fig. 11.39). The distalis produces lactotropin (LTH) involved in larval growth, whereas somatotropin (STH) regulates growth of adults. Thyrotropin (TSH) is responsible for metamorphosis, its secretion depending on a releasing factor from the hypothalamus (TRF) and there is also evidence for ACTH, controlled by a hypothalamic ACTH-RF. Two gonadotropins (FSH and LH) appear to be produced.

The pars intermedia produces a melanophore stimulating hormone (MSH) under inhibitory control by fibres from the hypothalamus. The neurohypophysis produces arginine vasotocin and mesotocin which, in anurans, accelerate water uptake through the skin and urinary bladder and, in both anurans and urodeles, reduce glomerula filtration. These responses are greater

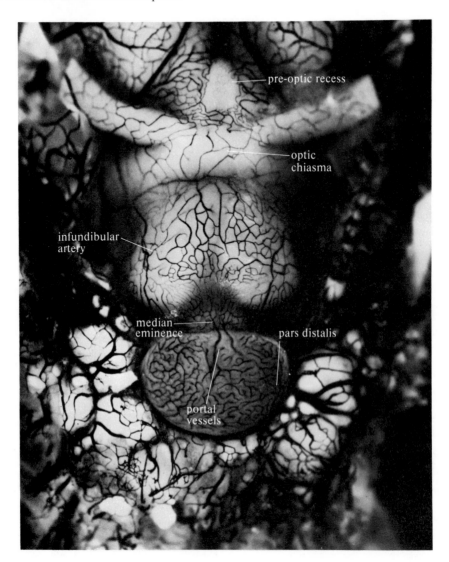

Fig. 11.39. Base of brain of a toad *Bufo bufo*. Vessels injected with Indian ink gelatin suspension. (Kindly provided by K.G. Wingstrand.)

in terrestrial than larval or aquatic forms and their evolution must have been one of the secrets of life on land. The drive to seek water for reproduction is dependent on prolactin (in newts).

The thyroids are paired. Their secretion controls metamorphosis and moulting but does not seem to stimulate oxygen consumption in the adult. Parathyroids, derived from the pharyngeal pouches, appear for the first time in amphibians as regulators of mineral metabolism by secretion of parathormone. After their removal from a frog the blood calcium falls and phosphorus rises. They have perhaps evolved by modification of the ionocytes (chloride-secreting cells) of the

gills of fishes (p. 165). The ultimobranchial bodies develop from the floor of the last pharyngeal pouch. They secrete calcitonin which has the opposite action to parathormone, mobilizing calcium from the endolymphatic sacs, especially at the time of metamorphosis.

The amphibians are the first vertebrates in which there is a close anatomical association of the ectodermal, adrenaline-secreting chromaffin tissue of the suprarenals with the mesodermal steroid-secreting interrenals. The reason for the association is no more clear in amphibia than in other tetrapods. In urodeles the adrenals form an extended series, whereas in the frog they are compact orange masses on the kidneys.

Adrenaline has sympathomimetic actions similar to those in mammals. It accelerates the heart, raises blood pressure and dilates the lungs. It causes hypergylycemia and dilates the pupils of the eyes and causes contraction of the melanophores. The adrenal cortex (cortical and chromaffin tissues) plays a part in control of water balance and carbohydrate metabolism but little is known of it in detail (see Barrington 1975).

The islets of Langerhans of the pancreas contain α- and β-cells and are presumed to secrete glucagon and insulin, respectively acting to raise and lower the blood sugar.

29. Skin receptors

Lateral-line organs are present in the skin of all aquatic amphibian larvae and in some aquatic adults, such as those of the anuran family Pipidae. They are of simple form, consisting of groups of cells in an open pit. In newts they are present in the larvae, which are aquatic, but are covered by epidermal layers during the first post-larval stage during which the newt lives on land. In the final aquatic adult stage the organs reappear.

The skin, of course, also contains tactile organs, and in addition is often sensitive to chemical stimuli and to temperature changes. This chemical sense is mediated by fibres running in the spinal nerves, not by special elements such as the taste buds found spread out over the body in fishes. The skin is also sensitive to heat and cold, and there is some evidence that these senses are served by fibres different from those that mediate touch, pain, or the chemical senses. Histologically, however, there is little sign of the development of the special sensory corpuscles that are so conspicuous in the skin of birds and mammals. All the nerve endings are of the type known as 'free nerve endings', except for a few touch corpuscles on special regions such as the head and feet. In this the Amphibia again resemble the fishes and show less differentiation than do the higher animals.

The taste buds on the tongue and palate are probably able to respond to the presence of only two of the four types of substance that are discriminated by mammals. Applications to the tongue of the frog and recordings of the impulses in its nerves show that there are chemoreceptors present able to respond to salt and sour substances, but that no reaction is given to substances that in mammals are classed as sweet or bitter.

The olfactory organ functions both on land and in the water, special mucous glands being present to keep it moist when in air. A continual circulation of water or air is maintained over the olfactory epithelium by cilia or the movements of respiration. The internal nostril may have originally developed from the double nostril of fishes in order to make a circulation around the

olfactory receptors possible. Jacobson's organ is a special diverticulum of the olfactory chamber, serving to test the 'smell' of food in the mouth. It is another organ characteristic of terrestrial life and absent in aquatic amphibians.

The Apoda, being blind, have a great development of the sense of smell, including a hollow tentacle or olfactory tube.

30. The eyes

Provided that certain requirements are met the air gives more scope for the use of photoreceptors than does the water. Light is transported with less disturbance through the air and image formation is facilitated by the refraction of the air–corneal surface. The Amphibia have exploited these advantages and sight has become the dominant sense of most of the species. For clear vision it is essential that the surface of the eye be protected and kept moist and free of particles, and these purposes are served by the eyelids and lachrymal glands the ducts of which lead to the nose. The upper lid is fixed, but the lower is very mobile and folded to make a transparent structure, the nictitating membrane, able to move rapidly across the surface of the eye.

The eyeball is almost spherical, with a rounded cornea. The lens is farther from the cornea than in fishes and is nearly spherical and unable to change shape. There is an iris, with a rapidly moving aperture, operated by powerful circular (sphincter) and radial (dilator) muscles. Although these muscles are partly actuated by a nervous mechanism they are also directly sensitive to light, and the pupil of the isolated eye of the frog shows wide excursions with change of illumination (Campenhausen 1963).

Accommodation is effected by protractor lentis muscles, attached to the fibres by which the lens is supported (Fig. 11.40). These muscles move the lens forward for near vision whereas in teleostean fishes the muscles move the lens backwards. Other fibres, the musculus tensor chorioideae, run radially and around the lens. They may help the protractors, and are probably the ancestors of the ciliary muscles of higher forms.

In amphibia living in the water the eye is based much more on the fish plan and the lens is rounder. There are then no lids or lachrymal glands and the eye is enabled to make an image, in spite of the absence of the air–corneal interface, by a thickening of the inside of the cornea.

Rods and cones are present in the retina, the former containing visual purple, which may be red or greenish. The two sets of receptor are apparently found throughout the retina in urodeles, but in *Rana* there is a macular region in which the cones are in excess and this is still

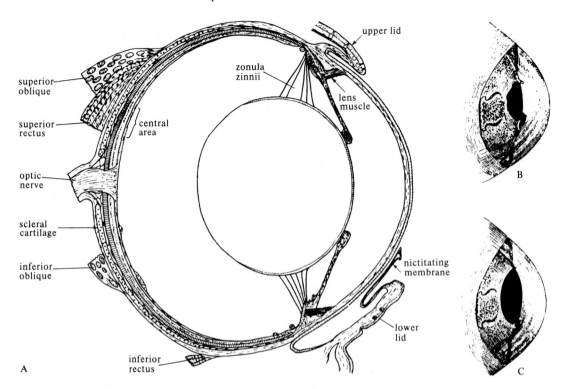

FIG. 11.40. The amphibian eye and its accommodation. A, anuran eye in vertical section. B, anterior segment of *Bufo* in relaxation. C, in accommodation; note forward movement of lens. (From Walls 1942.)

further developed in *Bufo*. Study of the impulses in the optic nerves of *Rana* shows that there is much coding in the retina of the information provided by the rods and cones. (1) Contrast detectors give a sustained response when a sharp edge moves into the visual field; (2) convexity detectors respond to objects that are curved, the discharge being greater the more curved (smaller) they are. These two types together may be called 'on' fibres; (3) moving-edge detectors ('on/off' fibres) respond with a frequency proportional to the velocity of movement; (4) dimming detectors respond on reduction of illumination ('off' fibres); (5) darkness detectors fire with frequency inversely proportional to illumination. These types of fibre project to different depths in the tectum as sheets of endings, and the arrangement of the retina is accurately reproduced there although the fibres are interwoven in the nerve (perhaps to prevent 'cross-talk'). Moreover, if the nerve is severed the fibres regenerate in such a way as to reconstitute the map. A sixth type of fibre is sensitive to blue light and is connected with the thalamus.

This coding serves to provide reports of the types of change relevant to the animal. Thus the second type might be called 'insect detectors', responding when a small dark object enters the field and moves about

intermittently. More complex visual discriminations are also possible, for example toads can distinguish between shapes. Frogs can learn to discriminate by sight between insects that are palatable or distasteful. They gaze for a long time before jumping and seizing with the arms and tongue. The prey is out of sight during the jump, which must be timed 'ballistically' perhaps by the cerebellum (see Horridge 1968). The pathway to the thalamus mediates a positive phototactic response which is specific for blue wavelengths irrespective of either intensity or saturation. It is not elicited by green light and is probably responsible for the reaction by which frogs jump into the water away from vegetation (Muntz 1962).

Pineal and parapineal organs are present in anurans and contain rod-like cells sensitive to light. They may be concerned with endogenous rhythms and other internal activities. The skin is also sensitive to light in amphibians: frogs are positively phototactic even after removal of the eyes and cerebral hemispheres. This skin sense is especially developed in certain cave-living urodeles, *Proteus*, in which the eyes are not functional.

31. The ear

The middle ear of the adult frog consists of a funnel-shaped tympanic cavity communicating with the pharynx and closed externally by a tympanum supported by a tympanic ring. The tympanum occupies an otic notch, which corresponds to the spiracular slit of a fish. Sound waves are received by the large drum and transmitted across the cavity by a rod, the columella or stapes. This fits by an expanded foot into a small hole in the wall of the auditory capsule, the fenestra ovalis. Here there is also a small plate, the operculum, attached by a muscle to the scapula. The columella is generally held to be the homologue of the hyomandibula, attached to the skull, but no longer to the jaw.

The tadpole uses the lungs as 'ear drums', connected to the *round* window by a 'branchial columella'. This degenerates at metamorphosis and a tympanic cavity is developed from the spiracular cleft, by a series of changes. The original cleft degenerates six days after hatching but about six of its lining cells persist and at the end of the tadpole stage form a tympanic vesicle, which becomes connected with the pharynx by a rod of cells. This rod then degenerates again and an open air passage to the vesicle of the drum is not established until some thirty days after emergence from the water, when a pouch from the pharynx joins the tympanic cavity. These events show the complexities that may result from the modification of developmental processes, and they emphasize the difficulty in assigning 'homologies'. It is still debated to what extent the middle ear of the frog can be compared with that of amniotes. The hyomandibular nerve, which divides above the middle ear of amniotes (and above the spiracle of the dogfish) lies behind the tympanic cavity of the frog and branches below it. Some workers conclude that the tympanic membrane and stapes of amphibians (including labyrinthodonts) have been developed independently and are not ancestral to those of amniotes (Lombard and Bolt 1979).

The arrangement for conveying vibrations to the ear varies considerably among amphibians. In urodeles there is no tympanum or middle ear. In some of them the columella is attached to the squamosal, perhaps to receive ground vibrations in a semi-aquatic or burrowing habit. A similar arrangement may have been present in the earliest amphibians, which have a columella but no oval window. In other urodeles (Plethodontidae) the columella is attached to the quadrate and there may be a second ossicle, the operculum, working in parallel, with its inner end in the oval window caudal to the columella and its outer end attached by a muscle to the scapula. In terrestrial forms the columella becomes fused with the window at metamorphosis and its function is taken over by the operculum, probably receiving vibrations from the fore-legs. The opercular muscle is attached to the suprascapula. The more aquatic forms (*Crypto-*

branchus) retain the larval condition and never develop an operculum. The tympanum and columella are also reduced in some terrestrial anurans (*Bombinator*) but in the aquatic *Xenopus* and *Pipa* the operculum and its muscle are absent.

The inner ear includes six vestibular organs and two auditory organs. The vestibular system includes the sacculus, utriculus, and lagena sensitive to airborne vibration and gravity and the semicircular canals registering angular accelerations. The amphibians are unique among vertebrates in making discrimination between sounds by means of two auditory organs, the basilar papilla and amphibian papilla. They lie respectively in the floor and roof of the sacculus and each consists of hair cells with a tectorial membrane. The pathway to the basilar papilla is much longer and only passes higher frequencies (Fig. 11.41). Experiments with synthetic mating calls show that bull-frogs only respond to calls containing both low frequency (200 Hz) and high frequency (1400 Hz) with little energy between. The two receptors show maximum sensitivity at the frequencies of the mating call. The basilar papilla contains only about 50 hair cells, tuned to high frequency and with no efferent innervation and not inhibitable by other sounds, as are the low frequency units of the amphibian papilla, which receives efferents. The intervention of environmental noise can thus be suppressed during reproductive chorusing (Lombard and Straughan 1974).

The central pathway for hearing passes from the medullary auditory nucleus to the superior olive, here found for the first time in vertebrates, lying at the upper end of the medulla. Its cells receive the inputs from both ears and are probably responsible for sound localization. From this the lateral bulbotectal tract passes to the torus semicircularis in the tegmentum in the floor of the midbrain. Pathways onwards from this are not well known but are said to reach the dorsal thalamus and the midbrain tectum. Some cells of the torus are sensitive to both high- and low-frequency

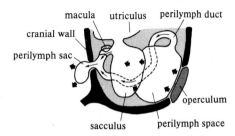

Fig. 11.41. Amphibian ear. Directions of displacements resulting from inward movement of the operculum. Broad arrows, low-amplitude, high-pressure displacements; fine arrows, high-amplitude, low-pressure displacements. (Smith, J.J.B. (1968), *Journal of experimental Biology* **48**, 191–205.)

tones, but it is not clear how the convergence initiates responses. The high-frequency sensitivities are matched to specific features of the mating calls in some species. The control of sound production is by the hypothalamus and tegmentum.

The auditory system must be capable of considerable discrimination since anurans may emit up to seven distinct signals including mating calls, territorial calls, release calls, rain calls, distress calls, and warning calls (much fewer, of course, than in birds). Some species produce one only, but others several of these. The prey may also be located by sound.

A peculiar feature of many Anura is an immense backward development of the perilymphatic space of the inner ear, forming a sac extending above the brain and on either side of the spinal cord as far back as the sacrum. Portions of this sac emerge between the vertebrae, showing as whitish masses on account of the granules of chalk they contain. The calcium salts in these sacs diminish greatly during metamorphosis and they then refill. The system may serve as a calcium reserve also for the adult.

32. Behaviour

The habits of Amphibia, like their special structures, enable them to deal with the various emergencies that threaten the continuation of life on land. Frogs, toads, and newts have a strong sense of place and they show distinct 'homing' reactions. The newt *Tarichia* was found to return home from as far as 13 km, but the power was lost after section of the olfactory nerves. *Rana clamitans* displaced from their breeding site orientate towards the nearest pond where there is a chorus of its species. Frogs are able to learn to find their way out of mazes and to remember the way for periods of at least thirty days.

Complex migrations are made by many species; nearly all migrate to the water in spring. In this migration the males usually precede the females, then attract the latter by their calling. The receptors for orientation towards the water are known to be the osmoreceptors in the mouth of the frog. This orientation is particularly clear in urodeles, in which sound plays no part in the migration. The power to find water is obviously of first importance for any animal living on land. Especially important is behaviour that avoids dessication. For instance tree frogs rarely ascend tall trees. In periods of drought many amphibians shelter under logs often in congregations, or they may dig in the earth, especially the spadefoot toads (*Scaphiopus*).

The search for food and the avoidance of enemies are not in principle more difficult on land than in the water, but they probably demand new mechanisms. For example, the greater range of visibility in air than in water can be a disadvantage if it is exploited by one's successful and predatory descendants. A hawk, owl or heron makes fuller use of its opportunities in this respect than does the frog, who can only remain safe from them by behaviour that keeps it concealed. Similarly there are dangers in certain situations, for instance of dessication, which are additional to those that are met by an animal in the water.

In the emergence of the first land vertebrates we thus see a conspicuous example of the invasion by living things of a medium far different from themselves. This produces a situation that calls forth all the powers of the race to produce new types of individual, and necessitates that the individuals make full use of their capacities. New patterns of structure and behaviour are developed as the various possible situations emerge. The types of organization that at first manage to survive gradually give place to others, still more complex or 'higher'. Some traces of the organization of the early venturers can still be seen in the Amphibia, which today exploit the damper situations on the earth, at least for part of their lives.

12 Evolution and adaptive radiation of Amphibia

1. Classification

Class AMPHIBIA
*Subclass 1: Labyrinthodontia (folded teeth)
  *Order 1: Ichthyostegalia (fish vertebrae). Upper Devonian–Carboniferous
    *Ichthyostega; *Elpistostege
  *Order 2: Temnospondyli (divided vertebrae). Carboniferous–Triassic
    *Suborder 1: Rhachitomi (stem animals). Carboniferous–Triassic
      *Loxomma; *Eryops; *Cacops; *Archegosaurus
    *Suborder 2: Stereospondyli (ring vertebrae). Permian–Triassic
      *Capitosaurus; *Buettneria; *Mastodonsaurus
  *Order 3: Anthracosauria (coal lizards). Carboniferous–Permian
    *Palaeogyrinus; *Seymouria; *Pteroplax
*Subclass 2: Lepospondyli (scale vertebrae). Carboniferous–Permian
  *Diplocaulus; *Ophiderpeton; *Microbrachis; *Sauropleura
*Subclass 3: Lissamphibia (smooth amphibia)
  Order 1: Anura (no tails). Carboniferous–Recent
    *Protobatrachus; Leiopelma; Rana; Bufo; Hyla; Pipa
  Order 2: Urodela (tails) (= Caudata). Jurassic–Recent
    Molge; Salamandra; Triton; Ambystoma; Necturus
  Order 3: Apoda (no limbs) (= Gymnophiona = Caecilia). Recent
    Ichthyophis; Typhlonectes

2. The earliest Amphibia. Labyrinthodontia

THERE are such close resemblances between the skulls of the earliest amphibians and those of the Devonian crossopterygian fishes that there can be no doubt of the relationship (Fig. 11.22). At present there is, however, no detailed fossil evidence of the stages of transition from the one type to the other. The fossils that appear to be closest to the possible tetrapod ancestor are the osteolepids of the Lower and Middle Devonian periods, about 375 million years ago. These were definitely fishes, though they may have used air as well as water for obtaining oxygen, as do modern lungfishes. *Elpistostege is a single Upper Devonian skull intermediate in proportions between such fishes and the earliest undoubted tetrapods, *Ichthyostega and similar forms, found in freshwater beds of Greenland (see Fig. 11.22). These are dated as very late Devonian or early Carboniferous, that is to say about 350 million years ago. They

are the oldest members yet found of the great group of Labyrinthodontia (folded teeth). As their name implies, the enamel and dentine surrounding the pulp cavity at the base of the tooth (as seen in cross-section) was folded into a labyrinthine pattern. All were fish eating. This condition was also present in their crossopterygian ancestors. Throughout the succeeding 100 million years of the Carboniferous and Permian periods they flourished and developed many different lines, one giving rise to the reptiles. They were in the main aquatic or semi-aquatic forms, and only a few seem to have been completely terrestrial.

The earliest labyrinthodonts were definitely tetrapods and already showed sharp changes from the fish type. Some had scales, while others had a dry leathery skin. The underside of the body was sometimes protected by ventral plates. In *Ichthyostega the skull (Figs. 11.22 and 12.1) shows all the characteristic amphibian features, but retains traces of fish ancestry in its shape, with

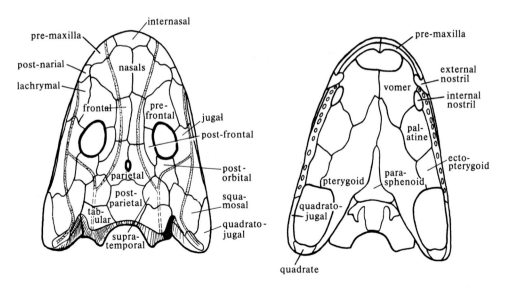

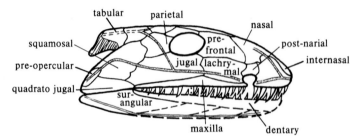

FIG. 12.1. Skull of *Ichthyostega*. (From Westoll, after Säve-Söderbergh.)

a short, wide snout and long posterior region, and presence of a preopercular bone. There was some sign of separation of the two parts of the brain case. The nostril lies on the very edge of the upper lip, apparently partly divided by a flange of the maxilla into internal and external openings. There being no gills the operculum had been lost and the second gill arch (spiracle) lay in an otic notch between the tabular and squamosal bones and presumably carried a tympanum (ear drum). These creatures were about 100 cm long and possessed a long tail with a dorsal fin, but had strong legs and strong, well-ossified 'rhachitomous' vertebrae. They must have been not very far from the ancestors of all tetrapods.

3. The vertebrae of Palaeozoic Amphibia

The Palaeozoic labyrinthodonts other than the Ichthyostegalia are of two main types grouped as the orders Temnospondyli and Anthracosauria, the latter containing forms that produced the reptiles. Both groups contain some more or less fully terrestrial types, but

many of the later temnospondyls became secondarily aquatic.

The main structural difference between the two orders is in the vertebrae, and the strength and structure of these give us clues to the habits of the animals. In the earliest amphibia (Ichthyostegalia) the vertebrae were like those of crossopterygians, composed of three parts, a dorsal neural arch and a centrum of two parts, anterior intercentrum and posterior pleurocentrum, the former carrying a ventral arch and rib. In the Temnospondyli the pleurocentrum became small, a type of vertebra known as rhachitomous, and in later members of the group only the intercentrum remained ('stereospondylous'). In the anthracosaurs on the other hand the pleurocentrum became more developed. In some of them both centra remained so that each vertebra consisted of two rings ('embolomerous'). In the line leading to the reptiles, however, the intercentra were reduced to little wedges between the main centra, which are formed from the pleurocentra.

In many other aquatic amphibians the vertebrae are

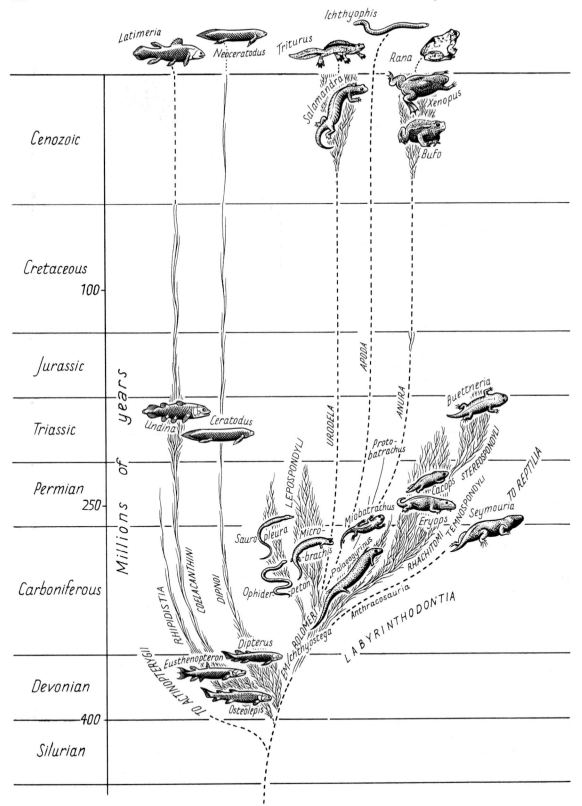

FIG. 12.2. Chart of evolution of Amphibia.

reduced to simple cylinders of bone surrounding the persistent notochord, a condition called 'lepospondylous', which is found also in modern forms.

4. Order Temnospondyli

This large group of Palaeozoic amphibians includes many lines, not easy to separate or classify. The earliest of the Carboniferous Rhachitomi with rather high skulls and rounded bodies, such as *Loxomma*, were probably largely aquatic. Later members such as *Eryops* and *Cacops* from the early Permian had quite stout legs and were probably able to live in swamps or on firm ground (Fig. 12.2). There was a tendency in these Permian and Triassic forms for the skull to become flattened, while the bony palate became incomplete, perhaps for reducing weight. Numerous tadpoles of these Rhachitomi are found and were long thought to be a separate genus called 'Branchiosaurus', because of their gills, and given their own order 'Phyllospondyli' (leaf-vertebrae). Some rhachitomes became fish-eaters, with long snouts (*Archegosaurus*).

The Stereospondyli, such as *Buettneria*, derived from some creatures like *Eryops* (Fig. 11.1), carried the flattening of the skull further in the Permian and Triassic. The intercentra became reduced and finally lost, as in the vertebrae of the giant *Mastodonsaurus* with its skull of 120 cm length.

5. Order Anthracosauria

These were the amphibians whose descendants became reptiles, and hence they include the ancestors of all later tetrapods, including ourselves. Some of them are so like reptiles that they are often classified with them. The earliest members such as *Palaeogyrinus* of the Carboniferous were close to the ancestry of all Amphibia. The vertebrae retained both pleurocentrum and intercentrum, the two-ringed condition from which the early Amphibia derive the name Embolomeri. These were largely aquatic animals, with small legs and long tail, perhaps they were fish-eaters. The later anthracosaurs

of the Permian, such as *Seymouria* were apparently wholly terrestrial as adults, although the young bore lateral line canals, indicating an aquatic larval stage, justifying the placing of these forms in the Amphibia (Figs. 12.3 and 12.4). They had strong limbs and must have been able to walk well. The intercentra are smaller than in the early anthracosaurs, but larger than in reptiles. The skull roof has a pattern like that of

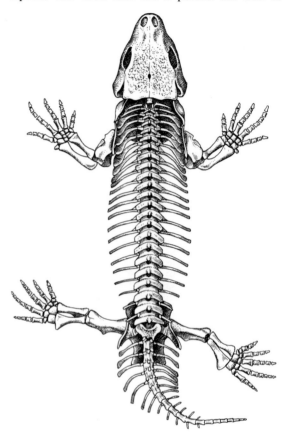

Fig. 12.3. Skeleton of *Seymouria*. Actual length 51 cm. (From Williston 1925.)

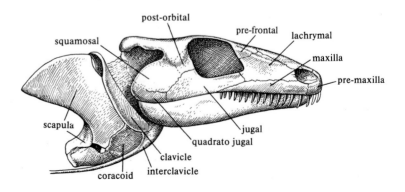

Fig. 12.4. Skull and pectoral girdle of *Seymouria*. (From Williston 1925.)

anthracosaurs including an intertemporal bone, absent in reptiles. The limb girdles resemble those of reptiles, with a coracoid bone and broad ilium, presumably connected with walking. The phalangeal formula was the reptilian one of 2.3.4.5.3.(or 4), not usual in Amphibia. There was a very large otic notch, carrying an ear drum. *Seymouria* is clearly intermediate.

6. Subclass Lepospondyli

This includes three or more lines of Carboniferous and Permian amphibians in which the vertebrae were single rings around the continuous notochord. *Ophiderpeton* and other forms from the Carboniferous were like snakes and had lost their limbs. *Diplocaulus* from the Permian possessed remarkable horned skulls. These nectridians with broad flat heads and upward-looking eyes and small limbs were secondarily aquatic carnivores, with long bodies and vertebrae adapted for bending movements. A third group placed here, the microsaurs, such as *Microbrachis*, had more normal limbs and bodies; some were aquatic, others fully terrestrial, some used burrows and might have been the ancestors of the Apoda.

7. The origin of modern Amphibia

The modern forms are generally considered to be a single monophyletic group, but this is not certain and their relationship to the Palaeozoic Amphibia is unclear. They show some characters typical of later temnospondyls, for instance flattening of the skull and loss of ossification in the roof and the palate. The vertebrae, on the other hand, are simple rings like those of lepospondyls. The earliest known anuran *Protobatrachus* (= *Triadobatrachus*) comes from the lower Triassic and has a frog-like skull but many vertebrae with ribs, and a segmented tail. The ilia and tarsal bones are long. The urodeles are superficially more like early amphibians. Fossil salamanders are known back to the Jurassic, but resemble modern forms. An independent origin has been suggested from the fish *Porolepis*, supposed to have an internal nostril (Jarvik 1980). No fossil Apoda are known, but their retention of scales suggests an early separation from the rest, and they may be derived from the lepospondyls rather than from the labyrinthodonts.

8. Tendencies in the evolution of fossil Amphibia

The changes in the structure of Amphibia can be followed from the beginning of the Carboniferous to the end of the Triassic period, and, indeed, in the form of their reptilian descendants far beyond. There were many distinct lines (as we should expect), and it is not possible to trace details of the history or fate of particular populations. The earlier amphibians continued to get their food in the water. It was not possible for them to become large and fully terrestrial until there were animals on land suitable as prey. Only a few types really succeeded in this such as the small *Cacops* and larger *Seymouria* (Fig. 12.3). Most Palaeozoic Amphibia were aquatic or semi-aquatic. Two distinct tendencies thus appear over this period: (1) to become fully terrestrial, (2) to remain in or return to the water. The terrestrial forms became gradually shorter in body and stronger in leg. The skull remained fairly high and domed and the otic notch became deeper, as a more effective tympanum developed. In the vertebrae the intercentrum became reduced as muscles developed attached to the pleurocentrum.

Return to the water led to animals of two distinct types, (a) snake-like or (b) flattened, but in both there was a reduction of limbs and a secondary lengthening of the body, with return to the sinuous movements of fish-like locomotion. In the bottom-living forms, such as some Stereospondyli and *Diplocaulus* among Lepospondyli, the skull became flattened, with the eyes looking upwards, the otic notch being shallow. The snake-like *Dolichosoma* retained the more normal skull shape but became immensely elongated and lost the limbs altogether.

9. Newts and salamanders: Order Urodela (= Caudata)

The urodeles show less deviation from the general form and habitats of the amphibia as a whole than do the specialized anurans (Fig. 12.5). The adult and larval urodeles differ little from each other, and characters suitable for aquatic life are frequently found in the adult. Indeed, all stages of suitability for land occur, from the terrestrial salamanders, such as *Salamandra maculosa*, the European salamander, which is viviparous, to the fully aquatic forms, for instance *Necturus*, the mud-puppy of North America. In many of the aquatic animals there is a tendency to retain in the adult characters usually found in larvae. This process of paedomorphosis (neoteny) has developed to various extents, and independently in several groups. Thus the giant salamander *Megalobatrachus*, 170 cm in length, in China and Japan, has no eyelids, but loses its gills in the adult. In *Cryptobranchus*, the hell bender of the United States, the spiracle remains open and is used for the outlet of water during respiration.

Amphiuma, also from the southern USA, is a very elongate form, with absurdly small legs, no eyelids, and four branchial arches. In the still more modified forms, such as *Necturus*, external gills are present and the lung is so reduced that the animals can live permanently

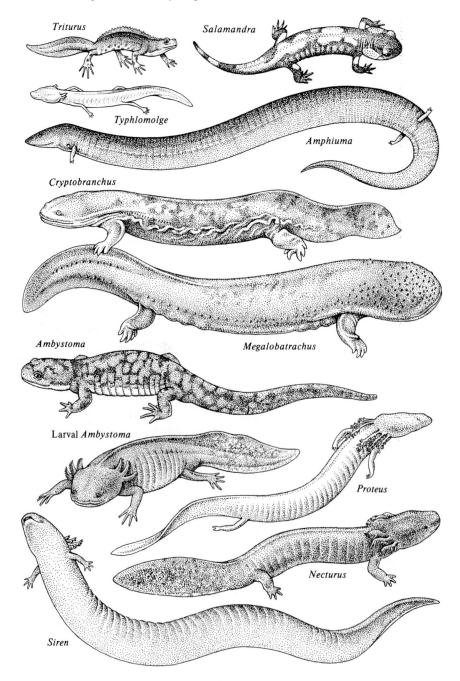

FIG. 12.5. Various urodele amphibians, not all to same scale (mostly from life.)

submerged, walking along the bottom. *Proteus* from European caves is a blind urodele with external gills and no pigment. *Siren* shows almost entirely larval characteristics and has no hind limbs (Fig. 12.5).

The more typical terrestrial newts are of several sorts.

In North America the common genus *Ambystoma* (often written *Amblystoma*) has eleven species, many adapted to special habits, including *A. mexicanum*, in which some races become mature without metamorphosis, because of lack of iodine in the water, whereas others, the axolotls of Mexico are genetically neotenous.

The common British newt *Triturus vulgaris* is a typical example of the more definitely terrestrial urodeles, though it is not able to live in very dry situations. However, the limbs support much of the weight of the body, and their soles are applied to the ground and turned forwards. The tail shows various degrees of reduction to a rod-like organ, but in the breeding season, when both sexes return to the water, it develops a large fin, especially in the male. The common newts of America form a distinct family, including *Plethodon* and many specialized forms, such as the blind *Typhlomolge*, inhabiting the waters of caves (Fig. 12.5).

10. Frogs and toads: Order Anura (= Salientia)

Among the frogs and toads are very many suited for special modes of life, and it must again be emphasized that this is far from being a static and precariously surviving group. We have already mentioned the frog *Ascaphus*, which lives in mountain streams in the northwest of the United States and has reduced lungs. It shows a combination of specialized and primitive features. Internal fertilization is assured by a penis-like extension of the cloaca. In this genus and the New Zealand *Leiopelma* (Fig. 12.6) there are other primitive features, including tail muscles (absent in all other anurans), amphicoelous vertebrae, free ribs, abdominal ribs, and persistent posterior cardinal veins.

In *Alytes*, the midwife toad of Europe, the male carries the eggs wrapped round the legs. *Pipa* is a related and still more specialized aquatic frog from South America; it has no tongue, and, curiously enough, has developed an elaborate arrangement by which the young are carried in pits on the back. *Xenopus* of Africa is related to *Pipa*, but without the habit of carrying its young (Fig. 12.6).

The bufonid toads are among the most successful of all amphibian groups and are more fully adapted than most for a terrestrial life, but nearly always return to the water to breed. *Bufo* itself is found in almost all possible parts except in Australia and Madagascar; related genera, many of them with special features, are found all over the world. Curiously enough only one genus, *Nectophrynoides* from West Africa, is viviparous, the young being in that case provided with a long vascular tail, by means of which they maintain contact with the wall of the 'uterus', even though embedded in a mass of embryos.

Hyla and other tree-frogs, very widely distributed, are similar to the bufonids but have pads on the toes by which they climb, and many other adaptations to arboreal life. Many tropical frogs have invented me-

thods of avoiding having to return to water to breed. *Gastrotheca* (= *Nototrema*), the marsupial frog, is a genus in which the young develop in a sac on the back of the female, this sac being in one species protected by special calcareous plates. *Rana* and its allies, the true frogs, are also cosmopolitan. A number of frogs related to *Rana* have taken to a tree-living habit, developing pads on the toes. *Polypedates* is a widespread genus and there are several others, each independently derived from ranids. This is therefore a striking illustration of parallel evolution – the hylid tree frogs having arisen from bufonids and probably several sorts of polypedatids from ranids.

Burrowing with the legs has also been evolved several times by anurans. In *Breviceps* (Fig. 12.6), which digs for ants, there is a snout, as in other anteaters.

11. Order Apoda (= Gymnophiona = Caecilia)

These (such as *Ichthyophis*) live in the tropics and are burrowing, limbless creatures, like earthworms. They show several interesting primitive features, often including the retention of small scales in the skin. They are specialized, however, in having a very short tail and some features suited to their usually terrestrial life, such as copulatory organs. There is a solidly built bony skull, probably secondarily evolved for burrowing. The animals are blind, but carry special sensory tentacles. The eggs are large and yolky and cleavage is meroblastic; they are laid on land and the embryos develop around the yolk sac, but often have long, plumed gills. Viviparity is common, as in the aquatic form *Typhlonectes*.

12. Adaptive radiation and parallel evolution in modern Amphibia

Even this superficial study of the 250 genera and about 2000 species of modern amphibians shows that the features we have already recognized in fish evolution are found also in evolution on land. It is difficult in a short time to gain an impression of the very great variety that is characteristic of any group of animals when closely studied. Besides the main types that can be distinguished, countless lesser variations will be found, and one realizes that the characteristics of the populations are still today in process of continual and perhaps rapid change. Anyone trying to discover the relationships of the various derivatives of ranids or bufonids must be impressed by the presence of series of parallel lines of evolution, so that it is impossible to disentangle the relationships. Evolution viewed at close quarters by the student of abundant modern animals looks very different from the picture that may be seen by the collector of a few rare fossils, who tries to arrange his types in

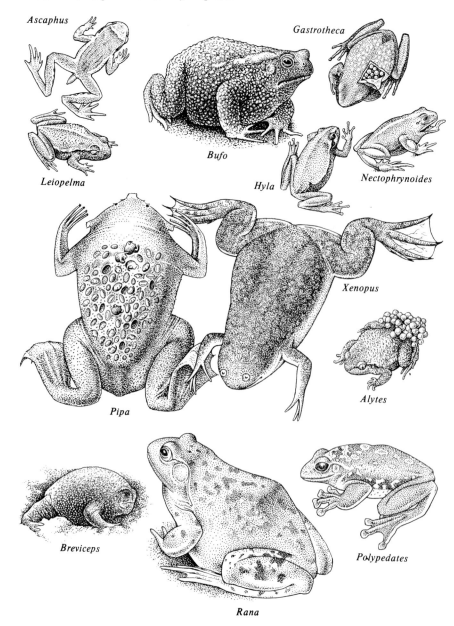

Fig. 12.6. Various anuran amphibians. (Not all to same scale.)

genealogical trees although they represent only an infinitesimally small sample of numerous and varied populations.

We can perhaps find certain tendencies in the modern amphibian populations that are similar to the tendencies of the fossil series. Many return to the water, especially among the urodeles. Others become more fully terrestrial, either by climbing trees or by burrowing into the earth. Both these habits have been independently adopted many times by recently evolved lines and, no doubt, still more often in the past by creatures that have died out, leaving no trace.

13. Can Amphibia be said to be higher animals than fishes?

It is not easy to decide whether there is a clear sense in which amphibians can be said to have advanced over their fish ancestors. They have moved from the water into environments that are in a sense less suitable for life. In order to maintain a watery system, such as a frog or toad, outside the water, various special structures and methods of behaviour have been evolved. The presence of such additional systems can be said to add complexity to the organization. It is difficult to make a count of the number of 'parts' involved in the organization of any animal. Amphibia possess many special devices, for instance, for respiration without loss of moisture, for control of water intake and water loss, for return to water to breed, and so on. Even without making a proper quantitative computation it seems reasonable to say that these add up to make an organization more complicated than that of a fish. The integration of the action of so many parts requires an elaborate nervous system, and there is evidently some connection between the increased size and importance of the nervous system and the development of this more complicated organization that enables life to continue in a different environment.

Considering the matter in this way it is hardly sensible to ask the question 'Are the amphibians more efficient than the fishes?' The work that they do in maintaining life is so different that a comparison of 'efficiency' is fallacious. One method of assessing living efficiency might be to judge each animal organization by the extent to which it maintains its constancy – by its power of homeostasis. Data about the fluctuations of the internal environment are so scanty among lower vertebrates that we cannot proceed very far on these lines. It is probable that the blood of fishes shows greater fluctuations, for instance in osmotic pressure or lactic acid content, than does that of amphibians, such fluctuations being perhaps even an advantage in allowing life in waters of differing salinity. In fact, to say that the whole mechanism of homeostasis becomes more complicated in land animals is only to say over again that they are 'higher' because they have more special work to do to maintain themselves in a difficult environment. Almost every part of the body shows signs of this greater complexity; the central nervous system becomes larger, the autonomic nervous system develops more elaborate control of the viscera. The endocrine glands become more numerous and differentiated, the muscular system shows more distinct parts, enabling the animal to act in new ways.

However difficult such comparisons may be it is hardly possible to deny them some validity. Amphibian organization differs from that of fishes and may be said to be 'higher' in the sense that it is more elaborate and allows life in conditions that the fish organization cannot tolerate.

13 Life on land: the reptiles

1. Reptilia (= crawlers)

Towards the end of the Devonian period, say 350 million years ago, the vertebrate organization produced a population of amphibian creatures and from this has been derived not only various modern groups classed as Amphibia but also the more fully terrestrial populations that do not need to breed in water – the Amniota. Since that time many divergent lines have evolved from this stock, including the birds and the mammals, and it is therefore difficult to specify what is meant by a reptile, as distinct from an amphibian or a bird or a mammal. The term does not define a single vertical line of development or branch of an evolutionary tree, but is rather a horizontal division, marking a band on the evolutionary bush, specifying a level of organization beyond that of an amphibian but before that of either bird or mammal. Attempts have been made to divide the reptiles vertically into sauropsidan (bird-like) and theropsidan (mammal-like) lines, but such a division, although it has some foundation, obscures the fact that their bush-like evolutionary radiation has produced not two but many lines.

The existing reptiles belong to four out of the dozen or more main lines that have existed. The most successful modern forms are placed in the order Squamata, the lizards and snakes, the latter being of relatively recent appearance in their present state. Secondly, the tuatara, *Sphenodon*, of New Zealand is a relic surviving with little change from the Triassic beginnings of this group. Thirdly, the crocodiles are an older offshoot from the stock from which the modern birds were derived. Finally the tortoises and turtles (Chelonia) have retained in some respects the organization of still earlier times, perhaps through the special protection of their shells. Though they are much modified in some ways, they still show us several characteristics of the earliest Permian reptiles.

These four modern types are all that remain of the reptiles that flourished throughout the Mesozoic, culminating in the giant dinosaurs of the Jurassic and Cretaceous. Evidently a profound change affected the world including the populations of land and sea, between the end of the Cretaceous and the Eocene. This was considered in Chapter 1, but we must briefly discuss here the possible relation of the decline of the reptile populations to the rise of their descendants, the birds and mammals. It can hardly have been only the more efficient organization due to the warm blood that gave these their opportunity, for there were forms in the Triassic so similar to mammals in their skeletons that we may reasonably (though not certainly) suppose them to have been warm-blooded. There were birds with feathers in the Jurassic, and it is probable that they also already had warm blood. However, as a working hypothesis, we may suppose that the climate, which had been suitable for reptiles in the Mesozoic, became less so in the early Tertiary, and the most obvious suggestion is that colder conditions developed all over the earth's surface. The modern reptiles for the most part live in the temperate and tropical zones, indeed they flourish only in the latter. However, it must be remembered that climate fluctuates continually (p. 23); it is dangerous to make generalizations about conditions over such long periods as the Cretaceous.

2. Temperature of reptiles

The organization we call reptilian is, generally speaking, suitable for life in warm countries, though two species, the common lizard (*Lacerta vivipara*) and the adder (*Vipera berus*), are found as far north as the Arctic Circle. No doubt the distribution of reptiles is limited largely by the fact that they cannot usually maintain a temperature above that of the surroundings by production of heat from within. The widespread idea that reptiles have no means of regulating their body temperature, however, has been overemphasized. In the wild (though not as a rule under laboratory conditions) reptiles are often able by suitable behaviour to maintain their body temperatures at a remarkably high and constant level throughout much of the day, by varying their exposure to the available sources of heat. When they get cold they bask in the sun or rest on warm rocks;

when they get too hot they shelter under vegetation or in holes. In some species, such as monitor lizards, colour change plays an important part in temperature control, the animals becoming darker or paler in colour, according to whether heat absorption or reflection is the appropriate response.

It has also been shown that each species of reptile has an optimum range of temperature, below which the animals become inactive and above which they quickly die since they have no sweat glands and though they may pant, this is expensive in water. In some desert lizards the upper limit is above 40 °C. The range tends, as one would expect, to be higher in diurnal than nocturnal forms, and is in general higher in lizards than it is in snakes or alligators. Turtles must live at the temperature of the surrounding water.

The reptilian method of temperature control differs essentially from that of mammals in that it depends on the availability of external sources of heat such as the sun, rather than on the ability to conserve or lose heat generated within the body. For this reason reptiles are sometimes termed 'ectothermic' and mammals 'endothermic'. These terms have a somewhat different meaning to 'poikilothermic' (having a variable temperature) and 'homoiothermic' (having a constant temperature), and are much used by herpetologists.

The ectothermic method of temperature control presupposes some sensitive mechanism for registering slight changes in the temperature of the surroundings. There is evidence that the pineal complex is a receptor in some reptiles and that the hypothalamus may be involved in thermal homeostasis. Yet the tuatara, *Sphenodon*, which has the most perfect pineal eye, seems to be tolerant to surprisingly low temperatures and to be active mostly at night (p. 289).

It remains true to say, however, that no existing reptile can retain an independent body temperature for a long period. For this reason, reptiles living in temperate climates must hibernate during the winter, while in warm countries some, conversely, aestivate during the hottest months.

There is, however, some evidence of endothermy among extinct reptiles. The organization of the blood supply of bone is characteristically different in birds and mammals from that of cold-blooded vertebrates (Desmond 1975). The difference is connected with the more active metabolism of the endotherms. It is claimed that a similar pattern is found in thecodonts, both groups of dinosaurs and in pterosaurs as well as in the therapsid ancestors of the mammals. Other evidence that these animals were endothermic is the absence of annual growth lines (p. 315) and indeed their great size, which could only have been reached after several hundred

years at known reptilian growth rates! (see Bakker 1977).

3. Skin

The skin is characteristically dry and waterproofed, unlike that of amphibians. It contains few or no glands. The Malphigian layer of the epidermis produces a thick covering of keratin and the horny scales (scutes), which are periodically shed in flakes, or, as in snakes, cast as a single slough. Beneath the horny scales many reptiles (some lizards, crocodiles, some dinosaurs) develop bony plates in the dermis (called osteoderms). These may be restricted to the head, where they lie superficial to the skull bones, or may cover most of the body. The tortoise's shell contains both horny (epidermal) and bony (dermal) components (p. 296). The horny scales of many reptiles are modified to form crests, spines, and other appendages.

Many reptiles, particularly lizards and snakes, have bold and elaborate colour patterns, mostly cryptic. Local races of lizards differ in colour to match light or dark soil (Porter 1972). The poisonous *Heloderma* has warning colours (p. 308). In some forms, especially lizards, there are marked colour differences between the sexes (see p. 287). Colour change is marked in chameleons and some other lizards but not in snakes and alligators. There are melanophores lying deep in the dermis and iridocytes and lipophores more superficially, as in Amphibia (p. 227). The mechanism of colour change varies and is discussed on page 308. The internal organs of diurnal reptiles are often protected by black pigment.

Various snakes deceive their enemies by 'head mimicry'. The tail has a red underside and is displayed to invite an attack, which the real head can then defeat.

4. Posture, locomotion, and skeleton

The elongated body and small laterally projecting legs of many reptiles recall those of a urodele, and the method of locomotion is in general similar in the two groups. Many retain the primitive five digits in both hand and foot. With the similarity of movement goes a general similarity in plan of the skeleton: there are, however, certain most significant features, characteristic of the reptiles. The head is usually carried off the ground, on a well developed neck. The two first cervical vertebrae are modified to form the atlas and axis. The atlas is a ring of bone without centrum, but with a facet in front for the occipital condyle and one behind for the odontoid process, a peg attached to the front of the axis but derived in development from the centrum of the atlas segment.

The vertebrae articulate with each other by a system

of interlocking processes much more elaborate than that found in fish-like vertebrates and presumably serving to allow the column to carry weight. As a rule in modern reptiles, each centrum is concave in front, covering the convex hind end of the vertebra next to it, a condition known as procoelous. In aquatic vertebrates the centra articulate by flat surfaces and this condition is found in the acoelous vertebrae of some primitive and a few modern reptiles. Besides the articulation of the centra the vertebrae are also united by the zygapophyses, facets on the neural arches, so arranged that the upwardly facing surfaces of the anterior zygapophyses slide over the down-facing surfaces of the posterior zygapophyses, an arrangement that is found throughout the amniotes.

Ribs are found on most of the vertebrae but are well developed in the middle or trunk region; each articulates with the body of the vertebra usually by a single facet. They are attached to a sternum in the thoracolumbar segments (Fig. 13.1). The ribs of the two sacral vertebrae are short and broad and articulate with the ilia. The numerous caudal vertebrae show reduction of all parts, especially towards the tip of the tail. The chevron bones are ossicles attached to the caudal centra and representing the reduced intercentra (p. 115). The gastralia are bony rods in the ventral body wall of crocodiles and some other reptiles. They may be remnants of the bony scales of crossopterygians.

The girdles and limbs (Figs. 11.9, 11.14, and 13.1) show the same general structural and functional fea-

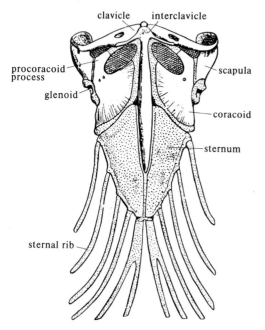

FIG. 13.1. Shoulder girdle and sternum of a lizard (*Iguana*).

tures as those of Amphibia. The limbs form the main locomotor system, the metachronal contraction of the myotomes playing a lesser part than in urodele amphibians. The humerus and femur are normally held in such a position that their outer ends lie higher than the inner, that is to say, in a position of abduction. The radius and ulna and tibia and fibula proceed downwards towards the ground (at right angles to the proximal bones) and the hand and foot are turned outwards at right angles, to rest on the ground. The main muscles thus draw the humerus and femur backwards and forwards as well as downwards, and the ventral regions of the girdles are large and flattened to receive these muscles. In mammals, with a different system of progression, the more dorsal parts of the girdles have become developed.

The pectoral girdle (Figs. 11.10 and 13.1) consists of a dorsal scapula and a large ventral coracoid, which may be fenestrated. Distinct pro- and post-coracoid elements are found in the extinct mammal-like reptiles. The dermal components are represented by the paired clavicles and median interclavicle. A cleithrum is present in a few very primitive forms.

In the pelvic girdle (Fig. 11.12) the usual dorsal ilium, anterior pubis, and posterior ischium are found, the last two meeting their fellows in midline symphysis.

The characteristic modifications of the reptilian skull are discussed on p. 293. The general plan is similar to that of primitive amphibians, but in all except the most primitive reptiles there is a development of holes (fossae) in the temporal region to provide space for the bulging temporal muscles. The skull roof is in some respects more primitive than that in modern amphibians (Figs. 13.2 and 13.3). It is made up of a large series of dermal bones, including the nasals, prefrontals, frontals, post-orbitals, and parietals. The side of the skull is usually less complete, composed of the tooth-bearing premaxilla and maxilla, lachrymal, jugal, post-orbital, squamosal, supratemporal and quadrate. The naming of some of the smaller bones round the orbit and above the quadrate is a matter of controversy.

The margins of the palate are formed by flanges of the premaxillae and maxillae and the small ectopterygoids. The internal nostrils usually lie forwards between the maxillae, vomers, and palatines. More posteriorly the floor of the skull is made up mainly by pterygoid bones and the parasphenoid, which is partly fused with the lower surface of the basisphenoid. Occipital bones surround the foramen magnum and make up the single occipital condyle, which in some forms is indented to form three partly distinct lobes. In many reptiles there is an epipterygoid bone on either side of the brain case behind the orbits; this is regarded as an ossification in

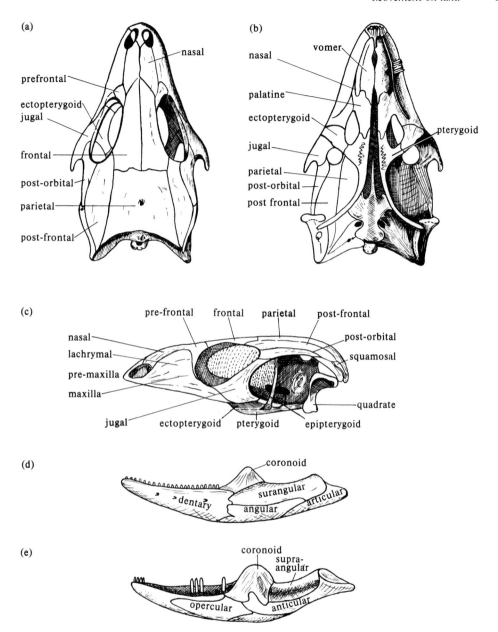

Fig. 13.2. Skull and lower jaw of *Lacerta*. (a) dorsal view; (b) ventral view; (c) from left side; (d) right half of lower jaw from outside and (e) from the inside, showing the pleurodont arrangement of the teeth. The regions of persistent cartilage are not shown in detail. (After Gadow, H. (1901). *Cambridge Natural History*, Vol. 8. Macmillan, London.)

the ascending process of the palato-quadrate. The lower jaw usually consists of six bones, the articular forming the joint with the quadrate, and the dentary carrying teeth (Fig. 13.2).

The anterior part of the chondrocranium, surrounding the front of the brain, and the nasal capsule, remain more or less unossified, and in places may be membranous. There may, however, be small ossified orbitosphenoids and farther back pleuro- or laterosphenoids, which develop in the pila pro-otica uniting the orbital cartilage with the otic capsules. Between the eyes there is in most reptiles a thin sheet of cartilage known as the interorbital septum, which is seldom if ever ossified. The posterior part of the chondrocranium ossifies to form

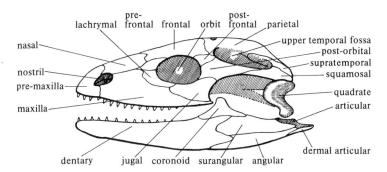

FIG. 13.3. Diagram of the skull of a lizard to show the temporal fossa. The upper temporal fossa is shown diagrammatically, as it occurs in many lizards; in *Lacerta* it is largely covered by an extension of the post-frontal—see Fig. 13.2. (From Goodrich 1930.)

the occipital complex, basisphenoid, and the ossifications in the otic capsule (pro-otic, opisthotic, etc.).

In many reptiles the upper jaw and front part of the skull can move to some extent in relation to the occipital region and cranial base, such movement being termed kinesis (p. 346). This is often associated with mobility of the quadrate, as in lizards, snakes, and certain dinosaurs and is thought to permit adjustment between the movements of the upper and lower jaws during snapping, and perhaps also helps to increase the gape.

The postmandibular visceral arches play no part in jaw support but are incorporated into the ear and hyoid apparatus. There is a rod-like columella auris with a small cartilaginous element (extracolumella) at its outer end. The columellar system usually conducts vibrations from the tympanum, lying behind the quadrate, to the fenestra ovalis and inner ear. In some forms, e.g. snakes, however, the tympanum is absent and the outer end of the columella is applied to the quadrate. These animals may be sensitive to ground vibrations, transmitted through the bones of the jaw (p. 416).

The hyoid apparatus consists of a basal process, which projects into the tongue, and often three pairs of ascending horns. These represent the remains of the hyoid and branchial arches.

5. Feeding and digestion

Food is seized either by the teeth or, in some specialized lizards such as the chameleon, with the elongated tongue. The teeth are situated along the edges of the jaws and often also on some of the bones of the palate. Typically, they are all of the same conical shape, but may be slightly serrated, or modified to form crushing plates, poison fangs, and other devices. As a rule, tooth succession is continuous throughout life, though exceptions to this are found among the lizards. Salivary glands are well developed in some forms; in snakes and one genus of lizards (*Heloderma*) some of them are modified to form poison-glands. The tongue is very variable, being hardly movable in some reptiles (e.g. crocodiles) but long, forked, and highly mobile in others

(e.g. snakes). In about one-third of all reptiles the salivary glands produce some form of venom toxic to predators or prey, these are mostly the snakes (Gans 1978). The secretion is often found at the base of enlarged teeth and may be injected by running down a deep groove or hollow tube. Immobilization of the prey is an obvious aid to swallowing it whole. Snake venom contains a complex mixture of enzymes, other proteins, and carbohydrates, not all poisonous. The enzymes of some crotaline and viperine venoms break up the tissues of the prey. Some components of the secretions may have been evolved first for lubrication of the prey and cleaning of the mouth and teeth and later as venoms and aids to the digestion. The fact that snakes are venomous is an important deterrent to predators.

Digestion proper begins in the stomach which is divided into a main part, the corpus or fundic region, and a pars pylorica. The branched fundic glands contain clear neck-cells near the surface, secreting mucus, and deeper-lying dark cells which secrete both hydrochloric acid and pepsin. These may, therefore, be called chief cells or main gastric glandular cells. The glands of the pyloric part are shorter and less branched and contain only mucous cells. The venom of snakes assists in digestion.

The intestine is short and its epithelium folded into simple crypts of Lieberkühn. The columnar epithelium may be simple or stratified and it includes goblet cells, Paneth cells and enterochromaffin (endocrine) cells. The large urodaeum (cloaca) is lined by columnar epithelium. Cloacal glands are present in crocodiles and many other forms and are probably scent glands. Smaller reptiles are mostly insectivores and make very efficient use of the food (80 per cent of the calorific value). Larger animals have to rely on the more abundant supplies provided by plant foods, but they use them less efficiently than mammals because of their inability to digest cellulose. Herbivorous lizards extract about 50 per cent of the calorific value of their food, against 71 per cent for cattle and 64 per cent for rabbits (Bennet and Dawson 1976).

There is a well-marked cloacal chamber in all reptiles, subdivided into a coprodaeum for the faeces, and a urodaeum for the products of the kidneys and genital organs. These two chambers open into a final common proctodaeum, closed by a cloacal sphincter. This division of the cloaca is associated with the necessity for the retention of water, the cloacal chambers serving for water resorption from both the faeces and urinary excreta (p. 284).

6. Respiration, circulation, and metabolism

The typical method of respiration is a backward movement of the ribs, produced by the muscles attached to them. Reptiles breathe by aspiration, not by the positive pressure pumping used by amphibians. Many reptiles make rhythmic gular movements but these serve mainly for smell, not air pumping. The respiratory movements are performed by the intercostal muscles, which are active during both inspiration and expiration. Variable periods of apnea intervene between respiratory cycles. In most reptiles the pleural and peritoneal cavities communicate but in crocodiles they are separated by a kind of diaphragm, and a diaphragmaticus muscle running from the liver to the pelvis assists in respiration. The glottis is a slit at the back of the mouth and leads into a larynx with supporting cricoid and arytenoid cartilages. Many reptiles are able to produce small sounds, but the voice box is less developed than in either Amphibia or birds.

The lungs are sacs whose walls are folded into ridges, separating a number of chambers or bronchioles. The volume of the lungs is relatively larger than in mammals but the surface area is sometimes as much as 100 times smaller (in proportion to body weight). This arrangement is not so unsuitable as it seems. The large volume provides a reservoir of air, useful in diving species, but also for the long periods of holding the breath when startled and so remaining still. In aquatic species the lungs also provide buoyancy and are then often increased by smooth avascular air sacs.

Reptiles have a relatively low need for oxygen. Their standard metabolic rate (that of a fasting animal measured at rest in the dark) is only 10–20 per cent of that in homeotherms (Bennet and Dawson 1976). The concentration of mitochondria and specific activity of their aerobic enzymes in lizards is one-fifth of that in rats. Most reptiles are therefore unable to sustain long periods of activity. They tend to move in short bursts, during which their muscles contract anaerobically degrading glucose and glycogen to lactic acid. This provides a rapid source of energy but the lactic acid accumulates in the blood, lowering its pH and hence impairing oxygen transport through the Bohr effect on

haemoglobin (see *Life of Mammals*). However, reptiles are able to tolerate much greater changes in the circulatory components of the blood than mammals. This advantage is put to use in their capacity to exist for long periods in low oxygen conditions. Lizards, snakes, and crocodiles can all survive for 30 min in pure nitrogen and turtles for several hours. The lactate appears in the blood mainly *after* emergence, the blood supply to the muscles being cut off under water and contraction wholly anaerobic. Recovery after activity is slow. Turtles require 5 h to remove half of the lactate accumulated in a one hour dive.

In the heart of lizards and other reptiles except crocodiles there is a partial separation of venous and arterial blood (Fig. 13.4). There are two auricles, but only one ventricle, this being partly divided by a septum into right and left sides. Three arterial trunks arise directly from the ventricle, these being the right and left aortae, and the pulmonary trunk twisted so that the opening of the left aorta lies opposite the right side of the ventricle. It used to be thought that this arrangement led to mixing of venous and arterial blood, but radiographic studies have shown that when a reptile is breathing normally there is an almost complete separation. The conditions within the ventricle and the lower resistance in the lungs ensure that the first blood ejected passes to the pulmonary arch. The ventricle is partly divided into three chambers (Fig. 13.5) (White 1976). The cavum pulmonale (or ventrale) is the right-hand part, separated by horizontal and smaller vertical septa from the cavum venosum receiving blood from the right auricle and cavum arteriosum from the left. When the auricles contract the atrioventricular valves meet the septum dividing the cavum arteriosum from the combined cavum venosum and pulmonale (Fig. 13.5) which fill with venous blood. In the first part of systole this passes to the pulmonary artery, but with increasing pressure the septa move over and nearly meet the ventral wall, obliterating the cavum venosum (Fig. 13.6). The difference in pressure in the two main chambers can be shown by pulling a catheter from one to the other (Fig. 13.7).

These events explain the long-known fact that the left arch contains red blood (i.e. oxygenated). Mixing does occur, however, if the pulmonary pressure is raised, as in turtles when diving and perhaps in other conditions such as thermoregulation. The left arch is probably retained for its functional value, it is not a useless relic. It may indeed provide means of adjustment that are no longer possible in the separated systems of birds and mammals. In crocodiles the septum completely divides the ventricles. The left aorta arises from the right ventricle but the two aortae are joined by an opening, the foramen of Panizza. The higher pressure of the left

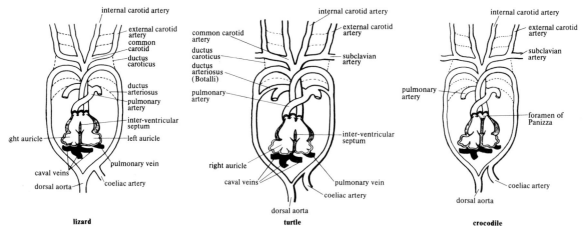

Fig. 13.4. Diagrams of the heart and arteries (Goodrich 1916). Holmes (1975) comments that 'the usually reliable E. S. Goodrich published a wholly inaccurate diagram of the reptilian heart'. The errors are in the twisting of the arches and especially the single 'interventricular septum'. Goodrich himself in a later publication said, 'I willingly admit that my description of the reptilian heart is very general, and in some respects incomplete, and that the diagram given may be somewhat misleading when applied to the Lacertilia and Ophidia (though correct for the Chelonia)' (1919). The structure of the reptilian ventricle is still controversial even today.

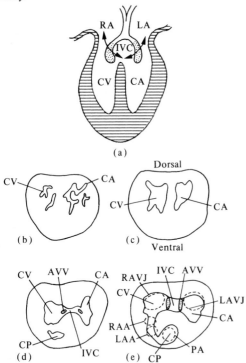

Fig. 13.5. Diagrams of a squamate heart (a) in frontal section dorsal of the cavum pulmonale. (b–f) Cross-sections, progressing from near the apex (b) towards the base of the heart in (e). AVV atrioventricular valve; CA cavum arteriosum; CP cavum pulmonale; CV cavum venosum; IVC interventricular canal; LA left atrium; LAA left aortic arch; LAVJ left atrioventricular junction; PA pulmonary arch; RAA right aortic arch; RAVJ right atrioventricular junction. (From White 1968.)

ventricle ensures that no deoxygenated blood enters the left aorta, except when the crocodile dives and the pressure in the lungs increases. This conserves oxygen by reducing the blood flow in the lungs. The brain still receives oxygenated blood through the right aorta, the mixed blood in the left passing only to the hinder part of the body.

The arterial blood pressure is around 70 mm Hg, varying with temperature and activity. The heart receives sympathetic accelerator and vagal depressor fibres, similar to those of mammals (Fig. 13.8). There are glomus bodies of unknown function on the right aortic arch. The arteries and chief veins receive an abundant adrenergic innervation. There is a well developed control of the circulation, and electrical stimulation of the hypothalamus produces changes in heart-rate. Heating and cooling of the hypothalamus produces changes in arterial pressure and peripheral circulation of iguanas. These animals have a preferred temperature of 36°C and in midday heat adopt a posture that limits exposure to the sun and perhaps allows heat-loss to the ground (Fig. 13.9). It seems likely that temperature regulation from the hypothalamus was developed long before endothermy.

Other changes in the circulation are produced during diving. In turtles bradycardia (slowing of the heart rate) occurs within the first beat or two after submergence and disappears with the first breath of air, or even slightly before emergence. The control is apparently produced by vagal inhibition.

7. Blood

Variations that would be fatal to a mammal are

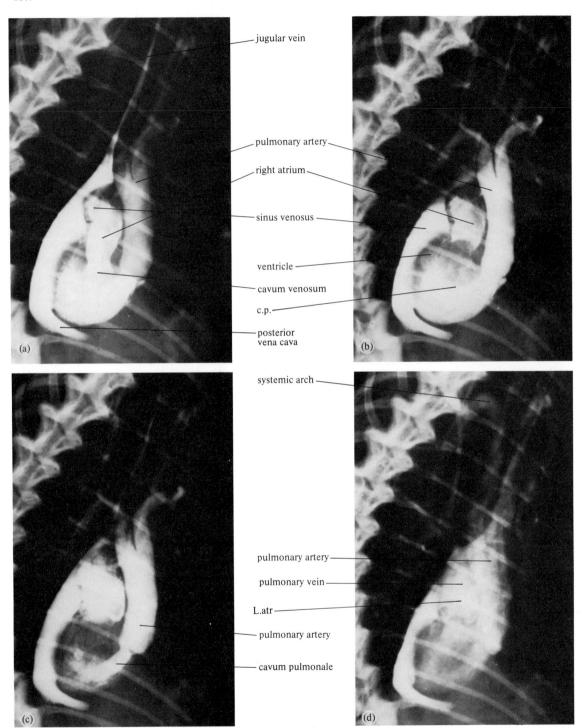

Fig. 13.6. The passage of a radiopaque medium (injected into the right jugular vein) passing through the heart of *Varanus niloticus* followed by cineradiography. Part of two cardiac cycles, the right frame follows the left. In the left frame contrast can be seen in the cavum venosum and pulmonary artery from the preceding ventricular systole. In the right frame the venosum is obliterated and the cavum pulmonale (c.p) and cavum arteriosum are full. No labelled blood has yet entered the aortic arches. L.atr., left atrium. (From Johansen 1977.)

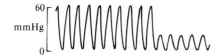

FIG. 13.7. *Varanus varius*. Recording of pressure in the heart as a catheter is pulled from the cavum venosum (left) into the cavum pulmonale (right). (After Harrison, J.M. (1965). M.Sc. Thesis, University of Sydney.)

2 min

(a)

(b)

(c)

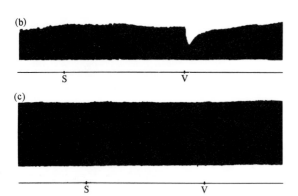

FIG. 13.8. Lizard. Responses of isolated auricles to stimulation of the sympathetic (S) and vagal (V) nerves (10 Hz for 10 s) at the points marked on the baseline.
(a) control responses; (b) 60 min after addition of bretylium (2×10^{-6} g/ml) to the organ bath, the response to sympathetic stimulation is abolished; (c) 25 min after addition of hyoscine (10^{-7} g/ml), the vagal response is also abolished and the force of the beat is increased. (From Berger, P.J. (1971). *Journal of experimental Biology and Medical Science* **49**, 297–304.)

continually occurring in a reptile, particularly with the temperature changes characteristic of an ectotherm. Glucose concentration in the blood varies widely with conditions. It is raised during anaerobic activity up to 1200 mg% after 24 h in nitrogen (in a turtle). The plasma osmotic pressure in turtles may vary from 150–450 mOs/l. Calcium may reach 200 mg% in snakes. All the chloride disappears from the blood of a crocodile after the demand for hydrochloric acid imposed by feeding. Bicarbonate enters instead and the pH may reach 8.1. The pH of the blood of reptiles may vary from this level down to 6.5 due to accumulation of lactic acid. They can maintain the pH of their blood at preferred

levels, although the tissues are able to tolerate the much lower pH values produced by lactic acid after activity. Evidently the constancy of the blood is not a prerequisite condition for the free life of reptiles as it is of mammals.

The red cells of reptiles are nucleated and survive for much longer than those of mammals (up to 800 days in tortoises). The total haemoglobin concentrations and oxygen carrying power of the blood are only half those of mammals. Moreover, there is little capacity to increase erythropoiesis at high altitudes. Reptiles tend to meet changed conditions by their tolerance of variation rather than by special mechanisms to maintain constancy.

The white cells of reptiles include the same types as in other vertebrates. The lymphatic system is exceptionally well developed, with larger vessels than in mammals. There are no lymph nodes but large cisterns occur at the sites where nodes are found in birds and mammals. Lymph is pumped by paired lymphatic hearts to a large cisterna chyli in the abdomen and from there passes through thoracic ducts, forming sheaths around the aortae, to enter the base of the subclavian and jugular veins.

The venous system is based on the same plan as that of the frog, with pelvic veins receiving blood from the tail and hind legs and returning it to the heart through either an anterior abdominal vein or renal portals, and the inferior vena cava.

8. Excretion

In the urinogenital system is seen another feature characteristic of amniotes, the development of a posterior region, the metanephros, concerned solely with excretion, leaving the mesonephric (Wolffian) duct to function as the vas deferens in the male. There is sometimes an endodermal (allantoic) bladder.

The waste nitrogen is largely excreted as uric acid in Squamata, and this allows the resorption of much of the water in the urodaeum, with precipitation of the organic matter as a chalky white mass of urates. The advantage of this uricotelic method of excretion is that it allows for a greater economy of water than would be possible if the end product was the more soluble urea (ureotelic). There is, however, great variation in the mode of excretion, depending on the manner of life of the species and the necessity for water conservation. Thus among Chelonia the more aquatic forms (*Emys*) produce considerable amounts of ammonia and urea, but relatively little uric acid, whereas the last is the main excretory product of the fully terrestrial types, such as the Greek tortoise (*Testudo graeca*), which can live under almost desert conditions. However, under some conditions tortoises

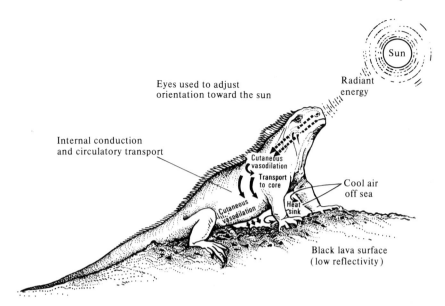

Eyes used to adjust
orientation toward the sun

Radiant
energy

Sun

Internal conduction
and circulatory transport

Cutaneous
vasodilation

Transport
to core

Cool air
off sea

Cutaneous
vasodilation

Heat
sink

Black lava surface
(low reflectivity)

FIG. 13.9. *Amblyrhynchus*, the marine iguana as a regulated heat shunt. The model emphasizes the interplay of behaviour (elevated basking) with reference to the sun, creation of a heat sink (microenvironment), and the graded cutaneous vascular response to thermal gradients. The result is relative stability of core temperature under heavy thermal loading. (From White, F.M. (1973). *Comparative Biochemistry and Physiology* **45A**, 503–13.)

excrete urea and a thin watery urine. Crocodiles excrete much ammonia. The bladder and cloaca of reptiles often assist the renal tubules in the regulation of the water level of the blood.

Salt glands have been evolved independently in several types of reptile, because they live either in the sea or in deserts. In turtles there are orbital glands (Harderian glands) and the secretion is washed away by abundant tears. Sea snakes have sublingual salt glands and in various lizards the nasal glands have developed this function and are perhaps homologous with those of birds.

9. Reproduction

Some of the most serious difficulties in the colonization of the land are concerned with reproduction, and these problems have been largely solved in the reptiles, allowing the animals to reproduce without returning, as many Amphibia must do, to the water. Fertilization has become internal, and in all modern reptiles except *Sphenodon* special organs of copulation derived from the cloacal wall are developed in the male. In crocodiles and tortoises there is a single median penis, but in lizards and snakes there is a pair of these structures, though only one is inserted at a time. The mechanism of erection involves both muscular action and vascular engorgement converting the cloacal pocket or groove into a tube for the sperm. The sperms pass from the vasa deferentia

into the urodaeum, and after traversing this region they are carried into a groove along each penis. In turtles and snakes the sperms may survive within the female for long periods, and instances are known of isolated individuals laying fertile eggs after months, sometimes even years, in captivity.

The eggs of oviparous reptiles are always laid on land. They therefore require a shell for physical support and protection against desiccation, as well as an adequate supply of food and special means of gaseous exchange and storage of waste products. These requirements are met by the development of a shell, secreted by the walls of the oviduct and often hardened by lime impregnation, by the formation of special embryonic membranes, the amnion and allantois, and by the provision of a large quantity of yolk enclosed in a bag, the yolk-sac (Fig. 13.10). The embryonic cleavage is affected by the great amount of yolk, and as in birds is only partial. An albumen or egg-white layer is present in the eggs of crocodiles and tortoises, and presumably serves as a reservoir of water; but in the eggs of lizards and snakes is poorly developed or absent.

The formation of the amnion and allantois is one of the most remarkable features of the development of reptiles and is characteristic of all higher vertebrates, distinguishing them sharply from the lower types. The amnion is developed from folds, which cover the embryo and enclose a sac filled with fluid, where

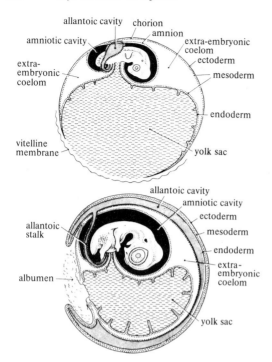

allantoic cavity   chorion
amniotic cavity   amnion
                              extra-embryonic
                              coelom
extra-   ectoderm
embryonic
coelom   mesoderm

                              endoderm

vitelline
membrane
                      yolk sac

allantoic cavity
       amniotic cavity
              ectoderm
allantoic   mesoderm
stalk
              endoderm
                      extra-
                      embryonic
albumen   coelom

                      yolk sac

FIG. 13.10. Diagrams showing the relations between extra-
embryonic membranes and the chick embryo at fourth (above)
and ninth day (below) of incubation.

development can proceed in the absence of the pond
that was necessary for the earlier vertebrates. The
allantois began as an enlarged bladder, serving for the
reception of the waste products during the life within the
shell. It is therefore lined by endoderm and covered by
well vascularized mesoderm. Coming close to the surface and fusing with the chorion, it then becomes the
vehicle for the transport of oxygen to the embryo. The
allantois is also essential for the storage of the waste
produced during development. The final product of
nitrogen metabolism in the chick is only to a small extent
ammonia, which might become toxic, or urea, which is
soluble and would increase osmotic pressure, but 80 per
cent is insoluble uric acid or urates. In the black snake,
however, excretion is at first ureotelic and uric acid is
produced only in later stages. In the alligator egg
excretion is almost wholly of ammonia and urea.

The evolution of these eggs and embryonic membranes must have been an event of critical importance in
tetrapod history. They are called cleidoic or closed box
eggs. This advance may have taken place under climatic
conditions of alternate drought and flooding, so that
eggs laid above the high-water mark had the best chance
of survival. Since many of the early reptiles are thought
to have spent much of their time in the water, it is

possible that the egg preceded the adult in the process of
adaptation to terrestrial life.

Most reptiles lay their eggs, but in many lizards and
snakes these are retained within the oviduct until the
young are ready or nearly ready to hatch (e.g. *Lacerta
vivipara, Anguis fragilis, Vipera berus*). The eggs are
always large, however, and the method of reproduction
is termed ovoviviparous. In forms that practise it the
eggshell is reduced to a thin membrane or is lost
altogether. In some species (e.g. certain skinks and other
lizards, sea-snakes) a placenta is developed from the
chorio-allantois or the yolk sac or both. The placenta
may, as in *Lacerta vivipara*, serve only for transfer of
water and dissolved gases, but in more advanced forms
it probably provides a means of transport for food
(supplementing the yolk) and excretory products. In
ovoviviparous species corpora lutea are formed from
the discharged ovarian follicles, and they produce
progesterone.

Young born alive are perhaps less susceptible to the
hazards of weather than those left to hatch in the sun or
among rotting vegetation, and it is interesting that all of
the few reptiles that live in places where the climate is
really severe are ovoviviparous.

In several different species of lizards and geckos there
are races that consist only of females, reproducing
parthenogenetically (e.g. *Lacerta saxicola* in the Cau-
cacus and *Cnemidophorus* in North America).

Young reptiles have special devices to assist their
escape from the egg. In *Sphenodon*, Chelonia, and
Crocodilia, as in birds, there is a horny epidermal egg-
breaker on top of the snout tip, called the egg-caruncle.
In the Squamata, a true egg-tooth, projecting from the
front of the upper jaw, has the same function. The egg-
tooth is present, though sometimes rudimentary, in
ovoviviparous forms.

Some reptiles make a simple nest but the group is not
noted for maternal care, usually abandoning their new-
laid eggs or newborn young. There are, however, some
exceptions to this; female pythons and certain other
snakes and lizards brood their eggs, and female croco-
dilians guard their nests, often help to liberate the young
from beneath the impacted sand, and may remain with
the young for some time after hatching. These reptiles
demonstrate far more elaborate forms of parental care
(in which even the male may participate) than was
formerly believed.

Many reptiles exhibit well marked courtship and
display phenomena during the breeding season. The
males fight and display ritually either to intimidate each
other or to evoke a suitable response from the female.
This is particularly striking in certain lizards, notably
those of the iguanid and agamid groups, where the

males are often brightly coloured and may be adorned with crests and highly coloured distensible fans under the throat. In these lizards bobbing movements of the head and front part of the body, often accompanied by colour change, form an important part of the display. As in birds, courtship may be associated with territory, a male holding an area of ground on which females, but not rival males, are tolerated. The formalized communication system of the lizard *Anolis* has been investigated in detail (Greenberg 1977). When a male enters another's territory they exchange a sequence of signals. Fighting occurs only if the invader does not answer the threats by retreat. Male snakes also show elaborate male combat rituals, twining round each other and seeking dominance by pressing down the rival (Carpenter 1977) (Fig. 13.11). The communication systems of crocodiles are visual and auditory, those of tortoises mainly tactile and chemical. The newly-hatched young of snakes, turtles, iguanas and other reptiles show various social signals leading to collaborative aggregations and groups, behaviour that may promote survival in the absence of parental care (Burghardt 1977). Breeding behaviour and sexual coloration are, of course, under the control of the endocrine system, especially the anterior pituitary and the gonads, and may be modified by castration. The onset of the breeding season is also influenced by climatic conditions; most reptiles breed only once or twice a year, but a few species living in warm stable climates may breed at intervals nearly all the year round.

10. The brain

All the modifications of structure that fit the reptiles for life on land would be useless without the development of appropriate behaviour. This in turn depends on suitable structure and function of the nervous system, and the brain shows some interesting developments, though it is relatively small, being, at most, 1 per cent of the body weight. In the dinosaurs of up to 30 500 kg the brain probably weighed only about 100 g. The cerebral hemispheres are relatively larger in reptiles than in amphibians (Fig. 13.12). The increased bulk lies mainly in the basal parts of the hemisphere (the corpus striatum), as in birds (Fig. 13.13). The roof (pallium) is little developed and lacks the elaborate cortical differentiation found in mammals. The pallium is, however, moderately well developed in turtles and could be regarded as 'ancestral' to the mammalian neocortex. This agrees with the evidence that the Chelonia are surviving anapsids. The condition in lizards and crocodiles is more similar to that in birds (p. 360).

The midbrain receives visual and other sensory inputs and is largely involved in the guidance of movement. The retinal projection is somatotopically organized on

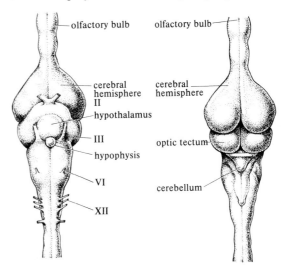

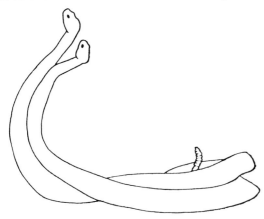

FIG. 13.11. The mutual display during combat between two male *Crotalus atrax*. Note high vertical stance, neck flexion, and posterior trunk contact. (Taken from ciné frames, Carpenter 1977.)

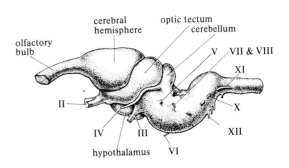

FIG. 13.12. Three views of the brain of a lizard. (After Frederikse.)

Fig. 13.13. Transverse section through forebrain of *Lacerta*. (After de Lange and Kappers.)

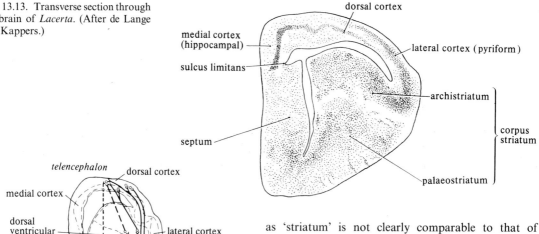

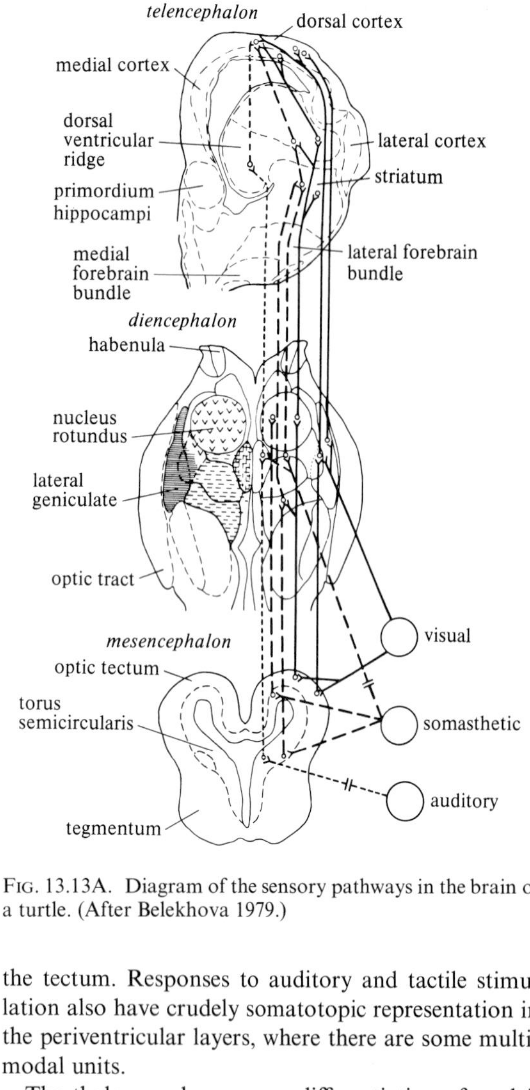

dorsal cortex

medial cortex (hippocampal)

sulcus limitans

lateral cortex (pyriform)

archistriatum

corpus striatum

septum

palaeostriatum

telencephalon

dorsal cortex

medial cortex

dorsal ventricular ridge

lateral cortex

striatum

primordium hippocampi

medial forebrain bundle

lateral forebrain bundle

diencephalon

habenula

nucleus rotundus

lateral geniculate

optic tract

visual

mesencephalon

optic tectum

somasthetic

torus semicircularis

auditory

tegmentum

Fig. 13.13A. Diagram of the sensory pathways in the brain of a turtle. (After Belekhova 1979.)

the tectum. Responses to auditory and tactile stimulation also have crudely somatotopic representation in the periventricular layers, where there are some multimodal units.

The thalamus shows some differentiation of nuclei, especially in turtles (Fig. 13.13A) (Belekhova 1979). Visual projections reach the thalamus both from the tectum to a large nucleus rotundus and directly to the lateral thalamus. The large ventrolateral region known as 'striatum' is not clearly comparable to that of mammals. It receives visual and somatosensory projections through the thalamus. The pallium is differentiated into medial, dorsal and lateral divisions and a large subpallium with characteristic dorsal ventricular ridge (Fig. 13.13A).

Olfactory stimulation produces evoked potentials in all parts of the forebrain except a small area of the dorsal cortex. This area receives visual, somatosensory and auditory projections through thalamic relays. Many of its cells respond to several modalities. The visual impulses reach the cortex and dorsal ventricular ridge by two pathways, from the tectum through the nucleus rotundus and more directly through the lateral thalamic nucleus. The two visual pathways to the cortex are more distinct in reptiles than in anamniotes but less so than in either birds or mammals. There is a crude retinotopic organization in the dorsal cortex of turtles. The cells respond maximally to moving stimuli and may show discrimination of intensity, velocity of movement and size of stimulus. The majority of cells have very large receptive fields, often covering the entire visual field of both eyes. Some units have restricted fields with directional sensitivity (see Belekhova 1979). Cells of the subpallium and striatum show similar properties.

Many types of visual response are still possible after ablation of the cortex or even the whole forebrain in turtles or lizards including conditioned spatial discrimination. Only combined cortical and tectal lesions produce large deficits in visually guided behaviour. However, more trials are needed to learn habits after cortical ablation and the capacity to make delayed responses disappears. The forebrain thus appears to add to the effectiveness of the tectum in discrimination and learning rather than providing a qualitatively different behavioural capacity. Lizards with lesions in the striatum were found to fail to respond to the challenge display of another male (Greenberg 1977). They were

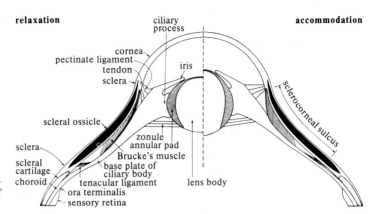

FIG. 13.14. Diagram to show the mechanism of accommodation in the eye of reptiles. (From Walls 1942.)

not otherwise blind but seemed not to perceive the challenger as an antagonist. However, some behaviour patterns are altered by telencephalic ablation. Further research is needed to improve our understanding of the localization in the brain of the programs that regulate behaviour (Young 1979).

Reptilian brains show electroencephalograms with frequencies similar to those of mammals. There is evidence of behavioural and cerebral sleep in turtles, as distinct from the periods of reduced mobility found in some fishes and amphibians. They do not show clear fast-wave ('paradoxical') sleep.

11. Receptors

The eyes are the main exteroceptive sense-organs of most reptiles. The cornea is curved and provides the main refracting surface. It is protected by movable eyelids, including a third eyelid or nictitating membrane which sweeps backwards rapidly to clean the cornea. In snakes and some lizards the cornea is protected by a transparent skin, the spectacle, and there is no nictitating membrane. The lachrymal and Harderian glands provide secretions that keep the surface of the cornea moist and in lizards a lachrymal duct carries tears away to the mouth while in crocodiles it enters the nasal cavity. The eye is supported in most reptiles by a scleral cartilage and a ring of bony scleral plates. Accommodation for near vision is usually produced by the striated ciliary muscles, so arranged that they cause the ciliary process to squeeze the lens, making its anterior surface more rounded (Fig. 13.14). In many reptiles the retina possesses both rods and cones, the latter predominating in diurnal types including most lizards and many turtles, which can discriminate between colours. The double cones of turtles and lizards may serve to detect polarized light (Underwood 1970). The conus papillaris is a vascular rod projecting into the vitreous (Fig. 14.19). Probably it provides nutrition, like the pecten of birds (p. 368).

In *Sphenodon* and many lizards a 'pineal' or parietal eye is present with lens-like and retina-like components. In such forms there is a pineal foramen in the parietal bone near the frontoparietal suture. Similar foramina are found in many fossil reptiles, especially the more primitive types. The function of the reptilian pineal is still rather obscure, but there is evidence that in lizards it registers solar radiation, and, perhaps by the secretion of melatonin, influences the animal's thermoregulatory behaviour in exposing itself to sunlight. It is also possible that the pineal complex plays some part in the control of reproduction.

The nose is a sac opening by external nostrils on the head and internally through the palate. Not all the epithelium is olfactory, the remainder being well vascularized and perhaps concerned in temperature control. The organ of Jacobson (vomeronasal organ), a specialized and sometimes separate region of the nose, innervated by a separate branch of the olfactory nerve, is present in *Sphenodon*, and well developed in most Squamata (see p. 306).

The tympanum when present lies behind the jaws, sunk a little below the surface. Vibrations are carried across the middle-ear cavity by a rod, the stapes or columella, whose inner end forms a footplate attached to the fenestra ovalis of the inner ear (Fig. 13.15). In snakes and many lizards there is no tympanum and the stapes is attached to the quadrate, detecting vibrations of the ground, through the jaw.

The inner ear contains the usual parts, with a large endolymphatic duct and sac, containing calcareous matter and sometimes extending outside the skull. Hearing is probably performed by the lagena or cochlear duct, an uncoiled tube leading from the saccule. Its wall, known as the limbus, is strengthened by a peculiar form of connective tissue. The receptor hairs, (organ of Corti) are carried on a basilar membrane and are in contact with a tectorial membrane attached to the limbus. Hearing is good in some lizards which also produce

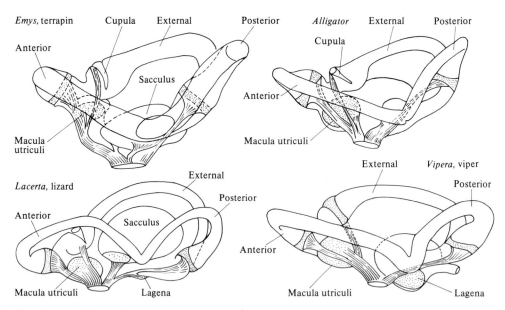

FIG. 13.15. The shape and form of the semicircular canals of several reptiles. (Retzius, M.C. (1881). *Das Gehörgan der Wirbelthiere. Morphologische-histologische Studien.* Stockholm.)

sounds for communication. Crocodiles also hear well but chelonians only hear up to 1000 Hz and snakes respond mainly to earth-borne vibrations through the quadrate (p. 416), although they are also quite sensitive to a narrow waveband of low-frequency air-borne vibrations (see Wever 1978).

12. Endocrine glands

The pituitary contains the usual parts, with a portal system that carries blood from capillaries in the median eminence to the distal lobe. There are five different types of chromophilic cells in the distal lobe similar to those of amphibians and including lactrotropes and somatotropes, basophil thyrotropes, corticotropes, and gonadotropes. Reptiles, like other tetrapods, secrete two distinct gonadotropins, FSH and LH. The pars intermedia produces MSH and is large in reptiles that change colour (p. 308) but very small in burrowing forms. The neurohypophysis contains arginine vasotocin, and mesotocin, the former being the main antidiuretic hormone. The neurohypophysis also regulates the oviduct and injection of extracts to *Lacerta vivipara* leads to premature oviposition (Fox 1977).

An interesting feature of the thyroid is that injection of T_4 has no influence on oxygen consumption in lizards (*Lacerta*) at 20 °C but causes a marked rise at 30 °C, which is the preferred temperature. After removal of the thyroid, moulting ceases in *Lacerta* but increases in snakes.

The parathyroids are similar to those of mammals.

After their removal from lizards the blood calcium falls to half its normal level and the phosphate rises. The animals become hyperexcitable and have tetanic convulsions after exercise. Parathyroid injections return them to normal. There is therefore a definite mechanism for calcium regulation, in spite of variations in its level (p. 365).

The adrenal glands usually lie close to the gonads except in turtles. The parts secreting steroids and catecholamines are mixed up together, with the former often inside the layers of chromaffin tissue. The steroid tissue ('interrenal') atrophies after hypophysectomy.

Reptiles have well developed thymus glands and they react against foreign material by cellular proliferation and antibody production. There is an immunological memory and a heightened response to a second challenge by foreign material and these are at least partly dependent on the thymus. Removal of it from young individuals reduces the capacity for adaptive response. The thymus atrophies in adults.

13. Behaviour

Many species have elaborate courtship behaviour and some are social (p. 286). Their communications are by means of stereotyped sets of visual or auditory signals, probably largely innate. Thus many lizards give nods or 'bobs' with the head. The display of a hybrid *Anolis* was found to be intermediate between those of the parent species. Vocal displays are mostly by geckos such as the barking gecko *Ptenopus garrulus* of the Kalahari Desert.

Males exhibit courtship patterns. In some species these are variants of aggression displays used to hold territory or in aggregations of males. Females may give non-receptivity signals when gravid.

14 Evolution of the reptiles

1. Classification

Class REPTILIA
  Sublcass 1: Anapsida
    Order 1: *Cotylosauria. Carboniferous–Triassic
      *Romeriscus; *Solenodonsaurus; *Captorhinus; *Limnoscelis; *Labidosaurus; *Milleretta; *Diadectes
    Order 2: *Mesosauria. Late Permian
      *Mesosaurus
    Order 3: Chelonia. Permian–Recent
      *Eunotosaurus; *Proganochelys; *Archelon; Chelus; Emys; Chelonia; Testudo
  Subclass 2: *Synaptosauria (=Euryapsida)
    Order 1: *Protorosauria. Permian–Triassic
      *Araeoscelis; *Tanystrophaeus
    Order 2: *Sauropterygia. Triassic–Cretaceous
      *Lariosaurus; *Pliosaurus; *Plesiosaurus; *Placodus
    Order 3: *Placodontia. Triassic
      *Henodus
  Subclass 3: *Ichthyopterygia
    Order 1: *Ichthyosauridae. Triassic–Cretaceous
      *Mixosaurus; *Ichthyosaurus
  Subclass 4: Lepidosauria
    Order 1: *Eosuchia. Permian–Eocene
      *Youngina; *Prolacerta
    Order 2: Rhynchocephalia. Triassic–Recent
      *Homoeosaurus; *Rhynchosaurus; Sphenodon (=Hatteria); *Scaphonyx
    Order 3: Squamata. Triassic–Recent
      Suborder 1: Lacertilia (=Sauria). Triassic–Recent
        Infraorder 1: Gekkota. Mainly Recent
          Gekko; Pygopus; Hemidactylus; Ptychozoon
        Infraorder 2: Iguania. Cretaceous–Recent
          Iguana; Anolis; Phrynosoma; Draco; Lyrocephalus; Agama; Chamaeleo; Amblyrhynchus; Calotes
        Infraorder 3: Scincomorpha. Eocene–Recent
          Lacerta; Scincus
        Infraorder 4: Anguimorpha.
         Cretaceous–Recent
          *Dolichosaurus; *Aigialosaurus; *Tylosaurus; *Mesosaurus; Varanus; *Lanthanotus; Anguis
      Suborder 2: Ophidia (=Serpentes). Cretaceous–Recent
        *Dinilysia *Palaeophis; Python; Natrix; Naja; Vipera
      Suborder 3: Amphisbaenia. Recent
        Amphisbaena
  Subclass 5: Archosauria
    Order 1: *Thecodontia. Triassic

*Euparkeria; *Saltoposuchus; *Phytosaurus; *Mystriosuchus
　Order 2: Crocodilia. Triassic–Recent
　　*Protosuchus; Crocodylus; Alligator; Caiman; Gavialis
　Order 3: *Saurischia. Triassic–Cretaceous
　　Suborder 1: *Theropoda
　　　*Compsognathus; *Ornitholestes; *Allosaurus; *Tyrannosaurus; *Struthiomimus; *Plateosaurus
　　Suborder 2: *Sauropoda
　　　*Apatosaurus; *Brontosaurus; *Diplodocus; *Yaleosaurus; *Plateosaurus; *Brachiosaurus
　Order 4: *Ornithischia. Triassic–Cretaceous
　　Suborder 1: *Ornithopoda
　　　*Camptosaurus; *Iguanodon; *Hadrosaurus
　　Suborder 2: *Stegosauria
　　　*Stegosaurus
　　Suborder 3: *Ankylosauria
　　　*Ankylosaurus; *Nodosaurus
　　Suborder 4: *Ceratopsia
　　　*Triceratops
　Order 5: *Pterosauria. Jurassic–Cretaceous
　　*Rhamphorhynchus; *Pteranodon
Subclass 6: *Synapsida. Carboniferous–Permian
　Order 1: *Pelycosauria (=*Theromorpha)
　　*Varanosaurus; *Edaphosaurus; *Dimetrodon
　Order 2: *Therapsida. Permian–Jurassic.
　　*Scymnognathus; *Cynognathus; *Bauria; *Dromatherium; *Dicynodon; *Gorgonops; *Lystrosaurus;
　　*Kannemeyeria

2. Skull types among the reptiles

SINCE our knowledge of reptiles depends mainly on fossil remains it is convenient to classify them by means of the skull into four great groups (Fig. 14.1). Such a classification is in some ways artificial, but it serves to indicate in a broad way the main lines of evolution within the class.

In the cotylosaurs the dermal bones of the temporal region of the skull presented an unbroken surface and there were no temporal fossae. There were therefore no arches or 'apses' of bone in the temporal region. Such

forms are placed in the subclass Anapsida. The jaw muscles took origin from the deep surface of the temporal side wall, between it and the brain case, and they passed down through holes in the palate to be inserted on the lower jaw. This represents the most primitive condition found in reptiles, and resembles that in the early amphibians. It is still seen today, though often in a modified form, in the Chelonia, which are hence placed in the anapsid subclass.

In more advanced groups of reptiles fossae bounded

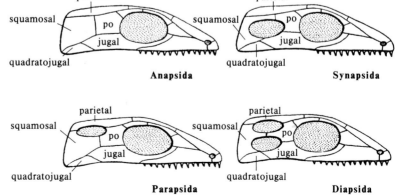

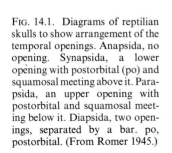

FIG. 14.1. Diagrams of reptilian skulls to show arrangement of the temporal openings. Anapsida, no opening. Synapsida, a lower opening with postorbital (po) and squamosal meeting above it. Parapsida, an upper opening with postorbital and squamosal meeting below it. Diapsida, two openings, separated by a bar. po, postorbital. (From Romer 1945.)

by bony arches appear in the temple region, enabling the jaw muscles to bulge into and so facilitate their action.

In many reptiles, two such fossae appeared, the condition being termed diapsid. This is seen in the subclasses Lepidosauria and Archosauria, perhaps the most successful groups of reptiles. In lepidosaurs of the order Squamata, however, the lower temporal arch is always incomplete, having no quadratojugal bone and the jugal separated from the squamosal. In some lizards and in snakes the upper arch is also lost.

In other groups only a single fossa and arch is present. When this is situated high on the skull the condition is known as parapsid. Parapsid skulls are seen in the subclasses Ichthyopterygia (icthyosaurs) and Synaptosauria (plesiosaurs, etc.). Formerly, these two subclasses were placed together in a group known as the Parapsida, as is shown in Fig. 14.2, but this classification is now regarded as artificial, since the ichthyosaurs and sauropterygians are not closely related; in fact a careful analysis shows that the bony relationships of their single temporal fossae were rather different.

In the remaining subclass, the Synapsida, there is also a single fossa, but in the earlier forms at least it is placed low down and the parietal and squamosal meet above it, while it is bounded below by the jugal and squamosal. The term synapsid, meaning 'fused arch', is actually a misnomer, due to the fact that early workers believed,

wrongly, that the single arch was formed from the fusion of the two seen in diapsids.

The synapsids comprise the mammal-like reptiles, but in the later members of the group, such as *Cynognathus*, and in their descendants the mammals, the temporal fossa has greatly enlarged, and has lost its primitive relationships.

3. The earliest reptile populations. Subclass Anapsida (without arches)

The organization of a reptile is well suited to maintain life on land. Many features show a considerable advance in this respect over the Amphibia, for example, the dryness of the skin, the method of reproduction, and the devices for economizing in the use of water. The immense radiation of the reptiles into every sort of land habitat during the Mesozoic period shows the efficiency of these mechanisms, which were probably present, at least in imperfect form, in the Carboniferous and Permian offshoots from the ancestral Labyrinthodontia (p. 242).

3.1. Order Cotylosauria (= stem reptiles)

The earliest fossil generally agreed to be a reptile is the cotylosaur *Romeriscus*, a single specimen from the lower Pennsylvanian. Of similar age is *Solenodonsaurus* (Fig. 14.3) a form often classified as an anthraco-

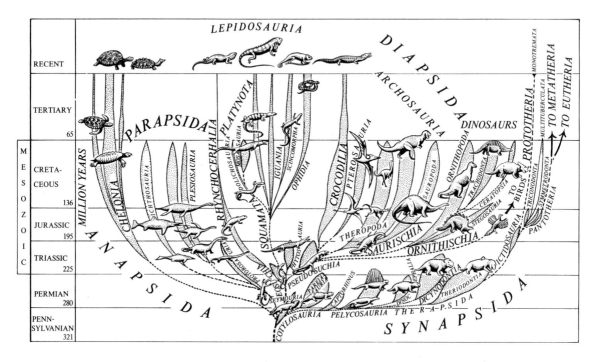

Fig. 14.2. Chart to show the course of evolution in the reptiles. For *Seymouria* see p. 271.

saurian amphibian (p. 270). It could have been close to the ancestry of all reptiles and perhaps also to the amphibian *Seymouria* (Figs. 14.2 and 12.3) (p. 270). It possessed large pleurocentra but also small intercentra. There were scales and reptilian limbs and girdles. The skull was long and high with large orbits. The otic notches were present but smaller than in *Seymouria*. There was an intertemporal bone as in anthracosaurs but the tabular and supratemporal were reptilian, as was the palate (no teeth) and pterygoid (with a flange). The teeth in the maxilla ('canines') were very like those of definite cotylosaurs. *Solenodonsaurus* (Fig. 14.3) and some similar fossils are almost precisely intermediate between the conditions that are arbitrarily called amphibian and reptilian. The name Cotylosauria (usually

translated as 'stem reptiles') is now used by some for *Solenodonsaurus*, *Seymouria*, *Diadectes*, and some allies. These cotylosaurs are considered to be an early egg-laying offshoot of an amphibian group *Batrachosauria*, which gave rise independently to the captorhinomorphs and other true reptiles (see Panchen 1980; Heaton 1980). There evidently remains much uncertainty and confusion about the classification of these early forms intermediate between amphibians and reptiles.

Better known cotylosaurs from the beginning of the Permian (the Red Beds of Texas) are *Limnoscelis*, *Captorhinus*, and *Labidosaurus*, all with the rather high narrow skulls and pointed nose characteristic of reptiles rather than amphibians, and differing from the

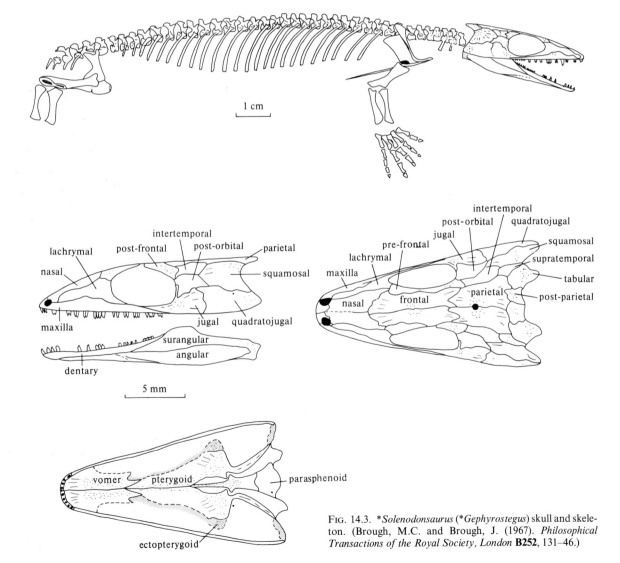

FIG. 14.3. *Solenodonsaurus* (*Gephyrostegus*) skull and skeleton. (Brough, M.C. and Brough, J. (1967). *Philosophical Transactions of the Royal Society, London* **B252**, 131–46.)

latter in the absence of the otic notch. This and other features suggest that these animals were related to the early mammal-like reptiles (p. 405). *Limnoscelis*, however, was a particularly primitive form, and may have been partly aquatic in habits (Fig. 14.4).

Solenodonsaurus, *Diadectes* and *Pareiasaurus* (Figs. 14.3 and 14.4) with specialized teeth including cropping 'incisors' and grinding 'molars' were reptiles. There was an enormous otic notch, which carried an ossified tympanic membrane attached to a huge stapes. The function of this is not known but the large notch and other features have led Romer (1966) to place *Diadectes* as an amphibian, perhaps close to *Seymouria*.

Bradysaurus and other 'pareiasaurs' from the Permian and Triassic of Europe, Africa, and America were up to 3 m long and probably carried the body well off the ground, the limbs being held underneath the body and showing some reduction of specialized digits. This, together with the large size of the animals, suggests that with *Diadectes* they may have been the first of the many types of large herbivore to appear on the land (see p. 321). In some of them the skull developed grotesque protective protuberances, recalling similar later developments in reptiles (*Ceratopsia, p. 318) and mammals (*Amblypoda, p. 529).

Milleretta from the Upper Permian was a small form with a short gap between the jugal and squamosal. It may have been close to the ancestry of the lepidosaurs (p. 303).

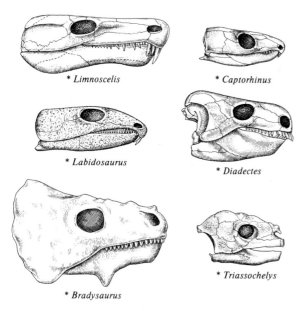

* *Limnoscelis*

* *Captorhinus*

* *Labidosaurus*

* *Diadectes*

* *Bradysaurus*

* *Triassochelys*

FIG. 14.4. Skulls of various early reptiles.

The early reptiles multiplied and became very diversified throughout the 45 Ma of the Permian period, by which time the main reptilian types had appeared. The individual reptilian orders nearly all became established during the subsequent 45 million years of the Triassic and most of them reached their maximum development in the Jurassic and Cretaceous.

3.2. Order Chelonia (turtles)

Shut away in their boxes the tortoises and turtles have retained some of the features of the earliest anapsid reptiles. Even today they are a not unsuccessful and quite varied and widespread group, with more than 200 species. These include terrestrial animals, such as *Testudo graeca* and *T. imarginata* tortoises of southern Europe, which are herbivorous; the freshwater tortoises, such as *Chrysemys* and other American terrapins; and *Emys* the European water-tortoise, all of which are carnivorous. The marine Chelonia, usually known as turtles, are often very large. *Dermochelys*, the leathery turtle, which has no horny shell, is over 2 m long and weighs over 500 kg. *Chelonia mydas*, the green or edible turtle, is about 1 m long (Fig. 14.5).

The characteristic of chelonian organization is the shortening and broadening of the body, together with the development of bony plates, forming a box into which the head and limbs can be withdrawn (Fig. 14.6). The total number of segments is only about 8 in the neck, 10 in the trunk, and a series of reduced caudals; the body is therefore morphologically shorter than in any other vertebrate except the frog. Probably this shortening and broadening is the result of some quite simple change in morphogenesis.

The shell is usually considered to include a dorsal carapace and ventral plastron. Each of these is made up of inner plates of bone, covered by separate outer plates of horny material, comparable to the scales of other reptiles. The carapace includes five rows of bony plates, namely median neurals, and paired costals and marginals (Fig. 14.7). These plates are ossifications in the dermis, attached to the vertebrae and ribs, but not actually formed from the latter. The plastron is developed from the expanded dermal bones of the pectoral girdle, together with dermal ossifications comparable to the abdominal ribs found in crocodiles and other reptiles. The whole is covered in most chelonians by rows of special smooth epidermal plates forming the 'tortoise' shell (Fig. 14.7). A new and larger layer is added to each of these plates each year, the old one remaining above it, thus making a number of 'growth rings' from which some indication of the age of the tortoise can be calculated, though the outer members often become rubbed off.

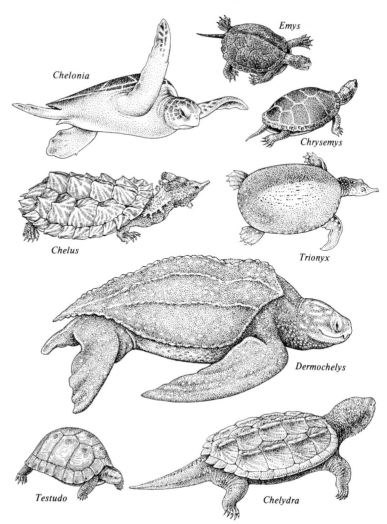

FIG. 14.5. Various chelonians, not all to same scale. (*Emys* and *Chelys* after Gadow, *Dermochelys* after Deraniyagala.)

In order to support this box the limb girdles have become much modified and lie inside the encircling ribs (Fig. 14.6). The pectoral girdle has three prongs, a scapula that meets the carapace dorsally and carries a long 'acromial process' and a backwardly directed coracoid, the two last being attached by ligaments to the plastron. The ilia are attached to two sacral vertebrae and the ischia and pubis are broad. The limbs are stout, but otherwise typically reptilian, with five digits in each. In the marine turtles they are transformed into paddles.

The interpretation of the skull is still somewhat doubtful, but it seems not unlikely that the turtle, *Chelonia*, shows the simplest case, namely, the original anapsid condition (Fig. 14.8). Here the roofing is complete, the dermal bones being widely separated from the brain case, forming a tunnel for the jaw muscles and those producing retraction of the neck. The tympanum is stretched across a sort of otic notch, bounded by the squamosal, quadratojugal and quadrate, the columella auris articulating with the latter. In other groups of Chelonia, however, the dermal roofing has been reduced or 'emarginated', presumably to give still better attachment for the powerful adductor mandibuli of the jaws and for the neck muscles (Fig. 14.8). It has been argued that the condition in *Chelonia* is secondary, but there is no evidence of true temporal fossae in any chelonian, and since the early form *Triassochelys* also had a fully roofed skull there seems no reason for denying that we have here essentially an anapsid condition.

A peculiarity of recent Chelonia is the entire absence of teeth, both in the herbivorous and carnivorous forms. The edges of the jaws form sharp ridges, covered with a formidable horny beak.

The similarity of the soft parts of Chelonia to those of other surviving reptiles (which are all diapsids) suggests that the general organization of the group has changed

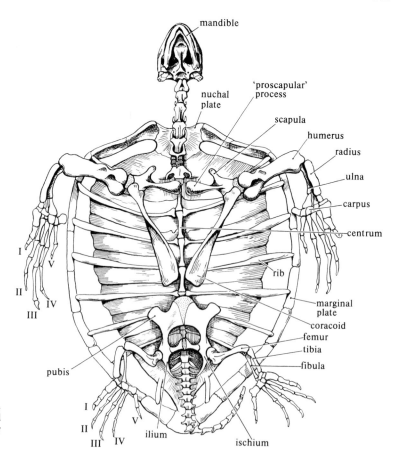

FIG. 14.6. Skeleton of turtle (*Chelone*).
Note hook-shaped 5th metacarpal and
foreshortened scapula. (After Shipley
and McBride and Reynolds.)

little since the Permian. The heart has a muscular sinus
venosus, separated auricles, a partly divided ventricle
and two equal aortic arches (Fig. 13.4). Respiration is
modified by the rigidity of the body wall; the lungs are

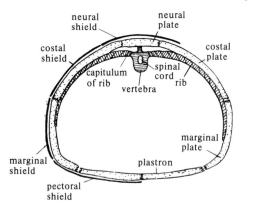

FIG. 14.7. Diagram of the arrangement of the shell of the
tortoise (*Testudo*). The horny shields are shown only on the
left. (After Gadow.)

spongy structures attached to the dorsal surface of the
shell, sometimes enclosed in a separate pleural cavity
(*Testudo*). Breathing is mainly by a suction pump type of
action, operated by the contraction of muscles at the
openings at front and back, which are probably partly
modified abdominal muscles. They function in a manner
comparable with that of the mammalian diaphragm. In
Testudo the pectoral girdle also moves the fore limbs in
and out. There may be pumping by the pharynx. Some
aquatic forms (*Emys*) also respire by taking water into
special vascularized diverticuli of the urodaeum. The
metabolism of tortoises is low and they can remain for
long periods without breathing. The leathery turtle,
Trionyx, of the Nile can stay underwater for as much as
6 h by taking up oxygen from the water through the skin
and papillae in the pharynx. Long stay underwater is
assisted by a capacity to obtain energy by anaerobic
glycolysis. Further, the tissues including the brain are
unusually tolerant of anoxia. In temperate climates all
species hibernate regularly.

The kidney is metanephric and the nitrogenous
excreta vary between urea and uric acid, there being a

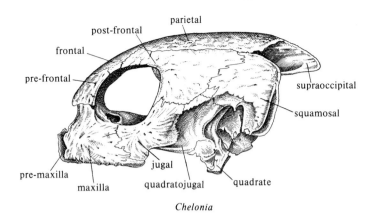

Chelonia

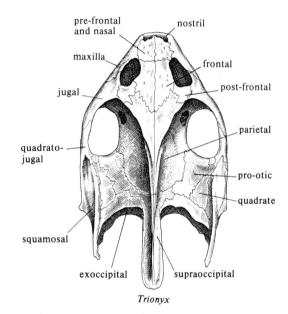

FIG. 14.8. Turtle skulls. (After Goodrich from
Parker and Haswell, and Zittel.)

Trionyx

typical subdivision of the cloaca and reabsorption of
water to form a solid whitish excretory product (see
p. 284). There is a single copulatory organ and the male
mounts on the back of the female after courting her by
following until she will accept him. The eggshells have
mineral and fibrous layers and are usually white. A
female may lay 10 clutches of 100 eggs in a year. Like
other aquatic reptiles that lay eggs, turtles all come
ashore to breed. Thus the marine *Chelonia mydas*, the
green turtle, may make migrations of more than
1500 km, perhaps by sun compass. The eggs are carefully placed in holes made by scooping away the sand
with the hind feet. The traces of the nest are then
covered, often with considerable success. Some species

have well defined breeding seasons and lay large clutches, but these suffer much predation. Species laying
few eggs at irregular intervals and with no defined nest
area avoid the danger of training the predators to follow
the cycles. *Geochelone gigas*, the giant tortoise of the
Aldabra Islands also digs a nest for its eggs. The number
laid is inversely proportional to the density of the
population. Emergence after hatching is facilitated by
mutual stimulation so that all the young wriggle together and open the nest. The young have many enemies,
from crabs to mammals, the principal one being man,
both eggs and young being used as human food. The
adults are also caught as they come ashore, ending as
turtle soup. Strenuous attempts are now being made to

save turtles by conservation of the beaches and farming (see Solomon and Baird 1979).

The proverbial slowness of the tortoise is a necessary corollary of its heavy armour, but the nervous organization and behaviour is more complex than is sometimes supposed. The brain shows well developed cerebral hemispheres, with not only the basal regions (striatum) but also the pallium quite large. This was therefore probably true also of the earliest reptiles, as it is of amphibians. In the mammals, also derived from cotylosaur ancestors, there has been still further development of the dorsal regions of the hemispheres to form the cerebral cortex, whereas in the remaining reptile groups and in the birds, the ventral portion has become large, the dorsal thin. Only part of the hemispheres of *Chelonia* receives olfactory fibres. There are also visual, auditory, and somatic sensory inputs from the thalamus.

The eyes are probably the chief receptors of chelonians. The retina consists largely of cones, many containing oil droplets of various colours. There are three photopigments in turtles, with different absorption maxima. The olfactory system is well developed and *Pseudemys* can be trained to avoid olfactory (or visual) cues. Some turtles produce liquid with a strong smell when disturbed. The musk glands also play a part in courtship. Turtles return to their birthplace to breed. Perhaps they are imprinted on the odour of its water (as in salmon), but the homing mechanism has not been identified (Harless and Morlock 1979). The tympanum is often covered with ordinary skin and hearing is probably only for low tones. In some species there are mating calls as part of their elaborate courtship. Chelonians are capable of learning considerable feats of maze learning, discrimination, and operant conditioning (Harless and Morlock 1979). They show several forms of homing behaviour, some, including time-compensated solar orientation, presumably involving an internal clock.

Chelonia occupy many different habits, some of them involving complicated and ingenious behaviour, especially among the aquatic forms (Fig. 14.5). For instance, the snapping turtles (*Chelydra*) and alligator turtles (*Macroclemys*) of North America and *Emys* in Europe show considerable care and skill in stalking and capturing not only fish but also young ducks and other birds. Similarly the smaller turtles, such as *Chrysemys picta*, the painted terrapin, with bright yellow, black, and red colours, feed not only on insect larvae, but also on flies, which they catch near the water surface. *Chelus*, the metamata of South America, breathes through a long proboscis. It lies below the water and fishes tempted by a filamentous lure are sucked in by opening the large mouth, as by an angler fish (p. 202).

Testudo and other land tortoises of Europe and Asia are mainly herbivorous but will eat meat. The giant tortoises (*Geochelone*) of Galapagos and Aldabra reach 1.5m and 250kg. Much is being done by conservationists to ensure the future of these tortoises.

Our knowledge of the geological history of the Chelonia extends back to the Triassic. *_Proganochelys_ (=*_Triassochelys_) (Fig. 14.4) was an early turtle, with a shell like that of modern forms, but still possessing teeth on the palate. The skull was anapsid and the pectoral girdle contained interclavicles, clavicles, and perhaps cleithra; these dermal bones were already somewhat enlarged and incorporated in the plastron. The head, tail, and limbs could not be withdrawn into the shell and were protected by spines. In the later evolution of the Chelonia retraction of the head became possible by one of two methods. In the suborder Pleurodira, 'side-neck turtles', the neck is folded sideways. The group was world-wide in the Cretaceous and survives today in tropical Africa (*Chelus*), South America, and Australia (Fig. 14.5). The more successful group is the suborder Cryptodira, in which the neck is curved in a dorsoventral plane. This type is also known from the Jurassic and Cretaceous and includes most of the modern types. The aquatic chelonians show various modifications and it is probable that several lines have independently returned to the water. As a result of this habit the bony shell is often reduced, presumably in the interests of lightness and because of absence of enemies. This had occurred already in *_Archelon_ of the Cretaceous, which is very similar to the modern *Chelonia*. *Dermochelys*, the leathery turtle, has a curious 'carapace' consisting only of a mosaic of small bony plates beneath its leathery skin, and *Trionyx* is a freshwater turtle with a soft shell and no horny plates. All of these forms are best considered as aberrant cryptodirans.

The origins of Chelonia are obscure but the group presumably originated from cotylosaurs of some kind. *_Eunotosaurus_ from the Permian of South Africa had a small number of vertebrae, with very broad, expanded ribs. This perhaps suggests some affinity with Chelonia, though in the latter the ribs themselves are not expanded. There is therefore little to tell us how, when, or why one of the early reptilian populations shortened its body and covered it with armour for protection against the hazards of the land it had recently invaded.

4. Subclass *Synaptosauria (connecting lizards)

4.1. Order *Protorosauria (dawn lizards)

All the Synaptosauria characteristically possessed a single temporal fossa in the upper or parapsid portion.

The earliest forms were small terrestrial lizard-like creatures such as *Araeoscelis* from the lower Permian (Fig. 14.9) quite like a cotylosaur except for the temporal fossa. The remarkable Triassic *Tanystropheus* had a long neck and short body. The protorosaurs were never an important element of the early reptilian fauna but it is not unlikely that they gave rise to the sauropterygians. The theory that the Squamata were derived from protorosaurs by the emargination of the lower temporal region is now held to be unlikely.

4.2. Order *Sauropterygia (lizard fins)

This was a very successful line of marine reptiles, extending from the Triassic to the end of the Cretaceous. The earlier nothosaurs, such as *Lariosaurus* (Fig. 14.9) from marine Triassic deposits, were small (90 cm long) and had a long neck, and limbs partly converted into paddles. The upper temporal fossa was enlarged and the

nostrils lay rather far back, as in many water reptiles.

All of these features were further developed in the plesiosaurs of the Jurassic and Cretaceous, such as *Muraenosaurus* (Fig. 14.9). In some the neck became very long, presumably for catching fish; 76 cervical vertebrae have been recorded. In others the neck was shorter and the skull longer. The limbs were developed into huge paddles, as much as half as long as the whole animal. The ventral portions of the girdles were large for the attachment of muscle masses inserted on the flattened humerus and femur. The dorsal portion of the girdles, so well developed in terrestrial reptiles, was here small: the ilium hardly articulates with the sacral vertebrae. The hands and feet were enlarged by increase in number of joints (hyperphalangy), but there was no increase of digits (hyperdactyly) such as is seen in ichthyosaurs. The skull was as in nothosaurs, but with the nostril still further displaced on to the upper surface.

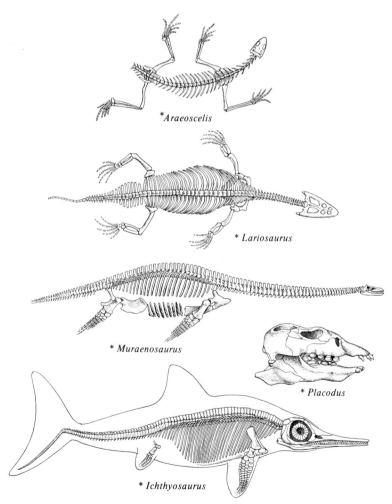

FIG. 14.9.   Ichthyosaurs, plesiosaurs, and their allies. (Partly after Romer.)

These animals were numerous in Jurassic and Cretaceous seas and some of them reached 15 m in length. They were obviously fish-eaters, but little is known of their habits. It is not known whether they were viviparous or came ashore to lay eggs.

4.3. Order Placodontia (plate-like teeth)

Placodonts were related to the plesiosaurs and were specialized for mollusc-eating by the development of large grinding teeth on the jaws and palate (Fig. 14.9).

5. Subclass Ichthyopterygia (fish fins)

These animals, found chiefly in the Triassic and Jurassic seas, were even more modified for aquatic life than the plesiosaurs (*Ichthyosaurus*, Fig. 14.9). They reached lengths of up to 9 m. They occupied a position comparable with that of the dolphins and whales during the Tertiary period. The body possessed a streamlined fish shape and swimming was by lateral undulatory movements. The vertebrae were amphicoelous discs and there were large dorsal and caudal fins, with the vertebral column apparently continued into the lower lobe of the latter (reversed heterocercal, p. 115). The paired fins

were small and presumably used as hydrofoils as stabilizing and steering agents. The pelvic girdle did not articulate with the backbone. In the limbs the number of digits was often greater or less than the usual five, and there was often hyperphalangy. Evidently this type of skeleton gives better support for a fish-like paddle than does the pentadactyl tetrapod type and it is interesting to find it evolved again in vertebrate stocks that returned to the water. This seems in a sense to be a case of reversal of evolution.

The head was much modified for aquatic life, with a very long snout armed with sharp teeth, and nostrils set far back. The eyes were large and surrounded by a ring of sclerotic bony plates. The temporal fossa, though in the parapsid position, had boundaries different from that in the synaptosaurians.

The Triassic ichthyosaurs were already greatly modified and we have no trace of the origin of the group. The ichthyosaurs were more highly adapted to aquatic life than any other reptiles known. They seem to have been viviparous, since the skeletons of small specimens have been found within larger ones. Like the plesiosaurs they developed a special mollusc-eating type, *Omphalosaurus*, in the Triassic. Ichthyosaurs were most nu-

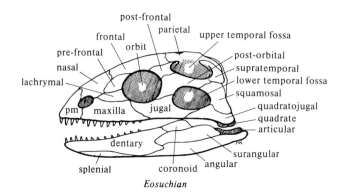

Eosuchian

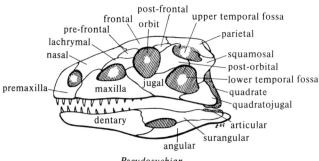

Pseudosuchian

merous in the Jurassic and declined in the Cretaceous, before the reduction of other marine reptiles and long before the appearance of the whales.

6. Subclass Lepidosauria (scaly lizards)

Most of the animals popularly considered as characteristic of the period of reptilian dominance have a two-arched or diapsid skull. This condition, or some modification of it, is found in all the surviving reptiles except the Chelonia, and in the birds. Formerly all the two-arched reptiles were placed in a single subclass, the Diapsida, but it is now customary to divide them into two subclasses, the Lepidosauria, which includes *Sphenodon*, the lizards and snakes (Fig. 14.15) and the Archosauria, including the crocodiles, dinosaurs, pterosaurs, and the ancestors of birds. It is not known if the archosaurs were derived from primitive lepidosaurs such as *Youngina*, or whether the two groups arose independently from cotylosaurian ancestors. *Milleretta* (p. 296) was a cotylosaur with a small lower opening; the addition of an upper one could have produced a lepidosaur.

6.1. Order *Eosuchia (dawn reptiles)

The best known of these earliest lepidosaurs *Youngina*, was a lizard-like creature, found in the Upper Permian of South Africa, and retaining many cotylosaurian features, for instance, teeth on the palate as well as on the margins of the jaws, and no opening between the bones of the snout (antorbital vacuity) (Figs. 14.10 and

14.11). The two fossae at the back of the skull immediately show the affinity with other diapsids. Little is known of the post-cranial skeleton. The fifth metatarsal does not show the hooked shape that is found in other diapsids and also in Chelonia. It is difficult, however, to ascribe very great weight to this single point, as against the general features of the skull, which indicate that *Youngina* could have given rise to the later two-arched reptiles and, by loss of the lower margin of the lower temporal fossa, also to the lizards and snakes. *Prolacerta*, from the Lower Triassic, shows how this may have come about; it is so like *Youngina* that it is classed as an eosuchian, but there is a gap in the lower temporal arch, suggesting that the animal may have been near the ancestry of lizards in which the quadrate has a movable joint with the squamosal, allowing greater mobility of the jaws.

6.2. Order Rhynchocephalia (beak-headed)

Sphenodon (= *Hatteria*), the tuatara of New Zealand, is the oldest surviving lepidosaurian reptile (Fig. 14.15); it still remains in essentially the eosuchian condition. Very similar Mesozoic fossils (e.g. *Homoeosaurus* from the Jurassic) show the continuity of the type (Fig. 14.2). Among the many primitive features that this race has preserved unchanged for 200 Ma are the two complete temporal fossae (Fig. 14.12), the well developed pineal eye (the pineal foramen is marked in the early diapsid fossil skulls), and the amphicoelous vertebrae with intercentra. *Sphenodon*, alone of surviving reptiles, has

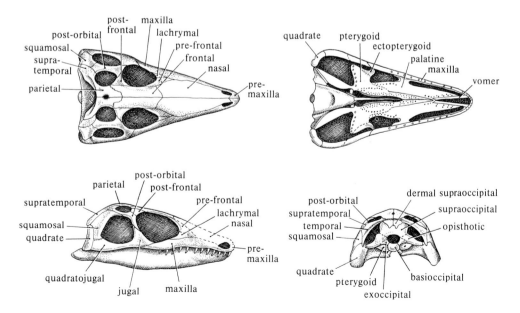

Fig. 14.11. *Youngina*. (After Romer 1945.)

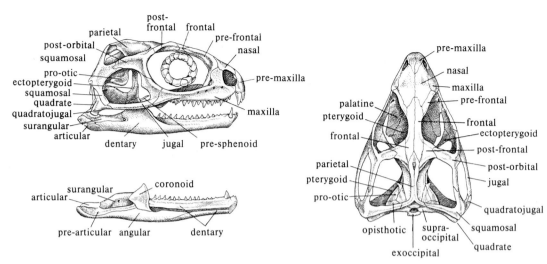

FIG. 14.12. *Sphenodon*. (After Romer 1945.)

no copulatory organ. As in some lizards the teeth are of acrodont type, fused with the jaws. They are not replaced and become much worn down in later life. The name *Sphenodon* means 'wedge-toothed' and refers to the shape of the large premaxillary teeth. There is a beak on the upper jaw.

The tuatara was once widespread throughout New Zealand but became much reduced and in danger of extinction. Recently rigid conservation measures seem to have allowed it to recover in numbers on some small northern islands. The animals are up to 60 cm long and are insectivorous and carnivorous. They are mostly nocturnal and may be active with a body temperature of 8 °C, at which most reptiles would be immobile. They live in burrows, often in association with petrels. The eggs take over a year to hatch, much longer than other reptiles.

Sphenodon evidently shows us a type that has departed relatively little from the condition of diapsids in the late Permian. yet its appearance, habits, and soft parts are very like those of lizards, and provide us with evidence that these animals remain in essential rather close to the original amniote populations. The rhynchosaurs such as *Scaphonyx were herbivorous rhynchocephalians very common in the Middle Triassic. They were up to 2 m long with a huge beak and sharp edge to the lower jaw which cut into a groove between rows of teeth on the upper jaw (Fig. 14.13).

6.3. Order Squamata (scaly ones)

The lizards and snakes are the most successful of modern reptiles, numbering between them nearly 6000 species. Probably the groups arose from eosuchians related to *Prolacerta. Such forms would also have been close to the Rhynchocephalia, differing from them, however, in a tendency to lose the lower temporal arch and to develop a movable quadrate.

It has now been shown that the lizards are a more ancient group than was formerly supposed, and had appeared by the end of the Triassic. Furthermore, some of the early forms were already considerably specialized. Our knowledge of the early lacertilian radiation is still incomplete, however, and none of the existing lizard families is known much before the Cretaceous. The earliest undoubted snake occurs in the upper Cretaceous and the group does not seem to have become abundant until the Oligocene.

Although typical lizards preserve a number of primitive reptilian features, the Squamata as a group show several interesting specializations that are absent in *Sphenodon*, and to which their success may be partly attributed. In the majority of forms, especially in the snakes, the skull is highly kinetic, having a freely movable quadrate, which imparts its motion to the bones of the upper jaw. Paired copulatory organs of a unique type are present in the male. There is a widespread tendency towards limb reduction, which has apparently occurred independently in members of about half the existing families of lizards, and in snakes.

The paired organs of Jacobson are highly elaborated and usually of great functional importance. In snakes and in most lizards these organs (Fig. 14.14) are hollow domed structures above the front of the palate, each opening into the mouth by means of a slender duct. The lachrymal duct opens into or near the duct of Jacobson's organ, instead of into the nose, suggesting that the secretions of the eye glands may have some special function related to that of the organ. Odorous particles

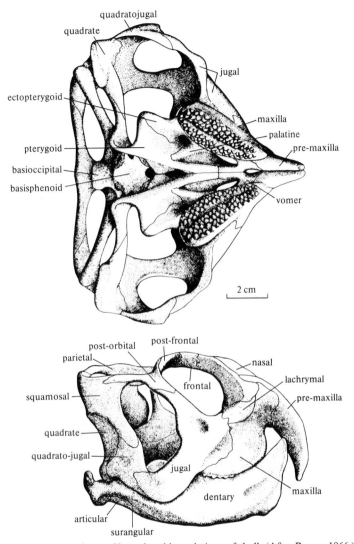

FIG. 14.13. *Scaphonyx.* Ventral and lateral views of skull. (After Romer 1966.)

are carried to the ducts of Jacobson's organs, or to the immediate neighbourhood of them, by the tongue tip, which is forked in snakes and many lizards. The lumen of the organ is partly lined by sensory epithelium, supplied by a separate branch of the olfactory nerve. Experiments have shown that the organs assist in such functions as sex recognition and following trails left by prey.

6.3.1. Suborder Lacertilia (lizards)

The modern lizards show extensive adaptive radiation (Fig. 14.15) and include terrestrial, arboreal, burrowing, and aquatic forms. The majority are carnivorous but there are some herbivores. It is difficult to say which of the twenty or so living families is the most primitive, and

the grouping of these into infraorders is a matter of some difficulty.

The Gekkota contains the geckos and a small group of Australasian limbless forms, the pygopodids. Geckos are mainly small nocturnal and arboreal insectivorous lizards of warm climates, with ridged pads on the toes furnished with minute bristles. These together with the sharp claws enable them to climb an almost smooth surface. Some species have taken to living in houses (*Hemidactylus*, the house lizard). The tree gecko *Ptychozoon* has webs of skin on the limbs and along the sides of the body, which perhaps act as a parachute to break its fall. Many geckos live in colonies and unlike most lizards are extremely vocal, making clicking and cheeping sounds. Their hearing is probably acute. The

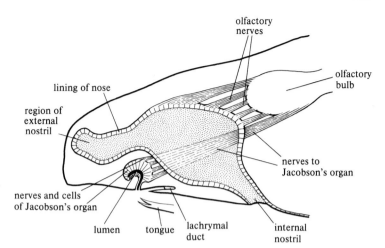

FIG. 14.14. Jacobson's organ. Diagram of reconstructed longitudinal section of snout of lizard showing nose and organ of Jacobson, together with its nerves and lumen from the side.

endolymphatic ducts of the inner ear are greatly expanded to form sacs in the neck containing calcareous deposits. These perhaps serve as reservoirs of calcium, since geckos are the only squamata that lay hard-shelled eggs; those of other forms being leathery in texture. Their eyes are, as a rule, covered by a transparent spectacle (fused eyelids) as in snakes.

The infraorder Iguania is a large group comprising the agamids (Fig. 14.16), iguanids (Fig. 14.15), *Iguana, Anolis, Phrynosoma*, and chameleons. The first two include terrestrial, arboreal, and amphibious types, sometimes of large size, and often furnished with crests, dewlaps, expansile throat fans, and other appendages that play a part in rivalry and courtship. The males are often brightly coloured and the whole group is characterized by a visually dominant behaviour pattern. In some arboreal forms, as in the chameleons, the sensory parts of the nose and the organ of Jacobson are reduced.

The agamids are found in the Old World and Australasia and include such well known types as the oriental 'blood-sucker' (*Calotes*), so-called because of the red colour of the throat, the spiny lizards (*Uromastix*) of the north African and Indian deserts, and the Australian frilled lizard (*Chlamydosaurus*) (Fig. 14.16). The Indo–Malayan genus *Draco* has a large lateral web, supported by the ribs, which can be spread out and used for gliding, though it is not moved in true flight. *Lyriocephalus* from Ceylon is an agamid with a remarkable convergent similarity to the chameleons. Some agamids (and other lizards) can run on their hind legs when they are in a hurry (Fig. 14.16). In agamids most of the teeth are set squarely on the summit of the jaw, as in *Sphenodon*; this condition is termed acrodont (= top teeth). In most other lizards the teeth are attached obliquely to the inner side of the jaw (pleurodont = side teeth).

The iguanids are found mainly in the New World and parallel the agamids in many ways. *Anolis* is a small, common North American form. *Iguana* from south and central America reaches 180 cm in length. *Amblyrhynchus*, the marine iguana, found only in the Galapagos Islands, is remarkable as the only existing marine lizard, though it spends much of its time on shore, basking and feeding on the sea-weed. *Phrynosoma*, the horned 'toad', with spikes on the head and back, is found in the deserts of North America and burrows in the sand (Fig. 14.15). This is one of the few ovoviviparous iguanids.

The chameleons are highly modified arboreal lizards from Africa, Madagascar, and India (Fig. 14.17). Some species have casques on the head, or one or more horns on the snout. The tail is prehensile and the digits are arranged in groups of two and three so as to be opposable and allow the grasping of branches. Chameleons live on insects, caught by means of the very long tongue (Fig. 14.17(c)), which has an adhesive clubbed tip and is projected by a remarkable muscular mechanism. Their movements are slow and deliberate, but they show considerable care in stalking their prey. As they approach it their eyes, which normally move independently, converge so as to bring the prey into binocular vision and they can judge distance by use of the mechanism of accommodation (Harkness 1977). Their powers of colour change are described on p. 308.

The infraorder Scincomorpha is another large assemblage, including *Lacerta* (Fig. 14.15), of Europe and North Africa. *Lacerta vivipara* is common in Britain. Its northerly extension is probably helped by its viviparous habit, unusual among lacertids. Another group, the skinks, form a large family with world-wide distribution. Many are modified for burrowing, sometimes in the sand. Many skinks have well developed limbs, but

FIG. 14.15. Various Squamata.

others show all degrees of limb reduction, either the fore or hind limbs, or both, being lost.

The infraorder Anguimorpha contains the anguids (Fig. 14.15), of which the European slow-worm *Anguis* is a familiar limbless example. The monitor lizards (*Varanus*) (Fig. 14.15) and their allies are placed in the superfamily Platynota, sometimes as a separate infraorder, since they have been distinct for a very long time. The monitors of the Old World and Australia include the largest of existing lizards, one species, the Komodo Dragon, of the island of Komodo east of Java, growing to at least 3 m long. They are ferocious

FIG. 14.16. Three frilled lizards (*Chlamydosaurus*) and a *Grammatophora* (at right) to show the bipedal habit. (Drawings made by Heilmann from photographs of the lizards running at full speed, taken by Saville Kent.)

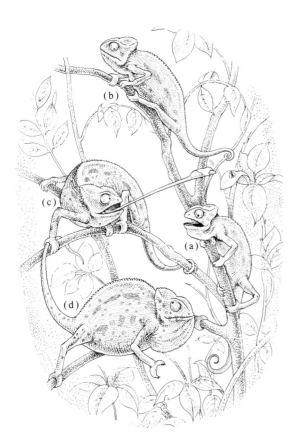

FIG. 14.17. Chameleon catching a fly, showing its changes in colour. A, cream with yellow patches, the usual night colour. B, Grey–green with darker patches. C, Dark brown patches and yellow spots. D, Reaction produced by pinching tail, inflation and darkening of all spots. (After Gadow.)

carnivores, killing vertebrates as well as insects, and are often semiaquatic. The quadrate is mobile and there is an extra joint in the lower jaw to increase the gape. Three related groups, now extinct, occurred in the Cretaceous. The aigialosaurs and dolichosaurs were amphibious lizards of moderate size, but the later mosasaurs, such as *Tylosaurus*, were huge creatures, sometimes 9 m long, and were highly adapted for marine life, with long jaws and paddle-like limbs showing some hyperphalangy. The strikingly coloured Gila monster, *Heloderma suspecta*, from the deserts of western North America, is also a platynotid; it and a closely related Mexican form are the only known poisonous lizards. Another allied form is the rare earless monitor, *Lanthanotus*, from Borneo, which seems to be a survivor of the primitive platynotid stock, perhaps related to the ancestors of the snakes. It lives partly underground but can swim well.

Many lizards are able to change colour, the chameleons, *Anolis* (often called 'American chameleons'), and certain agamids being the most notable for this. The colour may change with the environment, serving the obvious purpose of concealment. Special colour patterns are displayed in courtship or threat, and colour change may also occur in response to temperature and other environmental changes. The physiological mechanism of colour control varies in different reptiles. In *Anolis* there are probably no nerves to the melanophores, which are controlled by the pituitary hormone MSH causing darkening, and by adrenaline, producing paling. In chameleons, however, the melanophores are controlled partly or entirely by the sympathetic nervous system (Fig. 14.18). Colour is largely controlled through the eyes but in chameleons there is some local mechanism and a covered area of skin becomes pale.

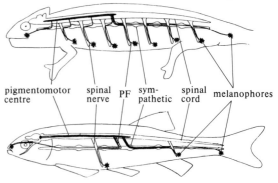

pigmentomotor centre   spinal nerve   PF   sympathetic   spinal cord   melanophores

FIG. 14.18. Diagram of the nervous control of the melanophores in the chameleon (above) and minnow (below). PF, Pathway of pigmentomotor fibres (the synapse in the sympathetic ganglia is omitted). (From Sand, minnow after v. Frisch.)

Many lizards are able to break off their tails when threatened or seized by a predator; this ability, known as autotomy, is due to the presence of special planes of weakness through the bodies of the caudal vertebrae. Such fracture planes are also found in *Sphenodon*, but not in snakes. After autotomy the tail regenerates, but the new member is not a replica of the normal. The vertebrae, for instance, are not regenerated, their place being taken by an unsegmented tube of cartilage. The new sensory innervation of the skin is derived entirely from the surviving dorsal roots, whose cells become greatly enlarged.

6.3.2. Suborder Amphisbaenia (both ways)

These 'worm-lizards' are tropical and subtropical creatures which have probably been distinct for a long time. In most of them the limbs have been lost (Fig. 14.15). The eyes are minute and the skull bones interlock to make a firm block for burrowing. They live in damp soil in underground burrows which they patrol in search of termites, earthworms, etc. The scales form rings round the body. The tail is short and the animals can move in either direction – hence the name 'both ways'. The extra stapes is usually attached to the outside of the lower jaw and probably provides acute hearing by which ants and other prey are detected underground.

6.3.3. Suborder Ophidia (snakes)

The snakes are obviously descended from lizards of some kind, but their precise mode or origin is obscure. Some workers believe that their nearest living relatives are the platynotid lizards (monitors, etc.). There is evidence that the snakes passed through a burrowing stage in their early history. A burrowing ancestry is particularly suggested by the structure of the eye, which, as Walls (1942) has pointed out, differs widely from that in typical lizards (Fig. 14.19). Thus there are no scleral ossicles or cartilages in snakes, and accommodation is brought about in a manner unusual for reptiles, involving displacement of the lens. The visual cells include cones of a peculiar type, which have apparently been derived from rods. The yellow retinal droplets that serve in lacertilian retinae to filter off blue light and so reduce chromatic aberration are absent, and instead some diurnal snakes protect their retinas by a yellow-tinted lens. These features can all be interpreted on the supposition that the ophidian eye was once drastically reduced, but has subsequently been refurbished in response to the needs of life above ground. Other characters that seem to point in the same direction include the structure of the ears, which are reduced as in many burrowing lizards, (but also in some that live above ground). The ear drums, tympanic cavities, and

Eustachian tubes are reduced or absent, and the columella auris articulates with the quadrate. There is evidence that snakes can hear air-borne sounds only of low frequency. But they respond at very low amplitude to vibrations at 300 Hz transmitted through the bones of the jaw.

The snakes show many other interesting peculiarities, the most obvious being the complete absence of limbs. Only in a few of the more primitive forms such as the boas and pythons can vestiges of the hind limbs and their girdles be found; in these snakes claws may be present externally on either side of the cloaca and are said to play a part in coitus.

Locomotion is produced by the lateral undulation of the body, which exerts pressure on surrounding objects and pushes the snake forwards; the enlarged transverse ventral scales of most species help to prevent slipping. A few snakes (e.g. some boas and vipers) can also progress by muscular movements of the ventral scales, with their bodies stretched out almost in a straight line. The spine is strengthened by additional intervertebral articulations known as the zygantra and zygosphenes, which are also found in some lizards.

The skull is highly modified, permitting, in all except a few burrowing forms, an enormous gape and the swallowing whole of large prey. The premaxilla is small and usually toothless, and the bones of the upper jaw are loosely attached to the rest of the skull. The upper arch is lost and the quadrate jointed to a bone sometimes called squamosal, but probably strictly a supratemporal (Fig. 14.20). There are movable joints between the frontals behind and prefrontals and nasals in front and also between several other bones of the brain case, palate, and jaws. These joints have loose ligaments and allow movement in several directions and so permit a huge gape. The two halves of the lower jaw are not firmly united. The sharp recurved teeth are carried on the palatal bones as well as on the maxilla and dentary. The brain case is strong and compact, the brain being protected from mechanical injury during swallowing by the massive parasphenoid and by flanges of the frontals and parietals, which lie between the orbits, so that there is no interorbital septum.

In the normal ophidian kinetic mechanism the upper jaw as a whole is raised as the result of forward rotation of the lower end of the freely mobile quadrate, which is loosely attached to the back of the pterygoid. The well developed protractor muscles of the pterygoid and quadrate play an important part in the process. In the viperid snakes a further elaboration of this mechanism is seen, the maxilla being very short and able to rotate on the prefrontal so that the fangs can be erected (Fig. 14.21). A slip from one of the muscles is attached to

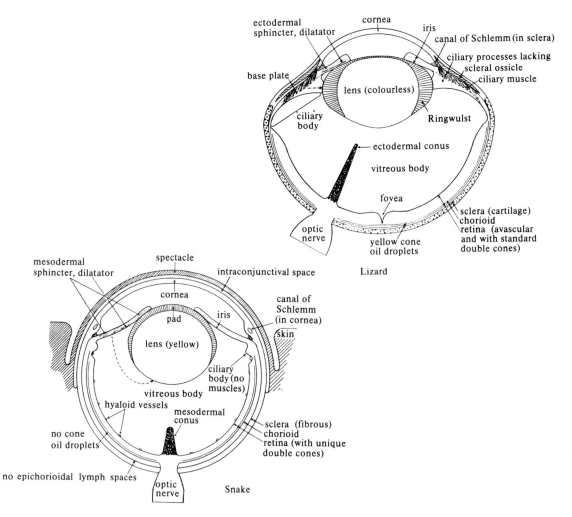

ectodermal sphincter, dilatator
base plate
ciliary body
cornea
iris
canal of Schlemm (in sclera)
ciliary processes lacking
scleral ossicle
ciliary muscle
lens (colourless)
Ringwulst
ectodermal conus
vitreous body
fovea
optic nerve
yellow cone oil droplets
sclera (cartilage)
chorioid
retina (avascular and with standard double cones)

Lizard

mesodermal sphincter, dilatator
spectacle
intraconjunctival space
cornea
iris
canal of Schlemm (in cornea)
pad
skin
lens (yellow)
ciliary body (no muscles)
vitreous body
hyaloid vessels
mesodermal conus
no cone oil droplets
sclera (fibrous)
chorioid
retina (with unique double cones)
no epichorioidal lymph spaces
optic nerve
Snake

FIG. 14.19. Diagrams of the eyes of a lizard and a snake showing the marked contrasts between them. The presence of a spectacle in snakes of all habitats indicates that their ancestors lived underground or were nocturnal. The dotted arrows show the direction of application of force during accommodation. (From Walls 1942.)

the poison gland and helps to expel the venom as the snake bites. Snake venoms vary in composition, most are mixtures. They often include polypeptides that poison the respiratory muscles of the prey. Many include proteases, which help in digestion. Some venoms are haemolytic. After a boa constrictor has swallowed a rat, the bones of the head have disappeared in two days and the whole skeleton in five days.

The respiratory system and viscera of snakes are also much modified. The glottis can be protruded so as to keep the airway clear while prey is being swallowed, and in some forms a part of the trachea is specialized for respiration as a tracheal lung. The left of the two paired lungs is usually reduced, often to a rudiment, as in some limbless lizards, and the other paired viscera tend to lie

at different levels on the two sides. The heart usually lies a quarter to a third of the way down the body, and the carotid arches are asymmetrical, the right common carotid artery tending to be suppressed.

The snakes show nearly as much adaptive radiation as the lizards, though there is less structural variation among them. The more primitive forms, with pelvic rudiments, include a number of small burrowers such as *Typhlops*, as well as the large boas and pythons of the family Boidae, which tend to be arboreal and amphibious in habits and kill their prey by constriction. In general, the pythons lay eggs, whereas the boas are ovoviviparous.

The majority of living snakes belong to the family Colubridae, which contains many medium-sized harmless snakes such as those found in England, the common grass snake (*Natrix* (= *Tropidonotus*)) and the rare smooth snake, *Coronella*. Some colubrids are moderately poisonous, with grooved fangs at the back of the maxillae; these are known as back-fang snakes or opisthoglyphs. The South African boomslang (*Dispholidus*) is one of the few whose bite may be lethal in man. *Dasypeltis*, the egg-eating snake, is also a member of this group; it swallows the eggs whole and crushes them with special tooth-like processes of the neck vertebrae.

The family Elapidae contains the cobras, kraits, and coral snakes, all highly poisonous with quite small and relatively non-movable fangs at the front of the maxilla,

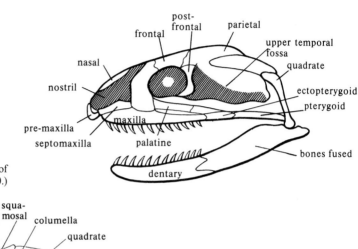

FIG. 14.20. Diagram of skull and lower jaw of an ophidian. (Modified from Goodrich 1930.)

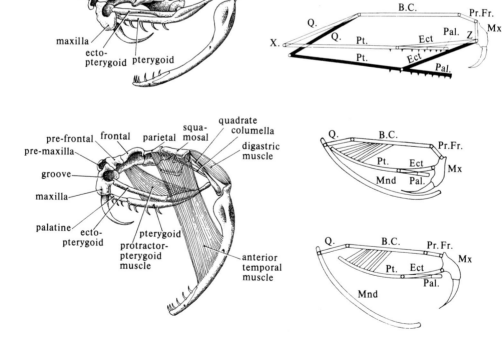

FIG. 14.21. Skull of rattle-snake (*Crotalus*) with jaws partly and fully opened. The protractor-pterygoid muscle pulls the pterygoid forward, causing it to push the ectopterygoid, which rotates the maxilla and erects the fang. The digastric muscle assists in opening the jaw. The depressor mandibulae shuts the mouth. The diagrams show the mechanism for lowering the pterygoid teeth; and the points at which movement occurs. (Boltt and Ewer 1964.)

and a venom predominantly neurotoxic in action. All the poisonous Australian snakes belong to this group. The hood of the cobra is expanded by the long cervical ribs and probably has a warning (sematic) function. The king cobra (*Hamadryas*) is the largest poisonous snake and reaches 5.5 m in length.

The very poisonous sea-snakes (Hydrophiidae) are related to the elapids. Their tails are vertically compressed for swimming; some species can hardly move on land. Like many freshwater snakes they are mostly viviparous.

The family Viperidae are the most specialized of all poisonous snakes. It includes the vipers of the Old World and the rattle-snakes and pit-vipers, mainly from the New World. The latter (subfamily Crotalinae) are distinguished by the presence of a remarkable sensory pit on each side of the head between eye and nostril (Fig. 14.22). The two cavities contain membranes with numerous nerve endings sensitive to changes in the flux of infrared radiation or temperature change (Figs. 14.23 and 14.24) (Bullock and Fox 1957; Cordier 1964; Barrett 1970; Hartline 1974). These provide directional sensitivity sufficient to allow the snake to strike at a small mammal 1 m away (Hartline, Kass, and Loop 1978). Changes of 0.003 °C can be detected. Similar thermometers can be found in the labial pits of Boidae. The rattle-snakes carry caudal appendages composed of articulated rings and modified skin. One ring is formed at each moult, though the older and most posterior ones break off periodically. The rattle is vibrated voluntarily as a warning and perhaps prevents the snake from being trodden on by large mammals.

The fangs of vipers are canalized, the canal having apparently being evolved by the progressive deepening of a groove until its margins have come into apposition. The fact that the fangs are erected when the snake strikes and can be folded back along the roof of the mouth when not in use, makes it possible for these

structures to be very long, about 25 mm in the case of a large puff adder. The bite of the European adder (*Vipera berus*) is seldom fatal to man. Some of the American pit vipers are very large, the dreaded bush-master (*Lachesis*) reaching about 3 m. The majority of the Viperidae bear their young alive and the finding of late embryos within the bodies of female adders and rattle-snakes may have given rise to the tale that these reptiles temporarily hide their young by swallowing them in the face of danger.

7. Subclass Archosauria (ruling reptiles)

The archosaurs were the dominant land animals of the late Mesozoic, and they include the dinosaurs and pterosaurs. Crocodiles are the only living descendants of the group that have remained at the reptilian level. The birds, which are also undoubtedly descendants of this archosaurian group, give us in some ways a better idea of the characteristic structure than do the crocodiles. The archosaurs have diapsid skulls and have usually been considered as descendants of the eosuchians. They may, however, have arisen directly from cotylosaurs and evolved the diapsid condition independently.

Some lines of archosaurs are characterized by a tendency to walk on the hind legs. These were much longer than the front and the acetabulum formed a cup, open below, so that the legs were held vertically below the body (propubic pelvis). At the same time the ischium and pubis became elongated to allow for the attachment of muscles producing a fore-and-aft movement (see p. 514). In later forms the ilium became fused with several sacral vertebrae. The femur has a head and neck, so that its shaft is vertical and the legs move fore-and-aft. The tibia becomes long and strong and sometimes fused with the proximal tarsals; the distal tarsals may fuse with the metatarsals as in birds, and the digits are reduced, usually to three long ones turned forward while

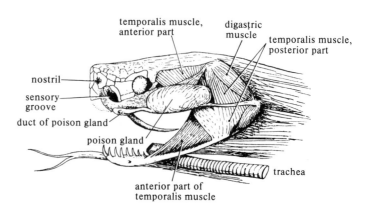

Fig. 14.22. Head of crotaline snake (*Lachesis*) after removal of skin. (From Gadow.)

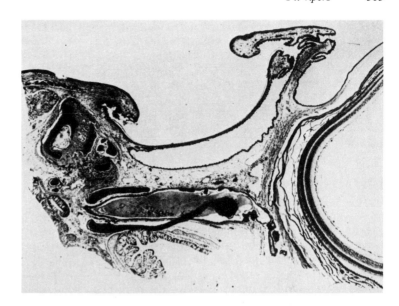

FIG. 14.23. *Agkistrodon.* Cross-section of facial pit organ showing membrane dividing it into anterior and posterior chambers. (Cordier 1964.)

the first is turned back. The skull is typically diapsid, but tends to have certain modifications, such as the development of antorbital vacuities behind the nostrils, and other spaces in the palate, presumably serving to give lightness without loss of strength.

7.1. Order *Thecodontia (socketed teeth)

The earliest archosaurs were the Triassic pseudosuchians such as *Saltoposuchus.* They can be visualized as lizards that sometimes ran on their hind legs which is indeed possible in some modern forms (Fig. 14.16). They were small and carnivorous, having sharp teeth set in sockets along the edges of the jaws (hence 'thecodont'). The skeleton showed all the archosaur characters in a most interesting incipient form. Thus the bones of the pelvis were still plate-like, but arranged in the characteristic propubic manner. The front legs were already much shorter than the hind. Antorbital va-

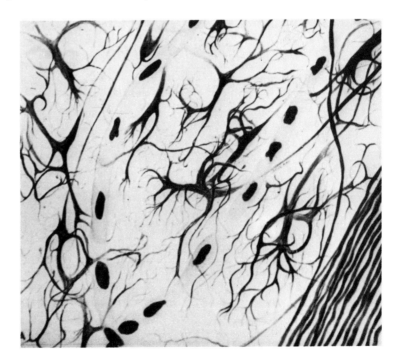

FIG. 14.24. *Crotalis viridis.* Membrane from the facial pit showing the nerve endings. Silver-stained whole mount. (Bullock and Fox 1957.)

cuities were present and there was no pineal foramen.

Even in the Triassic at least one line of thecodonts, the phytosaurs, abandoned the bipedal habit, becoming amphibious. These creatures such as *Mystriosuchus* were not actually ancestral to the crocodiles, but show remarkable parallelism to them in the elongated jaws and general build (Fig. 14.25). However, the nostrils were set far back and there was no palate. The hind legs were still longer than the front, although they could not have been used alone.

7.2. Order Crocodilia (crocodiles)

The true crocodiles first appeared in the late Triassic and are modified for amphibious life. (Fig. 14.26). The nostrils are at the tip of the snout and the air is carried back in a long tube, the maxillae, palatines, and pterygoids forming a bony secondary palate, as in mammals. There is a flap behind the tongue, which, with a fold of the palate, enables the mouth to be closed off from the respiratory passage and hence kept open under water. The nostrils can also be closed by a special set of muscles, and the ear drums are protected by scaly movable flaps. The Eustachian tubes are very complicated and there is a large lagena. They have acute hearing and eyesight. Parts of the skull are pneumatized

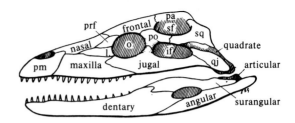

FIG. 14.26. Diagram of the skull and lower jaw of Crocodilia. (From Goodrich 1930.)

by extensions from the middle-ear cavity, as in birds. The teeth are sharp but can only hold the prey and not cut or chew (Fig. 14.26). They are continually replaced. The oesophagus can be distended to store food. The stomach consists of a thick-walled muscular gizzard containing stones (gastroliths) and a glandular pyloric region. There are canals leading from the peritoneal cavity to the exterior at the cloaca ('abdominal pores', p. 280).

Crocodiles have a social organization in which the males defend territories by fighting and roaring. There is a complex courtship and fertilization is in the water by insertion of the penis. The female has a clitoris. The Nile crocodile buries her eggs in a nest, then stays near them throughout incubation, helps the young to hatch, and escorts them to the water.

The crocodiles use all four limbs in walking, but the front are shorter than the hind, suggesting bipedal ancestry (but this is controversial). The pelvis of the crocodiles shows signs of the typical propubic structure, but there are only two sacral vertebrae. Rapid swimming is produced by lateral movements of the tail, but when moving slowly the partly webbed feet are used to push the animal along. The ribs (Fig. 14.27) are two-headed and there is a proatlas element between the skull and atlas. As in the earlier thecodonts there are dermal plates (osteoderms) along the back and belly underneath horny scales. There are well developed abdominal ribs (gastralia).

The soft parts of the crocodiles are of special interest because crocodiles are, except the birds, the only living creatures closely related to the great group of dinosaurs.

The lungs are well developed, having a system of tubes ending in sacs. A transverse partition separates off a thoracic from the main abdominal cavity. This 'diaphragm' is not itself muscular, but is continued into a diaphragmatic muscle attached to the abdominal sternal plates. This muscle, innervated by abdominal spinal nerves, presumably assists in respiration. It is a development for this purpose quite distinct from the mammalian diaphragm. It is not impossible that the dinosaurs possessed further developments of this arrange-

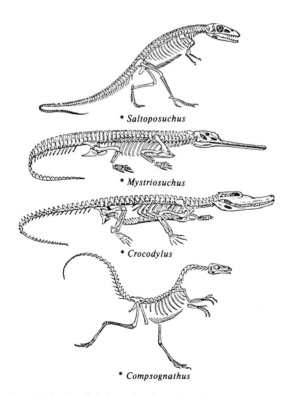

* *Saltoposuchus*

* *Mystriosuchus*

* *Crocodylus*

* *Compsognathus*

FIG. 14.25. The skeletons of various diapsids.

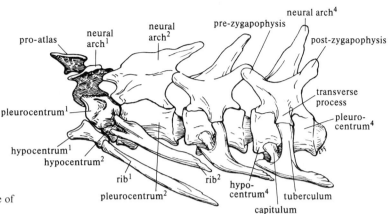

Fig. 14.27. Anterior cervical vertebrae of *Crocodilus*. (From Goodrich 1930.)

ment of the heart and lungs, and that they owed some of their success to this mechanism.

The nine genera of the modern crocodiles represent only the survivors of a once much more abundant group. *Crocodylus* is the most widespread genus, occurring in Central America, Africa and Asia, Malaya and East Indies, and North Australia. *Alligator*, with each fourth lower tooth penetrating into a hole in the maxilla, is found in North America and in China. *Caiman* of Central and South America is related to *Alligator*. The length of the snout varies considerably in different species, and is extremely long and slender in the fish-eating *Gavialis*, the Indian gharial, and *Tomistoma* of the East Indies. Crocodiles lay hard shelled eggs in large clutches, depositing them in the sand or in nests composed of vegetation.

The crocodiles have changed considerably since they first appeared in the late Triassic, perhaps 190 Ma ago. *Protosuchus* of that time had a pelvis like that of crocodiles but was otherwise very like a pseudosuchian. There were numerous types of crocodile in the Jurassic and Cretaceous, living both in fresh water and in the sea. *Deinosuchus* of the Cretaceous was up to 12 m long. In earlier forms the palate was closed only as far back as the palatine bones; the addition of flanges of the pterygoids took place only in the Eocene crocodiles, which were numerous in many parts of the world, including northern regions such as England that today are too cold for such animals.

In spite of their specializations for aquatic life, the crocodiles show us many features that were present in the earliest archosaurs and they, therefore, give some idea of the characteristics of the ancestors of the pterodactyls, dinosaurs, and birds. Today many species of crocodiles are seriously threatened by the demands of the skin trade. Conservation measures, in association with farming, which can provide a good commercial yield, should be strongly encouraged.

7.3. Order *Saurischia (lizard pelvis)

7.3.1. The 'terrible lizards', dinosaurs

At the end of the Triassic some of the descendants of the pseudosuchians became very successful and numerous and many of them very large (see Charig 1979). The large size was not a characteristic only of one but of two quite distinct lines, each with several sub-divisions. The term dinosaur is applied to all of them, but the two main lines have little in common beyond the characters common to all archosaurs. The desire to explain this extraordinary exuberance of reptiles has attracted much attention to these giants. It has been argued that some of them, at least, may have been endothermic, mainly on the ground that the bone structure shows the well developed Haversian systems characteristic of high metabolic activity (see *Life of mammals*). Tuna fish show the same structure. However, this bone structure may have been needed to support the weight. In any case these large creatures may have retained ectothermic heat for long periods thus permitting maintenance of a high metabolic rate as 'inertial homoiotherms'.

It has been argued that they must have had an active metabolism (and high temperature) in order to produce sufficient muscular energy to stand or walk. Further evidence on the question is the finding of growth rings in the dentine of dinosaurs. These are found also in the teeth of crocodiles and are due to the effect of seasonal variations on an ectotherm. There are no rings in the dentine of modern mammals except those of the Arctic and some temperate regions that are subject to marked seasonal contrasts. Nor are there rings on the teeth of late Cretaceous mammals such as *Protungulatum* (p. 517) (Johnston 1979).

There is also evidence that large dinosaurs possessed special arrangements for cooling the brain and spinal cord. As in some mammals cooled blood from the nasal turbinals may have passed to a rete mirabile in the cavernous sinuses. The nasal crests of the hadrosaurs (p. 318) probably served to provide large nasal surfaces, either for special olfactory epithelia or for cooling surfaces (or both). There were also sinuses in the bony plates along the back, for instance in *Stegosaurus (p. 318). The need to cool the central nervous system could of course arise in either an ectotherm or en-dothermous animal (Wheeler 1978).

Another mystery is that very few skeletons of *young* dinosaurs have been found. A possible explanation is that the eggs were laid and young 'kept' in uplands, where they would rarely become fossilized. This view is supported by a find of fifteen skeletons of young hadrosaurs 1 m long and jumbled together with many shell fragments, in a shallow nest (Horner and Makela 1979). Their batteries of teeth show some wear, so they must have been feeding for some time, presumably kept together by their parent(s). This gives some evidence of their social life.

7.3.2. Suborder *Theropoda

These include forms with a propubic pelvis, very like that of the pseudosuchians. The earlier types, like their ancestors, were bipedal carnivores of no great size, such as *Compsognathus from the Jurassic of Europe and *Ornitholestes from that of North America (Fig. 14.25). The front legs were short, with 4 or 3 digits, provided with claws; the pectoral girdle was reduced to scapula and small coracoid, with no trace of clavicles. Some members of this line, the theropods, soon developed into large carnivores, such as *Allosaurus (Figs. 14.28 and 14.29), over 9 m long (Jurassic, North America). These animals apparently swallowed their food whole and to help with this the quadrate was movable and there was a joint between the frontals and parietals, as in many lizards. In other respects the skull was very similar to that of the pseudosuchians.

At the end of the Cretaceous this theropod line produced the largest carnivores that have appeared on the earth, such as *Tyrannosaurus rex, nearly 15 m long and 6 m high, from North America. They must have weighed over 8000 kg. All the previously mentioned tendencies were here accentuated, producing creatures with bipedal habit, very powerful head and jaws with some teeth up to 15 cm long, and much-reduced fore limbs. They presumably preyed upon the large herbi-vorous dinosaurs of the Cretaceous and became extinct with their prey, either from a common inability to meet the rigours of the climate or in competition with the mammals and birds. Throughout most of the Jurassic and Cretaceous the theropods were the dominant carnivores of the world, taking the place occupied earlier by the synapsid reptiles (p. 405) and later again by

descendants of the synapsids, the carnivorous mammals.

In the Cretaceous the organization of this saurischian line also produced some exceedingly bird-like forms, *Struthiomimus and *Ornithomimus, walking on three toes and having three opposable, clawed digits in the hand. The skull became very lightly built and the teeth disappeared, possibly in connection with the egg-eating habit (Fig. 14.28). *Deinonychus walked only on *two* digits and the third carried a powerful 13 cm long claw, used presumably while it stood on one leg, balancing by the fused vertebra in the tail! (Desmond 1975).

All these carnivorous, bipedal saurischians may be grouped into a suborder Theropoda. Another line of organization, starting perhaps from some pseudo-suchian forebears were herbivorous and quadripedal. Perhaps they never had fully bipedal ancestors. These animals, the suborder Sauropoda, culminated in the immense Jurassic forms *Apatosaurus *Brontosaurus (Fig. 14.28), and *Diplodocus, the largest of all ter-restrial vertebrates. *Yaleosaurus from the Triassic was a partly bipedal creature 2 m long, with rather long front and short hind legs. *Plateosaurus, also of the Triassic, was 6 m long, but also partly bipedal. In later forms the neck became immensely elongated and the head was very small with a lightly built skull. The nostrils lay on the top of the head and in *Diplodocus formed a single opening. This seems to indicate that the animals were aquatic or amphibious. The front legs were rather larger than the hind ones. The question of their mode of life remains undecided. *Diplodocus and *Brachiosaurus were over 24 m long and the weight of the latter must have been nearly 50 800 kg. However, the structure of the vertebral column shows that much weight was carried on the legs, for the vertebrae are strong, though hollowed in places. Footprints of the animals have been found, apparently made under water. One or more of the digits bore claws. The skull became relatively short and broad, and among the many puzzling features of these giant animals is the weakness of the jaws and small size of the teeth, mostly crowded towards the front of the mouth. These teeth would have served well enough for cropping, but there are no teeth on the hind part of the jaws and no provision for grinding the food. Animals of large size can only have been supported by this feeble apparatus if some very nutritious food was readily available. This perhaps agrees with the small size of the brain case, which was several times smaller than the lumbar enlargement (though this latter may have been full of storage tissue, as in some birds). The mode of life of these giants remains uncertain. If they walked along the bottom some special mechanism was needed to allow breathing by a body 6 m below the surface.

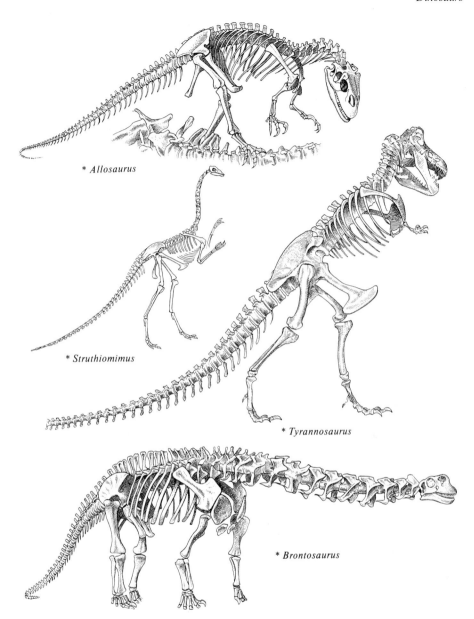

FIG. 14.28. The skeletons of various saurischian dinosaurs.

Living partly on land and feeding on high trees like a giraffe is perhaps the most likely solution.

7.4. Order *Ornithischia (bird pelvis)

The second main group of dinosaurs appeared later than the sauropods and possessed a 4-radiate pelvis, with the pubis directed backwards and an extra prepubic bone pointing forwards. The teeth were restricted to the hind part of the jaws, the front bearing a beak. At the front end of the lower jaw there was an extra bone (predentary). These were herbivorous forms and they appeared in the Jurassic and achieved their maximum in the Cretaceous, by which time the sauropods had become less common. The earliest of the ornithischians were bipedal animals, included in a suborder Ornithopoda, from the Jurassic and Cretaceous. These animals, such as *Iguanodon* (Fig. 14.30), were built on the same general lines as the pseudosuchians, from which they were presumably derived. The skull was heavily built and adapted for a herbivorous diet, with

FIG. 14.29. The first sauropod tracks were located in Glen Rose in 1938 and then R.T. Bird uncovered this spectacular trail of 1 m long brontosaur footprints together with those of an *Allosaurus*-like flesh-eater. The latter perhaps chased the former. (With permission National Geographic Society.)

powerful muscles attached to a coronoid process of the lower jaw. The bipedalism was less marked than in saurischians and the fore limbs less reduced. Several separate lines then reverted to a quadrupedal habit. The trachodonts (*Hadrosaurus*) were a very successful group of amphibious forms in the Upper Cretaceous. Some mummified skins have been found, showing that they had a covering of small scales, and webbed feet. The teeth were suited for grinding, forming parallel rows, making as many as 2000 teeth in one animal. In several types of hadrosaur the top of the head was prolonged in various ways, giving a structure with a complicated set of channels, perhaps increasing the area of the turbinals (p. 461). These animals reached 9 m in length and may have supplanted the sauropods as marsh-living forms. possibly when the soft foods gave place to tougher plants.

Other lines of ornithischians became more fully terrestrial and quadrupedal and were mostly heavily armoured. Thus the stegosaurs of the Jurassic carried immense spines on the back and the tail bore sharp spikes. The hind legs were much longer than the front, a relic of bipedal ancestry. The feet carried hoof-like structures. The skull was very small and the brain much smaller than the lumbar swelling of the cord. The teeth were in a single row and small. The ankylosaurs of the Cretaceous were covered all over with bony plates, somewhat in the manner of the mammalian glyptodonts (*Nodosaurus*, Fig. 14.30). Finally, the horned dinosaurs, the ceratopsians, such as *Triceratops* of the late Cretaceous, developed enormous heads, with huge horns and a large bony frill, formed by extension of the parietals and squamosals to cover the neck. These later Cretaceous animals appear to have lived on dry land and to have walked on all fours, in spite of the shortness of the front legs. There are several indications that the climate at the close of the Cretaceous was becoming drier and the organization of the giant reptiles became modified accordingly. They survived successfully for a while, but were ultimately replaced by the mammals, perhaps as a result of still further change in the climate (p. 573).

7.5. Order *Pterosauria (wing lizards)

The Triassic archosaurian reptiles gave rise to two independent stocks that took to the air, the pterodactyls and the birds. Both of these appear first in the Jurassic as animals already well equipped for flight, although obviously basically of archosaurian structure. We cannot therefore say anything about the steps by which their flight was evolved and can only speculate about the influences that drove them to take to the air. The early archosaurs were bipedal animals, and the fore limbs were therefore free and available for use as wings. There has been much speculation about the intermediate stages by which flight was produced. Other reptiles, such as *Draco* (Fig. 14.15), the flying lizard (p. 306), develop a membrane between the limbs and the body to assist them in making soaring jumps. The flight of pterodactyls and birds may have originated thus, or (Nopcsa 1907, 1923) by the flapping of the fore limbs during rapid running on the ground, the animals then becoming airborne for longer and longer periods (see p. 339).

The stages of the evolution of flight may have been different in the two cases, for whereas the birds are fully bipedal animals and the similarity to such reptiles as *Struthiomimus* and *Ornithomimus* is obvious, the pterodactyls probably could not walk on their hind legs and may have used the wing more for soaring than for flapping flight. In spite of great differences there are

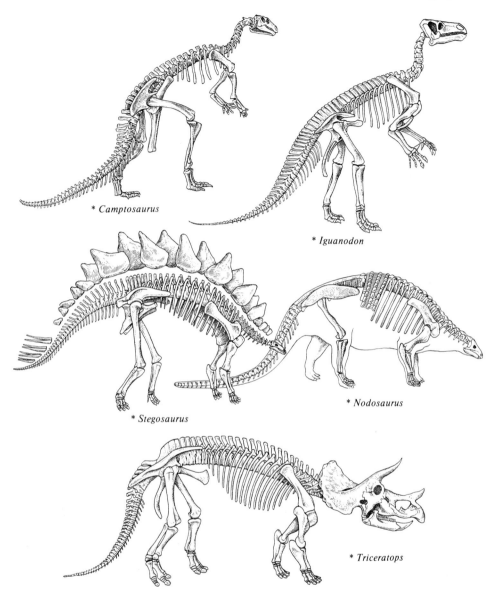

Fig. 14.30. The skeletons of various ornithischian dinosaurs.

interesting parallelisms in the structure of the fully evolved fliers of the two groups, for instance the limb bones became hollowed and light, the skull bones fused, and the jaws toothless and beaked. This parallelism in lines known to be distinct, although of remote common origin, is similar to that which we have noticed before in aquatic animals. Populations with similar genotypes will respond to similar environmental stimuli in the same way.

The pterodactyls (Fig. 14.31) are most commonly found in the Jurassic strata, less often in the Cretaceous.

Many specimens have been found in marine deposits and seem to have been fish-eaters. The characteristic features that have produced the pterodactyl structure from a thecodont ancestry may be described as a lengthening of the head and neck, shortening of the body and ultimately of the tail, lengthening of the arms and especially of the fourth digit, shortening of the hind limbs, and development of the ventral parts of the limb girdles. These are the changes that can be recognized in the bony parts available for study; no doubt there were many others in the soft parts also paralleling the

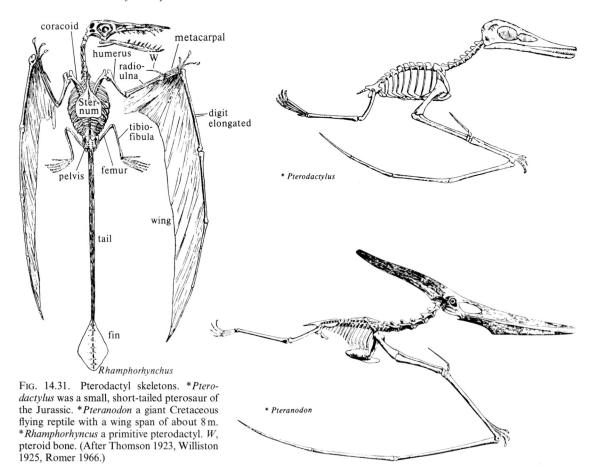

FIG. 14.31. Pterodactyl skeletons. *Ptero-dactylus* was a small, short-tailed pterosaur of the Jurassic. *Pteranodon* a giant Cretaceous flying reptile with a wing span of about 8 m. *Rhamphorhyncus* a primitive pterodactyl. *W*, pteroid bone. (After Thomson 1923, Williston 1925, Romer 1966.)

evolution of birds, for instance there is evidence from the bone structure that pterodactyls were warm-blooded (p. 323). Casts have been found showing that there was fur on the legs and the wings. However, there is no evidence that they possessed feathers; the wing was a membrane (patagium). Casts of the skull show reduced olfactory lobes but large cerebral hemispheres, optic lobes and cerebellum, as in birds.

Rhamphorhynchus of the Jurassic is still recognizably of archosaurian structure, especially in the skull, which has two fossae and large forward-sloping teeth (Fig. 14.31). The humerus was short and strong, radius and ulna long, carpus rather short but with an extra 'pteroid' bone in front. The first three digits were short and hooked, the fourth long, supporting the wing, and the fifth absent. The hind limb was slender, with five hooked digits. There was a long tail, ending in an expanded 'fin'. Both girdles had well developed ventral regions and there was a large 'sternum', keeled in front. The scapula articulated directly with the vertebral column.

Pteranodon of the Cretaceous, showed further modifications (Fig. 14.31). The trunk became shortened to ten or fewer segments and the fore limb further lengthened, the carpus being long and the fourth digit much longer than the other three. The hind limb remained small and the tail became very short. The very large and elongated head gradually lost its teeth, presumably acquiring a horny beak. In the latest forms the skull was drawn out backwards into an extraordinary process. Some earlier related forms were only a few cm long, but *Pteranodon* itself, of the late Cretaceous, had a wing-span of 8 m, though it probably weighed only 17 kg. Remains of an even larger pterodactyl, said to have a wing span of 15.5 m, have been found in deposits that were probably non-marine. Flapping flight cannot have been possible for such animals. They could carry enough muscle only if very lightly built and weighing perhaps 5 kg, of which all except 6 g would be muscle. However, the wing-loading was perhaps only a quarter that of an albatross, so that such creatures could fly without stalling at 24 km/h (Bramwell 1971; Bramwell and

Whitfield 1974). The flexible joints allowed variation of the aspect ratio to ensure that the boundary layer over the wings and body was kept laminar (see Tucker 1977).

Zoologists have not yet succeeded in reconstructing the life of these animals, and it is hard to see how they could have walked on land. The membrane, which stretched between both legs and the body, and perhaps also included the head, must have been easily torn. The feathers of birds can be ruffled without breaking and the loss of a few does no great harm: the bat's wing can be torn, but at least it is supported by many digits, whereas that of the pterodactyl was a huge continuous membrane supported by a single finger. Again it is difficult to see how the animals can have perched; if they hung, as the claws suggest, was it with the front or with the hind legs? And how can they have staged a take-off, which in birds is greatly helped by the jump of the hind legs? It is possible that they always came to rest hanging from cliffs, which they could leave by soaring. Even the flight itself presents many difficulties. Although there is a sternum and a strong humerus, neither suggests the presence of muscles sufficiently strong to carry a creature as large as *Pteranodon*. We cannot solve any of these mysteries, but one clue is that the biggest pterodactyls were mostly, if not all, marine. The largest flying birds alive today are the albatrosses, which use their great weight to gain height with the increasing velocity of the wind a few metres or so above the sea (p. 345). It is possible that the pterodactyls used a similar method of soaring. They were presumably unable to compete with the birds, however, and died out at the end of the Cretaceous, along with so many other reptiles.

8. Conclusions from study of evolution of the reptiles

Many of the conclusions that have been drawn from study of vertebrate evolution in the water also apply to the forms that have come on land. The fossil record leaves no doubt that almost all the populations have changed very markedly. Few forms of reptile alive today are closely similar to any found in the Permian or Triassic periods. *Sphenodon* has shown relatively less change than most others; it may be significant that it is found in an isolated island region (but see p. 580).

The data are not sufficient to show the rate of evolutionary change. We cannot be sure whether it has been constant or even continuous, but particular types are found only from a limited range of strata and there is little evidence that any terrestrial form remains unchanged for more than a few million years, at most. Each type is successful for a while and then the niche that it fills becomes occupied by another type, either descended from the first or, more usually, from some

related stock. Thus the earliest large land herbivores were probably the pareiasaurs; these were replaced by other reptilian types such as the herbivorous mammal-like reptiles, and later the sauropods (in so far as these were terrestrial) and various types of ornithischians; then perhaps by the hadrosaurs in the more watery habitats and the stegosaurs, ankylosaurs, and ceratopsians on drier ground. Finally, all these gave place to the earliest mammalian herbivores, which were in turn replaced by others (p. 581).

Throughout early tetrapod evolution there is a tendency to return to the water, perhaps under some pressure of competition from descendants on land. This is marked among reptiles, where besides the chelonians and ichthyosaurs and plesiosaurs there are the phytosaurs and crocodiles, mosasaurs and several types of dinosaur, not to mention the sea-snakes.

The large size of many reptiles has been one of their most striking features, but it is, of course, not true to say that there is a strong tendency for size to increase in all reptile groups. While many have become enormous, others, such as the lizards, have produced probably as great a biomass spread over a large number of small individuals. Large size in a reptile may help to conserve heat (p. 276), but could also endanger the animal from overheating, since the ratio of surface area to volume decreases as the absolute size increases, and heat cannot be lost so readily through the skin. Up to a point size may be a protection, but it involves the dangers of those who place all the eggs in one basket: incidentally, the actual eggs of these large animals must have provided formidable physical problems for their support.

Parallel evolution of several lines descended from a single stock is as common among reptiles as among other groups of vertebrates. Thus the bipedal habit, with hind legs longer than the front ones, has been adopted independently by a number of diapsids; again, elongated jaws are found among fish-eaters, whether ichthyosaurs, plesiosaurs, phytosaurs, crocodiles, or mosasaurs.

Although it is difficult to see in all this any persistent tendency except to change, yet the very fact that each type is so rapidly replaced suggests that descendants in some way more efficient are continually appearing. In the case of the reptiles the more interesting of these are the birds and mammals, and we shall therefore leave the problem of serial replacement among amniotes for later discussion (p. 581). Meanwhile we may note once again that the reptiles surviving today, although not of larger size nor obviously better suited for life than their Mesozoic ancestors, yet exist in considerable numbers alongside and even in competition with the birds and the mammals.

Throughout the 130 Ma between the late Triassic and

the end of the Cretaceous the reptiles were the dominant land vertebrates. They continued to diversify throughout, for instance the hadrosaurs and ceratopsians appeared only in the late Cretaceous. Evidently the reptiles then still remained a very active and versatile type. Yet many became extinct at the end of the Cretaceous. There is evidence that there were great earth changes at that time, leading to mountain building, temperature instability, and probably to an overall colder climate (p. 25). The conifers, cycads, and ferns gave place to higher flowering plants and deciduous trees. Yet it is hard to see why the reptiles could not adapt to such changes. In any case the numerous marine forms also disappeared at the same time. The question remains unsolved (see Bellairs and Attridge 1975).

15 Life in the air: the birds

1. Features of bird life

THE quality we define as 'life' is perhaps more fully represented in birds than in other vertebrates, or indeed in any animals whatsoever. It is difficult to find units by which accurate comparisons can be made of such matters, but there is a meaning in the statement that the life of a bird is more intense that that of, say, a reptile or a fish. Following out our definition of the life of a species as the total of the activities by which that particular type of organization is preserved, we shall find that the birds have many and very varied activities, by means of which a great deal of matter is collected into the bird type of organization. Moreover, this is achieved under conditions remote from those in which life first arose; the birds get a living by moving in the air, which is at once the most difficult medium of all and also the most rewarding for freedom of movement.

Flight is of course the characteristic that gives us most fully the feeling that the birds are active animals; it impresses us as a technical marvel and as a means by which the animals obtain a most enviable and valuable freedom, enabling them to avoid their enemies and to seek new habitats. The air provides the optimum medium for movement, as man also has discovered, and the whole organization of birds is devoted to exploiting it. Like any flying machine they have sacrificed everything to low weight and high power. The essential lightness and strength are achieved by using thin sheets and tubes of bone and extending air sacs into them. There are no teeth or heavy jaws and the food eaten is usually quickly digested and of high energy value. There is little accumulation of fat and the young develop outside the body. By such means birds achieve a very low relative density (specific gravity, only 0.6 in a heavy duck). Life in the air also demands a powerful propellant and for this they have high temperatures and metabolic rates, rapid heart-beats, and high-performance respiratory and muscular systems. To take advantage of the freedom of movement there has been great development of the brain, though on lines different from those of mammals. The high and constant temperature allows full use of the possibilities of delicate balance of activities within large masses of nervous tissue. In homoiothermic birds and mammals we find larger brains and more elaborate social and family habits than in any other animals.

2. Numbers and variety

Flight necessitates a high surface-weight ratio, therefore birds do not become so large as some mammals; nevertheless, an immense biomass is produced by their very great numbers. Any attempt to enumerate the bird population is largely guess-work, but the density of breeding birds in different habitats in Britain has been estimated, and varies from 200 per 4 hectares in woodland to 20 on agricultural land and 10 or less on moorland. Calculating from such figures, estimates indicate that there may be 100 million land birds in Great Britain and 100 000 million birds in the world altogether, including sea birds. This is perhaps a low estimate; it would represent a total biomass of the same order as that of 3000 million human beings. With all their activity, therefore, the birds organize no more matter into themselves than do the mammals.

Some 8600 living species are recognized and although the basic organization remains fairly constant differing types show a great variety of special features, fitting them for numerous habitats. Survival under various conditions has led to differences in behaviour, in body form and in powers of flight, in the shape of the bill, and hence in food habits, and in the details of many other parts, such as the feet. All these make fascinating studies in adaptation to environment. Birds are the most easily studied of all vertebrates, they are neither too small nor too large, and usually they are conspicuous. They have therefore provided the data for much recent theory on such matters as ethology and natural selection.

3. The skin and feathers

The skin of birds differs from that of mammals in being thin, loose, and dry. There are no sweat glands, indeed the only large cutaneous gland present is the uropygial gland or preen gland at the base of the tail. The bird cleans its feathers with its beak, obtaining oil from this gland, which is specially well developed in aquatic birds. Sebaceous secretions have also been found in other parts of the skin (Stettenheim 1972).

The keratin-producing powers of the skin are of course mostly devoted to producing feathers, but scales like those of reptiles are present on the legs and feet and sometimes elsewhere. The horny covering of the bill (p. 346) and claws are also specialized scale-like structures and are sometimes moulted.

The feathers of modern birds provide a covering whose uses vary from heat insulation and flight to protective coloration and sexual display. It is likely that in evolutionary history the function of heat regulation came first. The two main functions, of heat conservation and flight, are indeed today performed by feathers of different types. All feathers are based on the same plan (Fig. 15.1) but it is useful to distinguish between the large flight feathers or pennae of the wings and tail and the numerous small contour feathers covering the body. These latter provide warmth and the smooth air flow, without turbulence. In the pennae there is a central rachis and rows of barbs and barbules, linked to form a vane. In the down feathers or plumules of the nestling the barbs are all separate and soft. The semiplumes are downy feathers of the adult, lying beneath the contour feathers and giving warmth and shape. Filoplumes are hair-like structures accompanying the contour feathers and probably sensory. Bristles consisting mainly or wholly of a rachis occur around the bill and eyes and sometimes elsewhere.

Usually several generations of feathers are produced; first the nestling feathers (neoptiles), then one or more

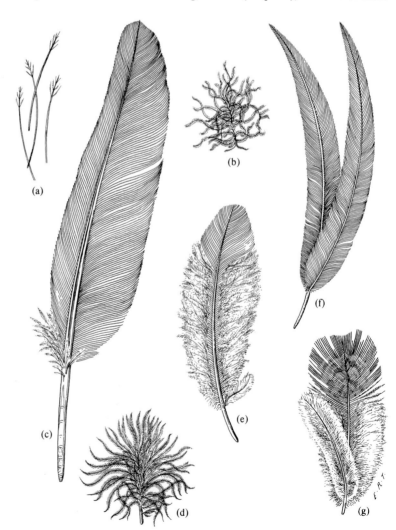

Fig. 15.1. Various types of feather. A, filoplumes; B, nestling down-feather; C, primary wing feather of pigeon; D, permanent down feather, semiplume; E, feather with free barbs; F, emu's feather with long aftershaft; G, contour feather of pheasant with aftershaft. (Partly after Thomson 1923.)

generations of juvenile feathers, which may be of various types, prefiloplumes, preplumules, and prepennae; finally, the adult feathers (teleoptiles).

Each feather, of whatever type, is formed from a dermal papilla or follicle, over whose surface keratin is produced (Fig. 15.2). In down-feathers the surface of the papilla is ridged all round and the result is to produce a number of fine threads or barbs of keratin, covering the body with a coat of fluff, which acts as a heat insulator by preventing air circulation.

Feathers, like other epidermal structures, are moulted, either at a certain stage in the life-cycle or seasonally, a new generation being produced from the old papillae. There is an elaborate hormonal control of the moult, basically by the thyroid. The flight feathers are usually not all moulted at once but in a regular order, but some heavy birds such as ducks and geese, whose flight would be impaired by loss of a few feathers, retreat to safe places near water. Here many birds gather together to moult all their flight feathers at once, becoming temporarily flightless. Ducks cover themselves with an obscure 'eclipse plumage' while the new feathers are growing.

The down-feathers of the nestling are partly replaced by contour feathers; the follicle, instead of producing equal barbs, now forms two large ones at one side, which together become the central axis (rachis), carrying a series of further barbs that spread at right angles to it to form the vane, or vexillum (Fig. 15.2) (Stettenheim 1972). Each feather thus consists of a central rachis, forming the hollow calamus or quill below and carrying the barbs, which make the vane (Fig. 15.3). The calamus opens at the base by the inferior umbilicus, the entrance of the mesodermal papilla, and at the beginning of the vane there is a second hole, the superior umbilicus. At this point there is often a loose tuft of barbs or an extra shaft, the aftershaft, perhaps in some way representing the down-feather.

In the flight feathers the barbs or rami make up the vane and are held together by rows of barbules (radii) running nearly at right angles to the barbs and carrying hooks (hamuli) by which the barbules of one radius become fixed to grooves in those of the next (Fig. 15.3). Anyone who has played with a feather knows that these connections can be broken down so that the barbs become separate, but can be joined again by 'preening' the whole feather. This ingenious arrangement allows for rapid repair even of the non-living feather, which would be impossible by new growth of a torn membrane.

The feathers are provided with non-striated muscles at the base, with an autonomic innervation. In addition there are striated subcutaneous muscles, which move groups of feathers. Control of feather position is important for the regulation of heat loss, for flight, and in several other activities, for instance sexual display. Like the hairs of mammals the feathers are also used as

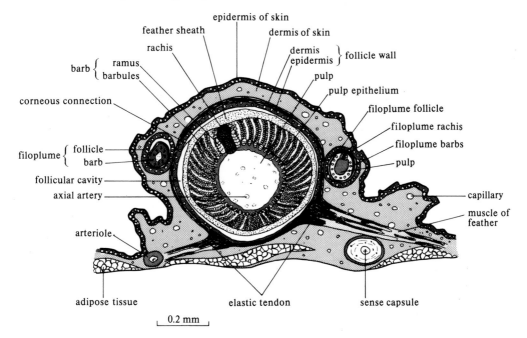

FIG. 15.2. Contour feather of 34-day-old domestic chicken. Transverse section of growing contour feather and filoplumes from crural tract. (Stettenheim 1972.)

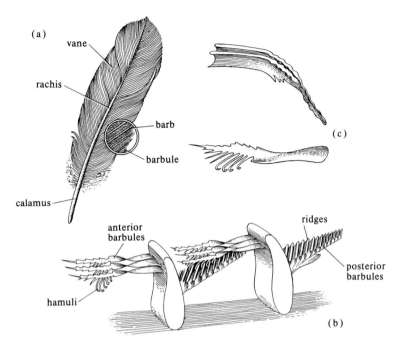

FIG. 15.3. The structure of a feather. (a) whole feather showing calamus (quill). On the right side a small area is shown as it appears under a lens. (b) A section cut at right angles to two barbs in the plane of the barbules of the anterior series. Note how the hamuli of the anterior barbs interlock with ridges on the posterior barbules. (c) Shows one anterior and two posterior barbules isolated. (After Pycraft 1910.)

organs for the sensation of touch. Nerve fibres are wound round the base of the papillae and the filoplumes transmit pressure changes from the contour feathers to encapsulated lamellar corpuscles at their base (Fig. 15.2). These Herbst's corpuscles are selectively sensitive to certain vibration frequencies. In owls and other night-birds vibrissae, analogous to those of mammals, are present. In some birds patches of special feathers without a rachis break up to make a greasy 'powder down'.

The feathers are not spread uniformly over the body but are arranged in rows in certain tracts, the pterylae, separated by bare areas, apteria. Among the contour feathers it is usual to recognize the remiges (remex, singular) of the wing and rectrices (rectrix, singular) of the 'tail'. Each large feather, whether in wing or tail, is usually covered above and below by several rows of upper and under coverts. The remiges are divided into ten primaries on the hand and about 10–20 secondaries on the ulna of the forearm (Fig. 15.9). In many birds there is a peculiar gap in the secondary feathers of the wing, the fifth remex feather being absent (diastataxis); the condition in which this feather is present is called eutaxis. The feathers have a remarkably flexible structure, so that they adopt different shapes with different positions of the wing. The shape of the quill and barbs varies between feathers and parts of a feather, for instance the barbs at the tip of the primary feathers provide a streamlined cross-section, like that of certain aeroplane propellers (p. 337). The small covert feathers

at the front of the wing stand up vertically, but have a right-angle bend, thus providing the wing camber.

The rectrices vary greatly, being very short or absent in some ground-living birds (e.g. quails, tinamous), but very large in fast-moving birds that change direction quickly (swallows). In these latter the outer rectrices are enlarged for steering purposes. The rectrices may be put to special uses, as in the woodpeckers, where they make a rigid brace, or in the peacocks, whose display feathers are the tail coverts.

4. Colours

Birds possess colour patterns more vivid than those of any other vertebrates, using them not only for concealment but also as the chief means of recognition and sexual stimulation and hence as the basis of their social life. Like other animals that live far from the ground and move fast (primates) the birds have a poor sense of smell, often none at all, but they have very good vision, and in many species the turning of discriminating eyes by one sex upon the other has led to the development of a very gorgeous covering. The feathers alter the appearance of the bird so completely that it is not fantastic to compare their effect with that of clothing in man.

As in other animal groups the colours are produced partly by pigments and partly by reflection and diffraction effects (structural coloration). The most common pigments are melanins, ranging from black through brown to yellow, and laid down in the feathers by special cells in the papilla. The processes of these amoeboid chromatophores convey granules of pigment into the barbs and barbules, where they are laid down in layers between those of keratin (Fig. 15.4). Carotenoid pigments (soluble in organic solvents) are also found, such as the yellow zooxanthin of the canary and the beaks and feet of the duck, and the red astaxanthin of pheasant wattles. White is usually given by reflection. In blue colours incident light is reflected from a layer of spongy keratin, pierced by holes $1-2\,\mu m$ in diameter which absorbs the red and reflects blue light. In iridescent feathers interference of light in thin surface films gives colours like those of soap bubbles. The more specialized iridescent feathers produce Newton's rings, with colours of the second and even third orders. The turacos or plantain-eaters of Africa contain two very peculiar pigments, a red copper-containing porphyrin, turacin and a green iron-containing turacoverdin.

The actual colour patterns vary with the habits of the bird. Concealing (cryptic) coloration is very common; even the brighter colours may serve this purpose, by breaking up the outline of the bird when at rest or in motion. Most birds are dark above and pale below. The feathers often show mottled or speckled patterns rather than a homogeneous colour. Finches and other birds living in the sunlit upper branches show bright yellow, yellow–green, and blue colours, either singly or combined. Birds living in thickets, such as the thrush and blackbird, are usually duller brown or black. An ex-

ample of disruptive coloration that is easy to observe is the white patch on the throat of a thrush if the nest is approached while the bird is sitting. The head is held rigidly still, with the beak upwards; the white mark on the neck breaks the outline and instead of an obvious bird's head there appear only the meaningless shapes of the sides of the jaws. In most species coloration is a compromise between concealment and display. Sometimes selection has acted so that the female is cryptic, the male conspicuous (e.g. ducks). In hole-nesting shelducks both sexes are conspicuous. In other birds bright colours are concealed most of the time (e.g. the robin's red breast is underneath, many waders have colours under their wings).

Some colour patterns that make the bird conspicuous may be a warning of a distasteful quality. The black and white pattern shown by the magpie may be an example of such aposematic coloration; certainly this bird is seldom preyed upon, no doubt partly because of its large size. The conspicuous black of rooks and starlings may be connected with their social life, making it desirable that the birds should easily follow each other, the group being protected by the combined receptors of its many members and the quick response of all to escape movements by any one.

5. The skeleton. Sacral and sternal girders

The arrangement of the whole locomotor apparatus is based on the plan of the bipedal archosaurian reptiles which has allowed the formation of two separate locomotor mechanisms for use in the air or on land or water. The bones are very light and often of tubular form, but sometimes with internal strutting well suited to the stresses they must bear (Fig. 15.5). Many of the bones contain extensions of the air-sacs; even the wing and leg bones are pneumatized in this way in very good fliers, such as some birds of prey and the albatross. Fusion of bones has proceeded so far that the skeleton consists of a few hollow girders and large plates of special shape (p. 329). This result is achieved by limiting the joints at which movement occurs and simplifying the muscular system. The long bones ossify from a single diaphysis, there are no epiphyses at the ends.

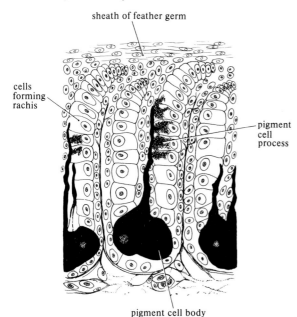

FIG. 15.4. Deposition of pigment in feather germ. Transverse section through a developing arm feather. (After Strong, from Stresemann.)

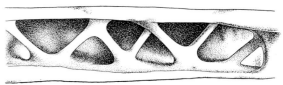

FIG. 15.5. Metacarpal bone from the wing of a vulture, sectioned to show the arrangement of the struts similar to that known to the engineer as a Warren's truss, such as is often used in aeroplane wings. (After Prochnow and D'Arcy Thompson.)

The skeleton of the backbone and limb girdles is so modified as to allow the weight of the body to be carried in two quite distinct ways, on the wings or on the legs. For this purpose there are two plate-like girders, the sternum and the synsacrum, curved in opposite directions (Fig. 15.7). The muscles around shoulder and hip joints balance the weight on these girders and produce propulsion. The main thrusts come from the pectoralis major in flying and from the leg retractors in walking. Perhaps no other animals are suited so perfectly for locomotion by two distinct means, and of course many birds can swim as well as fly and walk.

The whole axis of a bird is morphologically shorter than that of any other vertebrate except a frog or a tortoise (Fig. 15.6): only the neck remains a long and mobile structure. The number of cervical vertebrae varies and is greater in the birds with longer necks; there are fourteen in the pigeon, if we include two that bear ribs not articulating with the sternum. The cervical centra have saddle-shaped surfaces, the concavity running from side to side on the front and up and down behind, allowing turning in any direction, compensating for the fixture of the eyes in the orbits.

There are four or five thoracic vertebrae, all except the

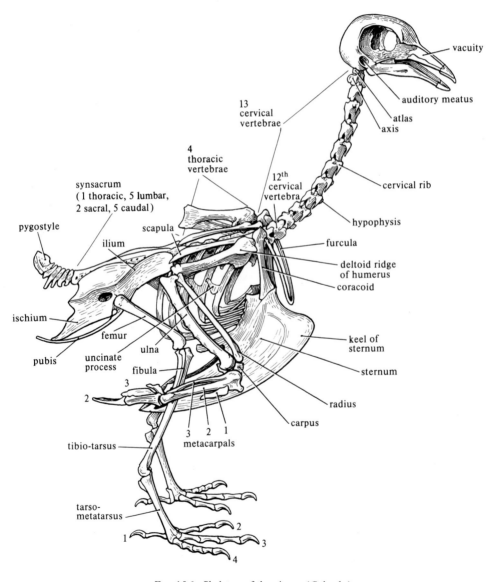

FIG. 15.6. Skeleton of the pigeon (*Columba*).

last united, in a pigeon, into a single mass. The ribs are large, double-headed, and jointed to the vertebrae. They bear uncinate processes on their vertebral portions, hook-like projections overlapping the rib next behind, thus strengthening the whole thoracic cage. There is a well-marked joint between the vertebral and sternal portions of the ribs. The latter are bony, not cartilaginous as in mammals, and are jointed to the sternum, which is a very large keeled structure in all flying birds, serving to carry the weight of the body to the wings by the attachment of the main wing muscles (Fig. 15.7). The pectoralis major, which depresses the wing in flight, is attached to the edge of the sternum and the great depth of the keel serves to increase the length and mechanical advantage of the fibres of the muscle and also, by its shape, to strengthen the sternum. When the bird is in the air the sternum is carrying a large part of the weight. By this arrangement the centre of gravity is kept well below the centre of air pressure on the wings, giving greater stability.

The last thoracic (rib-bearing) vertebra is united with about five that can be regarded as lumbars, two sacrals and five caudals to make a synsacrum, which is also fused with the ilium. This produces a very thin plate-like structure, whose ridged shape gives it sufficient strength to carry the bird's weight. Finally, there is a short bony tail of about six free caudal vertebrae, carrying four that are fused together to form the upturned pygostyle, supporting the tail feathers.

The joints of the vertebral column are therefore reduced so as to allow movement only in the cervical region, between the thorax and synsacrum and in the tail. The axial muscles have been correspondingly reduced. Those of the neck are large and the hinder cervical and the thoracic vertebrae have special ventral hypapophyses for attachment of the flexor muscles of the neck. The other back muscles, except those of the tail, are reduced and the whole back forms a single rigid strut, carrying the weight of the breast and viscera through the ribs and the abdominal muscles either to the pelvic girdle or to the sternum. In flying this weight is suspended on the wings and there is therefore a compression stress throughout the ribs, and this no doubt accounts for the ossification of their ventral parts. The weight of the bird when resting on its wings (Fig. 15.7) is thus carried by the pectoralis major as a tension member, through the plate-like sternum; the ribs, and especially the coracoid, act as compression members. The last-named bone lies nearly in the plane of the pectoralis major and is very strong.

6. The sacral girder and legs

In standing, perching, and walking the weight is balanced on two legs. To achieve this posture the type of girder found in the vertebral column of other terrestrial vertebrates has been abandoned, and with it the system of braces (back muscles) holding up the weight of the forepart of the body. Instead, the whole axis is so shortened that the centre of gravity lies far back, low, and over the feet. (This is not apparent from Fig. 15.8, which is not in a normal perched position.) Birds whose feet are placed far back for swimming must hold the body nearly upright to achieve a stable position with the centre of gravity over the feet (auks, penguins). The ribs and abdominal muscles transfer the weight to the greatly elongated ilia, which are fused to the vertebrae, making a long girder of approximately parabolic form. Though this is composed of bone of almost paper thinness, it is strengthened by longitudinal ridges (Fig. 15.7). Its strength, like that of the sternum, lies not

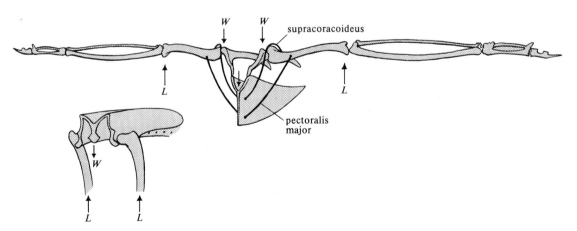

FIG. 15.7. Diagrams of the pectoral and pelvic girdles of an eagle, to show the methods of support in flying and walking. In each case the weight is carried on an arch, the strength of which is obtained by the peculiar kinked shape of the thin sheets of bone.

in its arched shape in the transverse plane, but in the distribution of weight that is achieved by its longitudinal curve and peculiar kinked shape. The whole pelvic girdle is modified to allow this arrangement. The ischium and pubis are directed backwards and do not meet in a symphysis, which would prevent the underslinging of the viscera.

The legs are used for balance and walking or hopping in ways that show interesting similarities and differences from those of man. The femur is turned under the body and articulates with the acetabulum in such a way that movement is almost restricted to the anteroposterior direction. The bird balances on its hips only in the sagittal plane; there are no movements of abduction and adduction such as are found in man. Abduction of the leg, or the falling medially of the bird's body when standing on one leg, is prevented by the fact that besides the ball-and-socket articulation of the femoral head there is also a second joint surface between the trochanter and an antitrochanter of the ilium. The ligaments across the top of this joint are very strong and they limit abduction movements, while movements of adduction are restricted by a strong ligamentum teres attached to the femoral head.

In life the femur is held nearly horizontal, bringing the legs well forward. The bird replaces the movements of abduction and adduction, which we make at the hip during walking in order to prevent falling over while only one leg is on the ground, by movements of rotation at the knee. The muscles around the hip joint form a system of braces allowing balancing and locomotion much as in man, but they are well developed only anteriorly and posteriorly; the lateral and medial (abductor and adductor) elements are weak (Figs. 15.8–15.11). The anterior group (protractors) includes a sartorius (iliotibialis internus) running from the ilium to the tibia, an iliofemoral, and a large anterior iliotibial inserted through a patella to a ridge on the front of the tibia. Associated with this muscle, which crosses both hip and knee joints, there are also, as in man, femorotibial muscles, making up with the longer muscles the extensor system of the knee.

The lateral side of the hip joint is supported by rather small abductor braces, the iliotrochanteric muscles,

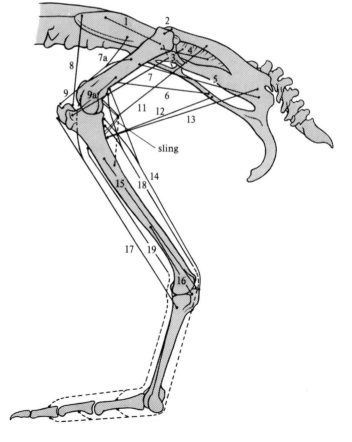

FIG. 15.8. Diagram of the muscles of the hind leg of a bird. Tendons are shown dotted. The sling is for tendon of iliofibularis. 1, iliotrochantericus; 2, iliofemoralis; 3, obturator; 4, ischiofemoralis; 5, caudiliofemoralis; 6, pubischiofemoralis; 7, iliotibialis posterior; 7a, iliotibialis anterior; 8, sartorius (iliotibialis internus); 9, femorotibialis medius; 9a, femorotibialis externus; 11, iliofibularis; 12, ischioflexorius; 13, caudilioflexorius; 14, gastronemius; 15, peroneus superficialis; 16, peroneus profundus; 17, tibialis anterior; 18, flexores digitorum; 19, extensores digitorum. (After Stolpe.)

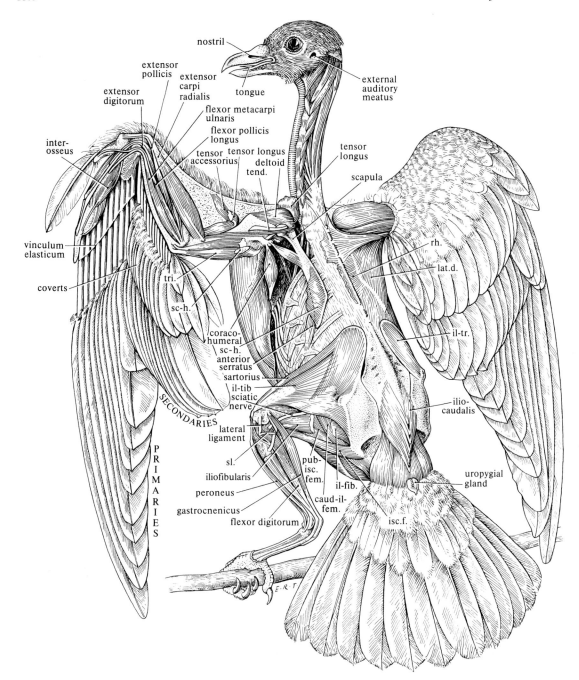

FIG. 15.9. Dissection of pigeon from back. *caud.-il-fem.,* caudiliofemoralis; *il-fib.* iliofibularis (cut); *il-tib.* iliotibialis; *il-tr.* iliotrochantericus; *isc.f.* ischiofemoralis; *lat.d.* latissimus dorsi; *pub-isc. fem.,* pubischiofemoralis; *rh.* rhomboid; *sc.h.* scapulohumeral (cut); *sl.* sling for tendon of iliofibularis; *tend.* tendon of pectoralis minor; *tri.* triceps.

corresponding to our glutei, and acting mainly as medial rotators, opposed by obturator and ischiofemoral muscles, which work as lateral rotators. The main locomotor muscles are the posterior braces or retractors, lying behind the hip joint and including muscles known as the posterior iliotibial, iliofibular, caudilioflexorius, pubischiofemoral, ischiofemoral, and caudischiofemoral. Some of these also act with the obturator muscle as

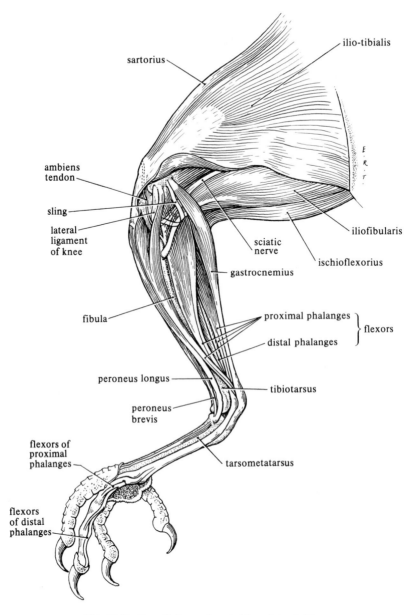

FIG. 15.10. Leg of pigeon dissected from lateral side.

lateral rotators, and those placed more medially function as adductors or medial braces, so far as such are required.

The femur articulates with both tibia and fibula at the knee. The fibula is distinct at its upper end, fused with the tibia below. The joints of the foot are greatly simplified by the union of the proximal tarsals with the tibia to make a tibiotarsus, articulating at an intertarsal joint with the remaining three tarsal and metatarsal bones, fused to make a single tarsometatarsus. There are

usually four digits articulating with the tarsometatarsus; three directed forwards and one backwards. In standing, the weight is usually balanced in tripod fashion on three of the four points provided by the front and back portions of the feet.

The knee joint has some remarkable similarities to that of man. It is stabilized by lateral, medial, and cruciate ligaments and contains a pair of lunate cartilages or 'menisci'. The joint allows movements of flexion and extension and the femur as it extends on the

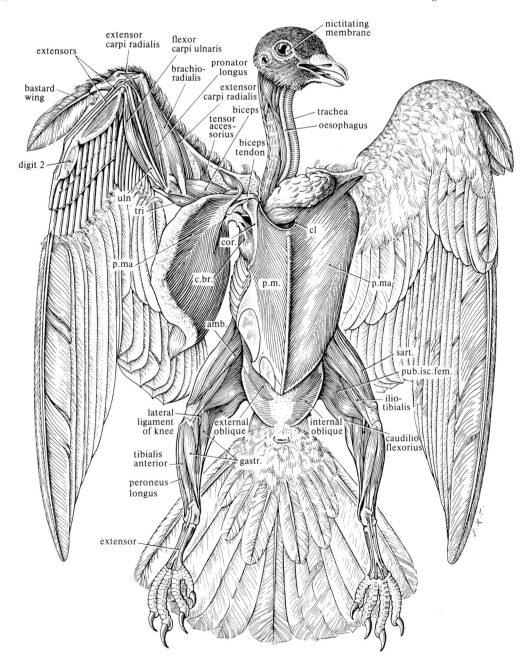

FIG. 15.11. Pigeon dissected from ventral surface. *amb.* ambiens; *c.br.* coracobrachialis; *cl.* clavicle; *cor.* coracoid; *gastr.* gastrocnemius; *p.ma.* pectoralis major; *p.m.* pectoralis minor (supracoracoideus); *pub-ise-fem.* pubischiofemoralis; *sart.* sartorius; *tri.* triceps; *uln.* ulna.

tibia in walking rotates laterally because of the arrangement of the joint surfaces. The bird thus balances in the mediolateral plane by rotation at the knee, somewhat as we do by abduction–adduction at the hip (Fig. 15.12). When it makes a step forward the weight is brought by

this rotation at the knee over the leg that remains on the ground.

The intertarsal joint allows mainly movements of flexion and extension. It is largely supported by ligaments and has a very strong capsule and lateral and

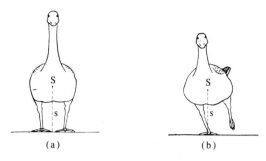

FIG. 15.12. Drawings from photographs of a goose, (a) standing; (b) stepping. The centre of gravity S is brought over the foot on the ground by lateral rotation of the femur on the tibia. Note the position of the tail in (b). (After Heinroth, from Stolpe.)

cruciate ligaments rather like those of the knee; there is even a meniscus on the lateral side. The back of the tibia is occupied by the gastrocnemius and the flexor muscles of the toes and at the front there is a tibialis anterior acting across the intertarsal joint, and also extensors of the toes. The calf muscles are mainly concerned with producing flexion of the toes in the act of perching and they form an elaborate system of tendons attached to the phalanges. These tendons often act as a single unit, and there is an arrangement by which the flexion is passively maintained by the weight of the body, even during sleep (automatic perching mechanism). Many of the muscles are specially arranged to allow support of the joint whether in the flexed or extended position. The iliofibular muscle passes through a conspicuous sling for this purpose (Figs. 15.8 and 15.10). The flexor muscles of the toes are inserted largely above the knee and thus tend to tighten as the bird sinks. In this they may be assisted by the ambiens, a muscle found in reptiles and some birds, which takes origin from the ilium. The muscle belly lies on the medial side of the thigh, and its tendon runs beneath the patella on to the lateral surface of the lower leg, where it is attached to the upper end of the muscles that flex the toes. This arrangement provides a single string crossing hip, knee, and ankle and allows the weight of the body to flex the toes as the joints bend.

The second mechanism for maintaining the bird on its perch is a locking device that holds the toes flexed. The under surface of the flexor tendon is ridged at the metatarsophalangeal joint, where the weight of the body presses it against a branch. The upper side of the tendon sheath is also ribbed and as the bird settles on its perch the two sets of ridges interlock.

The feet show a wide variety of adaptations for special habitats (Fig. 15.13). In the cursorial and walking birds there are often long digits in front and behind

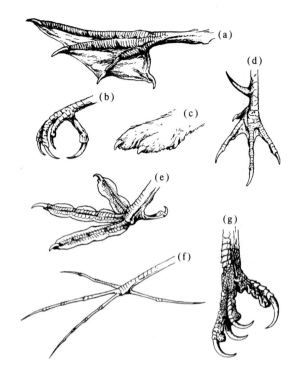

FIG. 15.13. Various types of feet in birds. (a) shag (swimming); (b) crow (perching, lifting); (c) ptarmigan (stockinged by feathers); (d) jungle fowl (walking, scraping); (e) coot (lobate, swimming); (f) jacana (suited for walking on floating plants); (g) sea eagle (raptorial). (From Thomson 1923.)

to give a good base for balance, but the number may be reduced – to two in running birds, such as the ostrich. Hopping is used by small birds on the ground and in the trees and produces quick movement. It is expensive because of the large displacements of the centre of gravity, and for long distances or large animals walking is more efficient. Many different groups of birds have acquired webbed feet for swimming. In birds exposed to cold the digits may be enclosed in a coat of feathers. Birds of prey develop long raptorial talons. Throughout the great group of perching birds one digit is directed backwards, allowing firm grasp of a branch. In climbing birds the fourth digit is often directed backwards as well as the first, so that the foot forms a sort of pincer, with long curved claws.

7. Skeleton of the wings

The wing is designed to have a minimum moment of inertia about an axis parallel to the sagittal plane and passing through the shoulder joint. Movements are produced by muscles lying either outside the arm or in its proximal part, with long tendons. The wing feathers are carried along the post-axial border of the humerus, ulna, and hand, and the shape of the wing depends on the position in which the feathers are held by their muscles, as well as on membranes, the pre- and post-patagia, developed where the limb joins the body. The active movements of flight are produced mainly by the pectoral muscles. The joints and muscles of the wing itself serve to spread the wing and to adjust its shape during each beat. The humerus is short and broad with a large head and an expanded surface for attachment of the pectoral muscles (Fig. 15.6). Radius and ulna are both large, especially the latter. There are only two free proximal carpals and the remainder of the wrist is formed of three metacarpals, one short and two long and fused. Only one digit, probably representing the second, is well developed, having two broad phalanges. The third and the first digits consist of single rods, the latter, standing somewhat apart at the front of the base of the hand, is capable of independent movement; it carries the bastard wing (alula or ala spuria).

The glenoid cavity is formed at the union of a blade-like scapula and a stout coracoid. The former lies horizontally and is attached by muscles to the vertebral column and ribs. The coracoid resists the compression produced by the wing muscles, holding the wing away from the sternum, with which it makes a joint (Fig. 15.6). The furcula (wish-bone), probably consist-ing of the combined clavicles and interclavicles, is loosely attached to the sternum and carries the origin of muscles that rotate the humerus about its long axis.

8. Wing muscles

Depression of the wing is produced mainly by a single mass of muscle, the huge pectoralis major, making up as much as one-fifth of the whole weight of the body. It runs from the sternum and furcula to the underside of the humerus, to which it is attached, at some distance from the joint, by a complicated tendon of insertion. The fibres of this muscle are very red in strongly flying birds and often contain numerous lipoid inclusions. In the fowl the fibres are white and contain glycogen but little lipoid. Elevation of the wing is produced by a muscle also attached to the sternum, lying deep to the pectoralis major and often called the pectoralis minor, but more properly supracoracoideus. Its tendon passes through the foramen triosseum acting as a pulley, between the furcula, scapula, and coracoid, to be inserted on the upper side of the humerus (Fig. 15.7). It is assisted by latissimus dorsi and deltoid muscles.

The chief muscles of the shoulder are thus a massive set serving to raise and lower the wing. There is little development of the muscles present in other vertebrates for the purpose of balance and drawing the limb

backwards and forwards for standing and locomotion. Such a system of braces all round the joint is un-necessary; the bird balances on its wings mainly by the action of the pectoralis major as the chief brace, between the sternum and the humerus, with the coracoid as a compression member between. Stresses must of course arise in other directions besides those tending to pro-duce a vertical force and these are met by the muscles that produce rotation of the humerus and various other movements of the wing, especially a pronation, depress-ing the leading edge. The muscles used in other tet-rapods to sling the weight of the body to the pectoral girdle and fore limb are little developed. The scapula is held to the vertebral column by small rhomboid muscles and there is a short series of slips attached to the ribs, the serratus anterior.

Other muscles running from the body to the humerus produce rotation of the humerus at the glenoid and adjustments of the patagia, movements that are very important in flight. From the outer surface of the scapula arises a scapulohumeral muscle, inserted in such a way as to produce adduction and lateral rotation of the humerus, raising the hinder edge of the wing. The coracohumeral muscle is a compact bundle attached near to the last and producing the opposite effect of abduction and medial rotation, lowering the hinder aspect of the wing.

The deltoid muscle is divided into several parts and besides its main abductor action on the wing also has slips inserted into the skin of the anterior patagium, muscles known as the long and short tensors of that membrane. There is also a tensor accessorius, running from the surface of the biceps to the skin of the leading edge of the wing.

The muscles within the arm itself serve to extend or fold the whole wing and to alter the positions of the parts, especially by pronation and supination during flight. Large triceps and smaller biceps muscles act at the elbow. In the forearm there is a large extensor carpi radialis and an extensor carpi ulnaris, serving to keep the wing extended at the wrist. Flexor carpi ulnaris folds the wing. There are also two large pronators, brevis and longus (brachioradialis), rotating the radius medially and lifting the back of the wing. A system of digital flexors and extensors, inserted into the distal phalanx of the main digit, keeps the wing tip spread out or folds it. The position of individual feathers is controlled by an elaborate system of tendons and muscles along the back of the hand. The first digit is moved independently by abductor and adductor pollicis muscles, controlling the position of the bastard wing, which increases the angle of stall and thus allows slow flying speeds in take-off and landing.

9. Principles of flight

A plane surface moved through the air in a direction inclined at an angle to its plane is known as an aerofoil. The forces generated can be resolved into a lift force acting upwards and a drag force tending to stop the motion. On this fact depends the power of supporting weight in the air that is possessed by birds and human heavier-than-air machines. Both lift and drag forces are proportional to the square of the speed, and the requirement for sustained flight in still air is that the object shall have sufficient speed to generate a lift force equal to its weight (Lighthill 1975; Pennycuick 1972, 1975).

The flow of air over the upper surface of the wing reduces the pressure there and provides the main portion of the lift (Fig. 15.14). By tilting the wing (increasing the 'angle of attack') the pressure on the underside can be increased, but the air flow now tends not to follow the upper surface but to become turbulent, especially at the hind edge, destroying the lift (Fig. 15.14). When an aerofoil falls below the critical speed it stalls; that is to say, drops suddenly, being no longer supported. The smooth flow of air over the wing tends to be especially disturbed at its hinder ('trailing') edge and by eddies round the end ('tip vortex'). The proportion of length to breadth (aspect ratio) in the wing suitable for a particular type of flight depends on the need to provide a sufficiently large undisturbed area.

10. Wing shape

The shape of the aerofoil is of critical importance in determining its aerodynamic capacities. For birds, as for aeroplanes, there are differing shapes, suitable for various types of flight. To understand them we must classify the means by which birds attain the necessary forward velocity. First and most obvious is flapping flight. Though the details of this are varied and not fully understood, it can be regarded as a screw-like motion of the wings, providing forward and upward components (p. 339). In still air the only alternative to flapping flight is to glide downwards, which obviously cannot continue indefinitely. Yet some birds, such as the gulls, and especially the albatrosses and the buzzards, condors, and other birds of prey, can be seen to soar for many minutes, gaining height without flapping the wings.

A wing of the shape that allows an albatross or swift to make its superb manoeuvres would stall immediately at the low speed of flight adopted by a crow. In discussing wing shape the chief factors to be considered are (1) the wing area, (2) the aspect ratio (wing length/breadth), (3) the wing outline and taper, (4) the presence of holes or slots, (5) the camber or curvature of the wing.

11. Wing area and loading

A small wing area is necessary for fast flight, since the drag α area $\times$ speed^2, at least for high speeds. For this

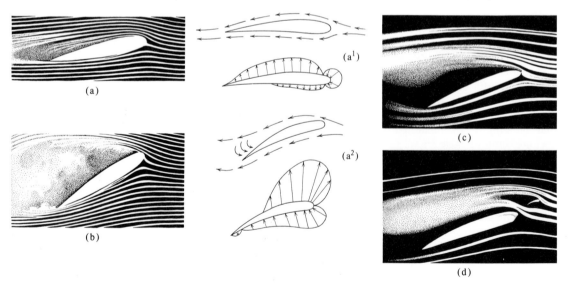

FIG. 14.15. Effect of an air-stream flowing past a wing under various conditions in a wind tunnel. The white lines are jets of smoke. In (a) they are close together below the wing and farther apart above it. These increased and reduced pressures produce the lift and the arrows in (a^1) and (a^2) show the actual pressures registered by gauges, with different angles of attack. In (b) and (c) the angle of attack is so large that the stream does not follow the upper surface, turbulence develops, the stream produces drag instead of lift, and the wing stalls. (d) shows how this is prevented by slotting. (After Storer from official photographs of the National Advisory Committee for Aeronautics, and from Jones.)

reason fast aeroplanes and birds have small wings, but in the bird the fact that the wing provides the forward momentum as well as the lift greatly complicates matters. For flapping flight the wings must be moved relatively fast and for this small size is an advantage. On the other hand, a large wing area allows slow flight (lift $\propto$ area $\times$ speed^2) and is found in hawks, vultures, storks, and other birds that fly slowly to hunt, especially if they soar on thermal currents, for which a large lift is necessary (p. 341).

The loading of the wing varies considerably. Since the weight increases with the cube but the wing area only with the square of the linear dimensions, it follows that large birds must have relatively larger wings than small. However, the larger birds usually have a heavier loading of the wing, for instance, 10 kg/m^2 in the duck (*Anas*), 20 in the swan (*Cygnus*), 1 in the goldcrest (*Regulus*), 3 in the crow (*Corvus*). A considerable 'safety margin' remains in most birds; for instance, pigeons were found to be able to fly until as much as 45 per cent of the wing surface was removed; hawks and owls have an especially high safety margin and they can carry prey almost as heavy as themselves.

The power required from unit mass of muscle to fly at a speed sufficient to provide lift increases with increasing body weight but the power available becomes relatively less. This deficiency of power sets a limit to the maximum weight at about 12 kg. The albatross, the bird with the largest wing span weighs up to 12.0 kg (mean 9.8 kg). Mute swans reach 13.5 kg (mean 11.9 kg) and great bustards have been reported up to 16 kg, but their mean is also 12.0 kg. For the same reason it is unlikely that *Pteranodon*, weighing about 18 kg, used flapping flight (p. 389) (see Pennycuick 1972, 1975; Bramwell and Whitfield 1974).

12. Aspect ratio

Although a small wing area reduces drag, many fast-flying birds have a large wing-span. The aerodynamic advantages of this allow a low rate of descent when gliding, reducing the expenditure of energy necessary to sustain flight. High aspect ratio is therefore found in birds that fly fast by flapping flight (swifts and swallows) and especially in those, such as the albatross, that glide fast in order to obtain sufficient kinetic energy to convert into altitude. However, these wings with very high aspect ratio stall at relatively high speeds and the birds that soar slowly on thermal up-currents over the land mostly have a low aspect ratio. Some figures for aspect ratios are:

Albatross (*Diomedea*)	25
Gull (*Larus*)	11
Swift (*Apus*)	11
Shearwater (*Puffinus*)	10
Vulture (*Neophron*)	6
Rook (*Corvus*)	6
Sparrow (*Passer*)	5

13. Wing tips, slots, and camber

A pointed wing tends to stall first at its tip and is therefore only suitable for fast fliers. Such birds show great development of the hand feathers, producing a long narrow wing, whereas birds built for slower flight and manoeuvre have a shorter broader wing with long arm feathers (Fig. 15.15).

The condition of the air around the wing is of first importance for the maintenance of lift; if there is not a smooth stream over the upper and under surfaces the air becomes turbulent, and the aerofoil stalls (Fig. 15.14). This tends to happen either if the speed falls too low or if the angle of the wing relative to the line of motion increases above about 20°. Turbulence is mitigated, however, by the provision of openings, known as slots, which let through part of the air and provide the necessary smooth stream. The spaces that occur between the feathers, especially towards the wing tip, almost certainly function as slots. Probably the arrangement provides a series of such apertures, giving a very efficient high-lift device. Such slots are conspicuous in slow fliers (rooks) and especially in those that soar on

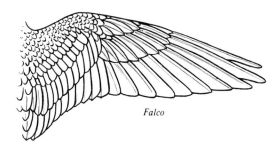

Falco

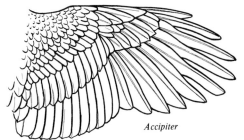

Accipiter

FIG. 15.15. Wings built for speed (falcon, *Falco*) and for manoeuvring (hawk, *Accipiter*). The former is long and narrow, with relatively large hand feathers. The latter is short and broad, the arm feathers being long and the primaries arranged to make slots. (After Fuertes.)

thermal up-currents (vultures). The feathers of such birds are often individually tapered (Fig. 15.15). Slots are also found in the wings of large birds that are fast fliers (pheasants), the wing being liable to stall in certain phases of the down strokes. It is possible that the bastard wing acts as a slotting device; indeed, consideration of it played a part in development of the theory of turbulence and slotting.

The shape of the wings have a very important influence on the air stream. In most birds there is a thick leading edge and a thinner trailing edge. Nearly all wings are cambered, that is to say, they are convexly curved in cross-section, especially in the region of the forearm (Fig. 15.14). This arrangement directs the air stream over the upper surface of the wing in such a way as to provide an extra lift by creating a 'suction zone' of reduced pressure. However, high camber, like low aspect ratio, reduces the speed of the bird.

The feathers are held by an elaborate system of tendons and in some birds they are allowed to twist only when the wing is being raised and the barbs of the feathers themselves are so arranged that they open like the vanes of a blind when under pressure from above, but close when the pressure is from below (Figs. 15.16 and 15.17). In other birds, especially those that fly fast with a slow wing beat, such as gulls and swans, the wing is probably rigid on the up as well as on the down strokes, and is twisted so as to produce forward and upward components on the upstroke.

The whole upward movement is usually faster than the downward one. Before the wing tip has reached its highest point the upper arm is already beginning to descend and in this way the line of flight is maintained

FIG. 15.16. Bill-fisher leaving its hole in a bank, showing wings half-way through the upstroke. The upper arm has reached its highest point and the forearm is just starting upwards, its primary feathers have opened on the right wing, reducing resistance. (Drawn from photograph by Aymar 1935.)

almost straight and does not follow a wavy path as it would do if the parts of the wing vibrated together. In small birds the wing works more nearly as a whole and the flight differs in several respects from that of larger birds. In general the wing is a very labile system and regulates itself automatically with changes in the aero-dynamical forces. This regulation is produced partly by feather plasticity and joint mobility, with participation of reflex muscular adjustments that are little understood.

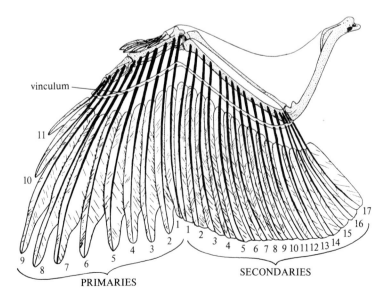

FIG. 15.17. Diagram of wing to show arrangement of the flight feathers. (From Pycraft 1910.)

14. Flapping flight

Flapping flight involves a complex, screw-like motion of the wing, downwards and forwards then upwards and backwards, more rapid upwards than downwards (Fig. 15.18). The action of the wings differs during take off or landing from that in sustained flight (Brown 1953). In the flight of a pigeon at a medium speed of 10 m/s the wings beat at a frequency of about 5 Hz and the sequence is shown in Fig. 15.19. The body is nearly horizontal. At the top of the downstroke the wings are fully extended and then move downwards through an arc of 90° (Fig. 15.19 (a)–(c)). Before the upstroke the elbow is first flexed and the wrist and the manus supinated. The primary feathers are then turned backwards (Fig. 15.19 (d) and (e)) producing a 'feathered' upstroke with a large positive pitch angle to the direction of locomotion and a thrust as the primaries are swung backwards. At the end of the upstroke the elbow and wrist are extended and the wrist pronated, spreading the primaries for the next downstroke. These feathers separate as the wing descends, an arrangement that increases their lift in the same way as the leading edge slots used in aeroplanes.

15. Hovering flight

Support of weight without horizontal movement requires very much greater consumption of energy than

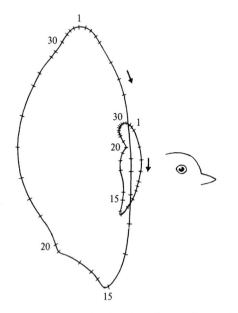

FIG. 15.18. Pathway of the wing tip and wrist joint relative to the body during free flapping flight of a gull. Equal time-intervals are shown. Note the great speed of the upward beat and that the forearm is raised before the wing tip. (After Demoll.)

forward flight and the largest hummingbirds weigh only 20 g. The principle involved is also that used for short times by larger birds in take off and landing (Weiss-Fogh 1973). Lift for weight support requires horizontal movements and since the wings beat in a plane perpendicular to the body, hovering is achieved essentially by holding the body vertically (Fig. 15.20). In this way the hummingbirds (Trochilidae) can remain in one place in the air or even move backwards. The wings beat backwards and forwards, often as fast as 200 times a second (Fig. 15.20). The supracoracoideus is about as large as the pectoralis major (p. 329).

16. Take off and landing

At take off the bird has to acquire sufficient forward momentum to provide lift, and yet must leave the air sufficiently undisturbed for subsequent beats to be effective. In many birds the jump provided by the legs is adequate for the take off (Fig. 15.21). Heavier birds have first to acquire speed by running, and some of the heaviest, such as condors, probably cannot take off at all in this way and do not inhabit flat country. Great bustards living in flat country must run very fast for take off. Similarly large bats and probably pterosaurs, whose wings involve the feet, must start by diving from a tree or hill. Eagles are said to be unable to rise without a long run, and many large birds nest on a cliff or tree, which gives them an up-current for the take off (Fig. 15.22). Swifts usually come to rest high up and can only rise off the ground with difficulty. The albatross is unable to take off from the sea surface in a dead calm.

The first beats are usually very large, beginning with the wings above the back and held at such an angle as to produce a large forward component. The wings may be heard actually clapping together in pigeons (Fig. 15.23). During rapid ascent, as in the larks, the body is nearly vertical and the wing changes its angle at the shoulder very sharply between the downward and recovery strokes. The bastard wing is held in such a position that the beat provides extra forward momentum.

In take off of a pigeon the body is held vertically. The action of the wing on the downstroke is as in sustained flight but on the upstroke the wing is first adducted, folded and flexed and supinated at the wrist, by the actions of supracoracoideus and other muscles. A very rapid flick then follows, produced by upward and forward rotation of the humerus, extension of the wing and pronation of the humerus. These movements, produced largely by the triceps and other extensors, result in a forward component. After jumping the bird is thus able to acquire sufficient forward velocity to take off from flat ground.

Landing is also a delicate operation, especially since it

FIG. 15.19. Cycle of movements of the wing of a pigeon flying at 32 km/h.
(a) Beginning of the downstroke, with wing fully extended.
(b) Middle of downstroke showing twisting of the wing tip to give forward drive.
(c) Near the end of the downstroke, when both velocity and angle of attack are falling.
(d) Beginning of upstroke. Flexure and retractions of wrist tip feathers, unstressed wrist rising.
(e) Middle of upstroke. Primaries are being swung violently backwards, giving a forward drive.
(f) End of upstroke. After the flick the primaries remain unstressed until the wing extends for the next downstroke. (Brown 1953.)

(a)

(b)

(c)

(d)

(e)

(f)

often involves coming to rest suddenly on a branch (Fig. 15.24). This is achieved by lowering and fanning out the tail, which thus acts as a flap, providing both lift and braking. The legs are then lowered; often one further wing stroke is given to bring the bird forward to drop onto the perch. The adjustment of braking in such a way as to prevent stalling involves a very special system of co-ordination (Fig. 15.25). Other methods of landing are possible for instance rooks may make a roll and side-slip to the ground.

17. Soaring flight

Many birds economize in the energy needed for flapping flight by making skilful use of the possibilities presented by movement of the air. Theoretically the bird can use three types of air movement: (1) ascending currents,

usually thermal; (2) variations in the wind velocity at any one level (gusts); (3) differences in wind velocity at different levels. The first method is that used by human sail-planes and is certainly adopted by many soaring land birds. The gustiness of the wind is probably turned to advantage by gulls, rooks, and many other birds, and the variations in wind velocity near the sea surface are used by marine soaring birds, notably the albatross.

All birds glide for short distances, some small birds with wings folded, others with wings outstretched. The higher the speed of a glide the smaller is the optimum wing area (least drag for adequate lift), and it is easy to see that birds fold their wings as they glide fast. Sustained gliding and soaring upwards without flapping the wings is found only in large birds, probably because considerable weight is necessary to provide kinetic

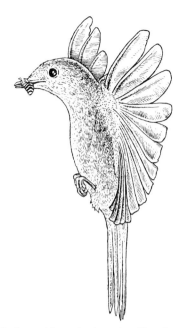

FIG. 15.20. Spotted flycatcher hovering. The wings are passing backwards and there are spaces between the feathers. (Drawn from a photograph by Hosking, with permission.)

energy sufficient to ensure continuous flight and efficient use of wind variations. As has been suggested, there are two distinct types of soaring birds: (1) land birds using thermal up-currents, (2) marine birds using variations in wind above sea level.

18. Soaring on up-currents

Up-currents of air arise in the neighbourhood of large objects on the ground (cliffs, tall buildings, or even a ship) and particularly from variations in the rate of heating of the earth's surface in the sun, over rocks, vegetation, mountain shadows, etc. some of which can be anticipated by observing the nature of the ground. Birds using such currents usually proceed upwards in a series of small circles gliding slowly, a behaviour seen in buzzards and other hawks and especially characteristic of vultures, which may ascend in this way above 300 m (Figs. 15.26–15.28). The characteristic features of such thermal soarers are large wing area, low aspect ratio, and wings broad at the tip and usually provided with well marked slots. Soaring economizes fuel but still makes some demands for energy to hold the wings down in their horizontal position. In all soaring birds the pectoralis muscle is divided and its deep portion is

FIG. 15.21. Heron (*Ardea*) leaving its perch. The legs have been used to make a jump and the wings are fully spread. (Drawn from photograph by Aymar 1935.)

FIG. 15.22. Drawings from four photographs from film of a take-off by an eagle (*Aquila*). (a) The legs are jumping and the upper arms nearly vertical before the wrist joint is extended. (b) Wing fully extended with bastard wing spread out, some pronation. (c) Marked pronation as the upper arm reaches its lowest point. (d) The upper arm is proceeding upwards, although the hand has not yet reached its lowest point. (Drawn from photographs by Knight, from Stresemann 1934.)

probably a tonic muscle serving this purpose. Storks (*Ciconia*) take advantage of the economy of soaring for their migrations. Because it is not possible over the sea they go all the way round the east end of the Mediterranean and Suez, taking about 23 days for the journey of 7000 km. They can carry enough fat to do this without stopping to renew their supplies, as they would have to do for flapping flight. Small passerines travelling between Europe and tropical Africa migrate by a straight line of about 5500 km, taking both the sea and the Sahara in single stages. But they must stop two or three times for several days to refuel.

FIG. 15.23. Pigeons (*Columba*), photographed during take-off, with exposures of 1/825 second. (a) front, and
(b) rear view, with wings together. (c)
nearly, and (d) quite at bottom of
downstroke; note pronation and forward movement of wing. (e) and (f)
wings during the upstroke; in (f) the
primary feathers have opened; note
that the wing moves backwards and
that the motion is faster than on the
downstroke. (From photographs by
Aymar 1935.)

FIG. 15.24. Jackdaw about to land. The
wings are fully extended on the downbeat,
and the tail is fanned out and bent downwards. (From photograph by Aymar
1935.)

FIG. 15.25. Hawk striking at a dummy owl. Note long legs and
the method of braking. The wings are broad and rounded,
giving a large safety factor. (From photograph by Aymar
1935.)

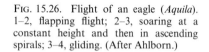

FIG. 15.26. Flight of an eagle (*Aquila*).
1–2, flapping flight; 2–3, soaring at a
constant height and then in ascending
spirals; 3–4, gliding. (After Ahlborn.)

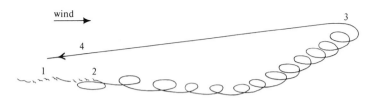

FIG. 15.27. Soaring flight of kite
(*Milvus*). Losing height downwind
and gaining it upwind. Time marks
one second. (After Hankin.)

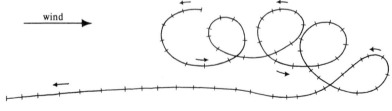

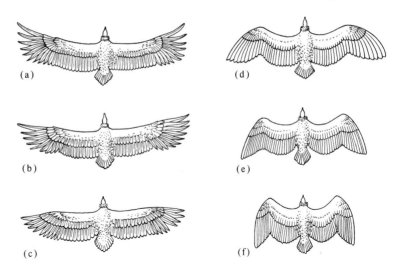

FIG. 15.28. Wings of vulture (*Gyps*). (a) soaring upwind, gaining height; (b) upwind on level; (c) downwind, losing height; (d)–(e) gliding flight. (After Ahlborn.)

19. Use of vertical wind variations

The decreasing effect of surface friction causes the wind to blow faster at greater heights and this phenomenon is used by some birds at sea, where conditions are presumably more uniform than over the land. The albatross (*Diomedea*) (Fig. 15.29) is the classic example of this type of bird, proceeding in a regular series of movements, without flapping the wings, downwind losing height and gaining speed and then upwind gaining height and passing into a faster-moving layer.

Each downwind tack is longer than the upwind one. During the upwind tack the wings are spread forwards, downwind backwards. The albatross remains all the time within 15 m of the sea surface, because the variations in wind velocity are marked only at low levels. The albatross shows the characteristics suitable for the rapid gliding needed for this type of flight, namely, large size and the greatest wing span of all birds (3.35 m), high aspect ratio (25), and pointed wing tips, without slots.

Other sea birds, such as the gulls, though not so highly

FIG. 15.29. Wings of the albatross (*Diomedea*), used for soaring flight. Wing very narrow, with long upper arm region. (Drawn from photograph by Aymar 1935.)

specialized, can take advantage of variations in horizontal wind velocity, including gusts at any one level. The bird moves upwards as it meets an accelerating gust and turns when the wind decelerates. Gulls also use up-currents at cliff faces and, no doubt, air movements of all sorts are widely used, especially by large birds. However, it is evident that the wing equipment only allows the bird a limited range of choice and probably even the slightly different wing shapes of related species depend on the differing conditions they are called upon to meet. A vulture could no more zoom backwards and forwards over the waves than an albatross could circle slowly on a gentle thermal up-current. A pigeon cannot equal a gull at steady gliding and soaring, but the pigeon can rise more steeply or descend more rapidly without stalling. The different capacities are very obvious when these two species come to town.

20. Speed of flight

Estimation of the speed of flight involves distinguishing between air and ground speed. The speed relative to the ground may be very high; there are records of birds covering more than 160 km in an hour, with the help of the wind. Racing pigeons can average 64 km an hour or more for considerable periods. Air speeds of 50–80 km/h can certainly be reached by many birds: swifts are said to reach 160 km/h in still air.

21. The skull

The arrangement of the parts of the bird's skull is similar to that of archosaurian reptiles (Fig. 14.25). Individual bones can be recognized in the young, but they mostly become united in the adult to form a continuous thin-walled structure that encloses the brain and sense organs and supports the beak (Fig. 15.30). Most birds are microsmatic; the nasal passages are simple and the turbinals reduced. There is seldom a complete bony secondary palate, such as there is in mammals, instead the internal nostril opens into the mouth relatively far forward. The large size of the brain and reduction of its olfactory portions are responsible for the rounded form of the top of the head, and there are very large orbits at the sides, separated by an ossified septum. The base of the skull is formed by a basioccipital behind, carrying a single occipital condyle. There is a large basisphenoid, covered ventrally by a pair of basitemporals, probably representing the parasphenoid, the front part of which makes a 'basisphenoid rostrum', as in archosaurs.

The jaws are characteristically slender and elongated. In the more advanced birds they have a very special form of support. The upper part of the front of the skull is composed of the enlarged premaxillae, the nostrils lying very far back and the nasal bones being small. The palatines are long and fused far forward with the maxillae, but they articulate movably behind with the pterygoids and base of the skull (Fig. 15.31) (Bock 1964). The pterygoid is a slender rod, itself movably articulated with the skull and with the quadrate, which is a triangular bone with clearly separate otic and basal articular processes. The upper jaw is thus a long thin bar composed of maxillae, quadratojugal, and jugal. As in many reptiles it is capable of considerable movement ('kinesis'). When the jaw opens the quadrate moves forward and pushes the upper jaw forwards and upwards. This mechanism is particularly well developed in parrots, where the beak is freely hinged on the skull. This type of palatal arrangement is known as neognathous. In some birds, such as the flightless ratites, the palatines are shorter, the vomer larger, and the pterygoids less movable, a condition called palaeognathous (p. 391). The lower jaw, also elongated, consists of the articular bone and four membrane bones.

22. The jaws, beak, and feeding mechansims

There is a complete lower temporal bar, composed of jugal and quadratojugal bones. The temporal region is hard to interpret, but has presumably been derived from the diapsid archosaurian condition. Typically, there is a single large fossa, communicating with the orbit, but this is often partly subdivided by bony processes; occasionally, there is a complete post-orbital bar (parrots). There are moderately large temporal and ptery-goid muscles, but the jaws are not usually very powerful, though, of course, formidable in carnivores. Having completely lost the teeth, the birds must rely largely on internal processes to break up the food. The outer and part of the inner surface of the beak has a light but hard, horny covering, the rhamphotheca. At the proximal end of the upper jaw this is modified as a soft, specially sensitive region, the cere. Lamellated sense organs, the corpuscles of Grandry and Herbst, occur in the rhamphotheca of ducks and many other birds (Figs. 15.64 and 15.65). These corpuscles presumably provide information about touch or vibration, perhaps especially for structures under water or otherwise out of sight.

The beak is characteristically modified according to the food habits (Fig. 15.32). There is very great variety in the feeding, as in so much of the life of birds, and though many species keep strictly to one diet others are able to adapt themselves to the food available. The ingenuity and persistence with which birds seek and collect food must be a main factor in their success.

Many birds with a moderately long bill, such as the song-thrush (*Turdus*) or starling (*Sturnus*), can eat either flesh (snails, earthworms, or caterpillars) or fruit. The snail's shell is cracked open to obtain the food, by

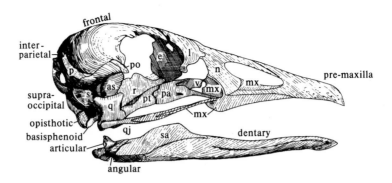

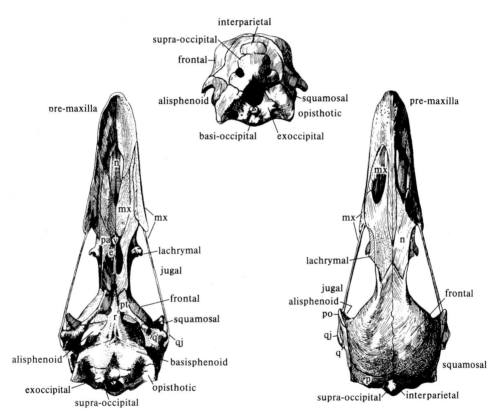

Fig. 15.30. Skull of young gosling (*Anser*). *as.* alisphenoid; *e.* ethmoid; *l.* lachrymal; *mx.* maxilla; *n.* nasal; *p.* parietal; *pa.* palatine; *po.* postorbital; *pt.* pterygoid; *q.* quadrate; *qj.* quadratojugal; *r.* rostrum of basisphenoid; *s.* squamosal; *sa.* surangular; *v.* vomer. (From Heilmann 1926.)

beating it against a stone (anvil). Birds that mainly eat seeds, such as the finches, usually have short, thick, strongly ridged bills. Some open the seeds by crushing or cutting with the sharp edge of the bill. Nuthatches wedge seeds in the bark before breaking them. Crossbills use their beaks to extract the seeds from pine cones. Large strong bills are present in the hornbills and toucans and are used to pluck fruit while perched on a thick branch. In parrots the beak is moved on the skull, pushed up by the jaw when the latter is pulled forward by the digastric muscle, giving a large gape.

The carnivorous birds, such as most eagles and owls, have short and sharp hooked beaks, whereas fish-eating, as in other vertebrates, results in long jaws. Another

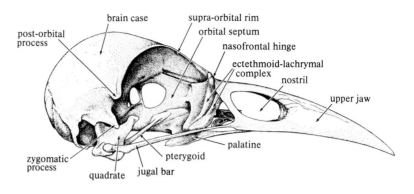

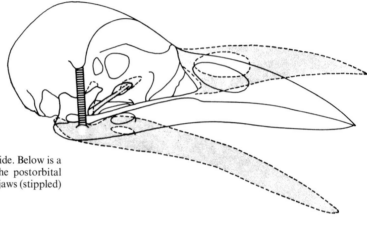

FIG. 15.31. Crow (*Corvus*) skull seen from the side. Below is a schematic figure showing the mandible and the postorbital ligament in place, and the positions of the open jaws (stippled) relative to the closed jaws. (From Bock 1964.)

widely found arrangement is the flattened bill of some ducks, similar to those of some sturgeons and of the platypus, which also sift out food from water or mud. The long, thin beaks of the waders select food from mud in a different way, mostly worms, crustaceans and small molluscs, sometimes stabbing inside the open shells of mussels. Flamingoes feed on shrimps and other small organisms by the piston-like action of the thick, fleshy tongue which pushes the water out of the mouth and through the filtering lamellae lining the inner surface of the curved mandibles (Fig. 15.32b). There are many special adaptations for insect feeding. Flycatchers snap their prey by special ligaments that contract as the mouth gapes. Tree-creepers have thin, pointed beaks to

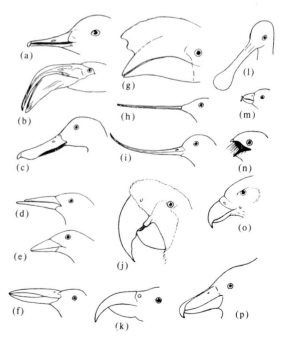

FIG. 15.32. Variety found in the beaks of birds. (a) merganser; (b) flamingo; (c) shoveller; (d) scissor-bill (adult); (e) scissor-bill (young); (f) *Anastomus*; (g) hornbill; (h) humming bird; (i) avocet; (j) parrot; (k) parrot; (l) spoon-bill; (m) crossbill; (n) nightjar; (o) eagle; (p) *Balaeniceps*. (From Pycraft 1910.)

take insects from cracks in the bark. The woodpeckers have a strong beak like a pick-axe for excavating in wood, and an elaborate apparatus for licking up insects by an enormously long, flexible, sticky tongue and modified hyoid. The woodpecker finch (*Camarhynchus pallidus*) on the Galapagos Islands probes insects from the bark by means of a cactus spine, a remarkable case of the use of a tool by a bird (Fig. 15.33). Sapsuckers dig holes in trees and may get so drunk on the fermented sap that they can no longer fly. Among the most specialized feeders are the hummingbirds, of which there are some 330 species. Many of them are adapted to feed on the nectar of particular flowers, the beak being long or short according to the type of flower visited, and the tongue provided with a special tubular tip. The tongue of most birds has no intrinsic musculature but is moved by the hyoid apparatus. It is covered by a special cornified epithelium, forming papillae on the dorsal side.

23. Digestive system

Once the food is in the mouth it is manipulated by the long, thin tongue, moistened with saliva, which is mainly mucus and rarely if ever contains amylase. In swifts the large salivary glands produce a glycoprotein used to glue together the nest. A North American species of jay uses the saliva to form balls of food which it hides for future use.

The food is usually swallowed entire down the oesophagus, which is often dilated below to make a large receptacle, the crop, found especially in birds that eat grain and fish (Fig. 15.34). This is mainly a storage chamber, grain may macerate there, but there is at most only a little enzymatic digestion. The crop is also used to store food for the young and in pigeons of both sexes it produces milk for them. This is formed by desquamated epithelium under the influence of pituitary prolactin, which also causes broodiness in the parents.

The true stomach is divided into two parts, a glandular proventriculus where food is mixed with peptic enzymes and a muscular gizzard where it is ground up. This has a hard lining of a peculiar material koilin, produced by the secretion of tubular glands, mixed with layers of cellular debris. It is coloured green or yellow by regurgitated bile. In birds that eat grain or other hard foods the koilin may form ridged plates and its action is assisted in many species by small stones. The gizzard is often said to be a substitute for the lack of teeth. Its smooth muscles make rhythmic contractions at 1–4 per min in the domestic fowl. The gizzard has very great grinding power. Lazzaro Spallanzani reported in 1784 that a turkey ground 12 steel needles to pieces in 36 hours.

The peptic juice has powerful digestive properties and many carnivorous and fish-eating birds dissolve even the bones of their prey, though in owls, hawks, and kingfishers these are regurgitated with fur or feathers, making characteristic pellets.

The duodenum and coiled intestine are of characteristic vertebrate type, relatively rather short, though somewhat longer in grain-eating birds. The bile and pancreatic ducts usually open into the distal limb of the duodenum; in pigeons the left bile-duct enters close to the pylorus (Fig. 15.34). There is a pair of blind caeca at the junction of rectum and intestine. The food enters these caeca, but it is not clear what function they perform, possibly it is related to the absorption of water

FIG. 15.33. The Galapagos woodpecker finch (*Camarhynchus pallidus*) using its stick. (From Lack 1947, drawn by R. Green from photograph by R. Leacock.)

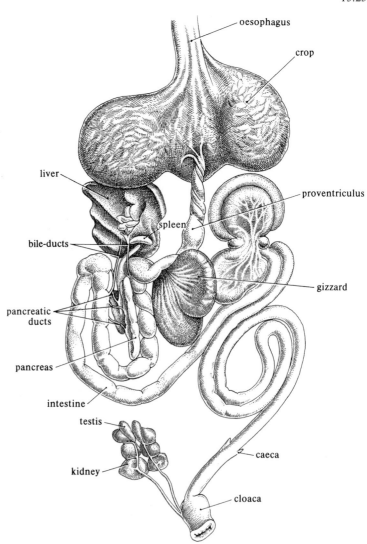

Fig. 15.34.   Dissection of pigeon. (After Schimkewitsch and Stresemann 1934.)

and there may also be microbial digestion here; the caeca are large in herbivorous birds. The cloaca is largely concerned with water conservation (Fig. 15.35). The rectum opens into a coprodaeum and this in turn receives a urodaeum, which is the terminal portion of the urinary and genital ducts. A final chamber, the proctodaeum, opens at the anus. The urinary products are made solid by absorption of water in the urodaeum and the walls of the other chambers serve a similar purpose. The bursa Fabricii is a blind sac with much lymphoid tissue, opening into the proctodaeum; its function is both to protect locally against infection and to produce lymphocytes for the bloodstream, hence it has been called a 'cloacal thymus'. Like the thymus, it is prominent in young animals and usually much reduced in the adult (see p. 290).

The large surface area, high temperature, and great activity of birds necessitate a high food intake, especially in the smaller types. This is made possible by rapid passage of food through the gut. Berries may pass through a warbler (*Sylvia*) in 12 minutes and a shrike (*Lanius*) is said to digest a mouse in 3 hours. Hens take only 12–24 hours over the most resistant grain. The amount of food taken per day may reach nearly 30 per cent of the body weight (6 g) in the very small goldcrest (*Regulus*) but is about 12 per cent in a starling (*Sturnus*) weighing 75 g. The food is very efficiently utilised in spite of the rapid passage. A young stork gains 1 kg for every 3 kg of fish eaten, whereas a young mammal would need 10 kg. Broiler chickens can put on 1 kg for every 2 kg of standard diet.

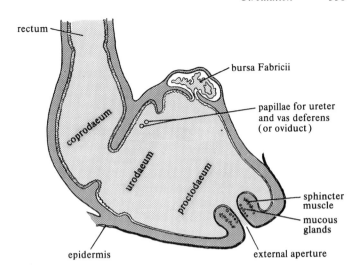

FIG. 15.35. Diagrammatic section through cloaca of pigeon. (After Clara, from Stresemann 1934.)

24. Circulatory system

Many of the features characteristic of birds depend on an efficient circulation, allowing of a high rate of metabolism, and hence a high and constant temperature. It is significant that the birds and mammals are the only vertebrates that have achieved complete separation of the respiratory and systemic circulations, making possible a high arteriolar pressure, which allows materials to reach the tissues rapidly.

The heart shows its sauropsidan characteristics clearly in that the ventral aorta is split to its base into aortic and pulmonary trunks. The former, arising from the left ventricle, curls round the pulmonary trunk to form a single right aortic arch. The heart has lost the sinus venosus; as in mammals no such extra chamber is necessary to step up the venous return pressure. The ventricles are large, especially the left. The right auricle and ventricle are separated by a single flap-like valve, the left side has bicuspid valves with chordae tendinae, somewhat as in mammals.

The embryonic ventral aorta becomes a short aortic bulb. The first and second aortic arches disappear and the third form the carotid arteries. The fourth right arch forms the aorta and the left disappears as do both fifth arches. The sixth form the pulmonary arches, losing connection with the aorta when the ductus arteriosus obliterates after hatching.

In the venous system there are renal portal veins but also a shunt, guarded by a muscular valve (Fig. 15.36) (Jones and Johansen 1972). When this is open the resistance of the kidneys and liver sends blood directly back to the heart allowing rapid circulation during exercise. The arteries and veins of the legs contain retia mirabilia allowing heat exchange with cooled venous

blood and so avoiding heat loss in the feet, especially in aquatic birds.

The size of the heart and rate of heart-beat vary with the size and activity of the bird, larger birds having in general relatively smaller and less rapid hearts. In a turkey the resting rate of beat may be less than 100 per minute, in a hen about 300, and in a sparrow nearly 500. Rates increase with exercise and a maximum of 1020/min has been recorded in a hummingbird. The large heart allows a greater stroke output and higher pressure than in mammals. In a turkey the systolic pressure reaches 400 mm Hg, one of the highest of all vertebrates, sometimes causing death by aortic rupture. The cardiovascular system is regulated much as in mammals. There are sympathetic accelerator and vagal

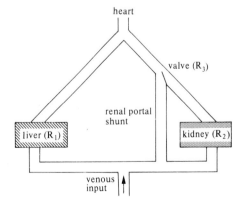

FIG. 15.36. Diagram to show the renal portal shunt in a bird. R_1, R_2, and R_3 represent the resistance to blood flow of the liver, kidneys, and renal portal valve, respectively. (Jones and Johansen 1972.)

depressor nerves and a complex sympathetic control of peripheral circulation.

The red corpuscles of birds differ from those of mammals in being oval and nucleated. They are large, $10–15\,\mu m \times 5–8\,\mu m$ and numerous $2.5–3.5$ million/mm^3. They carry a large amount of a haemoglobin that gives up its oxygen suddenly at a relatively high oxygen tension. The red corpuscles are smaller in actively flying birds than in the larger flightless ratites. Haemopoietic tissue is widespread in the young, but is restricted mainly to the marrow in the adult, although it may also be found in the liver and spleen. The white corpuscles are more numerous than in mammals. They include neutrophils laden with crystals, and thrombocytes, as well as the mammalian types of granulocytes, lymphocytes, and monocytes. Lymphatic tissue is dispersed rather than aggregated into nodes in most birds. Lymph glands are found in some aquatic and wading birds. There is a pair of lymph hearts in the sacral region of the embryo and these may persist in the adult.

There is a high basal metabolic rate and a temperature considerably higher than that of mammals, usually about 42 °C, reaching nearly 45 °C in some birds. The means by which this is kept constant in the absence of sweat glands are not known certainly. Heat loss is minimized by the absence of vascularized extremities, the feet being little more than keratin and collagen. The formation of the wing from large avascular surfaces has no doubt been a large part of the secret of the success of birds.

The air sacs may serve to conserve heat by providing an air cushion for the viscera, with perhaps the alternative possibility of losing heat in this way, by ventilation, when necessary. There is a system of direct arteriovenous connections in the feet, and elsewhere. These anastomotic regions have powerful muscles, whose contraction closes them and forces the blood through the capillary system. There must be a whole system of nervous pathways for the control of upward and downward temperature regulation, evolved independently of that found in mammals. At least one species (a nightjar) is known to hibernate, and certain hummingbirds and sunbirds, whose small size renders heat loss a serious problem, become temporarily poikilothermic at night.

25. Larynx, syrinx, and song

The larynx of birds is a small structure guarding the entrance to the trachea. The latter is often long and coiled and the tracheal rings are bony and complete. The voice is produced in the syrinx, a slight enlargement at the *lower* end of the trachea, containing a pair of semilunar tympaniform membranes with muscles that alter the tension on them and so the pitch of the sound. The apparatus is simple in many birds, but the muscles are very complicated in the singing birds and are especially large in the males. Many varieties of sound are produced, from simple cries appropriate to each sex to elaborate songs. The sounds produced are of two sorts, call notes used for warning and other social communication and song connected with territory, pair formation and nest building. In the song sparrow (*Melospiza melodia*) and other species the song is given in its full complexity by individuals that have had no opportunity of hearing others sing (Thorpe 1961). More usually some experience is necessary (Fig. 15.37). Exposure to song during the autumn, before the bird itself

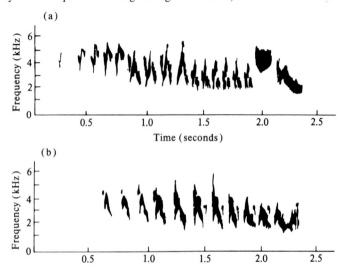

FIG. 15.37. Song of chaffinch. (a) Normal song. (b) An individual reared in isolation. (After Thorpe 1961.)

sings, is sufficient to allow later development of full song. When the song is largely learned by the young it may show considerable local variation. The voice is used for communication in various ways, including, in social birds such as rooks, the giving of warning and the frightening away of intruders. The language may include as many as fifteen sounds, used under different circumstances, in a chaffinch. The calls may provide information about such matters as intention to move, presence of food, avoidance of danger and, of course, sexual and parental themes. The type of sound and song varies to provide optimum propagation for the particular habitat, for instance the complex cries of mynah birds serve for social recognition among dense foliage, those of rooks are adequate in open country. Calls are an important factor in individual recognition, for instance in sea bird colonies of mates or parents. The more elaborate song of male birds is used to advertise the presence of an unmated male and as a threat to other birds invading the chosen territory (p. 382).

26. Respiration

The lungs use quite different principles from those of mammals, consisting of a series of passages through which the air sweeps. From these, gases pass by diffusion through air capillaries, which are not closed like the alveoli. The air sacs function as bellows blowing the air through the lungs. At inspiration air is drawn largely directly to the posterior sacs and then at expiration through the lungs to the anterior sacs, from which it is expelled at the next expiration (Fig. 15.38) (Schmidt-Nielsen 1979).

The lungs are rather small spongy organs, with little

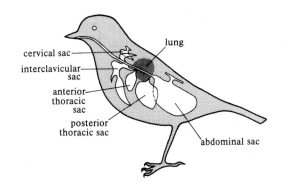

FIG. 15.38. A bird has several large, thin-walled air sacs. The paired lungs are small and lie close to the vertebral column. The main bronchus runs through the lung and has connections to the air sacs as well as to the lung. (Schmidt-Nielsen, K. (1972). *How animals work*. Cambridge University Press.)

elasticity. The air passes backwards in a large main bronchus running through the lungs and giving off branches to the lung substance, but continuing beyond to the inspiratory air sacs (Fig. 15.39). Each primary bronchus as it enters the lung gives rise to a number of secondary and tertiary bronchi which re-join posteriorly. Along the tertiaries are many openings, the atria, leading to vascularized air capillaries where the exchange occurs (Fig. 15.40). These air capillaries anastomose freely so that there are no end sacs.

The air sacs are thin-walled chambers, divided into two sets, the posterior, inspiratory and anterior, expiratory. The posterior air sacs are the abdominal and posterior thoracic, and they are filled by the air rushing into them through the primary bronchus. The anterior or expiratory air sacs include an anterior thoracic,

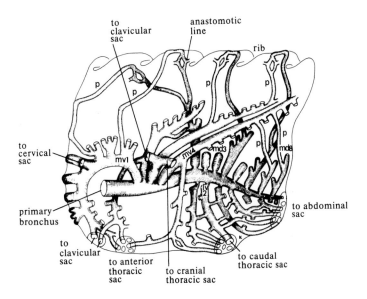

FIG. 15.39. Medial view of the right lung of the adult domestic chicken. lv2, lateroventral secondary bronchus; md3–8, mediodorsal secondary bronchi; mvl–4, medioventral secondary bronchi; p, parabronchus. (King 1966.)

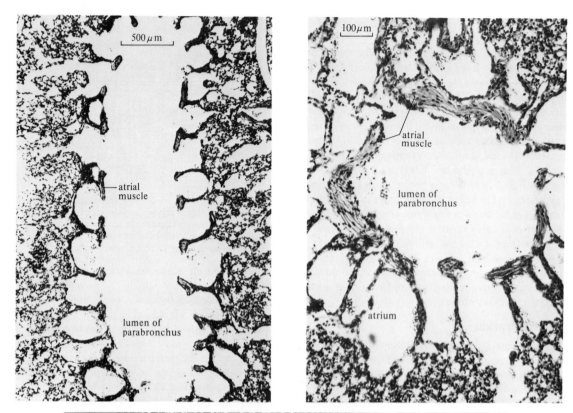

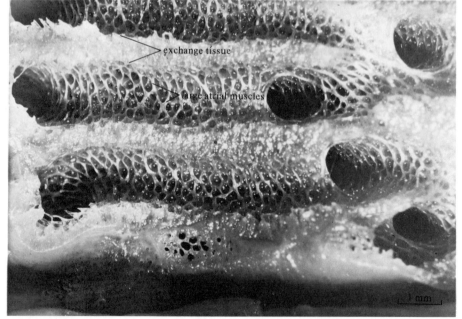

FIG. 15.40. Structure of tertiary bronchi of domestic chicken. Top left, longitudinal section of parabronchus (5-week chick). Top right, transverse section of tertiary bronchus. Below, scanning electron micrograph of tertiary bronchi, of adult chicken, showing cylindrical tubes that allow unidirectional air flow. (Photography by T.S. Fleming, University of Liverpool. From King, A.S. and Cowie, A.F. (1969). *Journal of Anatomy, London* **105**, 323–36.

median interclavicular, and cervical, and these often continue as spaces in the bones. (They communicate with the anterior secondary bronchi, but the sequence of the air flow is still somewhat uncertain.) A breath of pure oxygen appears immediately in the caudal air sacs but in the anterior ones only after a full respiratory cycle. Probably at inspiration some air passes directly to the posterior sacs, another part through the posterior secondaries to the tertiaries and indirectly to the air sacs. Then at expiration it again passes from posterior secondaries to the tertiaries but now to the anterior sacs and trachea (Fig. 15.41). The posterior sacs contain only 4 per cent of carbon dioxide but 17 per cent of oxygen, whereas in the anterior the figures are 7 per cent and 14 per cent. The interchanges with the blood therefore take place mainly as the gases pass forwards and the blood vessels are so arranged that the air richest in oxygen meets the blood just before it leaves the lungs (Fig. 15.42). This is not a full counter-current system but a system of cross-currents, and it allows the blood to become very fully oxygenated.

The metabolic rate is not in general higher in birds than mammals, though that of passerines is twice that of others. This respiratory mechanism is not necessary for flight, for bats manage without it (and even migrate). The advantage for birds is probably that the continuous flow and cross current allows the extraction of more oxygen, especially at high altitudes. Mice exposed to a pressure corresponding to 6000 m altitude can hardly crawl, but sparrows can fly. Birds have been seen flying in the Himalayas at heights where a man can hardly walk without oxygen. In some conditions especially in diving birds, the air may be passed backwards and

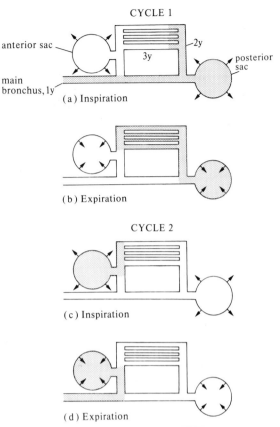

FIG. 15.41. Passage of a single inhaled volume of gas (stipple) through the avian respiratory system. It takes two full cycles to move the gas through its complete path. (Bretz, W.L. and Schmidt-Nielsen, K. (1972). *Journal of experimental Biology* **56**, 57–65.)

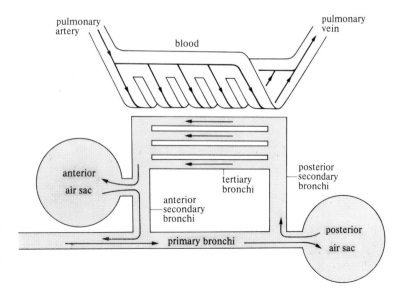

FIG. 15.42. Diagram of the probable pathway of air-flow in the lung of a bird and of its relationship to the cross-current flow of blood. (After drawing kindly supplied by K. Schmidt-Nielsen.)

forwards through the lungs several times, until all its oxygen has been used.

The mechanism by which the ventilation is produced is complicated and depends largely on the movements produced during locomotion. The upper surface of the lung adheres to the ribs, its lower surface is covered by a special membrane derived from the peritoneum and known as the pulmonary aponeurosis (Fig. 15.43). This is connected with the ribs by costopulmonary muscles. The floor of the thoracic air sacs, which lie below the lungs, is also covered by a fibrous membrane, the oblique septum, but the walls of the remaining air sacs are very thin. Quiet respiratory movements are produced by the intercostal (inspiratory) and abdominal (expiratory) muscles, acting upon the thoracic and abdominal cavities so as to enlarge and contract the thorax, drawing air in and out of the air sacs, through the lungs. During flight the movements of the pectoral muscles provide the ventilation, the sternum moving towards and away from the vertebral column.

The respiratory system occupies as much as 20 per cent of the volume of the body (5 per cent in man). The air sacs mostly lie dorsally, allowing a low centre of gravity. Besides their respiratory functions they ensure cooling and regulation of water loss by panting. At low temperatures there is increased respiration to provide heat but water loss is avoided by a counter-current heat exchange in the nasal turbinals producing re-condensation of water and saving of heat (Schmidt-Nielsen, Hainsworth, and Murrish 1970; see Lasiewski 1972). Birds have various ways of keeping warm, including special winter plumage, ruffling of the feathers, heat exchangers along the legs (p. 351), and clustering together in flocks (starlings). Young birds are poikilothermic and those of some species can survive in a lethargic state for several days of bad weather at little above air temperature (swifts, *Apus*).

27. Excretory system

The kidneys are, of course, metanephric and are relatively large, elongated, and lobulated (Fig. 15.44). They are provided with venous blood by the renal portal veins and arterial blood from the renal arteries. The arrangement is essentially as in amphibia and reptiles, with the renal arteries supplying the glomeruli while the tubules are supplied both by efferent glomerular arterioles and branches of the portal veins, which break up into inter-lobular vessels, whence the blood is collected into a central intra-lobular vein. It is not certain, however, exactly how the system operates, and it is possible that much of the blood flow is directly from the renal portal to the renal veins, making little contact with the tubule walls.

The excretory system is highly specialized for water-saving. For this purpose the end product of nitrogenous metabolism is the relatively insoluble uric acid, synthesized in the liver, probably from ammonium lactate. After excretion by the kidney the uric acid precipitates in the collecting tubules as whitish granules. There is no urinary bladder in the adult bird. More soluble excretory end substances, such as urea, would reach toxic concentrations. The glomeruli are numerous and small with few loops. The urinary tubules effect a considerable concentration of the urine by means of long loops of Henle. The viscous fluid that enters the urodaeum then passes up into the coprodaeum and large intestine, where further water is resorbed, and the mixed faeces and urinary products are then excreted as the characteristic semi-solid white guano. The water-conservation system is certainly very effective, and some desert-living

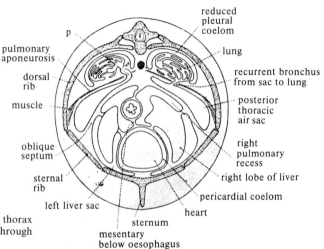

Fig. 15.43. Diagram of transverse section through the thorax of a bird. *p.* ex-current passage from lung to air-sac through pulmonary aponeurosis. (From Goodrich.)

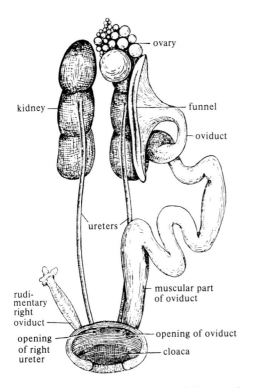

FIG. 15.44. Female reproductive organs of domestic fowl. (From Thomson 1923.)

season than it is in the non-breeding period, when it contains only spermatogonia.

The testis consists of coiled tubules of the usual type, joining to form a long epididymis and vas deferens, opening into the urodaeum by an erectile papilla that is the only copulatory organ of most birds. During copulation the proctodaea of male and female are everted and pressed together, so that the sperm is ejaculated direct into the female urodaeum and finds its way up the oviduct. A definite penis (and also clitoris) is found in ratites, anseriformes, and a few other birds.

The provision of material sufficient for the development of a warm-blooded creature is, of course, made possible in birds by the extremely yolky eggs, so large that they allow room for development of only one ovary, nearly always the left (Fig. 15.44). The right ovary remains present as a rudiment and if the left is destroyed by operation or disease the right is able to differentiate, but then forms not an ovary but a testis. Complete sex reversal can thus occur, at least in some races of domestic fowl, and the transformed bird may acquire cock plumage and tread and fertilize hens (Fig. 15.45). Sex reversal rarely, if ever, takes place in the opposite direction. We must suppose that there is some switch over in the balance of male and female determin-

birds are said to be able to survive for many weeks without water. In this respect the birds have freed themselves from the original aquatic environment to a remarkable degree. The control of urine production is as in mammals by variation both of the glomerular filtration rate and of tubular resorption, producing up to a 100-fold volume range. Sodium and chloride are also resorbed from the tubules. Production of a urine hyperosmotic to the blood is facilitated by a counter-current system as in mammals. Solutes leave the water-impermeable ascending loop of Henle and accumulate in the collecting tubules lying parallel to them. This provides a gradient throughout the medulla by which water is withdrawn maximally at the distal ends of the tubules. Birds also excrete salt by nasal glands, especially well developed in marine forms.

28. Reproductive system

The gonads of birds produce steroid hormones under the influence of two gonadotropic hormones (p. 364). Reproduction is seasonal, the gonads of both sexes becoming active only for part of the year, usually under regulation by changes in illumination. The weight of the testis is as much as 1000 times greater in the breeding

FIG. 15.45. Secondary sexual characters of the fowl (*Gallus*). (a) normal; (b) castrated; (c) cock with implanted ovary and hen with implanted testis. (After Zawadowsky.)

ing processes, taking place relatively early in the case of normal definitive males but later on in life also in 'females', so that all birds become potentially 'male' at the end of their life.

Of the large number of oocytes only few ripen to make the enormous follicles. After each follicle has burst it quickly regresses; there is no corpus luteum. A bird's ovary produces oestrogens, progesterone, and androgens probably from the follicle cells in the early stages of maturation. The lipids and proteins of the yolk are mainly synthesized in the liver and transferred to the egg through the follicle cells (Bellairs 1971). Oestrogens stimulate this synthesis but the uptake is dependent on pituitary gonadotropin. Ovulation depends on the cyclic release of luteinizing hormone (LH).

The egg is taken up by the ciliated and muscular funnel of the left oviduct, and passes down a tube with circular and longitudinal muscles and a glandular, ciliated mucosa. The albumen of the egg is produced by long tubular glands, opening to the lumen. The oviduct has various parts, the upper secreting mainly albumen, the lower producing the shell, and the lowest mucus, to assist the act of laying. The blue background colour of the egg (oocyanin) is produced during shell-formation in the upper part of the tube; spots of red–brown ooporphyrin are added lower down. The pigments are derived from the bile, ultimately from haemoglobin.

As much as a third of the weight of calcium in the whole skeleton is needed for the shells of the two eggs laid by a pigeon. A reserve is collected as the ovarian follicles mature. The oestrogen they produce increases the uptake of calcium from the food and stimulates its deposition in the bones. After ovulation the oestrogen level falls, the calcium is mobilized from the bones, probably by parathyroid hormone (p. 365), and its concentration in the blood becomes very high, until used by the eggs.

29. The brain

The brain is larger relative to the body in birds than in any other vertebrates except mammals, and there is no doubt that one result of the high temperature has been to allow opportunity for an elaborate nervous organization and complicated behaviour. There are considerable differences in the development of the parts in various birds, for instance, the forebrain is especially large in the rooks and crows (*Corvus*) and in the parrots, whose behaviour also shows signs of outstanding 'intelligence'.

In the spinal cord the most characteristic feature is the relatively small size of the dorsal funiculi, and their nuclei in the medulla are also small but ascending pathways pass on to the tectum and the thalamus (Karten 1963). However, the sense of touch is less well developed over the body than it is in mammals, perhaps less than in reptiles. No doubt movement of the feathers provides impulses leading to reflex actions, but it is not surprising that the loose covering does not allow elaborate organization of the sense of touch. The finer senses of birds are restricted to the eyes, ears, and bill. On the other hand, there are large spinocerebellar tracts, presumably proprioceptive and concerned with the delicate adjustments necessary for flight. The spinal cord is controlled by large efferent tracts from the brain, including cerebellospinal, vestibulospinal, and tectospinal pathways. There is a small tract running from the forebrain directly to the spinal cord. This invites comparison with the mammalian pyramidal tract, but it seems likely that tracts from the midbrain are of greater functional importance. Certainly the mechanisms of flying and walking are unimpaired by removal of the forebrain. After transection of the spinal cord a pigeon cannot stand but will make walking movements if supported in a wheeled carriage and can even fly.

The cerebellum is large and folded, with a central body and lateral flocculi (Fig. 15.46), and is probably involved in the precise timing and control of movement in all planes of space during flight. Signals from various parts of the head and body activate cells in different parts of the cerebellum in a somatotopic manner.

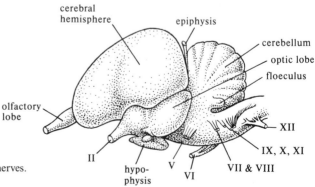

FIG. 15.46. Brain of a duck (*Anser*). II–XII, cranial nerves. (After Butschli and Ihle.)

Besides large spinocerebellar and vestibulocerebellar pathways there are also tectocerebellar and striocerebellar tracts, the latter perhaps conducting in both directions. The effect of the cerebellum on other parts of the brain is exercised through cerebellar nuclei, the cells of which give origin to the cerebellospinal tract.

The optic tracts are completely crossed and end partly in the midbrain tectum, as in lower vertebrates. However, a considerable portion of the optic tract fibres pass to the thalamus and hypothalamus. The tectum has an elaborate layered structure on which the retina and visual field are topographically projected. Impulses pass back from the tectum to the retina via the isthmo-optic nucleus at the base of the midbrain and a special part of the optic tract (see Crossland 1979). The optic lobes also receive ascending fibres from the trigeminal nuclei, the spinal cord and cerebellum and descending ones from the cerebrum (topographically organized). Their efferent pathways run to the oculomotor nuclei, to the cerebrum via a topographically organized path to the nucleus rotundus of the thalamus, and to the medulla and spinal cord. Evidently they play a large part in correlating visual with other afferent impulses.

The tectal cells are arranged in columns and the field sizes increase with depth. They are sensitive to movement. Edge detectors, which are common in the retina are rare in the tectum. The cells of the isthmo-optic nucleus have small receptive fields and wide inhibitory surrounds. They respond to small moving stimuli and their axons end on the amacrine cells of the retina. Their action may either increase or suppress the responses of retinal ganglion cells. The pathway may serve to minimize disruption of sensory processing by the animal's own movements (von Holst's 'efference copy', 1973), especially turning off one eye. Ablation of both isthmo-optic nuclei impairs the ability of a pigeon to peck at grain.

The thalamus is large and its dorsal part well differentiated into nuclei. Its dorsolateral region receives fibres from the optic tract and sends them to the 'Wulst' (German for swelling or ridge) of the hyperstriatum. Its cells show responses to small visual fields. It may therefore perhaps be compared to the mammalian lateral geniculate body. Other thalamic nuclear projections come from tactile, pain, temperature, and perhaps auditory sources. The large tracts from the thalamus to the forebrain, probably conduct in both directions.

The cerebral hemispheres are much larger than any other part of the brain and show an exaggeration of the condition found in reptiles (Fig. 15.47A). The ventrolateral portions are enormously developed, and usually called the 'corpus striatum', whereas the medial ventral walls are thin and the pallium is quite small, thin, and not folded. The whole plan of the cerebral hemispheres is thus different from that of mammals, and consists of large masses of tissue rather than extended sheets. Nevertheless, there has been the same tendency for the visual, auditory, and somatosensory systems to be projected forwards into the cerebrum: each has its own subdivision in the outer part of the cerebrum. These same regions, which comprise the bulk of the cerebrum, in turn send projections back to the thalamus, midbrain, medulla, cerebellum, and spinal cord. The whole arrangement calls into question the wisdom of referring to them collectively as the 'corpus striatum' (Karten 1979; Webster 1980). It seems likely that only the deepest parts (the 'palaeostriatum') of the cerebral hemispheres warrant this name. The more superficial parts are, in spite of their appearance, probably better thought of as 'cortex' (Fig. 15.47B). Unfortunately, these outer subdivisions still bear names that betray their origin in the older and now largely discarded way of looking at the avian forebrain: the region dealing with auditory information is known as the 'neostriatum', and those related to vision the 'hyperstriatum' and 'ectostriatum'.

There are two distinct visual pathways to the cerebrum (Fig. 15.47C). One goes topologically via the tectum and wide-field cells of the nucleus rotundus to the ectostriatum. The other passes through the narrow-field cells of the dorsolateral thalamus to the hyperstriatum (Maxwell and Granda 1979). The dorsal part of the hyperstriatum, called the 'Wulst', receives the thalamic fibres in a retinotopic order. Its cells are arranged in six layers, remarkably like those of the mammalian striate cortex. In pigeons and owls many Wulst cells have been found to be activated by a moving stimulus, especially a bar or edge with a particular orientation. They are thus very like mammalian simple and complex cortical cells.

The efferent projections of the Wulst, like those of the cortex, pass to the thalamus and tectum. They also run to a region around the ectostriatum known as the belt (Fig. 15.47D). This also receives fibres from the tectofugal pathway and the two visual pathways thus converge here. The axons of the belt cells proceed to the neostriatum (mainly an auditory area) and also to the archistriatum, an area of efferent cells to the tectum and brainstem, held to be similar in some ways to those of the deeper layer (V and VI) of the mammalian cortex (Fig. 15.47A).

There are thus many similarities between the avian and mammalian visual systems. Neither is fully understood but we begin to see how the two pathways may combine to produce higher-order 'perceptual' events. The ectostriatal belt cells may be tuned to respond to different combinations of signals. The wide-field tecto-

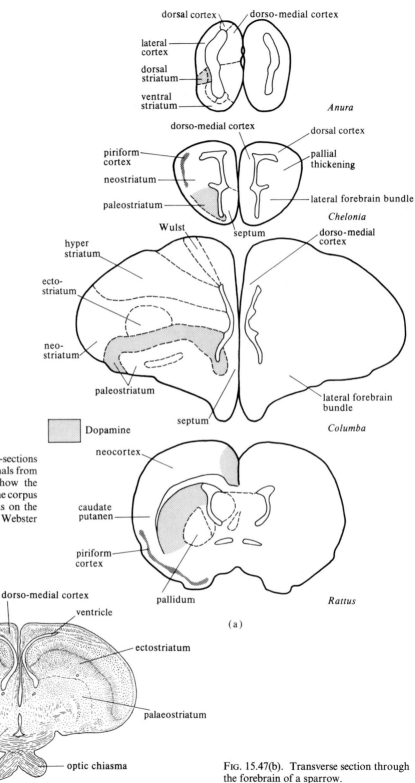

FIG. 15.47(a). Diagrams of cross-sections through the telencephalon of animals from four groups of vertebrates to show the probable relationship of parts of the corpus striatum. The finely stippled areas on the left contain dopamine. (After Webster 1980.)

FIG. 15.47(b). Transverse section through the forebrain of a sparrow.

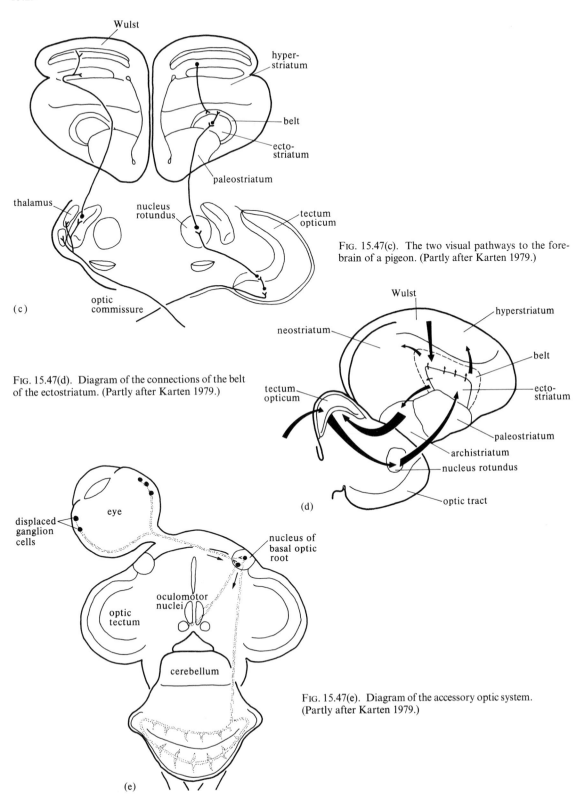

FIG. 15.47(c). The two visual pathways to the fore-brain of a pigeon. (Partly after Karten 1979.)

FIG. 15.47(d). Diagram of the connections of the belt of the ectostriatum. (Partly after Karten 1979.)

FIG. 15.47(e). Diagram of the accessory optic system. (Partly after Karten 1979.)

cortical pathway may encode such dimensions as wavelength and orientation, while the narrow-field thalamofugal units indicate the location, contrast, orientation and other features of edges or bars. Thus a cell might respond maximally to small blue bars in a certain part of the field. Such coding would efficiently allow for specificity to some dimensions and generalization to others. A few wide-field units could combine with many spatially specific ones (Maxwell and Granda 1979).

A further visual pathway is the accessory optic system, concerned with rapid eye movements (Fig. 15.47E) (Karten 1979). Large 'displaced' ganglion cells of the retina, with wide receptive fields sensitive to movement and direction, send large medullated axons to the nucleus of the basal optic root, whose cells project to the oculomotor nuclei and to the cerebellum. This ancient system, present in all vertebrates, is one of the first central visual structures to develop in a bird (Karten 1979). The whole visual system of birds seems to be organized for response to movement, not surprisingly since they move so rapidly themselves. The 'problems' that a bird's brain mostly has to resolve involve quick decisions as to where to go next.

30. Functioning of the brain

There is only a small direct pathway from the forebrain to the spinal cord, corresponding to the pyramidal tract of the mammals. Nevertheless, the forebrain is shown by extirpation experiments to contain areas responsible for distinct functions such as eating, locomotion, copulation, nest building, and care of the young. For example electrical stimulation of the hyperstriatum of a pigeon produces a stereotyped sequence of raising the head and then walking about, bowing and cooing. This region is especially concerned with reproductive behaviour and after implantation of a pellet of testosterone here a male pigeon attacks others and shows courtship and copulation with females. Lesions of either the tectum or nucleus rotundus or 'Wulst' of the cerebrum may impair the capacity of a pigeon to learn a visual discrimination. Impulses from the head and especially the beak reach via the trigeminal nerve and quintofrontal tract to the forebrain, which is important in the regulation of feeding.

Complete removal of both hemispheres does not reduce a pigeon to a helpless state. The animal can still maintain its temperature and its balance and can feed itself if the food is placed near to it. It never flies spontaneously but if thrown in the air will fly and land successfully. It may show a lack of activity, remaining inert for long periods, and then become aimlessly restless for a while. Evidently the normal balance of excitation and inhibition has been upset. Such birds are not blind but show some deficiencies of vision. They cannot fight, court, mate, build a nest, or feed the young.

These observations on the functions of the brain of pigeons may not be applicable to birds in general. The forebrain is larger in many other birds than in the pigeon, and there is some evidence that in the parrots movements and even 'phonation' can be elicited by electrical stimulation of the cerebrum, which is especially large. Removal of one particular area is said to lead to disturbances of 'speech'.

It seems, therefore, that the large masses of nerve cells comprising the cerebrum are concerned in some way with the elaboration of the more complex programs of behaviour. This is a very vague statement, but is the best that we can give at present. It may be that further investigation of the reciprocal actions of the cerebrum and thalamus will show whether the essentials of their action consist in some reverberating or scanning systems and whether these actions are at all similar to those in the forebrain of mammals. The fact that the cerebrum consists of solid masses of tissue suggests that the arrangement does not depend, as does the mammalian cerebral cortex, on the projection of patterns of excitation onto an extended surface. It is interesting that electroencephalograms recorded from a bird's brain (like those of fishes and amphibia) are rather similar to those of mammals in spite of the differences in basic structure. They are small and irregular at 20–40 Hz when the bird is alert, but large and rhythmic in deep sleep at 2–4 Hz, with bursts at 30–60 Hz for a few seconds, probably representing paradoxical sleep.

31. Learning in birds

Birds are usually said to show more stereotyped patterns of instinctive behaviour than primitive mammals ('fixed action patterns'). Once they have embarked on a line of action, even a complex one like nest-building, they are supposed to pursue it in a given manner, without ability to adapt themselves to unusual happenings. Watching the exploratory behaviour of a robin or a tit it is difficult to feel that the existence of this difference is adequately proved, but whatever distinction exists is presumably a reflection of the differences between the functions of large masses and spread out sheets of nervous tissue. The organization of the brain of the bird allows for rapid responses that perhaps involve 'knowledge' of abstract general features of space and time. Their performance in mazes and puzzle boxes is at least comparable to that of most mammals, but having learned one maze they can only slowly change to another.

The wide ranging life and variety of conditions encountered make special demands for adaptability and the capacities for learning are greater than in any other vertebrates except mammals (see Thorpe 1963). By operant conditioning, by pecking at a key in a Skinner box, pigeons can learn to discriminate between a wide variety of visual and auditory stimuli (see Delius and Emmerton 1979). They even show a capacity to learn to distinguish between figures on the basis of whether they are symmetrical (Fig. 15.47F). This suggested to the experimenters that 'when attempting to understand the function of central visual structures we must unfortunately also consider what, by human standards, are cognitive processes' – one may ask 'Why unfortunately?'

The homing abilities of birds involve very elaborate forms of learning as do the development of song (p. 352). Crows and jackdaws can learn quite complex tasks such as the distinction between short sequences of notes in spite of changes of pitch, tempo, or timbre. They can also learn to choose between boxes on a basis of the number of objects lying in front of them – so they can be said to count at least as well as rodents, that is up to five or six.

32. Imprinting

Newly-hatched nidifugous birds such as hens, geese, or partridges, learn to follow the first object seen or heard, usually of course the parent species (Sluckin 1972). This capacity is absent from ducks after removal of a particular small part of the hyperstriatum and during imprinting there is increased incorporation of uric acid with RNA in this region (Horn, McCabe, and Bateson 1979). Development often consists in increasing specificity of response. Chicks at first peck at anything, but later only at objects whose colour and shape has been associated with food. The capacity to imprint is at a maximum a short time after hatching and then wanes and may be replaced by a fear response of retreat from any strange object or sound. Imprinting differs from other forms of learning in several ways. It occurs at a short 'critical period', does not require any reinforcing reward, and it is largely irreversible. The 'imprint' may

in fact produce a 'courtship fixation'. A peacock in a zoo would only copulate with Galapagos tortoises in whose cage it had been raised. Phenomena similar to imprinting are found in precocial mammals such as ungulates or dogs and occur also in altricial species including man.

33. Releasing stimuli

There are many signs that a suitable initial stimulus sets off in the bird a whole train of behaviour, organized from within. On the other hand, inappropriate stimuli may sometimes set off a reaction, as if the 'keys' to these cerebral 'locks' were not very elaborate or specific. A robin held in the hand may burst into song; cock ostriches frightened by an aeroplane fall to the ground in their characteristic sexual display. Birds frightened or disturbed may proceed to the actions of bathing, preening, feeding, or drinking, performed in a ritual and cursory manner for a long time. Such 'displacement activities' show that the organization of the bird's nervous system, like that of a mammal, provides for some strange deviations, whose study may reveal much about the method of working of the brain.

Many complex forms of behaviour are responses to only limited parts of the natural stimulus situation. Thus when the models shown in Fig. 15.48 were towed above certain young birds, only the models marked with a plus induced escape reactions: apparently the configuration of the short neck is the essential feature. Much of the elaborate social life of birds depends on such 'sign stimuli' displayed by one bird (the 'actor') and serving as 'releasers' setting off particular actions or trains of action in another bird (the 'reactor'). Many of the elaborate forms of display evolved by birds (p. 379) are releasers of this sort (Fig. 15.49), and structures and actions on the part of the young release the appropriate behaviour of the parent. The red breast of the robin (*Erithacus*) is the agent that releases attacks by other birds (Fig. 15.50). Similar phenomena are known in fishes and other vertebrates (p. 186), and it remains to be shown whether they can be attributed to any single or

(a)

(b)

FIG. 15.48. A, Models used by Lorenz and Tinbergen. Small birds reacted with escape movements to the models marked +. B, The model induced escape reactions when towed to the right ('hawk') but not when towed to left ('goose'). (From Tinbergen.)

FIG. 15.49. Pintail ducks (*Anas acuta*). Males displaying dark brown feathers of neck with white bands on either side. (From Tinbergen, after K. Lorenz.)

particular neural basis. Releasers are sounds, postures, or structures that have been evolved for the purpose of communication. Some of them are ritualized developments of previously functional acts such as movements preliminary to flight or attack or copulation. There may be between 5 and 20 distinct signals and responses in the repertoire. They may be combined and the responses produced vary considerably. Bird communication is not by any means always characterized by agreement.

34. Endocrine organs

34.1. Pituitary

In the pituitary the pars distalis lies away from the brain and is divided into cranial and caudal parts. (Fig. 15.51) (Kobayashi and Wada 1973). The median eminence is a morphologically separate entity, also divided into two parts, only the anterior receiving fibres of the supraoptic–hypophysial tract, whose neurosecretory granules stain with Gomori's aldehyde–acid fuchsin (Fig. 15.52). These fibres also reach to the distinct pars

nervosa and the terminals of this system release two hormones into the systemic circulation, arginine vasotocin and mesotocin, a form of oxytocin. These assist in control of water-balance, heart-rate, and reproduction. Arginine vasotocin is antidiuretic and vasodepressor, mesotocin has the opposite actions, diuretic and vasopressor. Both cause contraction of the oviduct but vasotocin is the more active oviposition hormone. Mesotocin also has other actions, for instance increasing fatty acids and sugars in the blood.

Both parts of the median eminence receive Gomorinegative fibres from nuclei in the hypothalamus. They contain granules, smaller than those in the supra-optic hypophysial tract, presumably of pituitary releasing factors. Their terminals lie near capillaries of the primary plexus of a system of hypophysial portal vessels. These proceed over the outside of the eminence to the distal lobe where they have been shown to control secretion of some of the seven types of hormone. The distal lobe contains the following proteins: somato-

FIG. 15.50. The red tuft of feathers is attacked by male robins (*Erithacus*) holding territory, but the complete juvenile bird (without any red feathers) is left alone. (After Lack, D. (1943). *The life of the robin.* Witherby, London.)

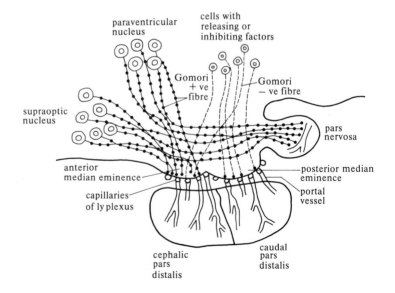

FIG. 15.51. Diagram of sagittal section of the median eminence–hypophysial region of a bird. (Kobayashi and Wada 1973.)

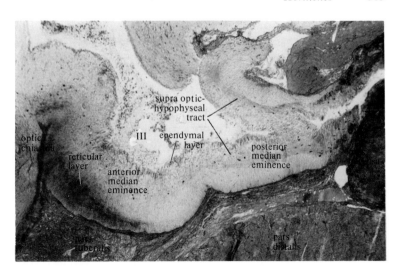

FIG. 15.52. Pigeon, *Columba*. Sagittal section of the hypothalamo–hypophysial region. (Kobayashi, H., Matsui, T., and Ishii, S. (1970). *International Review of Cytology* **29**, 281–381.)

tropin (growth hormone STH) and prolactin, thyrotropin (TSH), and two gonadotropins, luteinizing hormone (LH) and follicle-stimulating hormone (FSH). There are also two polypeptides, adrenocorticotropic hormone (ACTH) and melanotropic hormone (MSH), which in other vertebrates is produced by the pars intermedia. The function of the MSH is uncertain, but it is the first pituitary hormone to appear, on the fifth day of incubation. The other hormones have essentially the same actions as in other vertebrates. Prolactin has the effect of producing brood patches and broodiness, including in phalaropes even the males, who alone incubate the eggs. Prolactin also stimulates the production of pigeon's milk by the crop in both sexes.

34.2. Adrenals

In the adrenal glands the medullary cells may be mingled with the cortical ones or lie in separate islands. The cortical tissue does not show the characteristic layers seen in mammals. Corticosterone is the main adrenal corticoid in birds and there is some aldosterone and small amounts of others. The adrenals of marine birds are very large and their corticosterone is the chief agent controlling the supraorbital salt gland (p. 357).

34.3. Thyroids

The paired thyroids of birds produce tri-iodothyronine and thyroxine. There are several signs of great activity. The content of total ^{127}I and of T_4 ^{127}I after isotope injection is higher than in mammals. Iodine uptake is rapid and storage is high. On the other hand the plasma proteins have a low binding power for the hormones, whose half-life in the blood is therefore short. This rapid turnover is no doubt connected with the high tempera-

ture and metabolism and there is a rise of temperature after injection of thyroxine. The gland also controls growth and moulting (p. 325) and perhaps influences migration and reproduction.

34.4. Parathyroids

The parathyroids are large, their parathormone is very active and probably plays a major part in control of release of calcium from the skeleton during egg-laying. The ultimobranchial glands develop from the fifth and sixth branchial pouches and in many birds remain outside the thyroid. This situation has been much used to investigate the origin and properties of their hypocalcaemic polypeptide hormone, calcitonin.

34.5. Pancreas

The pancreas produces relatively less of its hypoglycaemic hormone, insulin, than in mammals, but up to ten times more glucagon, which mobilizes sugar from the liver and raises its level in the blood. The removal of the pancreas therefore has complicated effects, not that of classical diabetes. Glucagon also has a powerful lipolytic effect in birds and these actions of the pancreatic hormones are no doubt another feature of the need for very active metabolic interchange.

34.6. Thymus, bursa of Fabricius, and immunity

The two types of immune responses of birds are largely separately controlled, cellular immunity by the thymus, humoral by the cloacal bursa (p. 350). After neonatal thymectomy there is a reduction of small lymphocytes but not of immunoglobulins. These cannot be properly produced however after bursectomy. Both organs no doubt direct the differentiation of lymphocytes into immunologically competent cells, able to react to antigens in the lymph nodules and spleen.

35. The eyes

Birds depend more on their eyes than on the other senses; they are perhaps more fully visual than are any other animals. The eyes are extremely large: those of hawks and owls, for instance, may be absolutely larger than in man. The shape is not spherical, the lens and cornea bulge forwards in front of the posterior chamber, this form being maintained by a ring of bony sclerotic plates (Fig. 15.53). In most birds the whole eye is thus broader than it is deep, but in those with very acute sight

it is longer, and in some eagles and crows becomes almost tubular. The great distance between lens and retina allows broadening of the image, thus improving the fine two-point discrimination that is needed by these diurnal birds.

The shape of the back of the eye is such that 'the retina lies almost wholly in the image plane, so that all distant objects within the visual angle are sharply focused on the photosensitive cells, whereas in the human eye this is only true of objects lying close to the optic axis' (Pumphrey 1948a; Fig. 15.54). Accommodation is effected by changes of both the lens and the cornea. The shape of the lens, and especially the curvature of its anterior surface, are altered by the pressure upon it of the ciliary muscles behind. These, like the iris muscles, are striated, presumably allowing for the quick accommodation necessary in a rapidly moving bird, though it must not be forgotten that these muscles are also striated in lizards. The ciliary muscle is characteristically divided into 'anterior' and 'posterior' portions, the muscles of Crampton and Brücke (Fig. 15.53). The latter draws the lens forward into the anterior chamber so that since the shape of the eye is fixed by the sclerotic plates, the lens becomes more curved and hence accommodated for near vision; contraction of the iris sphincter assists in the process. Crampton's muscle is so arranged as to pull on the cornea, shortening its radius and further assisting in accommodation.

This double method of active accommodation for near vision is most fully developed in diurnal predators, such as the hawks, less so in night-birds. In aquatic birds Crampton's muscle is reduced, and the cornea is of little importance in image formation. Special arrangements are found in diving birds, for instance in the cormorants

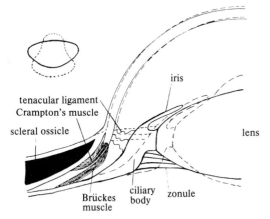

FIG. 15.53. The mechanism of accommodation in a bird's eye; the positions during near vision are shown dotted. Inset. The lens of the cormorant's eye at rest (full line) and fully accommodated (dotted line). (From Pumphrey 1948a, after Franz and Hess.)

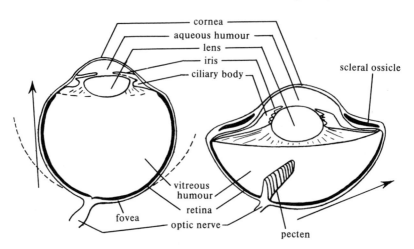

FIG. 15.54. Diagrams of right eye of man and left eye of swan to show the difference in shape. The position of the image plane in man is shown dotted; it lies behind the retina except near the centre. The arrows point forward. (From Pumphrey 1948a.)

Brücke's muscle is large and there is a very powerful iris muscle, which assists the ciliary muscles to give the great change in shape of the soft lens, allowing accommodation of 40–50 diopters (about 10 in man). The kingfishers are said to possess an amazing arrangement of double foveas, placed at different distances from the lens, so that as the bird dives under water the image is transferred from one fovea to the other without any change in the dioptric apparatus. These details of the visual system show, like so many other features of bird anatomy, how readily the structure conforms to special habits of life.

The retina of day-birds consists largely but not wholly of cones (Fig. 15.55); these animals are more fully diurnal than is man. The high resolving power and hence high powers of discrimination and of movement detection depend on the great density of the cones, as many as 1 million to each square millimetre in the fovea of a hawk, three times denser than in man. Nocturnal birds, on the other hand, have retinas composed mainly or completely of rods, and the differences between the behaviour of these two types of eye, found in birds as in mammals, have been a powerful support for the duplicity theory of vision. There are usually one or more areae, regions of the retina consisting of tightly packed receptors. In birds that live on the sea, in the desert, or other open spaces the area often has the form of an elongated horizontal band (Fig. 15.56), whereas in tree-living birds it is circular. Some birds have two areae, a central one in the optic axis and a second placed on the temporal surface of the eye, so that the image of objects in front of the head falls on the temporal areae of both eyes. This arrangement is common in birds that follow moving prey (shrike, *Lanius*) or for some other reason require accurate perception of distance (swallows, hummingbirds; the former feed their young on insects caught on the wing). The density of the cones is so high in diurnal birds, even outside the areae, that they probably obtain a good detailed picture in all directions. They do not, therefore, scan the world with the central area of the retina as we do; indeed, the eyes move relatively little. Instead the bird is able to detect very small movements anywhere in its surroundings. The bird's-eye view usually lacks stereoscopic solidity and it is possible that in compensation for this the animals appreciate distance by movements of the striated intrinsic eye muscles. The familiar cocking of the head of a bird before pecking may be its means of judging distance.

As in man there is often within the central area of the eye a fovea or pit, and in many birds the sides of this pit are steeply curved (Fig. 15.57). Walls (1937, 1942) has suggested that since the vitreous humour and retina differ in refractive index this curvature serves to magnify

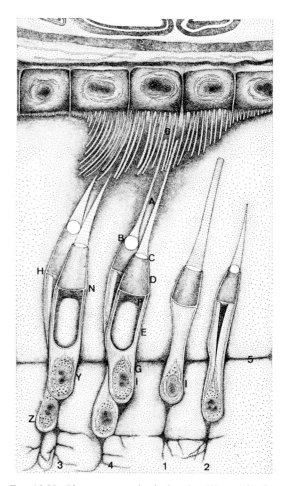

FIG. 15.55. Photoreceptors in the bunting (*Zonotrichia leucophrys gambelii*); 1, rod; 2, single cone; 3, double cone; 4, double cone with rodlike outer segment; 5, external limiting membrane; 6, pigment epithelium; A, outer segments; B, oil droplet; C, connecting cilium; D, ellipsoid; E, paraboloid; F, basophilic inclusion bodies; G, basophilic perinuclear granules; H, chief cone; N, accessory cone; I, Z, Y, nuclei. (After Hartwig, H.G. (1968). *Zeitschrift für Zellforschung Mikroskopische Anatomie* **91**, 411–28.)

the image and increase acuity which is often said to be very high in birds. It has been pointed out, however, by Pumphrey (1948b) that the visual acuity of birds is not much better than our own. What they are good at is detecting minute movements. The 'convexiclivate' fovea may in fact help this by actually introducing an aberration that disturbs the picture (Figs. 15.58 and 15.59). As the image moves across the retina it would be maintained in focus on the extrafoveal retina, as it passed across the fovea it would be out of focus and by keeping it so the bird can lock on it. The birds thus seem to have better vision than us because they can find objects better and keep them in sight. Foveas with steep

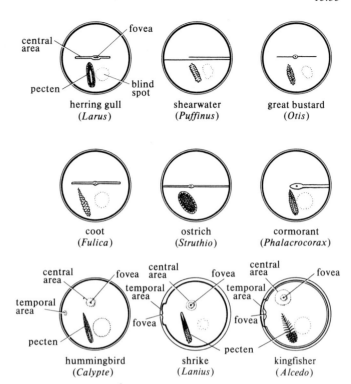

FIG. 15.56. The appearance of the retina of various birds as seen with an ophthalmoscope through the pupil. (From Pumphrey 1948*a*, after Wood.)

sides are found in birds of prey, kingfishers, and others that have very high powers of detecting movement; a similar but less pronounced arrangement is found in fishes and reptiles. It is probable that the primitive functions of the eyes were fixation and detection of movement, rather than resolution of detail and recognition of patterns. Some birds have one convexiclivate and one flatter fovea (Figs. 15.56 and 15.57), the latter being on the temporal surface of the retina and used in binocular vision. The fovea is also flat in the retinae of primates with binocular vision; evidently the optical errors of a curved fovea cannot be tolerated where there is fusion of the two retinal images. However, the temporal fovea is very deep in eagles and it is uncertain whether Wall's theory or Pumphrey's is correct.

The retina is thicker than in mammals and a great deal of computation goes on there. The inner nuclear and plexiform layers are especially thick in diurnal birds because each of the numerous cones has one correspond-

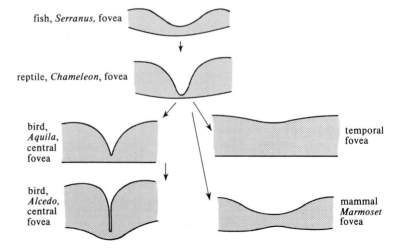

FIG. 15.57. Forms of the fovea in various vertebrates, showing the development from moderately to sharply convexiclivate types in foveas adapted for detection of movement, but flattening of the fovea where there is binocular vision. (After Pumphrey 1948*b*.)

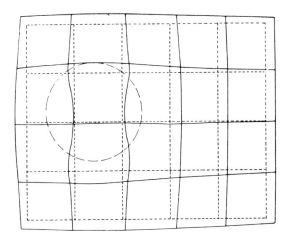

FIG. 15.58. Effect of refraction produced by the curvature of the fovea of the golden eagle. An image of the form shown with dotted lines is distorted by the fovea to the form shown in solid lines. The circle represents a radius of 10 μm at the centre of the fovea. (From Pumphrey 1948*b*.)

ing bipolar cell and there are also many amacrine and horizontal cells. The information passed on to the brain is already coded into five types of ganglion cell, each indicating one feature, such as verticality, horizontality, edges, moving edges, or convex edges (Maturana 1962). Such retinal coding occurs also in amphibians and reptiles, but in mammals it is mostly done in the brain.

Birds undoubtedly discriminate colours, apparently on a trichromatic basis similar to that of mammals. Studies by microspectrophotometry in various birds

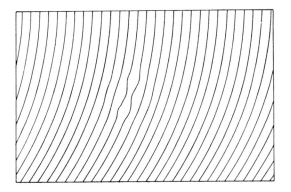

FIG. 15.59. Distortion by the fovea. The lines represent the successive images at equal time-intervals of the boundary of a regular object when the object moves steadily across the visual field. If this picture is viewed at 7 m the area of irregularity subtends an angle about equal to the angle subtended by the central part of the hawk fovea. It will be found that the irregularity is very evident to the human eye at this distance though the lines are resolvable with difficulty.

have identified three groups of cone pigments, absorbing maximally in the yellow, green, and blue (Bowmaker 1979). No other animals, except perhaps primates, show such responsiveness to colour in their surroundings, including the food and other members of the species. In animals that move so freely recognition and attraction of the sexes is more efficiently performed in this way than by touch or odour.

The cones of birds often contain carotenoid oil droplets, which are also found in frogs, turtles, and marsupials, but not in placental mammals. In diurnal birds they are red, yellow, orange, or colourless, but in nocturnal ones pale yellow or colourless.

The function of the colour droplets is not certain but may be to produce 'narrow-band sensitivity channels for the mediation of colour discrimination'. Quails raised on a diet deficient in carotenoids differentiated colour gratings on a basis of luminance rather than wavelength as in normal animals (Wallman 1979). Sometimes the droplets are so arranged as to allow accentuation of different contrasts in the parts of the visual field. The dorsoposterior part of the pigeon's retina contains red, the rest yellow filters. The red part may perhaps provide finer colour discrimination of objects in feeding.

Although the eyes of some birds are directed forwards, so that their fields overlap, they are said not to have binocular vision and decussation of the optic tracts is complete. Perception of distance, a very important function for the bird, must be performed in some other manner. In many birds the eyes are directed sideways, and the fields of view may even overlap behind the head, for example in waders. This may serve to give warning of predators.

The most enigmatic organ of the bird's eye is the pecten, a pleated highly vascular fold, projecting from the retina into the vitreous humour. Probably its main function is to bring oxygen and nourishment to the retina, which in birds has no capillary circulation. It is possible that the irregular shadow cast by this organ provides, as it were, numerous small blind spots and hence by a stroboscopic action increases the number of on-and-off effects produced by a small object in the visual field, increasing contrast and allowing detection of its movement. This, however, is only one of the numerous suggestions about the function of the pecten, and the only real support for the idea is that the body is large and much pleated in predatory birds, which detect minute movements at great distances, and is small and smooth in nocturnal birds (Fig. 15.60). The pecten may be in some way connected with accommodation; it is not likely that it actually assists in focussing, for instance, by pressing forward the lens, and no changes have been seen in it during accommodation. However, it might

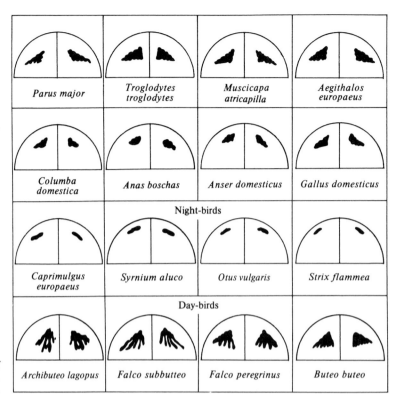

Parus major   *Troglodytes troglodytes*   *Muscicapa atricapilla*   *Aegithalos europaeus*

Columba domestica   *Anas boschas*   *Anser domesticus*   *Gallus domesticus*

Night-birds

Caprimulgus europaeus   *Syrnium aluco*   *Otus vulgaris*   *Strix flammea*

Day-birds

Archibuteo lagopus   *Falco subbutteo*   *Falco peregrinus*   *Buteo buteo*

FIG. 15.60. Tracing of the shadow of the pecten on the retina in various birds. (After Pumphrey 1948*a*, and Menner.)

possibly assist by adjusting the intraocular pressure, which must be increased by the extensive changes in the lens during accommodation.

36. The ear and hearing

The macula of the utriculus is horizontal, that of the sacculus vertical (Fig. 15.61) (Schwartzkopf 1973). The macula lagena is curved and separated from the other maculae by a long cochlear duct. It is partly innervated by fibres connected with the cochlear nucleus and may be concerned with the hearing of low tones. The maculae and semicircular canals have a continual tonic effect on the musculature. After removal of both labyrinths a pigeon is hardly able to stand, and after unilateral operation it falls to the lesioned side. The gravity receptors of the maculae ensure maintenance of the natural position of the body and head (Fig. 15.62).

The apparatus for hearing is surprisingly like that of mammals. The cochlear duct is a slightly curved tube, surrounded by a perilymphatic space within a bony tube (Fig. 15.61). Sound is transmitted from the tympanum by the columella auris (stapes), derived from the hyoid arch. There are several small pieces at its outer end, serving to reduce the amplitude and increase the force of vibrations to produce an impedance match with the fluids of the inner ear. There is a single middle ear muscle attached to the columella and tympanum and innervated by the facial nerve. Acoustic vibrations are transmitted to an oval window and so around the cochlea to a round window, as in mammals.

The cochlea has a basilar membrane, its fibres increase in length towards the tip. It carries rows of up to 40 hair cells on which rests a tectorial membrane. Different segments of the membrane oscillate according to the frequency. The endolymph contains a high concentration of potassium, as in mammals, maintained by the activity of a tegmentum vasculosum, which also absorbs sodium. There is therefore a small standing DC potential (20 mV) across the hair cells, which presumably contributes to their functions.

The central auditory pathways pass through several medullary cochlear nuclei to the torus semicircularis of the midbrain and from here to the thalamus and neostriatum (Fig. 15.63). Single cells here respond to sounds from either ear and serve to discriminate time patterns rather than frequencies (down to differences of 7 ms). Auditory pathways also reach to the cerebellum. Sounds are produced by electrical stimulation close to the afferent centres in both the midbrain and striatum.

The range of frequencies detected by birds is rather narrower than that of most mammals (0.3–7.5 kHz in pigeons, up to 21 kHz in song birds). Their capacity for

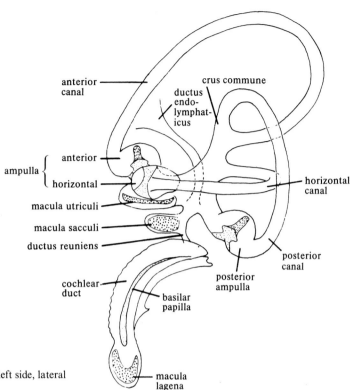

Fig. 15.61. Membranous labyrinth of a bird (left side, lateral view). (After Schwartzkopf 1973.)

frequency discrimination is not much greater than ours but their capacity for time resolution is ten times better. We cannot follow some of the rapid changes that make up a bird's song.

Ability to localize sound is high and owls and other nightbirds probably find their prey largely by ear. For the purpose of direction-finding they have developed very long cochleas and an asymmetrical arrangement of the ear cavities (*Strix*) or asymmetrical external ears (*Asio*). The detection of the sound source is partly by time differences between the ears but monaural localization may be achieved by the time difference between sound entering the auditory canal directly and after reflection from the contours of the pinna (which incidentally is probably also the function of the funny

shaped ears of man). A few birds that live in caves have the power of avoiding obstacles by echolocation, *Steatornis*, the oil bird, *Collocalia*, swiftlet. The swiftlets emit up to about 20 clicks a second at 4 to 5 Kc. Oilbirds emit lower-pitched clicks and their orientation by sonar is much less highly developed. The rate varies inversely with the amount of light and increases when obstacles are met. They are only one-tenth as good at avoiding obstacles as are bats, but the time resolution is similar. Penguins are also said to catch fish by sonar (see Schwartzkopf 1973).

37. Other receptors

The corpuscles of Grandry in the bill of ducks and other birds are probably touch receptors. They are composed of pairs of cells with a flattened nerve ending between (Fig. 15.64), and are comparable to Meissner's corpuscles in mammals. Merkel's corpuscles are similar but smaller, and found in various situations in many birds. The corpuscles of Herbst, found in the dermis elsewhere in the body, resemble Pacinian corpuscles but are bilaterally organized (Fig. 15.65). They are vibration receptors, sensitive to mechanical deformation by rapid pressure changes (500 Hz), acting as a band pass to suppress lower frequencies. They are found in great numbers in the beaks of all birds, and also in the feather

Fig. 15.62. Postural reflex in the duck; note position of the head, independent of trunk. (After Huxley 1914.)

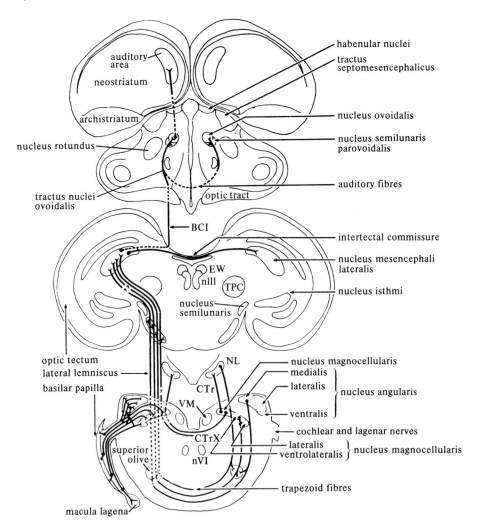

Fig. 15.63. Pigeon. Diagram of composite transverse sections through selected levels of the brain showing the ascending auditory connections. BCI, auditory fibres in brachium of inferior colliculus; CTr, uncrossed dorsal cochlear tract; CTrX, crossed dorsal cochlear tract; EW, Edinger–Westphal nucleus; NL, nucleus laminaris; nIII, oculomotor nucleus; nIV, abducens nucleus; TPC, nucleu tegmenti pedunculopontinus pars compacta; VM, medial vestibular nucleus. (By courtesy of Boord, R.L. (1969). *Annals of the New York Academy of Science* **167**, 186–98.)

follicles, between the tibia and fibula, and in the tip of the tongue of a woodpecker.

Chemoreceptors for taste and smell are little developed. There are only a few taste-buds on the tongue, 24 in a hen's, compared to 10 000 in man. The nasal cavity is large but the olfactory epithelium is restricted. It is doubtful whether most birds use the nose as a distance receptor; they may use it to test air coming from the internal nostril. In kiwis, however, which are nocturnal and terrestrial, the olfactory sense is well developed.

Muscle spindles and other proprioceptors are abundant and must be very important to birds, but have been little studied.

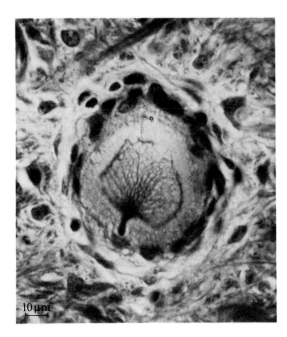

FIG. 15.64. Grandry corpuscle from bill of domestic duck. (Kindly provided by T.A. Quilliam.)

FIG. 15.65. Duck. Herbst corpuscle from the beak region. (a) Semischematic presentation showing the central core composed of receptor axon and sheath of palisade cells forming lamellae. An unmyelinated nerve fibre intruding into capsular space, branches within the palisade lamellae, forming synaptic contacts. (b) Transverse section across the axis of the central core (approximately × 10 000), showing axon, palisade lamellae, and synaptic contacts of a possibly vegetative nerve fibre. The lamellae are connected with each other and with the axon membrane by desmosomes. (Schwartzkopf 1973.)

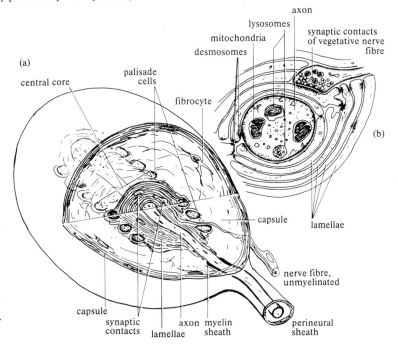

16 Bird behaviour

1. Habitat selection

THE success of bird life has been largely due to the great variety and ingenuity displayed in finding situations suitable for providing food and allowing reproduction. Their powers of habitat selection allow diversification of species by reducing interspecific competition. On finding itself in a new environment a bird actively explores it for food or enemies. It is often said that birds are creatures of stereotyped habits, yet they have certainly exploited their mobility to the full; they obtain the means of life in most various ways. This mobility makes it difficult to specify the 'environment' of a bird. For instance, a swallow may pass part of its life in the tropics, part near the Arctic Circle. A gull may nest on a rock, eat grain in a field, and then fish in the sea, all within a few hours. Observation of the familiar birds of town and country soon shows that they are at home in a much greater variety of situations than could be tolerated by most animals, and that within limits they can adapt their behaviour to each situation. Birds show, therefore, in a marked degree, two of the features most commonly used as criteria for the recognition of a higher animal, namely, freedom to move to different conditions and ability to obtain a living in unpromising circumstances.

Nevertheless, each species nests in a limited variety of conditions and eats a limited variety of food. There is evidence that the appropriate habitat is recognized by all members of the species by a relatively small number of conspicuous, characteristic features, independently of learning.

2. Food selection

Where the food is very specific substitutes will only be accepted under unusual conditions of starvation, but many birds are more catholic in tastes and some of these are among the most common, for instance rooks, starlings, thrushes, blackbirds, and gulls. Both field observation and experiment suggest that birds quickly learn from experience where and how food may be obtained. They will remember to visit an abundant source of supply and there are numerous stories of the ingenuity of such birds as the jackdaws in obtaining it. This type of memory is probably of great importance in allowing the bird to establish an effective routine for each part of its life, especially as it is combined with the power to explore elsewhere when conditions change. These polyphagous birds mostly have a reputation for intelligence.

Food selection thus depends on species-characteristic motor patterns and structures, whose use varies to suit the circumstances. There is usually little competition between species over food, even when they live close together (Gause's principle). Thus different species of finch eat seeds of different sizes or at different levels in the trees (Hinde 1973). There is an initial responsiveness to a wide range of stimuli, later modified by learning. In young birds these actions are not necessarily related to appropriate objects or situations. Thus young chaffinches or tits peck at spots of many sizes, but only when they are *not* hungry. When they are, they beg from the parents. Young kestrels 'play' at hunting pine cones, even after obtaining food by real hunting (see Hinde 1959).

This range of response is then narrowed by learning. Objects that provide food are pecked again, those that do not or are distasteful are avoided. The range of learning that is possible must, however, be influenced by the hereditary equipment. Thus chaffinches never use the foot to hold objects but tits (*Parus*) do so and can learn to pull over a grass stem and peck off insects otherwise out of reach, or indeed to open milk bottles!

3. Responses to predators and reliability of communication

Birds are especially vulnerable when not in flight and many species have very effective cryptic colours and postures. The frogmouths, *Podargus*, of Asia resemble a broken branch while perching. Many birds have innate responses to the shapes of owls, hawks, or snakes (Fig. 15.48). Passerines and other small birds will mob predators not actually attacking them. Members of gregarious species emit warning calls, even if these increase the danger to themselves. This altruism is also very characteristic of distraction displays, such as that of the peewit. These may have evolved from conflicting tendencies to flee and to attack. In such warning and distraction displays, as in any form of communication, the individuals concerned, here prey and predator, have a common interest in the avoidance of *deceit*. The transmitter must establish the reliability of his signals by exposing himself to a genuine risk. This has been called the handicap principle and it is believed by some workers to apply to a wide range of communication (Zahavi 1977). The prey informs the predator by the display that he has been seen and that he may be mobbed. He proves his sincerity by leaving cover and exposing himself. Similarly non-palatable species show by complicated colours that they are not afraid of being attacked, because they really are nasty. Again in sexual selection a male advertises his superior strength by very bright coloration, risking a higher rate of predation. It is not proved, however, that this is a general principle.

4. Recognition and social behaviour

Great mobility has made it necessary for birds to develop specific means for recognizing their fellows, their enemies, and their competitors. From this power an elaborate social life has developed in many species. In spite of their freedom many birds are not individualists, for much of their lives they remain together in flocks; the unit of life is larger than the 'individual' body. Indeed flocks often contain members of several species. However, many birds have characteristic flight movements and call sounds, these being adjusted for audibility over distances not greater than the habitat requires. Separation from the flock sets up a pattern of exploratory search and calling. The 'social facilitation' of the flock is so strong that a sated bird will start eating if it sees others doing so but will not feed even if hungry unless the others are eating too.

Species feeding on the ground, such as rooks, starlings, and partridges, commonly move about in groups during the winter and obtain the advantage that the alertness of each single bird serves to warn for many. The lack of procryptic coloration in some of these social birds is a measure of the effectiveness of the protection afforded by the society; indeed, it may be advantageous that the birds should be conspicuous to their fellows. Starlings carry the communal life farther by collecting together in large numbers each night to roost. As many as 100 000 may be found in one roost, the birds flying home from their feeding-grounds every night for distances of many miles. Possibly in this way the disadvantage of the conspicuous outline is minimized while roosting. This coming together may also serve for the regulation of the numbers of the species according to the theory proposed by Wynne-Edwards (1962). Rooks show somewhat similar behaviour, but it is not found in the protectively coloured partridge.

Many different means are adopted by birds for recognition of other members of the species and of the same and opposite sex; bird-life contains elaborate social and sexual rituals for many occasions. For example, relief of the one bird by the other at the nest is accompanied by a peculiar wing-flapping ceremony in herons and other birds. Greeting ceremonies are common in many species, and there is a host of sexual recognition and courtship rituals, to be considered later. These serve the immediate function of regulating the aggressiveness that otherwise arises when two individuals approach each other closely. Such ceremonies prevent attempted copulation with the same sex and ensure it with a member of the opposite sex, and of the right species.

Apart from sexual behaviour birds show many complex mutual and social reactions. Thus in communities of hens or pigeons there is quickly established a rank of 'pecking order', such that each bird is submissive to the one above it. The order changes, however, when age, moulting, or experiment (e.g. sex-hormone injection) alters the state of the birds. More pleasing communal habits are the dances and corporate flights, which are well known in cranes and other species.

5. Bird migration and homing

Among the remarkable devices of birds is their habit of seasonal movements to obtain the advantage of the favourable conditions offered in more northerly regions only during the summer. More than two-thirds of the species breeding in the northern United States move south for the winter, making trips of up to 6000 km each way (Emlen 1975). Some shore birds migrate for more than 10 000 km in each direction between breeding grounds in the Arctic and wintering areas in the southern hemisphere. It is calculated that 5000 million land birds migrate from Europe to Africa each autumn, and half of them succeed in returning next spring (Moreau 1972). Some tropical birds migrate to breed in the rainy season in the outer tropics, removing to the central tropics in the dry season. Marine birds also may make extensive migrations. Thus the great shearwater (*Puffinus*) breeds on Tristan da Cunha, but comes as far north as Greenland or Iceland in May, returning again after months of wandering at sea, apparently without making a landfall. The Arctic tern (*Sterna*) breeds in the north temperate zone, and migrates to the Antarctic along both sides of the Atlantic. Penguins make migrations by swimming.

Small song birds mostly migrate singly and at night, navigating partly by the stars, and covering up to 600 km a night, choosing only favourable weather. Diurnal migrants mostly fly in flocks, following topographical cues, Some birds make very long continuous flights to get across inhospitable zones, for instance European migrants crossing the Sahara on their way to or from tropical Africa. Adult birds return to the same general and even detailed places at both ends of the journeys. Young birds mostly do not learn from their elders, indeed may leave before them, flying off in a direction that is presumably genetically determined.

Hand-reared warblers orient correctly by the stars without previous experience.

Birds probably make use of several of the possible cues available for navigation in different parts of the journey, depending on the conditions. This redundancy has made it difficult to evaluate suggested mechanisms. Thus several workers showed that the homing of pigeons is not prevented by attaching magnets to them, but the experiments were done on clear days. In fact birds probably can use magnetic information – if necessary (p. 378).

It is convenient to consider three types of orientation ability (Griffin 1955). Type I is navigating by visual or

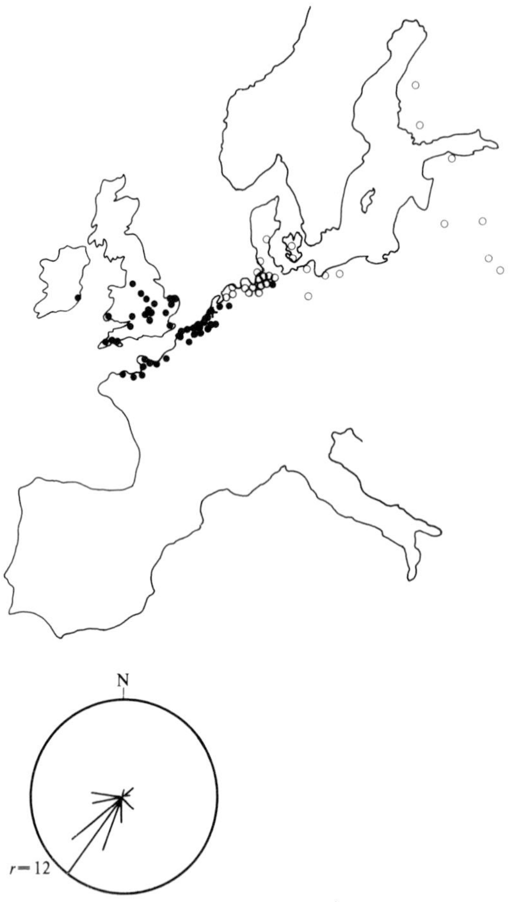

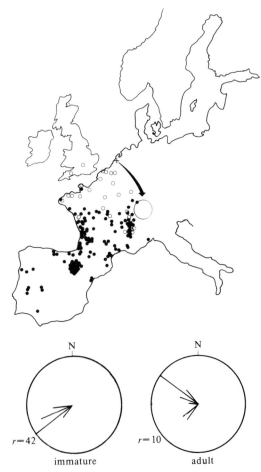

Fig. 16.1. European starlings. Locations where breeding (o) and wintering (●) starlings were recaptured after being banded during their autumn passage through the Hague (+) (seen better in Fig. 16.2). Below is a vector diagram of the directions of winter receptive locations relative to the Hague. The data are grouped into 15° sectors and plotted on a proportionality basis with the radius equalling the greater number of recaptures in any one sector. The number that this represents is presented at the lower left of the diagram. (After Emlen 1975.)

Fig. 16.2. Top: locations where adult (o) and immature (●) European starlings were recaptured after being displaced from the Hague to three locations in Switzerland during their autumn migration. Bottom: vector diagram of the directions of winter recaptures of these birds, plotted relative to the release sites in Switzerland. (After Emlen 1975.)

other land marks, Type II flying in a particular compass direction, and Type III is true navigation, the capacity to orientate to a specific goal from an unfamiliar area. This third type is rare and most migrants use combinations of the first two.

Topographic cues can give long range guidance in daylight. Migrants fly high and at 610 m the horizon is 30 km in either direction. Landmarks may be learned. Ducks and geese travelling in family groups transmit the information from generation to generation. Most birds make little use of landmarks except for the last stages of homing. They navigate mainly by the sun and stars. Pigeons fitted with translucent lenses that allow localization of the sun but not recognition of landmarks can home to the loft from up to 130 km.

If landmarks are little used migration and homing must depend largely on type II navigation, by the sun or stars, and there is now much evidence of this. Displacement of starlings (*Sturnus*) during migration has shown that adults can alter course and arrive at their 'planned' destination, whereas juveniles continue as before (Figs. 16.1 and 16.2) (Emlen 1975). The mechanism is evidently of type II. Such redirection would be useful to compensate for wind drift. But it is probable that much migration in fact makes use of the wind. The economy of energy moving down wind has obvious selective value and this has made birds into very good meteorologists. Migration occurs only on a small proportion of days, when conditions are favourable. Thus autumn movements occur on the east side of a high-pressure cell behind a cold front and spring flights on the west of a high-pressure area ahead of the cold front of a low. Birds often fly at the height where the wind is most favourable.

The importance of orientation by the sun was first shown in experiments by Kramer (1951). Caged starlings show a restlessness ('Zugunruhe') at times when migration is due and flutter in the appropriate direction in relation to the sun (Fig. 16.3). This would explain starting in the right direction but navigation on a long flight implies allowing for the sun's motion. Kramer showed that birds have a clock that does this. He trained them to orientate in a particular compass direction by the morning sun, and found they maintained this so long as they could see it – or even a lamp bulb. If the bulb was kept stationary, however, they changed their angle *as if it had moved*. The bird's clock can be shifted by an artificial day/night cycle and starlings trained to find food in a particular compass direction showed the expected shift of 15°/h which is the rate of the sun's motion.

Such a sun compass can keep a bird on course but to set its course after displacement is a more difficult matter involving very delicate bico-ordinate navigation. It has been satisfactorily proved that pigeons return home after release from a distant point even if they have never been there before, nor have had any previous training in returning from situations out of sight from the loft. Matthews (1955) and others have shown that upon release the birds fly off towards home provided that they can see the sun. Capacity to navigate by the sun would depend upon determination of position on two coordinates by observations of the sun's altitude, azimuth, and/or movement. This would imply the use of a very accurate chronometer as a means of determining the difference between home and local time. It is hard to believe that all this could be achieved in the few seconds following release, and probably only the azimuth is used, together with magnetic or other clues. However, some experiments designed to upset the 'chronometer' by altering the period of daylight have been claimed to be effective in altering the direction of flight of birds upon release.

Navigation by starlight has been studied by following the direction of the 'Zugunruhe' movements of birds in a planetarium. Warblers and others altered their directions when the stars were shifted. The requirements for navigation by the stars are more demanding than for solar observation. The stars change and presumably it is a pattern that is recognized. Stars appear to move at different rates, how does the chronometer deal with the

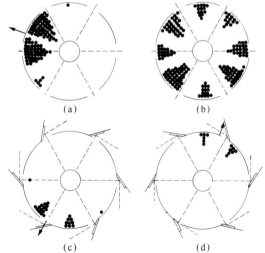

FIG. 16.3. European starling. Orientation of spontaneous diurnal migratory activity under various conditions. Bird was tested outdoors in a pavilion with six windows during the spring migration season. Behaviour (a) under clear sky and (b) overcast sky, sun not visible. When the image of the sun was deflected by mirrors 90° counterclockwise (c) and (d) 90° clockwise. Each dot represents 10 seconds of fluttering activity. Dotted lines, incidence of light from the sky. Solid arrow, direction of activity. (After Emlen 1975.)

fact that only the pole star is fixed? Emlen found that major groups can be removed without altering the orientation of buntings. Covering whole sections of the sky produced considerable deterioration of performance, and the northern sky within 35° of Polaris is of particular importance. It seems, however, that birds do not use time in star-compass orientation. Matthews found that time-shifted mallards make the expected errors when navigating by day, but not at night (Fig. 16.4).

In spite of the evidence for navigation by the sun and stars much remains uncertain. It is not clear what determines the initial compass bearing and in particular what is responsible for the reversal of migration between spring and autumn. These questions are especially acute in the case of birds that cross the equator. A further problem is that although birds usually avoid flying in clouds they seem to have some ability to maintain a course when forced to move under overcast skies or even in fog, though they may twist and loop along the way. Probably they use wind direction and other meteorological information but there is some evidence that birds can use magnetic cues. Pigeons released under conditions of total overcast 50 km from home take up an initial bearing towards home, but not if magnets are attached to their backs (Fig. 16.5). Producing a magnetic field by a miniature Helmholtz coil has a disturbing effect on homing under overcast but not sunny conditions. Robins (*Erithacus*) orientated their nocturnal activity in the migratory direction even without visual cues, but not if they were placed in a large steel chamber that reduced the magnetic field from 0.41 G to 0.14 G (Wiltschko 1972). They altered their direction appropriately if the direction of the field was changed by Helmholtz coils. These effects are not very large and

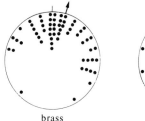

brass   magnet

FIG. 16.5. Departure bearings (pooled) of experienced homing pigeons when released under conditions of total overcast at distances 27 to 50 km from the home loft. Home direction = 360°. Left: results from birds with brass bars glued to their backs. Mean direction = 15° from correct, with some scatter. Right: results from birds with a magnetic bar attached. Mean direction = 31° from correct with much greater scatter. (After Emlen 1975.)

only show by statistical treatment of large samples of movements and few individual birds show clear cut orientation. Particles of magnetite have been found in the brain, and muscles of the head and trunk, of pigeons and sparrows. Magnetic forces may influence the cells and/or receptors in the muscles (Presti and Pettigrew 1980).

Whatever mechanisms are adopted some birds can return home from spectacular distances, an instance being the Manx shearwater removed from its nest (burrow) on Skokholm Island off the Welsh coast and sent to Boston by air: it returned in 12 days, the distance being 4940 km across the Atlantic. In another experiment three out of ten untrained terns returned to their nests from a distance of 1370 km, in about 6 days. There are many peculiar and unexplained features about such long journeys.

6. The stimulus to migration

Migrations are obviously correlated with the seasons and changes in day-length certainly play a central part in initiating them. It was first shown by Rowan in 1925 that juncos (*Junco*) or crows submitted to artificially increased day-length in winter show precocious migratory movements if released. The longer days also caused growth of the gonads (which regress in the autumn) and Rowan supposed that the secretion of sex hormones was the stimulus to spring migration and that its reduction produces the autumn one. Later work has shown that this is only partly true and that other endocrine influences are also at work (see Berthold 1972). Many nocturnal migrants make characteristic calls while in flight and these noises overhead may stimulate birds on the ground to join in the movement.

However, studies with several migrant species have shown that artificially prolonged day length can pro-

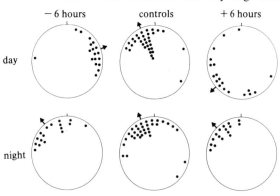

        − 6 hours   controls   + 6 hours

day

night

FIG. 16.4. Effect of clock-shifting upon the departure directions selected by Mallard ducks (*Anas platyrhynchos*) when released during the day (top) and night (bottom) near Slimbridge, England. Arrows show mean bearings. (From Matthews 1963.)

duce hyperphagia, increased body weight, deposition of fat and restlessness even in the late autumn. Non-migrants showed no such effects. Manipulation of the photoperiod can be made to produce the tendency to move either north or south by buntings (*Passerina*). The increase of food availability in spring and its decrease in autumn may be the actual triggers to the migration change. However, birds of several species kept in constant daylengths have been shown to develop restlessness and to lay down fat. Some were kept constant from less than a year after birth so the circannual rhythmicity is at least partly inborn. The temporal program of migration is an inherited species-specific character. Birds that migrate over a longer distance show earlier and greater restlessness. Each species has enough reserves of fat and migration activity to reach its particular destination along its standard route.

The patterns of migration of many species must have been evolved quite recently. At the end of the last glaciation 18 000 years ago there can have been hardly any breeding territories north of the Mediterranean. The maximum of northern breeding was probably about 5000 years ago and human clearance has probably reduced it by more than half. However swallows (*Hirundo*), house martins (*Delichon*), and swifts (*Apus*) have taken advantage of human habitation.

7. The breeding habits of birds

The complexity and variety of bird behaviour show especially in their breeding; perhaps in no other creatures except men is such elaborate behaviour involved in bringing the two sexes together and caring for the young. In order to ensure adequate provision for the development of a warm-blooded animal and its nourishment until it can fend for itself it is necessary either to keep it within the mother or to provide a means of incubating the eggs. In either case a long period of care is necessary after birth; the young animals, having a large surface area, require a great amount of food to metabolize to keep warm. Thus young starlings and crows may eat as much as their own weight of food each day.

Mammals and birds set about providing for this warmth and food in different ways. Since a female mammal can move about and get food while pregnant the father can desert her altogether, though frequently he does not do so. In birds the eggs and young cannot be left cold for long and it is therefore especially desirable that the father should help. In birds, therefore, perhaps even more than in mammals, the breeding habits involve the development of elaborate systems of mutual relations, serving not only to bring the parents together but also to keep them together throughout the period of incubation and while feeding the young. The actual

building of the nest may be an intricate business in which both birds collaborate and a further factor is that the pair occupies a territory around the nest, which they defend against others of the same species.

The type of association of the sexes varies greatly. The wren (*Troglodytes*) and other birds are polygamous, each male forming continuous association with several females. The great majority of birds form pairs throughout a single season, occasionally they change mates for the second brood. The same pair may mate in successive years (crows, swifts) and a few birds stay together through the year (ducks). In the ruff and certain game birds (blackcock) there is no pair formation: display and copulation occur at communal display grounds.

The fulmars (*Fulmarus glacialis*) are albatrosses where a successful slow breeding strategy has allowed a remarkable increase of numbers in recent years. They are nine years old when they breed for the first time, having spent the previous years prospecting for nest sites and a mate. The pairs produce only one egg and remain together for several years. Their average life is 44 years. Dunnet and his colleagues have followed individual birds for 30 years and shown that the breeding success of inexperienced birds is low and increases to a maximum of more than 0.6 and then declines to about 0.4. Incidentally, this is one of the first demonstrations of senescence in a wild bird. The fulmar has thus evolved a strategy that includes delayed maturity, large size, and therefore presumably large brain, allowing careful choice of nest site, compatibility with mate, low reproductive effort achieving a high success rate by careful attention of the parents until the young reaches maximum weight. In spite of their slow breeding fulmars have increased dramatically in Britain and Ireland from 50 breeding pairs recorded in 1879 to 300 000 in 1969 (Dunnet and Ollason 1979).

8. Courtship and display

The breeding of birds is nearly always seasonal, even in the tropics where conditions are apparently almost uniform throughout the year. In temperate latitudes breeding begins in spring as the gonads develop, probably under the influence of increasing illumination. The changes in behaviour with ripening of the gonads vary, of course, greatly with the species. Birds that have been social through the winter, for instance buntings, begin to leave their flocks, and the voice of the male changes from the simple winter notes to the more complex breeding song. The production of the elaborate secondary sexual characters of the plumage and other features used in display is controlled partly by direct genetic effects on the tissues, partly through hormones. Injections of male or female sex hormones or anterior pituitary extracts influence the production of some characters but not others, according to the species. In the majority of birds there is a breeding season, initiated, at least in many, by the effect of increasing length of day in the spring, acting through the pituitary on the gonads. Other factors such as degree of activity and food taken play their parts, especially near the equator, where there is little seasonal variation.

The song is one feature of the elaborate business of display and courtship. It has somewhat different functions from species to species. In the simplest case the display serves to bring the sexes together, to enable recognition, and at a later stage as a stimulus to copulation. Moreover, in some birds (doves, budgerigars, and canaries), the display serves as part of the stimulus to ovulation. Involved in these actions, there is, however, often a threat to other males. The aggressive displays are usually different from those of courtship, though in many species the male's first response to a potential mate is an aggressive one. When the female does not flee or fight back, as a male would do, he gradually changes over to courtship display. Finally, some forms of courtship, especially those that are mutual, seem to serve to keep the partners together for the period of incubation and feeding.

We may recognize in courtship, therefore, three elements, first threat to other males or to the female if unreceptive, secondly sexual stimulation, and thirdly mutual stimulation while rearing a family. Each individual has characteristics and actions that may produce attack, retreat, or sexual behaviour in others. The sequence of events depends upon the relative strengths of the ambivalent tendencies of the participants, depending on hormonal and other factors. The various types of display are combined in so many different ways that any classification or analysis is bound to be arbitrary and only a few examples that have been thoroughly studied can be given. There are, of course, also begging displays, given by young to parents, various displays given to potential predators, and others.

Song and displays that bring the sexes together are responsible in part at least for many of the pronounced secondary sexual characters in which the male and female differ. The beautiful plumage of the cock pheasant or peacock and the more bizarre combs and wattles of turkeys are displayed before the female in a manner that is clearly an excitant to copulation. The secondary sexual characters by which the sexes are differentiated may affect features as different as the colour, length, and structure of the feathers, the colour of the iris and size of the pupil, the shape and size of the body, the voice, the ornamentation of the head, and the spurs on the feet.

The importance of species recognition in leading to differentiation of male plumage is shown by Darwin's finches (p. 395) which, in the isolation of the Galapagos Islands, where there are few other passerine birds, have not developed the bright male plumage found in other finches. The recognition of individuals of the same species in this case is probably based on the characteristics of the beak; a male will begin to attack an intruder only when the face is seen.

Display takes place either by one bird to the other or mutually, and has the effect of bringing the animals together and keeping them together for periods varying from a few minutes to several years. Long unions are common in the large birds of prey and are often a result as much of mutual association with the nest as with display by the other bird. Some birds pair in the winter long before the gonads are ripe (ducks), but more usually after two birds pair off the display produces a gradual heightening of tension, leading to nest-building, copulation, and ovulation within a few days. Probably the process of bringing the birds together and ensuring coition requires even more elaborate stimulation in birds than other animals because of their great mobility and the fact that the male cannot grasp the female. Before he can treat her in such a way as to ensure coition she must be brought into a suitably receptive state. The function of the display is certainly largely to induce this state in the female, though much more is involved in addition. The processes of sexual stimulation, nest-building, incubation, and caring for the young necessitate a particular state of excitability that must last a long time, and the display serves to provide this condition in both birds. The reproductive process may be interrupted at any time if the stimuli are inadequate. Thus ovulation may depend upon courtship and copulation (pigeons) and eggs are often deserted if the birds are interfered with in any way, or if they fail to stimulate each other.

The actual procedure of display is as varied as any other feature of bird life. When the sexes are alike in colour and shape the performance is usually mutual, as in the great crested grebes (*Podiceps*), which approach each other over the water and go through various actions such as the head-shaking ceremony (Fig. 16.6). Where the male is more strongly coloured, provided with a special comb, etc., he often displays before the female, who adopts a more passive role. A very common element in the procedure of courtship is the sudden revelation of some feature or pattern, serving as it were to arrest and awaken the female and at the same time almost to reduce her to passiveness. The actual movements involved are very various. An excellent example is the peacock, who approaches showing his dull-coloured

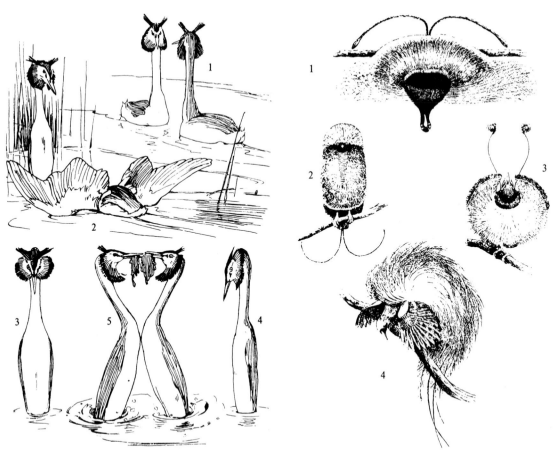

FIG. 16.6. Incidents in the courtship of the great crested grebe. 1. Mutual head shaking. 2. The female is displaying before the male who has dived and shoots out of the water in front of her. 3 and 4. Further views of the male rising from the water. 5. Both birds have dived and brought up weeds. (From Huxley 1914.)

FIG. 16.7. Display of various birds of paradise. 1 and 4, *Paradisea*; 2, *Diphyllodes*; 3, *Cicinnurus*. (From Stresemann 1934.)

back and then suddenly turns on his hen, revealing the pattern of spots (which we have noted elsewhere to be an arresting shape), shaking his 'tail' with a rustling noise and himself emitting a scream, in a way that can easily

be believed both arresting and fascinating. It is a common characteristic of all bird displays that they involve sights and actions that have a peculiar and exaggerated quality for us and probably also for the partner (Figs. 16.7 and 16.8). Watching the effects of such appearances on the female one has the impression

FIG. 16.8. Display of the pheasant, *Centrocercus*. Left: beginning of inflation. Right: complete display with cervical air-sacs inflated. (From Stresemann 1934.)

that this vision acts as it were as a key or 'releaser', initiating in her the appropriate program (Young 1978) of behaviour. Under its influence she acts as if 'mechanically', moving towards the male and adopting the receptive position, while his display activity passes over into that of mounting and coition. This sort of behaviour we have seen elsewhere in bird life; there must be in the nervous organization a receptive matrix ('the lock') ready to be actuated by the appropriate 'key', which may be the display of structures as bizarre as the wattles and tippets of a turkey.

Many displays are modified versions of everyday actions of the birds and some at least of these seem to be symbolic. Thus, in many birds courtship and coition include ritual feeding of the female by the male. Involved in this is a reversion by the female to an infantile condition. She may beg for food, often with actions similar to those she used as a nestling. The prime significance of this feeding is not in the nourishment provided but, in many cases at least, in the fact that it is the male who provides it. This is proved by the fact that the female (robin, for instance) will not feed herself when food is all around, but will beg the male to give it to her. Female herring gulls bringing back fish may even beg to be fed by the male, who has not left the rock! There are, of course, many cases of practical feeding of the sitting female by the male, but it is possible that these have been derived from the ritual feeding, rather than vice versa.

The various forms of billing and gaping ceremony probably represent a further degree of abstraction from ritual feeding. A variety of birds touch bills during courtship, for instance, gulls, ravens, great crested grebes, and some finches. The inside of the mouth is sometimes brilliantly coloured in adults, as it is in so many nestlings; it is green in some birds of paradise, yellow in many birds. During display the mouth may be opened suddenly, producing an obvious effect on the mate, who approaches fascinated into a state of passive acceptance by this surprising revelation, perhaps recalling a possibility of satisfaction remembered from childhood.

Other aspects of the courtship may show this reversion to infantile behaviour. Thus, in female sparrows and many other birds one or both wings are held drooping and fluttering during display, as they are by the chick craving food or by the frightened adult; a behaviour known as injury feigning or distraction display. The female hedge-sparrow may quiver with one wing and open her bill to the male at the same time.

The effects of bird display are by no means restricted to ensuring mutual recognition of males and females and stimulating them to coition, especially important

though such functions must be in 'flighty' creatures. The element of threat and even fighting with other males is very common. It is seen in its purest form in such birds as the blackcock and ruff, which are promiscuous. The male ruffs congregate on a chosen 'courting ground' and go through an elaborate series of ritual fights. The females ('reeves') do not take part in this procedure, but at intervals one of them will 'select' a male by fondling him with her bill and then adopt the receptive attitude for copulation; males with large ruffs were chosen especially often. Such selection of males by females was the basis of Darwin's theory of sexual selection, the supposition being that the males chosen would be the most gorgeous, victorious, and hence most vigorous and effective breeders. In one area, during a period of $3\frac{1}{2}$ hours, 12 copulations were recorded, 10 of them with a single male. It must be very difficult when observing birds to establish exactly the actions and relationships of males and females and hence to distinguish between the 'stimulation' of female by male and 'selection' of male by female. Perhaps there is only a verbal difference between the two.

In many of these forms of sexual communication the colours and behaviour seem excessive, and must involve heavy costs of material, time, and energy. These apparent exaggerations may serve to establish the 'sincerity' of the communications, involving the 'handicap principle' already referred to. In any case they emphasize the importance of sex for survival.

9. Bird territory

The element of threat in singing and display has a further importance in connection with the territories that many birds establish around their nests. An amateur ornithologist Eliot Howard (1920) first developed the concept of bird territory as a result of observation mainly of warblers and buntings. In the warblers (Sylviidae) the males, returning from migration some days earlier than the females, establish themselves on a certain area, singing often from a tall tree or other headquarters near its centre. As other males arrive boundaries develop, so that the region becomes divided up into a number of areas, at first each of about 1 hectare, later reducing to 0.5 hectare. When the females arrive they pair off with the males and throughout the whole season the two birds occupy a single territory, driving off other birds that encroach and in this way establishing quite definite boundaries to their area.

Howard supposed that this arrangement was widespread in birds and that it has four desirable effects for the species. 1. Uniform distribution over the habitable area is ensured. 2. Females are assisted to find unmated males. 3. The two birds are kept together and are not distracted by wanderings far from home. 4. It is possible to find adequate food without travelling far from the nest, this being especially important during the period of incubation and rearing of the young. There is no doubt that many birds do remain mostly in the area around their nest and that they may resist invasion. It is probable, however, that the territory is usually less rigid than Howard implied, and there is certainly much variety between different species. According to Lack the territory is often mainly associated with the sexual display of the male; it is his area, part, as it were, of the method he adopts to stimulate the female and to keep the pair together (1968). By establishing a territory he ensures the opportunity to display and copulate without disturbance, a very necessary precaution since he is vulnerable at these times and other individuals may attack a copulating male, trying to displace him. The complicated song, characteristic of so many male birds at the breeding season, is, on this view, partly an attraction to the female, but largely also a threat to warn off other males. We cannot exclude that it serves as a stimulus to the male himself. The impression is strong when one hears a thrush 'singing for joy' at his headquarters or a lark soaring above his patch of ground.

Territory is therefore put to various uses in different species. It may be either (1) a mating arena only, as in the ruff and blackcock mentioned above, or (2) it may be a mating station and nest as in the plover and swallow, which birds will not allow others near the nest, although all mix freely for feeding. (3) In sparrows and herring gulls the nest is likewise defended, but is not a mating station. (4) In the warblers investigated by Howard the territory, besides being a mating station, is also a feeding ground. The significance of this in spacing out the birds remains, however, doubtful. It may limit the effect of predators, by ensuring dispersal. (5) In still other birds, such as the robin, it is a feeding ground mainly, and therefore it is kept throughout the winter. Examination of the territory concept thus shows that birds have a strong sense of place and that they associate this in various ways with their life, especially during the breeding season. It is certain that the occupation of territory helps in the initiation and maintenance of the pair, but not yet proved that it serves to limit the breeding density and ensure a food supply for the young. Territories may extend for several square kilometres in birds of prey or be limited to the distance that a cliff-nesting bird can peck. The owner will attack a conspecific entering his territory or flee from one upon whose area he is trespassing, the 'combats' being largely a matter of exchange of ritual inherited signals of threat and appeasement.

10. Mutual courtship

In many birds courtship displays do not necessarily end in coition and may continue long after the eggs have been laid. A classic example of this sort is the great crested grebe, a water-bird watched by J. S. Huxley (1914) (Fig. 16.6). The male and female birds do not differ greatly and the ceremonies are mutual. One bird may dive and come up close to the other and they then approach with necks stretched out on the water, giving a curious ripple pattern that Huxley called the plesiosaur appearance. When they meet the birds come together neck to neck and a period of swaying ensues and may sometimes end by the mounting of one bird by the other, not necessarily the female by the male. At other times the diving bird comes up with pieces of nest-building material and elaborately presents them to its mate. This is apparently a symbolic act; the material is not actually used to make the nest and it is not far-fetched to suppose that such behaviour is an expression of the mutual activity in which the birds are engaged. In so far as it has a biological function it serves, like the rest of the ritual including post-ovulatory copulation, to keep the two individuals together while rearing the young.

As already mentioned, courtship may include ritual feeding of the other sex during display and often before coition, for instance, in pigeons and gulls, and this again may have a symbolic function. It is perhaps not fantastic to find analogies between behaviour of this sort and the elaborate and prolonged courtship and frequent copulation of man, continuing without cyclical breeding seasons for as long as the pair remain together and rear their young. In birds, as in man, the 'procreation of children' is not fully accomplished by a single act of fertilization.

11. Nest building

As in every other aspect of their life we find the nest of birds varied in many ways to suit different modes of life. It is suggested that the habit of making a nest may have arisen from the 'sex-fidgeting' that is commonly seen before, during, and after copulation. This fidgeting may take various forms, including making a 'scrape' in the soil or picking up pieces of grass, etc., after copulation. There is certainly, in many birds, a close connection between nest building and copulation. Ritual offering of nest material is an important element in many courtship displays and in some species (e.g. magpie) male birds may build extra nests.

The nest is therefore often at first a sex site and its position may be chosen by the male. The actual building of the nest is done very variously. Sometimes the male brings the material and the female uses it. She may do the fetching as well, perhaps accompanied by the lazy male; or he may have nothing to do with the whole business. The nest is built by means of a limited number of stereotyped movements, which are characteristic of the species. The integration of these movements into a functional sequence of behaviour depends, however, on experience. Nest building and copulation occur at about the same stage of the reproductive cycle and both can be induced by oestrogen (in canaries).

The complicated forms of nest are found mainly in passerine birds; in others it is usually simply a hollow in the ground or a heap of sticks. The more elaborate nests show many protective device. In temperate regions, where predators come largely from below, the nests are often open. Where there are many snakes they are mostly domed or hung from branches, or provided with a long tubular entrance (weavers).

In building the nest the bird follows a set pattern, laid down in some way by the method of working of its brain and showing no sign of foreknowledge of the result. Young birds, however, build rougher nests than mature ones. Weaver birds reared by hand for four generations made perfect nests of a type which, of course, they had never seen. On the other hand, it has been claimed that canaries deprived of nest-building materials for some generations then build clumsily at first, though they quickly improve.

The methods used, of course, vary with the materials. Many of them involve most elaborate tying of grasses to the branches, which is done with the beak or the feet or both (Fig. 16.9). The shape and construction of the nest varies with the habits of the bird. Many sea birds,

nesting safely on the cliffs, make only very simple nests or none at all. In such birds as the plovers or larks the nest is a cup in the earth, the eggs being procryptically coloured with a blotched green and brown pattern, so that is is very difficult to see them on the spring ploughland or partly green earth. Nests constructed in trees vary from the simple sticks of rooks or pigeons to the elaborate domed and lined nests of many passerines. The nest is woven from materials brought in with the beak and often lined with hair, or in ducks with feathers from the breast of the bird. The thrush lines its nest with mud moistened with saliva and some swifts make nearly the whole nest of saliva. Many birds build roughly and often use the old nests of others, for instance, kestrels are often found using the nests of crows. Others build in burrows, either accidental or, in kingfishers, made by the bird itself. Similarly, holes already in trees may be used (tit-mice), but the woodpeckers drill their own holes. The female hornbill walls herself into a hole in a tree with the help of mud brought by the male and for several weeks (up to 18 in very large species) she is fed by the male through a small aperture. In the brush turkeys (Megapodes) the eggs are not incubated by the bird but are buried in a mound of decaying material, which provides the necessary heat; the young bird is independent from the moment it hatches.

The bower-birds of Australia and New Guinea are passerines in which the male builds an elaborate structure of plants, ornamented with bright objects. Here he displays to the females and sings. A bower that is left may be destroyed by another male and its decorations stolen. The male constantly refurbishes his bower and

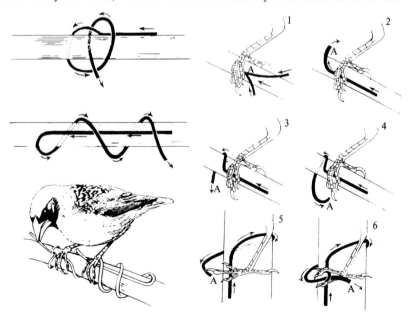

FIG. 16.9. Process of nest-building by a weaver bird (*Quelea*). 1–7. The arrows show directions in which the piece of grass is pulled. 8. Co-operative weaving by the foot and beak. A, the points of holding by the beak. (From Stresemann 1934.)

gyrates round it, tossing the decorations violently. This seems to be a form of displacement activity. When a female ultimately becomes receptive copulation occurs. The eggs are laid in a nest nearby.

12. Shape and colour of the eggs

Eggs are as varied as the rest of bird structures. The shape is determined by the pressure of the oviducal wall and the blunt end usually emerges first. The shape may serve to prevent the eggs rolling away (guillemot) or to help the eggs to fit together (plover). The coloration is usually procryptic in eggs laid on the ground, whereas those laid in holes are white and birds like the pigeons that do not attempt to protect themselves or their nests by concealment also have light-coloured eggs. The significance of the varied colours of eggs that are not procryptic is obscure. It is likely that they serve as a stimulus to the brooding bird, who will sometimes leave a nest when a wrong-coloured egg is inserted. A further sign of this is that there are various races of cuckoo, each laying eggs appropriate to the nest it parasitizes; but the genetics and behaviour of cuckoos are still very obscure subjects.

13. Brooding and care of the young

Usually it is the hen who broods the eggs, but the cock may assist and in a very few species he does all the brooding (e.g. phalaropes). Brooding is not a mere sitting on the eggs, but depends on the development of a vascular and oedematous response with a local moult of the feathers in a part of the skin, the brood spots, under the joint influence of oestrogen and prolactin. At hatching the chick breaks the shell by an egg tooth (caruncle) and the parents may assist. The care of the nestlings is a very elaborate business in most birds, involving many separate actions. The young are warmed, fed, and occasionally watered. In many species the nest is kept clean by careful removal of the faeces, which may be produced as pellets enclosed in a skin of mucus; these, being shining white, are eaten by the parents. In warm climates the parents may shield the nestlings from the sun during the heat of the day.

The work of caring for the young birds is often performed by both parents and there are various adaptations to ensure this. The young react strongly to the return of the parents to the nest, usually by opening the beak and displaying the coloured inside of the mouth, an action that strongly stimulates the parent, releasing the feeding behaviour, which varies with the species. The young pigeon thrusts its bill into the throat of the adult to collect the milk secreted by the crop. It is probable that this careful attention by the parents is ensured by a series of somewhat simple stimulus reactions. For example, Eliot Howard showed that a female linnet responds to its own nest rather than to its own young. When its young were put into an abandoned nest and the latter placed near to its own the hen usually returned to its nest and neglected the young, though the male gave them some food. Cuckoos similarly make use of this undiscriminating 'instinctive' behaviour; birds will feed any young that provide the appropriate stimulus and if no young appear they will make little effort to find them.

The rate of development after hatching varies greatly. In the gallinaceous birds, in many ways a primitive group, the young are well developed at hatching and soon fend for themselves (nidifugal). In nidicolous species, on the other hand, the young is naked and helpless, it is a growing machine, with a large liver and digestive system but little developed nervous system. Yet the birds of these species ultimately have much larger brains and are more 'intelligent' than those that leave the nest soon after hatching.

17 The origin and evolution of birds

1. Classification

Class AVES
    *Subclass 1: Archaeornithes (ancient birds). Jurassic
      *Archaeopteryx
    Subclass 2: Neornithes (new birds). Cretaceous–Recent
      *Superorder 1: Odontognathae (toothed jaws). Cretaceous
        *Hesperornis
      Superorder 2: Neognathae (new jaws)
        Order 1: Tinamiformes
          Tinamus, tinamous
        Order 2: Rheiformes
          Rhea, nandus
        Order 3: Struthioniformes
          Struthio, ostrich
        Order 4: Casuariformes
          Dromicieus, emu; *Casuarius*, cassowary
        *Order 5: Aepyornithiformes
          *Aepyornis, elephant birds
        Order 6: Dinornithiformes
          *Dinornis, moa; *Apteryx*, kiwi
        Order 7: Podicipediformes
          Podiceps, grebe
         Order 8: Sphenisciformes. Eocene–Recent
          Spheniscus, penguin; *Apenodytes*, Emperor and King penguins
        Order 9: Procellariiformes
          Fulmarus, petrel; *Puffinus*, shearwater; *Diomedea*, albatross
        Order 10: Pelecaniformes
          Phalacrocorax, cormorant; *Pelecanus*, pelican; *Sula*, gannet
        Order 11: Ciconiiformes
          Ciconia, stork; *Ardea*, heron; *Phoenicopterus*, flamingo
        Order 12: Anseriformes
          Anas, duck; *Cygnus*, swan
         Order 13: Falconiformes
          Falco, kestrel; *Aquila*, eagle; *Buteo*, buzzard; *Neophron*, vulture; *Milvus*, kite
        Order 14: Galliformes
          Gallus, fowl; *Phasianus*, pheasant; *Perdix*, partridge; *Lagopus*, grouse; *Meleagris*, turkey; *Numida*, guinea fowl; *Pavo*, peacock
        Order 15: Gruiformes
          Fulica, coot; *Gallinula*, moorhen; *Crex*, corn-crake; *Grus*, crane; *Phororhacos; *Diatryma
        Order 16: Charadriiformes
          Numenius, curlew; *Capella*, snipe; *Calidris*, sandpiper; *Vanellus*, lapwing; *Scolopax*, woodcock; *Larus*, gull; *Uria*, guillemot; *Plautus*, little auk

Order 17: Gaviiformes
  Gavia, diver (loon)
Order 18: Columbiformes
  Columba, pigeon; **Raphus*, dodo
Order 19: Cuculiformes
  Cuculus, cuckoo; *Opisthocomus*, hoatzin
Order 20: Psittaciformes
  Psittacus, parrot; *Probosciger*, cockatoo; *Trichoglossus*, lorikeet
Order 21: Strigiformes
  Athene, little owl; *Tyto*, barn owl; *Strix*, tawny owl; *Bubo*, eagle owl
Order 22: Caprimulgiformes
  Caprimulgus, nightjar
Order 23: Apodiformes
  Apus, swift; *Trochilus*, hummingbird
Order 24: Coliiformes
  Colius, mouse bird
Order 25: Trogoniformes
  Trogon, trogon
Order 26: Coraciiformes
  Merops, bee-eater; *Alcedo*, kingfisher
Order 27: Piciformes
  Picus, woodpecker
Order 28: Passeriformes (perching birds)
  Corvus, rook; *Sturnus*, starling; *Fringilla*, finch; *Passer*, house-sparrow; *Alauda*, lark; *Anthus*, pipit; *Motacilla*, wagtail; *Certhia*, tree-creeper; *Parus*, tit; *Lanius*, shrike; *Sylvia*, warbler; *Turdus*, blackbird, thrush, American robin; *Erithacus*, British robin; *Luscinia*, nightingale; *Prunella*, hedge-sparrow; *Troglodytes*, wren; *Hirundo*, swallow

2. Origin of the birds

MANY characteristics of birds show close resemblance to those of reptiles and in particular to the archosaurian diapsids. Already in the early Triassic period the small pseudosuchians such as **Saltoposuchus* (p. 313) showed the essential characteristics of the bird group, especially those associated with a bipedal habit. From some such form the birds have almost certainly been derived, by a series of changes parallel in many cases to those found in other descendants of the pseudosuchians, such as the crocodiles, dinosaurs, and pterosaurs.

3. Jurassic birds and the origin of flight

We have no detailed evidence of the stages by which cold-blooded terrestrial reptiles were transformed into warm-blooded flying birds, but five fossil specimens from the upper Jurassic rocks of Bavaria show us one intermediate stage on the way (Fig. 17.1). These **Archaeopteryx* certainly had achieved some powers of flight or gliding, but they were less specialized for the purpose than are modern birds. The whole body axis was still elongated and lizard-like. The vertebrae articulated by simple concave facets as in reptiles, without the saddle-shaped articular facets of the centrum seen in birds. The dorsal vertebrae were not fixed and only about five went to make up the sacrum. There was a long tail, with feathers arranged in parallel rows along its sides, probably an important organ, as in other animals that live in trees and jump and glide. The fore limb ended in three clawed digits, with separate metacarpals and phalanges, the hallux being opposable. The limb was used as a wing, for the fossils show feathers on the back of the ulna and hand, but the wing area was small and the shape rounded, like that of the wing of birds that fly for short distances only. There was a furculum and a small sternum. The ribs were slender and had no uncinate processes. The pelvic girdle and hind limb resembled those of archosaurs, with elongated ilium and backwardly directed pubis. Only six vertebrae were fused to form the sacrum (at least eleven in birds). The fibula was complete and the proximal tarsals were free, but the distal ones were united with the metatarsals.

In the skull of **Archaeopteryx* (Fig. 17.2) there were sharp teeth in both jaws. The shape was more reptilian than bird-like, with rather small eyes and brain, and premaxillae and frontals much smaller than in modern

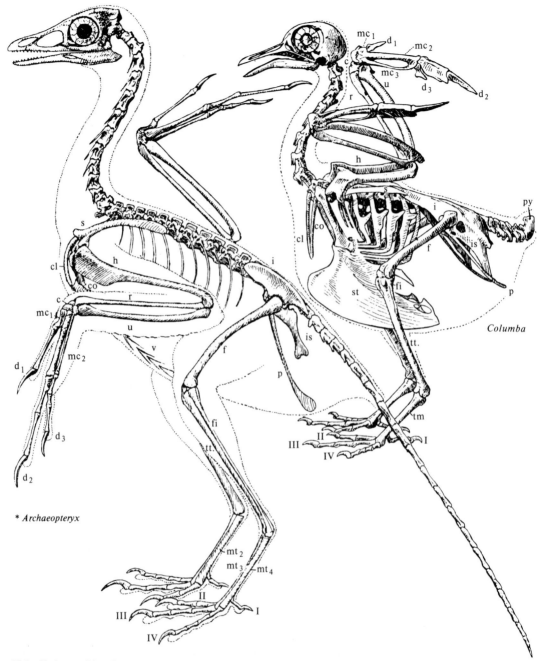

FIG. 17.1. Skeleton of *Archaeopteryx* (restored), compared with that of a pigeon drawn on a more reduced scale. *c*. carpal; *cl*. clavicle; *co*. coracoid; *d*. digit; *f*. femur; *fi*. fibula; *h*. humerus; *i*. ilium; *is*. ischium; *mc*. metacarpal; *mt*. metatarsal; *p*. pubis; *py*. pygostyle; *r*. radius; *s*. scapula; *st*. sternum; *tm*. tarsometatarsus; *tt*. tibiotarsus; *u*. ulna; *v*. ventral ribs; *I–IV*, toes. (From Heilmann 1926.)

birds. There was a large vacuity in front of the eye and probably there were post-frontal and post-orbital bones. The condition of the temporal region is unfortunately not clear on account of the crushing of the material. The brain case was large and many of the bones were united, as in modern birds. The bones were not pneumatized. The cerebral hemispheres were elongated as in reptiles and the cerebellum was small.

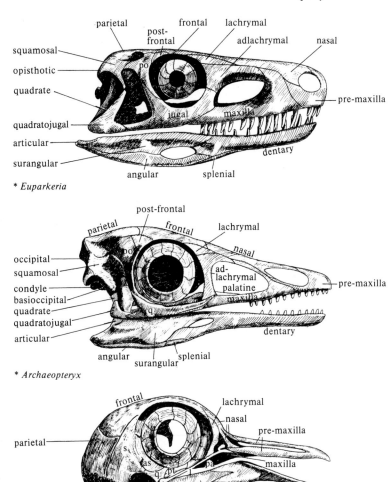

FIG. 17.2. Skulls of *Euparkeria* and the birds *Archaeopteryx* and *Columba*. (From Heilmann 1926.)

These very interesting fossils suggest that the birds arose from a race of bipedal arboreal reptiles, living in forests and accustomed to running, jumping, and gliding among the branches (Fig. 17.3). There has been much controversy about the origin of flight, some maintaining that the earliest birds were terrestrial and used the wings to assist in running, leading eventually to a take-off, perhaps at first for short distances. A recent theory is that the enlarged feathers on the back of the hands may have served as nets to assist in catching insects (Fig. 17.4) (Ostrom 1979). The tail would have provided a stabilizer for quick manoeuvring in the chase. The proto-birds may have used the wings to augment short leaps but could hardly have achieved genuine powered flight in this way since the thrust of the legs would be lost after leaving the ground. The theory assumes that they adopted an arboreal life and then used the wings for gliding. Thus the two theories that flight came 'from the ground up' and 'from the trees down' may both be true.

4. Cretaceous birds

These Jurassic fossils are so distinct from other birds that they are placed in a distinct subclass *Archaeornithes. All other known living and extinct birds have a short tail, reduced hand, a sternum, and other characteristics of the subclass Neornithes. A few fossils are known from the upper Cretaceous in which certain reptilian characteristics are still preserved. *Hesperornis* was a diver, swimming with the legs and with reduced wings, and it possessed teeth, the jaws being modified to catch fish. Other reported toothed Cretaceous birds (*Ichthyornis*) were probably mosasaurs.

Already in the Cretaceous there were some birds that

FIG. 17.3. Restoration of hypothetical proavian. (From Heilmann 1926.)

had lost the teeth and can be referred to orders found alive today. Among the earliest fossil birds were flamingoes, loons, grebes, cormorants, and rails. It may be significant that many of these are aquatic and marine animals and hence likely to be preserved. Birds are not commonly found as fossils, for obvious reasons, and it is

5 cm

Proto-Archaeopteryx

5 cm

*Archaeopteryx

Fig. 17.4. *Proto-Archaeopteryx* shows a hypothetical early stage in the enlargement of feathers on the hands and arms as aids in catching insects. **Archaeopteryx* presumably was at or just past the threshold of powered flight, but is drawn here in a similar predaceous pose. (Ostrom 1979.)

not possible to give a detailed history of the evolution of the various orders that are recognized. We do not even know whether bird life first became abundant after the Cretaceous, at the same time as the mammals began to be numerous.

5. Modern birds. Neornithes

Nearly all the orders of birds now living had appeared by the end of the Eocene, including the passerines. It is estimated that the total number of bird species that have existed is 154 000, of which 9600 are alive today (Brodkorb 1971). The Pleistocene seems to have been a time of rapid formation of new species, each living for about half a million years. During the Tertiary as a whole evolution may have been slower, with an average species life of 3 million years. Such estimates are very hazardous but they show an attempt to make evolutionary study quantitative. The great variety of modern birds are classed into 29 orders arranged here in the form provided for the survey *Check-list of birds of the world* (Peters 1931). We can only describe some of the more important orders.

6. Flightless birds. Ratites

The flightless birds or 'ratites', such as the ostrich, cassowary, and kiwi, with reduced wings and no sternal keel, long legs and curly feathers, have in the past been placed in a distinct group and regarded as primitive. Indeed, it has even been suggested that they diverged so early from the ancestral avian stock that they never passed through a flying stage. Because of the presumed primitive structure of the palate they were lumped together in a superorder Palaeognathae. However, de Beer (1956) and others have pointed out that these characters may be regarded as manifestations of neoteny and do not indicate a truly primitive condition. Some 'neognathous' birds pass through a 'palaeognathous stage' during development. The evidence strongly suggests that the 'ratites' have descended from flying birds and are not a natural group, but represent several different evolutionary lines. On zoogeographic grounds it seems certain that the 'ratite' type must have been evolved from flying birds at least four times, probably more. They are arranged together at the beginning of our classification but their relationship to other orders is unknown.

The ostriches (*Struthio*) are the largest living birds, found on steppes and now limited to parts of Africa. The rhea (*Rhea*) occupies the same ecological position in South America and the emu (*Dromiceius*) and cassowary (*Casuarius*) in Australasia. The moas (**Dinornis*) were another type; several species lived in New Zealand until recent times. The elephant-birds (**Aepyornis*) were similar, with several species in Madagascar in the Pleistocene. Some were larger than ostriches, with eggs estimated to weigh more than 10 kg, presumably the largest single cells that have existed!

The kiwis (*Apteryx*) of New Zealand are smaller, terrestrial birds whose relationship to the other ratites is doubtful. They are nocturnal and insectivorous or worm-eating, with a long beak and small eyes. The sense of smell and the parts of the brain related to it are better developed than in other birds; it is not clear whether this is the retention of a primitive feature. The palate shows large basipterygoid processes. There is a penis, as also in other ratites.

Still more doubtful is the position of the tinamous (*Tinamus*), terrestrial birds rather like hens, of which about fifty species are found throughout South America. They show similarities to the ratites in the palate and other features. Perhaps all these ratite types are early offshoots that have developed independently for a long time.

7. Some orders of modern birds

7.1. Order 7: Podicipediformes. Grebes

The grebes, *Podiceps* (Fig. 16.7) are aquatic birds, almost unable to walk on land. They resemble the loons in some ways, but are perhaps not closely related to them. They nest on lakes, laying a small number of white eggs in a floating nest.

7.2. Order 8: Sphenisciformes. Penguins

The penguins (*Spheniscus*) are birds that early lost the power of flight and became specialized for aquatic life. They may have a common ancestry with the petrels. Unlike most other water birds they swim chiefly by means of the fore limbs, modified into flippers; the feet are webbed. The penguins are mainly confined to the southern hemisphere. They come ashore to breed; many make no nests, but sometimes carry the one or two eggs on the feet throughout the incubation period. The emperor penguin breeds in winter on the Antarctic ice and is the only bird that never comes on land. The egg is supported on the feet.

7.3. Order 9: Procellariiformes. Petrels

The petrels (*Fulmarus*), shearwaters (*Puffinus*), and albatrosses (*Diomedea*) are birds highly modified for oceanic pelagic life, some of them very large. They lay one white egg, often in burrows. Their long narrow wings are specialized for soaring flight (Fig. 15.29).

7.4. Order 10: Pelecaniformes. Cormorants, pelicans, and gannets

This is another order of aquatic birds, much modified for diving and fishing and including the cormorants (*Phalacrocorax*), pelicans (*Pelecanus*), and gannets (*Sula*). They nest in colonies on rocks or trees; the eggs are usually unspotted and covered with a rough chalky substance. These birds make spectacular dives when fishing; gannets may plunge from more than 15 m.

7.5. Order 11: Ciconiiformes. Storks and herons

The storks (*Ciconia*), herons (*Ardea*) (Fig. 15.21), and flamingoes (*Phoenicopterus*) are large, long-legged birds, living mostly in marshes. Herons feed mainly on fish, storks on a mixed diet. Flamingoes are filter feeders (p. 348). All are strong flyers and some of them perform extensive migrations. Nests are usually in colonies and may be used year after year; there are elaborate display ceremonies. Eggs are few and unspotted.

7.6. Order 12: Anseriformes. Ducks

The ducks (*Anas*) (Fig. 15.49) and swans (*Cygnus*) represent yet another group of birds specialized for aquatic life. The characteristic flattened bill is used to feed on various diets. Some are vegetarians, a few filter-feeders; some eat molluscs, others fish. The numerous eggs are usually white or pale and the nest is usually built on the ground.

7.7. Order 13: Falconiformes. Hawks

This order includes the birds of prey that hunt by day, having sharp, strong, curved bills and powerful feet and claws (Fig. 15.25). The retina contains mainly cones. Many different types are found throughout the world. Most feed on birds or mammals, some on carrion, and a few on fish or reptiles. Typical examples are the kestrel (*Falco*), eagle (*Aquila* Fig. 15.22), buzzard (*Buteo*), and vulture (*Neophron*). The eggs, few in number, are usually spotted and the nests are generally made on cliffs, tree-tops, or other inaccessible places; some, however, are on the ground.

7.8. Order 14: Galliformes. Game birds

These are mainly terrestrial, grain-eating birds, capable only of short, rapid flights; some of their structural characters and habits are certainly primitive. The palate differs from both that of ratites and of most modern birds, suggesting an early divergence. There is often a marked difference in plumage, and sometimes in size, between the sexes. The nest, usually made on the ground, is simple and the eggs numerous, plain, or spotted. The young develop very quickly after birth. The order contains many successful types and is of world-wide distribution. It includes *Gallus*, the jungle-fowl of India, and all its domesticated descendants, also *Pha-*

sianus and other pheasants, *Perdix* (partridge), *Lagopus* (grouse), *Meleagris* (turkey), *Numida* (guinea-fowl), and *Pavo* (peacock). The Megapodes or mound-builders of the Australasian and east Indian regions lay their eggs in mounds of decaying leaves and earth.

7.9. Order 15: Gruiformes. Rails and cranes

Rails are mostly secretive, terrestrial birds, compressed laterally and often living in marshy country and having an omnivorous diet; common British members are the coots (*Fulica*) and moorhens (*Gallinula*). They run, swim, and dive easily, but are poor flyers; they build rather simple nests and lay numerous, often dark-spotted eggs. *Crex* (the corncrake) and other landrails are of more terrestrial habit. The cranes (*Grus*) are long-legged birds found in swamps and probably allied to the rails rather than to the waders as is still often supposed.

Possibly related to the rails are the cariamas of South America, carnivorous wading birds with very long legs, hardly able to fly, living largely on reptiles. The Miocene 'terror cranes' *Phororhacos* were similar birds reaching 180 cm high; evidently the group was successful in a region free of mammalian carnivores. *Diatryma* was an even larger flightless carnivorous bird, found in the Palaeocene–Eocene of Europe and North America and perhaps also related to the early ancestors of the Gruiformes, though usually classified in a separate order, or near the herons.

7.10. Order 16: Charadriiformes. Waders and gulls

This is a large order including the wading birds and the gulls, terns, and auks, which have evolved from them. The typical waders are birds that live mainly on the ground, often inhabiting open watery places or marshes. They are usually gregarious out of the breeding season and are often very numerous on the sea-shore. They often have long legs and long bills and feed chiefly on small invertebrates. The curlews (*Numenius*), snipe (*Capella*), and sandpipers (*Calidris*) are well known examples. The lapwings (*Vanellus*) and related plovers are birds found on drier land than is usual among other waders; the woodcocks (*Scolopax*) inhabit swampy woods.

The gulls (e.g. *Larus*) are a very important group of birds derived from the waders and adapted to life by and on the sea. Usually they have a grey or white colour, often with black head and wing-tips. The young are usually darker than the adults and mottled with brown. The guillemots (*Uria*) and little auks (*Plautus*) are more fully marine animals, breeding in very large colonies on the cliffs.

7.11. Order 17: Gaviiformes. Divers or loons

The divers are aquatic birds retaining some primitive characteristics. They are birds of open waters, feeding mainly on fishes. Various species of *Gavia* live mostly on the sea, but breed by lakes throughout the holarctic region.

7.12. Order 18: Columbiformes. Pigeons

The pigeons are tree-living, grain- or fruit-eating birds, mostly good flyers but retaining some primitive features (Fig. 15.23). They are of world-wide distribution. There is little sexual dimorphism; the nest is usually simple and the eggs normally one or two and white. The young are born very little developed and are nourished by the 'milk' secreted by the crop (p. 365). The dodo (*Raphus* = *Didus*) was a pigeon that adopted a terrestrial habit in the island of Mauritius and grew to a large size, but was exterminated by man in the seventeenth century.

7.13. Order 19: Cuculiformes. Cuckoos

The cuckoos include some species that build nests but many lay their eggs in those of other birds. In the common cuckoo (*Cuculus*), any one individual female lays mostly in the nests of a single foster species, in England often the meadow-pipit or hedge-sparrow. She watches the building of the nest and lays her egg on the same day as the foster parent, removing one of the clutch before she does so. Often about twelve eggs are laid in this way, each in a different nest; even more have been recorded. The eggs are usually strongly mimetic with those of the host, variable in colour, more so when varied host nests are available. The young hatch before the host eggs, which are then ejected from the nest by the young cuckoo.

The hoatzins, *Opisthocomus*, of tropical South America, have been shown by protein tests and anatomy to be related to the cuckoos. The young possess well marked claws on the digits of the wing (Fig. 17.5), which they use for climbing (see Attenborough 1979). These claws are usually considered to be a secondary development; their resemblance to the claws of *Archaeopteryx* is remarkable.

7.14. Order 20: Psittaciformes. Parrots

The parrots are birds found mainly in warm climates, living among the trees and having many special characteristics. With the crows, they are usually reckoned to be the most 'intelligent' birds and certainly have considerable powers of memory. They are predominantly vegetarian and some, though by no means all, make use of

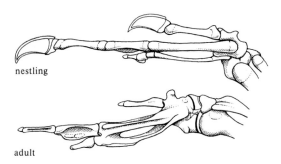

nestling

adult

FIG. 17.5. The hand of the hoatzin (*Opisthocomus*) showing claws in young. (After Heilmann 1926.)

the beak for breaking open hard shells. The eggs are usually laid in holes and are white and round. The period of parental care after hatching is unusually long (2–3 months).

7.15. Order 21: Strigiformes. Owls

The owls, specialized for hunting at night, resemble the hawks, by convergence, in their beaks, claws, and in other ways. The food is swallowed whole. They probably detect their prey mainly by sound, and show various specializations in the ears. The eyes contain mostly rods and are directed forwards; they are very large and they cannot be moved in the orbits, the movements of the neck compensating for this restriction. The feathers are so arranged as to make very little noise in flight. The eggs are white and laid in holes or in the old nests of other birds, some on the ground. Many genera are recognized from all parts of the world, examples being the barn owls (*Tyto*) and the eared owls (*Asio*).

7.16. Order 22: Caprimulgiformes. Nightjars

The nightjars (*Caprimulgus*) are a rather isolated group of crepuscular birds, feeding on insects taken on the wing. Two mottled eggs are laid on the bare ground.

7.17. Order 23: Apodiformes. Swifts and humming-birds

The swifts (*Apus*) and humming-birds (*Trochilus*) are perhaps more fully adapted to the air than are any other birds. The wings are very long, composed of a short humerus and long distal segments. The swifts are insectivorous and have very large mouths, adapted for feeding on the wing. The nests are often made in holes as by the only British species (*Apus apus*). Other species attach bracket, sleeve, or pocket-like structures to trees, cliffs, or buildings. The eggs are white, and the young helpless at birth.

7.18. Order 26: Coraciiformes. Bee-eaters and kingfishers

This is a large group of birds, including the bee-eaters (*Merops*), mainly tropical and often brightly coloured. The three anterior toes are united (syndactyly). The nests are usually made in holes and the eggs are white. The kingfishers (*Alcedo*) are modified for diving into the water to catch fish.

7.19. Order 27: Piciformes. Woodpeckers

The woodpeckers (*Picus*) are highly specialized climbing, insectivorous, and wood-boring birds. The bill is very hard and powerful and the tongue long and protrusible and used for removing insects from beneath bark. The tail feathers are used to support the bird as it climbs the tree-trunk. The nest is made in a hole in a tree and the eggs are white.

7.20. Order 28: Passeriformes. Perching birds

The great order of perching birds contains about half of all the known species. They often live close to the ground, though there are many tree-top species in the tropics. They are rather small, and of very varied habits. There are always four toes arranged to allow the gripping of the perch. The display and nesting behaviour is usually complicated, with a well developed song in the male. Many species build very complicated nests and the eggs are often brightly coloured and elaborately marked. The young are helpless at birth. Only a few of the many and varied types can be mentioned here.

The rooks and jackdaws (*Corvus*) (Fig. 15.24) are the largest passerines and perhaps 'highest' of all birds; they are mostly colonial. The starlings (*Sturnus*) are also partly colonial and nest in holes. The finches (*Fringilla*, etc.) are seed-eating birds with a short, stout, conical bill. The house-sparrows (*Passer*) are closely related to the finches and have become commensals of man all over the world. The larks (*Alauda*) make their nests on the ground. The pipits (*Anthus*) and wagtails (*Motacilla*) are somewhat like the larks, largely terrestrial birds with slender bills. The tree-creepers (*Certhia*) are tree-living, insectivorous birds with long bills, showing some convergent resemblance to woodpeckers. The tits (*Parus*, etc.) are a large group of woodland birds; they chiefly eat insects, also buds and fruits. The shrikes (*Lanius*) are peculiar among passerines in being mainly carnivorous, using their strong bills to eat other birds, amphibia, reptiles, and large insects.

The warblers (*Sylvia*, etc.) make a very large group of woodland birds, living in trees or scrub. Related to them are the thrushes and blackbirds (*Turdus*), the British robins (*Erithacus*), and nightingales (*Luscinia*), mainly eating small invertebrates, also fruits. They have a very wide distribution and are among the most recently evolved and successful members of the whole class. The hedge-sparrows (*Prunella*) are small omnivorous passerines, possibly related to the thrush group. The wrens (*Troglodytes*) are small and mainly insectivorous. The swallows (*Hirundo*) form a very distinct family of passerines, suited for powerful flight and feeding on insects caught in the air. The very long pointed wings and 'forked' tail allow rapid manœuvring in the air and the insects are taken in a wide mouth. These features, together with the elaborate migrations, mark the swallows as among the most specialized of all birds.

8. Tendencies in the evolution of birds

The bird plan of structure, originating in the Jurassic period, perhaps 150 Ma ago, has become modified to produce the great variety of modern birds. In trying to discover the factors that have influenced this modification we are handicapped by the poverty of fossil remains; it is not possible to trace out individual lines as it is in other vertebrate groups. It is clear that the process of change has been radical, the later types often completely replacing the earlier ones: for instance no long-tailed or toothed birds remain today.

Our knowledge of direction of the change is largely dependent on study of the variety of birds existing today, which is perhaps more thoroughly known than in any other group of animals. In the reports of those who have studied this variation there are two distinct, indeed opposite, tendencies. Many have observed that adaptive radiation has occurred; birds are found occupying a wide variety of habitats, with modifications appropriate to each way of life. Other workers, recording minor differences between races occurring in different areas, have found difficulty in believing that these have adaptive significance. The existence of such 'subspecies' with a geographical limitation is a striking characteristic, especially conspicuous in widely distributed species such as the chaffinch (*Fringilla*). On continental areas the subspecies usually grade into each other (making 'clines') and are interfertile at the areas where they meet. On the other hand, where a group of individuals becomes isolated on an island, or by some other geographical barrier, it may become infertile with the 'parent' species. If the two groups again come to occupy a common area, then either one eliminates the other or slight modifications of habits enable the two to survive side by side as two distinct species.

Definite cases of formation of new species in this way have been recorded in birds, which are especially suitable for such study. Thus in the Canary Islands, besides a local form of the European chaffinch (*Fringilla coelebs*) there is the blue chaffinch (*F. teydea*), which was probably originally an offshoot from the European form. On the mainland *F. coelebs* inhabits both broad-leaved and coniferous forests, but in Grand Canary *F. teydea* occupies the pine woods, *F. coelebs* the chestnut and other woods. On the island of Palma, however, the blue chaffinch is absent and the European form occupies both habitats. It is presumed that the blue form is more suited to the coniferous woods, but this could not be deduced from its specific characters as recorded by a systematist. The adaptive significance of the differences between groups of animals may not be easy to discover, but it is most unwise to assume that it does not exist until a very thorough study has been made. Detailed observation generally shows small differences in habits and behaviour between animals occupying what seems at first to be a single 'habitat'. Lack, who studied this question, wrote 'A quick walk through the English countryside might suggest that there is a wide ecological overlap between the various song-birds. In fact, close analysis shows that there are extremely few cases in which two species with similar feeding habits are found in the same habitat' (1933). The differences may be in the breeding habitat, as in the case of the meadow, tree, and rock pipits (*Anthus pratensis*, *A. trivialis*, and *A. spinoletta*). The spotted and pied flycatchers (*Muscicapa striata* and *M. hypoleuca*) catch their food in slightly different ways and the chiff-chaff (*Phylloscopus collybita*) feeds higher in the trees than the willow warbler (*P. trochilus*).

A complicating factor is that birds that occupy similar habitats for one part of the year may migrate to different regions, for instance, the tree and meadow pipits and the chiff-chaff and willow warbler.

Birds are able to get their food and to breed in many different ways, and when a race finds a situation occupied it perhaps often survives by a slight change of habits, creating a new 'habitat' not previously occupied. This presumably results from the action of certain individuals whose constitution differs from the mean of the race, enabling them to pioneer. It is not certain to what extent individual birds are able to 'adapt themselves' in this way to new habitats. Probably the majority of the members of any population are limited by their structure and behaviour pattern to a rather narrow habitat range.

9. Darwin's finches

A remarkable example of evolution and adaptive radiation is provided by the birds of the Galapagos Islands. It was these birds and the giant tortoises (after which the islands are named) that started Darwin on his study of evolution. 'In July opened first note-book on "Transmutation of Species". Had been greatly struck from about month of previous March on character of South American fossils – and species on Galapagos Archipelago – These facts origin (especially latter) of all my views' (Darwin's Diary 1837).

These islands are volcanic and probably of Tertiary (? Miocene) date. They lie on the equator, 960 km west of Ecuador, the only other nearby land being the island of Cocos, 960 km north-west (Fig. 17.6). It is no longer suggested that the islands were connected by a land bridge with South America. Their limited stock of plants and animals has arrived across the sea. Of hundreds of species of land birds on the mainland, descendants of only seven species are found in the Galapagos. The only land mammals are a rat and a bat. The land reptiles include giant tortoises, iguanas, a snake, one lizard, and one gecko. There are no amphibians and only a limited number of land insects and molluscs. There are large gaps in the flora; for instance, no conifers, palms, aroids, or Liliaceae. This fragmentary flora and fauna strongly suggest that the islands have been colonized by chance transportation across the sea, and that, once arrived, the animals and plants have proceeded to settle not only in the habitats they occupied on the mainland but also in others, not filled, as in their homeland, by rivals. Thus the tortoises and iguanas, arriving presumably by chance, have grown to large size, to occupy the ecological position usually taken in other faunas by mammalian herbivores. The composite plant *Scalesia* and the prickly pear, *Opuntia*, have become tall trees in the Galapagos. We have here, therefore, an example of the results of evolution over a relatively limited period of

time (perhaps less than 5 million years) from a limited number of initial creatures, and this provides an excellent opportunity for trying to discern the forces that have been at work.

There are thirteen larger islands in the Galapagos group, the largest 128 km long. They are separated by distances of up to 160 km and several of the peculiar Galapagos animals have formed island races. The land birds present perhaps the most interesting features of the whole strange fauna. Besides two species of owls and a hawk they consist of five passerine types and a cuckoo, all very close to others found on the South American mainland, and a group of fourteen species of finches, placed in a distinct subfamily, Geospizinae. These finches are related to the family Fringillidae, which is represented in South America, but they cannot be derived from any single species now existing there. Like the other animals in the islands the birds tend to become differentiated into distinct island races, but this process has gone to varying extents. The cuckoo, warbler, martin, and tyrant flycatcher are similar in all the islands and it is significant that they are all very close to species occurring in South America. Presumably they are recent arrivals. The vermilion flycatcher, also a South American species, has three island races. The mocking-bird, *Nesomimus*, is placed in a genus distinct from that on the mainland and has different races on each island, some of them being reckoned as separate species. The extreme of island differentiation is shown by the finches, now reckoned to belong to fourteen species, classed in four genera (Fig. 17.7), one of these being found on the distant Cocos Island.

Evidently these finches have been in the archipelago

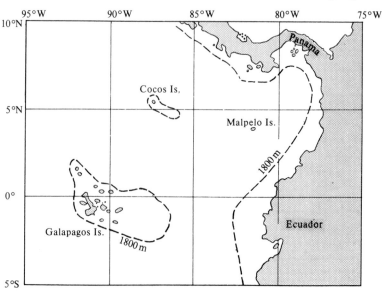

Fig. 17.6. Position of the Galapagos Islands. (From Lack 1947.)

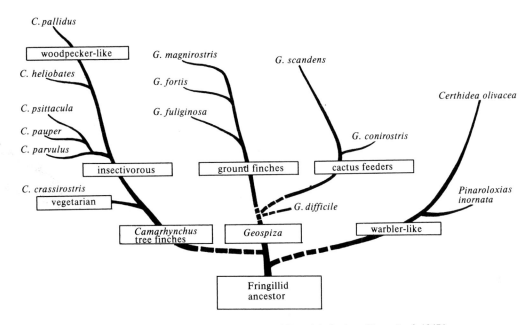

FIG. 17.7. Suggested evolutionary tree of Darwin's finches. (From Lack 1947.)

for a considerable time and the specially interesting feature is that not only have they formed races recognizably distinct, but they have radiated to form a series of birds that have quite varied habits, many of them very un-finchlike. The main differences are in the form of the beak, which varies greatly with the food habits (Fig. 17.8). The central species, which are also the nearest to the presumed Fringillid ancestor, are the ground-finches, *Geospiza*, of which there are five species, feeding mainly on seeds. Two further species of *Geospiza* have left the ground and taken to eating cactus plants. The tree-finches, placed in a distinct genus *Camarhynchus*, include one vegetarian and five mainly insectivorous species, one of the latter, *C. pallidus*, having acquired the habit of climbing up the trees like a woodpecker and excavating insects with a stick (Fig. 15.33). A third group of this remarkable subfamily has acquired a convergent likeness to warblers. *Certhidea* has a long slender beak and eats small soft insects; by many it has been regarded as distinct from the other Galapagos finches, but it has now been shown to resemble them, not only in structure but also in breeding habits. Nevertheless, it probably diverged some time ago and is found on all the islands. Presumably its success is due to the absence of other warbler-like birds, since the true Galapagos warbler is a recent arrival. The fourth genus of the Geospizinae is *Pinaroloxias*, the Cocos-finch, found on Cocos Island 960 km away, and also having warbler-like characteristics.

These birds provide a remarkable example of adap-

tive radiation 'with seed-eaters, fruit-eaters, cactus-feeders, wood-borers and eaters of small insects. Some feed on the ground, others in the trees – originally finch-like, they have become like tits, like woodpeckers and like warblers' (Lack 1947). This is interesting enough, but more can be learned from this extraordinary natural experiment than from other examples of adaptive radiation. Another most striking feature is that the birds are very variable, and are by no means uniformly

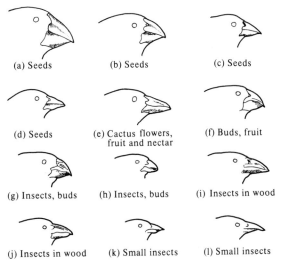

(a) Seeds   (b) Seeds   (c) Seeds

(d) Seeds   (e) Cactus flowers,   (f) Buds, fruit
                 fruit and nectar

(g) Insects, buds   (h) Insects, buds   (i) Insects in wood

(j) Insects in wood   (k) Small insects   (l) Small insects

FIG. 17.8. Beaks of Darwin's finches and their food. (From Lack 1947, after H.S. Swarth.)

distributed over the islands. The outlying islands lack certain species and their place is then taken by variants of others, with corresponding modification of the beak. In some such cases the local subspecies can be clearly seen to have an adaptive significance, but this is by no means always so. Some races found on one or more islands differ in minor features of colour or size. It remains to be shown whether these are themselves significant or connected with other significant factors.

The striking feature is that so many distinct races should appear, even in birds, which could easily cross the distances, mostly less than 80 km, between the islands. Evidently the birds tend to remain at home, and there is no doubt that degree of geographical separation is a main factor in producing the races. The central islands of the group, lying close together, have no endemic subspecies, whereas the forms in the outlying islands are mostly distinct (Fig. 17.9). It is evident that isolation is tending to break up the population into a number of distinct units and it is remarkable that, in spite of the large amount of variation, intermediates are rare between species and even between subspecies. The species, of course, do not cross in nature, and such as have been tested in captivity mate only with their own species.

It is still not possible to give an entirely clear idea of the factors responsible for the development of varied animals, even in this much simplified case. There is no clear line between subspecific differences, say of the beak, that are not obviously adaptive and those with clearly adaptive character. Since favourable mutations will spread through a population the easiest assumption is that all these differences are in fact adaptive, or linked with unseen adaptive characters. The easiest assumption is not necessarily correct, but the demonstrably adaptive character of many of these differences certainly constitutes a case for considering that the others are also of this nature. In very small populations (< 1000) unfavourable characteristics may become established by chance ('genetic drift', Wright 1968), and this factor may be responsible for some of the island races (Chapter 1).

It is especially interesting that in the Galapagos animals special conditions have made it possible for variety to arise. If, as seems likely, the earliest *Geospiza* species were among the first birds to arrive in the archipelago, they found there neither competitors nor enemies. Because of the distance from the mainland this condition has remained with little change ever since. Even today the six passerines not belonging to the

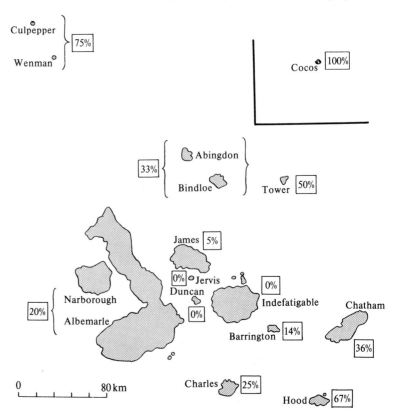

FIG. 17.9. Percentages of endemic forms of Darwin's finches on each island, showing the effects of isolation. (From Lack 1947.)

Geospizinae do not provide serious rivals, and the local owls and hawks are apparently not serious predators on the finches.

The other great factor that has led to differentiation is the splitting up of the area into a number of isolated units. This has allowed slightly different races to emerge, and we may suppose that if these again came into contact with each other they would find slightly different optimal conditions, and therefore, with partial or complete intersterility, would continue as distinct species. This has almost certainly happened with the Canary chaffinch (p. 395), and in Galapagos there is a similar case in that two species of the large insectivorous tree-finch *Camarhynchus* occur together on Charles Island. Both are derived from a single population, offshoots from which have independently colonized the island (Fig. 17.10).

10. Birds on other oceanic islands

While the case of Darwin's finches is very striking it is important to recognize that similar radiation from a few species is not found on all oceanic islands. For instance,

there are numerous birds on the Azores and they differ little from those of Europe. Whereas migrants are frequent in the Azores they are rare in the Galapagos; evidently there are special factors producing the isolation of the latter.

A radiation similar to that of the Galapagos has, however, occurred in Hawaii, where there are only five passerine forms and one of these, a finch-like bird, the sicklebill, has produced an even greater variety than Darwin's finches (see Lack 1971, 1976) (Figs. 17.11 and 17.12). Some of these sicklebills feed on insects, others on nectar, fruit, or seeds, and the beaks have developed accordingly. One species climbs like a woodpecker, digs for beetles with a short lower mandible, and probes them out with a long curved upper one.

11. The development of variety of bird life

This and similar evidence from the study of the relatively recent evolution of birds might be summarized by saying that the production of the variety of animal types is largely due to the interplay of biotic factors. The particular characteristics that each population acquires

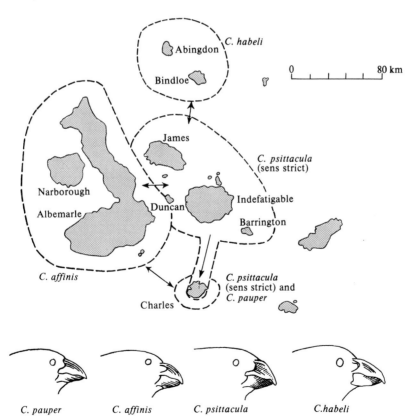

FIG. 17.10. Development of distinct forms from a single one in the Galapagos finch *Camarhynchus*. Charles Island has two forms derived independently from the central form *C. affinis*. (From Lack 1947, partly after H.S. Swarth.)

FIG. 17.11. Adaptive radiation of sicklebills in Hawaii. (From
Lack 1947, after J.G. Keulemanns.)

depend partly on physical and geographical conditions, but the stimulus to change, or the check on change, comes from the interaction of the animals and plants with each other.

It has already been suggested that the tendencies to increase and to vary are important among these factors influencing animal evolution and there is evidence that both have been at work in the development of the population of Galapagos finches and other animals. It is safe to say that there are more and varied finches, tortoises, iguanas, and mocking-birds than there were when each arrived. The other factors that we are now able to isolate, by their absence in this case, are the competitors and the predators. In more fully developed continental populations these probably tend to limit the development of variety, allowing only those individuals of a population showing the mean or 'normal' structure

and behaviour to survive. Extreme variants that venture to brave the enemies and seek new habitats are eliminated. Variation will arise when these checks are weak. In a crowded habitat this may occur as a result of some peculiar swing of the elaborately balanced interacting system of biotic factors, or by some external physical change. We still do not know which of these is the more important in producing the changes of these complicated mainland populations, but the evolutionary laboratories provided by the volcanoes of the Galapagos and other islands suggest that evolutionary change does not follow only on climatic or other physical changes. A single population will become divided into several distinct ones by its own tendencies to growth and variation, given absence of competitors and predators and some means of isolating the animals in different parts of the range.

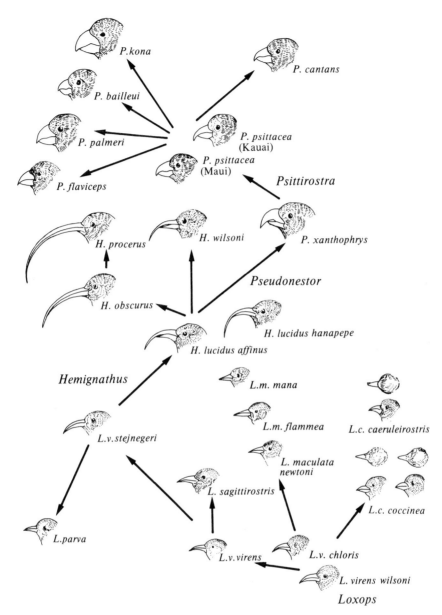

FIG. 17.12. Phylogenetic relations inferred for living species of Hawaiian sicklebills of the family Psittirostrinae. A species of the genus *Loxops* founded the psittirostrines, and radiated into at least five species including *Hemignathus lucidus*. The latter is not greatly different from its ancestral subspecies, *Loxops virens stejnegeri*, but has itself radiated to produce three species with which it forms a rather distinctive adaptive group. Accordingly, this group is recognized as a genus, *Hemignathus*. Another species derived from *H. lucidus* is placed in a separate, monotypic genus, *Pseudonestor xanthophrys*. Six living species have radiated from a rather compact group of their own, *Psittirostra*. (Modified after Dobzhansky, Ayala, Stebbins, and Valentine 1977.)

18   The origin of mammals

1. Classification

Class REPTILIA
  Subclass *Synapsida
    Order 1: *Pelycosauria (= Theromorpha). Carboniferous–Permian
      *Varanosaurus; *Edaphosaurus; *Dimetrodon
    Order 2: *Therapsida. Permian–Jurassic
      Suborder 1: *Anomodontia
        *Galepus; *Moschops; *Dicynodon; *Kannemeyeria
      Suborder 2: *Theriodontia
        *Cynognathus; *Scymnognathus; *Bauria; *Dromatherium; *Tritylodon; *Oligokyphus; *Diarthrognathus
Class MAMMALIA
  Subclass 1: Prototheria
    *Order 1: Docodonta. Triassic–Jurassic
      *Morganucodon, Triassic, Europe; *Docodon, Jurassic, North America and Europe
    Order 2: *Triconodonta. Triassic–Jurassic. Europe and North America
      *Amphilestes, *Triconodon, *Megazostrodon, *Eozostrodon
    *Order 3: Multituberculata. Jurassic–Eocene. Europe and North America
      *Plagiaulax, Jurassic, Europe; *Ptilodus, Palaeocene, North America; *Kamptobaatar, Upper Cretaceous, Mongolia.
    Order 4: Monotremata. Pleistocene–Recent. Australasia
      Tachyglossus (=Echidna), spiny anteater, Australia and New Guinea; Zaglossus (=Proechidna), New Guinea; *Obdurodon, Miocene, Australia; Ornithorhynchus, platypus, Australia
  Subclass 2: Theria
    *Infraclass 1: Pantotheria
      *Order 1: Eupantotheria. Jurassic. Europe and North America
        *Amphitherium; *Kuehneotherium
      *Order 2: Symmetrodonta. Jurassic. Europe and North America
        *Spalacotherium
    Infraclass 2: Metatheria. Cretaceous–Recent
      Order Marsupialia
    Infraclass 3: Eutheria (=Placentalia). Cretaceous–Recent

2. The characteristics of mammals

The idea that the mammals include the highest of animals can easily be ridiculed as a product of human vanity; we shall find, however, that in many aspects of their structure and activities they do indeed stand apart as animals that are 'higher' than others, in the sense in which we have used the word throughout this study. The mammals and the birds are the vertebrates that have become most fully suited for life on land; among them

are many species in which the processes of life are carried on under conditions far remote from those in which life first arose. The information in their DNA (deoxyribonucleic acid) provides them with numerous special adaptive devices, not the least being brain capacities that enable them to collect further information during their individual lifetimes. The mammalian organization includes a great number of special features that together enable life to be supported under conditions that seem to be extravagantly improbable or 'difficult'. For example, the surface of the body is waterproofed, and elaborate devices for obtaining water are developed. A camel and the man he is carrying through the desert may perhaps contain more water than is to be found in the air and sandy wastes around. This is only an extreme example of the 'improbability' of mammalian life, which is one of its most characteristic features.

The faculty of maintaining a high and constant temperature has opened to the birds and mammals many habitats that were closed to the reptiles. Besides making life possible under extremely cold conditions, such as those of the polar regions, the warm blood vastly extends the opportunities for life in more temperate climates. Mammalian populations, which can feed at night and all through the winter, can of course expand more rapidly than their reptilian cousins, which must hibernate for much of the year, during which period they consume rather than produce living matter.

The success of the mammals in maintaining life in strange environments is largely due to the remarkable powers they possess of keeping their own composition constant. All living things tend to do this, but it seems probable that the mammals maintain a greater constancy than any other animals, except perhaps the birds. Claude Bernard's famous dictum 'La fixité du milieu intérieur c'est la condition de la vie libre' (1878) may be doubtful of vertebrates in general, but it can certainly be applied to mammals. They have a life that is more free than that of other groups, in the sense that they can exist and grow in circumstances that other forms of life would not tolerate, and they can do this because of the elaborate mechanisms by which their composition is kept constant. Besides the regulation of temperature there is also a regulation of nearly all the components of the blood, which are kept constant within narrow limits. Barcroft (1932) has pointed out that the achievement of this constancy has enabled the mammals to develop some parts of their organization in ways not possible in lower forms. For instance, an elaborate pattern of cerebral activities requires that there shall be no disturbances by sudden fluctuations in the pressure and composition of the blood.

We shall expect to find in the mammals, therefore, even more devices for correcting the possible effects of external change than are found in other groups. Besides means for regulating such features as those mentioned above we shall find that the receptors are especially sensitive and the motor mechanisms able to produce remarkable adjustments of the environment to suit the organism, culminating in man with his astonishing perception of the 'World' around him and his powers of altering the whole fabric of the surface of large parts of the earth to suit his needs.

Such devices for maintaining stability often take peculiar and specialized forms in particular cases. The activity and 'enterprise' of mammals has led many of them to make use of particular structures and tendencies in order to develop very odd specializations, which enable them to occupy peculiar niches. What could be more bizarre than the development of the muscles of the nose until a huge mobile trunk appears, so that the heaviest of four-footed beasts, while using its legs for support, can also 'handle' objects more delicately than almost any other animal?

The mammals have developed along many special lines and many of these have already become extinct; others, especially among rodents and primates, remain among the dominant land-animals today. Warmth, enterprise, ingenuity, and care of the young have been the basis of mammalian success throughout their history. The most characteristic features of the modern mammals are thus seen to be largely in their behaviour and soft structures. Mammalian life is above all else active and exploratory. Mammals might well be defined as highly percipient and mobile animals, with large brains, spiral cochlea, warm blood, left aortic arch, and a waterproofed, usually hairy skin, whose young are born alive, and are nourished by milk. Since it is difficult to recognize such characters as these in fossils we cannot say exactly when they arose, and our technical definition of a mammal must be made on the basis of hard parts. These include tribosphenic molars, a dentary – squamosal articulation and three ossicles in the middle ear. The side wall of the skull is formed by the alisphenoid bone. There is an axis–atlas complex and seven cervical vertebrae. The pelvis has an anterior iliac blade and small pubis. Some of these features were not yet present in the earliest mammals, the Prototheria (p. 409).

The whole series of mammal-like forms from Carboniferous anapsids onwards forms a natural unit, and it is only by an arbitrary convention that we separate the reptilian subclass Synapsida from the class Mammalia. The present day mammals form a distinct group of animals, which we identify superficially by their possession of hair. For instance, the duck-billed platypus is immediately referred because of its hair to the mammals. We class it apart from reptiles or birds, even though its internal organization and the fact that it lays eggs show it to have many similarities with reptiles, and its bill is like that of a duck. The technical characteristic of the class Mammalia is conventionally given by the presence of a single dentary bone in the lower jaw, the articular and other bones forming, with the quadrate, part of the mechanism of the middle ear. However, fossils showing intermediate conditions are now known from the upper Triassic, say 180 million years ago, and all stages can thus be traced in the jaws. The full mammal-like condition was established by the middle Jurassic period, about 150 million years ago. The reduction of the jaw bones may perhaps have been associated with the habit of chewing the food, in order to obtain the large amounts necessary to maintain a high temperature.

It would not be very rash to suggest that by Jurassic times the synapsids had developed the other mammalian characters, such as active habits, large brain, warm blood, hair, and perhaps also a diaphragm, four-chambered heart, and single left aortic arch. They may even have been viviparous, for the surviving monotremes, which lay eggs, probably diverged at a still earlier period, perhaps in the Triassic (p. 422).

3. Mammals of the Mesozoic

In spite of all the uncertainties of the fossil record it is now possible to follow the history of the Mammalia back to their origin from cotylosaurian reptiles of the Permian, more than 225 Ma ago. Sufficient information is available for us to be quite sure that some population of early anapsid reptiles, such as *Solenodonsaurus* of the Carboniferous and Permian times, besides giving rise to all the modern reptiles, to the dinosaurs, and to the birds, also produced the mammals. The evidence for this connection rests on a most interesting series of fossils, together with some 'living fossils', the monotremes of Australia. The fossil history is not at all times equally clear. The mammalian stock first became distinct in late Carboniferous times, as a special type of cotylosaurian reptile, with a tympanum placed behind the jaw and later a lateral temporal aperture. This stock quickly became very abundant and successful as the pelycosaurs and therapsids of Permian and Triassic times, but then nearly died out in the Jurassic, during which period we know of the mammals only from fragmentary remains of a few small animals. Then, in the Cretaceous, some of these small forms became more numerous and from them arose, before the Eocene, a variety of different types of mammal (see Lillegraven, Kielan-Jaworowska, and Clemens 1979), from which the histories of the modern orders can be followed in some detail.

We have, therefore, abundant evidence of the earliest stages of mammalian evolution, say 200 Ma ago from fossils in the Permian Red Beds of Texas. For the next following 40 Ma or so we have also rich material from the Upper Permian and Triassic Karroo beds of South Africa. In these early times the mammal line was a flourishing one, more so indeed than the diapsids or any

other of the descendants of the cotylosaurs. The animals were mostly carnivorous, though there were also herbivorous types. Many of these early mammal-like forms became quite large and numerous, in fact this stock dominated the land scene in the Permian. In the later Triassic, however, if our fossil evidence is a safe guide, their numbers became reduced, perhaps the carnivorous dinosaurs took their place. Already at this time many of the essential features of mammalian organization had been developed, so far as these can be judged from the bones. Possibly the soft parts of these Permian forms were also mammal-like, the animals may have been active and 'intelligent', have possessed hair and warm blood. But their brains were small and reptilian in structure.

Throughout the succeeding 100 Ma of Jurassic and Cretaceous times our knowledge of the mammalian stock is dim. Mammals of a somewhat rodent-like type, the multituberculates, became quite numerous and produced some large forms, but then became extinct in the Eocene. Unless we have been singularly unlucky in the preservation of mammalian remains, no other mammal-like animals larger than a polecat existed throughout this long period. Since we possess detailed information, based on numerous fossils, about scores of large and small diapsid reptilian types, it can hardly be only an accident that mammal-like fossils are so rare.

We must conclude that the mammalian organization, after an initial success in the Permian and Triassic, was almost supplanted in the Jurassic and Cretaceous by the various diapsid creatures. Only a few are fossils, mostly of teeth and lower jaw bones, to give us some idea of the nature of the animals that carried our type of organization through the Jurassic period. Then from the top of the Cretaceous, about 70 Ma ago come the first fossil remains of mammals similar to those alive today; rare, insectivorous creatures, not widely different from modern hedgehogs. This was the time of the beginning of an astonishing revolution, which completely altered the life on the face of the earth. The descendants of these shrew-like animals multiplied considerably in the new conditions, and by the earliest Palaeocene times, 65 million years ago, they had produced recognizable ancestors of most of the modern orders of mammals.

Our knowledge of the origins of mammals derived from fossils is supplemented by certain surviving mammals, whose structure shows them to have diverged rather early from the main stock, especially the monotremes (duck-billed platypus and spiny anteater) and the marsupials (kangaroos, opossums, etc.). Unfortunately it is still not clear exactly how these survivors are related to the main stocks as revealed by the fossils. The monotremes are almost unknown except as Pleistocene fossils; so, tantalizingly, we know only little about the affinities of this ancient group. The characteristics of their bony skeleton show that they must have diverged from other mammals well back in Mesozoic times. For instance, the pelvic and pectoral girdles are very 'reptilian' in structure. It is therefore not wild speculation to use the characters of the soft parts of monotremes to deduce those of the mammalian stock in late Triassic or early Jurassic times, say, 195 Ma ago. The marsupials appear as fossils in the late Cretaceous. Some of the earliest forms are very like modern opossums, and we may conclude that the condition of modern marsupials throws some light on the probable condition of the soft parts of mammals 70 or 80 Ma ago. However, it has been realized in recent years that many features in marsupial organization are special developments, not 'primitive' characteristics carried over from the ancestral condition.

The piecing together of all this evidence to give a reliable picture of the history of mammals is a valuable exercise in zoological and geological method, as well as a means of becoming familiar with a fascinating story. We may now return to the Carboniferous times to examine the evidence more in detail (see Fig. 14.4).

4. Mammal-like reptiles, Synapsida (joined arches)

All our evidence about the origin of mammals must of course be based upon the study of hard parts, which can become fossilized, and in particular on the skull. The characteristic feature of the skull in the populations that led to the mammals is the development of a hole in the roof, low down on the side of the temporal region, the lower temporal fossa (Fig. 14.3). This hole was at first bounded above by the post-orbital and squamosal bones, below by the jugal and squamosal. It was at one time supposed that this single fossa was formed by union of the two present in diapsid reptiles, and thus the whole group acquired the inappropriate name Synapsida. Animals having this characteristic appear in the rocks at the same time as the cotylosaurs, with no hole in the skull roof (Anapsida), from which they were presumably derived. We have only fragmentary knowledge of anapsids from the Carboniferous; most of our information comes from lower Permian forms, such as

Solenodonsaurus, yet there are quite well preserved synapsids of Upper Carboniferous date. Therefore we have no complete series of successive types to show how the earliest mammal-like types arose; nevertheless we can see something of the probable stages of this progress within the Permian Anapsida. One of the characteristics of the skull of amphibians was the presence of an 'otic-notch', in which lay the tympanic membrane, at the back of the skull (Fig. 12.4). In *Captorhinus* and similar anapsid fossils from the Lower Permian this notch had disappeared (Fig. 14.2). The tympanum apparently lay behind the quadrate, leaving the whole side of the skull as a rigid support for the jaw. A long series of subsequent evolutionary changes led to one of the most characteristic features of the skeleton of mammals, namely, the conversion of the hinder jaw bones (quadrate and articular) into ossicles for the transmission of vibrations to the ear (p. 416).

Captorhinus and its allies were 30 cm or more in length and showed some development of the limbs towards the mammalian condition, in that the bones were moderately elongated and slender and the limbs perhaps tended to be held vertically under the body. However, there was no great development of a backwardly directed elbow and forward knee. The teeth were considerably specialized, possibly for eating molluscs, and there were several rows of crushing teeth on the edge of the jaw, and long overhanging ones in the premaxillae.

5. Order *Pelycosauria (= *Theromorpha, mammal-like)

The synapsid line began when an early offshoot from some such anapsid as *Captorhinus* developed a temporal fossa. This must have occurred in the middle of the Carboniferous, nearly 300 Ma ago, producing in the later part of that period, and in various Pennsylvanian and Lower Permian strata, especially in North America, these earliest synapsid populations, classified together as pelycosaurs or theromorphs. *Varanosaurus* (Fig. 18.1), from the Texas Red Beds, is a typical example, a carnivore, about 90 cm long, showing a lizard-like appearance little different from that of anapsids or primitive diapsids. Intercentra were found all along the vertebral column, as in anapsids, and there were abdo-

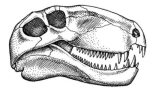

FIG. 18.1. Skulls of two early synapsids, that of *Dimetrodon* was about 40 cm long and of *Varanosaurus* 23 cm. (After Romer 1945.)

Dimetrodon   Varanosaurus

minal ribs. The teeth showed the beginnings of the mammalian differentiation in that one or more near the front of the series were elongated as 'canines'. The skull of pelycosaurs also showed tendencies in the mammalian direction in having a long anterior region and a relatively high posterior part, giving a large brain case and deep jaw.

Other pelycosaurs developed long neural spines, in more than one line, *Edaphosaurus* and its allies being herbivores, some of them as much as 360 cm long, whereas *Dimetrodon* (Fig. 18.1) was a carnivore. The spines presumably supported a web, perhaps used in temperature regulation to absorb or radiate heat, which would explain that it was relatively bigger in some of the larger animals. These animals were important in the late Carboniferous and early Permian fauna, the carnivores preying on the other early reptiles and on rhachitomous amphibia. They did not survive long, however, being apparently replaced by their descendants. From some form similar to *Dimetrodon* arose a whole series of lines classed together as Therapsida and leading on to the mammals.

6. Order *Therapsida (= mammal-arched)

In these animals there was never any trace of the bipedalism that developed in the Archosauria; the fore limbs were always at least as long as the hind, and tended to be turned under the body and to carry it off the ground. The skull became deeper and its brain case enlarged; a bony secondary palate developed from flanges of the premaxillae, maxillae, palatines, and pterygoids. The teeth became differentiated for various functions and the bones of the lower jaw, except the dentary, were gradually reduced.

These features seem to have developed in several different lines descended from theromorph ancestors; the sorting out of the various genealogies is not yet complete. It is therefore still difficult to decide for certain the interesting question whether parallel evolution occurred, and especially whether similar mammalian features appeared independently in animals of different habits. The fossils are nearly all found in South Africa and have been studied in great detail by Broom (1932).

Much of the surface of the southern part of Africa is covered by rocks known as the Karroo system. These consist partly of shales and mud-stones formed by the matter brought down by a large Mesozoic river, and the remainder are sandstones composed of blown sand. Both sorts of rock were particularly favourable for the preservation of the remains of terrestrial animals. Unfortunately the absence of marine fossils makes it

difficult to give dates for these rocks. Altogether the strata present a thickness of some 4500 m, laid down over a period corresponding probably to that from the Middle Permian to the Upper Triassic in Europe, that is to say, from 250 to 180 Ma ago.

The therapsids fall into two groups, the mainly herbivorous Anomodontia and the carnivorous Theriodontia, probably preying upon the former. *Galepus* (Fig. 18.2) shows a stage of evolution of anomodonts. The temporal opening was small, the teeth all alike and the bones at the hind end of the jaw large. The early anomodonts include the Dinocephalia (= terrible-heads), such as *Moschops*, which retained many primitive features, but the legs were turned under the body and the phalanges became reduced to the mammalian formula of 2.3.3.3.3. The roof of the head was expanded into a large dome, giving the name to the group. The later anomodonts were a still more specialized and successful group, the first of the many great tetrapod herbivores. They became the commonest of all reptiles in the later Permian. They were large creatures (*Kannemeyeria* and *Dicynodon*, Fig. 18.2), some probably living in marshes. The margins of the jaws were covered with a horny beak and the teeth reduced to a single pair of upper 'canine' tusks (from which they get their name two-tuskers), and even these were absent from some.

The most interesting of the therapsid reptiles are, however, those placed in the suborder *Theriodontia, which, by the early Triassic, had produced a very mammal-like type of organization, apparently in several independent lines. The mammals themselves very likely arose from one of them, though we do not know which. The temporal opening became progressively wider, presumably for the accommodation of larger jaw muscles, so that the parietal bone entered into the margin of the fossa and post-orbital and squamosal bones no longer met above it. Eventually the post-orbital bar itself became incomplete, leading to the typical mammalian condition, with the orbit and temporal fossa confluent. The jaw elements other than the dentary became reduced, but never wholly lost. There was a large columella, articulating broadly with the quadrate and perhaps serving to brace the latter as well as to transmit vibrations to the inner ear. A bone of this size could act as an efficient vibrator (Hotton 1959). The brain case was high and large and the cerebellum and cerebral hemispheres better developed than in modern reptiles. The occipital condyle became double as in mammals. A secondary palate became developed, allowing breathing to continue while holding the prey.

The ribs were well developed and seem to have formed a cage, which may have been used, with a diaphragm, for respiration. The limbs came to support the body off

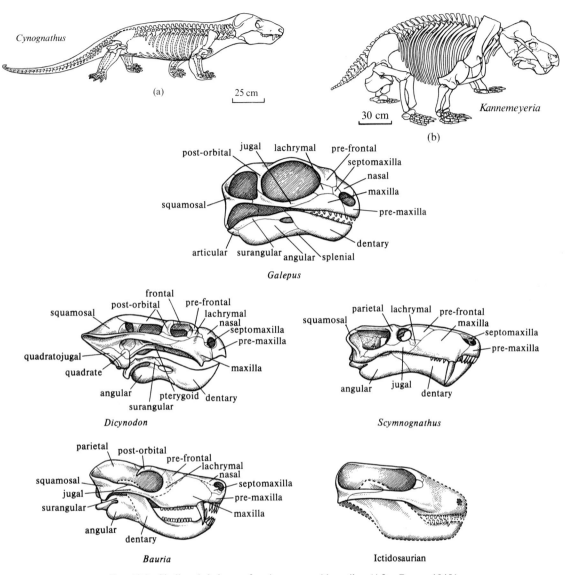

FIG. 18.2. Skull and skeleton of various synapsid reptiles. (After Romer 1945.)

the ground, and the dorsal parts of the girdles were developed accordingly, the scapula becoming large and bearing an acromial spine for muscle attachments, the coracoids being reduced. The anterior portion of the ilium became large and a hole appeared between the pubis and ischium. The head of the femur lay at the side, and a special knob, the great trochanter, appeared on it for the attachment of gluteal muscles running from the front part of the ilium and producing a backward thrust. In the hands and feet the phalanges show a reduction in several distinct evolutionary lines to 2.3.3.3.3. The teeth became differentiated into incisors, canines, and cheek teeth, the latter having several cusps

in place of the single cones of a typical reptile tooth. In most forms the tooth replacement was serial as in reptiles but in some it was more limited.

These features were little developed in the earliest theriodonts, such as *Scymnognathus* and other Permian forms (Fig. 18.2). The temporal fossa still resembled that of pelycosaurs, there was a single condyle, no secondary palate, and a phalangeal formula of 2.3.4.5.3. *Cynognathus* and other Triassic cynodonts (= dog-toothed) were typical theriodonts, showing the above 'mammalian' features. They were carnivores, of distinctly dog-like appearance, although they still retained many signs of a heavy reptilian build. *Bauria*

was of another type, still more advanced than *Cynognathus in that the orbit and temporal fossa were
confluent (Fig. 18.2). The foramina of the maxilla of
some of these animals suggest the passage of nerves and
blood-vessels for a facial musculature, which is characteristic of mammals but absent in other vertebrates.
Other types of theriodont were more specialized, with
rodent-like features (*Tritylodon, *Oligokyphus).

These *Theriodontia were the dominant carnivores of
the early Triassic, but by the end of that period they had
almost disappeared. The latest synapsids include in
some classifications the *Ictidosauria, such as *Diarth-
rognathus of the Upper Triassic. They are rare fossils,
showing almost completely mammal-like structure.
There was no post-orbital bar and no prefrontal, post-
frontal, or post-orbital bones. A well developed secondary palate was present. The bones at the hind end of the
lower jaw had become very small (Fig. 18.3). In at least
one form the dentary probably articulated with the
squamosal, though there was also a quadrate-articular
joint. This fossil has been appropriately named *Diarthro-
gnathus. The double articulation by more medial
quadrate-articular and lateral squamosal-dentary may
have served to resist the forces produced by a shearing
dentition. It is found again in the triconodonts
In the fully mammalian condition the medial articulation is lost (Kermack 1972).

Study of the synapsids, mainly from the Karroo
system, shows us, therefore, a series of types, of which
the earliest were very like the first reptiles and the latest
very like true mammals. There can be no doubt of the
general tendency, but the series is not complete enough
to enable us to follow the details of the evolution of the
populations. These fossils are revealed by denudation
and can be given only approximate dates. It is certain

that some of the mammal-like features appeared independently in lines whose evolution proceeded separately
from a common ancestor. Thus the dicynodonts and
later theriodonts all had the mammalian phalangeal
numbers, but each of these lines had certainly evolved
independently from pelycosaur-like ancestors having a
greater number of phalanges.

The influences that produced the evolution of these
populations must have been quite complex, since they
did not affect all parts of the body at once. For instance,
some early therapsids, in spite of their mammalian
phalangeal formula, still showed pelycosaur features in
the absence of a secondary palate, and presence of a
single occipital condyle and small dentary. If the
presence of a squamodentary articulation is taken as the
criterion of 'a mammal' this condition was almost
certainly reached independently by several different
lines (see Simpson 1959). We still know too little to be
able to specify clearly the conditions controlling such
evolutionary changes, but it seems possible that the
gradual appearance of terrestrial life and of large
herbivores led various animals of a suitable structure
and disposition to a carnivorous life. For this purpose
certain changes of the ancestral structure would be
needed, leading to parallel evolution in related stocks.
However, at present we can hardly do more than pose
questions about such matters and resolve to be rigorous
in interpretation of the available evidence.

7. Mammals from the Triassic to the Cretaceous

The types classified as synapsid reptiles, which we have
been considering, are not found later than the early
Jurassic, 170 Ma ago; mammals of approximately the
modern type appear in the late Cretaceous. For the

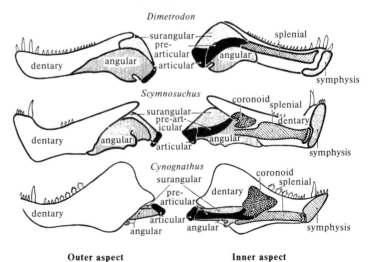

FIG. 18.3. Reduction of the articular and
other bones and increase in the dentary in
therapsid reptiles. (From Neal, H.V. and
Rand, H.W. (1936). *Comparative anatomy.*
Blakiston.)

Outer aspect   **Inner aspect**

enormous time of more than 90 Ma between these dates the mammalian organization maintained itself in the form mostly of rather small animals, many probably arboreal and nocturnal (Kielan-Jaworowska 1971; Jenkins and Parrington 1976). The fossil material of these early mammals is still scarce, and authors disagree as to the best interpretation and classification of them. It is often held that they include several different lines, which independently crossed the border between reptiles and mammals (Simpson 1961). As more is found out about the skull structure it seems likely that they are in fact all related and that we can recognize the Mammalia as a monophyletic class with two subclasses Prototheria and Theria (Kermack 1967; Hopson 1970). The subclass Prototheria includes the order *Multituberculata, which were the most abundant forms and are quite well known (Fig. 18.4) and the surviving order Monotremata. We shall also put here the orders *Triconodonta and *Docodonta about which there is much controversy. The subclass Theria includes all the modern mammals, with which we put the infraclass *Symmetrodonta and the other especially controversial group the infraclass *Pantotheria (= Trituberculata).

The differences between the two subclasses Prototheria and Theria are largely in the structure of the teeth and wall of the skull. In therian mammals this is formed by a large squamosal and alisphenoid, which are small in Prototheria. The two subclasses have diverged from Upper Triassic forms, the triconodonts and docodonts, which are put in one group or the other rather arbitrarily. This simplified classification will not be generally agreed and may be upset by further discoveries, but seems to give a synthesis of recent views (see Simpson 1961; Crompton 1980).

8. Subclass *Prototheria

8.1. Order *Multituberculata

The multituberculates were very numerous and long lasting Mesozoic mammals, surviving for over 100 Ma from the early Jurassic to the lower Eocene, longer than any other mammalian order, and radiating into six distinct families. They were herbivorous and later ones were relatively large. They had rodent-like chisel incisors and a diastema as in other mammals that chew large amounts of vegetable matter. The cheek teeth carried longitudinally arranged rows of cusps, presumably used for grinding the food. The arrangement of the muscles can be deduced from the jaws. The temporal muscle was small and there was a small coronoid process on the mandible for its attachment. There was therefore no wide sweeping up and down movement of the jaw as in carnivores. On the other hand, there was a large masseter, pulling anteriorly, and a pterygoid

pulling transversely, and a shallow glenoid fossa. All these features will be found again in placental herbivores. The skull shows similarity to that of *Morganucodon and Monotremata with small squamosal and alisphenoid and a large anterior lamina of the petrosal (Fig. 18.4). The jaw articulation was between squamosal and dentary. Epipubic bones were present, suggesting that they had a pouch. Moreover the birth passage seems to have been much too narrow for any cleidoic egg but would have allowed the birth of a very small young of about 1 g as in marsupials (Kielan-Jaworowska 1979).

8.2. Orders *Triconodonta and *Docodonta

The *Triconodonta including *Triconodon, *Morganucodon, and *Amphilestes were probably mainly carnivores and the molars carry three cusps all in one line (Fig. 18.5). There was division into premolars and molars with limited replacement as in mammals. This suggests that the young were fed by milk (p. 402). The jaw articulation included medial quadrate-articular and lateral squamosal dentary facets, perhaps connected with shearing action of the teeth (p. 414). The brain was probably larger than in reptiles. Already in the Triassic these animals had 'achieved nearly all the major adaptations for movement and stability characteristic of later mammals' (Fig. 18.6) (Jenkins and Parrington 1976). They were perhaps rather like modern shrews. There was a division between thoracic and lumbar vertebrae, characteristic of mammals. This allows a dorsoventral flexure, useful for tree-life and for running. The shoulder girdle was intermediate between that of cynodonts and monotremes, and the pelvic girdle fully mammalian. The digits carried claws and the animals were probably at least partly arboreal. Possibly they were homeotherms and invaded a nocturnal habitat not available to reptiles. The Docodonta include controversial fossils of uncertain affinities, probably close to the Triconodonta (Fig. 18.5).

8.3. Therians of the Mesozoic

The remaining Mesozoic mammals are closer to modern mammals and are placed with them in the subclass Theria, forming the infraclass *Pantotheria. The molar cusps are not all in one line but form triangles as described below. This is the basic condition from which all modern mammalian teeth have been derived.

The symmetrodonts such as *Spalacotherium (= mole-beast) were Jurassic and Cretaceous animals with three-cusped molars but no talonid on the lower ones. *Kuenotherium is a Triassic form known only from teeth, but these are two rooted and similar to those of triconodonts, which is further evidence for Osborn's

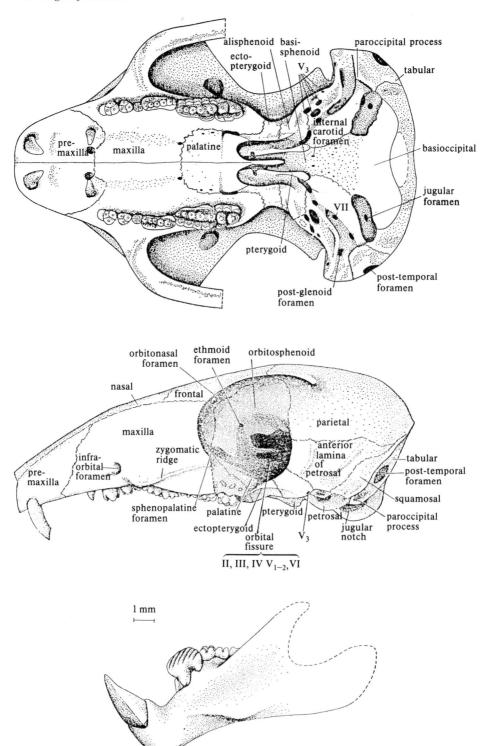

Fig. 18.4. *Kamptobaatar kuczynskii*. Reconstruction of a multituberculate skull (lateral and palatal views) and dentary (labial view). (After Kielan-Jaworowska (1971.)

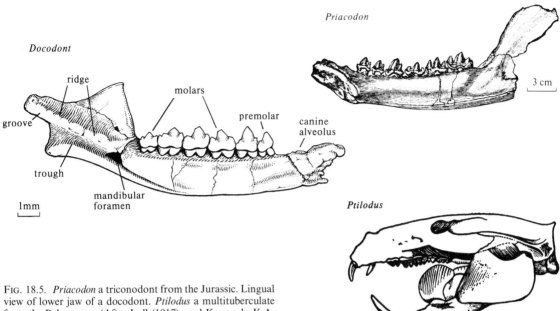

Priacodon

Docodont

ridge

molars

premolar

canine
alveolus

groove

trough

mandibular
foramen

1mm

3 cm

Ptilodus

FIG. 18.5. *Priacodon* a triconodont from the Jurassic. Lingual view of lower jaw of a docodont. *Ptilodus* a multituberculate from the Palaeocene. (After Lull (1917), and Kermack, K.A. and Mussett, F. (1958). *Proceedings of the Royal Society, London* **B148**, 204–15.)

theory that trituberculates evolved from tricondonts (Parrington 1978) (Fig. 18.7). In *Amphitherium* there were 4 pairs of incisors on each side of the lower jaw, one canine and 11 cheek teeth, 4 of these being preceded by milk teeth and hence classed as premolars. Several different sorts of eupantotherian are known from the Jurassic and could have given rise to the earliest placental insectivores, which appear in the late Cretaceous (p. 408).

9. Mammalian teeth

The teeth of reptiles prevent the escape of prey but are not usually capable of grinding food. In early mammals the upper and lower teeth began to meet in 'occlusion'. The tooth surfaces then gradually became developed for functions appropriate to the way of life; sharp in carnivores, rough in animals that grind grasses and so on. Whatever their present condition the patterns of the cusps on the teeth of the different mammalian orders can all be traced back to a common triangular form, or trigon, which was derived from the single cones of reptilian teeth. Teeth with these triangles are called tribosphenic (= three-cusped) or sometimes tuberculosectorial. They are suited to an omnivorous diet combining puncturing, shearing, and crushing. The American

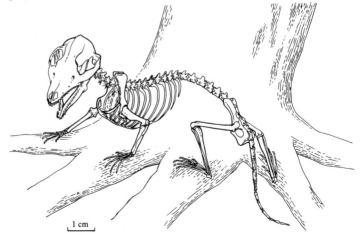

FIG. 18.6. Reconstruction of the skeleton of a Triassic triconodont, largely based on *Megazostrodon rudnerae.* (Jenkins and Parrington 1976.)

1 cm

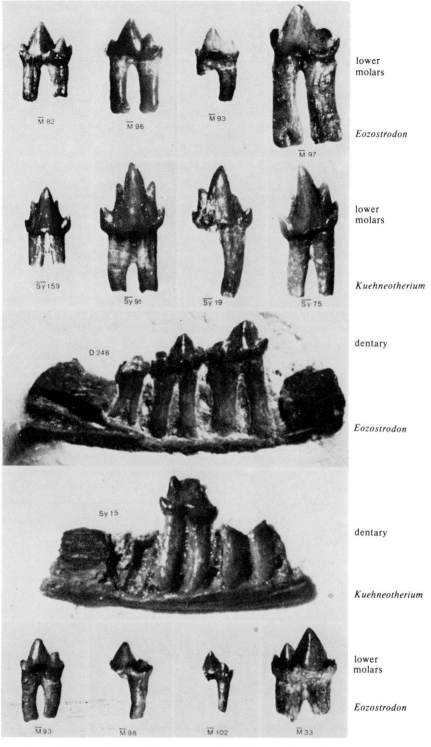

FIG. 18.7. Lower molars and dissected dentary from a triconodont, *Eozostrodon*, and a symmetrodont, *Kuehneotherium*. (Parrington 1978.)

palaeontologist Osborn (1888, 1907) suggested that the three cusps were derived from those of supposed ancestors of mammals such as triconodonts, where they are in line, by an evolutionary process of 'rotation' (Fig. 18.8) (Butler 1978). Osborn named the cusps by the order in which he thought they had appeared in phylogeny, the inner cusp on the lingual side of the upper molar being called the protocone (= first cusp) and the other members of the trigon were the paracone in front and metacone behind. Later work has shown that two triangles are involved (Fig. 18.8). The original reptilian cusp represents the paracone of the tribosphenic molar and the 'protocone' was a later addition (see Butler 1978). Several workers have therefore suggested alternative names for the cusps, but Osborn's system is well established and will be used here. The original cusp of the pretribosphenic molar can be called the eocone.

The surfaces of the molars are often complicated by further cusps. A hypocone behind the protocone turns the pattern into a square (Fig. 18.9). In front of the protocone is a protoconule and behind the metacone a metaconule. The outer border of the tooth of more primitive mammals often bears a ridge, the cingulum, and there may also be ridges along the front and back of the tooth (Fig. 18.10) (Butler 1978).

The lower molars of early mammals are also based on a triangular pattern, but here the supposed original cusp (protoconid) lies on the outer (labial) side. Osborn used similar names for the lower cusps, attaching to them the ending -id. The lower molar develops a low posterior extension, the talonid or heel, which occludes with the middle of the upper trigon. On this heel there are two or three further cusps, the hypoconid externally, entoconid internally and a smaller hypoconulid. The outer side may also have a cingulum.

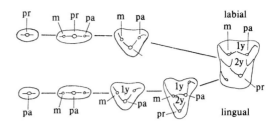

FIG. 18.8. The origin of the tribosphenic upper molar according to Osborn (above) and current ideas that two triangles are involved, labelled 1y and 2y (below).
m, metacone; *pa*, paracone; *pr*, protocone. (Butler 1978.)

10. Development of teeth and serial homology

Each tooth develops from a clump of ectomesenchymal cells, the dental papilla, derived from the neural crest. This is capped by a layer of ectoderm, the enamel organ (Fig. 18.11) (Osborn 1978). The shape of the surface of the tooth and hence the cusp pattern is determined by folding of the interface between the enamel organ and the papilla. In nearly all mammals there are gradients in tooth shape and size, with incisors in front, a single large canine on either side with premolars and molars behind. The premolars are the teeth that are preceded by milk teeth. They are often simpler in structure than the molars and may approach more closely to the ancestral tribosphenic pattern. The gradation along the tooth row strongly suggests that the form and folding of the enamel organ is controlled by some system of morphogenetic gradients (Butler 1939; Osborn 1978).

11. The chewing cycle

Cinematic recording has shown that in most modern mammals there are two phases of mastication (Kay and Hiiemae 1974; Hiiemae 1978). The initial phase is called

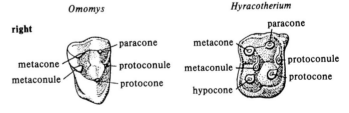

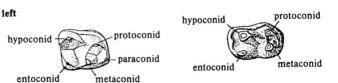

FIG. 18.9. Molar teeth of primitive placentals. *Omomys*, an Eocene tarsioid, with a tritubercular pattern; *Hyracotherium*, a Lower Eocene horse, showing the evolved quadritubercular pattern. (After Romer 1966.)

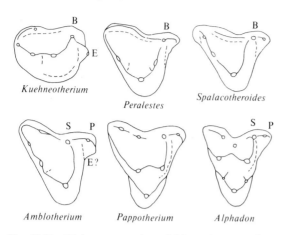

FIG. 18.10. Right upper molars of Mesozoic mammals to show possible homology of cusp B of *Kuehneotherium*. S = stylocone; P = parastyle. (After Butler 1978.)

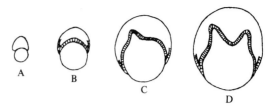

FIG. 18.11. Tooth development. A, Tooth primordium which consists of a roughly spherical clump of ectomesenchymal cells (dental papilla) surmounted by ectoderm (enamel organ). B, Later the inner cells of the enamel organ are arranged as a regular layer of columnar cells, the internal enamel epithelium. The enamel organ becomes a cap on the papilla (cap stage). C, The enamel organ continues to surround the papilla producing a bell shape (bell stage). The internal enamel epithelium now buckles in a region which later becomes the first cusp to develop. D, Subsequent growth leads to further folding of the internal enamel epithelium into a pattern very like the fully developed crown of the tooth. (Osborn 1978.)

puncture-crushing, with more vertical opposition of the cusps and basins. This produces a preliminary reduction of the food ingested, by tooth-food-tooth contact. Chewing cycles of regular mastication follow, with more transverse movement allowing the grinding and cutting edges of the molars to reduce the food so that the teeth finally reach full occlusion. The two types of lemur shown in Figure 18.12 have teeth adapted to eating somewhat different plant food (Seligsohn and Szalay

1978). *Hapalemur* divides its diet of young bamboo shoots by puncturing and crushing and has sharp conical cusps, using especially PM $\frac{4}{4}$–M$\frac{3}{3}$ which are alike. *Lepilemur* feeds on tough, leafy material of low nutritive value and chews these fibrous sheets by squat, elongated crests, with long curved cutting edges.

12. Functioning of the teeth

The mode of functioning of the cusps of the teeth has

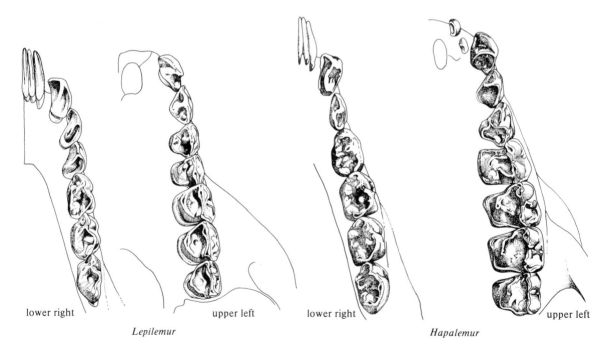

lower right   upper left   lower right   upper left

Lepilemur   *Hapalemur*

FIG. 18.12. Occlusal view of the relatively unworn dentition of *Lepilemur* and *Hapalemur*. (Seligsohn and Szalay 1978.)

been determined by cinematic observations of living forms. Fortunately the conclusions can be extended to fossils because teeth carry a record of the way they have been used by the signs of wear of the facets. Two forms of wear are recognized (Butler 1972). Abrasion is a general wear of the surface and tips of the cusps due to contact with food during the power stroke (chewing). Attrition takes place where the teeth are in contact with each other during the crushing, puncturing stroke and leads to the formation of shiny facets, where striations show the actual direction of movement between the teeth.

In most of the Jurassic mammals the upper and lower teeth pass each other with a purely shearing action, in none is there any opposition between teeth (Butler 1972) (Fig. 18.13). This was true not only of triconodonts but also of the symmetrodonts and pantotheres at that time. The only Jurassic mammals showing opposition were the herbivorous multituberculates. Then in the later Cretaceous the full tribosphenic pattern appeared in the Pantotheria (p. 411) and the teeth came into opposition, allowing not only puncturing and shearing but also crushing and grinding. The protocone was added on the lingual side of the upper molars and the trigon crushed

the food against the talonid basin. The shearing facets were retained. The new mode of functioning involved a change not only in the cusps but also in the jaw articulation, the muscles of mastication, and the brain programs for using these (Young 1978). The lower teeth were now moved medially across the upper, as can be seen from the wear facets (Crompton and Kielan-Jaworowska 1978). This stage is retained today in the teeth of the opossum (*Didelphis*). The capacity to eat a variety of food allowed the therian mammals to enter new adaptive zones and may have been a major factor in their radiation at the end of the Cretaceous. The addition of the hypocone made a further crushing surface with the trigonid of the molar behind. This is the state of the late Cretaceous *Protungulatum* and *Purgatorius* (p. 458) and from something like this the grinding teeth of the various ungulate types can be derived, the shearing function being then lost.

There have been various theories about how the triangular plan was reached. The original trituercular theory supposed that there was a 'rotation' into a triangular position from the three cusps in line of a triconodont. Even if this could be shown to have happened in a series of fossil teeth, we should still require a knowledge of the change of morphogenetic process by which the 'rotation' was produced. There have also been attempts to explain the many-cusped mammalian tooth as due to the 'fusion' of a number of reptilian tooth germs, either those making one series on the gum or the teeth of successive series. It is plausible that changes in relative time and/or place of tooth development could occur in this way, leading to a partial 'fusion' and the production of many-cusped structures. At present there is too little information to decide how the reptilian became changed into the mammalian tooth, but there can be little doubt that the trituercular theory shows us approximately the nature of the earliest mammalian cusp patterns. Indeed it was originally put forward because nearly all the Eocene representatives of the various mammalian orders show signs of a triangular cusp pattern.

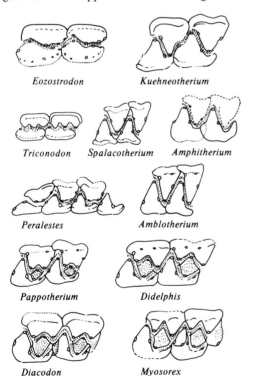

Eozostrodon   *Kuehneotherium*

Triconodon   *Spalacotherium*   *Amphitherium*

Peralestes   *Amblotherium*

Pappotherium   *Didelphis*

Diacodon   *Myosorex*

Fig. 18.13. Upper and lower molars of various 'primitive' mammals fitted together, as seen in the direction of relative movement. (Butler, P.M. (1972). *Evolution* **26**, 474–83.)

13. Subclass Prototheria

13.1. Egg-laying mammals. Order Monotremata (= one aperture)

These include the duck-billed platypus (*Ornithorhynchus*) and one species of echidna or spiny anteater, *Tachyglossus* (swift tongue) from Australia and two species of *Zaglossus* (mostly tongue) from New Guinea (Figs. 18.14 and 18.15) basically similar to each other, but so different from other mammals that it is certain that they left the main stock far back in the

FIG. 18.14.   Duck-billed platypus (*Ornithorhynchus*). (From photographs.)

Mesozoic. Their organization possibly shows us many of the characteristics of mammalian populations at that time.

13.2. The skull

Since we are comparing the platypus and echidna chiefly with Mesozoic reptiles we shall deal first with their hard parts, examining the living animals as if they were fossils. The lower jaw consists of a single dentary bone. The quadrate, articular, and tympanic have entered the ear, but the malleus is large, the incus small, and the stapes elongated (Fig. 18.16) like the columella of reptiles (as it is also in marsupials). The tympanic bone

FIG. 18.15.   Five-toed   echidna   (*Tachyglossus*).   (From photographs.)

forms a partial ring around the tympanum, and the whole apparatus is not enclosed in a bony 'bulla' as it is in modern mammals. The ossicles of *Tachyglossus* are fused together and to the periotic bone, presumably allowing detection of bone-conducted sounds of low frequencies (Fig. 18.17).

Neither type of monotreme possesses true teeth in the adult. The platypus has a flattened bill covered with soft skin and used for 'paddling' for the small bottom-living aquatic animals, especially mussels and snails, on which it lives. *Tachyglossus* and *Zaglossus* have long, strong beaks and tiny round mouths. However, in the young platypus flattened, ridged teeth are present, unlike those of any other mammals (Fig. 18.18). These true teeth are replaced by horny structures, formed by an ingrowth of epidermis beneath them and apparently used for breaking the shells of the molluscs. It is particularly unfortunate, since most of our knowledge of early mammalian affinities comes from their teeth, that we can deduce so little from those of the monotremes. Fossil teeth from the Miocene platypus named **Obdurodon* are larger than those of the modern form and have two parallel transverse lophs.

The skull (Fig. 18.19) is specialized in both genera, particularly at the front end, and many of the bones fuse early. There is a wide communication between the orbit and temporal fossa. There are many 'reptilian' features; for instance, separate pterygoid bones. The 'dumb-bell shaped bones' are perhaps the remains of the prevomer but may be vestiges of the palatine processes of the premaxillae. Small prefrontal and postfrontal bones are present. In the temporal region there is a narrow canal open at each end between the squamosal and periotic and this may represent the posterior temporal fossa of Therapsida. The lateral wall of the brain case is formed by an anterior extension of the periotic, a condition found also in multituberculates (p. 404), and not by the alisphenoid, as it is in therian mammals.

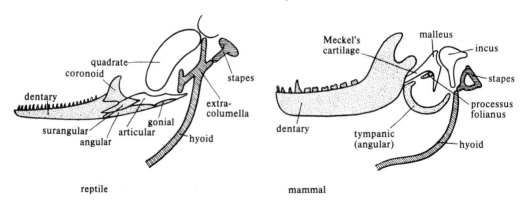

FIG. 18.16.   Diagram of arrangement of jaw and auditory ossicles in a reptile and a mammal.

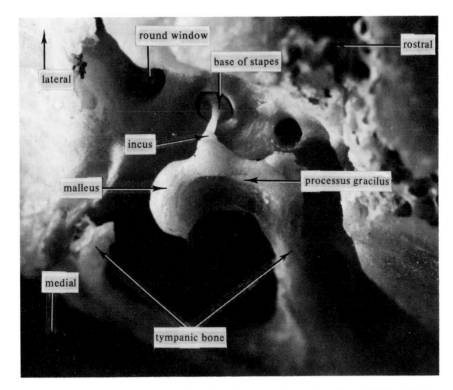

FIG. 18.17. *Tachyglossus*. Tympanic cavity with ear ossicles *in situ*. (From Griffiths 1968.)

13.3. Vertebral column and limbs

The vertebrae are very reptile-like, especially the cervicals, which bear separate ribs, as in the synapsid reptiles. There is an atlas-axis complex and seven cervical vertebrae. In the dorsal region differentiation has proceeded less far than in other mammals, there being 16 thoracic ribs in the spiny ant-eaters and 17 in the platypus, with 3 or 4 lumbars in the former and only 2 in

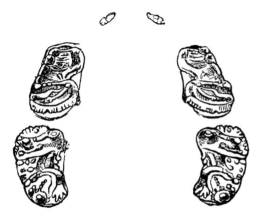

FIG. 18.18. The temporary upper teeth of the duck-billed platypus. (From British Museum Guide.)

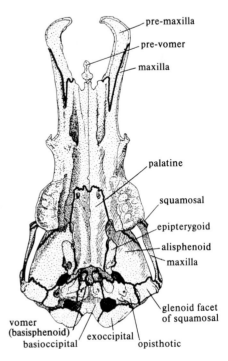

FIG. 18.19. Skull of platypus seen from below. (From Ihle.)

the latter. The ribs articulate only with the bodies of the vertebrae, not with the transverse processes. The tail is vestigial in *Zaglossus*, but forms a flattened swimming organ in the platypus. The limbs and their muscles and girdles are remarkably reptilian (Fig. 18.20). They tend to be held laterally rather than beneath the body and in general the ventral parts of the girdles are far better developed than in modern mammals, and this is sometimes spoken of as a 'plate-like' condition. It may be related to the use of the limbs for burrowing. In the pectoral girdle there are separate clavicles and a median interclavicle. The coracoid region includes two quite large and separate bones, the coracoid and 'precoracoid'. There is no spine on the scapula. The ventral portion of the pelvic girdle is enlarged by the development of epipubic bones, presumably partly for the support of the marsupial pouch. There is a large and broad humerus in both animals, held horizontally. In

Ornithorhynchus the fibula is expanded at its upper end like the ulna of other mammals, for the attachment of the large muscles that produce the swimming action.

The condition of the skeleton is therefore quite sufficient to establish the early divergence of the monotremes from other mammals and we are justified when looking at the soft parts in supposing that many of the characters were possessed by the synapsid reptiles. However, it must always be borne in mind that many of the 'mammalian' characters could conceivably have been produced by parallel evolution, subsequent to the divergence of the two lines.

13.4. Reproduction

Perhaps the outstanding non-skeletal feature is the egg-laying habit. The large yolky eggs undergo partial cleavage while still in the uterus, probably remaining there for several days while forming a blastoderm. There is rapid extension of the extra-embryonic ectoderm, which absorbs uterine gland secretions through the porous shell, formed of protein. The embryo has 19 somites and is 17 mm long at laying (in *Tachyglossus*). There is no such precocious extra-embryonic ectoderm in reptiles or birds. The monotremes thus show 'the beginnings of the process of substitution of uterine for ovarian nutriment' (Hill 1910). The eggs have a whitish shell and in the spiny ant-eater they are transferred by the mother to a special marsupial pouch in which she incubates the egg for 10 days (Fig. 18.21) (Griffiths 1978). The female platypus makes a burrow up to 15 m long terminating in a nest for her two or three eggs. She remains curled around them probably continuously until after hatching. Monotremes are unique in possessing a caruncle on the head as well as the egg-tooth, suggesting that both were present in the ancestor of amniotes as means of breaking out of the shell (Fig. 18.22). After incubation and hatching the young enter the pouch and are fed by milk. The postnatal care of the young therefore developed before the egg laying habit was lost. Both genera produce milk from about 120 large specialized sweat glands on the ventral abdominal wall of the female, but the ducts of these are not united to open on central nipples. Milk secretion and 'let down' are controlled by oxytocin and the milk contains lactose and casein.

13.5. Hair, and temperature regulation

The presence of hair again gives us a valuable clue. Unless this feature has been separately evolved on several lines we may conclude that the Mesozoic mammals and perhaps even the synapsids had made some progress in temperature regulation. The platypus has fur, dark brown in colour, of fine short hairs; in the

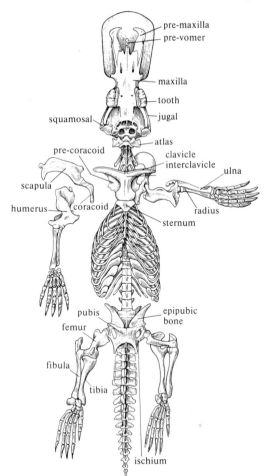

Fig. 18.20. Skull and skeleton of a female platypus. (Modified after Owen.)

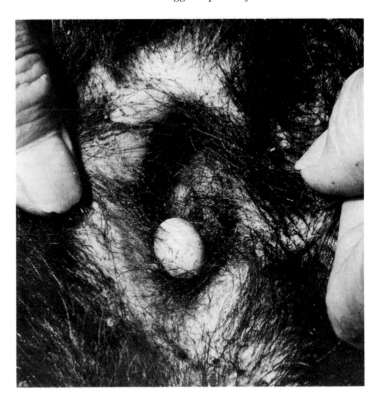

Fig. 18.21. An egg in the pouch of an echidna (eighth day of incubation) with tumescent lips, a sign that egg laying is imminent. (From Griffiths 1978.)

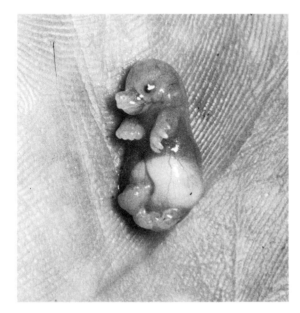

Fig. 18.22. Newly hatched echidna, showing caruncle and remains of foetal membranes. As in marsupials the forelimbs are large. Ingested milk shows through the transparent abdominal wall. (From Griffiths 1978.)

spiny ant-eater the back carries a mixture of spines and hairs, the belly carries hairs alone.

Early observations suggesting that temperature regulation was imperfect in monotremes were probably due to use of stressed animals. They do indeed have a rather low temperature, but can regulate as well as many eutherians (Griffiths 1978). Platypuses in good health maintained temperatures of about 32 °C in ambient temperatures of either air or water from 5–30 °C. There is some evidence that they cannot tolerate higher temperatures, although they have sweat glands. *Tachyglossus* also maintains a body temperature of 29–32 °C even when held for 12 h at 5 °C. In the winter, however, they hibernate and the body temperature falls to a few-tenths of a degree above ambient. Echidnas have few or no functional sweat glands and above 30 °C the body temperature rises and they die at 38 °C. In hot conditions they burrow or retire to caves.

13.6. Urinogenital system

The rectum and urinogenital system open to a common cloaca, a 'reptilian' feature from which the group gets its name, although it is found also in marsupials (Fig. 18.23). The testes are undescended. The penis of the male is a groove with erectile tissue in the cloacal floor

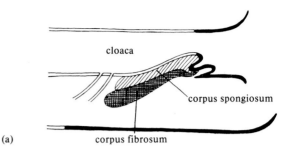

cloaca

corpus spongiosum

(a)   corpus fibrosum

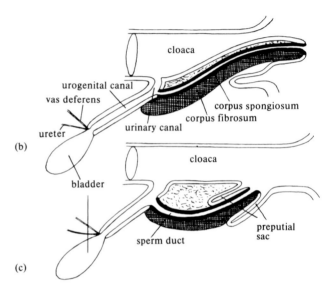

cloaca

urogenital canal
vas deferens

corpus spongiosum
corpus fibrosum

ureter   urinary canal

(b)

cloaca

bladder

preputial
sac

sperm duct

FIG. 18.23. Comparison of the cloaca and penis of
tortoise (a) and a monotreme (b). The penis is
shown erect in (b), withdrawn in (c). (From Ihle.)   (c)

and is used only for the passage of sperm, the urine
entering the cloaca by a special urinary canal. A curious
feature found in both monotremes but in no other
mammals is a grooved erectile poison spine on the tarsus
of the male, served by a gland in the thigh. The gland
increases during the breeding season and the spur and
poison are probably used in territorial conflicts, the pain
inflicted ensuring spatial separation. It is possible that it
is used to immobilize a female during coition. The
poison is agonizing to man and can kill a dog.

13.7. Receptors

The olfactory system of the platypus is reduced and the
eyes and ears are closed under water. The main sensory
organs are the mechanoreceptors of the bill. These
include many specialized organs and Pacinian corpuscles. The trigeminal nerve is much larger than the
optic or auditory nerves. There are large swellings where
it enters the brain and it is represented somatotopically
over a very large area of the surface of the cerebral
cortex (Fig. 18.24).

In *Tachyglossus* the olfactory nerves are well de-

veloped and vision is good. The trigeminal nerve is large
but its cortical representation smaller than in the
platypus.

In the ear of both monotremes the cochlea is only
partly coiled but contains an organ of Corti. As in
reptiles it carries a lagenar macula at its tip.

13.8. The brain

This is relatively large and arranged essentially on the
mammalian plan (Fig. 18.25). The pallial portion of the
cerebral hemispheres is large, not the striatal portion as
in reptiles, and as in marsupials there is no corpus
callosum. The surface is smooth in the platypus but in
echidnas the neopallium is exceptionally large and
convoluted, with a pattern different from that of
eutherians. The distribution of sensory and motor areas
is entirely different from that of other mammals
(Fig. 18.26). The visual cortex lies dorsally and the
auditory one ventrally and posteriorly. There is a large
area devoted to both somatic sensory and motor
functions, another is purely motor and in front of this is
'relatively more frontal cortex than in any other mam-

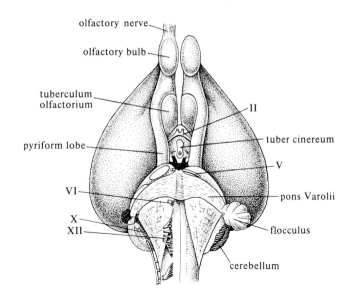

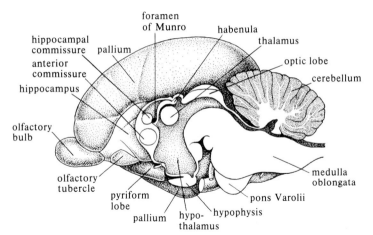

FIG. 18.24. Brain of the platypus. *II–XII*. cranial nerves. (From Kingsley, J.S. (1917). *Outlines of comparative anatomy of vertebrates*. Blakiston. After Elliot Smith.)

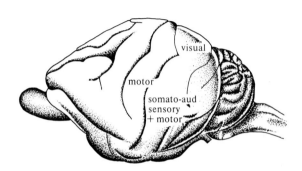

FIG. 18.25. *Tachyglossus*. Sensory and motor areas of the cortex. (From Lende, R.A. (1969). *Annals of the New York Academy of Science* **167**, 262–76.)

mal, including man, the function of which remains unexplored' (Lende 1964). A corticospinal tract passes in the opposite doral column as far as the 24th segment of the spinal cord.

Platypuses are solitary out of the breeding season. Courtship and copulation occur in the water. The young animals play.

13.9. Other features

The diaphragm is fully developed and the heart and single left aortic arch resemble those of other mammals. There are no renal portal veins. These animals have therefore advanced in their circulatory system beyond the anapsid condition, such as is probably shown today in Chelonia (p. 293). The larynx is mammalian and platypuses make growling sounds and echidnas coo like a ringdove! In the adrenals of echidnas, 'cortical', and

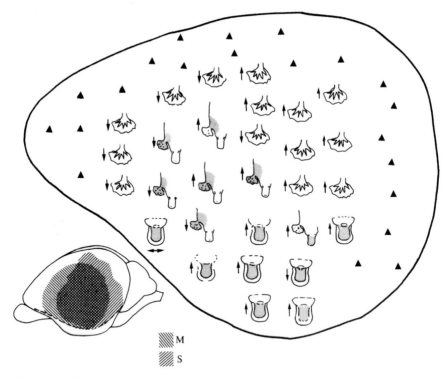

FIG. 18.26. *Ornithorhynchus.* The figures show the areas of the body from which movements were elicited by stimulation of the cortex. Flexion of forelimb or muzzle opening (↑), forelimb extension or muzzle closings (↓) and movements of mandible (↔). Triangles mark points from which no movements could be elicited. The solid inset shows the relative positions of motor (M) and somatosensory (S) cortical areas. View of hemisphere from dorsolateral aspect. (From Bohringer, R.C. and Rowe, M.J. (1977). *Journal of comparative Neurology* **174**, 1–14.)

'medullary' tissues are distinct compact masses but are not arranged one within the other. The chromosomes include microsomes, which are present in Sauropsida but absent from other mammals. The male is heterogametic but there is no Y chromosome, present in all other mammals.

13.10. Specialized features

With all their archaic features the monotremes also show many specializations. The platypus is highly modified for aquatic life. Apart from its bill there are the webbed feet, dorsal nostrils, long palate, short fur, thick tail, and absence of external ears. The animals burrow in the banks, making nests in which the young are reared. They are not uncommon in the rivers of southern and eastern Australia and Tasmania and fortunately are difficult to catch and therefore in no danger of extinction, though their fur and flesh are both useful.

The spiny ant-eaters (*Tachyglossus*) show specializations for eating ants and termites similar to those of *Myrmecophaga* (p. 447). There is a long snout, long erectile sticky tongue, and large salivary glands. The insects are ground between spines on the base of the

tongue and palate. The lining of the stomach is horny and has no glands. The clawed feet are used to make burrows as well as for digging up the nests of ants. It is perhaps significant that both surviving types of very early mammal burrow and make nests to assist in the protection of their young. The 'ant-eater' of New Guinea, *Zaglossus*, probably lives mainly on earthworms, which it sucks up a groove in the tongue (Griffiths 1978).

The Monotremata thus show a peculiar mixture of mammalian and reptilian characteristics. In their brain, hair, warm blood, heart, larynx, and diaphragm they are mammalian, but in skeleton and egg-laying habit they resemble reptiles. A large part of their interest is that they suggest an intermediate stage, in many features, between the two groups. Thus the pectoral girdle appears to be that of a reptile partly changed to a mammal. There are certainly many things to be discovered from these extraordinary creatures, which have perhaps remained with little change in fundamental organization for nearly 195 Ma from Jurassic times. The characters they show literally provide us with a view of the past.

19 Marsupials

1. Marsupial characteristics

THE pouched mammals including some 250 living species are essentially very similar to placentals, other than in their reproduction, though they undoubtedly diverged from some early stage of the main mammalian stocks (Tyndale-Biscoe 1973). They parallel, in the isolation of Australasia, the adaptive radiation accomplished elsewhere by the placentals though they never produced forms like bats, whales or seals (Stonehouse and Gilmore 1977). Many of their features are specialized, so that they represent not a stage on the way to placental evolution but a specialized side-branch. Today some 172 species are found in Australasia (Ride 1970), and there are successful representatives in North and South America. Marsupials are often supposed to be in some sense 'imperfect' mammals, though opossums and some others have survived in America in competition with placentals. Temperature regulation and perhaps other homeostatic processes may be imperfect but other studies have shown performance equal to that of placentals, for example in learning visual discrimination and problem solving (Kirkby 1977).

2. The skull

The skull shows many characteristics found also in Insectivora and other early mammalian groups (Figs. 19.1 and 19.2). The brain case is small and the top of the skull therefore rather flat. The orbit and temporal fossa remain fully confluent and there is no post-orbital bar. The bony palate is incomplete posteriorly, there

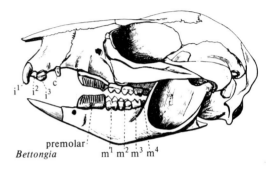

FIG. 19.2. Skull of the rat kangaroo. *c.* canine; *i.* incisor; *m.* molar. (From Flower and Lydekker 1891.)

being large holes in the palatine portion of it. The jugal bone always reaches back to the glenoid articulation of the jaw. The lower jaw, consisting of course of a single dentary, has a characteristically inturned or 'inflected' inner 'angle' to which the large lateral pterygoid muscle is attached. Other special features of the skull, not usually found in placentals, are that the foramen for the optic nerve and that for the eye-muscle nerves and trigeminal ophthalmic nerves are not separated from each other and that the lachrymal bone extends outside the orbit. More interesting than these apparently trivial and unconnected diagnostic features is the fact that the auditory region is not protected by the formation of a bulla of the petrous bone as it is in other mammals; instead the alisphenoid bone sends a wing over the middle ear.

The teeth are not easy to interpret (Fig. 19.3). The

FIG. 19.1. Skull of the Tasmanian wolf. (From Flower and Lydekker 1891.)

FIG. 19.3. Teeth of the upper jaw of the opossum (*Didelphys*) showing the last premolar, whose place is occupied in the young by a molariform tooth. (From Flower and Lydekker 1891.)

basic dentition is $I_4^5 - C_1^1 - P_3^3 - M_4^4$, found today only in opossums. The incisors are thus more numerous than in placentals. Of the cheek teeth only one, the third of the series, is replaced in both jaws in modern forms, and if this is regarded as the last premolar we have a dentition of 3 premolars and 4 molars, as against the 4 and 3 of a typical placental. However, fossil marsupials with three replacing teeth have been found (p. 409) and the significance of the peculiar condition of the modern forms remains obscure; it may be connected with the specialization of the mouth of the young for life in the pouch. The cusps of the teeth of many marsupials (e.g. opossum) show a close approach to the presumed primitive tribosphenic plan (p. 411), but in addition to the main triangle the upper molars often carry an outer row of 'stylar' cusps (Fig. 19.4), which is also present in the molars of some primitive placentals. As in placentals, in herbivorous types of marsupial the pattern of cusps on the cheek teeth becomes square, producing flat grinding surfaces (p. 413).

3. Skeletal features

The general plan of the muscles and backbone is essentially that found in placentals and has been greatly changed from the reptilian or monotreme condition. There are no cervical ribs. The thoracic region consists of about 13 rib-bearing vertebrae, as in placentals, and there are usually 7 lumbars. The pectoral girdle shows no interclavicle, but the clavicle remains large. The coracoid is reduced, as in placentals, and the scapula enlarged and provided with a spine. In fact all the developments of the dorsal region of the girdle that are typical of the mammalian method of locomotion have taken place. In the pelvic girdle there are epipubic bones (sometimes wrongly called 'marsupial bones'), reminiscent of those of monotremes; they are two large flat bones, curving forwards and outwards from the mid-

line as partial crescents, and only lightly joined to the other bones (Jones 1923). They take the special stresses produced by the abdominal muscles and pouch, but are reduced in the fully terrestrial and quadrupedal Tasmanian wolf. The hands usually carry five digits, armed with claws, but the number of toes is often reduced and they may bear hoof-like structures (Fig. 19.5).

4. Reproduction

Marsupial reproduction follows a pattern distinct from that of eutherians. Pregnancy is shorter than the oestrus cycle. The pituitary and the ovary can be removed without terminating pregnancy although they are both necessary for normal parturition (Tyndale-Biscoe, Hearn, and Renfree 1974; Hearn 1975a). There is no special immune suppressant system protecting the foetus from the mother's antibodies. A thin shell protects the ovum against the response of the uterus to the father's antigens. This protection can only be brief and the young is born within 10–12 days of the shell breaking down. The life of the mother is not interrupted by any dramatic birth – she appears to be unaware of its occurrence and does not recognize the new born, although she has cleaned the pouch in preparation for birth and continues to do so throughout the neonate pouch life. Thus although the young gain less from the mother before birth than in eutherians she risks less and

Fig. 19.4. Molar teeth, upper right (above) and lower left (below), of a fossil opossum of the Upper Cretaceous. The upper molar has a well developed row of outer cusps in addition to the normal trigon. (From Romer 1966, after Osborn.)

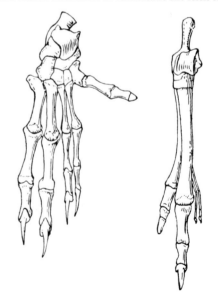

(a) *Trichosaurus*   (b) *Macropus*

Fig. 19.5. Hind foot of *Trichosurus* (a) and *Macropus* (b), showing elongation of the toes of the latter and reduction of the 1st, 2nd, and 3rd toes. (From Zittel, C.A. v. (1920–5). *Textbook of Palaeontology*. Macmillan, London. After Dollo.)

retains more freedom – which may be an equally or more effective reproductive pattern. After birth the young enjoy long protection in the pouch but the mother does not look after them and they may be lost easily if she is stressed or chased.

The Müllerian ducts are paired and differentiated into upper 'uterine' and lower 'vaginal' portions. In many species the latter are provided with median diverticula, the two meeting in the midline as a sinus vaginalis (Fig. 19.6). This ends blindly until the young are about to be born, when an opening is formed through the tissues – the pseudovagina or birth canal. This usually then closes until the next parturition but in some species, e.g. most macropids, forms a permanent median vagina. The lateral vaginae serve only for entry of sperm and may be enlarged enormously after copulation; attaining a far greater volume than does the gravid uterus. The median birth canal has been developed in marsupials to avoid the ureters, which pass dorsal to the Müllerian ducts. In eutherians the ureters run ventrally, allowing complete fusion of the female ducts. As in monotremes the rectum and urinogenital sinus open together at a common cloaca, though this is not very long, longer in the female than the male (Fig. 19.7). There is a well developed penis, bifid at the tip in several species, in which case the clitoris is also double, the arrangements

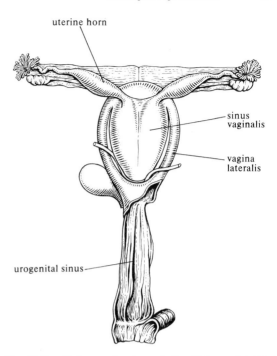

Fig. 19.7. *Phalanger*, female genitalia. Note caudal blind ending of the sinus vaginalis. (After Brass and Ottow.)

presumably ensuring fertilization of both oviducal tubes. The testes descend to a scrotum.

The reproduction of marsupials shows a viviparous condition that is not closely similar to any found in placentals. The egg is rather yolky and covered with albumen and a shell membrane; cleavage is very unequal. In some forms (*Dasyurus*) there is a contact of the vascular wall of the yolk sac with the somewhat hypertrophied uterine wall (yolk sac or omphaloidean placenta). Only in the bandicoot, *Perameles*, does the allantois develop a nutritive function to some extent. In many marsupials there is only a very limited placental development, and instead, uterine milk may be taken up by the yolk sac. The embryos are born very young, as little as 8 days from conception in the opossum (Fig. 19.8). They crawl to the pouch and become attached to the teats. This is not by any means a primitive plan and it involves many specializations. To make the journey to the pouch the fore limbs and their nervous centres are precociously formed being fully functional at birth, when the hind limbs are mere buds. The mouth and the olfactory system are also precociously developed. The method of suckling also involves special developments of both mother and foetus, so that milk is injected into the latter without choking it. The sides of the lips grow together round the teat, which is thrust far back in the pharynx, the larynx

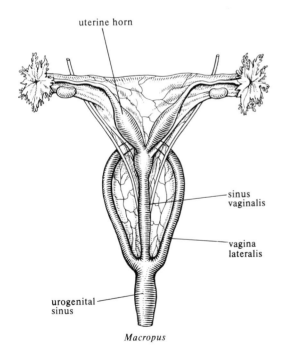

Macropus

Fig. 19.6. Kangaroo, female genitalia. Note that the sinus vaginalis opens directly into the urogenital sinus. (After Brass and Ottow.)

FIG. 19.8. Pouch young of *Dasyurus* about 10 days old. (After Hill and Osman Hill.)

extending forwards into the nasal passage. Milk is pressed out by a special muscle (homologue of the male cremaster) attached to the epipubic bones. During the 8–9 months that a young wallaby stays in the pouch the mother remains in lactationally induced anoestrus. In the tammar and Bennetts wallabies there is also a seasonally induced anoestrus which lasts from when the joey stops suckling until about the longest day (Hearn 1975*b*).

Many marsupials have a system of delayed implantation. Copulation occurs a day or two after birth but an 80-cell blastocyst in its shell can remain dormant for up to 11 months, controlled in turn by lactational and seasonal quiescence. Suckling probably stimulates release of oxytocin and there are high circulating levels of prolactin which may inhibit the quiescent corpus luteum. When the joey begins to be weaned (or if it is lost) the inhibition is removed and the embryo develops quickly and is born about 30 days later (in kangaroos). The first offspring may still suckle low protein, high fat milk from one teat while the new one gets high protein, low fat milk from another one (Bailey and Lemon 1966; Lemon and Barker 1967).

The marsupial mouse *Antechinus*, common near Sydney, has a curious form of population control. All the males die one week after mating. The death is associated with increased adrenocortical activity in the last weeks of life leading to greater activity and aggression and more frequent encounters with others. Males captured before mating and kept in isolation survive beyond the natural time of death, but not if they are caged together. Females survive for two or even

three seasons. This seems to be the only case of 'semelparity' or single reproduction, among mammals, although there is often a greater mortality among rodents after synchronous reproduction (e.g. in lemmings and voles).

5. The brain

The brain shows some reptilian features. The cerebral hemispheres are small for a mammal, and the olfactory bulbs large. The hemispheres are not prolonged backwards over the cerebellum, which is itself small and simple. An obvious difference from eutherians is that there is no corpus callosum, but large dorsal (hippocampal) and ventral (anterior) commissures the latter including a special bundle, the fasciculus aberrans, carrying fibres that connect all parts of the neocortex. The sensory and motor areas of the cortex are organized as in eutherians. The cochlea of the ear is spirally coiled.

6. Characteristic features

Marsupials are often divided into two suborders, the more primitive insectivorous or carnivorous polypro-

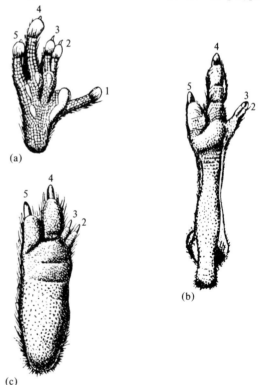

FIG. 19.9. Hind feet of some marsupials. (a) The arboreal opossum has grasping hallux. (b) The cursorial kangaroo is without hallux and digits II and III are syndactyl. (c) The secondarily arboreal tree kangaroo has no hallux. (After Lull 1917.)

todonts, found outside as well as within Australasia, and the diprotodonts, more specialized and restricted. The distinction is based on the presence of more than three pairs of incisors in each jaw in the first group while in the other only two remain in the lower jaw and protrude. A further interesting feature is that in most diprotodonts the second and third digits of the hind limb are fused to make a comb for cleaning the hair (Fig. 19.9). This character is absent in nearly all polyprotodonts, which are hence said to be didactylous, and this is no doubt the primitive condition. However, the comb-like condition (syndactyly) is found not only in diprotodonts but also in the polyprotodont bandicoots; conversely the curious South American opossum-rat (*Caenolestes*), though didactylous, has diprotodont teeth.

The fact is that marsupials, having radiated perhaps over 60 Ma ago, present us today with a number of distinct types of organization. Simpson (1945) groups them into six superfamilies and it is perhaps better not to attempt any higher grouping of these (see Kirsch and Calaby 1977). The absence of Australian Tertiary deposits before the Pleistocene increases the difficulties of study of marsupial phylogenesis.

7. Classification

Infraclass 2: Metatheria
  Order Marsupialia
    Superfamily 1: Didelphoidea. Upper Cretaceous–Recent. Europe and America
      Eodelphis, Upper Cretaceous, North America; *Didelphis*, opossum, Pliocene–Recent, North and South America; *Chironectes*, water opossum, Central and South America
    *Superfamily 2: Borhyaenoidea. Palaeocene–Pliocene. South America
      Thylacosmilus; *Borhyaena*
    Superfamily 3: Dasyuroidea. Pleistocene–Recent. Australasia
      Dasyurus, native cat; *Sarcophilus*, Tasmanian devil; *Thylacinus*, Tasmanian wolf; *Myrmecobius*, banded ant-eater; *Notoryctes*, marsupial mole; *Sminthopsis*, pouched mouse
    Superfamily 4: Perameloidea. Pleistocene–Recent. Australasia
      Perameles, bandicoot
    Superfamily 5: Caenolestoidea. Eocene–Recent. South America
      Palaeothentes (= *Epanorthus*); *Caenolestes*, opossum-rat, Ecuador, Colombia, Peru
    Superfamily 6: Phalangeroidea. Pliocene–Recent. Australasia
      Trichosurus, Australian opossum; *Petaurus*, flying opossum; *Phascolarctos*, koala bear; *Vombatus*, wombat; *Macropus*, kangaroo; *Bettongia*, rat kangaroo; *Diprotodon*; *Thylacoleo*

8. Opossums

The opossums (Didelphoidea) were the earliest group to appear and probably originated in America. The other families have probably evolved from them, with syndactyly appearing twice. They are arboreal, mainly nocturnal and omnivorous or insectivorous animals, with a prehensile tail, occurring over the southern United States, Central and South America (Fig. 19.10). The pouch is generally absent. Similar forms are found back to the Upper Cretaceous (*Eodelphis*) and the American opossums (*Didelphys*) are certainly the closest of living marsupials to the ancestors of the group, perhaps they are the least modified of all therian mammals. *Chironectes* is a related South and Central American otter-like form, with webbed feet.

The bandicoots (*Perameles*) with several species in Australia and New Guinea are burrowing animals, rather rabbit-like but mainly insectivorous. They have a polyprotodont dentition, quadrituberculous grinding molars, syndactyly of the hind toes, and an allantoic placenta. Their affinities are uncertain.

FIG. 19.10. Azara's opossum, South America (*Didelphis*). (From photographs.)

9. Carnivorous marsupials

The Dasyuroidea include some nocturnal carnivorous polyprotodonts that show remarkable convergence with placental carnivores, the teeth being modified for cutting flesh in a way similar to the carnivore carnassial. *Thylacinus* (Fig. 19.11), the Tasmanian wolf, is now nearly or quite extinct. *Sarcophilus*, the Tasmanian devil (Fig. 19.12), is still common there. *Dasyurus* (native cat) (Figs. 19.13 and 19.14) includes several species of cat-like creatures still fairly common in Australia, Tasmania, and New Guinea. The Borhyaenoidea were successful carnivorous marsupials living in South America, while it was not colonized by early placental carnivores. They became extinct when the latter arrived from North America in the Pliocene. They were perhaps related to the Dasyuridae; *Borhyaena* shows many similarities to *Thylacinus* and this suggests (like *Caenolestes*) that marsupials reached Australia from South America, via Antarctica. *Thylacosmilus* was a Miocene and Pliocene sabretooth, the size of a panther; its huge stabbing upper canine and other features closely parallel the placental *Smilodon* (p. 413). It is said that in some of these borhyaenoids, two or more milk teeth were replaced; if true this suggests that the condition in modern marsupials is secondary.

10. Marsupial ant-eaters and other types

Myrmecobius is an ant-eating form, with elongated snout and as many as 52 teeth (Fig. 19.15). *Notoryctes*, the marsupial mole (Fig. 19.16), from South Australia, has reduced eyes, well developed fore limbs, fused cervical vertebrae, and many features suiting it for burrowing and feeding upon ants. Its pouch opens backwards. *Sminthopsis*, the pouched mice (Fig. 19.17), are small marsupials occupying the niche taken in other parts of the world by the shrews.

Caenolestes, the opossum rat of the forests of the Andes, is an interesting shrew-like creature. It has four upper incisors like the polyprotodonts but three or four lower incisors the middle ones being strong, elongate, and procumbent, which resemble those of diprotodonts. There is no syndactyly. It is the survivor of a group formerly abundant in South America, some with teeth similar to those of multituberculates.

11. Kangaroos and phalangers

The diprotodont marsupials form a compact group (leaving out *Caenolestes*) of Australian forms, here included as the superfamily Phalangeroidea. Their fossil history is little known. They have become specialized for various modes of life, mainly as herbivores, in Australia and the neighbouring islands. The kangaroos and wallabies (Macropodidae) (Fig. 19.18) have become

Fig. 19.11.   Tasmanian   wolf   (*Thylacinus*).   (From photographs.)

Fig. 19.12.   Tasmanian   devil   (*Sarcophilus*).   (From photographs.)

Fig. 19.13.   Tasmanian   tiger   cat   (*Dasyurus*).   (From photographs.)

Fig. 19.14.   Eastern   native   cat   (*Dasyurus*).   (From photographs.)

Fig. 19.15.   Banded   ant-eater   (*Myrmecobius*).   (From photographs.)

mostly terrestrial and developed a bipedal method of progression, involving modification of the ilia and thigh muscles, for whose attachment the tibia bears a marked anterior crest. The foot gains increased leverage by elongation of the metatarsal of digit 4. Digits 2 and 3 are very small and syndactylous. There are several modifications for a herbivorous diet; the single pair of lower incisors is directed forwards and their sharp inner edges can be moved in such a way as to cut grass like shears. This condition recalls that of Rodentia (p. 488) and, as in that group, a special transverse muscle is developed (m. orbicularis oris), but in this case it is part of the facial musculature and innervated from the seventh nerve, whereas the analogous muscle of the rodents is a part of the mylohyoid and innervated from the trigeminal. The molar teeth are modified for grinding, by the fusion of the cusps to make two transverse ridges, recalling those of ruminants. The stomach has a large sacculated non-glandular chamber like a rumen, allowing digestion by protozoan and bacterial symbionts. *Bettongia* and other 'rat-kangaroos' are terrestrial and bipedal jumpers. They also have a prehensile tail (Fig. 19.19). In the tree-kangaroos (*Dendrolagus*) the forelegs have become almost as long as the hind and provided with claws.

The Australian opossums or phalangers are less modified than the kangaroos and are arboreal animals, with a prehensile tail and various special modifications, mostly for herbivorous diets; *Trichosurus* is the common phalanger or possum found over much of Australia. The pygmy possum (*Cercatretus*) is a mouse-like creature, living in holes in trees. *Tarsipes*, the honey possum has a long snout and tongue for feeding on nectar. *Phalanger*, the cuscus (Fig. 19.20), eats mainly leaves. The koala or native bear, *Phascolarctos* (Fig. 19.21), lives on the leaves of *Eucalyptus*; it has cheek-pouches, an enlarged caecum, and a reduced tail. The first two fingers are opposable to the rest. Three distinct genera of phalangers have developed extensions of the skin for purposes of soaring; *Petaurus* (Fig. 19.22) is the best known of these flying phalangers. *Vombatus*, the wombat (Fig. 19.23), is a large, burrowing, tailless animal, with rodent-like grinding teeth; it eats roots. *Diprotodon* was a very large marsupial, of the size and form of a rhinoceros, which lived in Australia in the Pleistocene. *Thylacoleo* was a marsupial lion in which the incisors were developed as fangs.

12. Significance of marsupial isolation

The knowledge we now have of continental drift explains the facts about the distribution of fossil and modern marsupials, which used to puzzle zoologists. The first marsupial opossum-like fossils of the upper Cretaceous and Eocene, about 100 million years ago,

FIG. 19.16.   Marsupial   mole   (*Notoryctes*).   (From photographs.)

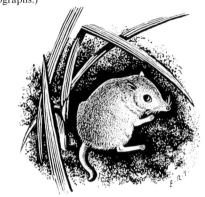

FIG. 19.17.   Pouched   mouse   (*Sminthopsis*).   (From photographs.)

FIG. 19.18. Kangaroo (*Macropus*). (From photographs.)

FIG. 19.19. Rat kangaroo (*Bettongia*). (From photographs.)

FIG. 19.20. Cuscus (*Phalanger*). (From photographs.)

are found in Europe and North and South America and
these continents were probably then connected (p. 19).
By the time placentals appeared Australia and South
America were no longer connected with the other
continents and the marsupials became extinct in Europe
and North America, which were still connected. South
America remained isolated until the formation of the
Panama bridge in the Pliocene and most of its marsupial
fauna became extinct after this time. In North America
also there were no marsupials between the middle
Miocene and the Pleistocene, when opossums (*Didelphis*) appeared, probably from South America. Australian marsupials only became threatened by the
advent of Man from Europe and the other placentals he
brought with him. However, the large macropids have
thrived with the spread of agriculture in marginal
climatic areas of their range.

The opossums show that marsupial life can continue
effectively in competition with placentals. But it can
hardly be an accident that the diversification of mar-

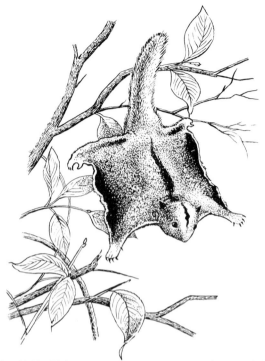

FIG. 19.22. Flying opossum (*Petaurus*). (From photographs.)

supials in Australia has been accomplished in isolation.
It is true that there are 108 species of placentals in
Australia, as against 124 marsupials, but the placentals
are almost all bats (50 species) and murid rodents (48
species). The marsupials, on the other hand, have
become differentiated into numerous types, arboreal,
fruit-eating, grazing, gnawing, digging, burrowing, ant-
eating, insectivorous or carnivorous, in each case with
appropriate structure. It will be interesting to see how
this assemblage stands up to competition with placentals in the future. Carnivores, ruminants, lagomorphs,
rodents, and primates have recently become firmly
established in Australia and it can hardly be an accident
that some of the corresponding marsupial types are
already becoming rare or extinct.

FIG. 19.21. Koala bear (*Phascolarctos*). (From photographs.)

FIG. 19.23. Wombat (*Vombatus*). (From photographs.)

20 Evolution and classification of placental mammals

1. Eutherians at the end of the Mesozoic

SEVERAL different lines of evidence converge to show that all the eutherians (placentals) have been derived from small, perhaps nocturnal, insectivorous or omnivorous animals, living in the Cretaceous period about 100 million years ago. Many features of marsupials and placentals alike suggest origin from a small Cretaceous shrew-like form, perhaps itself descended from some animal like the Jurassic pantotheres. It is especially interesting, therefore, that fossil evidence is now available to show that both opossums (p. 427) and placental insectivores (p. 436) existed in the Cretaceous. We may be reasonably sure that the population from which those groups were derived resembled these animals, which are indeed basically similar. At this Cretaceous period the arrangements for nourishing the young were presumably not yet fully developed and in the marsupials (Metatheria) they have remained at a simple level, little above ovoviviparity, but the condition in *Perameles* makes it possible though not likely that an allantoic placenta has been lost by the other marsupials.

The stock that was to give rise to the eutherians was therefore already differentiated in the Cretaceous. As the revolution proceeded these animals began to flourish and to develop into several divergent populations. Some very simple 'ungulate' animals are now known from the upper Cretaceous (p. 517) and by the very beginning of the Cenozoic period, in the early Palaeocene, several different types of placental are found.

2. The earliest eutherians

Fossil placentals found in the Cretaceous period have nearly all been insectivorans (p. 438). Those of the Palaeocene include also some that can be referred to the primates and to the Carnivora, but they are very unlike modern members of those groups and could almost equally well be classed as Insectivora. Similarly, the ungulates and various other types that appeared in the late Cretaceous and Palaeocene were very like insectivores. Evidently, therefore, during these times the original placental stock was branching out into various

habitats, and this occurred relatively rapidly. Simpson's (1945) careful classification recognizes twenty-six orders of placentals (p. 433), and many of these were distinct by the Eocene and ten since that time have become extinct. Evidently the great period of mammalian expansion was in the earlier part of the Cenozoic and the group may be considered to have passed its peak for the present. Only the bats, rodents, lagomorphs, and perhaps the primates and carnivores can be considered really successful land animals at the present time; to these we may add the whales in the sea. The Artiodactyla are also abundant, but most of the remaining placental orders are today poorly represented in numbers.

In order to gain a comprehensive general understanding of the great placental history during its 80 million or so years duration, we will first give a technical definition of a placental, then the characteristics of the earlier types, and finally try to list some of the tendencies to change that are widely found in different groups. For simplification we can attempt some grouping of the great list of orders, before dealing with them individually.

3. Definition of a placental mammal (Eutheria)

3.1. Placentation

The blastocyst early develops an outer layer of cuboidal or squamous cells surrounding a small group, the inner cell mass, from which the embryo develops. This trophoblast protects the embryo from immunological attack and provides the earliest nourishment from the mother. The embryo is then retained for a considerable time in the uterus and nourished by means of an allantoic placenta. This was at first a simple apposition of foetal and maternal tissues without invasion (epitheliochorial placenta). This primitive condition is still found in Strepsirhini, Artiodactyla, Perissodactyla, Pholidota, and Cetacea. In these the allantois remains large and the amnion is usually formed by folds. In other mammals there are varying degrees of interpenetration of foetal and maternal tissues producing the endotheliochorial placenta of carnivores or the haemochorial one of insectivores, bats, rodents, edentates, and anthropoid primates. Since many of these latter orders are in other respects primitive several workers would read this evolutionary sequence in the reverse direction. However, many of these forms also show more advanced conditions such as small allantois and an amnion formed by cavitation (see *Life of Mammals*; Luckett and Szalay 1975).

3.2. Skeleton

There is no pouch or epipubic bones. The face is long and the skull tubular. There is usually a separate optic foramen, no palatal vacuities, and no in-turned angle of the jaw. The tympanic bone is either ring-like or forms a bulla, there is never an alisphenoid bulla. The legs are short with few digits and a plantigrade foot. The dental formula is $\frac{3.1.4.3}{3.1.4.3}$ or some number reduced from this and the molars are basically tribosphenic.

Many of these are obviously small points of formal definition, artificially abstracted for the purpose of classification. They are not really satisfactory as a definition of the life of a placental, such as we may hope to have in a more developed biology. The early population of Cretaceous insectivores presumably possessed most of these features, and showed characteristics that are common to all the earlier mammals.

4. Evolutionary trends of eutherians

In the descendants of these early mammals we can recognize modifications in each of these sets of characters; changes occurring, independently, in some members at least of all the later lines. (1) Many of the mammals became larger. Increase in size seems to be advantageous to many animal types and may be connected with the possession of a large brain storing much information during the life of the animal and so allowing slow reproduction (p. 452). Large size is especially important for warm-blooded animals in cold climates, since it reduces the relative area of heat loss, though also introducing new problems of obtaining adequate amounts of food. This may have to be finely ground by tooth surfaces whose increase with size is less rapid than that of the weight of tissue they must support. (2) The limbs became longer and specialized in various ways for locomotion, often by raising the heel off the ground, so that the animals came to walk on the digits instead of the sole of the foot (Fig. 20.1) and the number of toes became reduced. (3) Teeth were reduced in number and their shape specialized, often by the addition of cusps and their fusion to make transverse or longitudinal grinding ridges in herbivorous animals or cutting blades in carnivores. (4) The brain of the earlier mammals resembled that of reptiles (Fig. 20.2); later

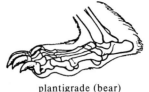

plantigrade (bear)

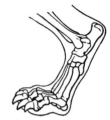

digitigrade (hyaena)

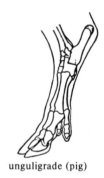

unguligrade (pig)

FIG. 20.1. Postures of the foot in various mammals. (From Lull 1929.)

forms showed increasing development of the non-olfactory part of the cortex, increase of the frontal lobes, and other changes probably correlated with more complicated behaviour and better memory. It has often been supposed that the brain becomes relatively larger in later forms, but in fossil horses as the body size increased the brain/body weight ratio decreased (Campbell 1975).

Confident statements are often made about evolution of the brain within the Mammalia, assuming that features such as a larger pyramidal tract are 'more advanced'. Unfortunately too few species have been studied to allow such confidence. Much of the life pattern of mammals centres around their well developed nervous organization. The layered cerebral cortex is common to all of them and is found in no other vertebrates. No doubt it provides special opportunities for the display of complex patterns of innate and learned behaviour but we do not really understand in what respects it is superior to the forebrain of other vertebrates, such as sharks or birds which also receive projections from the thalamus. Other parts of

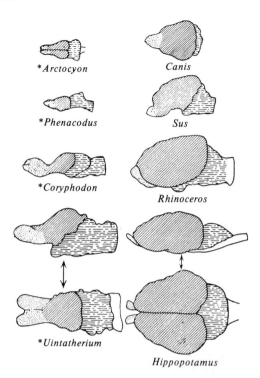

Arctocyon

Canis

Phenacodus

Sus

Coryphodon

Rhinoceros

Uintatherium

Hippopotamus

FIG. 20.2. Comparison of pairs of brains of archaic and modern mammals of similar size. Olfactory lobes dotted, cerebral hemispheres oblique lines, cerebellum and medulla dashes. (From Lull 1929.)

the nervous system presumably also have unique mammalian features and vary between orders. For instance the hypothalamus, as the guardian of homeostasis, may provide mammals with standards that are more rigorous than those of other vertebrates and for each peculiar to its way of life. But we are still too ignorant to make precise comparisons.

Understanding of the anatomy and functioning of any brain is only possible after long study by special techniques of its neurons and fibre tracts. Until recently evolutionary comparisons rested upon a few easily obtained species such as opossum, rat, cat, monkey, and man, which have all too often been considered to form a phyletic sequence from lower to higher. Study of further animals usually shows examples of functional convergence that upset established stereotypes (Campbell 1975). For instance the presence of a corticospinal (pyramidal) tract with direct connections to motor-neurons was thought to be a primate characteristic responsible for fine finger movements. However, it is also present in the racoon (*Procyon*), a carnivore which holds its food in its hands, though cats and dogs have no such connections.

Again the laminated dorsal lateral geniculate nucleus

is not a characteristic of 'higher' mammals as was once thought but is present in many placental orders and even in the arboreal marsupial phalanger (*Trichosurus*) and in kangaroos. It seems to be a feature of arboreal or rapidly moving animals. These are examples of brain features that are 'homoplastic' rather than 'homologous'. In fact we are still too ignorant to be able to say much about evolutionary tendencies in the brain, interesting though the subject is.

5. Conservative eutherians

Changes in the directions outlined above have taken place in many separate mammalian lines, but the evidence contradicts the thesis that they are the result of some force of orthogenesis, driving the animals infallibly along. In nearly every group there are examples of some animals that have remained nearly unchanged for long periods, e.g. opossums and shrews since the Cretaceous (80 Ma), lemurs and tarsiers since the Eocene (50 Ma), pigs and tapirs since the Oligocene (35 Ma), and deer since the Miocene (20 Ma), to name only a few. Several of these examples of 'bradytely' (slow evolution) live in tropical forests, which may provide a relatively conservative niche. But of course all have changed somewhat and each shows some special features of structure and behaviour.

It is important to study such animals in which there has been little change; they form the 'controls', and may enable us to recognize the factors inducing change when it does occur. Moreover, in many further lines there has been change in some but not all of the above directions; for instance, many of the most successful mammals have remained small. There may even be changes in the directions opposite to those listed, for instance, some edentates and whales have more than the original number of teeth. Mammals do not commonly decrease in size during evolution (but they may do so), and they probably never reacquire lost digits, though in a few claws have reappeared after they had been lost.

6. Divisions and classification of Eutheria

Infraclass 3: Eutheria
   Cohort 1: Unguiculata
      Order 1: Insectivora
      Order 2: Chiroptera
      Order 3: Dermoptera
      *Order 4: Taeniodontia
      *Order 5: Tillodontia
      Order 6: Edentata
      Order 7: Pholidota
      Order 8: Primates
   Cohort 2: Glires
      Order 1: Rodentia

Order 2: Lagomorpha
Cohort 3: Mutica
   Order Cetacea
Cohort 4: Ferungulata
   Superorder 1: Ferae
      Order Carnivora
   Superorder 2: Protungulata
      *Order 1: Condylarthra
      *Order 2: Notoungulata
      *Order 3: Litopterna
      *Order 4: Astrapotheria
      Order 5: Tubulidentata
   Superorder 3: Paenungulata
      Order 1: Hyracoidea
      Order 2: Proboscidea
      *Order 3: Pantodonta
      *Order 4: Dinocerata
      *Order 5: Pyrotheria
      *Order 6: Embrithopoda
      Order 7: Sirenia
   Superorder 4: Mesaxonia
      Order Perissodactyla
   Superorder 5: Paraxonia
      Order Artiodactyla

For purposes of phylogenetic study as well as classificatory convenience it is desirable to attempt to discover how the original eutherian population became divided at its Cretaceous origin, and whether there were main trunks of the placental tree. By Eocene times most of the existing orders were already well established (Fig. 20.3)

and it is often stated that the branching of the population occurred relatively rapidly, though it is doubtful if we may use this term for a process occupying perhaps 30 Ma during the late Cretaceous and Palaeocene! This early expansion into varied branches, occurring at a time when few fossils were being formed, makes it difficult to discover the outlines of the main divisions. An attempt has been made by some workers to provide a cladistic classification of mammals, that is to say one following strictly and only the lines of phyletic subdivision (McKenna 1975). It is too soon to say whether zoologists will accept the many new names and ranks such as Superlegions and Magnorders that are needed to provide the best possible approach to a cladistic classification for mammals of the type suggested by the entomologist Hennig (1950). There are many difficulties in deciding what weight to give to characters that seem to be primitive (now called 'plesiomorphic') rather than derived or 'distinctive and independently acquired by the later stock', otherwise called 'advanced' or 'apomorphic'. The very problem of finding good words shows the difficulty of the whole enterprise. Some workers at the opposite extreme believe that classifications should be purely practical and empirical, depending only on resemblance and not considering phylogeny. We think this is a feeble approach and prefer to use the classification proposed by the American palaeontologist Simpson (1945, 1975). After careful piecing together of evidence he suggested grouping of the twenty-six eutherian orders into four main cohorts,

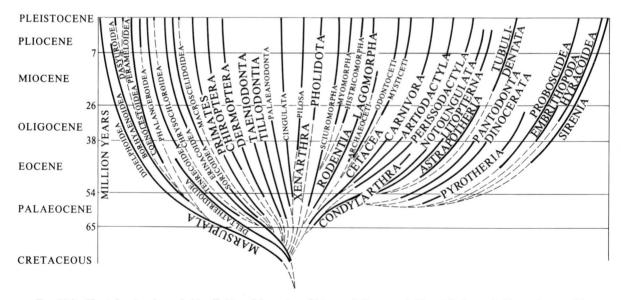

FIG. 20.3. Chart showing the probable affinities of the orders of Mammals (large capitals) and the lesser divisions of some of the more primitive orders (small capitals).

and their division corresponds in the main to the classification originally proposed by Linnaeus in 1766 on a basis of the foot structure.

1. The **Unguiculata** include orders in which the original characteristics of the mammalian type have been largely preserved, the Insectivora themselves and the Chiroptera (bats), the Primates and the Edentata (sloths, ant-eaters, and armadilloes), all of which can be easily and directly derived from the insectivores. Here also probably belong the scaly ant-eater, *Manis* (Pholidota), The *Taeniodontia and *Tillodonta were small groups of herbivorous animals derived from insectivoran ancestors in the Palaeocene and Eocene and can be included here though their affinities are obscure. Some became as large as bears. The teeth of taeniodonts became high crowned pegs, persistently growing, with enamel only at the sides, but earlier forms were very like insectivores, with tribosphenic dentition.

2. The **Rodentia** (rats) and the **Lagomorpha** (rabbits and hares) are often also considered to belong with the insectivores but they appear fully developed in the Eocene and must have diverged very early and are therefore placed alone in a separate cohort **Glires**.

3. Similarly the **Cetacea** (whales) have certainly been distinct since the Eocene. Their affinities are quite obscure, and they are classed as a separate cohort named, by Linnaeus, Mutica (most inappropriately since they communicate by sounds).

4. All the remaining mammals have been held to show signs of a common origin and Simpson (1945) suggests grouping them together as a cohort **Ferungulata**. It has long been realized that the hoofed animals include two distinct types, those with an uneven number of toes, Perissodactyla, and those with even toes, Artiodactyla. These lines have been separate at least since the Eocene. The former can be derived from the Palaeocene and Eocene animals known as *Condylarthra which may also include the ancestors of the artiodactyls. The Carnivora seem at first sight to have no similarity to either of the ungulate types, but it has been suspected that the ancestral Carnivora, the *Creodonta, resembled the earliest artiodactyls. Creodonts and condylarths are often so alike as to be hardly separable, but they are also very like Insectivora, and many palaeontologists feel that there is no real basis for major grouping (Romer 1966). However, Simpson's (1945) suggested cohort Ferungulata recognizes the existence of a common creodont-condylarth stock, perhaps in early Palaeocene times. The cohort may then conveniently be subdivided into five superorders; the first **Ferae** for the Carnivora, the second **Protungulata** (= 'first ungulates') for the condylarths, the extinct South American ungulates (*Litopterna and *Notoungulata), and the obscure *Astrapotheria and Tubulidentata (*Orycteropus*, the Cape ant-eater). The third superorder **Paenungulata** (= 'near ungulates') includes the elephants (Proboscidea), hyraxes (Hyracoidea), and sea cows (Sirenia), as well as the extinct *Pantodonta, *Dinocerata, *Pyrotheria, and *Embrithopoda. The fourth superorder **Perissodactyla** is then made to include only the horses, tapirs, and rhinoceroses; and the fifth superorder **Artiodactyla** the pigs, camels, and ruminants. This system gives us a means of grouping that is phylogenetically reasonably accurate and also conveniently close to the usually accepted uses of familiar names.

21 Insectivores, bats, and edentates

1. Insectivores

1.1. Classification

Order: INSECTIVORA

    Suborder 1: Proteutheria. Upper Cretaceous–Recent
        Deltatheridium, Upper Cretaceous, Asia; *Didelphodus*, Eocene, North America; *Zalambdalestes*, Upper Cretaceous, Asia; *Leptictis*, Oligocene, North America; *Anagale*, Oligocene, Asia; *Tupaia*, tree-shrew, Asia; *Ptilocercus*, pen-tailed tree shrew, Asia

    Suborder 2: Macroscelidea. Oligocene–Recent
        Macroscelides, elephant shrew, Africa

    Suborder 3: Dermoptera. Palaeocene–Recent
        Cynocephalus (= *Galeopithecus*), colugo or flying lemur, East Asia

    Suborder 4: Lipotyphla. Lower Eocene–Recent
        Superfamily 1: Erinaceoidea. Palaeocene–Recent. Holarctic, Oriental, Africa, North America
            Echinosorex, moonrat, Asia; *Erinaceus*, hedgehog, Old World
        Superfamily 2: Soricoidea. Palaeocene–Recent
            Family Soricidae: Oligocene–Recent. Holarctic, Africa
                Sorex, shrew, Holarctic; *Neomys*, water shrew, Eurasia; *Crocidura*, Africa, Asia, Europe
            Family Talpidae: Upper Eocene–Recent. Holarctic
                Talpa, mole; *Desmana*, desman, water mole
            Family Solenodontidae: Recent. West Indies
                Solenodon, alamiqui
            Family Tenrecidae: Miocene–Recent. Africa, Madagascar
                Potamogale, otter shrew, West Africa; *Tenrec* (= *Centetes*), tenrec, Madagascar
            Family Chrysochloridae: Miocene–Recent. Africa
                Chrysochloris, golden mole, South Africa

The order Insectivora includes the ancestors of all eutherian mammals and a considerable number of modern descendants who we hardly recognize as our relatives. It therefore includes four rather different types. Firstly the typical living forms such as shrews, moles, and hedgehogs. These are included in the suborder Lipotyphla whereas all the rest used to be put together as 'Menotyphla'. It is now better to drop that term in favour of three further suborders. For the most primitive insectivores, the basic eutherian stock, Romer suggests the very descriptive name Proteutheria (1966). This suborder includes the living tree shrews, perhaps the most primitive of living eutherians, and fossils so generalized that they cannot be put in any other placental order. The other two suborders include small divergent offshoots from the insectivoran stock, the elephant shrews, Macroscelididea, and flying lemurs, Dermoptera.

1.2. Modern insectivores. Suborder Lipotyphla

These are mostly small, nocturnal animals, maintaining many of the earliest mammalian features, possibly because of their special habits. The full dentition of $\frac{3.1.4.3}{3.1.4.3}$ is usually preserved, and the cusps have diverged little from the tribosphenic pattern but two extra cusps are often added on the outer side of the tooth to make a W pattern ('dilambdodont'). The skull (Fig. 21.1) shows many primitive features. The orbit is broadly continuous with the temporal fossa (except in

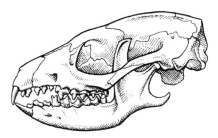

FIG. 21.1. Skull of hedgehog (*Erinaceus*).

FIG. 21.3. *Solenodon*, alamiqui. (After *Cambridge Natural History*.)

tree-shrews). There is an incomplete bony palate and an open tympanic cavity with no ossified bulla, in which the tympanic bone (the old angular) forms a partial ring. In the post-cranial skeleton there is usually found a clavicle, five digits with claws in both limbs, and the method of locomotion is plantigrade. A specialization is the reduction or elimination of the pubic symphysis. In the soft parts the primitive characters again predominate. The stomach is simple. The brain has large olfactory bulbs and small cerebral hemispheres, composed mainly of large pyriform lobes (rhinopallium), usually not covering the corpora quadrigemina or cerebellum, and with little convolution. The neopallium and corpus callosum are small. Besides the nasal receptors insectivores have a sensitive snout often drawn out into a short trunk. They have vibrissae and acute hearing (especially moles). The eyes are small and in many species the retina contains only rods but there are cones in some species of *Sorex* and in the tree shrews, *Tupaia*. Some insectivores retain the cloaca. The uterus is bicornuate and the testes are never fully descended into a scrotum. The placenta is discoidal and haemochorial (p. 432), that is to say of a type not obviously close to the presumed ancestral mammalian condition. Numerous young are produced (up to 32 in *Tenrec*). Many insectivores hibernate in winter and are provided with special reserves of fat for this purpose. They use various strategies to meet the large demands for energy and water due to their small size (see Schmidt-Nielsen, Bolis, and Taylor 1980). They find microclimates that protect them from heat or cold, and may become

hypothermic if food is short. Most insectivores are solitary but some have social habits, exchanging auditory and olfactory signals (*Solenodon*). They may make simple nests.

Tenrec (Fig. 21.2) from Madagascar, and *Solenodon*, the alamiqui (Fig. 21.3), from the West Indies are remarkably similar animals, showing in their dentition, brain, and other features characters more primitive even than those of other insectivores. The teeth have a tritubercular V pattern, by which they are sometimes distinguished as 'zalambdodont' from the remaining or 'dilambdodont' insectivores. The resemblance of the alamiqui and the tenrec has often been cited as evidence of a land bridge, but is probably a result of retention of primitive features. *Potamogale* (the otter shrew) is a related aquatic African form, feeding on fish.

The golden moles (*Chrysochloris*) of Africa are burrowing animals, with interesting features of similarity to the marsupial and true moles. Hedgehogs (*Erinaceus*) (Fig. 21.4) are mainly nocturnal creatures, feeding on a mixed diet of insects, slugs, small birds, amphibians, and snakes or even fruit. They have some immunity to snake-bite and indeed to bacterial and other toxins. They increase the irritation caused by their spines by smearing them with a froth made by chewing the skin of toads, or other toxic substances. This habit of self-anointing is inherited and manifested by the young before the eyes open at 20 days (Brodie 1977). Related genera in south-east Asia, such as *Echinosorex*, are more primitive in that the hairs of the back are normal, and not converted into spines as in the hedgehogs. In the Oligocene and Miocene of Europe both types were equally common. The shrews (Soricidae) are mouse-

FIG. 21.2. *Tenrec*. (From photographs.)

FIG. 21.4. *Erinaceus*, hedgehog. (From photographs.)

FIG. 21.5. *Sorex*, common shrew. (From photographs.)

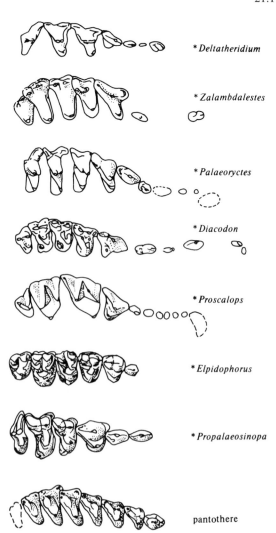

FIG. 21.7. The upper cheek teeth of primitive placentals. (Romer 1966.)

like, insectivorous and omnivorous animals of various types, some terrestrial, others aquatic, found throughout the world. The incisors are specialized as pincers. *Sorex* (Fig. 21.5) is a very ancient genus, hardly changed since the Oligocene. Shrews emit ultrasounds at 30–60 kHz in pulses of 5–30 ms and use these to find their way. They are disorientated if the ears are plugged (Gould, Negus, and Novick 1964). Moles (*Talpa*) (Fig. 21.6) are related to shrews and are found throughout the Holarctic region. They are highly specialized for burrowing, with eyes that are either very small or covered with opaque skin, no external ears, smooth fur, fused cervical vertebrae, massive pectoral girdle, including a procoracoid, and broad digging claws on the hands. They feed mainly on earthworms. *Desmana*, the desman, of south Europe in an aquatic mole, with webbed feet.

1.3. Primitive insectivores. Suborder Proteutheria

These are obviously of especial interest since they include the oldest of all eutherian mammals. *Delta-theridium* and *Zalambdalestes* are rather different types found together in upper Cretaceous deposits in Mongolia. In the former there are partly separated cusps near the centre of the tooth, which some believe to represent the partly divided central cusp of ancestral pantotheres (p. 411) (Fig. 21.7). In *Zalambdalestes* and related forms such as *Leptictis* the paracone and metacone lie in a more typical placental position on the outer border of the tooth (Fig. 21.7).

The modern tree shrews are classed with these early forms by their primitive features, and *Anagale* of the Oligocene may be an intermediate type. *Tupaia* and

Ptilocercus are diurnal, arboreal, squirrel-like creatures with a long tail (Fig. 21.8). They feed on insects or fruit and show many lemuroid characters. For instance, there is a complete post-orbital bar, and the brain has relatively larger hemispheres than in other insectivores, and less development of the olfactory regions. The eyes, lateral geniculate body, and visual cortex are well developed. They have three incisors rather than the two of primates and the cheek teeth have a very primitive tribosphenic pattern. They have claws, not nails. These animals, though they are like insectivores, have a life very like that of lemurs and they are often classified with the primates; they show how narrow is the gap between the two groups, but we need not worry unduly whether they 'really' belong in one group or another.

FIG. 21.6. *Talpa*, common mole.

FIG. 21.8. *Tupaia*, tree shrew. (From photographs.)

FIG. 21.9. *Macroscelides*, elephant shrew. (From a photograph.)

1.4. Suborder Macroscelidea

The elephants shrews (Fig. 21.9) *Macroscelides* of Africa are omnivorous creatures in which the molars are square, like those of ungulates. They are probably a distinct line with no special relationship to tree shrews.

1.5. Suborder Dermoptera

The colugo or flying lemur of the orient, correctly called *Cynocephalus* (= *Galeopithecus*), was probably an early offshoot from the insectivoran stock, with a patagium developed for parachuting. The wing differs from that of bats in that the fingers are not elongated and the wing is not moved in flight. The animals are nocturnal and feed on leaves and fruit. A peculiarity is the forwardly projecting lower incisors with tips divided to form a comb, as in lemurs. A related Palaeocene form shows that this line has been separate for more than 50 million years.

2. Chiroptera

2.1. Classification

Order Chiroptera. Lower Eocene–Recent
  Suborder 1: Megachiroptera. Oligocene–Recent
    Family: Pteropidae: Asia, Australia, Africa
      Pteropus, fruit bat
  Suborder 2: Microchiroptera. Eocene–Recent
    19 recent families, including:
    Family Rhinolophidae
      Rhinolophus, horseshoe bats, Europe, Asia, Australasia
    Family Phyllostomatidae
      Desmodus, vampire bats, South America
    Family Vespertilionidae
      Vespertilio, European bats, Palaearctic

2.2. Bats

Except for their specializations for flight the bats stand very close to the insectivores. They diverged early, however, and their characteristics were already developed in the early Eocene. They are the only mammals that truly fly, by flapping the wings, as distinct from the soaring of flying phalangers, colugos, and others. In acquiring the power of flight they have evolved many features in parallel with the birds, such as economy of weight in the skeleton and gut, and active metabolism. However, the differences of anatomy and behaviour are striking. The method of flight of many forms is specialized to give the great manœuvrability needed for catching insects by echolocation at short distances rather than by vision. The wings vary greatly but are thin aerofoils, often with high camber, giving high lift at low speed (p. 336). The wing is a patagium or membrane of skin that can be folded, involving all the digits of the hands, even the first in some species, and extending also along the sides of the body to include the legs (but not the feet) and, usually, the tail. The chief skeletal modification is therefore a great elongation of the arm, and especially of its more distal bones (Fig. 21.10). The sternum carries a keel for the attachment of the large pectoral muscles but this is not so large as in birds. The scapula is movable, not fixed as in birds, allowing control of the angle of attack of the wing. The downstroke is the main action in flight and is achieved by the pectoralis major, serratus anterior and subscapularis muscles, which together make over 15 per cent of the weight of the bat. The distal muscles of the arm are

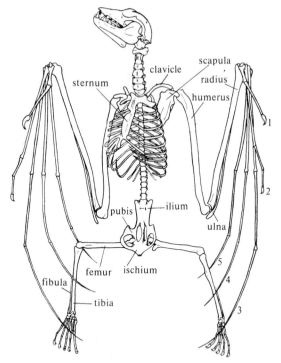

FIG. 21.10. Skeleton of fruit bat (*Pteropus*). (From Reynolds 1897.)

slender and serve to transfer distally the forces generated by the more proximal muscles. The weight is thus concentrated near the centre of gravity, as in birds.

Since the thorax is used as a fixation point for the flight muscles the ribs move relatively little on each other and respiration is mainly by the diaphragm. The ribs are flattened and some may indeed be fused together and with the vertebrae. The characteristics of the arms and thorax are rather similar in these flying animals to those found in brachiating arboreal creatures, such as gibbons, which also have long arms and large, fixed thoracic cages.

The humerus is very long and carries a large greater tuberosity, which may acquire a special articulation with the scapula. The flight movements occur mostly at the shoulder, with the rest of the limb held stiff. The radius is long and the ulna reduced and fused with the radius; the elbow joint allows only flexion and extension. The carpus is much specialized by fusion of bones, allowing flexion-extension and spreading of the digits. Of the five fingers the first is stout and usually free of the wing; it bears a claw in Microchiroptera, as does the second also in most fruit bats. The remaining metacarpals and proximal phalanges are enormously elongated to support the wing, the distal phalanges being relatively short. As in birds the wings are short and broad in the

slower fliers (horseshoe bats), long and narrow in those that fly faster, with long rapid beats (noctules). On landing horseshoe bats turn a somersault forwards and catch on with the hind legs. Others land on all fours and can walk and crawl reasonably well.

The pelvis is rotated so that the acetabulum lies dorsally and the limb is held outwards and upwards. The ventral portions of the girdle are thus drawn apart and are often not united in a symphysis. The hind legs are weak and carry five clawed digits, by which the animal is suspended upside down when at rest, the tendons providing a catch mechanism so that no muscular effort is needed. Even a large fruit bat remains suspended if shot while hanging.

Bats usually hang head downwards when not flying and to excrete they turn and hang by the claw of the pollex, so that the wing is not soiled. There is no upward temperature regulation when a bat is hanging. The animal becomes cold every time it rests (Fig. 21.11). This cooling provides a great economy of food. When first wakened the bat can walk, open its mouth, and bite or cry out, but can only fly after a period of some minutes of warming up by jerking the legs and shivering (for biology of bats see Wimsatt 1970–72; Yalden and Morris 1975).

Hibernation is common in bats of the temperate zones (Vespertilionidae and Rhinolophidae). It is an accentuated form of the daily sleep but species that hibernate store a special reserve of subcutaneous fat. Some bats can stand freezing temperatures for a short while, but they usually hibernate in caves at moderate temperatures around 10–15 °C and may waken and fly on warm days. Some aggregate in dense clusters in caves in the winter and can then survive in temperatures as low as −15 °C. This clustering serves to stabilize temperatures near the optimum and minimize water loss, which is an important factor.

2.3. Microchiroptera

The typical microchiropteran is insectivorous, often with molars of the ancestral tritubercular type arranged in a dilambdodont W pattern. They are tuberculosectorial, with cusps and ridges that chop or crush the food. A bat catches insects in its large mouth whilst in flight and then bites the wings off neatly. A bat can fill its stomach with insects up to one-third of its body weight in half an hour. The gut is very short, as in birds, and in some food passes through in half an hour. Diets differ considerably: some bats lick nectar from flowers, which they thus pollinate. The vampires *Desmodus* and *Diphylla* of South America drink blood and have the upper incisors modified into cutting blades. Other Microchiroptera eat fruit, fish, or flesh.

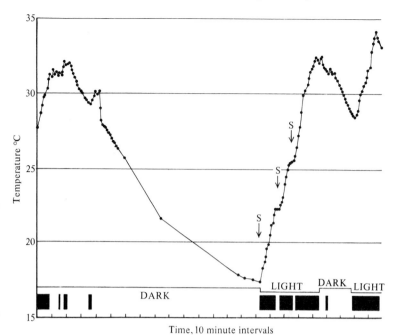

FIG. 21.11. Temperature chart of greater horseshoe bat. Manipulation during attachment of thermocouple to the back has caused warming at beginning of experiment. Bat stimulated at points marked S. The black rectangles mark periods during which the animal was shivering or moving. Room temperature 15.5°C. (From Burbank, R.C. and Young, J.Z. (1934). *Journal of Physiology, London* **82**, 459–67).

These bats obtain the major part of their information through the ears, and the reflection of the sound waves that they themselves emit. The cerebral hemispheres are small and the olfactory portions reduced, but the inferior collicula (concerned with hearing) and cerebellum are large. The eyes are often moderately large and presumably used in twilight; the retina contains mainly rods. The touch receptors are well developed, especially on the wings but are not used as sensors in flight.

The special feature of their life, apart from flight, is the capacity to be active in the dark. This has allowed many to live safely in deep caves and to find their food at night. The capacity to catch insects in the dark may be a secondary use of the sonar system.

The echolocation is performed by discrete pulses of high intensity and up to 150 kHz frequency emitted through the nose and/or mouth. These are produced by the very large larynx, whose cartilages are ossified to make a rigid framework. The strong cricothyroid muscles put great tension on the light vocal cords. In many types such as the horseshoe bats there are special resonating chambers and the face is elaborately modified forming a nose-leaf to beam the sound forwards (Fig. 21.12).

The ears of bats are greatly specialized, some have very large pinnae varying in shape. They are small in bats that fly fast and emit loud sounds, enormous in those that hunt insects on the ground or vegetation, using faint pulses. The pinnae provide directional sensitivity, especially in the vertical direction (where it is

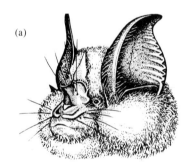

FIG. 21.12. Bats (a) *Rhinolophis*, (b) *Myotis*. (After Grassé.)

not needed in terrestrial mammals). The cochlea is large and the basilar membrane narrow and tightly stretched. The tensor tympani and stapedius muscles are large. The hearing is especially acute for high frequencies and greatest sensitivity is often limited to one or a few frequency bands.

The bat is often entirely dependent on echolocation for avoiding obstacles and catching insects. If the larynx is damaged or the ears blocked it blunders against even large obstacles. The normal animal can avoid wires less than 0.5 mm thick in complete darkness even if blinded. The presence of loud noise at high frequency disturbs the bat, but lower frequencies do not. There is evidence that the bat hunts by echolocation and not (usually) by listening to the sounds made by the insects. By use of its sonar system the bat derives information on the distance, direction, velocity, and nature of the reflecting object and makes decisions to avoid an obstacle, catch an insect, or dip for water. It can assess more than one target at a time and perhaps perceive the three-dimensional geometry of its surroundings.

The mechanism adopted is not fully understood and certainly is not always the same (see Henson 1970; Sales and Pye 1974; Novick 1977). Many different patterns of sound are used. In Vespertilionidae and Molossidae frequency modulated pulses of 1–4 ms duration are emitted by the mouth. The note falls through about an octave in each pulse. The pulse repetition rate varies from less than 10/s at rest to over 100/s when avoiding obstacles or hunting. The pinna of these bats is very large and below it lies a fold, the antitragus (Figs. 21.12 and 21.13). Vespertilionids commonly detect insects at 50 cm and may do so at 1 m.

In the horseshoe bats the pulse is much longer (40–100 ms), and of high and constant frequency (85–100 kHz). It is emitted through the nose and beamed by interference at the nostrils, which are set half a wavelength apart. The pulse repetition rate is low (<10/s). This mechanism is even more effective than the other and is said to detect insects even at 6 m (see Novick 1977).

It was first suggested by Hartridge (1920, 1945) by analogy with early audiolocation and radar devices that bats estimate distance by measuring the echo delay. The middle ear muscles and intra-aural reflexes do indeed allow a very rapid recovery of sensitivity after short loud sounds, as would be necessary, forming a sort of transmit–receive switch (Griffin 1958). Yet it hardly seems possible that the reflex can work fast enough to allow accuracy at short distances. An alternative hypothesis is that the bat measures the loudness of the echoes, especially the horseshoe bats, with their long pulses. The beam movements might give direction and searching movements the range by triangulation. This theory seems to require a very complete acoustic

FIG. 21.14. A bat chasing a moth in flight in darkness. The pulses of ultrasonic sound from the bat are reflected from the moth and enable the bat to keep the moth in its own flight path. The moth is sensitive to the high-frequency note of the bat, and takes avoiding action. Successive positions of the bat are at 1 to 4, of the moth a to f. (After Horridge 1968.)

FIG. 21.13. Big-eared bat (*Plecotus*). (After Hamilton 1902.)

separation of ear and nasopharynx and special cerebral capacities for calculation.

A third suggestion is that the difference in frequency between the outgoing and reflected notes is perceived as a low-beat note (Pye 1960). Since the sound is used only for location its absolute qualities are not important for a bat as they are for man. With this method any object within range will be located by the variation in the beat notes that are produced as its position changes. The auditory nerve carries information only about the differences, those of a few kHz could be readily recognized by the brain. Recordings by electrodes from the inferior colliculus of the midbrain show units whose response changes with the direction of stimulation. Some have very narrow response bandwidths tuned to the bat's own vocal frequency and responding only to a *faint* intensity range and to very slight changes of intensity. These faint pulse units may be facilitated by previous loud noise, acting as echoreceptors each measuring a particular delay in the range 1–5 ms. Young bats have to learn to perfect their locating system

so that they become able to construct 'images' of the world with auditory units as we do with visual ones (see Fig. 21.14) (Horridge 1968).

Recording of evoked potentials from the inferior colliculus shows that responses are produced not during the outgoing pulse but to the frequency-modulated sweep at the end of it and to the echo pulses (Fig. 21.15). Clearly there is 'protection' from the outgoing signal as required by Hartridge's theory. The middle-ear muscles do in fact contract during emission and relax in the frequency modulated and echo phases.

Horseshoe bats often seem to use echolocation to search around when they are stationary and there is some evidence that under these conditions they introduce an artificial velocity factor, and hence a Doppler shift, by rapid movements of the ears, which may occur at as much as 50/s. However, it is not clear exactly how this is achieved or how it is related to a second opening of the meatus that lies at the non-moving base of the pinna and leads by a groove to the nose-leaf, with its lance and shield (Fig. 21.12).

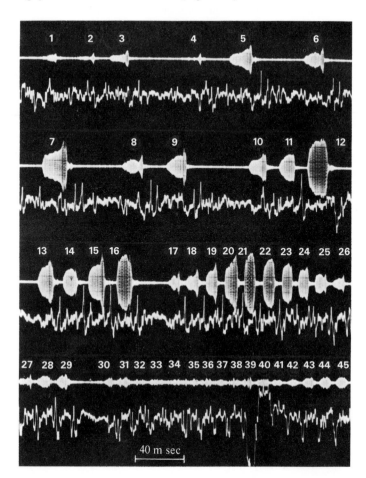

FIG. 21.15. Neurophysiological potentials elicited by orientation cries and echoes of *Chilonycteris* while flying. Upper trace pulse detected by microphone, lower trace evoked potentials in the inferior colliculus. Pulses 1–4 emitted as bat dropped from far wall of recording chamber. During emission of 5–29 bat was flying towards landing area (window); in pulses 30–38 it was turning, and then landing from 39–43 during which time (39) the bat made contact with the window. (Bar = 40 ms.) (Henson 1970.)

The placenta is of a discoidal form and haemochorial, at least in some bats. A peculiar feature in some families is that copulation and ovulation occur in the autumn and the sperms and ova remain alive (but presumably not active) within the female until delayed fertilization takes place in the spring. Gestation is long, from 50 days to 8 months, and post-natal development is also slow and the young may not fly until they are 10 weeks old. This dependence on the mother seems curious for flying animals. The young are well formed at birth and have then already cut the milk dentition of special teeth, with sharp, backwardly directed hooks, which, with the claws, enable the baby to remain attached to the mother in flight.

Microchiropterans characteristically live in caves, cracks, or crannies and the whole shape of the body is suited for this. Many bats live massed together in colonies during the day, with a considerable social organization. They spread out at night and home accurately. After artificial displacement marked bats return home from 100 km or more. The method of orientation used is not certain but is probably auditory or olfactory rather than visual. A few species hibernate in large colonies where there are suitable caves and then migrate for 2000 km or more and return to the same cave next winter.

The Microchiroptera is one of the most successful groups among modern mammals, including a large number of families, genera, and species, with differing habitats. As would be expected, the families often have wide geographical ranges, vespertilionids, for instance,

are found all over the world. It is interesting, however, that some genera have a rather restricted range. For instance, *Vespertilio* is limited to the Palaearctic. The vampire bats (Phyllostomatidae) of Central and South America are restricted by the 10 °C isotherm both north and south. There is evidence that their distribution has been greatly extended by the introduction of domesticated animals in quantity since European colonization in about 1500 AD. The fact that even flying mammals should be so restricted is good evidence that the simple problem of travel is one of the least of the difficulties standing in the way of the dispersal of an animal type.

2.4. Megachiroptera

Megachiropterans, fruit bats or flying foxes, are quite large animals, with wing span up to 1.5 m, living in Asia, the Pacific, Australia, and Africa. They are mainly fruit eaters and have flattened, grinding teeth. In spite of their diet they are in some ways the less specialized group, having snout, head, and ears of more usual mammalian form. The skull retains many primitive features and resembles that of a tree-shrew. There is an annular tympanic bone and no post-orbital bar. Flying foxes differ in many ways from the Microchiroptera. They roost in colonies in trees, emit characteristic loud cries, and may also feed in groups. They depend heavily on sight and smell and do not echolocate, at least in the way that microchiropterans do. Flying foxes also orientate visually, and have no need for great manœuvrability, but they can glide.

3. Taeniodontia (= band teeth) and Tillodontia (= tearing teeth)

*Order Taeniodonta. Palaeocene–Eocene. North America
   Psittacotherium; *Stylinodon*
*Order Tillodontia. Palaeocene–Eocene. Europe, North America
   Trogosus

These were short-lived Tertiary groups, outgrowths near the base of the placental stem, close to the primitive insectivoran stock.

Stylinodon, a Middle Eocene taeniodont, was quite large, the short, deep skull reaching 30 cm long had powerful jaws. The high-crowned teeth were simple rootless pegs, the enamel covering being restricted to bands along the side. The canines were well developed but only a single pair of incisors remained. They had

grinding teeth and were perhaps leaf-eaters.

Trogosus, a tillodont of the Middle Eocene, was the size of a bear with plantigrade, clawed, five-toed feet. The skull was about 30 cm long provided with a slim snout, and the brain case small. There were two pairs of incisors, the second enlarged and rootless, small canines and low crowned molars. They were perhaps herbivorous or omnivorous.

4. Edentates

4.1. Classification

Order 6: Edentata (= toothless). Upper Palaeocene–Recent
   Suborder 1: Palaeanodonta (= ancient toothless). Upper Palaeocene–Lower Oligocene. North America
      Metacheiromys

Suborder 2: Xenarthra (= extra-jointed). Palaeocene–Recent. Central and South America
   Infraorder 1: Cingulata (= belted). Palaeocene–Recent
      Superfamily 1: Dasypodoidea. Lower Eocene–Recent
        Dasypus, nine-banded armadillo
      *Superfamily 2: Glyptodontoidea. Upper Eocene–Pleistocene
        Glyptodon
   Infraorder 2: Pilosa (= hairy). Upper Eocene–Recent. Central and South America
      *Superfamily 1: Megalonychoidea. Ground sloths. Upper Eocene–Pleistocene
        Megatherium; *Mylodon*; *Nototherium*
      Superfamily 2: Myrmecophagoidea (= Ant-eaters). Pliocene–Recent
        Myrmecophaga, giant ant-eater; *Tamandua*, tamandua; *Cyclopes*, two-toed ant-eater
      Superfamily 3: Bradypodoidea (short-footed). Recent
        Bradypus, three-toed sloth; *Choloepus*, two-toed sloth (unau)

The reduction or loss of the teeth with adoption of a diet of invertebrates and especially ants has occurred independently at least five times among mammals; this habit is indeed to be expected, since the whole mammalian stock was at first insectivorous. We have already noticed the occurrence of ant-eating characteristics in the echidnas and in *Myrmecobius*, the marsupial ant-eater. Among eutherians the habit is well developed in animals of three different types, (1) the ant-eaters of South America, *Myrmecophaga* and its allies, (2) the pangolins of Africa and Asia, *Manis*, and (3) the aardvark or Cape ant-eater, *Orycteropus*. These ant-eating animals have many features in common. They all possess a long snout and tongue, very large salivary glands, and reduced teeth; because of these similarities they were for a long time classed together as Edentata. It has gradually become apparent, however, that the three groups of placental ant-eaters have evolved separately. The aardvark was probably an early offshoot from the ungulate stock (p. 520). The pangolins, placed in the Unguiculata, represent a separate line, diverging from the insectivoran stock very early (p. 449). The South American ant-eaters form a natural group with the armadillos and sloths having, like the South American ungulates and other animals, proceeded along several courses of evolution of their own during the long isolation of their continent throughout the Cenozoic period. The term Edentata is now reserved for this South American group.

In many ways the Edentata remain close to the basic eutherian condition. The characteristic feature has been a simplification of the teeth, which are absent altogether in the ant-eaters themselves. In sloths and armadillos the front teeth are absent and the hinder ones are rows of similar pegs, with no covering of enamel. Except in sloths, there is considerable elongation of the snout and the whole cranium is of tubular form, with a low brain case, containing a small brain with poorly developed hemispheres, having a large olfactory region. The jugal bar is often incomplete, but the hind end of the jugal carries a large downward extension in sloths and ground sloths. A characteristic common to all the Edentata is the presence of extra articulations between the lumbar vertebrae, a striking feature in view of many different modes of locomotion in the group. From these articulations the group gets its name, Xenarthra. Several other peculiar features of the skeleton are common to most or all of these animals, such as a fusion of the coracoid with the acromion to enclose a coracoscapular foramen and a union between the ischium and the caudal vertebrae. The feet have well developed claws, often used for digging, and the animals may walk on the outside of the claws, though some species are arboreal and use the claws for hanging.

Many of the characteristics of the group are obviously those of all generalized eutherians, the edentates having departed little from the original mammalian plan. For example, they all have rather low temperatures, fluctuating widely with the environment. Their features are mostly the result of special ways of life, often leading to bizarre external appearances, such as the long snout of the great ant-eater or the carapace of the armadillo.

The order Edentata is divided into two suborders, the first *Palaeanodonta for a few Palaeocene and Eocene types such as *Metacheiromys*, which had not yet acquired the structure of the vertebrae found in all the remaining edentates (suborder Xenarthra). The palaeanodonts are found in North America and are held by some to be survivors of the original stock, existing before the separation of the continents. The xenarthrous population itself split up early and we can recognize two main groups (infraorders), the Cingulata for the armadillos and glyptodons, and Pilosa for the ant-eaters, sloths, and extinct giant ground sloths.

4.2. Armadillos

The armadillos (Dasypodidae) (Fig. 21.16) have departed least from the ancestral plan and are a very ancient group, already differentiated in Palaeocene times. They are nocturnal and fossorial and obtain

FIG. 21.16. Hairy armadillo, *Dasypus*. (From photographs.)

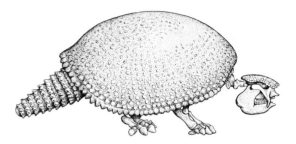

FIG. 21.17. *Glyptodon*. (From a reconstruction lent by the Trustees of the British Museum.)

protection by the development of bony plates in the skin, these being covered by horny scutes. The plates are usually arranged in rings round the body united by the skin muscles (panniculus carnosus) which in some genera allow the animal to roll up into a ball. The vertebrae tend to be fused to support the shield, and many vertebrae unite in the sacrum. The teeth are simple uniform pegs, without enamel, and with open roots and continuous growth. They are often more numerous than in other mammals (as many as twenty five in each jaw); with simplification of the system of tooth morphogenesis, repetition becomes possible, as we see also in whales.

The armadillos are abundant creatures, living as insectivores and omnivorous scavengers in the southern United States and Central and South America; there are many different genera and species. The nine-banded armadillo (*Dasypus novemcinctatus*) is a very active burrower and is spreading northwards in the United States with the destruction of its carnivore enemies by man. The haemochorial placenta, at first diffuse then discoidal, is modified as a result of the process of polyembryony. The ovum divides to form a blastocyst and remains dormant for 2–4 months. It then divides into up to twelve secondary embryos.

During the Pleistocene and earlier periods, besides the modern armadillos, there were also giant armadillos. The glyptodonts (Fig. 21.17) were a related type, diverging as early as the Upper Eocene, with a skull and carapace composed of many fused small pieces and sometimes the well known 'battle-axe' tail. They show a remarkable convergence with tortoises and some dinosaurs. They reached up to 2.7 m long and were numerous from the southern United States to Argentina until the late Pleistocene.

4.3. Ant-eaters and sloths

4.3.1. Ground sloths

The enormous extinct ground sloths were undoubtedly related to the ancestors of the modern soft-skinned edentates, the ant-eaters and sloths, but few fossils of the latter are available. The ground sloths were abundant in both North America and South America from the Oligocene to the late Pleistocene. The earlier ones were small and mainly arboreal, probably the ancestors of the modern sloths. From them evolved three distinct lines of large animals. They all had five, clawed, digits on hands and feet, the fore limbs being shorter than the hind. The skeleton of *Nototherium*, the last member of one line, has been found with skin and tendons still adhering. *Megatherium* (Fig. 21.18) was larger, 6 m long, bigger than an elephant. The last known representative of another line *Mylodon* found in a cave in Patagonia had been killed by man. It is uncertain how these giant creatures lived and why they became extinct. Nearly fifty genera of ground sloths have been recognized and they were evidently successful in the American forests.

FIG. 21.18. Giant South American edentates of the Pleistocene. *Megatherium*, the giant ground sloth, and glyptodonts related to the armadillos. (From a mural by C.R. Knight.)

FIG. 21.19. Great ant-eater, *Myrmeco-phaga*, from life.

4.3.2. Ant-eaters of the New World

The ant-eaters (Myrmecophagidae) have a characteristic elongated snout, without teeth. There are three genera, differing in size, and the larger species have relatively much the longer snouts. The great ant-eater *Myrmecophaga* (Fig. 21.19) has an enormously elongated face, but this is much shorter in the smaller *Tamandua* (Fig. 21.20), and the very small tree-living *Cyclopes* (Fig. 21.21) has a head of normal mammalian shape. Analysis shows that there is little difference between the relative rate of growth of the face in these three genera, and the differing final forms result mainly, though not wholly, from the differences in absolute size

(Figs. 21.22 and 21.23). In all ant-eaters the face becomes relatively longer as the animal increases in size, and the enormous snout of the great ant-eater is produced by a relative growth-rate only slightly higher

FIG. 21.21. Tree ant-eater, *Cyclopes*, showing the defence attitude adopted, perhaps to startle an attacker (dymantic posture). (From a photograph.)

FIG. 21.20. Lesser ant-eater, *Tamandua*. (From photographs.)

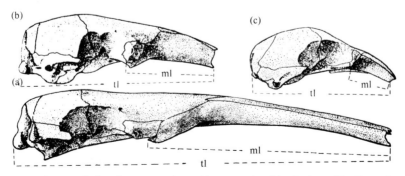

FIG. 21.22. Side views of adult skulls of *Myrmecophaga* (a); *Tamandua* (b); *Cyclopes* (c). *Tamandua* and *Cyclopes* are approximately $1\frac{1}{2}$ and 3 times the scale of *Myrmecophaga*. tl measurement of total length; ml of maxilla length. (From Reeve.)

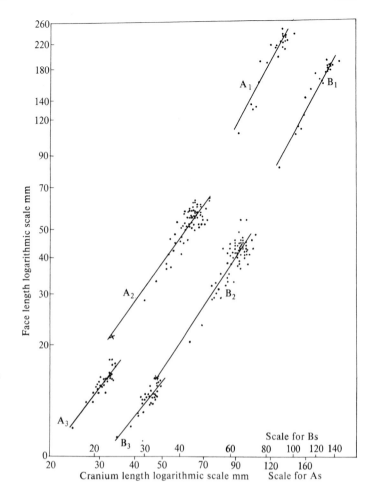

FIG. 21.23. Logarithmic plots of lengths of maxillae (A_{1-3}) and nasal bones (B_{1-3}) against cranium.

The suffixes represent: 1, *Myrmecophaga*; 2, *Tamandua*; 3, *Cyclopes*. Scale for Bs is shifted to the right compared with As. Crosses at bottom of lines for A_2 and B_1 represent a very young *Tamandua*. The lines were fitted to each sample by least squares. (From Reeve.)

than that found in *Tamandua* and *Cyclopes*. This is an excellent example of the way in which the proportions of an organ will vary in animals of different sizes if its growth is allometric, that is to say, relatively faster or slower than that of the body as a whole.

The hard palate is prolonged backwards in *Myrmecophaga* by union of the pterygoids, a condition found also in some armadillos (*Dasypus*).

The great ant-eater is a fine animal, 2 m long, with a long hairy coat, including a long bushy tail. It lives in forests and savannah country from Central America to Argentina. It has a long thin tongue for collecting ants and termites, and enormous submaxillary salivary glands. The claws of the front legs are very large and used for defence as well as for digging. *Tamandua* and *Cyclopes* differ from *Myrmecophaga* in other features besides the length of the snout. They are arboreal and can climb and cling by the hind legs and prehensile tail,

leaving the fore legs free to dig for ants or for defence (Figs. 21.20 and 21.21).

4.3.3. Sloths

The sloths (Bradypodidae) (Figs. 21.24 and 21.25) are fully adapted for arboreal life and cannot walk on the ground. They show, as do the bats, how the mammalian skeleton can be used with surprisingly little change to support weight by hanging, the limbs being used as tension members rather than as pillars. In marked contrast to the ant-eaters the face is short and the head rounded, with large frontal air sinuses. The neck is peculiar for the presence of nine or ten cervical vertebrae in the three-toed sloth, *Bradypus* (Fig. 21.25). This might be supposed to provide a flexible neck for an animal that must often face backward, were it not that in the two-toed sloth *Choloepus* (Fig. 21.24) there are but six cervical vertebrae.

FIG. 21.24. Two-toed sloth, *Choloepus*. (From a photograph in Scott 1913.)

The limbs are long, especially the fore limbs, and the digits carry hooked claws for hanging (Fig. 21.25). In the pectoral girdle the clavicle articulates with the coracoid, a unique condition among mammals. As in ant-eaters the acromion is connected with the coracoid, enclosing a coracoscapular foramen. The significance of these special features is not clear, but the habit of hanging upside down has produced some obvious modifications, for instance all the vertebral neural spines are low and the pelvis is short. Even here, however, we find peculiar similarities to the ant-eaters, in the union of the ischium with caudal vertebrae, a feature whose adaptive significance is obscure.

The sloths live on foliage, but this herbivorous diet is perhaps secondary to a long period of insectivorous life, during which there was a reduction of the teeth and loss of enamel. On adoption of the new way of life the enamel could not be restored, but a grinding surface is provided by the presence of cement and continuous growth of the teeth. The stomach is large and divided into several chambers, recalling those of ruminants. The rectum is enormous and the masses of faeces are retained for several days, intestinal peristalsis being as slow as all the other movements of these creatures. Interesting features connected with this slow life are the small size of the thyroid and adrenals. The body temperature is low and variable (24–37 °C) and the animals neither shiver when cold nor sweat when hot, dying of hyperthermia if held at 40 °C.

The sloths live in the rain forests of South and Central America, moving very slowly among the branches, which they come to resemble closely by the growth of blue–green algae in special grooves in the hairs. Their upside down posture has led to interesting changes from the typical mammalian organization, including, it is said, a reversal of the usual mechanism for maintaining posture. When a normal mammal is decerebrated its legs assume a pillar-like extensor rigidity, because of the overaction of the reflexes of standing. A decerebrate sloth is said to show the opposite, flexor rigidity.

5. Order Pholidota: pangolins

Order 7: Pholidota. Oligocene–Recent
   Family Manidae
      Manis, scaly ant-eater (pangolin). Asia, Africa

The pangolins or scaly ant-eaters, *Manis* (Fig. 21.26) of

FIG. 21.26. Black-bellied tree pangolin, *Manis*. (From photographs.)

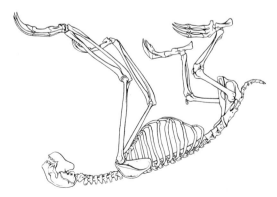

FIG. 21.25. Skeleton of three-toed sloth, *Bradypus*. (After Blainville.)

the Old World (Africa and Asia) have many features superficially like those of the New World ant-eaters and the groups may be remotely related. Unfortunately nothing is known of the fossil history of *Manis* and its position among the unguiculates is therefore provisional. The body is up to 1.5 m long, covered with horny epidermal scales, interspersed with hairs. The gait is plantigrade and there are five digits on hands and feet, armed with claws, used for opening nests. The absence of teeth, the elongated snout, long thin tongue, simple stomach, reduced ears, and long claws are all features found in other ant-eaters as is the habit of rolling into a ball. Rods of cartilage extending backwards from the xiphisternum have been compared with the abdominal ribs of reptiles, but are probably a special development, connected with the protrusion of the enormous tongue. This is carried in a special sac and operated by muscles attached to the xiphisternal processes. The animals are macrosmatic, with small eyes. The brain is very small, but the hemispheres are folded. The placenta is diffuse and epitheliochorial, with a large allantois and a yolk sac persisting until birth. Evidently the pangolins preserve many very ancient mammalian features. There are various species of *Manis*; some live in open savannah, others are able to climb trees and have prehensile tails. All are nocturnal and eat ants and termites.

22 Primates

1. Classification

Order 8. Primates
  Suborder 1: Prosimii. Palaeocene–Recent
    Infraorder 1: *Plesiadapiformes. Cretaceous–Eocene
      *Family 1: Paromomyidae. Cretaceous–Eocene
        Purgatorius, Cretaceous–Palaeocene
      *Family 2: Plesiadapidae. Palaeocene–Eocene. Europe and N. America
        Plesiadapis, Palaeocene
      *Family 3: Carpolestidae. Palaeocene–Eocene
        Carpolestes
      *Family 4: Picrodontidae. Palaeocene
        Picrodus
    Infraorder 2: Lemuriformes. Palaeocene–Recent
      *Family 1: Adapidae. Eocene. Europe and N. America
        Notharctus; *Adapis*
      Family 2: Lemuridae. Pleistocene–Recent. Madagascar
        Megaladapis, Pleistocene; *Lemur*, common lemur; *Lepilemur*, sportive lemur
      Family 3: Indridae. Pleistocene–Recent. Madagascar
        Indri, indris; *Propithecus*, sifaka
      Family 4: Daubentoniidae. Recent. Madagascar
        Daubentonia (= *Cheiromys*), aye-aye
    Infraorder 3: Lorisiformes. Miocene–Recent. Asia, India, and Africa
      Family. Lorisidae
        Loris, slender loris, S. E. Asia and India; *Galago*, bush baby, Africa; *Perodicticus*, potto, Africa;
          Progalago, Miocene
    Infraorder 4: Tarsiiformes. Palaeocene–Recent. Holarctic, Asia
      *Family 1: Anaptomorphidae. Eocene–Miocene
        Tetonius, *Omomys*
      Family 2: Tarsiidae. Eocene–Recent. E. Indies
        Necrolemur, Eocene, Europe; *Pseudoloris*, Eocene, Europe; *Tarsius*, tarsier
  Suborder 2: Anthropoidea. Oligocene–Recent
    Superfamily 1: Ceboidea. New World monkeys. Miocene–Recent. S. America
      Family 1: Callitrichidae. Recent
        Callithrix (= *Hapale*), marmoset
      Family 2: Cebidae. Miocene–Recent
        Homunculus, Miocene; *Cebus*, capuchin; *Ateles*, spider monkey; *Alouatta*, howler monkey; *Aotus*,
          night monkey
    Superfamily 2: Cercopithecoidea. Oligocene–Recent
      *Family 1: Parapithecidae. Oligocene. Africa
        Apidium *Parapithecus*
      Family 2: Cercopithecidae. Old World monkeys. Miocene–Recent. Africa and Asia.

Mesopithecus, Pliocene; *Macaca*, rhesus monkey, macaque, Asia, N. Africa; *Papio*, baboon, Africa; *Mandrillus*, mandrill, Africa; *Cercopithecus*, guenon, Africa; *Presbytis*, langur, E. Asia; *Colobus*, guereza, Africa

Superfamily 3: Hominoidea

    Family 1: Pongidae. Apes. Oligocene–Recent

        Oligopithecus, Lower Oligocene, Egypt; *Propliopithecus*, Lower Oligocene, Egypt; *Aeolopithecus*, Oligocene, Egypt; *Pliopithecus*, Lower Miocene, Europe, Africa; *Aegyptopithecus*, Oligocene, Egypt; *Dryopithecus*, Miocene, Africa and Asia; *Gigantopithecus*, Pliocene–Pleistocene, Asia; *Oreopithecus*, Pliocene, Europe; *Hylobates*, gibbon, S. E. Asia; *Pongo*, orang-utan, E. Indies; *Pan*, chimpanzee, Africa; *Gorilla*, gorilla, Africa

    Family 2: Hominidae. Man. Upper Miocene–Recent

        Ramapithecus (= *Kenyapithecus*), Upper Miocene–Pliocene, India, Asia, Africa, and Europe; *Australopithecus*, Pliocene–Pleistocene, south and east Africa; *Homo*, man (all living races), Pleistocene–Recent

2. Characters of primates

Linnaeus reserved his order Primates for the monkeys, apes, and men, distinguishing them thus from the other mammals, Secundates, and all other animals, Tertiates. The term primate carries with it the implication that the animals in the group are not only the nearest to ourselves but are also in some sense the first or most completely developed members of the animal world. We shall try to examine this belief in accordance with the principles adopted in this book and to inquire whether we and our relatives can be said to be the highest animals in the sense that we process more information and so possess a system of life able to survive under the most varied and unpromising conditions (Napier and Napier 1967; Walker 1964; Corbet and Hill 1980).

The earliest eutherians of the Cretaceous were probably arboreal; the primates have continued this habit and with it they retain many of the features present at the beginning of mammalian history, for instance the five fingers and toes and the clavicle. Primates already existed in the Palaeocene, 65 Ma ago and have a longer geological history than any other placentals except the insectivores and carnivores. It is not surprising, therefore, that it is difficult to separate the primates from the insectivores; the tree shrews, for instance (p. 438), have several times been transferred from the one order to the other.

The general plan of primate life has thus been to retain the original eutherian conditions with emphasis on those features important for tree life. Their characters are those of animals raised up from the ground. The opportunities offered in the trees for the use of hand and brain have no doubt been important influences in the shaping of man. In such an existence continual quick reaction to circumstances is likely to be necessary, the environment is varied, and the mechanical supports it offers are often precarious. Under these conditions safety is achieved by quick reactions rather than by

stability; thus primate more than any other life tends to be a matter of continual exploration and change. The information that ensures the life of the species is obtained by the individuals and stored in their brains, rather than by selection among large numbers of rapidly breeding individuals. The time taken for development increases in the primate series. Growth continues for about 3 years in larger prosimians, 7 in monkeys, 9 in gibbons, 12 in other apes, and 20 in man. Receptors are obviously of first importance in obtaining information, but in the tree-tops one cannot hunt by smell; the eyes and ears therefore became developed, at the expense of the nose. Later primates are microsmatic, with reduction of the number and length of the turbinal bones and hence of the long snout that houses them. Consequently the eyes come to face forwards, so that their fields overlap, binocular vision becomes possible, and in later forms central areas with cones appear in the retinas. Monkeys are certainly more dependent on vision than are most animals and for this reason they approach the birds in the adoption of colour patterns for sexual recognition and excitation.

The changes in the receptors were accompanied by conspicuous changes in the brain, which becomes very large in later primates, with cerebral hemispheres reaching far backwards. The olfactory bulbs and rhinopallium become relatively smaller and the neopallium very large, differentiated into areas and provided with a large corpus callosum. The occipital pole, concerned with vision, and the frontal areas, become especially well developed in the apes and man. Stereoscopic eyes with numerous cones would be of no value without a central analyser to allow the animal to discriminate shapes and colours, retain the impression of past situations, and otherwise make use of the available information. The marked differences in the rate of growth of the brain of different primates are shown in Fig. 22.1. At early stages of development all the primates studied have the same

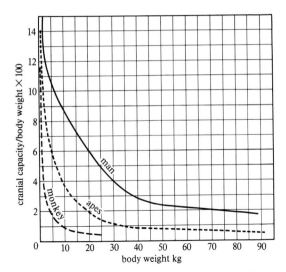

FIG. 22.1. The relative cranial capacity as a function of body-weight in various primates. The brain grows relatively faster in man than in either monkeys or apes. The curves are constructed by measuring cranial capacity and body-weight of individuals of differing ages. The monkeys included various Cercopithe-cidae, the apes only gorillas, chimpanzees, and orangs. (Modified after Schultz 1941.)

(high) relative brain weight, but in the adults the brain is relatively and absolutely larger in man than in monkeys or apes.

The special developments of the receptors and brain and mechanics of the jaws have marked effects on the skull, whose facial portion becomes shorter and the brain case relatively larger and rounder; the foramen magnum comes to face downwards, rather than backwards. As the eyes are directed forwards the orbits become closed off from the temporal fossae behind in tarsiers and Anthropoidea. The head is more clearly marked off from the body than is usual in mammals and the neck is very mobile, allowing the head to be turned in any direction for all round vision.

The skeletal and muscular systems of primates have become variously modified to provide locomotion in a three-dimensional environment. The generalized mammalian pentadactyl limb patterns are preserved to a greater extent than in most mammals. There is provision for a wide range of movements at the shoulder and hip joints. There is a strong clavicle and the scapula can be rotated. The radius and ulna remain separate and jointed to allow rotation to give pronation and supination. The tibia and fibula are separate except in tarsioids. The digits are mostly elongated and provided with sensitive pads and flat nails instead of claws though these are present in aye-ayes and callitrichids. The hallux and pollex are often opposable to allow grasping.

Many primates still retain a quadrupedal form of locomotion, modified to allow running along branches. All four limbs are then of equal length. Climbing is made possible by allowing the forelimbs to reach forwards and even in front of the head. The retractor muscles of both limbs then pull the body up, together with the distal flexors in the forelimb and extensors in the hind.

Some primates leap by rapid synchronous retraction and extension of the hind limbs, e.g. bush babies (*Galago*) and *Tarsius* (Figs. 22.11 and 22.13). Suspension and swinging of the body from the forearms leads to a form of locomotion loosely classed as brachiation (Fig. 23.8). This is found in different forms in a number of not closely related species, including the prehensile-tailed spider monkeys and the gibbons (Fig. 22.18). Suspension from the hind limbs is another form of arboreal locomotion, using especial mobility of the hip, as in some lorisines and again in spider monkeys (Fig. 22.2). Terrestrial locomotion on four limbs has been secondarily developed in baboons, with some restriction of hip movement and a partially digitigrade gait. Knuckle walking is another terrestrial adaptation, found in the African great apes. Several sorts of monkey can walk on the hind legs for short distances but man is the only specialized primate biped.

The primates early ceased to feed only on insects, and took to a mixed diet; the teeth have not become so specialized as in ungulate mammals. The hands are frequently and ingeniously used to obtain food. Omnivorous or frugivorous diets are common. There are only two incisors on each side of the upper and lower jaws, and the canines are generally small. The tooth row is short, with (usually) only two or three bicuspid molars (Fig. 22.3). The molars of the early forms were tribosphenic but have generally become quadritubercular, the upper adding a hypocone and the lower losing the

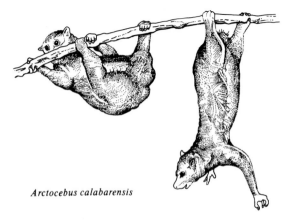

Arctocebus calabarensis

FIG. 22.2. Arboreal locomotion using four grasping extremities, the angwantibo. (After Oxnard 1975.)

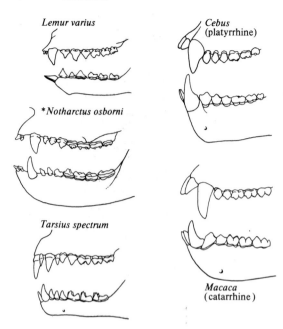

Lemur varius

Cebus
(platyrrhine)

Notharctus osborni

Tarsius spectrum

Macaca
(catarrhine)

Fig. 22.3. Upper and lower dentition of modern and fossil primates. (After Le Gros Clark 1934.)

young the primates also extend parental care for a long time after birth.

In many features of their life, therefore, the primates show to a high degree the adjustability and power to obtain information and hence sustenance from varying environments that is characteristic of all life. The receptors, brain, and hand provide means for doing this in more elaborate ways than are used by any other animals. The monkeys and apes have exploited these powers to a considerable extent and are successful animals, living, as we might say, by their wits, in a wide variety of circumstances. However, non-human primates are unable to adjust to conditions outside the tropical and subtropical regions. Man has made still better use of his talents; by creating his own environment managed in 1979 to support a population of more than 4100 million large individuals, scattered all over the globe.

3. Divisions of the primates

Fortunately many of the changes of habit characteristic of the various primates involved changes in the skull and these can be followed in the fossils. Our knowledge of the evolutionary development of primate life, though far from complete, is less so than might be expected from the rarity of preservation of skeletons of arboreal animals. During the 60 million years since the Palaeocene the various primate stocks have, of course, divided and subdivided many times, and invaded many special habitats. The forms at present known, living and as fossils, are placed by Simpson (1945) in 150 genera, two-thirds of them extinct, 70 of the total being prosimians. Most of primate evolution has occurred in the Old World, but numerous fossil primates are known from North America in the Eocene and Oligocene.

Bitter controversy still rages around the question of the best means of classification of Primates. Earlier zoologists tended to postulate a series of stages successively closer to man, the latest product of evolution. There has been increasing awareness of the unwisdom of this procedure. Recognition that many of the surviving stocks have been separate for a long time has led systematists to emphasize the distinctions between the groups more sharply. There is no general agreement about the best means of classification; the more traditional schemes, such as that adopted here, probably give an over-simplified idea of a progression of forms. A cladistic classification on more 'natural' or phyletic lines could be devised, but would necessitate the postulation of a large number of distinct categories, unless these were simplified by admitting speculations about the affinities of the lines (p. 11).

We shall, as usual, in the main follow Simpson (1945).

paraconid of the original pattern, leaving the metaconid and protoconid, while the hypoconid and entoconid become raised to make a posterior pair, sometimes with addition of a fifth cusp, the hypoconulid, posteriorly. The cusps are usually not of the sharp insectivorous type, but are low (bunodont) cones and extra ones may be added, or the cusps joined to make ridges. These changes are associated with the adoption by many primates of a diet of fruit or leaves, requiring treatment by biting and grinding, but the teeth are never hypsodont.

The method of reproduction is one of the most characteristic of primate features. The uterus retains signs of its double nature in the prosimians but has become a single chamber in monkeys, apes, and man. The number of young produced is small, as in other animals with large brains that learn well. There is often only a single pair of teats and in association with the arboreal habit these are pectoral. The arrangements for placentation involve elaborate changes in the uterine mucosa in preparation for reception of the embryo, followed by breakdown at regular intervals (menstruation) in Old World monkeys, apes, and man. This special nature of the uterine mucosa makes possible the efficient haemochorial form of placentation in these animals in which maternal and foetal bloodstreams are separated only by the walls of the foetal vessels themselves. Besides these arrangements for nutrition of the

His arrangement retains the order Primates and recognizes two great suborders, Prosimii and Anthropoidea. The division is 'horizontal' rather than 'vertical'; the two groups are not separate and divergent lines, they contain respectively the ancestral and the 'developed' forms. The Prosimii includes four sorts of primate, all 'primitive' in the sense of retaining insectivoran characters, such as long face, lateral eyes, and a brain smaller than in Anthropoidea; they are grouped here as four infraorders following Simons (1972). The *Plesiadapiformes include the earliest primates, Lemuriformes for the lemurs of Madagascar and their fossil allies; Lorisiformes the rather similar animals outside Madagascar; and Tarsiiformes the living tarsiers of the Malay Archipelago and its Eocene relatives. The suborder Anthropoidea includes three superfamilies: the New World monkeys, superfamily Ceboidea; the Old World monkeys Cercopithecoidea; and the apes and man Hominoidea.

4. Lemurs and lorises

The lemurs (Fig. 22.4) living in Madagascar today resemble certain fossils, the adapids, that existed in various parts of the world in Palaeocene and Eocene times (Tattersall and Sussman 1975; Martin, Doyle, and Walker 1974). We may, therefore, perhaps assume that they show us some of the characters of the primate stock more than 50 Ma ago. Madagascar has been an island since the late Cretaceous and in it the lemurs have radiated into a variety of niches. There are few other mammals there; apart from bats, only one insectivore, the tenrec, a few viverrid carnivores, some rodents, and bush pigs probably recently introduced. No other ungulates, catarrhines, rodents, or carnivores live there.

Lemurs show their 'primitive' nature in their habits and appearance, as well as in the details of their

FIG. 22.4. Ring-tailed lemur, *Lemur catta*. (From life.)

structure (Fig. 22.5). They are mostly nocturnal and the name means 'ghost'. The nose has numerous well-developed turbinal bones (Fig. 22.15). The brain (Fig. 22.6) has relatively small cerebral hemispheres, and olfactory regions large for a primate, though smaller than in insectivores or other primitive mammals. The cerebral sulci tend to run longitudinally, rather than transversely as in anthropoids.

The snout is long, with a cleft upper lip and moist rhinarium. The eyes are directed somewhat sideways, and the retina contains only rods except in the genus *Lemur*, which is diurnal and possesses cones. There is no

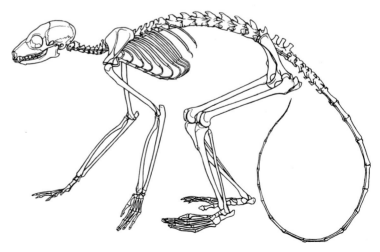

FIG. 22.5. Skeleton of ring-tailed lemur.

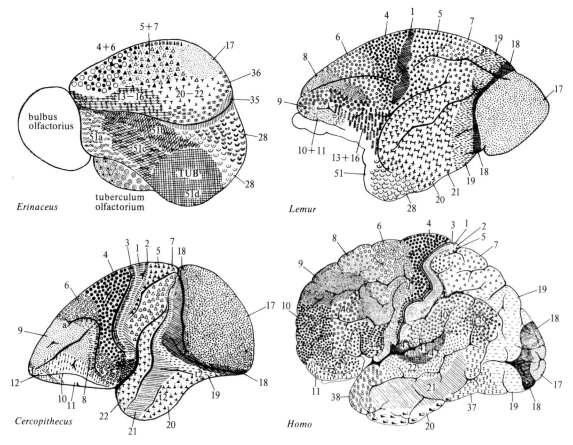

FIG. 22.6. Brains of hedgehog and various primates interpreted to show the relative development of various parts. The numbers refer to the areas recognized by Brodmann on a basis of their structure. 4 and 6 are the precentral motor areas, 8–12 the frontal and prefrontal areas, lacking in the earliest forms. 1–3 are the end station of skin sensations, and 5 and 7 are also concerned with these. 17 is visual end station, and 18 and 19 are also concerned with this sense. 22 is the auditory end station. (Later work has altered some of these interpretations.) (From Brodmann 1909.)

fovea and no binocular vision. The external ears may be large, as in other nocturnal animals. In the skull (Fig. 22.7) there is a postorbital bar, but the temporal fossa opens widely to the orbit. The tympanic region shows several peculiar features. The tympanic bone forms a ring, lying within a petrosal bulla, but not fused with it (Fig. 22.8), a condition not found in higher primates. The pollex and hallux are used for grasping; most of the digits have nails, but the second digit of the foot has a toilet claw. The fourth digit is usually the longest, whereas in anthropoids the whole symmetry of the hand and foot is arranged about a long third digit. The teeth show the typical primate number $\frac{2.1.3.3.}{2.1.3.3.}$, but the upper incisors are very small and the lower incisors and canines are procumbent, that is to say directed forwards and are used by the lemurs for grooming themselves or each other, and in some species also for

scraping bark to find insects or gum. The first lower premolar is caniniform. The molars are triangular in some genera, in others a hypocone gives them a square shape. The lower molars are of typical tuberculosectorial type, with a heel.

The reproduction shows several primitive features. There are marked breeding seasons and the females are polyoestrous. In *Lemur fulvus* there are great differences in the appearance of males and females, especially on the head. In the breeding season both sexes mark with glandular perianal skin and males have glands on the scrotum. The uterus is bicornuate and the placenta of a remarkably simple type, epitheliochorial and diffuse, with villi all over the surface of the chorion, which is vascularized directly by a large allantois, filled with fluid. The amnion arises as folds and not by cavitation as it does in higher primates.

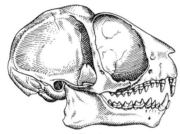

Tarsius

Lemur

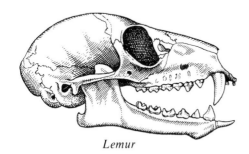

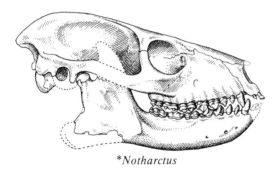

*Notharctus

FIG. 22.7. Skulls of relatively primitive primates. (After Flower and Lydekker 1891.)

Ten genera of lemurs occur today in Madagascar, where they have flourished in isolation throughout the Tertiary. The genus *Lemur* includes five species all arboreal, moving along horizontal branches, balancing with the tail and sometimes leaping. They feed on fruits,

flowers and leaves but not insects. They live in troops each holding a territory in some species. They spend much time in the canopy, but can travel on the ground. In *L. fulvus* it is said that there are no dominance hierarchies in the groups, but *L. catta* which is more often terrestrial live in troops of up to 24, organized around a core of adult females and their offspring, sometimes with one or more males. Subordinate males ('the Drones Club') follow the core group during progression and feed at the periphery. Dominance is maintained by frequent chasing, cuffing, scent marking, and calling, and females always dominate over males. Each troop defends a territory, most of the fighting being done by the dominant females. In the breeding season any male may copulate if the more dominant ones are otherwise occupied. Females spend their whole lives in the same troop but males may change troops.

Sportive lemurs, *Lepilemur*, are nocturnal and solitary, moving mainly by jumps. The mouse lemurs, *Microcebus*, are very small (60 g) and run like rodents, but they feed on insects in addition to fruit. *Indri* and *Propithecus* (sifakas) are large animals up to 1 m long. They travel through the forest by long leaps and on the ground usually hop on the hind legs. Some earlier lemurs became larger still, the skull of the Pleistocene **Megaladapis edwardsi* was nearly 30 cm long. These giant lemurs existed until about 2000 years ago.

The aye-aye, *Daubentonia* (Fig. 22.9) of Madagascar, is like other lemurs in some ways but with unique specializations and also some primitive and some anthropoid features. It is specialized for feeding on insect larvae buried in the bark of trees, filling the niche occupied in most parts of the world by woodpeckers (Cartmill 1975). It has large, continually growing upper and lower incisors, like a rodent, and a thin third finger with a claw, which it uses, with its teeth, to find the larvae. The other digits (except the hallux) also have claws, allowing it to cling to tree trunks. In New Guinea, where there are also no woodpeckers, the marsupial *Dactylospila* occupies the same niche, using an elongated fourth finger, it also has claws. Even the shortened skulls are similar in these unrelated animals, and are like

FIG. 22.8. Tympanic ring and tympanic bulla. (a) Primitive mammalian condition, floor of cavity unossified. (b) Lemuriformes, ring enclosed within bulla. (c) Lorisiformes and platyrrhines, ring part of bulla. (d) *Tarsius* and catarrhines, bony meatus. (After Le Gros Clark 1934.)

FIG. 22.9. Aye-Aye, *Daubentonia*. (From a photograph.)

those of woodpeckers. Apart from its specialization the aye-aye differs from lemurs in having inguinal mammae and other very 'primitive' features. It may have split off from the primate stock earlier than any other surviving form.

The Lorisiformes (Fig. 22.10) include two sorts of prosimian that are found outside Madagascar, the lorises, which are very slow moving animals, and the galagos which are very quick jumpers. The group is known back to the Miocene. The lorises of East Asia, India, and Sri Lanka (*Nycticebus* and *Loris*) are arboreal and nocturnal, proceeding by remarkably slow and deliberate movements and often hanging upside down. They occupy a curious niche, feeding largely upon prey unpalatable to other predators, such as urticant caterpillars, centipedes, ants, and bugs, detected mostly by smell. On the African mainland *Perodicticus*, the potto, is a similar slow creature catching its prey by stealth. *Galago*, the bush babies (Fig. 22.11), however, are very mobile animals, with several species throughout tropical Africa. They are omnivorous, catching active insects by leaping, they also eat small mammals and some fruit and plants. Using sight and sound for detection of their prey they move by vertical jumps with their very long legs and can reach heights of 2 m, which is 14 times their own height.

All of these lorisiform animals are very like lemurs in their basic features except that the tympanic ring is fused to the lateral wall of the petrosal bulla. In this and some other respects they approach the higher primates. Thus in some of them the face is shorter and the brain case rounder than in true lemurs. It is therefore possible that they are survivors of an earlier stock, closer to our own than are the lemurs. However, traces of very early features remain in lorises including a transverse skin fold on the abdomen of the female, which is considered by some to represent a marsupium.

FIG. 22.11. Bush-baby, *Galago*. (From a photograph.)

5. Fossil prosimians

The earliest known Primate remains consist only of teeth from the late Cretaceous and early Palaeocene and are named *Purgatorius* from Purgatory Hill in Montana where they were found together with at least six species of dinosaurs. These molars are tritubercular and very like those of early condylarths such as *Protungulatum* (p. 517) or insectivores. The animals were the size of shrews and were probably herbivores living on or close to the forest floor. They may be classified with some rather later forms in a family *Paromomyidae. The *Plesiadapidae were abundant primates from the Palaeocene and early Eocene of both Old and New Worlds with chisel-like upper and large procumbent lower incisors (Fig. 22.12) similar to those of

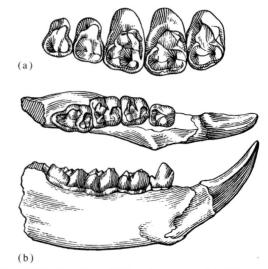

(a)

(b)

FIG. 22.12. Dentition of *Plesiadapis anceps*. (a) Right upper teeth showing P³P⁴ and 3 molars. (b) Lower dentition. (After Simpson.)

FIG. 22.10. Slender loris, *Loris*. (From life.)

the aye-aye, which were however a later parallel development. The skeleton was like that of a squirrel but more heavily built and probably used for vertical climbing and leaping, using the large compressed claws to hang from the bark (Simons 1972).

The *Carpolestidae and *Picrodontidae are further small groups of Palaeocene animals and these four groups of Palaeocene Primates may be classed together in a suborder *Plesiadapiformes.

The *Adapidae were Eocene animals, like lemurs in many ways but without procumbent incisors. The Old World members of the family (*Adapis* and *Pronycticebus*) could have given rise to the lemurs and lorises, which they probably resembled. Most were small but a few were large creatures with heads 10 cm or more long. The brain case was small but carried temporal crests (*Notharctus*, Fig. 22.7). There was typically a very full dentition ($\frac{2.1.4.3.}{2.1.4.3.}$). The incisors were not procumbent but the canines were incisiform in *Adapis*. The molars resembled those of some modern lemurs. The tympanic ring was included in the bulla. The cerebral hemispheres were smooth. These animals therefore showed many features common to other early mammals but they were remarkably similar to lemurs. The skeleton is well known and shows adaptations for grasping, leaping, and perching. *Progalago* is a Miocene fossil lorisid but no intermediate fossil lemurs are known.

6. Tarsiers

The fourth group of the Prosimii, the Tarsiiformes, includes one living genus, *Tarsius*, and a number of early Tertiary fossils. The whole group could be described by saying that its members show many characteristics similar to those of Insectivora and lemurs, but also others suggestive of the anthropoid primates. Yet there are present specializations that rule out the possibility that these animals are in the direct line of descent of the higher forms, and we must therefore regard them as an early offshoot, showing us something of the characteristics that were possessed by the anthropoid stock in Palaeocene or early Eocene times.

Tarsiers (Fig. 22.13) are arboreal, nocturnal, insectivorous creatures, the size of a small rat, living in the East Indian islands. They are vertical climbers and leapers, moving by jumps of as much as 4 m from trunk to trunk by extending the legs like a frog. Their long legs show considerable specialization for leaping, the fibula being fused with the tibia and the calcaneum and astragalus elongated (as in *Galago*) this provides an extra leg segment while retaining the grasping foot, which is not possible if the metatarsals are elongated as they are in kangaroos or ungulates (Fig. 22.14). Both first digits can

Fig. 22.13. Spectral tarsier, *Tarsius*. (From life.)

be used for grasping and the digits bear adhesive pads; all have nails except the second and third in the hind limb, which carry claws used for cleaning the fur. As in other jumping animals the ilium is very long. There is also a long tail. The dental formula is $\frac{2.1.3.3.}{1.1.3.3.}$. The molars retain a very simple tritubercular pattern and the lower incisors and canines do not show the specializations found in lemurs (Fig. 22.3). The head is more like that of a monkey than of a lemur; it is set on a mobile neck, indeed the animal has the uncanny power of turning its head through 180° so that it faces backwards, while the eyes, like those of owls, are so large that they can hardly move. The foramen magnum opens downwards. The eyes face more nearly forwards than in lemurs, the snout is shortened, and the turbinals of the nose reduced. The nose thus resembles that of a monkey (Fig. 22.15), and there is neither a cleft in the upper lip nor a moist rhinarium, such as is present in most mammals and in lemurs, but absent in anthropoids. This reduction of the snout as a tactile organ perhaps goes with the development of the hand for that purpose.

The eyes are enormous, relatively larger than in any other Primate, but suited for night vision, with the retina containing only rods, though, nevertheless, possessing a yellow macula and a small fovea. The external ears are large and mobile and the sense of hearing is keen. The

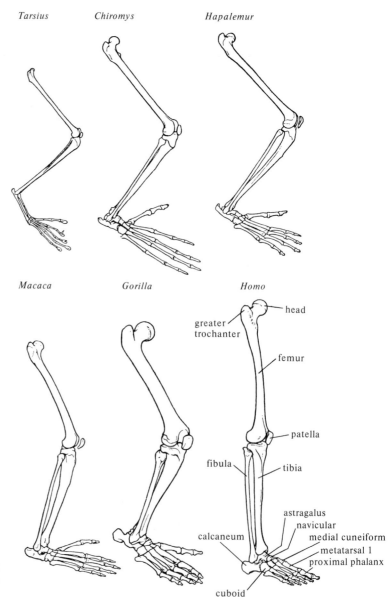

FIG. 22.14. Bones of hind leg and foot of various primates (not to the same scale).

orbit is partly divided off from the temporal fossa (Fig. 22.7). The tympanic bone is not only fused to the very large petrosal bulla but also somewhat drawn out into a tube, as in anthropoids (Fig. 22.8). The brain is relatively small but shows a curious mixture of early mammalian and advanced primate characters. The olfactory regions are small and the cerebral hemispheres large, though smooth. The visual (occipital) cortex shows remarkably well-differentiated layers. The corpus callosum is small and the anterior commissure large. The cerebellum is small and simple. The posterior corpora quadrigemina are large.

Tarsiers are said to live in pairs with only rudimentary social organization. Their reproduction shows some similarity to that of Anthropoidea. They breed throughout the year and the uterus is double, as in lemurs, but the placenta is of discoidal shape and haemochorial organization, with a much reduced allantois, almost like that of apes and humans, but the amnion is formed by folding.

Some of these characters indicate a type of organization so similar to that of anthropoids that the resemblance can hardly be entirely due to convergence. Many of the monkey-like features of the head could,

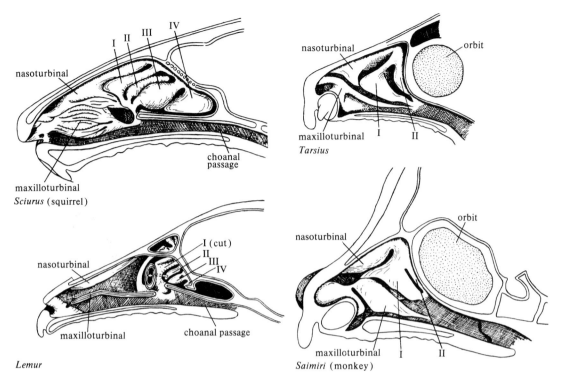

FIG. 22.15. The nasal passages of various mammals. *I–IV*, ethmoturbinals. (From Cave.)

however, be due to the large size of the eyes. Moreover, the reduction of the turbinals has taken place differently in *Tarsius* and anthropoids.

The *Anaptomorphidae (e.g. *Tetonius*, *Omomys*) are fossil tarsioids found from Palaeocene to Oligocene in Europe and America. At least 20 genera are known, mostly from skulls and teeth; where limb bones are found they already have indications of the tarsioid specializations. In most the eyes were large and the face short. Casts show that the brain was like that of *Tarsius* but with olfactory regions better developed. The temporal and occipital lobes were enlarged. Other Eocene fossils *Pseudoloris* and *Necrolemur* are so like modern tarsiers that they are placed in the family Tarsiidae. The tarsioids show more similarity to the Anthropoidea than to the lemurs. Immunodiffusion studies of serum proteins also show this similarity. In some classifications they are put with the anthropoids as Haplorhini (= whole lip) in contrast to the lemurs and lorises, which are Strepsirhini (= cleft lip). We are ignorant of tarsioid history from the Oligocene to recent times, but it seems likely that they have remained an isolated stock, their relationship to higher primates being one of common ancestry in early Tertiary times, when all these primates were so alike that it is best to class them together as Prosimii, some of which went on to develop into anthropoids.

7. Characteristics of Anthropoidea

Monkeys, apes, and humans form a natural group, almost certainly of common descent from some Eocene population. They first appear as fossils in the Oligocene and have flourished greatly since; Simpson (1945) lists 66 genera in the suborder, of which only 30 are extinct; evidently the type has been successful and is expanding. The outstanding characteristic of the anthropoids might be said to be their liveliness and exploratory activity, coming perhaps originally from life in the tree-tops, necessitating continual use of eye, brain, and limbs. With this is associated the development of an elaborate social life in many species, based not on smell, as in most mammals, including prosimians, but on sight. Monkeys show more bright colours than do other mammals, especially the curious reds and blues worn on the head and rear. The species may become subdivided into distinct races showing great differences of coat colour (Fig. 23.1). Communication between individuals is ensured by elaborate systems of vocal signals and the platysma muscle becomes differentiated into a set of facial muscles used to signal emotions.

Many of the characters of the group are those of *Tarsius*, listed already, but the Anthropoidea are mostly diurnal and microsmatic, with a short snout, large forwardly-directed eyes, many cones, a well marked central area in the retina and partial decussation in the optic tract, features that are associated with binocular vision and large powers of visual form discrimination. The external ears, no longer serving as tactile organs or for direction-finding, are small and the edge is usually rolled over. The orbit is closed off behind. The tympanic bone is fused to the petrosal and in later forms drawn out to a tube (Fig. 22.8). The tactile sense is greatly developed on the fingers and toes, which carry characteristic ridges in whorls. The brain (Fig. 22.6) is relatively much larger than in lemurs or *Tarsius* and its cerebral hemispheres are especially well developed, overhanging the cerebellum and medulla. The olfactory parts of the brain are reduced and the pyriform lobe becomes displaced on to the medial surface by the extension of the neopallium. The surface of the neopallium is highly fissured, showing a characteristic form of Sylvian fissure, and a well marked central sulcus, separating the motor and sensory areas. A parieto-occipital sulcus separates these lobes and a large lunate sulcus marks the visual area, especially in monkeys (simian fissure). The neocortex thus shows four distinct lobes, frontal, parietal, occipital, and temporal. The occipital (visual) and frontal regions are especially large.

The head is rounded to fit the brain, with the foramen magnum below, so that the head is carried up on a mobile neck. The gait of monkeys is typically quadrupedal and plantigrade when on the ground, with the fore-limbs somewhat longer than the hind. In the trees some may be described as low canopy runners (e.g. guenons) others are high canopy acrobatic types (spider monkeys). Only the lesser apes habitually swing along with the arms (brachiation). Some monkeys have become terrestrial (baboons). The pollex and hallux are opposable and the digits all carry nails. The hands and feet are used for feeding as well as for locomotion and many monkeys sit, freeing the hands for eating.

The characteristic of the tooth row is a tendency to shortening, presumably connected with the shortening of the face. The two mandibles are united by fusion of the symphysis. There are three premolars in the earlier Anthropoidea, later reduced to two (Fig. 22.3). In the line leading to man there is a tendency to still further reduction, with the last molar becoming smaller than the others. The cusp-pattern is tritubercular in earlier anthropoids, but later the molars become square and have four or more bunodont cusps in higher anthropoids. The incisors become spatulate, rather than pointed, and the premolars bicuspid (Figs. 22.16 and 22.17).

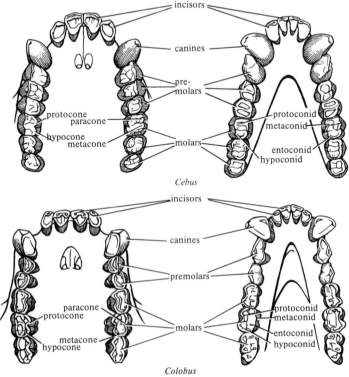

FIG. 22.16. Tooth rows of New World and Old World monkeys. (Top is of *Cebus*.) (After Buettner-Janusch 1966.)

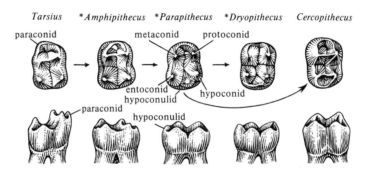

Tarsius   **Amphipithecus*   **Parapithecus*   **Dryopithecus*   *Cercopithecus*

FIG. 22.17. Diagram of the right lower molar cusp pattern in some primates, showing presumed evolutionary stages in the development of the cusp pattern. *Tarsius*, the primitive (tribosphenic) type, in which the paraconid is involved in the formation of the trigonid. In **Amphipithecus* the paraconid is undergoing reduction while the talonid and trigonid portions of the crown are at the same level. In **Parapithecus* the paraconid has completely disappeared, the talonid bears the hypoconid, the entoconid, and a relatively well developed hypoconulid; the trigonid portion bears the metaconid and protoconid. In **Dryopithecus* the five cusps are more or less equally developed and separated by a characteristic pattern of intervening grooves. *Cercopithecus* shows the characteristic bilophodont pattern, with transverse ridges. (After Le Gros Clark 1934.)

The reproduction is characterized by the presence of menstrual cycles, continuing throughout the year though these have not been clearly demonstrated in New World monkeys. Ovulation occurs once in each cycle, often accompanied by the development of sexual signals and behaviour patterns by the female. There is a discoidal, haemochorial placenta, with very early development of the extra-embryonic mesoderm and reduction of the yolk-sac, amniotic folds, and allantois. Usually a single offspring is produced and there is one pair of pectoral mammae. The young are looked after for a long period. The social life is often based upon families of one male and several females and young.

8. New World monkeys, Ceboidea

The continent of South America has a special type of monkey, as of so many other mammalian groups. These platyrrhine (flat-nosed) monkeys have presumably been isolated since Eocene times. They could not have been a later immigration from North America because, so far as we know, no cercopithecoids or hominoids reached that continent until man came. It has been suggested that they might have been derived from the Tertiary African stock by rafting, although the South Atlantic was already two-thirds as broad then as now. The differences from the Old World monkeys are not very profound, however; therefore either the characteristic monkey organization had appeared in the Eocene or the platyrrhines and catarrhines have evolved on parallel lines. The problem is unsolved but the primates from the Egyptian Oligocene (Fayum) are very much like ceboids in many characters.

In the teeth the second premolar is retained $\frac{2.1.3.3.}{2.1.3.3.}$ or $\frac{2.1.3.2.}{2.1.3.2.}$ in Callitrichidae, whereas it is lost in all Old World forms; the molars are quadritubercular

(Figs. 22.3 and 22.16). The brain is relatively larger in marmosets even than in man, but this results from the small size of the animal. The smaller ceboids show little fissuring, but this develops in the larger ones (*Ateles*), showing a pattern similar to that of Old World monkeys. The nasal apparatus, though smaller than in lemurs, is larger than in Old World monkeys, producing the wide separation of the nostrils from which the name platyrrhine derives. Facial vibrissae are present, but usually small. In the ear the tympanic bone is a ring fused with the petrosal, but is not drawn out into a tube as it is in catarrhines, and there is a large bulla, which is absent in the latter (Fig. 22.8). The caecum is relatively large. The reproductive system does not show the full 'anthropoid' pattern; for instance, there are at most only slight signs of menstrual bleeding at the end of the luteal phase of the oestrus cycle. Social life is sometimes well developed and individuals may live as small monogamous families in some species and in large troops in others. Social organization is probably more fluid than in Old World monkeys, with less fixed hierarchies of dominance, but further research is needed. The signalling system is also less complicated than in catarrhines. Thus the colour is seldom brilliant, and the facial musculature around the mouth relatively simple. The loud voice of the howler monkeys (*Alouatta*), which have special laryngeal sacs, is the most effective method of communication for these tree dwellers (Attenborough 1979). It is used in the assertion of territorial rights by the clan, which includes one fully mature male as well as females and young (age-graded male system). Co-operation is ensured by a language of at least nine distinct sounds with separate meanings. These New World monkeys are all adapted for arboreal life, with long limbs, delicate hands, and tail for balancing or seizing (Fig. 22.18). The thumb is not fully opposable. The tail pad, of species with prehensile tails, has special tactile sensitivity, with ridges like those on the digits and a large representation in the cerebral cortex.

The fourteen living genera of New World monkeys are very varied and are divided into two families. The marmosets and tamarins, Callitrichidae include five general of very small insect- and fruit-eating animals, of somewhat squirrel-like appearance and habits, living in tropical South America (Fig. 22.19). They have a thick fur and non-prehensile tail and there are claws on all the digits except the first of the foot, allowing the animals to run up trunks they cannot grasp. The pollex is not opposable. The incisors and canines are partly procumbent. Three premolars are present, but the molars are reduced to two, a condition found in no other anthropoid. The cusp-pattern is tritubercular. Unlike other anthropoids the marmosets give birth to two or three young and there are signs of ancient conditions in the placenta, where the yolk sac becomes larger than in most higher primates. They live in families of 3–8, but may also form larger groups.

The Cebidae are a larger family (11 genera) and include the owl monkeys, also called night monkeys or douroucoulis (*Aotus*) which are perhaps nearer than any

FIG. 22.19. Common marmoset, *Callithrix*. (From life.)

FIG. 22.18. Spider monkey, *Ateles*. (From life.)

other monkeys to the base of the anthropoid stock. They have a pure rod retina and are nocturnal. The fur is thick and they lack the prehensile tail present in some other cebids. The spider monkeys (*Ateles*) have very long limbs and reduced thumb (Fig. 22.18). They inhabit the canopy, either in small groups or in troops of up to 100, and they feed on fruit and nuts. They are basically quadrupedal but can show a form of brachiation, swinging with long jumps and downward drops.

Cebus includes several species of capuchins, widespread in Central and South America. They too live in the canopy, feeding on insects and fruit, using a partly opposable thumb and considerable manipulative skill. They produce a variety of sounds and visual signals and have an elaborate social structure. The howler monkeys (*Alouatta*) are leaf eaters with large molars. They often move by swinging from the tail, sometimes from the arms. Fossil cebids are known only back to the Oligocene and Miocene of South America (*Homunculus).

23 Monkeys, apes and men

1. Old World monkeys. Cercopithecoidea

The seventeen living genera of Old World monkeys, apes, and men are often classified together as Catarrhini because of the common characteristics in which they contrast with New World monkeys. Perhaps this union is justified and the Catarrhini is a monophyletic group, with a common ancestor in the late Eocene but the Old World monkeys, the superfamily Cercopithecoidea, diverged very early from the apes and men, Hominoidea.

The cercopithecid or Old World monkeys of Africa and Asia do not differ very strikingly in general habits and organization from monkeys of the New World, though they are mostly larger (p. 463). We must conclude either that the two groups have made many changes in parallel or that in the Eocene there were already animals with the good sensory systems, active brains, and skilled movements of the monkeys. The distinguishing features of the Old World types are rather trivial, for instance they sit upon ischial callosities, surrounded by naked and often highly coloured skin, which becomes enlarged in the female before ovulation. In cercopithecids there are cheek pouches in which to store food; usually there are complicated laryngeal sacs, a bony tympanic tube, and never a prehensile tail. The great reduction of the olfactory turbinals leaves the nostrils close together and pointing downwards hence the catarrhine or 'drop nose'. The dentition is reduced to $\frac{2.1.2.3}{2.1.2.3}$, the upper molars carry four bilophodont cusps, and the lower four except for the last, which has five (Figs. 22.3 and 22.16). The diet of the more generalized cercopithecids is omnivorous, including insects, lizards, eggs, and fruit, but many monkeys are specialized fruit-eaters, and in these the molar teeth are quadrangular, with the four cusps united to make two transverse ridges used for grinding, somewhat as in ungulates. These specialized molars make it unlikely that the modern cercopithecids could have been ancestral to the apes or man. The colon usually has a sigmoid flexure and a small caecum and appendix. The reproduction is similar to that of apes and man; there is menstrual bleeding and haemochorial placenta. The anogenital region (sexual skin) of the female may show marked signals (swelling and coloration) at the time of ovulation. The male shows continuous spermatogenesis in some species, without a breeding season. In some monkeys there are, however, seasonal fluctuations in the number of births.

The cerebral cortex is always large and fissured and its frontal regions well developed. Behaviour is exploratory and manipulative and learning powers high. Social behaviour is elaborate and often based upon polygamous families. The coat colour is often ornate (Fig. 23.1) and there are elaborate communication systems by the facial musculature and vocal apparatus.

The Cercopithecoidea includes thirteen living genera with various habits. Many are common and well known. *Macaca*, the macaque (Fig. 23.2), has several species in Asia and North Africa, one reaching Gibraltar. They have been introduced to Mauritius and the West Indies. They are carnivorous and partly terrestrial and may frequent cliffs, rocks, or coasts. The thumb is fully opposable. They live in groups of 20 or more with several males, females, and young. One male is dominant and disciplines the group. Over 30 sounds have been described and facial expressions and postures are

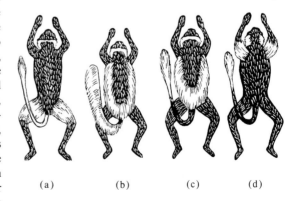

(a)   (b)   (c)   (d)

FIG. 23.1. Colour types of sub-species of *Colobus polykomos*. (a) *C.p. vellerosus* Graff. (b) *C.p. caudatus* Thomas. (c) *C.p. abyssinicus* Oken. (d) *C.p. angolensis* Schlater. (After Rodt.)

FIG. 23.2. Rhesus monkey, *Macaca*. (From life.)

FIG. 23.4. Mandrill, *Mandrillus*. (From life.)

complex. *Cercopithecus*, the guenons, are similar animals in Africa, many showing highly coloured flashes. The baboons, *Papio* (Fig. 23.3), of Africa and related forms in Arabia, and mandrills, *Mandrillus* (Fig. 23.4) of West Africa have become secondarily terrestrial and quadrupedal and the face has become elongated ('dog-face'), allowing for a long tooth-row of grinding molars. The Colobinae are fully arboreal, leaf-eating monkeys, without cheek pouches but with the stomach sacculated; the guerezas (*Colobus*) live in tropical Africa and the langurs or leaf monkeys (*Presbytis* = *Semnopithecus*) in south Asia.

2. The earliest monkeys. Parapithecidae

Parapithecus and *Apidium* are fossils from the Oligocene beds at Fayum in Egypt which may be close to the ancestry of the Old World monkeys (Fig. 23.5) (Simons and Delson 1978). However, some of their characters are so like those of condylarths that it was long doubted whether they were primates at all. *Apidium* was named 'little bull' because it was thought to be an artiodactyl (Fig. 23.6). The dental formula is $\frac{2.1.3.3}{2.1.3.3}$ and the face short, rather like that of a marmoset. The molars carry a quadrangle of four cusps and a hypoconulid behind. *Parapithecus* was an animal the size of a squirrel, which might have been derived from the anaptomorphids and led on to the catarrhines (Figs. 22.17 and 23.7). The two rami of the jaws diverge posteriorly, as in tarsioids,

rather than running parallel. However, the number of teeth is reduced to that of catarrhines, that is $\frac{2.1.2.3}{2.1.2.3}$. The molars carry five cusps, not united by ridges. The cercopithecid type was well established by the early Pliocene and abundant remains are available of *Mesopithecus*, with the molar cusps united to ridges. Species of the genus *Macaca* have been described from the late Miocene perhaps 7 Ma·ago.

3. The great apes: Pongidae

The question of the exact degree of affinity between the existing apes and man remains unsettled. There were plenty of fossil apes in Miocene times and man-like creatures are found in the late Pliocene and early Pleistocene. A few workers believe that it separated much earlier from the ancestral catarrhine stock, say, in the Oligocene. Evidence of definitely man-like creatures that can be placed with certainty in the family Hominidae is found back to about 5 million years ago. However the longest of the above estimates would say that our stock has been distinct and evolving separately for nearly 60 million years, without leaving any remains. The view that men have descended from apes is much the more widely held. We shall first survey the structure of the great apes by treating them as members of a separate family Pongidae.

The living apes include the gibbon, *Hylobates* (Fig. 23.8), and orang-utan, *Pongo* = *Simia* (Fig. 23.9), from east Asia and the chimpanzee, *Pan* (Fig. 23.10), and *Gorilla* (Fig. 23.11) from Africa. They and related fossil forms are marked off from the cercopithecids by their teeth and methods of locomotion. Many apes are rather large animals and this has made it impossible for them to walk along the branches as monkeys do. They therefore swing by the arms, which are longer than the legs and are provided with very powerful muscles, the hands and feet being efficient grasping organs. There is no tail. These brachiating habits have affected the entire skeleton. All the apes and men differ from the cercopithecids in having wider chests, longer necks, longer

FIG. 23.3. Sacred babbon, *Papio*. (From life.)

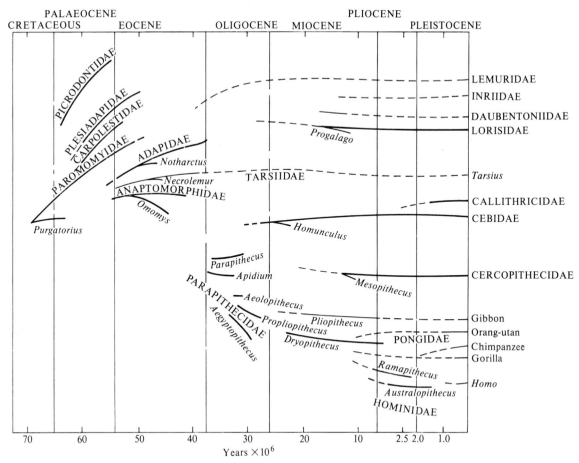

FIG. 23.5. Summary of the evolution and relationships of the main primate groups. The time scale changes at 2.5×10^5 years.

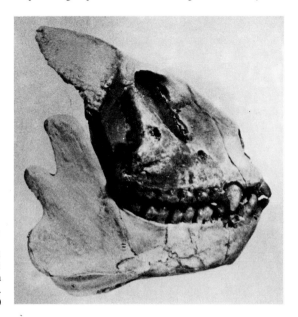

FIG. 23.6. *Apidium phiomense*, a primitive catarrhine from the Oligocene deposits of Egypt. Photographic reconstruction of the lateral view of the face based on the American Museum frontal, Cairo Museum maxilla, and mandible from Yale. (Photograph by A. H. Coleman, in Simons and Delson 1978.)

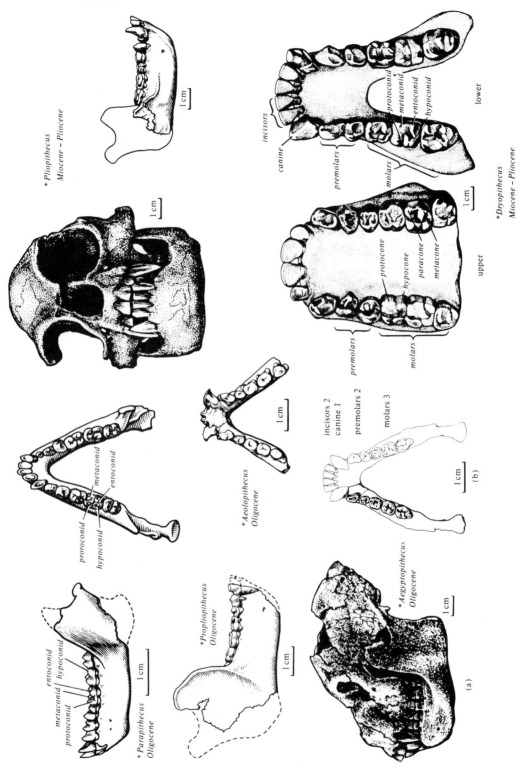

Fig. 23.7. Skulls, mandibles, and dental arches of early fossil primates.

FIG. 23.8. Gibbon, *Hylobates*. (From life.)

FIG. 23.10. Chimpanzee, *Pan*. (From life.)

limbs, and larger heads (Fig. 23.12). The cervical and sacral regions are longer in apes than in the monkeys and the lumbar region shorter. When proceeding on the ground the apes cannot balance on two legs for long; instead, the long forearms prop up the front of the body and produce a semi-erect position. The hands are specialized for brachiating, with a short thumb and long metacarpals and digits with curved phalanges parti-

FIG. 23.9. Orang-utan, *Pongo*. (From life.)

FIG. 23.11. Gorilla, *Gorilla*. (From life.)

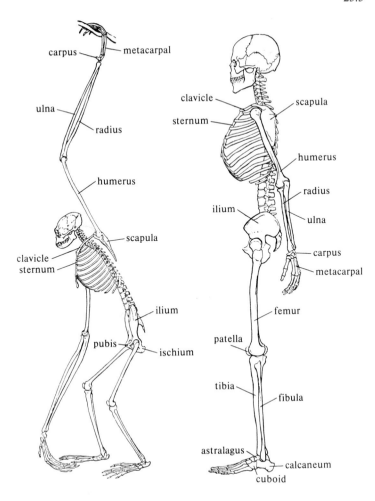

carpus — metacarpal

ulna

radius

humerus

clavicle
sternum

scapula

ilium

pubis

ischium

clavicle — scapula

sternum

humerus

radius

ilium

ulna

carpus

metacarpal

femur

patella

tibia

fibula

astralagus

calcaneum

cuboid

FIG. 23.12. Skeletons of gibbon and man.

cularly in *Hylobates* and *Pongo* (Figs. 23.13 and 24.2). The foot is in general similar but in the chimpanzee and gorilla it is more suited for walking, with broader sole and shorter toes. In the terrestrial *Gorilla gorilla beringei* the hallux lies parallel to the other toes almost as in man.

The teeth (Fig. 23.14) are of a rather generalized type. The canines are often large, especially in males, and the lower front premolar forms a sectorial blade. The molars carry grinding tubercles and often show a crenation of the enamel, which is characteristic of apes, though found as an abnormality in monkeys and man. The trigon is still present in the upper molar, the hypocone being small. The lower molars have a hypoconulid, making five cusps, as contrasted with the four of monkeys. All the apes are mainly vegetarians, but they may eat meat occasionally. The use of the teeth for grinding is associated with powerful masticatory muscles and the development of temporal and occipital crests. The supraorbital ridges are also large. In the

digestive system apes and man differ from other primates in the presence of a vermiform appendix.

The brain is much larger than in cercopithecids, and shows a pattern of convolutions similar to that of man, though less complex. The behaviour provides many signs of efficient memory, leading to the attitudes that we characterize as imitation and association of ideas. There are extensive powers of manipulation and for obtaining ends by indirect means. The communication system is highly developed except in the orang-utan, which is solitary. The young chimpanzee is said to be able to make at least thirty-two distinct sounds. The facial musculature is more highly differentiated than in monkeys and produces a wide range of expressions such as of rage, surprise, pleasure, and laughter.

Social organization is always well developed. The gibbons are monogamous, the family consisting of a pair and the young of current and previous years. Chimpanzees, and so far as is known gorillas, live in

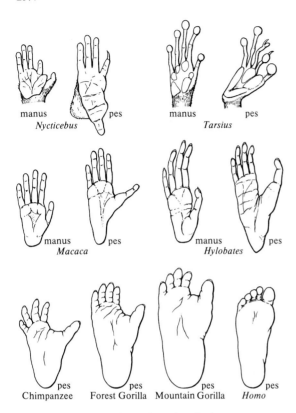

FIG. 23.13. Manus and pes of a series of primates.

bands, led by a dominant male (Lawick-Goodall 1971). The individuals of a band co-operate in helping each other and the groups show differing social traditions. The apes are diurnal animals, eating in the day. They make platforms on which they rest at night (except the gibbons).

Reproduction is based on a menstrual cycle of thirty-five days in chimpanzees, with a great development of the sexual skin at mid-cycle. Gestation is long, as is the growth period, seven to nine years in the gibbon, ten to twelve in the chimpanzee. The life span is also long, reaching forty years in chimpanzees, perhaps fifty in gorillas.

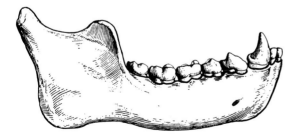

FIG. 23.14. Mandible of *Gorilla*.

The gibbons are fully arboreal animals, swinging rapidly with their very long arms (Fig. 23.8). This extreme brachiation may be a quite recent specialization. They eat mainly fruit and leaves, also insects and eggs. Gibbons are numerous throughout south Asia and the type is certainly the most successful among modern great apes. The characteristic cries are part of the defence of the territory that is occupied by each group. The orang-utan of Borneo and Sumatra is larger and also has very long arms. The chimpanzees (*Pan*) and gorillas (*Gorilla*), living in the forests of tropical Africa, are so alike that it is doubtful if the generic separation is justified. The chimpanzee is the smaller and less muscular animal, lacking, for example, the large parietal and occipital crests found in the male gorilla. There is a corresponding difference of temperament, the chimpanzee being lively and sometimes tameable, the gorilla gloomy, ferocious, and unafraid. Gorillas are mainly terrestrial, walking on all fours and sleeping either on the ground (males) or at a small height (females and young). Like the other apes they show much local variation but all may be referred to a single species, *G. gorilla*. A mountain form, *G. gorilla beringei*, is more fully terrestrial than the others (Fig. 23.13).

4. Fossil apes

The similarities and differences between modern apes and man will be discussed later. Their relationship with the other catarrhines is clearer. The lower jaw of *Parapithecus* contained teeth of a pattern that could have given rise to those of the Pongidae as well as the Cercopithecidae (Fig. 23.7). The Egyptian Fayum beds contain a rich fauna of animals that lived in lush tropical forest conditions 35–30 Ma ago. *Oligopithecus* from the lower zone is the oldest known hominoid (Fig. 24.8) (Simons, Andrews, and Pilbeam 1978). It shows some prosimian features but the dental formula was ?2.1.2.3., with a large canine. Rather later were *Propliopithecus* and *Aeolopithecus*, where the molars have a distinctly five-cusped pattern, with protoconid and metaconid in front and a large heel, carrying a hypoconid laterally and entoconid medially, and also a posterior hypoconulid (Fig. 23.7). These animals were small and like gibbons but could be close to the basal stock that gave rise to both pongids and hominids.

Aegyptopithecus also from Fayum was larger and could have led to *Dryopithecus* 10 million years later (Fig. 23.7). A robust ulna and vertebrae suggest that *Aegyptopithecus* was a high canopy dweller and possessed a tail. Great apes were found quite widely in the Old World during the Miocene and Pliocene. The earliest of these, *Dryopithecus* (=*Proconsul*), from the lower Miocene of Kenya, showed a combination of characters of cercopithecids, great apes, and man. The skull was more lightly built than in modern apes, with no brow ridges. The tooth rows converged anteriorly as in *Parapithecus* (Fig. 23.7). The incisors were small and like those of man, but the canines were large and the first premolar was sectorial as in apes. Its medial surface was sheared against the overlapping upper canine and kept it sharp. The molars have five cusps arranged in a Y pattern, which is also found in modern apes and (sometimes) in hominids. The limb bones suggest that the gait was arboreal – terrestrial and quadrupedal, and that the brachiating habit had not yet evolved. They were probably capable of running and leaping. Similar forms are known from the Miocene and Pliocene of Asia and Europe. Modern apes probably evolved from some dryopithecine ancestor. *Pliopithecus* (=*Limno-pithecus*) from the Miocene and Pliocene of Africa and Europe were perhaps ancestral to the gibbons (Fig. 23.7). Several other types are known and evidently the apes were widespread, varied, and successful animals in the Miocene and there are among them plenty of signs of the characteristics both of the modern apes and man.

Oreopithecus from the Pliocene of Italy shows a curious mixture of characters. It had very long arms but no tail. Its method of locomotion is not clear. The lower molars have four cusps arranged in pairs as in monkeys but not united by ridges. The upper molars resemble those of apes but the small canines, absence of diastema, and bicuspid first lower premolar have led some to place it close to man, but almost certainly it was an independent hominoid offshoot (Butler and Mills 1959). *Gigantopithecus*, known from huge teeth and jaws from the Pliocene and Pleistocene was a very large ape, probably feeding on grain. It may have survived to 500 000 years ago.

24   The origin of man

1. Human characteristics

IN order to discover the position of man in relationship
to the living and fossil ape populations we may try to
specify the characters distinctive of the family Hominidae and then discuss whether they could have been
derived from those of monkeys or apes. Schultz (1950),
who has made careful measurement of many features of
primates, lists the following as the chief specializations
of man: (1) elaboration of the brain and behaviour,
including communication by facial gestures and speech;
(2) the erect posture; (3) prolongation of post-natal
development; and (4) the great rise in population in
recent years. Others might make up the list differently,
but we may use it as a basis for discussion of the
differences between men and other creatures.

2. Brain of apes and man

The brain is much larger absolutely and relatively in
man than any living ape; Fig. 22.1 shows that man
stands farther apart from the apes in this respect than
they do from other anthropoids. The cranial capacity
for males of modern Caucasian man may be taken as
1375 ml whereas that of chimpanzees is given as 410,
gorillas as 510, and orangs as 450 (Table 24.1). The
general arrangement of function within the brain is
similar in man and apes, but the parts especially well
developed in man are the frontal and occipital lobes.
The latter are concerned with the sense of sight and are
related to our intensely visual life. The frontal lobes, so
far as is known, serve to maintain the balance between
caution or restraint and sustained active pursuit of
distant ends, which, above all else, ensures human
survival in such a variety of situations, and makes
possible the social life and use of language by which so
great a population is maintained. The difference of
behaviour between men and apes exceeds all the structural differences; our lives are so widely different from
theirs that any attempt to specify the divergences in
detail is apt to seem ridiculous. Perhaps the more
striking of them are related to the powers of communication by speech which, besides its obvious social

TABLE 24.1

*Rate of evolution of man. Change of 1/1000 part per 1000
years = 1 darwin. (After Campbell 1964.) Some of the
dates used in this table differ from those adopted
elsewhere in this chapter*

	Date BP $\times 10^6$	Cranial capacity (ml)	Rate in md
Australopithecus to	1.2	500	..
H. erectus (Java) to	0.5	900	280
H. erectus (Pekin) to	0.4	1000	351
H. sapiens steinheimensis to (Swanscombe)	0.15	1325	375
H. sapiens sapiens (modern)	0.0	1375	6

advantage, gives to man the power of abstract thought.
Whatever we may think about the consciousness of
animals there is no doubt that our own awareness of life,
being expressed in words, is widely different from that of
all other creatures. The speech system depends upon a
complex of features of the brain, larynx, tongue, mouth,
and auditory apparatus. The human brain is asym-
metric, with the parts responsible for speech located
mainly on one side, usually the left. In addition, the
facial musculature is more fully differentiated, especially
around the eyes and mouth, than in apes.

3. Human posture and gait

The gait of man differs from that of any ape in that the
body can be fully and continuously balanced on the two
legs. This involves considerable modifications throughout the skeleton and musculature (Fig. 23.2). The
backbone, instead of the single thoracic curve of
quadrupeds, has an S shape, being convex forward in
the lumbar, backward in the thoracic, and again forward in the cervical region. The thoracic curve develops

before birth, but the cervical only as the baby holds its head up and the lumbar as it begins to walk. The vertebral column, which in quadrupeds is a horizontal girder, in man becomes vertical, carrying bending and compression stresses along its length (see *Life of Mammals*). This entirely alters the arrangement of its secondary struts and ties. The bodies of the vertebrae carry much of the weight and are massive, tapering in size upwards. They are separated by well-developed intervertebral discs, acting as elastic cushions. The weight of the head is balanced on the backbone through the neck, and the thorax acts as a bracket from which the viscera are suspended. The muscles of the back, the ties of the vertebral girder, though arranged on the same general morphological plan as in quadrupeds, now carry very different stresses and no long neural spines or large transverse processes develop, since the girder is not now of cantilever type. For the same reason there is no sharp change in the direction of the neural spines at the hind end of the thoracic region; the girder is now one unit, with stressing by bending along its whole length.

The balancing of the body on the legs also involves many changes. The muscles around the hip joint achieve this balance, and the changes to allow this are that the iliac blade comes to face laterally, allowing the gluteus medius and minimus muscles to act as abductors, preventing the body from falling medially when the weight is on the opposite leg. The gluteus maximus becomes a powerful extensor able to lift the whole weight of the body into the upright position. The buttocks are therefore a characteristic human structure. The adoption of a bipedal position imposes entirely new requirements on the musculature of the limbs. In quadrupedal progression the retractor muscles are the main means of locomotion, drawing the leg backward at the hips while straightening the knee. In man the propulsive thrust is obtained mainly from the calf muscles and in particular from the soleus, which runs from the tibia to the heel, the gastrocnemius, since it tends also to bend the knee, being reduced. The quadriceps femoris becomes very large, serving to keep the knee extended both while the calf muscles develop their thrust and, as a check to the forward momentum, when the foot touches the ground. The ilium is therefore very broad in man, increasing the surfaces for attachment of the glutei, iliacus (a flexor of the hip), and for the abdominal muscles, which are attached along its crest and have an important part to play in carrying the weight of the viscera (Fig. 24.1).

4. The limbs of man

Many changes would be needed to convert an ape-like leg and foot to the human condition (Fig. 22.14). The

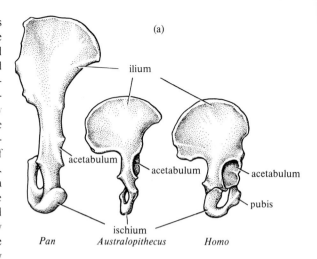

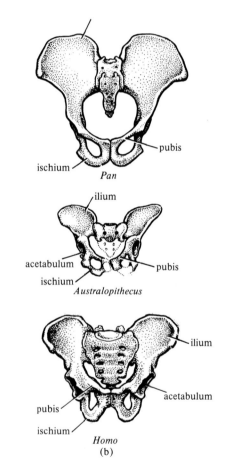

FIG. 24.1. Drawings of the pelvic bones of *Pan*, *Australopithecus, and *Homo* to show the changes in sacro-iliac articulation, acetabulum, and the length of the ischium. The ilium of *Australopithecus resembles that of man, but the ischium is much longer. (After Broom and Robinson, in Grassé 1955.)

femur of man is straight and the articular surface at its lower end set at an angle to the shaft. This allows the lower legs and feet to be as nearly as possible below the centre of gravity in standing, in other words, for the knees to be held together although the femoral heads are wide apart. At the ankle joint, on the other hand, the articular surface is at right angles to the tibia in man, at an oblique angle in apes, since in the latter the foot is turned outwards. In ourselves the weight is transferred from the tibia to the talus and then partly backwards to the calcaneum and partly forwards through the tarsus to the metatarsal heads (Fig. 24.2). The calcaneum is modified for this weight-bearing and the tarsus and digits even more so, the whole foot being converted into an arched system, no trace of which is found in apes. With this arrangement the hallux is not used for grasping and is very large. It is held in line with the other digits and the whole forms a compact wedge with a joint

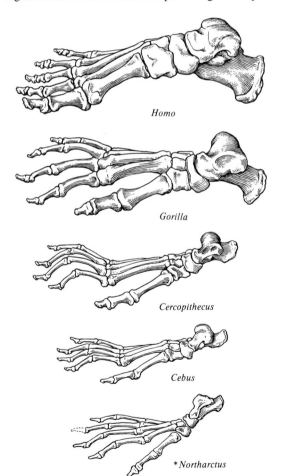

Homo

Gorilla

Cercopithecus

Cebus

* *Northarctus*

FIG. 24.2. Skeleton of foot from a series of primates. (After Gregory.)

at the metatarsal heads. In walking, when the heel is raised by the calf muscles, the toes remain on the ground, to prevent slipping forwards. The condition in which the first toe is the longest is peculiar to man, but in some monkeys and apes the axis tends to shift from the third digit medially and the human condition is an accentuation of this change, with the metatarsal and first phalanx of the first digit becoming long and strong. Even in modern human populations the second toe as a whole is often longer than the first; this condition was perhaps commoner in historical antiquity (the 'Grecian toe'), and may be a cause of foot trouble, the long second digit being unsuited to the stresses it is made to bear.

The differences between apes and men in the arms and hands (Figs. 23.12 and 23.13) are marked, though perhaps less striking than in the feet. The human fore limb is, of course, relatively much shorter than that of any ape and its muscles far less powerful. In order to carry the whole weight of the large body an ape needs enormous muscles all along the limb. Thus the serratus anterior, which pulls the body up on the scapula, is very large and the ribs to which it is attached have large flattened surfaces, are very long, and extend far cau- dally; the chest of man is much more lightly built. Similarly, the muscles of the shoulder and the flexor muscles of the elbow, wrist, and hand are all much larger in apes, as are the ridges to which they are attached, for instance on the palmar surfaces of the phalanges (Fig. 24.3). The human arm has specialized in mobility. The hand can be brought into almost any position in relation to the body by virtue of the wide range of movement at the shoulder, pronation and supination of the forearm and movements at the wrist.

In the hand itself the thumb is characteristically long in man and moved by powerful muscles. Only in humans can the thumb be in the fullest sense opposed to the other digits, so that the pads face each other. This is achieved by the special saddle shape of the joint between the first carpal and metacarpal. The third digit is the longest in apes, as in men, but the second digit (index) of man is generally at least as long as the fourth, often longer (the 'Napoleonic finger'). In lower primates the digits of the ulnar side are relatively much longer. Apart from proportions and skeletal features the human hand also has a very well developed sensory supply, which is essential for its use as a handling organ.

5. The human skull and jaws

Comparisons between the skulls of apes and men have attracted special attention because so many of the finds of early human types have been of skulls (Fig. 24.4). The differences are mainly referable to changes in the brain, dentition, and method of balancing the head upon the

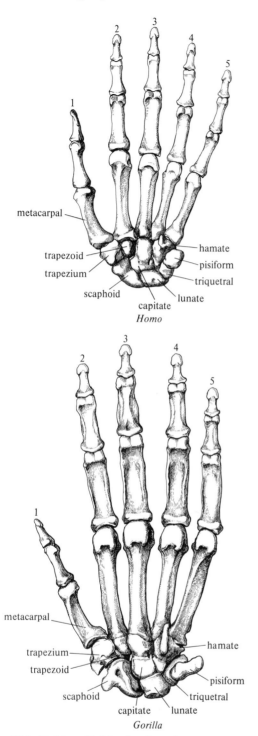

FIG. 24.3. Skeleton of left hand, palmar view.

FIG. 24.3A. Hominid footprint tracks 3.6 million years old from Laetoli, north Tanzania. The prints were made in volcanic ash by two individuals walking in tandem, with a smaller one to the left. To the right are the tracks of *Hipparion, an adult and a foal (Leakey 1981).
(Photographed by John Reader, with kind permission from Dr. M. Leakey and the National Geographic Society.)

neck. The enlargement of the brain has been in the occipital and especially in the frontal region (p. 473), giving a high forehead and the characteristic upright face. At the same time the jaws have receded, so that the human tooth row is unusually short. Moreover, the

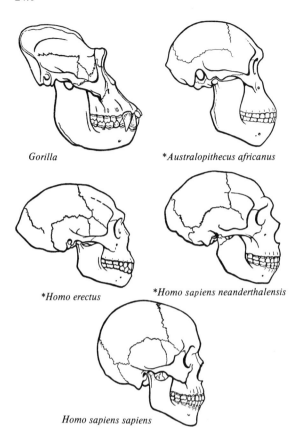

Gorilla

Australopithecus africanus

*Homo erectus

Homo sapiens neanderthalensis

Homo sapiens sapiens

FIG. 24.4. Skulls of apes and man. The face and back of jaw of *Homo erectus* has been restored. (Partly after Romer 1945.)

dental arcade is characteristically rounded in front, that of modern apes is rectangular, with large canines at the corners (Fig. 24.5). However, in Miocene apes and monkeys the jaws diverge posteriorly, so this is a primitive feature retained in the hominid line (see Simons 1978). In man the canines are small and incisiform; the first lower premolar is bicuspid, like the rest and not sectorial as in other catarrhines. The molars show a characteristic pattern that may be regarded as based on four cusps above and five below. The cusps are arranged roughly as a rectangle, so that the grooves between them make a + as compared with the Y pattern typical of the dryopithecine molar (Fig. 24.6). Thus the protoconid meets the entoconid in the human but not in the earlier type. However, there is great variation in these patterns, both in men and apes. Little can be said therefore about a single tooth, but the proportion of molars with four cusps and a + pattern is higher in man than in apes (Fig. 24.7). The last molar (wisdom tooth) is smaller than the others in man, but not in modern apes. There are, however, many signs of

possible ape-like ancestry in our teeth; for instance the canine has a long root and erupts late.

The lower jaw of man is less shortened than the upper. In apes it is strengthened by a 'simian shelf' of bone on its inner side, but in man this strengthening is on the outside, making the chin. The jaw is less massive in man than in apes, especially its posterior ramus; the muscles for moving it are less powerful. Correlated with this weakening of the jaw has been a rounding of the surface of the skull. Occipital and temporal crests for the attachment of the neck and jaw muscles are well developed in the male gorilla, suggested in other apes, but absent in man. The brow ridges, also characteristic of the apes, are large masses of bone above the eyes, probably produced to meet the compression stresses set up by the powerful action of the jaw muscles. Their absence, together with the large forehead, produces the human type of face. The large external nose is another corollary of the shortened face; it provides some extension of the nasal cavity for warming and filtering the air.

The balancing of the head on the neck is a result of the adoption of the upright position. Movement of the foramen magnum to a position beneath the skull has been noted as a primate characteristic and it reaches its extreme in man, allowing considerable reduction of the musculature at the back of the neck; the splenius and semispinalis capitis muscles are much smaller in man than in apes. The small size of the trapezius is partly a consequence of the good balance of the head, partly of the absence of brachiating habits. Reduction of these muscles leads to simplification of the bones at both ends of them. The area of their attachment to the occipital surface of the skull becomes much reduced and remains smooth, instead of being roughened and even raised into ridges as in apes. At the same time the spines of the cervical vertebrae, very long in the gorilla, are short and almost vestigial in man. When the head is properly balanced on the backbone it can be freely turned around, and for this purpose the sternomastoid muscles are well developed and the large mastoid ('breast-like') swellings where they are attached to the base of the skull provide a characteristic human feature.

6. Neoteny in the evolution of man

One of the most striking differences between man and apes is the slow rate of our own growth and development; there is a strong suspicion that many of our features are due to retardation of the time of onset of maturity. Schultz (1944, 1963) has shown that in the apes growth ceases between the ages of 10 and 12 and that the epiphyses finally close between 12 and 14. Many of the features of man, such as the large head set at right

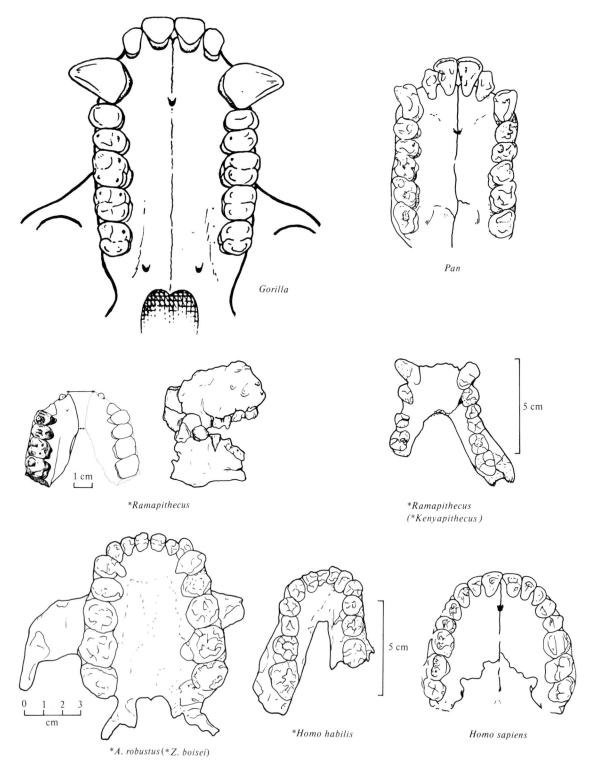

Gorilla

Pan

1 cm

Ramapithecus

Ramapithecus
(*Kenyapithecus*)

5 cm

0 1 2 3
cm

A. robustus (*Z. boisei*)

Homo habilis

5 cm

Homo sapiens

FIG. 24.5. Comparison of the jaws of apes and early hominids. (That of *Homo habilis* is now referred to *A. africanus.*) (After Day 1965. Tattersall 1975, Centre for Prehistory and Palaeontology, Nairobi.)

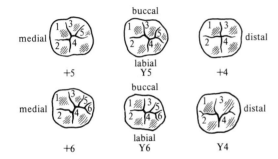

+5   Y5   +4

+6   Y6   Y4

FIG. 24.6. Mandibular molar patterns in chimpanzee and man. The chief distinguishing feature between Y and + patterns is the relationship of cusps 2 and 3 to each other. In the Y pattern they are in contact, in the + they are separated by cusps 1 and 4.

angles to the vertebral column, presence of a prepuce on the penis and hymen in the vagina, are those to be found in foetal apes, and it is therefore suggested that one of the main changes leading to our development has been delay in the rate of differentiation and onset of maturity, a process of paedomorphosis or neoteny. This might well depend on the endocrine balance, perhaps particularly on the action of the hypothalamus and anterior lobe of the pituitary. In mammals with seasonal reproduction the pineal produces a hormone (melatonin) which inhibits growth of the gonads. It is possible that the human pineal has this function in childhood since tumours of it lead to precocious maturity in boys. The circulating melatonin falls sharply in the early stages of puberty (Fig. 24.7A). The long period of childhood, whatever its origin, has profoundly influenced human social organization. Those family organizations were more efficient in which individuals developed late and were therefore better behaved and better learners. Families composed of such slow-developing and re-

strained individuals would therefore survive and the genetic factors involving delay of maturity be selected (see *Introduction to the study of man*).

7. Growth of human populations

This increase of the post-natal developmental period may well be connected with the appearance of the fourth outstanding feature of man noted by Schultz (p. 473), the great population increase in recent times. No exact figures are available, but it is probable that a first increase occurred when the Neolithic agricultural civilization developed, perhaps 10 000 years ago. This development presumably depended on factors making for orderly and restrained behaviour, such as we have been discussing; it is no accident that family customs are closely linked with those of tribes and nations in all stages of society. A further great increase of human population, probably at least a doubling, has occurred during the past 200 years, and we may associate this with the further extension of habits of thought and restraint in the conduct of affairs, making possible the development of logic and science and their application to human productivity.

8. The earliest hominids

Ramapithecus from Africa and Asia and related fossils from Africa, Hungary, and Greece are forms from the later Miocene and Pliocene (14–8 Ma ago) closer than any other known ape-like forms to the ancestry of the Hominidae (Fig. 24.8). Only jaws, teeth, and fragments of the skull have been found but these show several features more like those of man than ape (Fig. 24.5) (see Poirier 1977; Simons and Pilbeam 1978). The jaws are relatively short and delicate and the canines and premolars are smaller than in pongids with less extreme shearing adaptations. The dental arch is divergent

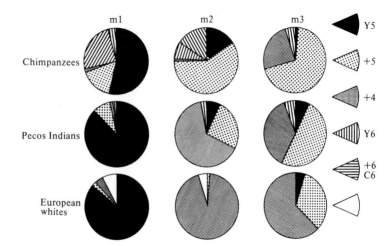

FIG. 24.7. Proportions of the various mandibular molar patterns found in chimpanzees, Pecos Indians, and white Europeans. (After Schuman and Brace.)

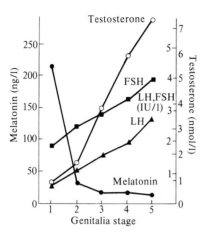

FIG. 24.7A. Decrease of melatonin and increase of pituitary and sex hormones related to the stages of puberty, assessed on a scale from 1 to 5 (Silman, Leone, Hooper, and Preece 1979).

posteriorly, but less so than in modern man. The molars, however, are like those of *Dryopithecus* (Fig. 23.12) and some workers consider that these fossils should be placed in that genus, or that they had evolved from it. The molars show considerable interstitial wear, and the massive jaws were used in powerful transverse mastication movements. These features suggest a tough herbivorous diet, perhaps of seeds or grasses as warm forests declined in a colder climate.

9. *Australopithecus* (= southern ape)

From about 4 Ma ago there is evidence from Africa of several distinct lines of creatures resembling man (see Howell 1978, Leakey 1981). The bones of *Australopithecus* probably represent three species none of which actually contained our ancestors but evolved parallel to them. The first specimen was a juvenile skull described by Dart in 1925 (Fig. 24.4) from a limestone cave at Taung in South Africa and named *A. africanus*. Other fossils referred to the genus have since been found in East

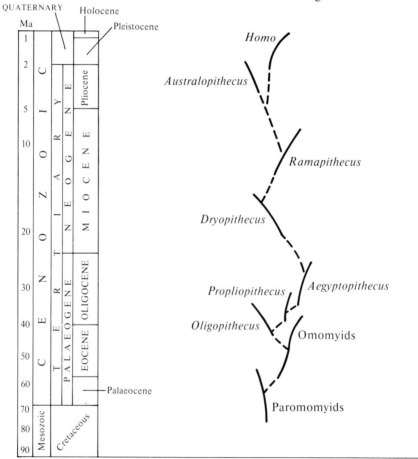

FIG. 24.8. Distribution and hypothetical phylogenetic tree of taxa leading to fossil hominids.

Africa in deposits dated between 3 and 1 Ma old. *Australopithecus* skulls have a cranial capacity estimated at 450–500 ml, little larger relative to body size than that of an ape (Fig. 24.4) (p. 473), yet they have been found in East Africa associated with stone tools. There is evidence that they walked on two legs, though perhaps with nearly the same rhythm as today (Fig. 24.3A). The area at the back of the skull for the attachment of neck muscles was smaller than in apes, while the mastoid process was larger. The foramen magnum is at the centre of the skull. These features mean that the head was *balanced* on the neck rather than hanging forward as in apes. The mandibles were larger than in modern man but the teeth form a rounded arcade. The incisors are vertical and the canines spatulate and smaller than in apes but larger than in man. The anterior lower premolars are bicuspid and did not form shearing facets with the canine. The molars are relatively larger than in man and the third is the largest, but the pattern of replacement was like that of man. Several pelvic bones have been found and they have a broad ilium to allow the gluteus maximus to act as an extensor, raising the body into an upright position on two legs. The foot was probably arched. Some workers hold that the *Australopithecines may have run rather than walked, because the iliac blade faces dorsally and the middle metatarsal was thickened rather than the first as in *Homo* showing that they did not use our striding gait (Fig. 24.1). Other workers believe the bones lie within the human range.

Efforts are being made to solve such questions of relationship by multivariate analysis. The results of eight logarithmically transformed measurements of the astragalus (talus) in various anthropoids are shown in Figure 24.9. The lengths of the lines are proportional to the mean distances between the specimens. If this analysis is sound *H. habilis* (see below) was an australopithecine since the talus is different from that of either apes or *Homo*. They probably had their own unique means of locomotion, perhaps a form of running on the ground and climbing trees (or cliffs?) with their arms (Oxnard 1975). The arms and hand were powerful but like those of man, with a fully opposable thumb, quite capable of making tools. The arms may have been used for locomotion, but not for knuckle walking.

Some australopithecine skulls have a heavier build than the Taung specimen, with large brow ridges and larger teeth. Most workers consider these as a separate species *A. robustus*, but some have supposed the differences to be only sexual (Fig. 24.5). A third species *A. boisei* covers the specimens found in East Africa and originally called *Zinjanthropus* by L. S. B. Leakey (Fig. 24.10).

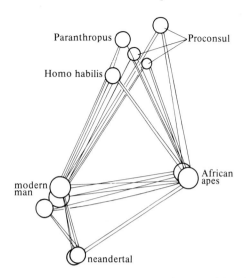

FIG. 24.9. Generalized distance analysis by Oxnard (1975) of measurements by Day (1974) of the talus of various anthropoid specimens.

Whatever interpretation is finally put upon these fossils they show that a considerable variety of almost human creatures existed in Africa. Discoveries in 1972 by R. E. Leakey have revealed some remains that are even more like modern man and may be near enough to our ancestry to be classified in the genus *Homo*. A fragmented cranium, about 1.9 Ma old, found at Koobi Fora in Kenya when assembled shows a capacity of 775 ml. A mandible as well as post-cranial remains found in the same place also resemble *Homo* rather than *Australopithecus* and there were stone tools and evidence of some social organization around a 'home base'. This skull (KNM-ER 1470) has been assigned to *Homo habilis*. This was the name given to specimens found in the lower levels at Olduvai Gorge by L. S. B. Leakey. The cranial capacity of 642 ml was considered to make them distinct from *Australopithecus*. Other workers, however, feel that these remains belong to *A. africanus* and that the small size of the brains of these fossils and of KNM-ER 1470 should exclude them from the genus *Homo*.

10. The earliest men?

Still more recent discoveries near Hadar in Ethiopia have revealed fossils from about 3.5 Ma ago, some with characteristics like those of *Homo* and others like *Australopithecus* (Johanson 1981). The evidence thus suggests that over the period from 4 to 1 Ma ago several distinct lines of creatures rather like man existed. Whether any of them should be called *Homo* is a matter of definition. None of these creatures had brains more

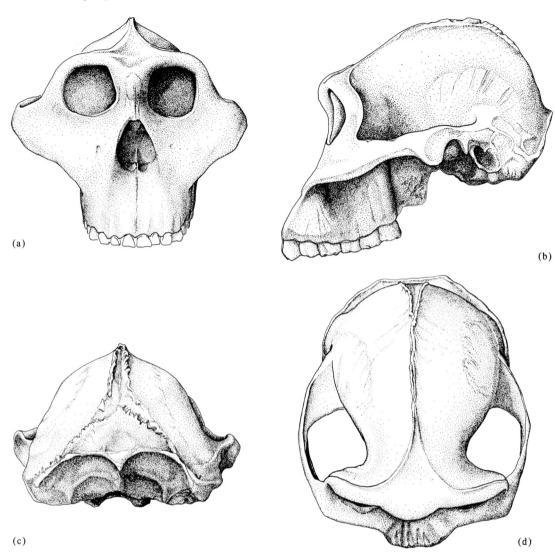

FIG. 24.10. *Australopithecus boisei*. Facial, lateral, posterior, and superior views of restored cranium. Two-fifths natural size. (Based on specimens from Olduvai Gorge and Koobi Fora.) (Howell 1978.)

than half the size of that of modern man (Table 24.1). However, their teeth were more like those of man than of ape (Fig. 24.5), they walked on two legs, had nearly human hands and feet and some of them made rough stone tools, the earliest of which are dated at 1.9 Ma ago from Omo, perhaps earlier from Ethiopia and the Koobi Fora Formation. The very difficulty in classifying the fossils is perhaps the most important fact of all. It shows firstly that they were truly in some sense missing links, intermediate between ape and man. Secondly their mixture of characters shows that the human features did not all evolve at the same time. This may lead us to look for the particular environmental features that led to the

adoption of the upright posture and manipulative hands and later the full enlargement of the brain.

11. *Homo erectus*

The problem of defining man is not made easier by the fact that Java and Pekin man, fossils from little later than those discussed, and with brains rather larger, are generally agreed to be called *Homo erectus* (though they were at first called *Pithecanthropus* and *Sinanthropus*). We therefore have a rather complete set of 'missing links'. Java man was first found in 1891 in gravel beds now given a radiometric date of 1.9 ± 0.4 Ma. Similar fossils have been found at Choukoutien near

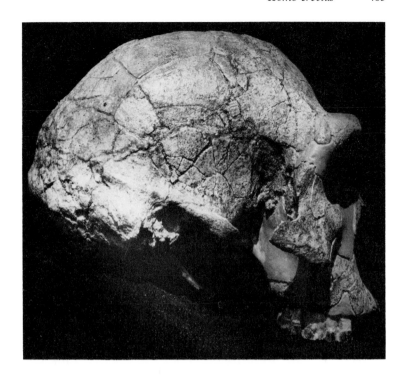

FIG. 24.11. *Homo erectus*, 1.5 Ma. Cranium KMN-ER 3733 from Koobi Fora. Right, lateral aspect. (Photograph kindly provided by R.E. Leakey.)

Pekin, including the remains of 50 individuals; also at Olduvai in Africa and in Hungary, some of these being as recent as 0.5 Ma. Moreover there is evidence that over this large range of time there was a considerable advance towards the modern human condition (Figs. 24.4 and 24.12).

The earliest cranial bones from Java are very thick, with a large brow ridge, low forehead and large nuchal area at the back for muscle attachments. The mean capacity of the cranium is estimated at 950 ml (varying from 800–1000 ml) (Table 24.1). The post-cranial skeleton of *H. erectus* was quite like that of modern man. The pelvis allowed an alternating tilt, the tibia had a strong soleus muscle attachment and the foot was probably arched (Day 1973).

The remains at Pekin were found in a cave with choppers and flakes retouched to make scrapers, and evidence of fire. The skulls are larger, 850–1300 ml, with a mean of 1075 ml. The forehead was steeper, the palate more rounded and the teeth smaller than in the earlier Java specimens. The age is uncertain but was perhaps 0.6 Ma BP. The *H. erectus* fossils in Africa also cover a long time span from about 1.6 (Koobi Fora) to 0.6 Ma ago (Ternifine in North Africa). Those in Europe (Mauer and Vértesszöllös) all seem to be of later date (0.5 Ma) and are the first hominids known there since *Ramapithecus* (p. 479). The cranial capacity of the skull from the latter site is estimated at as much as 1400 ml.

H. erectus brains therefore showed an increase in size from little more than that of *Australopithecus* to modern man. Yet all showed the characteristic features of thick bones, prominent supraorbital ridge and low forehead. It is possible but not certain that they included the ancestors of *H. sapiens*.

12. Presapiens

The earliest known skulls referred to our species were found at Swanscombe, near London, and Steinheim, in Germany, and are dated at about 0.5 Ma BP (Fig. 24.13). These 'presapiens' fossils are thicker and more heavily built than modern skulls but otherwise very like them. They may be classified in a subspecies *Homo sapiens steinheimensis*. Neanderthal men were of a later type nearly but not quite like modern man. They are found from sites dated as between 100 000 and 50 000 years ago. They show much variation but the skulls are generally long and low, with the sides bulging, large brow ridges, low forehead and projecting occiput (Fig. 24.4) (Wood 1978). The cranial capacity was large, often greater than today. The incisors were large but cheek teeth of about modern size. The post-cranial bones were large and strong and the Neanderthalers walked as we do, not with the stooping gait that is often shown in pictures that were based on an arthritic skeleton. The Neanderthal population lived in Europe and parts of Asia and north Africa. They may be classed

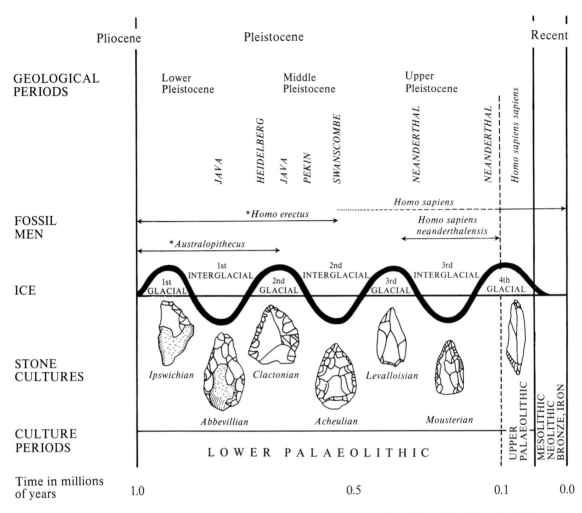

FIG. 24.12. Diagram of Pleistocene time and some stages of human cultural evolution. (After Howells 1944.)

as a subspecies *H.s. neanderthalensis*. They were as-sociated with a characteristic stone tool culture the Mousterian and were Palaeolithic hunters, living in caves, using fire and burying their dead with ceremony. It is probable that they contributed genes to the later population of the area, perhaps mixing with invaders from further east (Fig. 24.12).

13. *Homo sapiens sapiens*

Skulls of modern human type, *H.s. sapiens*, are nearly all less than 50 000 years old (Fig. 24.4). Humans today are so varied that it is difficult to know what characters should be taken to define the subspecies. Features generally used are greater breadth in the parietal region than at the base, steeply rising forehead, and prominent chin. The earliest of such skulls is from Omo in Kenya dated at about 130 000 years ago. Skulls from other

continents are all more recent. It is generally assumed that all modern men are of one stock, but we do not know when or where it arose.

14. Rate of human evolution

Even this superficial study of fossil hominids shows that there is strong evidence that man has reached his present condition by a gradual process of change, possibly with small quantal jumps (p. 8). 'When did the first man appear?' is therefore a non-question. There were evi-dently several parallel lines of pre-humans so we cannot even sensibly ask when the human line diverged from that of apes. Most anthropologists agree that our ancestors are to be found among the Miocene dryo-pithecines. Conversely studies of protein structure show remarkable similarity between man and gorilla and chimpanzee, suggesting separation only a few million

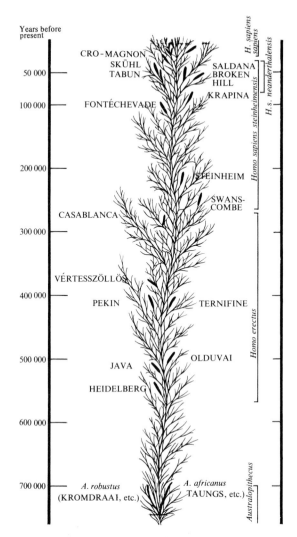

Fig. 24.13. A tentative scheme of human evolution. It shows the gradual change and also that there must have been many different branches and anastomoses of slightly separated lines. Note that there is no main 'trunk'. Some known fossil remains are shown in their relative positions but the dating of the early forms is uncertain. *H. s. steinheimensis* is shown surviving till 45 000 years ago to include such skulls as those of Fontéchevade and Skühl, but this is doubtfully correct. (From Young 1971.)

years ago (p. 4). There is only one difference in the sequence of 141 amino acids in the α-chains of human and gorilla haemoglobin. This problem has not been solved.

Another approach is to measure the rate of evolution using as a unit the darwin, the increase or decrease of a character by 1/1000 per 1000 years (p. 8) (Haldane 1949). Comparing the shape of the skull in *H. erectus* and modern man the rate was 620 millidarwins (md).

Rates for the brain are similar (Table 24.1) and suggest a rapid evolution compared with the teeth of horses of 78 md since the Eocene.

15. Human cultures

The earliest *Homo sapiens* were hunters, living apparently as small families, in caves. They used fire, but their only known instruments were of wood or stone. More definitely shaped flints, used as axes, are found during the second ice age, second interglacial, and third ice advance and are referred to the lower Palaeolithic (Chellean and Acheulian) stages. The third interglacial and last glacial period constitute the middle Palaeolithic (Mousterian), with well-made flints, probably produced by Neanderthal man. The recession of the last main glaciation began less than 100 000 years ago and after that time there was a series of upper Palaeolithic cultures (Aurignacian, Solutrean, and Magdalenian) in which besides wonderfully chipped flints there were also fine bone and ivory needles, blades, and other instruments. These were probably made by men of *H. sapiens sapiens* type, whose skulls first became abundant in caves and deposits of the last glacial period (Fig. 24.12).

The details of the replacement of these species and races by each other remain obscure. Probably invading races often took over parts of the cultures of their victims, as well as bringing in their own culture, so that the whole story becomes very confused. It is usually considered that *H.s. neanderthalensis* did not evolve into *H.s. sapiens*, at least in Europe, but was replaced by a wave of invaders coming from central Asia though the two groups may have interbred. Similarly, at the end of the Palaeolithic period, about 12 000 years ago, the hunters of that time gave place first to a series of Mesolithic invaders, builders of the lake dwellings found in various parts. These developed into the Neolithic people, who were farmers and city builders and from about 9000 BC onwards dominated the Middle East and south Mediterranean region and hence spread outwards, developing the series of cultures known as the Bronze and Iron Ages.

Study of the gradual mixing and changing of human populations, besides its personal interest, is of value to a zoologist in calling the process of evolution to our imagination and showing us its complexity and slowness. Without undue difficulty we can have in mind a picture of great populations of human beings, composed of individuals differing slightly in structure and habits. In this way we may get some faint indication of how slow and confused changes might constitute evolution. But the fossil record is so incomplete that we cannot be sure whether human evolution included any sudden jumps. The discovery of rapid climatic changes such as occurred 90 000 years ago (p. 28) prompts the thought that the modern human stock might have arisen suddenly from an isolated population of neotenous individuals (Young, Jope, and Oakley 1981).

In development of powers of obtaining information with the nervous system man is only showing an accentuation of the characteristics of mammals in general and especially primates. With this goes the low reproduction rate, slow development, and long period of post-natal care. The paedomorphosis in man, producing teachable, co-operative individuals may well have been the factor that has made possible the development of complex societies and their tools. By efficient communication man is able to produce a cumulative store of information outside his mortal body and pass it on not only to few individuals, as is the genetic store, or that passed by word of mouth, but to many. Each individual thus 'inherits' not from two or few parents but from the accumulated memory store of a large population. It is perhaps this 'multi-parental' or exosomatic inheritance of information that has changed man so rapidly in the past and is likely to do so even faster in the future. Success will clearly be for those populations whose individuals are able so to co-operate as to discover and transmit more and more information.

Summary of hominid evolution

Classification and dating of remains of early hominids is continually changing as new fossils are found and earlier discoveries evaluated. The data at present available are approximately as follow (see Curtis 1981; Leakey 1981):

			Ma B.P.
Ramapithecus			14–4
*Hominid	Lake Turkana	Jaws and teeth	?5.5
*Hominid	Laetoli, Tanzania	Footprints	3.75
'*Australopithecus afarensis*'? '*Homo*'?	Hadar, Ethiopia	Skull and skeleton ('Lucy')	3.5
*Australopithecus? *Homo?	Laetoli, Tanzania	Jaw etc.	3.5
*A. africanus	Makapansgat,		?3.0
	Sterkfontein,		?2.5
	Taung, S. Africa		?0.9
*A. robustus	Swartkrans, Kroomdrai, S. Africa		2.0–1.0
*A. africanus	Omo, Ethiopia		3.0–2.0
*A. boisei	Olduvai, Tanzania and Omo		2.0–1.4
*Homo habilis	Olduvai, Tanzania		1.9
*Homo habilis	Koobi Fora, Kenya	KNM-ER1470	1.8
*Homo habilis	E. Turkana		1.45
*Homo erectus	Olduvai		1.7
	Koobi Fora		1.6
	Java		1.9
	Pekin		?0.5
	Omo	Stone tools	1.9
*Homo sapiens steinheimensis	Swanscombe		0.25
*Homo sapiens neanderthalensis			0.10
Homo sapiens sapiens			0.05

25   Rodents and rabbits

1. Characteristics of rodent life

IN spite of the similarities of all the animals with gnawing teeth zoologists consider that the rabbits and hares to be sufficiently different from the others to be placed in a distinct order Lagomorpha, the order Rodentia being retained for all other 'rodents'. The two orders are not closely related. Fossil rodents (in the strict sense) certainly occurred in the late Palaeocene period and a probable lagomorph is reported from the same time. Both groups retain many primitive mammalian characters, for instance, a long, low skull, with small brain and small cerebral hemispheres, temporal fossa widely open to the orbit, pentadactyl limbs, separate radius and ulna, and so on. These features, being found in all early mammals, indicate no closer affinity of the two orders than depends on evolution from a common stock. It is not even clear exactly how the two groups are related to the ancestral eutherians, and we must be content to say that it is probable that animals with rodent specializations diverged from the insectivoran eutherian stock in the late Cretaceous or Palaeocene and then rapidly became differentiated into lagomorph and rodent types. The two orders are therefore placed together in an isolated cohort Glires. The animals loosely known as rodents are the most successful of modern mammals other than man. They live in all parts of the world, from the tropics nearly to the poles. Three thousand species are known, as many as are found in all other mammalian orders put together. They inhabit a considerable variety of ecological niches, mostly on the land, often in burrows, but many in the trees and some in the water. This most successful type of mammalian life is, however, in several ways atypical of the rest and indeed has been distinct since the early Tertiary. One striking point is that the animals have never become large in size, although such increase is a trend found in almost all other mammalian groups. The South American capybara, the largest living rodent, is the size of a small pig, and few fossil forms were much larger. Rodent life has specialized in rapid breeding and this system of production of large numbers of small animals has been very successful. The total rodent biomass today may well be greater than that of the whales, which are at the other extreme, and have all the advantages of aquatic life. The rapid reproduction presumably brings considerable evolutionary advantages, enabling the population to make the adjustments necessary to meet changing circumstances. One of the characteristics of rodent populations today is their great fluctuations (p. 495), notorious in the case of the voles, mice, and lemmings, but marked also in rats and other forms. The pressure of rodent life is such that no stable equilibrium is reached with the environment and extreme oscillations occur, often with results of great importance to man and to his crops.

2. Classification

Cohort 2: Glires
  Order 1: Rodentia (= Simplicidentata)
    Suborder 1: Sciuromorpha
      Thirteen families, including
      *Family Ischyromyidae: Palaeocene–Miocene. Eurasia, N. America
        Paramys
      Family Aplodontidae: Eocene–Recent
        Aplodontia, mountain beaver, N. America
      Family Sciuridae: Miocene–Recent
        Sciurus, squirrel, Holarctic; *Marmota*, marmot, woodchuck, Holarctic; *Tamias*, chipmunk,

N. America; *Petaurista*, flying squirrel, Eurasia; *Cynomys*, prairie dog, N. America; *Citellus*, ground squirrel, N. America

Family Geomyidae: Oligocene–Recent. N. America
 Geomys, pocket gopher; *Dipodomus*, kangaroo rat

Family Castoridae: Oligocene–Recent. Holarctic
 Castor, beaver

Family Anomaluridae:
 Anomalurus, flying scaly tails, Africa

Family Pedetidae:
 Pedetes, jumping hare, S. Africa

Suborder 2: Myomorpha
 Nine families, including

Family Dipodidae: Pliocene–Recent. Palaearctic
 Dipus, jerboa, Eurasia

Family Cricetidae: Oligocene–Recent. World-wide (except Australasia)
 Peromyscus, deer mouse, N. America; *Lemmus*, lemming, Holarctic; *Microtus*, vole, Holarctic; *Cricetus*, hamster, Holarctic; *Mesocricetus*, golden hamster, Asia Minor; *Gerbillus*, Africa, Arabia, Pakistan; *Reithrodontomys*. American harvest mice, N. and central America

Family Muridae: Pliocene–Recent. Native to Old World
 Apodemus, field mouse; *Rattus*, rat; *Mus*, house mouse; *Glis*, dormouse; *Notomys*, jerboa-rat

Family Spalacidae: Miocene–Recent
 Spalax, mole rat, S. E. Europe, N. Africa, Asia

Family Zapodidae: Oligocene–Recent. Holarctic
 Zapus, jumping mouse

Suborder 3: Hystricomorpha
 Nineteen families, including

Family Hystricidae: Oligocene–Recent. Palaearctic, Africa
 Hystrix, porcupine, Asia, Africa

Family Erethizontidae: Oligocene–Recent. N. America
 Erethizon, tree porcupine, N. America

Family Caviidae: Pliocene–Recent. S. America
 Cavia, guinea-pig; *Dolichotis*, Patagonian hare

Family Hydrochoeridae: Pliocene–Recent. S. America
 Hydrochoerus, capybara

Family Dasyproctidae: Recent. S. America
 Dasyprocta, agouti, pacas

Family Chinchillidae: Oligocene–Recent. S. America
 Lagostomus, vizcacha; *Chinchilla*, chinchilla; *Myocastor*, nutria, coypu, swamp beaver

Family Bathyergidae: Pleistocene–Recent. Africa
 Bathyergus, mole-rat

Order 2: Lagomorpha (= Duplicidentata)
 Family 1: *Eurymylidae. Palaeocene
  Eurymylus, Asia

 Family 2: Ochotonidae. Upper Oligocene–Recent
  Ochotona (= *Lagomys*), pika, cony, N. America, Asia

 Family 3: Leporidae. Upper Eocene–Recent
  Lepus, hare, Pleistocene–Recent, Palaearctic, N. Africa; *Oryctolagus*, rabbit, Pleistocene–Recent, Europe, N. Africa; *Sylvilagus*, cotton-tail, Pleistocene–Recent, N. and S. America

3. Order Rodentia

Rodents are mostly herbivorous and their most characteristic features are, of course, in their teeth, especially the incisors, one pair only of which persists in each jaw; hence they were the suborder 'Simplicidentata' of the older order Rodentia, which included also the rabbits and hares. These latter have a second pair of upper incisors, hence 'Duplicidentata'. The incisor has enamel only on its labial surface and thus maintains a cutting edge. It is worn away at the rate of perhaps several millimetres a week and is replaced by continual growth, for which it has a very wide open pulp cavity, or in the conventional term is said to be a 'rootless' tooth. The incisors are often very large and curved and their gnawing action against each other gives them chisel edges. If one incisor is lost the opposing one continues to grow round in a spiral, until it enters the skull.

The second incisors, canines, and anterior premolars are missing, leaving a large diastema in front of the cheek teeth. Folds of skin can be inserted into this gap to close off the front part of the mouth, so that material bitten off during gnawing is not necessarily swallowed. A distinct anterior chamber of the mouth is thus formed, and may be prolonged into deep pockets in which food is stored for transport to the hoards that are collected by many species. The teeth are not used only for obtaining food; rats will gnaw their way through a lead pipe.

The premolars are reduced to two above and one below in the more primitive squirrels and even fewer are present in the other rodents. The molars and premolars are usually alike in pattern and show modifications similar to those found in the grinders of other herbivorous mammals. The cusps of the original eutherian molar can still be recognized in the squirrels, but they are arranged in transverse rows, paracone and protocone in front, metacone and hypocone behind. The cusps become joined in pairs to form ridges, giving a bilophodont grinding tooth. In most of the rodents further ridges are then added in front, behind, and between the original ones, and these are also joined by cross-ridges, giving a multi-lophodont molar. Similar changes occur in the lower teeth. The teeth also become very high-crowned or 'hypsodont' and the enamel, dentine, and bone ('cement') wear at differing rates. The roots remain wide open and the teeth grow continually. All of these features are somewhat like those arrived at in ungulates by a convergent process of evolution. A further feature of the rodents is that the upper tooth rows are set closer together than the lower and bite inside the latter, often giving an oblique grinding surface. The milk teeth are shed very early and are not functional.

The lower jaw and its muscles show many modifications. The articulation is very long, and the lower jaw moves backwards and forwards on the upper; indeed the lower incisors are thrust so far forward that while gnawing the molar surfaces no longer occlude. A curious feature is that the mandibles are not united in a symphysis but are freely movable, with a joint cavity between them. A special portion of the mylohyoid muscle, known as the m. transversus mandibulae, draws the two mandibles together, causing the lower incisors

to separate (Fig. 25.1). The action of the lateral portions of the masseters then brings the two teeth together again with a scissor action.

The jaw muscles are very large and modified to produce backward and forward motion of the jaw. The masseter muscle is divided into three layers. In the earliest fossil rodents all three layers arose from the zygoma as they still do in the primitive sciuromorph *Aplodontia* (Fig. 25.2). The most superficial serves to close the jaw and draw it forward and arises far forward on the side of the face and runs backwards to insert at the angle. This part is the same in all rodents but the other two differ and have been made the basis of classification. In the Sciuromorpha the middle layer (known confusingly as the 'lateral masseter') spreads forward in a channel on the side of the skull. The deep layer ('medial masseter') retains the primitive condition. In the rats and mice (Myomorpha) besides this channel for the lateral masseter the medial masseter has also spread forwards through the enlarged infra-orbital canal. The South American rodents such as porcupines and some probably related Old World forms make the third group the Hystricomorpha. In these the lateral masseter retains the primitive position but the medial

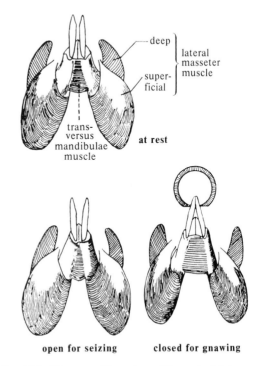

FIG. 25.1. Diagrams of lower jaws of the squirrel *Sciurus*, and the position of the teeth during various phases of feeding. In the closed position they are also used to prise open nuts (From Weber, after Krumbach.)

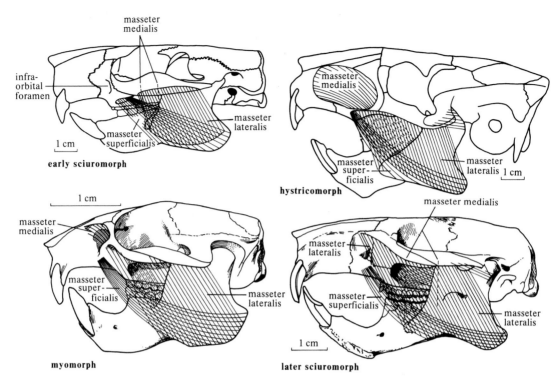

FIG. 25.2. The forms taken by the masseter muscle in rodents. The early sciuromorph type (*Paramys*, Middle Eocene). Hystricomorphous form (*Neoremys*, Middle Miocene). Sciuromorphous type (*Eutpomys*, Middle Oligocene). Myomorphous (*Eumys*, Middle Oligocene). (Wood 1974.)

layer spreads forwards through the infra-orbital canal. It remains uncertain, however, whether these are really three compact groups. Fossil evidence suggests that there has been much parallel evolution and that the hystricomorph condition evolved independently no less than eight times and the sciuromorph six times (Wood 1974). The lower jaw often carries a large flange, for the attachment of the masseter muscle. The pterygoid muscles and their attachments are also often large, but the temporal muscle is usually small.

Apart from their gnawing mechanism the rodents remain rather unspecialized mammals. The gait is plantigrade, the forelimbs often shorter than the hind and used for handling the food. In some this tendency is carried to the extent of producing a hopping, bipedal gait. There is a large caecum and food is passed twice through the gut (p. 494). A division of the stomach is found only in mice, which have a horny cardiac region. Rodents are macrosmatic, with large olfactory bulbs and relatively small, smooth neopallium. The eyes are often well developed, especially in arboreal rodents and those living in open country or steppe. Hearing is often good, and in some desert species the tympanic bulla is greatly dilated, perhaps to detect sounds made by the

widely separated individuals. Many rodents have well developed social habits, with olfactory and visual, as well as auditory and tactile signalling.

Rodents are mostly polyoestrous, often breeding throughout the year, at least in captivity. Numerous young are produced and are often cared for in a nest. Usually the uterus is double and the placentation of the discoidal and haemochorial type.

4. Suborder Sciuromorpha

No thoroughly satisfactory scheme for grouping the various types of rodent has been devised. We shall keep to the classical division of the order into three suborders, Sciuromorpha, Myomorpha, and Hystricomorpha. However, the Rodentia early radiated along many lines and are not easily divisible into major groups. Many authors prefer to use the term Caviomorpha for this last group, containing all the South American rodents. The Old World porcupines (Hystricidae) and some other families are then considered to be separate groups, not closely related to the porcupines of South America.

The Sciuromorpha have two upper premolars and one lower and no large infra-orbital canal for the masseter. The suborder includes the most primitive surviving rodent, *Aplodontia* the 'squirrel' or 'mountain beaver' of western North America. This can be regarded as a survivor of the oldest family of rodents the *Ischyromyidae, such as *Paramysis*, extending from the Palaeocene to the Oligocene. The suborder also includes the squirrels (Sciuridae) found in all major regions except Australasia. They are diurnal, with large eyes capable of colour vision, and often bright colouring. Squirrels, of course, build a nest, or drey, of twigs, or line hollow trunks with dried plants. They make numerous food stores for the winter, but find them again only by smell, not memory. The flying squirrels, *Petaurista*, glide for long distances by means of the patagium, whose muscles enable them to change direction in the air (Fig. 25.3). They are said to make use of thermals.

The marmots (*Marmota*, Fig. 25.4), ground squirrels (*Citellus*), and prairie dogs (*Cynomys*) are burrowers, with elaborate underground societies. Marmots can hibernate for as long as eight months. The Geomyidae (gophers) and kangaroo rats (*Dipodomys*) of North and Central America are also sometimes placed here. They are colonial burrowers and some of them are bipedal jumpers, paralleling with the jerboas, gerbils, and jerboa-rats.

Two other families are possibly related to sciuromorphs. The Anomaluridae, the scaly-tailed flying squirrels of tropical Africa have acquired a patagium and are skilful gliders. The Pedetidae include only *Pedetes*, the jumping hare or spring hare common in South Africa. They are like small kangaroos with elongated hind feet. Their burrows are a nuisance to farmers.

Possibly also in this group are the beavers, Castoridae, found throughout the Holarctic. These are aquatic rodents (Fig. 25.5) that show remarkable habits in preparing a house and store of food for the winter (Fig. 25.6). The house ('lodge') is built on a mass of

FIG. 25.4. Marmot, *Marmota*. (From photograph.)

debris so as to be surrounded by water. Sticks are built up to make a wall round the platform and the whole finally closed by a dome of sticks and mud, which is carried by the beavers with their fore paws. When this damp structure freezes it makes a strong protection against bears and other enemies, and the beavers keep warm inside it. Food is obtained from the bark of branches kept in a store under water and brought up to the lodge through a plunge hole. The dams are made by the beavers during the summer in order to deepen the streams; they may reach a height of 7 m and a length of 1000 m. The lodges and dams are the result of work by two or three families living together but although they are colonial they are not co-operative, each works independently for himself 'as if he owned the colony'. The beavers work compulsively, repairing the structures whether they need it or not. They will add sticks and mud to a concrete dam. These dams benefit the valleys in many ways. The beavers are therefore being successfully re-introduced in America and Russia, sometimes by dropping them from aircraft by parachute in cages that open automatically.

Like other rodents they mark their territory by scent, secreted by large anal oil glands. When an animal smells a deposit it visits it and adds its own 'castoreum'.

FIG. 25.3. Flying squirrel, *Petaurista*. (From a photograph.)

FIG. 25.5. Beaver, *Castor*. (From photographs.)

FIG. 25.6. Lodge and food store of the
beaver. (From Hamilton 1939.)

5. Suborder Myomorpha

The Myomorpha is a very large group, including the
rats, mice, voles, jerboas, and other types. All have the
medial portion of the masseter running through the
infra-orbital canal, but are probably not really closely
related. They include many families and genera, with
specializations for individual ecological niches. The
jerboas, Dipodidae (Fig. 25.7), are desert animals, with
limbs specialized for hopping, but in other respects with
somewhat primitive characteristics. There has been a
great elongation of the metatarsals of the three central
digits, which alone are well developed.

The Muridae (Fig. 25.8) are perhaps the most successful of all mammals, including 475 living species. It is an
ancient family originating in the Oligocene but radiating
explosively in the Pliocene and invading all parts of the
world. They reached South America and Australasia
only in relatively recent times, but they arrived in the
latter before man and have diversified into seven distinct
genera there. They are the only mammals indigenous to
New Zealand. The family includes an enormous variety
of mice, rats, dormice, field mice, hamsters and so on,
and also some special forms such as the jerboa-rat
(*Notomys*) of Australia, which has paralleled the true
jerboas in its jumping habits. The rats and mice may be
considered as parasites of man and are still changing

FIG. 25.8. Common brown rat, *Rattus*. (From photographs.)

their distribution. *Rattus* has been widely spread in
recent centuries. The black rat (*R. rattus*) prefers
warmer and drier conditions than the brown rat (*R.
norvegicus*) but the two often compete.

The mole-rats (Spalacidae) of south-east Europe,
north Africa, and Asia live permanently underground,
feeding on roots. They are the only blind rodents, with
vestigial eyes beneath the skin. The bamboo-rats (*Rhiz-
omys*) of Asia also live underground but are less
specialized.

The voles (Cricetidae) are a very large family, including partly aquatic as well as terrestrial forms. The field
vole (*Microtus agrestis*) is apt to increase greatly in
numbers, producing notoriously destructive plagues.
Lemmus, the lemming (Fig. 25.9), is a mouse-like form

FIG. 25.7. Jerboa, *Dipus*. (From life.)

FIG. 25.9. Lemming, *Lemmus*. (From a photograph.)

FIG. 25.10. Porcupine, *Hystrix*. (From a photograph.)

FIG. 25.12. Capybara, *Hydrochoerus*. (From a photograph.)

living on grasses and roots on the Norwegian mountains. At irregular intervals of 3–5 years, depending upon the conditions, the population grows greatly by increase in the young in each litter and in the number of litters per season. The population becomes too great for the area to support and large swarms emigrate to the lowlands and die of starvation or from predators. Others reaching the sea-shore swim out and are drowned.

The large house mice of the various Faeroe Islands (*Mus musculus*) are differentiated into six genetically distinct races, although one of the populations has been distinct only for 40 years and two others for less than 200 years. The differences are due to instant sub-speciation by the vagaries of colonization by small samples. This example shows that 'such stochastic change can be an important testing mechanism for new gene combinations' (Berry, Jakobson, and Peters 1978).

Cricetus, the hamster, is a burrower occurring from Europe to Siberia. The golden hamster, *Mesocricetus*, has spread throughout the world from the family of one female found in Syria in 1930. The American harvest mice *Reithrodontomys* include a great number of species superficially like the Old World rats and mice. The deer mice *Peromyscus* resemble European field-mice. The gerbils *Gerbillus* are jumping desert animals, with long hind legs.

6. Suborder Hystricomorpha

As already mentioned the histricoid condition has probably evolved independently several times, with the infra-orbital canal enlarged for the medial part of the masseter and the lateral part attached to the zygoma. The Old World porcupines of Africa and Asia (*Hystrix*) may then not be closely related to those of America (*Erethizon*). The former (Fig. 25.10) are burrowers and mainly vegetarian. They are slow moving, relying on the protection of the colour and noise of their quills, which give notice of the effect of sharp points as they charge backwards.

The New World hystricomorphs, perhaps better called caviomorphs, are the only rodents indigenous there and no fossils are found earlier than the Oligocene. Their niche was previously occupied by marsupials (p. 429). They were probably descended from North American paramyids (p. 491), crossing by a 'land bridge' route.

The American *Erethizon* is a tree-porcupine, feeding on leaves, buds, and bark. The male stimulates a female to copulation by spraying her with urine. The quills are barbed at the tips. The guinea pigs (Caviidae) are a large family of nocturnal burrowers. They are unusually precocious at birth and can then already eat solid food. The Patagonian hares (*Dolichotis*) are long-legged and digitigrade and run like a rabbit. The agoutis (*Dasyprocta*, Fig 25.11) are also digitigrade and cursorial. The capybaras (*Hydrochoerus*, Fig. 25.12) are the largest living rodents and are semi-aquatic. The vizcachas (*Lagostomus*, Fig. 25.13) burrow underground, often

FIG. 25.11. Agouti, *Dasyprocta*. (From life.)

FIG 25.13. Vizcacha, *Lagostomus*. (From a photograph.)

making large colonies. The chinchilla (*Chinchilla*) is a related mountain rodent probably now surviving only in domestication for its fur. The coypu (*Myocastor*) has been widely cultivated for its fur (nutria) and disliked as a pest when it escapes and digs burrows in the banks of rivers.

Possibly related to the hystricomorphs are animals still more modified for digging, the Bathyergidae or African mole-rats, which have lost most of the hair and developed long claws on the forelimbs.

7. Order Lagomorpha

The rabbits and hares are nowadays considered to be a very isolated offshoot of the early eutherian stock, whose similarities to the Rodentia may be only superficial. The two orders are kept together in one cohort Glires, more as a convenience than because of characters they have in common. The arrangements for gnawing found in the lagomorphs have indeed a superficial similarity to those of rodents, but on inspection the differences appear profound. Continually growing incisors are present in both groups, but in lagomorphs the upper pair is accompanied by a small second pair (hence 'Duplicidentata'). The diastema is common to the two groups, and both have cheek pouches with similar functions. The similarity of the molariform teeth is only superficial, however; in the lagomorphs three premolars remain in the upper jaw and two in the lower. The premolars and the molars all acquire sharp transverse ridges, usually two each, used for cutting rather than grinding, the upper teeth biting outside the lower ones, not inside as in rodents. The masseter is powerful, but simpler than in rodents, and it does not extend into the infra-orbital canal. The temporalis muscle is reduced in both groups, but the lagomorphs lack the power of movement between the two halves of the lower jaw.

Taken all together, therefore, the differences are as great as the similarities, even in the gnawing mechanism, which the rabbits share with the other rodents. In the remaining parts there are few points of close similarity, other than those due to the fact that neither set of animals has departed far from the original eutherian condition. Moreover, serological studies do not show any signs of closer affinity of the lagomorphs with the rodents than with other mammalian orders. If anything they are more like artiodactyls. However, the lagomorphs share with rodents the habit of passing food twice through the alimentary canal (caecotrophy). Dried faecal pellets are produced only during the day. At night soft pellets covered with mucus are formed in the caecum and are immediately taken from the anus by the lips. They are stored in the stomach and later mixed with further food taken (Fig. 25.14). The double passage

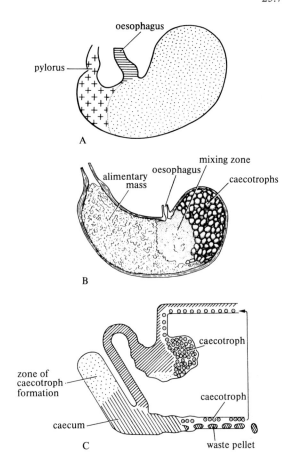

Fig. 25.14. A, Diagram of rabbit stomach. B, Longitudinal section of rabbit stomach. C, Diagram showing the passage of food and caecotrophs traversing the digestive tracts. In the stomach, under the mechanical action of peristalsis the caecotrophs are mixed with the food (mixing zone). In the lowest part of the caecum, the chyme under the action of bacteria is made into caecotrophs, which the animal eats soon after they are expelled from the anus. (B and C after Harder.)

of the food is necessary for the life of mice and guinea pigs as well as rabbits. The animals die in two to three weeks if prevented from reaching the anus. The moist pellets probably contain the metabolites that have been produced by breakdown of cellulose by the bacteria of the caecum, which cannot be absorbed by the organ itself.

Rabbits and hares (Leporidae) have characteristically developed the hind legs for a jumping method of locomotion and there has been a reduction of the tail. The rabbits (*Oryctolagus*) show a number of specializations for burrowing life. They are among the most successful of all mammals, especially in the Holarctic region, but have made relatively little progress in Africa or South America.

Hares (*Lepus*) do not burrow but lie hidden in a depression in the vegetation, the form. The young are born fully furred and with eyes open. The northern species moult to a white winter pelage (varying hares). The pikas (*Ochotona*), sometimes called conies, of Asia and western North America are mountain hares with the habit of storing hay, which they eat in the winter, rather than hibernating. They form colonies, but each animal guards a territory on which he makes a pile of cut vegetation, turning it to dry in the sun.

Fossil lagomorphs, quite like modern hares and rabbits, are found back to the Oligocene. Few remains are known from the Eocene, but there is evidence that the type was already distinct in the Palaeocene and has persisted with relatively little change ever since.

8. Fluctuations in numbers of mammals

The rapid spread of rabbits in Australia and their later reduction by disease show how rapidly the distribution of animal life can change. Domesticated hutch rabbits arrived with the first settlers in 1788 but the explosive spread came from 24 wild rabbits imported on the clipper *Lightning* in 1859 and liberated near Geelong. From here they spread north at 110 km a year and reached the Indian Ocean 1770 km away in 16 years and became a major pest. Their reduction was equally spectacular. Biological control had often been suggested and in 1888 Louis Pasteur had already sent his nephew to try to earn the reward offered by the government for control of rabbits, by introducing chicken cholera. The myxoma virus was known to be fatal to rabbits, carried by arthropod vectors. In 1950 experiments with it were in progress in the Murray River Valley and its accidental escape led to a spectacular epizootic with over 90 per cent deaths, spreading for 1600 km, probably carried by mosquitos in that very wet season. By the end of 1953 by natural spread and inoculation the disease had reached to all parts of the continent. Attenuated strains of the virus rapidly came to dominate the original lethal ones, which killed too many to allow it to persist over the winter. The population is now stable at perhaps 20 per cent of its original level (in Victoria) and has greater resistance.

Myxomatosis was introduced to France in 1952 and rapidly became an enzootic disease over the whole of Europe and North Africa. The carriers have probably been mainly mosquitoes in France, but fleas in Britain. The British rabbit population was reduced by 1955 to 10 per cent of the previous level but has since increased, though myxomatosis is still prevalent. The surviving rabbits have largely abandoned the habit of making fur-lined burrows in favour of nesting in hedges. Attenuated strains of virus are present, but the virulent ones also persist. The pathological manifestations of the diseases differ in various countries as much as does its virulence and means of transmission. This is an excellent example of the great significance of variety in animal life and its effect in producing rapid evolution (Fenner and Ratcliffe 1965).

Fluctuation in numbers is characteristic of many rodents and other small mammals. The phenomenon is usually first recorded as a 'plague' of the rats, mice, voles, or rabbits, and these may be cases of local and sporadic abundance. Study of some species has shown, however, that there are rather regular cyclical fluctuations in numbers, extending over many years. Thus figures for the furs collected by trappers for the Hudson Bay Company enable the variations in numbers of the varying hare to be followed back to 1850 (Fig. 25.15). The cycle is surprisingly regular, with a period of 9.7 years. During the periods of abundance of the hares the mammal and bird predators of these also increase and fall off a little later than the herbivore populations. Changes in food taken by the predators then produce all sorts of secondary effects on the animals of the area.

The existence of these fluctuations, like those of the rabbits in Australia, is striking evidence that animal populations are not maintained in any stable equilibrium. It is often suggested that the cycles depend on changes in the physical conditions, for instance, on the sunspot cycles, but no close correspondences have been discovered. The sunspot cycle has a period of 11.1 years, whereas rodent and lagomorph cycles varying from the 9.7 years for the hares to 5 years for lemmings and 3 for voles have been recorded. Changes in the amount of solar radiation no doubt produce considerable effects on the animals. There is some evidence for a direct effect of diet increasing the fertility. In South America plagues of rats have been observed to coincide with abundance of bamboos. But it seems likely that the cycles of numbers depend on the particular balances set up within the animal communities. These presumably depend on the interactions of the reproductive pressures of the plants, herbivores, carnivores, and parasites, and it is easy to believe that in some conditions these factors, either alone or in association with cycles of solar radiation or other climatic factors, would produce an oscillatory system.

When the numbers are at a maximum the animals show unusual behaviour patterns, including migration, and may enter into a pathological state, becoming cold and torpid and with a very low blood-sugar content. Probably the pressure of competition and lack of food operates on the hypophysio-adrenal system to produce a state of shock. It is this, rather than disease or predators, that finally reduces the numbers.

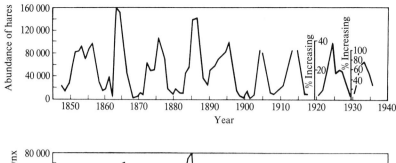

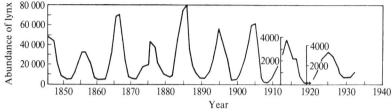

FIG. 25.15. Fluctuations in numbers of varying hares (above), and lynx (below). Estimated from the Hudson Bay Company's fur returns. (Note change of scales.) (From Hamilton 1939.)

Such phenomena are probably not peculiar to small mammals, but appear markedly in them because the rapid breeding and short life provide cycles of relatively brief duration. The facts are all consistent with the interpretation already reached that the control of animal populations depends very largely on the interaction of biotic factors. This continual pressure of the animals and plants on each other is probably a main factor responsible for the great variations in the characteristics of animal populations in different parts of their range, which is found wherever sufficiently careful study is undertaken. This variation in turn contributes largely to the phenomenon of organic evolution and the continual replacement of types by their descendants, which appears so clearly as one follows the history of the vertebrates (Chapters 1 and 33).

26 Whales

1. Classification

Cohort 3: Mutica (= silent ones)

  Order Cetacea (= whales)

    *Suborder 1: Archaeoceti (= ancient whales). Middle Eocene–Lower Miocene

      *Protocetus, Eocene; *Basilosaurus (= *Zeuglodon), Eocene

    Suborder 2: Odontoceti (= toothed whales) Eocene–Recent

      *Superfamily 1: Squalodontoidea (= shark-toothed) Eocene–Miocene

      *Squalodon, Miocene

    Superfamily 2: Platanistoidea. River dolphins. Miocene–Recent

      Platanista, Ganges dolphin, Recent

    Superfamily 3: Physeteroidea. Miocene–Recent

      Hyperoodon, bottle-nosed whale, North Atlantic, Antarctic; Physeter, sperm whale, all oceans

    Superfamily 4: Delphinoidea. Miocene–Recent

      Monodon (= one tooth), narwhal, Arctic; Delphinus, dolphin, all oceans; Phocaena, porpoise, all
        oceans; Tursiops, bottle-nosed dolphin, all oceans; Sousa, white dolphin, Old World Oceans

    Suborder 3: Mysticeti (= moustache whales). Baleen whales. Oligocene–Recent

      Balaenoptera, blue whale, fin whale, sei whale, minke whale, all oceans; Balaena, right whale,
        temperate and Arctic oceans; Eschrichtius, grey whale, North Pacific; Megaptera (= large-winged),
        humpback whale, all oceans

2. Characteristics of whales. Life in water

The whales are a successful set of populations of aquatic, carnivorous mammals, which diverged from the eutherian stock, possibly from the creodont branch, at a very remote date. Their specializations from the typical mammalian plan for aquatic life are greater than those found in any other order, and their organization shows in a remarkable way the effects of habit of life and environment (Matthews 1978). Whales spend their whole lives in the water, being conceived and born there. Large whales and dolphins sleep in the sea. In many respects the whales have reverted to the characteristics of a fish form of life, most noticeably in the body shape, with elongated head, no neck, and tapering, streamlined body. In terrestrial mammals the air resistance is not a serious factor and the body shape does not conform to it, but in water it becomes of first importance.

Dolphins can swim at up to 40 km/h for many hours. The power necessary to drag a wooden model through the water with the speed of the animal would necessitate a horse-power in the muscles several times greater than that of any known mammal. Underwater photographs show that the accommodation of the mobile skin produces stationary waves that absorb part of the turbulence energy, producing laminar flow and reducing drag. The surface is completely smoothed off by the loss of all hair, except for a few sensory bristles round the snout in some species. There is a thick layer of dermal fat (blubber), which, besides acting as a heat insulator, may also provide a reservoir of food and perhaps, when metabolized, of water. The fat also reduces the specific gravity of the animal and possibly provides an elastic covering to allow for changes in volume during deep diving. There are no glands in the skin.

The propulsive thrust comes largely from the horizontally placed tail flukes, which constitute a cambered aerofoil that is moved up and down by the tail muscles. These consist of upper and lower sets of longitudinal fibres inserted in the tail-stock vertebrae by means of long tendons from muscles originating from more forwardly situated vertebrae. More caudally placed muscles inserted on the hindmost vertebrae in the tail flukes allow upward movement of the flukes relative to the tail-stock to produce the forward thrust. Stability is provided by the paddle-like forelimbs acting as hydroplanes, and there is often a large dorsal fin, especially in fast swimmers such as the killer whales (*Orcinus*). The plasticity of animal form is shown by the fact that the flipper is a modification of the ambulatory forelimb, whereas the dorsal fins and tail flukes are 'neomorphs', folds of skin and hard connective tissue but no skeletal support, radical innovations indeed. It is not easy to imagine by what alterations of habit an early eutherian could come to develop fins out of such folds of skin. Equally remarkable has been the disappearance of the hindlimb, leaving no external trace and internally only paired pelvic vestiges with additional bony nodules in right whales representing femur and tibia. The pelvic bones serve as attachments for the corpora cavernosa of the penis and may therefore be regarded as ischia.

The vertebral column (Fig. 26.1) carries no weight except when the animal jumps out of the water. In the vertebral column the zygapophyses are reduced, and the centra are well developed to make a compression strut, as in fishes. The vertebral epiphyses remain separate to a late age. The neural spines and transverse processes are well developed for the attachment of muscles in a typically mammalian manner giving a dorsoventral movement of the body, this being in contrast with the lateral movement produced by the segmental myotome arrangement of fishes. The neck is very short and the cervical vertebrae partly or wholly fused together. The ribs, as in other aquatic vertebrates, are rounded and mobile; they are the chief agents of respiration, the diaphragm containing little muscle.

In the forelimb the humerus is short, the elbow-joint hardly mobile, and the hand increased in length and sometimes expanded (right whales, killer whales, river dolphins). The number of fingers is often reduced to four; the phalanges of some of the digits may be considerably increased in number (hyperphalangy). The scapula is flattened and there is no clavicle.

Some striking modifications are seen in the head. The skull shows a curious telescoping of the bones over each other. The premaxillae are elongated and the maxilla of toothed cetaceans extends above the frontal, combining with the latter to make a roof to the temporal fossa. Thus the maxilla almost reaches the supra-occipital. In baleen whales the backward prolongation of the maxilla is mainly *below* the supra-orbital process of the frontal, although there is a medial process extending dorsally towards the supra-occipital. This telescoping occurs only towards the end of foetal life. A further curious feature is that the skull in the toothed whales is asymmetrical, the vertex being shifted over to the left side; no satisfactory explanation for this phenomenon has yet been offered. The jaws are always greatly elongated, in the Greenland right whales they make up one-third of the total body length. The masticatory muscles and the coronoid process are reduced, the latter most extremely in *Balaena* in which it is distinguished only as an inconspicuous ridge (Fig. 26.1).

Whales are microsmatic or even in some species anosmatic (with no olfactory nerve). The brain case is therefore short and rounded while the nostrils have moved backwards, and open upwards. There are simple turbinals in baleen whales, none in odontocetes. The nasal bones have become reduced in length and no longer roof over the nasal cavity. The auditory region is much modified, the tympanic and periotic bones being surrounded by soft tissue and free from the rest of the skull. There is a large tympanic bulla, fused with the petrosal. Extensions of the middle ear cavity form pneumatic sacs, below the base of the skull, which serve to insulate sound and equilibrate the varying pressures experienced under water.

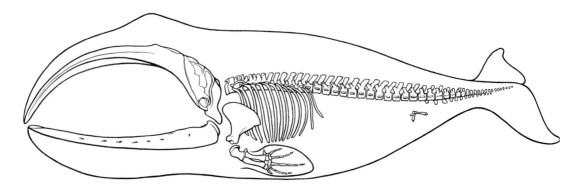

Fig. 26.1. Skeleton and outline of the right whale (*Balaena*). (After *British Museum Guide*.)

3. Feeding habits

The feeding arrangements provide further special features. Many of the toothed whales, such as the porpoises and dolphins, eat mainly fish and the teeth form a row of numerous (65/58), similar peg-like structures, usually in both jaws. With elongation of the jaws the masticatory function of the teeth has been reduced and they probably serve to hold the prey. *Orcinus*, the killer whale, has large powerful jaws and teeth and its diet includes dolphins, birds, seals, and the flesh of large whales. Squids are also often eaten, but in species that feed predominantly on these there is some tendency to a reduction in the number of teeth. For example, in the sperm whale functional teeth are confined to the lower jaw and in the bottle-nosed or beaked whales only one or two pairs of teeth are visible. The surfaces of the roof of the mouth and tongue are roughened to prevent the escape of squid.

In whalebone whales (Mysticeti) there are teeth only in the foetus. The food is plankton collected by the fringed baleen, which consists of rows of transverse plates of keratin. A fin whale feeds by a quick action lasting only 10 s in which it rolls on its right side close to the surface and scoops up the krill. Blue whales feed more continuously. The tongue of the right whales is powerfully muscular and forces the water from the mouth. In the rorquals the same purpose is achieved by the contraction of subcutaneous muscles associated with the external throat grooving. The shrimp-like krill (*Euphausia*) and other small organisms are then swallowed.

The stomach of whales has three or four parts. The fore-stomach is a muscular crop, often containing stones, lined by squamous epithelium and without glands. The main stomach has a folded mucosa and gastric glands. Thirdly there is a smooth pyloric stomach with few glands, leading to the duodenum. The intestine may be as much as sixteen times as long as the body.

Both types of nutrition are evidently efficient and the whales are abundant and of course very large, up to 24 times the weight of an adult elephant. It is more easy to see the advantage of large size for aquatic than for land animals. There are no problems of support of weight, and on the other hand a great premium is placed on large size by the fact that skin friction is thereby relatively reduced, and this, which forms but a small element in the work to be done by a land animal, must be important in the water. Further, the heat loss is greatly reduced by the size, and this may be a large factor in cold water, with its high thermal conductivity. However, it has also been claimed that downward temperature

regulation may be a problem and that the flippers and fins act as 'radiators'.

4. Respiration and metabolism

The respiratory system shows some special developments in the air passages, lungs, and nostrils. There are valves for closing the nostrils during diving, and special cartilaginous rings and muscles in the bronchioles. The epiglottis is extended as a tube inserted into the posterior narial cavity so that an uninterrupted air passage is provided from the blow-hole to the lungs. The very elastic and extensible lungs can thus be quickly filled with large volumes of air. In spite of the enlarged tracheobronchial tree the respiratory surface is small, but there are special arrangements of valves and venous plexuses to ensure economical distribution of the air and blood. Sperm whales have been timed to remain submerged for 82 min. One has been observed on sonar to reach a depth of 2250 m. Other whales dive to lesser depths, fin whales to 500 m and only for 10 min. There is relatively little experimental evidence about the means by which whales obtain oxygen and resist compression. The whole arrangement ensures the taking down of a maximum of oxygen but deep divers have small lungs and probably empty them before a dive. There is rapid ventilation while on the surface, followed by slower heart-beat and presumably reduced tissue respiration while below, so that the whale can take down enough oxygen to last throughout its dives. However, the heart rate slows only to one-half in the only cetacean fully investigated (*Tursiops*) and there is no evidence about retention in venous sinuses or other means of reducing circulation such as are found in seals (p. 512). The respiratory centre in the medulla has a lesser sensitivity to CO_2 than in land animals.

Besides the air in the lungs there may also be some provision for storage of extra oxygen in the large blood-volume of the retea mirabile, networks of blood-vessels, which abound throughout the body, especially in the thorax. The function of the retea mirabile is still uncertain but is probably connected with the accommodation of the animal to varying hydrostatic pressures and temperatures. They expand and contract to occupy the space in the thorax as the air in the lung is diminished or increased as the animal rises or descends while swimming. There is much myoglobin in the muscles, which can also act anaerobically for longer than in terrestrial mammals. The brain is supplied with blood entirely from meningeal arteries, which draw on the thoracic retia. The basilar artery and intracranial carotids close early.

No doubt the metabolism is also arranged to allow

accumulation of a high oxygen debt, but the special metabolic peculiarities that allow for this are not known. It is not necessary for whales to have a special defence against caisson sickness if they are using oxygen reserves. There is no continuous addition to the nitrogen dissolved in the blood such as would lead to the formation of the bubbles (bends) that occurs when miners or divers rise quickly after breathing air at great depths. The sudden expiration on surfacing produces a cloud of foetid vapour, the blow. This is generally supposed to be due to condensation, and it is very much less pronounced in hot air.

5. Brain and receptors

The brain is absolutely larger in whales than in any other animals (up to 9.2 kg), and the hemispheres are elaborately folded. The cerebellum is very large. The olfactory nerves and bulbs are very small in mysticetes and present only in the foetus in toothed whales. The inferior colliculi are as much as four times larger than the superior, no doubt in relation to the sense of hearing (as in bats). The amygdaloid complex is especially large, as it is also in bats, perhaps somehow in relation to echolocation. The cerebral cortex shows many differences from that of other mammals, especially the presence of a unique paralimbic lobe containing areas of both motor and granular (? somesthetic) type.

The eyes are small relative to the body (vestigial in *Platanista*) and in all whales much modified for aquatic life and diving. The cornea is more flattened than in subaerial mammals, and the lens rounded; in these respects the whales have returned to fish-like conditions. The whale eye is enclosed in a thickened sclera and further has special lid muscles. The tear glands and their duct are absent; instead, the surface of the eye is protected by a special fatty secretion of the Harderian glands.

There are few or no taste buds. The trigeminal and auditory are much the largest cranial nerves and provide the major receptor systems. The lateral lemniscus, superior olive, inferior colliculus, and medial geniculate are all very large. Presumably much of the cortex serves the sense of hearing. The apparatus concerned with reception of air-borne vibrations is reduced. The external opening is very small and the long meatus is often filled with a waxy plug. The tympanum is thick and ligamentous. It is to a normal ear drum as a closed umbrella is to an open one. The end of the tympanic 'ligament' is attached only to the tip of manubrium mallei. The distal end of the processus gracilis of the malleus is fused to the adjoining bone of the tympanic bulla. The ossicles have articulations with one another as in terrestrial mammals and the tip of the stapes is movable in the foramen ovale.

The petrotympanic bone or tympanic bulla is cowrie-shaped and made of very heavy bone which does not resonate. It is free from the skull and rests on a thick fibrous pad and is otherwise almost completely enveloped in a system of foam-filled air sinuses. The whole arrangement is believed to be designed to isolate the essential organ of hearing from vibrations extraneous to those reaching it by means of the meatus and auditory ossicles, and so to provide the means for directional hearing. Adjustments of amplitude and pressure to values normally experienced in the cochlea by terrestrial mammals are achieved in cetaceans by the modifications of the middle ear mechanism (Fraser and Purves 1960). Whales undoubtedly have a very acute sense of hearing and the sound waves are probably conducted by the wax plug (see Gaskin 1978).

6. Sound production and communication (whale song)

Whales emit sounds for various purposes. A bottle-nosed dolphin *Tursiops* in a tank recognizes the sex of a new arrival in another tank, out of sight. Several such social reactions have been reported, for example, between mother and young. *Tursiops* can also avoid obstacles in the dark. They scan by emitting intense trains of broad-band clicks. Some with low frequency (0.25–1 kHz) are orientation clicks providing a general profile of the environment, whereas discrimination clicks at 2–220 kHz identify details. The origin of the sounds is disputed. Some believe they arise from nasal diverticula around the blow-hole, others from the larynx, which lies far forward and has complex folds and muscles but no vocal chords.

Mysticete whales also produce sounds and blue whales and minke whales emit narrow-band pulses up to 30 kHz but these are not directional like those of delphinoids. Humpback whales (*Megaptera* = large wing) have a song lasting up to 35 min and repeated after surfacing (Winn and Winn 1978). This is produced day and night for months on end but only by unmated males. On the winter tropical calving banks the song serves to attract females and maintain contact, with a range of as much as 185 km. The songs consist of 15–20 syllables sung in a fixed sequence of six themes described as including (1) low frequency moans and mups, (2) yups, (3) higher frequency modulated notes, whos or wos, (4) long low frequency moans, with ees and oos, (5) even lower frequency snores, and finally (6) a ratchet cry on surfacing. Other sounds were included in each phrase and each was repeated several times. Differences were observed in consecutive years, suggesting dialects identifying herds and individuals. The high redundancy of the song perhaps avoids habituation and so ensures reception.

Fin whales (*Balaenoptera*) produce low-frequency sounds (20 Hz) of one second duration several times per minute. They are very loud for animal noises (75–80 dB). In the sea sounds below 100 Hz have very low attenuation and travel great distances, especially in the SOFAR Channel at 1000 m. In the pre-propeller ocean whale sound could spread spherically for 800 km or more. However, only sperm whales dive as deep as the SOFAR Channel. The baleen whales are able to make especially good use of sound. They have sufficient ear separation to check direction and are also reasonably predator-free or predators could home in on the signals. All whales of the same species probably use the same frequency and there is little individual communication. Herd cohesion is almost certainly much more important; for example suppose feeding whales vocalize and whales searching for food are silent – then this could bring them together in areas containing food. The actual usable range is likely to be from 72–800 km and 20 Hz is the ideal frequency for long-range signalling, it is very bad for use as sonar because of poor resolution. It lies just below storm-generated noise and no energy is lost on bottom reflection. It is the best frequency for polar oceans and under-ice conditions. No one knows for certain if fin whales can hear 20 Hz, but if they cannot then why do they make the noises? (Payne and Webb 1971).

7. Behaviour

The behaviour of whales is undoubtedly elaborate, involving social life, communication by sound, and probably much learning. There is co-operation between individuals in helping to keep a wounded companion or a new-born at the surface. Killer whales (*Orcinus*) hunt in packs and will tip up an ice floe to dislodge their prey. Play is common and rhythmical 'dancing' has been observed and also homosexual behaviour and masturbation in captivity.

Many species migrate, exploiting the seasonal productivity of waters in high latitudes but returning to breed in warm waters with lower but constant productivity. In grey whales (*Eschrichtius*) the orientation and navigation are achieved largely by following bottom topography parallel to coast lines. When arriving at an anomaly whales show 'spy-hopping', rising with the head clear of the water and looking around, perhaps for a coast line. The orientation of truly pelagic species (such as blue or sperm whales) may be by the sun (or moon) and there are suggestions that it involves learned behaviour transmitted from one generation to another (see Gaskin 1978).

8. Reproduction

The reproduction shows various modifications for aquatic life. The testes do not descend into sacs, but remain close to the kidneys. The penis is very long (up to 3 m), and curled when not erect. The uterus is bicornuate, but usually only one young is carried and is retained for a long time (more than a year in large whales), so that it is as much as a third of the length of the mother at birth. The placenta is diffuse but with a few villosities like the cotyledons of ungulates. Its structure is epitheliochorial and there is a large allantois. Birth takes place under water. There is a pair of teats in the inguinal region and the mammary glands are provided with many myo-epithelial cells so that milk is pumped into the mouth of the young. The milk contains 50 per cent of fat and is concentrated, economizing water and allowing rapid growth. The larger whales probably reproduce at about 12 years of age and may live to 60, porpoises at 15.

9. Evolution and diversity

It is clear that many factors have collaborated to concentrate the biomass of whale life into large units. Indeed, whales include the largest known animals, either fossil or recent. The blue whale, *Balaenoptera musculus* (Fig. 26.2), reaches nearly 120 tonnes with a length of 33.5 m. This is one of the whalebone whales, which are in general larger than the odontocetes, perhaps because of the immense sources of food directly available in the plankton; they have grown fat by eliminating the 'middle-men' upon which all toothed whales must feed. These mysticetes appeared in the Oligocene and radiated in the Miocene and since into a

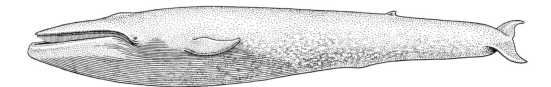

FIG. 26.2. Blue whale, *Balaenoptera*. (After Mackintosh, N.A. and Wheeler, J.F.G. (1929). *Discovery Report* **1**, 257–540.)

relatively small number of types, all of large size. *Balaena*, the right whale of the Arctic, is now very rare. The chief prey of the whalers during the 1970s have been sperm whales (*Physeter catodon*) in the Antarctic and both sperm and minke whale (*Balaenoptera acutorostrata*) elsewhere.

The odontocete whales are a more varied group; their history can be traced back to the late Eocene. The squalodonts of the Oligocene and Miocene were like the porpoises, but with small brains and as many as 180 triangular shark-like teeth. Most of them disappeared in the Miocene, but the river-porpoise *Platanista* of the Ganges and Indus, and related forms in the Amazon and Yangtse may be descendants. *Platanista* has over 50 slender pointed teeth and the eye is small with no lens. The cortex is simple but the cerebellum very large. They swim continually on the side, testing the bottom with one flipper and emitting pulsed clicks of up to 200 kHz whose echoes are used to explore the surroundings and to find fish in muddy water.

The modern porpoises and dolphins (Delphinoidea) are a very successful and numerous group of relatively small animals, with a dorsal fin and teeth in both jaws. They are nearly all active predators, but the habits vary from those of the killer whale *Orcinus* (Fig. 26.3), which is a fierce and cunning hunter, attacking even the largest whales, to the omnivorous porpoise *Phocaena* (Fig. 26.4), whose food includes crustacea as well as fishes and cephalopods. This is the commonest and smallest British cetacean, the largest individuals reaching 180 cm. The jaws are rather short especially the upper one, and the teeth are spade-shaped. The true dolphins, *Delphinus* (Fig. 26.5), are larger animals (240 cm), living mostly on fish and having long jaws with many conical teeth. The upper jaw forms a beak which

FIG. 26.4. Porpoise, *Phocaena*. (After *British Museum Guide*.)

distinguishes them from porpoises. They are very fast swimmers reaching 37 km/h. The narwhal, *Monodon*, sometimes placed in a separate superfamily (Fig. 26.8), is a large whale up to 600 cm long. There is a single tooth, usually the upper incisor (up to 300 cm) in the male. It grows continually to make the spirally twisted horn, up to 270 cm long; its use is not known. The female retains a pair of small incisors buried in the premaxillae. The long-backed dolphin (*Sousa*) feeds on vegetation in rivers in Senegal.

In the sperm whales, *Physeter catodon*, up to 20.7 m long, the rostrum is overlaid by an enormous reservoir containing spermaceti, a highly vascular oil-filled connective tissue (Fig. 26.6). The oil is an unusual mixture of fatty esters solidifying when cooled below 32 °C. There have been various suggestions for the function of the spermaceti. It may serve to focus sound waves during echolocation. A recent theory is that it provides control of buoyancy by cooling and hence decreasing its volume and increasing its density. A drop of 3 °C would allow neutral buoyancy down to depths of 2000 m or more to which these whales can dive. The heat loss may be through blood vessels from the spermaceti to the skin of the head, perhaps with a sort of blushing, but possibly by taking cool water into the right narial passage. The arteries and veins in the snout lie side by side, and this counter-current system allows arteries supplying blood to the spermaceti to be surrounded by a network of veins carrying cooler blood away. Sufficient heat to melt the

FIG. 26.3. Killer whale, *Orcinus*. (After *British Museum Guide*.)

FIG. 26.5. Dolphin, *Delphinus*. (After *British Museum Guide*.)

FIG. 26.6. Sperm whale. *Physeter*. (After Flower and Lydekker 1891.)

FIG. 26.7. Bottle-nosed whale. *Hyperoodon.* (After Norman and Fraser.)

FIG. 26.8. Narwhal, *Monodon.* (After Norman and Fraser.)

oil on rising is produced during a dive and stored in the muscles (Clarke 1978, 1979).

Sperm whales can remain submerged for over an hour and whalers have long known that they surface close to where they dived. They probably economize energy by moving little while they feed, perhaps largely on schools of spawning squid including the giant squid *Architeuthis*, whose suckers leave scars on them. They have functional teeth only in the lower jaw and vestigial ones in the upper. The jaws of squid combine with secretions in the intestine to make ambergris, a substance used as an absorbent in the manufacture of scent. The bottle-nosed whales, such as *Hyperoodon*, of polar seas, dives deeply, feeds upon cephalopods and also has oil in the head (Fig. 26.7). They have only a single pair of large conical teeth.

The whales probably arose from creodonts that took

to the water; *Protocetus* from the middle Eocene has a full eutherian dentition with sharp molars very like that of the creodonts (p. 513). These Eocene whales were different from either of the modern groups and are placed in a separate suborder *Archaeoceti*. The molars had sharp crenated edges. The hindlegs had already disappeared, though vestiges of their skeleton remained. The skull was long, and the nostril had moved some way back. The teeth were of the normal mammalian number, 44, and were heterodont. Casts of the brain show large olfactory centres but a small, little folded, cortex. The cerebellum was enormous (if it has been correctly reconstructed). In *Basilosaurus* (= *Zeuglodon*), from the upper Eocene (Fig. 26.9) the body was very long (up to 2130 cm) and apparently thinner than in modern whales, suggesting a sea-serpent form.

Evidently the whales had developed a long way from

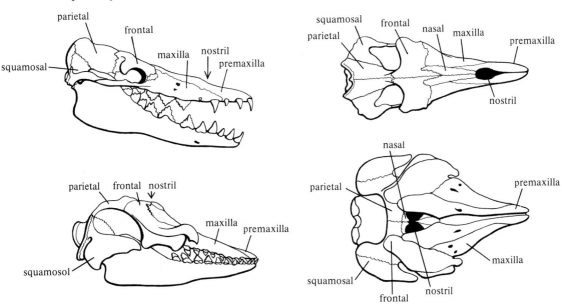

FIG. 26.9. Change in position of the blow-hole (nostril) during the evolution of whales. The two upper figures show the condition in the Eocene *Basilosaurus*, the lower figures a Miocene squalodont. (After Romer 1945.)

the main eutherian stock by middle Eocene times. This is an example of relatively rapid evolutionary change; it may be presumed that their ancestors had been in the condition of small insectivores not much later than the end of the Cretaceous, at the most 20 million years earlier. The basilosaur type persisted to the early Miocene, but the exact connections with the two modern sorts of whale are not clear. Whales have been abundant throughout the Tertiary, but there is not yet sufficient evidence available to reconstruct their full phylogenetic history. In broad outlines, however, it is clear that there has been a progressive adoption of features suitable for aquatic life, the long-bodied, heterodont basilosaurs giving place to the shorter, streamlined modern whales, provided with suitable stabilizing fins and with the mouth highly specialized for eating fish, cephalopods, or plankton. It is impossible to say whether the changing of the populations was due to indirect influences of climatic changes or to competition within the animal populations themselves. The course of

whale evolution, like that of teleosts, appears to have produced increasing efficiency within a single habitat, rather than a progressive colonization of new fields; but this appearance may be only a reflection of our ignorance and lack of knowledge of the varied and changing condition of the sea.

The depletion of the stocks of whales by man is notorious. The great blue whale (*Balaenoptera musculus*) may be down to 7000 individuals but since protection there are signs of an increase in the Southern Ocean. There are perhaps some 80 000 fin whales (*B. physalus*) but only 5000 humpbacks (*Megaptera*). The population of sperm whales (*Physeter*) is larger (500 000). It is uncertain whether they are in immediate danger (Gaskin 1978). Spermaceti is used for high quality engine oils (in gear boxes) and other purposes. Over 23 000 sperm whales were killed in 1961–2 but only 9921 were killed in 1978–9 in the Antarctic and 1979 elsewhere (see Allen 1980).

27 Carnivores

1. Affinities of carnivores and ungulates: Cohort Ferungulata

THE union of the modern carnivores and hoofed animals in a single cohort Ferungulata advocated by the American palaeontologist Simpson (1945) was based on the suggestion that both groups, together with some isolated surviving types such as the elephants and sea cows, and many other forms now extinct, all arose from a common population in Palaeocene times. The most ancient members of this group were the Condylarthra of the Palaeocene and Eocene, which gave rise to the Perissodactyla and probably also to the Artiodactyla. Very similar to the condylarths were a number of families classed together as Creodonta, whose evolution paralleled that of the carnivores (Fig. 27.1). It is a convenience to use these possible relationships as a basis for classification, but it must be recognized that modern carnivores have little more in common with ruminants than with, say, monkeys or rats. The three great groups that are joined in the Ferungulata diverged from each other only a short time after they had separated from the common stock of the other mammals; at that time all eutherians were so alike that we should probably place them in a single order if they had left no descendants.

The Ferungulata (if they are a real group) have become much more diversified than the other cohorts into which we have divided the Eutheria and we have to recognize no fewer than fifteen orders in the group. It is therefore convenient to make further subdivision into five superorders. The first of these, Ferae, makes the central group, including the Carnivora and the extinct Creodonta. The second superorder, Protoungulata, includes the earliest ungulates, the condylarths, and it is convenient to place here also certain early offshoots, such as the South American ungulates and one living possible survivor, the aardvark. The third superorder is known as Paenungulata ('near ungulates') and includes a group of orders with rather primitive organization, most of them extinct. The elephants, hyraxes, and sea cows remain as isolated vestiges of this great group, which included the huge pantodonts and dino-

cerates, formerly classed together as amblypods, also the pyrotheres, elephant-like animals from South America, and the embrithopods, large, horned animals from Africa. These large animals have only been grouped together because they are graviportal; they are probably not closely related. The fourth superorder, Mesaxonia, includes only the Perissodactyla, descended from the

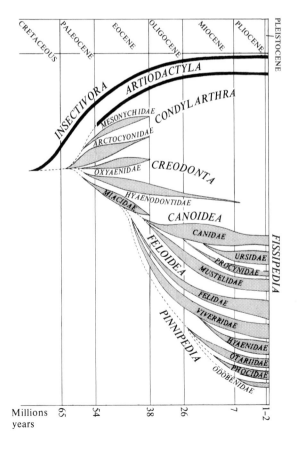

FIG. 27.1. Chart of the evolution of Carnivora.

condylarths, while the fifth superorder, Paraxonia, contains the artiodactyls, derived at the same time also from the condylarths.

Of all this assembly of various types only the car-

nivores and ruminant artiodactyls and some perissodactyls remain successful types at the present day abundant in species and individuals, the remaining orders are either extinct or are represented only by a few species.

2. Classification

Cohort 4: Ferungulata
  Superorder 1: Ferae
    Order 1: Carnivora
      Suborder 1: Fissipeda. Palaeocene–Recent
        *Superfamily 1: Miacoidea. Palaeocene–Eocene, Holarctic
          *Family Miacidae
            Didymictis, North America; *Miacis*, Holarctic
        Superfamily 2: Canoidea
          Family 1: Canidae. Dogs. Eocene–Recent
            Cynodictis, Europe; *Canis*, wolves, dogs, jackals, world-wide, but not originally wild in South America or Australia; *Vulpes*, fox, Holarctic, North Africa; *Dusicyon*, South American foxes; *Lycaon*, hunting dog, Africa
          Family 2: Ursidae. Miocene–Recent
            Ursus, bear, Holarctic
          Family 3: Procyonidae. Miocene–Recent
            Procyon, raccoon; North and South America; *Ailurus*, panda, Asia, *Ailuropoda*, giant panda, Asia; *Nasua*, coati, America
          Family 4: Mustelidae. Oligocene–Recent
            Mustela, weasel, ferret, stoat, Holarctic, South America, North Africa; *Meles*, badger, Eurasia; *Taxidea*, American badger, North America; *Mephitis*, skunk, Recent, North America; *Lutra*, otter, Holarctic, South America, Africa; *Martes*, marten, Holarctic
        Superfamily 3: Feloidea
          Family 1: Viverridae. Oligocene–Recent
            Viverra, civet, Asia; *Herpestes*, mongoose, Eurasia, Africa, and introduced to West Indies
          Family 2: Hyaenidae. Miocene–Recent. Eurasia, Africa
            Hyaena, hyena, Asia, Africa
          Family 3: Felidae. Upper Eocene–Recent
            Felis, cats, pumas, ocelots, leopards, lions, tigers, jaguars, world-wide except Australasia; *Hoplophoneus*, sabre-tooth; *Smilodon*, sabre-tooth
      Suborder 3: Pinnipedia
          Family 1: Otariidae. Miocene–Recent
            Eumetopias, sea lion, Atlantic and Pacific
          Family 2: Odobenidae. Walruses. Miocene–Recent
            Odobenus, walrus, Arctic
          Family 3: Phocidae. Seals. Miocene–Recent
            Phoca, seal, Atlantic and Pacific; *Halichoerus*, grey seal, North Atlantic
    Order 2: Creodonta. Palaeocene–Pliocene. Holarctic
        *Family 1: Oxyaenidae. Palaeocene–Eocene
           Oxyaena; *Patriofelis*; *Palaeonictis*
        *Family 2: Hyaenodontidae. Eocene–Pliocene
           Hyaenodon; *Apataelurus*; *Sinopa*

3. Order Carnivora

The earliest Cretaceous mammals were probably insectivorous and it is not therefore surprising that some of their descendants became flesh-eating; indeed it is curious that a single stock has provided nearly all the

hunters found among the mammals ever since, though the marsupials have produced carnivorous types in South America and Australasia. It is difficult to see why carnivores have not developed more often from the insectivoran or some other stock; that they have not done so may remind us that special circumstances are necessary for the origin even of a type for which a means of life would seem to be readily available.

The order is divided into distinct suborders, Fissipeda for the mainly land animals and Pinnipedia for the seals, sea lions, and walruses.

4. The cats

The changes that convert a mammal into an effective hunter occur in many parts of the body, without, as it were, radically distorting any (see Ewer 1973). We may illustrate this by considering the most specialized members of the group, the cats (Fig. 27.2). The head is large, with long ears, long whiskers, and nose with many turbinals. The brain is large, the cerebral hemispheres overlap the cerebellum; the olfactory centres are large. As is usual with carnivores, behaviour is complicated. In order to continue pursuit of prey that cannot be seen, or perhaps even smelt, the animals learn to associate the presence of food with obscure clues such as footmarks, and to make use of these clues they must lie in wait for the prey. All of this involves an elaborate balance of internal motivation with activity and restraint. This power of 'abstraction' of ultimate satisfaction from the immediate situation may perhaps be associated with the familiar play of the kitten or the less edifying treatment of a captured mouse by an adult cat.

Social or family groups are commonly well marked in carnivores and there are usually characteristic odours for recognition, often associated with large anal glands, especially well known as producers of civet and the 'poison' of the skunks.

The back of the head is enlarged to take the brain and there is a well developed snout for the nose, but the face is nevertheless short. It is characteristic of the specialized carnivores that the tooth-row is shortened, but developed especially at the front end, producing the incisors for piercing, canines for tearing, and premolars

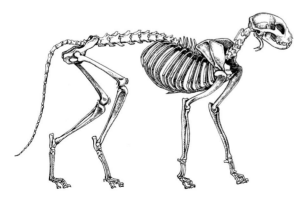

FIG. 27.2. Skeleton of the cat (*Felis*).

and anterior molars for cutting. In contrast to the ungulate type of dentition the hinder molars, not being needed for grinding, are reduced. In the cats, as in all modern carnivores except seals, the teeth most favourably placed for biting by their position relative to the jaw muscles, namely, the last upper premolar and first lower molar are specially developed into cutting blades, the carnassials. This is done by formation of a ridge along the outer side of the upper fourth premolar making a single cutting-edge. The protocone remains as an inwardly projecting ridge at the front of the tooth, which otherwise makes a single blade, shearing outside a similar blade formed by the paraconid and protoconid of the lower molar. This restriction to long sharp ridges also affects the teeth in front of the carnassials, but behind them the molars are so reduced that in true cats they are represented only by a single vestige in each jaw. More of the posterior molars remain in some carnivores for example the dogs, and in some, such as the bears, they may acquire a bunodont surface and hence the power of grinding.

The jaws are, of course, powerful in carnivores, the articulation being a tight, transverse hinge, allowing none of the rotatory movements found in other mammals. The jaw muscles include especially powerful temporals, for whose attachment there is a large coronoid process on the jaw and often large sagittal crests on the top of the skull. The temporal fossa is very wide and never closed off from the orbit, since there is no need for specially increased surfaces for the masseter, which is only moderately strong. The pterygoid muscles (and their fossa) are reduced, since the jaw has no rotary action.

The post-cranial skeleton shows a generalized mammalian build, with specializations for sudden leaping movements. The gait is often plantigrade but in the more cursorial canids it becomes digitigrade, like that of the ungulates that they chase. There are five digits in the hand and four in the foot of cats; in other carnivores the number is never less than four. The toes are armed with the characteristic claws, which are held drawn back by elastic ligaments and pulled out when needed, by the action of the flexor digitorum profundus muscles on the terminal phalanx, to which the claw is attached. The weight of the body is carried on special pads on the second interphalangeal joints and the metatarsal heads. The carpus of all modern carnivores is stabilized by fusion of the scaphoid and lunate bones. The arrangement of the limbs and backbone is that of a quadruped able to proceed over uneven surfaces and also to move steeply upwards and downwards, especially in the carnivores that are arboreal. Great speed is needed by those that hunt ungulates, the cheetah is said to be able to reach 112 km/h. Most carnivores have a long body, with much of the weight carried on the forelimbs; the thoracic neural spines are therefore high. On the other hand, the vertebral girder has to take the strain of powerful sacrospinalis muscles for leaping; the transverse processes are broad in the lumbar regions, and again in the neck for the muscles that move the head. The clavicle is reduced. The tail, well developed for the maintenance of balance in arboreal species is reduced in bears and pandas and under domestication, the extreme being the Manx cat, with only three caudal vertebrae.

As in so many carnivorous vertebrates the alimentary canal is short and the stomach never complex or the caecum large. The uterus retains the primitive mammalian bicornuate form. The chorio-vitelline placenta is important early in pregnancy. The chorio-allantoic placenta is of a type known as vasochorial with the villous portion of the chorion restricted to a characteristic band around the embryo (hence 'zonary' placenta).

5. Early carnivores. Superfamily Miacoidea

The creodonts who were for long considered to be ancestral to the carnivores are now thought to have been a parallel line (p. 505). The modern carnivores were probably separately derived from insectivoran ancestors and the earliest forms are included in the family Miacidae such as *Didymictis* from the middle Palaeocene and *Miacis* from the Eocene. They were long-bodied arboreal creatures with short legs and carnassial teeth formed by P^4 over M_1, these teeth being elongated, with cusps partly united to form ridges (Fig. 27.3). By the early Oligocene representatives of the modern fissiped families were present. *Cynodictis* was rather like a weasel with a long body and short legs. These animals are usually considered as ancestral dogs but may have been near the origin of all the modern families.

Fig. 27.4. Wolf, *Canis*. (From a photograph.)

6. Modern carnivores. Superfamily Canoidea

The dogs (Canidae) (Fig. 27.4) appeared very early and have changed relatively little since the Miocene (see Fox 1975). Numerous fossil dogs are known and the type has been very successful. The diet is varied and partly herbivorous. The European red fox, for instance, feeds on small mammals and birds, but also on snails, insects, and berries. The canids are not suited for climbing but for running in open country, for which purpose long legs and a digitigrade habit have been developed, the pollex and hallux being reduced (Fig. 20.1). The teeth have

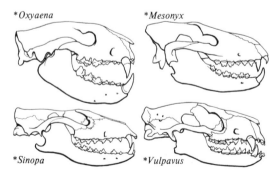

Fig. 27.3. The skulls of some early creodonts and carnivores. *Oxyaena*, *Mesonyx*, and *Sinopa* are creodonts and *Vulpavus* is a miacid fissipede. (After Romer, Scott, Wortmann, and Matthew.)

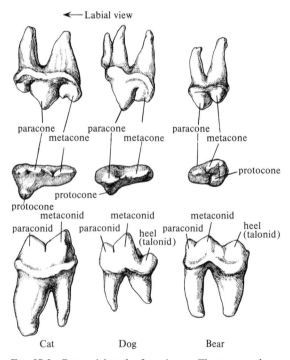

Fig. 27.5. Carnassial teeth of carnivores. The top row shows the last left upper premolar from the labial side; middle row the same from below; bottom row the first left lower molar from the labial side. (Partly after Flower and Lydekker 1891.)

Fɪɢ. 27.6. Brown bear, *Ursus*. (From life.)

remained unspecialized, with at least two post-carnassial grinding molars and still distinct signs of the triangular form in the carnassial (Fig. 27.5). Dogs, wolves, and foxes are found throughout the world, including South America, but not in Madagascar or New Zealand. So many types are known that the exact ancestry of the various wolves and foxes has not been fully disentangled.

The bears (Ursidae) (Fig. 27.6) were a Miocene offshoot from this dog stock and here the tendency to non-carnivorous diet became accentuated; there are no carnassials and the molars acquire bunodont grinding surfaces. The gait is plantigrade and the flat feet are suited for movement in rough or hilly country. Various types are found through the Holarctic region and South America but not in Africa. The raccoons (Procyonidae) (Fig. 27.7) are rather similar though smaller animals typically living in forests and often arboreal. The dentition is suited to an omnivorous diet, the upper carnassial having developed (? redeveloped) a hypocone. They are all American, except *Ailurus* and *Ailuropoda*, the panda and giant panda (Fig. 27.8),

Fɪɢ. 27.7. Raccoon, *Procyon*. (From a photograph.)

Fɪɢ. 27.8. Giant panda, *Ailuropoda*. (From a photograph.)

large, herbivorous creatures living in Asia. The latter has a special bone near the pollex, making a grasping organ for holding bamboo shoots.

Other carnivores that are included in the superfamily Canoidea are the stoats and weasels (Mustelidae) (Fig. 27.9). They can be recognized back to the Eocene. They have well developed carnassials and never more than one post-carnassial molar. Today they live mostly on rats and mice, rabbits, and other small herbivores. *Meles* the badger (Fig. 27.10), is omnivorous and has a hypocone; *Mephitis* the skunk (Fig. 27.11), a burrowing animal that ejects a stream of strong-smelling liquid from the anal glands. The otters, *Lutra* (Fig. 27.12), have webbed feet, short fur, small ears, and other features suited for life in water. In many mustelids there is delayed implantation of the blastocyst.

7. Superfamily Feloidea

These are the most modified carnivores. The civets and mongooses (Viverridae) (Fig. 27.13) are survivors that show us many of the characters possessed by feloids in the Oligocene. They are the small carnivores that

Fɪɢ. 27.9. Stoat, *Mustela*. (From a photograph.)

Fig. 27.10. Badger, *Meles*. (From photographs.)

Fig. 27.12. Otter, *Lutra*. (From photographs.)

occupy in the Old World tropics the position taken farther north by the weasels. In general they are like the ancestral miacids, with long skull, small brain, and short legs. *Herpestes*, the mongoose, is abundant throughout Africa and Asia.

The hyenas (Hyaenidae (Fig. 27.14) are large creatures, with massive teeth specialized for crushing bones and hence allowing a scavenging life. The aardwolf (*Proteles*) is related to the hyenas but has a reduced dentition and feeds on termites. The true cats (Felidae) were already differentiated from the miacid ancestry in the Oligocene, but at that early period they were mostly sabre-tooths with great development of the upper canines as cutting and piercing sabre-teeth (Figs. 27.15 and 27.16). There were numerous genera with this characteristic from the Eocene onwards until the Pleistocene, when they became extinct throughout the Holarctic region. Probably they attacked large thick-skinned herbivores. The jaw could be opened to a right angle to allow the fangs to strike, and there were such associated developments as large mastoid processes for the sternomastoid muscles that pulled the head downwards and forwards in the strike. The closing muscles of the jaw and the coronoid processes were, however, small in the sabre-tooths. The teeth were probably used to make penetrating wounds, especially in the throat, and then for slicing meat from the prey.

The modern cats, with smaller upper and larger lower canines, appeared in the Pliocene either from sabre-tooths or from unknown ancestors. They are very successful carnivores, partly arboreal and hence most common in tropical regions, where there are large forests. There is much difference in detailed habits between the many species of the family, as well as similarities in bony structure. Numerous attempts have been made to divide the group into genera and sub-genera, but alternatively they may be retained in a single genus *Felis*. Some authors separate the five big cats in a separate genus *Leo* (previously *Panthera*) (see Corbet and Hill 1980). Lions (*Felis leo*) have many distinct races in Africa and Asia and are mainly terrestrial, hunting in open country. Tigers (*F. tigris*) (Fig. 27.17) occur throughout Asia to Siberia, are usually solitary, and often frequent damp places. They also have many races, and in captivity lions and tigers can be crossed. Other cats are mostly smaller and more fully arboreal than the lion and tiger. The leopard (*F. pardus*) of Africa and Asia reaches 150 cm in body length. *F. silvestris*, the wild cat, formerly common, still exists in Britain and has a Palearctic distribution. The domestic cat (*F. catus*) may have first been bred in Egypt from the African subspecies *F. silvestris libyca*. The jaguar (*F. onca*), puma (*F. cougar*), and ocelot (*F. pardalis*) are large Central and North American cats, which have recently invaded South America.

Felis is thus one of the most widespread of all mammalian genera, occurring in all main parts of the

Fig. 27.11. Common skunk, *Mephitis*. (From a photograph.)

Fig. 27.13. Indian mongoose, *Herpestes*. (After a photograph by F.W. Bond.)

Fig. 27.14. Hyena, *Hyaena*. (From photographs.)

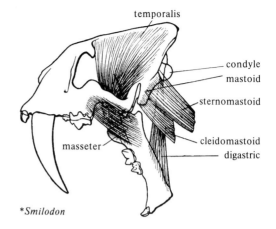

Smilodon

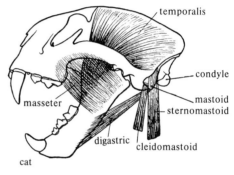

cat

Fig. 27.16. Neck and jaw muscles of sabre-tooth, *Smilodon, compared with those of a modern cat to show modifications for striking with the whole head and biting, respectively.

world except Australasia, Madagascar, and oceanic islands. Because of their striking and familiar characteristics we can form a vivid picture of all these slightly different sorts of cat, pursuing varying prey in the different regions. In order to visualize past evolution properly we should need to have an equally detailed knowledge of past populations. It is difficult enough to classify and describe a modern population of this sort and we need not be surprised that the palaeontologist, who has to consider also variations with time, finds insuperable problems of classification.

8. Suborder Pinnipedia

The seals, sea-lions, and walruses are marine carnivores that have existed since the Miocene. Their affinities with other carnivores are uncertain. The seals (Phocidae) may be a distinct group only distantly related to others. The sea-lions (Otariidae) and walruses (Odobenidae) show some similarity to bears (see Ronald and Mansfield 1975).

In the seals (*Phocidae*) the streamlined body is covered with thick fur, below which is a thin epidermis and a thick layer of blubber, making a quarter of the weight of the animal. Swimming is by means of the paddle-like limbs and flexion of the whole body, there is webbing between the digits, and the tail is reduced to a short rudiment. The basal segments of the limbs are shortened and some of the digits lengthened, without any increase in their number, though there are some

extra phalanges. The speed of swimming may reach 27 km/h if the seal is frightened. The cervical vertebrae are massive, with complex articulations, but the hinder ones are simplified and the column very flexible, so that it can be bent dorsally or laterally, allowing sudden

Fig. 27.15. Sabre-tooth, *Smilodon. (After Scott 1913.)

Fig. 27.17. Tiger, *Felis*. (From photographs.)

turns in the water and complicated balancing feats on land (Fig. 27.18). All the seals leave the water to breed and therefore need some support for their large bodies.

The teeth show a reduction in number and are rows of nearly similar, laterally compressed cones. They may carry three cusps in a row, a reversion to 'reptilian' conditions, which serves to prevent escape of the slippery prey. There are large canines. The milk dentition is lost very early, sometimes *in utero*. The intestine is long. Their supply of fluid is obtained from metabolic water.

The pinnipeds resemble the whales in being microsmatic and have good eyes, with flat cornea, round lens, and a muscular palpebral sphincter. The eyes are directed upwards and prey is often caught from below. The external ears are reduced but hearing is probably acute; the auditory ossicles are massive and there is a large cochlear aqueduct, perhaps transmitting sound energy. Seals hear better in water than humans do, the sound entering through an area ventral to the ear orifice. They probably do not use a sonar system for echolocation. There are numerous large vibrissae on the muzzle. The brain is large and rounded, with convoluted hemispheres and large midbrain and cerebellum. There is an elaborate vocal communication system, the calls varying to human ears from booming to chirping. Male walruses sing a repetitive song when courting.

Young seals can remain submerged for up to 25 minutes and have been shown to be able to stand a pressure equivalent to a dive to 95 m. Larger seals can remain submerged even longer and at greater depths. The nostrils are closed by special muscles. The lungs are large and the bronchi contain myoelastic valves. During a dive the heart slows from 120 to 4 beats a minute. There is no drop in blood pressure, because of a widespread reflex vasoconstriction, which prevents blood reaching the tissues, except the brain and heart muscle. Blood from the brain returns to the abdomen, by a large vessel above the spinal cord, and then accumulates there in extensive sinuses, including a huge dilatation of the vena cava above the liver, which is held shut by a sphincter of striated muscle above the diaphragm. There are few true rete mirabile but abundant venous plexuses. The blood can carry as much

as 35 ml of oxygen for 100 ml of blood (20 ml in man under the same conditions), and there is much myoglobin in the muscles. The respiratory centre tolerates a high CO_2 level. Lactic acid accumulates in the muscles, reducing metabolic levels. By such means the animal is provided with sufficient oxygen for the dive, without absorbing nitrogen and risking 'bends'.

Copulation takes place in the water in most seals and to aid this the penis bone is very large, especially in the walrus. The external genitalia, like the nipples, are withdrawn into folds of the surface. The eggs are fertilized shortly after parturition during a post-partum oestrus, the two horns of the uterus carrying alternate pregnancies. Implantation of the blastocyst is delayed for two months or more. As in other carnivores the placenta is zonary, with coloured margins due to the presence of bilirubin.

The Phocidae are the modern seals, found in all seas and fully aquatic, the hindlimbs being attached to the tail. They come ashore only for short periods in isolated places to breed, and can only just drag themselves along the beaches. Many seals migrate for long distances to particular breeding places, such as the Pribilof Islands in the case of the fur seals (*Callorhinus*). The males, arriving first, fight furiously with each other, the victors then collecting harems of twenty or more females, who give birth to their young and are soon afterwards impregnated again. The bulls remain on shore without feeding, guarding the family, while the females return to suckle the young at each tide for a period of about three weeks.

In the sea-lions (Otariidae) (Fig. 27.19) the legs can still be turned forward for use on land, and there are other unmodified features, including external ears. They are more mobile on land than are the seals and can even climb cliffs. The family dates back to the Miocene. The walrus, *Odobenus*, is a related form, highly specialized for eating oysters and other bottom-living molluscs, which it digs up with its enormous canines.

9. Order *Creodonta (= flesh tooth)

This order includes a number of extinct animals that in the early Tertiary evolved several types of flesh eaters similar to those that appeared later in the modern

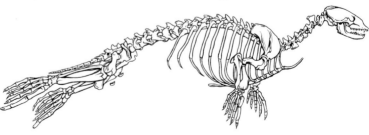

FIG. 27.18. Skeleton of the seal, *Phoca*. (After Blainville.)

FIG. 27.19. Sea-lion, *Eumetopias*. (With permission from the Zoological Society of London.)

Carnivora. Two related families, Oxyaenidae and Hyaenodontidae, now make up the group, two others having been removed to the Condylarthra (p. 517) (Fig. 27.1). They had carnassial teeth but these were formed from molar teeth, either M^1 over M_2 or M^2 over M_3. The creodonts lacked the characteristic ossified bulla and fused carpal bones of the carnivores. They had small brains.

*Oxyaena (Fig. 27.3) from the Eocene was like a large weasel and *Patriofelis* as big as a bear. *Paleonictis* was like a cat. The Hyaenodontidae appeared in the Eocene and some lasted to the Pliocene. *Sinopa* from the Eocene was weasel-like (Fig. 27.3), *Hyaenodon* and its allies were hyaena-like animals of various sizes. *Apataelurus* was an Eocene sabre-tooth. This array of early carnivore types constitutes a remarkable example of evolution parallel to that of the later carnivores. The creodonts flourished in the Eocene but then declined. Perhaps with their small brains and slow gait, though able to catch the cumbrous early herbivores, they were unable to make a living from the later, faster-moving ungulates.

28 Protoungulates

1. Origin of the ungulates

DURING the Palaeocene, or perhaps even earlier, a number of mammals abandoned the insectivorous habit and began to eat plants. The earliest of them are grouped together in the order Condylarthra. They rapidly radiated into numerous types, so that by the end of the Palaeocene several distinct orders descended from this stock can be recognized. In North America and elsewhere there appeared large, clumsy animals, the *Pantodonta (Amblypoda) and *Dinocerata, while in South America a special fauna, the *Notoungulata and *Litopterna, developed. Further types then arose in the Eocene, including the early elephants and Perissodactyla. The Artiodactyla first appeared in the lower Eocene, as rather pig-like creatures; their origin is uncertain but they may have come from some form such as an arctocyonid condylarth near *Tricentes.

During the Eocene and Oligocene there were, therefore, numerous large, heavy-bodied herbivorous mammals, perhaps mainly suited to forest life and living upon relatively soft green food, since their teeth were mostly not highly developed for grinding. They were, however, gradually replaced during the Miocene by swifter, grazing animals suited to the plains of that period.

Following Simpson (1945) we shall classify the numerous orders of herbivorous (ungulate) mammals into four superorders. The Protoungulata include the oldest forms and various early offshoots placed in the orders *Condylarthra, *Litopterna, *Notoungulata, and *Astrapotheria. One living creature, the aardvark or cape ant-eater, seems to retain some of the earliest characteristics and is included here in the order Tubulidentata (= tube-toothed). A second superorder, Paenungulata, or subungulates, includes a number of descendants of the condylarths that early achieved success, the *Pantodonta and *Dinocerata of the Holarctic region, *Pyrotheria of South America, and *Embrithopoda in Africa. With these are placed the Proboscidea, which succeeded them as large herbivores in the Miocene. The conies (Hyracoidea) are an isolated group that still shows some of the Eocene characteristics of this Paenungulate group, and the sea cows (Sirenia) are an early offshoot that took to aquatic life. This is obviously a very mixed group and it may be that its members are not really more closely related to each other than to other ungulate orders. Finally, the orders Perissodactyla and Artiodactyla occupy two separate superorders, Mesaxonia and Paraxonia.

2. Ungulate characters

When mammals adopt a herbivorous diet they assume certain characteristics, which it is convenient to recognize before dealing with the individual groups (Fig. 28.1). The animals often become large, but it must be remembered that, outside the ungulates, the rodents and lagomorphs include successful small herbivores, conversely among the ungulates the hyraxes are small. The skin is often thick and a variety of protective coloration schemes of spots and stripes appear, the underside usually being paler to eliminate shadows.

Defensive weapons like tusks or horns are common, but the problem of security often leads an unaggressive animal to the development of a swift gait. For this the limbs are lengthened by raising up on the toes (p. 432), producing first digitigrade and then unguligrade locomotion. When this happens the more lateral digits, failing to reach the ground, become reduced and may disappear, leaving finally the characteristic one or two. The movement of the limb becomes restricted to a fore-and-aft direction, and the joints assume a pulley-like form, especially characteristic in the trochlea of the talus, deeply grooved in artiodactyls but markedly so also in perissodactyls (Fig. 28.2). The carpal and tarsal bones of these swift-moving animals become arranged on the so-called interlocking plan, by which each elongated metapodial thrusts up against two carpals or tarsals. No movements of pronation occur, and the ulna and fibula tend to be reduced and fused with the radius and tibia. The hoofs themselves are a characteristic development, the terminal phalanx is broadened, and the claw becomes modified to surround it, while a pad

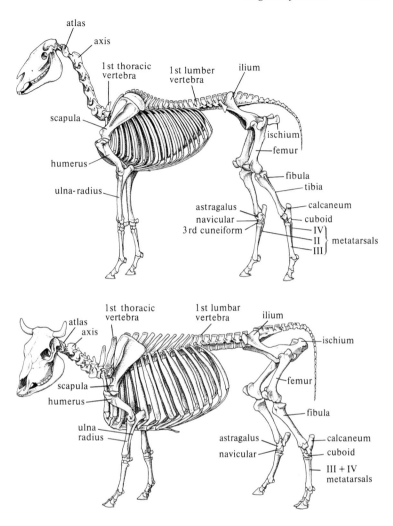

FIG. 28.1. Skeletons of horse and cow. (For alternative nomenclature see p. 236.) (After Ellenberger and Sisson.)

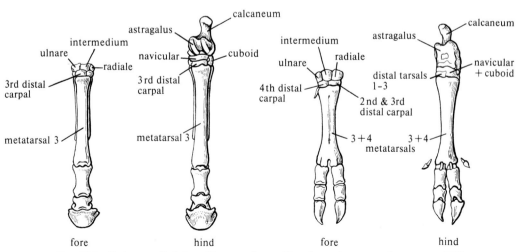

fore   hind   fore   hind

FIG. 28.2. Skeletons of left feet of horse and cow. (For alternative nomenclature see p. 236.)

forms below. The elongation of the limbs is mainly in the lower sections, the humerus and femur being short.

Locomotion is by moving of the whole limb by the action of its proximal muscles, the hindlimbs being the main propellents and the forelimbs weight-bearers, with corresponding modification of the vertebral girder (p. 535). The neural spines are very high above the forelegs and the ribs are numerous, so that the girder has large compression struts above and below and it balances largely on the forelegs and is pushed from behind. The ilium is broad and raised vertically, providing large attachments for the glutei, which are the important locomotor muscles, and for the abdominal muscles, which carry the weight of the belly. This arrangement of the column is essentially preserved even in the very large animals, such as elephants, rhinoceroses, and many extinct types, which are said to be 'graviportal'. But in these large animals the legs are arranged on a different plan, since the great weight can only be carried by very massive struts of large cross section. The proximal parts are therefore enlarged and several digits are retained to make broad supports for the pillars, as is well seen in the elephant's foot.

In many ungulates the neck becomes considerably lengthened, probably both in order to reach up or down for food and also allowing the head to keep a good look out. The ears are long and hearing acute, so that the direction of sounds may be easily detected. Sight is not especially developed, but the pupil is often horizontal in animals that live on the plains, giving a wide visual angle. The sense of smell is well developed and the animals often graze advancing up-wind, using for this purpose the receptors of the moist muzzle. The tongue is large and taste receptors sensitive.

The brain is large and the life of these herbivores is conducted with the use of much information learned during each lifetime. This enables them to range over large territories and to vary these with the seasons, in search of food and water. They show a remarkable alertness to changes of sound or scent.

Many herbivores are social animals, and information is shared among a large group. They have elaborate means of communication by scent glands, which are used to mark trails and territory, as well as for exchange of signals between individuals. Their sexual organization is complicated, involving elaborate interchange of visual, auditory, and olfactory signals and ritual combat between males. The establishment of a leader is apparently often needed to allow the advantages of social organization for protection and finding food. Gestation is long and relatively few young are produced (as in other large animals with efficient brains). The newborn is well developed and can soon run with the herd.

Among the most significant modifications are the means of obtaining and digesting the food. The triangular molar pattern gives place to a square one, by development of a hypocone on the posterior interior side of the upper molar. The lower molar also becomes square, by loss of the paraconid and raising the heel, whose outer hypoconid and inner entoconid make a pair, behind the metaconid and protoconid (Fig. 28.3). Even more characteristic is the change in the cusps themselves. Instead of the original sharp points they develop first low cones, giving so-called bunodont grinding surfaces. The cusps may become elongated and often crescentic ('selenodont' = moon-shaped) as in artiodactyls. Then, in many later evolutionary stages, ridges or lophs appear between the cusps; an ectoloph between paracone and metacone, transverse protoloph at the front of the tooth (between protocone and protoconule), and metaloph behind, between hypocone and metaconule. All sorts of further developments and cross connections may then take place in these lophodont molars; moreover, the whole tooth becomes surrounded by 'cement' (bone) which forms even between the cusps, so that the ridges are supported as they wear away and continually maintain a rough surface for grinding. Short (brachydont) molars, which would wear away too quickly, are replaced by deep (hypsodont) ones, which grow continually from open roots in extreme instances.

In these animals that need to increase the grinding surfaces the whole set of teeth is usually retained and the molar structure extends forwards to the premolars. This molarization may be said to be the opposite of the condition in carnivores, where the tooth row is shortened and some hinder teeth come to have cutting edges like the front ones. The incisors of ungulates become specialized for cropping the food; in artiodactyls the upper ones are lost and the lower work against a horny upper lip. The canine is often absent, leaving a diastema.

upper

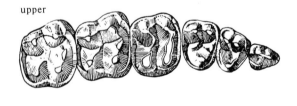

lower

Fig. 28.3. *Hyracotherium*, premolar and molar teeth. After Wortmann.)

The cropping and grinding mechanisms involve various modifications of the lips, palate, tongue, and, of course, the jaws and their muscles. The articulation of the jaw with the skull is usually made by a flattened facet, allowing rotatory action of the lower jaw. The pterygoid and masseter muscles are well developed, the temporal less so and the skull is flat and without a sagittal crest, in contrast with carnivores. To provide lateral attachment for these muscles there is a tendency for a redevelopment of the post-orbital bar.

In the digestive system of ungulates there is usually some chamber in which bacterial and protozoan action upon cellulose can take place, but this has evidently evolved independently in the different groups, being in the stomach of artiodactyls but in the caecum of perissodactyls.

This set of 'ungulate' characteristics has developed independently many times in descendants of the insectivoran eutherian ancestor, and shows strikingly how the adoption of a particular method of life leads to selection of variations of structure tending in similar directions. There is no special difficulty in understanding how this has happened if we imagine that each part varies genetically in its dimensions. A herbivorous diet will be easier for animals with ridged teeth and long legs, whereas those with sharper teeth can become carnivores. Types are selected that combine a nervous organization leading to certain habits with other features that make these habits successful. In the evolution of any population there is evidently an elaborate interplay between variation in different directions in various organ systems and changes in the environmental circumstances.

3. Classification

Superorder 2: Protoungulata
  *Order 1: Condylarthra. Cretaceous–Miocene
    *Family 1: Arctocyonidae. Cretaceous–Eocene
      Arctocyon, Europe; *Protungulatum*, North America
    *Family 2: Mesonychidae. Palaeocene–Eocene
      Mesonyx, North America
    *Family 3: Hyopsodontidae. Palaeocene–Eocene. North America
      Mioclaenus; *Hyopsodus*
    *Family 4: Phenacodontidae. Palaeocene–Eocene. Holarctic
      Tetraclaenodon; *Phenacodus*
    *Family 5: Didolodontidae. Palaeocene–Miocene. South America
      Didolodus
    *Family 6: Periptychidae. Palaeocene. North America
      Periptychus
    *Family 7: Meniscotheriidae. Palaeocene–Eocene
      Meniscotherium, Holarctic
  *Order 2: Notoungulata. Palaeocene–Pleistocene
    Palaeostylops, Asia; *Notostylops*, South America; *Toxodon*, South America; *Homalodotherium*, South America; *Hegetotherium*, South America; *Typotherium*, South America
  *Order 3: Litopterna. Palaeocene–Pleistocene. South America
    Thoatherium; *Macrauchenia*
  *Order 4: Astrapotheria. Eocene–Miocene. South America
    Astrapotherium
  Order 5: Tubulidentata. Pliocene–Recent
    Orycteropus, aardvark, Cape ant-eater, Africa

4. Superorder Protoungulata

*Order Condylarthra

This group includes animals so close to the central eutherian stock that it is still disputed whether some of them should be classified as insectivores, primates, or creodonts. Seven families are now recognized, nearly all from the late Cretaceous, Palaeocene, and Eocene periods. The first two families *Arctocyonidae and

*Mesonychidae were previously classified as Creodonta and believed to show that ungulates and carnivores were derived from a common stock (p. 505). No doubt there were many lines of separate evolution from early insectivoran stocks and it is very hard to decide which it is wise to group together, much as one may wish to do so for convenience.

The arctocyonids had already appeared in the Upper

Cretaceous, where *Protungulatum* of North America is the earliest known ungulate. These animals were numerous in the Palaeocene and *Arctocyon* and others were as large as bears and probably omnivorous. The teeth had sharp cusps but a square form with a hypocone. The limbs carried claws.

The *Mesonychidae were more like ungulates, especially in that they had hoofs. The dentition is curious with three blunt cusps on the upper molars and shearing talonids on the lower. It is not clear whether they ate flesh, molluscs, or plants, but they were able to grow very large, some skulls are up to 27 cm long.

The *Hyopsodontidae such as the Eocene *Hyopsodus*, had a complete row of bunodont, quadritubercular teeth, and short legs with clawed digits (Fig. 1.12). They were small (30 cm long) and perhaps arboreal, and therefore could be classified with lemurs or insectivores. *Mioclaenus* is an even older type, which possessed sharp-cusped teeth. *Phenacodus* (Fig. 28.4) is perhaps the best known condylarth. It was an Eocene form with ungulate characters already present, including hoofs and square bunodont molars. The build was, however, still that of a generalized carnivorous or insectivorous mammal, with a markedly curved spine, small brain case, sagittal crests, rather short limbs with slightly elongated metapodials, the central digit the longest, complete ulna and fibula, carpus and tarsus not interlocking, and a long tail. The animal was, however, rather large (120 cm long). Smaller Palaeocene phenacodonts, such as *Tetraclaenodon*, still possessed claws and may have been very close to the ancestry of all protoungulate types. Evidently some 10 million years of herbivorous life in the Palaeocene had produced only suggestions of the 'ungulate' facies. Other condylarths became more specialized in the Eocene. Thus *Meniscotherium* had lophodont grinders recalling the selenodont teeth of artiodactyls, but it retained the clawed digits.

Didolodus and similar forms from the Palaeocene of South America are condylarths that may perhaps have given rise to some of the characteristic South American ungulates though they themselves survived to the Miocene. The *Periptychidae were Eocene condylarths probably ancestral to the pantodonts. They also show similarities to the surviving *Orycteropus*, which may be descended from them.

5. South American ungulates
*Order Notoungulata

The ungulate fauna of South America provides a case of geographic isolation as striking as that of the marsupials in Australia (Simpson 1978). In Palaeocene times there were ungulates common to South America and the rest of the world. Besides the condylarths, considered above, there was *Palaeostylops* in the Palaeocene of Asia, and the similar *Notostylops* found in South America. These earliest notoungulates showed only a slight advance in size and other features from the basal condylarth condition. From some such beginnings there quickly developed, after the isolation of South America in the Eocene, a very rich fauna, including many large animals. Specimens of these peculiar fossils were first collected by Darwin during the voyage of the *Beagle* and were later described by Owen. Darwin records that their characteristics were among the earliest stimuli that turned his thoughts to evolution (see p. 396).

The teeth formed a complete series and often paralleled those of perissodactyls with persistently growing roots and high crowns and lophodont ridges. The toes became reduced to three, often with hoofs, but never in the full unguligrade position. A characteristic of the group was the very large tympanic bulla. The brain was small and especially the cerebral hemispheres.

As many as nine families of notoungulates can be recognized in the Oligocene; after this period they

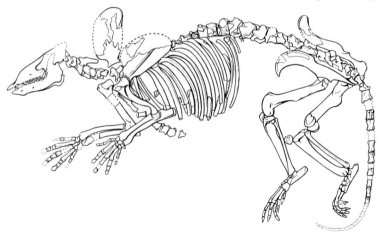

Fig. 28.4. Skeleton of *Phenacodus* as found in the rock. (Simplified after Woodward and Cope.)

became less numerous. Some of them persisted throughout the Tertiary, but all became extinct in the Pleistocene, after the connection with North America had been re-made in the Pliocene and competition was felt from more modern types, both ungulates and carnivores. The notoungulates known as toxodonts (= 'bow-teeth') were very large graviportal animals like rhinoceroses, the tooth row being curved to form a bow from which the group takes its name. The limbs were massive, with as few as three digits, the middle the longest, bearing hoofs. *Toxodon* itself (Fig. 28.5) survived to the Pleistocene and was a creature nearly 3 m long, with enormous head, short front and long hind legs. *Homalodotherium*, on the other hand, had front legs longer than the hind, and provided with claws. The incisors were small but formed rows suitable for cropping. Possibly the animals reared up on their hind legs to reach branches and the large ischia show that the muscles for this were strong (p. 473). Alternatively, they

may have dug for roots. The typotheres and hegetotheres were small rabbit-like creatures with gnawing incisors. The notoungulates thus radiated to form various types and for many millions of years they were the dominant herbivores in South American forests and perhaps other surroundings.

6. Order *Litopterna

Some descendants of the condylarths in South America developed along lines astonishingly similar to the horses. The didolodonts already show tendencies in this direction and are, indeed, sometimes removed from the condylarths and placed with the South American horse-like forms in the order *Litopterna. A series of fossils shows that members of this order became first digitigrade and then unguligrade, the central metapodials elongating and the lateral ones reducing, first to three and then, in *Thoatherium* (Fig. 28.5), to a single one, with splint bones even smaller than remain in modern

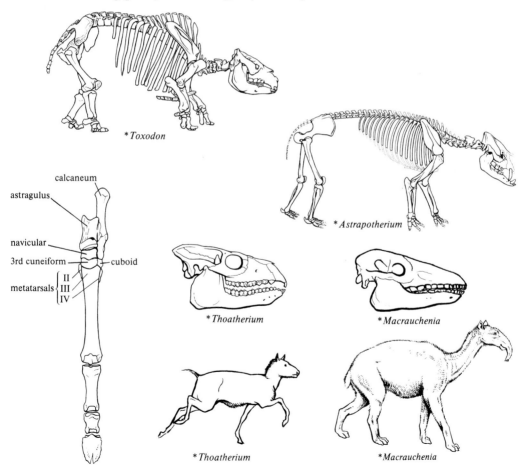

FIG. 28.5. Various South American ungulates. (After von Zittel, K.A. (1925). *Text-book of Palaeontology*. Macmillan, London; Romer 1945; Scott 1913.)

horses. The general appearance of the limbs was very horse-like, for instance in the grooved talus, but the carpus never became interlocking. Other respects in which these litopterns developed less far than the horses were that the tooth row remained nearly complete and the molars low-crowned, though provided with ridges. A post-orbital bar was developed. These differences from our horses are as interesting as the similarities, and they show that the features of the ungulate facies do not necessarily all evolve together. It is impossible to say what difference in conditions was responsible for forming horse feet on an animal whose head was only partly horse-like.

These ungulates were presumably evolved to meet conditions on the South American plains in the Miocene. It is interesting that the three-toed types outlived the one-toed. *Thoatherium*, lasted into the Pliocene but *Macrauchenia* (Fig. 28.5), a creature looking like a camel and perhaps living in swamps, was the only Pleistocene survivor.

7. Order *Astrapotheria

This order includes some Eocene, Oligocene and Miocene South American ungulates, with a short skull with long, curved, persistently growing canines but long lower jaw with cropping incisors and massive molar teeth (Fig. 28.5). There was probably a proboscis. The feet were small and perhaps rested on pads. The weak vertebral spines and transverse processes suggest that the animals may have been aquatic and the large lower canines, diverging in older animals, resemble those of a hippopotamus.

8. Order Tubulidentata (= tube-teeth)

The aardvark ('earth-pig') or Cape ant-eater, *Orycteropus* (Fig. 28.6), is a zoologically very isolated form, of unknown affinities, placed by Simpson with the Protoungulata, because it is possibly not very remote from the condylarths. It is the size of a small pig, with a highly curved back, and is much given to digging, both for protection and to obtain termites, which are its main food. It is quite common from South Africa to the Sudan. There is an elongated snout, round mouth, and long tongue, as in other ant-eaters. The peglike teeth are unlike those in any other mammal. They consist of numerous hexagonal columns of dentine, separated by

FIG. 28.6. *Orycteropus*, the aardvark. (From life.)

tubes of pulp. There is no enamel, though enamel organs are present in the tooth germs. In the adult, there are about five teeth in each jaw, but there is a full series of rudimentary milk teeth.

There are special arrangements in the mouth and throat to allow the animal to bury its snout in a mass of termites and then to swallow them while continuing to breathe. There are large salivary glands. The digits (4 in hand and 5 in foot) are covered by structures sometimes referred to as compressed nails, sometimes as hoofs. There is a strong clavicle and complete radius and ulna and tibia and fibula. The limbs are thus specialized for digging, which is done very fast. Otherwise aardvarks retain the characters of the earliest mammals. The head is long and the brain small and of an extremely primitive type, with extensive olfactory regions and very small neopallium. The olfactory turbinals are better developed than in any other mammal; the aardvarks find termites by their scent. The animals are nocturnal and the retina has only rods, and a tapetum. The ears are long and hearing acute and there are bristles on the long mobile snout. The uterus is paired and the placenta of a zonary type, somewhat like that of carnivores. There is a very large allantois.

Orycteropus occurs as fossils back to the Miocene. Its earlier history is unknown, but similar teeth have been reported from the Eocene, and many features of the skeleton are strikingly like those of condylarths. The animal obviously retains many characters that were present in the earliest eutherians, the fact that it is placed by some as an edentate or insectivore and by others close to the base of the ungulate stock suggests that it has diverged relatively little from the ancestor of all eutherians.

29 Elephants and related forms

1. Subungulates, superorder Paenungulata

FROM the Palaeocene ungulate stock, when it was yet hardly differentiated from that of other mammals, there diverged several lines of herbivorous animals and these rapidly increased and diversified in the Eocene, many of them becoming very large. Most of these lines declined in the Oligocene and only the huge elephants and tiny hyraxes remain today to show approximately the structure of this range of Eocene pantodonts, dinocerates, and other forms. The highly specialized Sirenia (seacows) were also an early offshoot of this type of animal. The various lines diverged so very long ago that we should hardly expect to find that they have much in common that they do not share with other ungulates, or indeed with all mammals, but it has long been recognized that there is a loose grouping of orders around the elephants and hyraxes. Simpson (1945) suggests the name Paenungulata ('near ungulates') for these forms that are all slightly, but not much, beyond the protungulate level. They are also often called 'subungulates'. The legs of all of them remain rather primitive, with long upper segments, complete ulna and fibula, and several digits, with nails but no well-marked hoofs. The incisors and canines often become reduced to single pairs of large tusks in each jaw and the molars are specialized for grinding, with the development of crossridges.

2. Classification

Superorder 3: Paenungulata
 Order 1: Hyracoidea. Oligocene–Recent. Palearctic, Africa
   Procavia (= *Hyrax*), hyrax, Africa, Asia; *Dendrohyrax*, tree hyrax, central and south Africa;
    *Megalohyrax, Oligocene, north Africa
 Order 2: Proboscidea. Eocene–Recent
  *Suborder 1: Moeritherioidea. Eocene–Oligocene. Africa
   *Moeritherium
  *Suborder 2: Deinotherioidea. Miocene–Pleistocene. Eurasia, Africa
   *Deinotherium
  Suborder 3: Gomphotherioidea
   *Family 1: Gomphotheriidae. Eocene–Sub-Recent
    *Palaeomastodon, Eocene–Oligocene, Africa; *Gomphotherium (= *Trilophodon), Miocene, Africa,
     Eurasia; *Anancus, Miocene, Africa, Eurasia; *Tetralophodon, Miocene, Africa, Eurasia;
     *Stegomastodon, Pliocene–Sub-Recent, North and South America; *Cuvieronius, Pleistocene,
     America
   Family 2: Elephantidae. Upper Miocene–Recent
    **Stegotetrabelodon*, Upper Miocene–Pleistocene, Africa, Eurasia; *Primelephas, Upper
     Miocene–Pliocene, Africa; *Mammuthus, mammoth, Pliocene–Sub-Recent, Africa, Eurasia, North
     America; *Loxodonta*, African elephant, Pleistocene–Recent, Africa; *Elephas*, Asian elephant,
     Pleistocene–Recent, Africa, Eurasia
  Suborder 4: Mammutoidea.
   *Family 1: Mammutidae. Lower Miocene–Sub-Recent
    *Mammut (= *Mastodon), Pleistocene–Sub-Recent, Africa, Holarctic; *Zygolophodon, Pleistocene,
     Africa, Holarctic
   *Family 2: Stegodontidae. Middle Miocene–Upper Pleistocene

Stegodon, Pliocene–Pleistocene, Asia
*Order 3: Pantodonta. Palaeocene–Eocene. Holarctic
    Pantolambda, Palaeocene; *Coryphodon*, Palaeocene–Eocene
*Order 4: Dinocerata. Palaeocene–Eocene. Holarctic
    Uintatherium, Eocene
*Order 5: Pyrotheria. Palaeocene–Oligocene. South America
    Pyrotherium, Oligocene
*Order 6: Embrithopoda. Oligocene. Africa
    Arsinoitherium
Order 7: Sirenia. Eocene–Recent
    Protosiren, Eocene; *Dugong* (= *Halicore*), sea-cow, Indian Ocean and Pacific; *Manatus*
    (= *Trichechus*), manatee, Atlantic

3. Order Hyracoidea

The hyraxes, the conies of the Bible, are abundant in Africa and neighbouring regions (Fig. 29.1). They have persisted throughout the Tertiary as small, gregarious, herbivorous creatures, occupying similar niches to rabbits, which they resemble superficially in some ways. Fossils are known in Africa back to the Oligocene and probably the group existed before that time and therefore shows us something of the appearance of smaller Eocene and Oligocene ungulates. The gait is plantigrade, with four anterior and three posterior digits, carrying somewhat hoof-like nails, except for a sharp bifid claw on the inner hind toe (Fig. 29.2), said by some to be used for grooming, but perhaps more for climbing. The palms and soles have soft pads, perhaps improving grip when climbing by physical means or by secretions. There is a single pair of continually growing incisors in the upper and two comb-like ones in the lower jaw. The upper are used for defence, the lower for grooming. Food is cropped by the molars and taken in laterally. There is a diastema and seven grinding molariform teeth of bunoselenodont type, with transverse ridges, looking like those of a tiny rhinoceros. The lower jaw is very deep, for the attachment of the masseter muscle, and, as is usual in ungulates, the post-orbital bar is nearly or quite complete. There is a serial carpus with a centrale, an unusually primitive feature for an ungulate. The intestine provides chambers for digestion by symbionts. In the large median caecum are found enormous ciliates (*Pycnothrix*), up to 5 mm long, and a flora of cellulose-splitting bacteria. Beyond this lies a further pair of caeca.

The brain is of macrosomatic type. The iris carries a lobe said to shade the retina when the cony basks in the sun. They often live in deserts and can survive with little water, producing a concentrated urine. As in the elephants, the testis fails to descend and remains close to the kidney, there being no sign of a scrotum or inguinal canal. The uterus is paired, and the placenta has an annular avascular allantois, drawn out into four sacs, rather as in elephants, but with a haemochorial struc-

Fig. 29.1. *Procavia*, hyrax. (From life.)

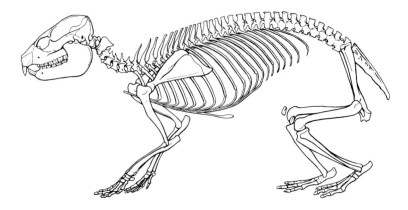

Fig. 29.2. Skeleton of *Procavia*.

ture with a resemblance to that of *Tarsius*. There is a single pair of pectoral mammae.

Various species of *Procavia* are common throughout Africa (not Madagascar), Arabia, Palestine, and Syria, living in desert regions in colonies under rocks and in holes. They do not dig burrows. They feed gregariously, eating quickly before bolting to their holes. At any sign of danger they produce an alarm note, the signal for all to flee. The related *Dendrohyrax* live in the trees. Earlier hyraxes were sometimes the size of small horses (*Megalohyrax from the Oligocene).

4. Elephants. Order Proboscidea

The two existing types of elephant, referred to distinct genera, live still in considerable numbers in Africa (*Loxodonta*) and Asia (*Elephas*). They are survivors of a much larger population, reaching its greatest variety in Pliocene and Pleistocene times. The essential feature of the type is the great size (the largest recorded *Loxodonta* had a shoulder height of 4 m) and the presence of a special food-collecting system able to gather enough raw material to support such a large living mass. Of the various factors influencing the optimum size for a given animal type, all those favouring increase must be present in the ingredients of elephant life. Elephants are larger than any other land animals, living or extinct, except perhaps the huge Oligocene rhinoceros *Baluchitherium and some of the largest dinosaurs (if these were indeed terrestrial). (For biology see Sikes 1971; for physiology Laws, Parker, and Johnstone 1975; for palaeontology Coppens, Maglio, Madden, and Beden 1978; for ecology Delany and Happold 1979.)

The significant features in the organization of elephants are that they are very large animals, with an efficient nervous organization for finding the food, efficient means of collecting it, and large well-organized surfaces for grinding it. When a suitable tree is found the elephant shakes it with his trunk for its fruits and then uses his weight to push it down with his forehead. He may scrape off the bark with the tusks. The trunk is the main means of collection – an enormously elongated nose and upper lip, with appropriate muscles and sensitive grasping tip. The muscles have been developed chiefly from the parts of the facial musculature that are responsible for moving the sides of the nose. It is innervated by the enlarged maxillary division of the trigeminal. Its complicated muscles allow its use in an astonishing variety of situations. It is an arm and 'hand' capable of movements of both great delicacy and power. It is sensitive to chemical and tactile stimuli and is used for finding and collecting single fruits or branches or leaves or other food and transferring it to the mouth. It is raised to test the wind and the direction of scent. As a

hose pipe it can be used to take up water for drinking or to spray it over the body. As a wind instrument its trumpet sounds are a major factor in social life and in courtship it caresses and tickles the partner.

The trunk probably evolved rather quickly, in late Miocene times, perhaps 10–15 Ma ago. The earlier elephants of the Miocene possessed very long lower jaws, which became shortened as the trunk developed. Any tall animal must have means of reaching the ground and the trunk is superior even to a very long neck for this purpose, because it can reach upwards and sideways as well as downwards.

Only one pair of continually growing upper incisors remains in modern elephants, forming the two enormous upcurved tusks, up to 3.2 m long in *Loxodonta*, composed of solid dentine except for a temporary cap of enamel at the tip. This mass of ivory is no doubt useful for defence and perhaps in food collection, but it seems to be a considerable waste of calcium and phosphorus, not to mention of the energy necessary to carry the 160 kg weight. This weight is balanced against that of the body, upon the pillar-like front legs, and it is perhaps not fantastic to suggest that the tusks serve partly as counterweights, for purposes of balance (see p. 514), extravagant though such an arrangement may be. However, the weight of the head is reduced by extensive development of air sinuses between the inner and outer tables of the bones of the skull. The tusks are smaller in females and in the Indian than in the African elephant; in the female Indian elephant they do not project beyond the lips. This difference may be connected with the relatively small size of these females.

The essential features of the grinding apparatus are the mechanisms of mastication and the immensely large hypsodont molars (Fig. 29.3) with numerous sharp transverse ridges. By these all sorts of plants, including hard grasses, are chopped into small fragments. The first three anterior teeth represent premolars, which are soon shed, and posteriorly three molars are then developed in a series and used one after the other. By a special arrangement of the palate (Fig. 29.4) the teeth are allowed to form high up in the skull, so that each tooth has a very great area, made up by the fusion of as many as twenty-seven separate 'plates', which develop as separate cones of dentine and enamel, each with its own pulp cavity, the cones being finally joined together by cement. The three elements of the teeth are exposed and they wear at slightly different rates, leaving a rough surface. Each tooth is worn away gradually from in front backwards (Fig. 29.3). The elephant molar is a horizontal shearing device, not a grinding one. During its evolution the number of plates has increased and they have become thinner and more closely packed providing

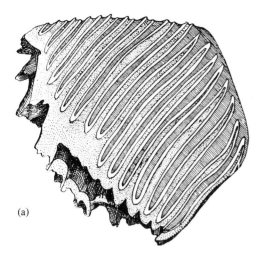

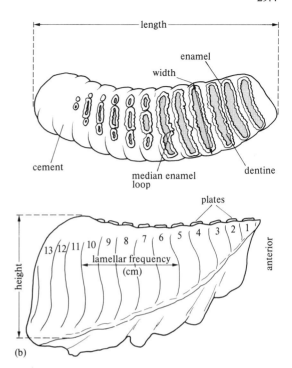

(a)

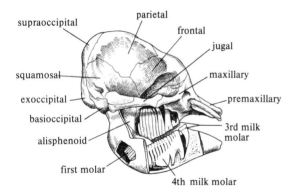

(b)

FIG.29.3. (a) Section of tooth of elephant. The front part of the crown (on the left) is already worn away. Notice the upstanding enamel lamellae, which reach to the base of the tooth. Dentine is shown dotted, cement by lines. (From Weber.)
(b) Molar$_2$ of *Elephas namadicus* showing structural features that can be measured. (Maglio 1973.)

FIG.29.4. Skull of young Indian elephant. The roots of the teeth have been exposed. (After Reynolds 1897.)

more shearing surfaces at each stroke. The movement of the lower jaw is mainly fore-and-aft. It is balanced in a sling by the temporalis muscle and the other masticatory muscles are almost all devoted to pulling it forward in the shearing stroke (Maglio 1973). The teeth are placed just above the ascending ramus of the mandible, so that the large jaw muscles work at maximum advantage. For their attachment the skull becomes extremely short and high, with the development of large air spaces between its tables. This shape also allows a large occipital region for the muscles that hold up the head.

With this head structure the elephants have been able

to grow to a size that must approach the limit possible for a fully terrestrial animal. The skeleton shows typical graviportal features. The backbone (Fig. 29.5) is based on a 'single girder' plan, with as many as twenty ribs, and high thoracic neural spines, forming together a huge beam that carries the weight of the abdomen and balances it on the forelegs against the weight of the head, the hindlegs acting as propellents. The ilium is nearly vertical and expanded transversely for the attachment of the large gluteal, iliacus, abdominal and sacrospinalis muscles. The acetabulum faces downwards.

As in other heavy animals the legs are enormous pillars, with long upper segments and no great extension of the lower. The ulna and fibula are complete and bear part of the weight, the ulna and radius being held permanently crossed in a fixed position of pronation. Walking is of a modified digitigrade type; all three of the short phalanges of each digit reach the ground, but the greater part of the weight is taken by a pad of elastic tissue at the back of the foot. There are five digits in each foot, united by a web to make a firm basis, and having small, flat, somewhat hoof-like nails at the tips. The ribs carry so much weight that respiration is almost wholly diaphragmatic and the lungs are fused to the walls of the thoracic cavity by elastic tissue.

The soft parts of elephants show some features retained unmodified from their early ungulate ancestry. Thus the cerebral hemispheres are relatively rather small

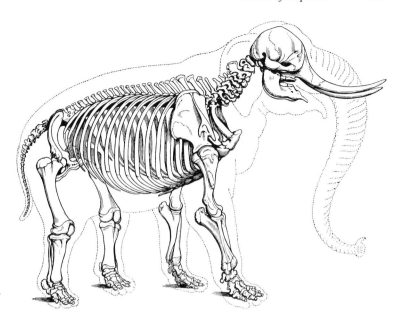

FIG.29.5. Skeleton of an Indian elephant. (From Owen 1866.)

and leave the cerebellum uncovered. In other respects the brain is well developed, it has a greater absolute size than that of any other land mammal (6700 ml). The proverbial intelligence and memory capacity have been verified by experiment. Smell, hearing, and the tactile organs of the trunk provide the main receptors, vision being less developed. Like many other animals with large brains there is a long period of post-natal growth and there is an elaborate social life. Much information is no doubt learned from other individuals, and it has been shown that elephants can learn to discriminate between upwards of 100 pairs of visual choices. There is good evidence that members of a group defend each other, and they may give assistance to the injured by putting the tusks under the body. They may prevent research workers from approaching an elephant immobilized by a drug.

Large size and large brain go with long life and by serial use of their molars female African elephants live up to 60 and males to 50 years in grasslands, perhaps to upwards of 80 years in montane areas (Fig. 29.6).

In spite of the specialization of the head for a herbivorous diet, the stomach and intestine remain simple. There is no special large fermentative chamber but the caecum is long and sacculated and there is an ileocaecal sphincter. The bacterial fauna has not been described.

The testes lie close to the kidneys, as in other paenungulates, and have made no movement of descent into a scrotum. The two horns of the uterus remain separate, though united externally. There are no stable relationships between the sexes and a cow in oestrus receives several males successively. Only one young is born at a time, after a gestation of 22 months. The placenta has a superficial similarity to the zonary arrangement of carnivores, but in structure resembles that of hyraxes and sirenians. At the poles are areas of diffuse, non-deciduate placenta while in an annular zone round the middle there is much invasion of the trophoblast. Development is slow and Asian elephants reach puberty at about 13–14 years, African elephants rather earlier. The basic social unit of African elephants is a mother–offspring family of about 12 ruled by a matriarch and including sometimes mature males and juveniles of both sexes. Males also associate as bull herds.

Moeritherium, from the Upper Eocene of Egypt, is usually considered to be an early member of the elephant line but may not be a proboscidean at all. Some place it in a separate order but we have left it isolated as a suborder. It was the size of a tapir and had a small snout. Perhaps it was partly aquatic with ears and eyes placed high on the head as in the hippopotamus. The skull was elongated and small tusks were present, but the dentition was nearly complete, $\frac{3.1.2.3}{2.0.3.3}$. The molars were bunodont and carried only two cross lophs, a condition easily derived from that of a condylarth quadritubercular tooth. From some such animal arose ultimately so great a collection of types that Osborn (1942) in his study of the group recognized 350 species, only two living at the present day. *Moeritherium* survived to the Lower Oligocene, where there is also *Palaeomastodon*, about twice as large, the first of the 'gomphotheres' (the long-jawed beasts). Both upper and

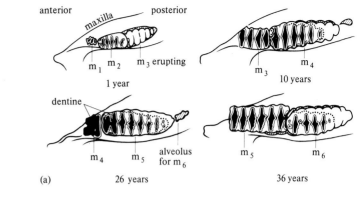

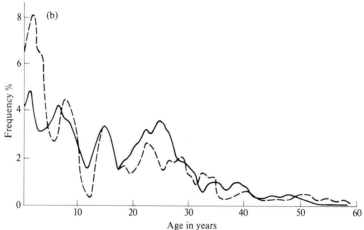

FIG.29.6. (a) Four stages in eruption, attrition, and elimination of the lower molar teeth of the elephant. (After Laws 1966.)
(b) Age structure of elephant populations in Kabalega Park North (solid line) and Tsavo Park (broken line) based on animals cropped between 1965 and 1967. (After Laws 1969.)

lower jaws carried tusks and the whole front of the head was greatly elongated, with formation of a long diastema. The wear of the lower incisors shows that they were used for digging. The molars were brachydont with three low ridges and were all used together, not successively. In the Miocene there was a considerable variety of these long-jawed animals, the various species of the genus *Gomphotherium*. In later members the premolar teeth tended to be reduced and the molars became covered with an increasing number of cusps, arranged to make a number of cross-ridges, seldom, however, more than five and always with a central cleft. These teeth are called bunomastodont (= breast cusped) and gave rise to those of later elephants by increase in the number and reduction in width of the cusps (Fig. 29.9). During the Pliocene and Pleistocene several lines of animals with these low-crowned molars changed in ways parallel to those of the true elephants. In *Tetralophodon* from the Pliocene of Europe and North America the lower jaw was short with a small tusk. There was a long upper tusk and probably a long proboscis. There was some tendency for the teeth to be used serially and they had a number of rather high cusps with some cement

between. *Anancus and *Stegomastodon had up to seven or eight cusps, forming crests and the latter persisted into the Pleistocene, perhaps to historical times in America. There were several other lines of bunomastodont elephants such as *Cuvieronius where the upper tusks were spirally twisted.

The line that produced the modern elephants (Family Elephantidae) can first be recognized in the Upper Miocene by the fact that the cusps of the cheek teeth are united into sharp ridges (Fig. 29.7) (see Maglio 1973; Coppens, Maglio, Madden, and Beden 1978). This allows feeding on hard grasses, which have to be cut, rather than on softer stalks, which can be ground. *Stegotetrabelodon living from the Miocene to the Pleistocene in Asia and Africa perhaps shows the transition from the bunolophodonts to this condition. It had 6–7 plates in the 3rd molars, arranged in transverse rows still with a cleft as in *Gomphotherium. In *Primelephas from the latest Miocene the molars were still low-crowned but the plates are triangular and not divided by a cleft. The true elephants appear only in the Pleistocene and are now grouped into three genera, the two living elephants (*Loxodonta* and *Elephas*) and the mammoths

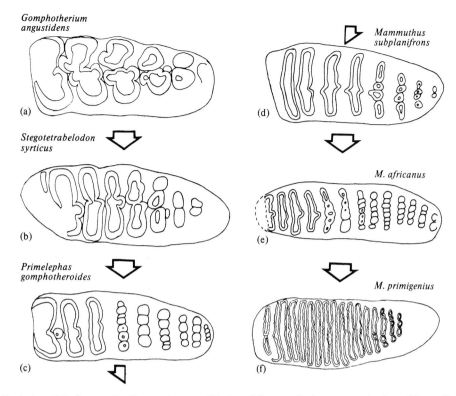

FIG.29.7. Evolution of elephant molar. Progressive consolidation of the gomphothere cone-pairs, loss of the median cleft, fusion of accessory columns, increase in plate number, and thinning of enamel is shown. Not to scale. (Maglio 1973.)

(*Mammuthus*). The mammoths had curved tusks up to 5 m long, turned upwards at the tips. There were several 'species', including such forms as *M. primigenius*, the woolly mammoth, common in Europe during the Pleistocene and surviving until at least the end of the last glaciation in Alaska and Siberia. The animals were adapted for the cold of the Ice Ages, with a thick pelage of brown hair and a layer of fat 9 cm thick under the skin. Mammoths were of course hunted by Palaeolithic man and many carcasses have been found in frozen soil and glaciers, allowing study of the soft parts and the contents of the stomach. *Loxodonta*, the African elephant, has straighter tusks and the surface of the molars wears to a diamond-shaped pattern. In *Elephas* the tusks are also nearly straight and the molar ridges are parallel. There are numerous other differences between the two modern elephants.

Knowledge of the later part of the phylogeny of the elephants is so nearly complete that it provides a particularly interesting opportunity to study the question whether new species arise by continuous change (phyletic evolution of chronospecies) or by splitting. The parent species *Primelephas gomphotherioides* has produced 21 groups recognized as species, of which 9

seem to have been formed by phyletic transition and 12 by branching (Fig. 29.8) (Maglio 1973; and see also Stanley 1980). Even the longest lasting lineage only produced two or three species in 3.5–4 Ma. The paradox is that nearly all the orders of Cenozoic mammals arose in an interval only three or four times longer than this.

The suborder Mammutoidea, the mastodons, includes a distinct line of evolution from some ancestor such as *Palaeomastodon*. There is great confusion about the naming and affinities of mastodons (see Maglio 1973). The genus *Mammut* was called *Mastodon* by Cuvier and French workers keep this name for the group (often including also the gomphotheres!). A further confusion is that *Mammuthus* is a true elephant, as we have seen. So the names mastodon and mammoth are likely to lead to confusion!

The members of this suborder always retained low-crowned, simple molars with no cement. *Mammut* survived in North America until 8000 years ago (or even less). The skull was short as in later bunomastodonts and elephants and only one or two teeth were in use at one time. The tusks of these mastodons curved upwards and out.

The stegodonts were related animals which paralleled

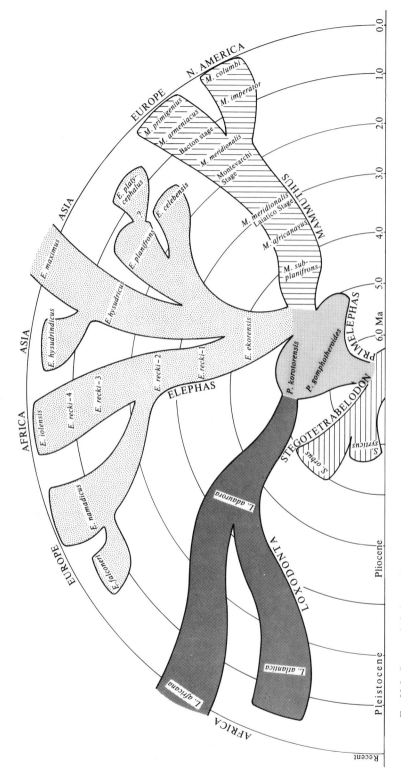

FIG. 29.8. Proposed phylogenetic relationships between all species of the family Elephantidae, recognized as valid at the present. (Maglio 1973.)

the elephants in developing continuous, transverse enamel ridges and a fore-and-aft shearing mechanism. It is not known where they originated but they were numerous in Asia in the Pliocene and Pleistocene and later invaded Africa (Fig. 29.9).

Thus from Miocene times onwards the lower jaw began to shorten and the skull to achieve an elephant-like form in at least four separate stocks (perhaps more). *Anancus and *Stegomastodon and the true elephantines evolved faster in this direction than *Mammut, which retained the 'mastodont' characters associated with browsing even into the Pleistocene, but the stegodonts also paralleled the elephants. The existence of parallel evolution may be regarded as established beyond reasonable doubt in this case; evidently there was some feature either in environmental change or internal 'tendency' or in both, leading all these stocks to change in similar ways, though at different times and rates (Fig. 29.10). *Loxodonta* has evolved much more slowly in number of tooth plates, thickness of enamel, and depth of tooth (hypsodonty) than *Mammuthus and *Elephas*, which expanded into numerous species, spreading over Europe, Asia, and North America in the Pleistocene. Their sudden reduction seems to require a special explanation, perhaps the advent of man (Maglio 1973).

The African elephant population is estimated to be now about 300 000, nearly 2/3 of these being in the Congo Basin and Tanzania. Elephants are being rapidly reduced both for the value of their ivory and because they compete with man on account of their high energy turnover. It is doubtful whether they can survive in the wild but they have increased in several national parks where their numbers are regulated. The killing of the best bull elephants for their tusks seems already to have led to a reduction in the size and quality of elephants and their ivory.

*Deinotherium was a distinct type similar to the elephants, but separate from all others from Miocene times or earlier and here placed in a separate suborder. There were down-turned lower tusks and probably also a long trunk. There were several brachydont molars simultaneously in the tooth row and the full elephant specializations did not develop. They probably fed on soft vegetation. The animals remained similar in structure for a long period, but became very large before they disappeared in the Pleistocene.

5. *Order Pantodonta (Amblypoda)

During the Palaeocene and Eocene the ungulate stock produced various large herbivores and these may be referred to the paenungulate group. The relationships of the numerous types discovered are still obscure and

classification is probably not yet final. The animals here placed (following Simpson 1945) in the order *Pantodonta were formerly, with others, known as amblypods ('blunt feet'). The Palaeocene *Pantolambda was about 90–120 cm long, with a long face and tricuspid molars. The limbs were short and broad, and the pelvis very like that of *Phenacodus.

Later members of the group, such as *Coryphodon (= ridge-toothed) of the Lower Eocene of North America and Europe were over 2.4 m long and heavily built, with some formation of ridges on the teeth, and feet with five digits; some had simple hoofs, others claws. The brain was small and evidently these were clumsy creatures, successful for a time in Europe, Asia, and America, but unable to compete with later herbivores.

6. *Order Dinocerata (= terrible horns)

These were even larger animals, of the same general graviportal build as the *Pantodonta, and were previously classed with the latter as Amblypoda. The two pairs of horns as well as nasal protuberances and very large dagger-like canines of the males provided weapons of defence. The molars showed folds and ridges and provided a reasonably efficient grinding battery. *Uintatherium (Fig. 29.11) was a typical Eocene form. Even though the brain was small and the gait clumsy, the animals were evidently successful at the time, and reached a size as great as that achieved by any other land mammals except the elephants.

7. *Order Pyrotheria (= fire beasts)

*Pyrotherium and its allies (Fig. 29.11) were Eocene and Oligocene South American ungulates and they are usually classed with Notoungulata, but more for geographical than phylogenetic reasons. They were remarkably similar to elephants, for instance in their large size and in the dorsal nostril, suggesting the presence of a trunk. The incisors were developed into tusks and the molar teeth carried two transverse rows of cusps, as in bilophodont early proboscidians. The similarities of the two groups are striking, but they probably indicate only common early ungulate derivation and provide another instance of convergence.

8. *Order Embrithopoda (= heavy feet)

*Arsinoitherium (called after Queen Arsinoe of Fayum in Egypt where they were found) was another large graviportal early ungulate creature, possibly related to the conies. Its limbs resembled those of elephants, with five semi-plantigrade digits. There was a pair of enormous nasal horns, with a keratinous covering like that of ruminants, also smaller frontal horns. There was a

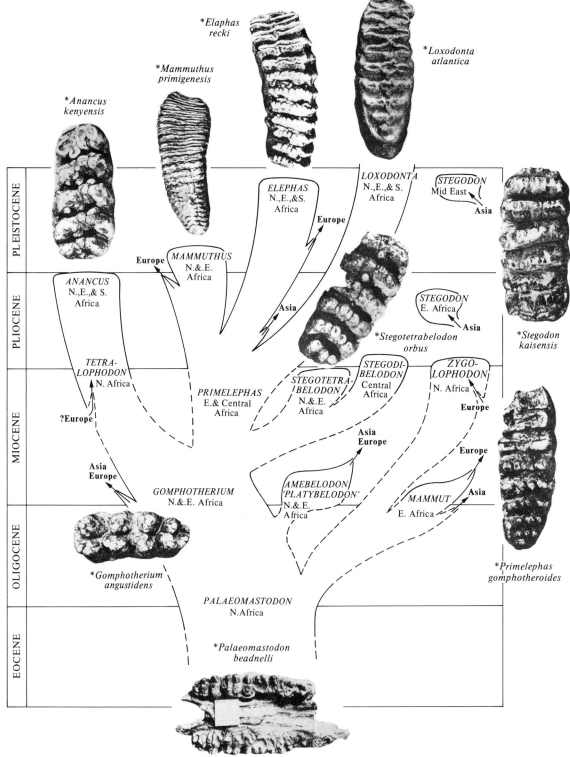

*Elaphas
recki

*Mammuthus
primigenesis

*Loxodonta
atlantica

*Anancus
kenyensis

LOXODONTA
N.,E.,& S.
Africa

STEGODON
Mid East
Asia

ELEPHAS
N.,E.,&S.
Africa

Europe

MAMMUTHUS
N.&.E.
Africa

STEGODON
E. Africa
Asia

Europe

ANANCUS
N.,E.,& S.
Africa

Asia

*Stegotetrabelodon
orbus

*Stegodon
kaisensis

TETRA-
LOPHODON
N. Africa

STEGODI-
BELODON
Central
Africa

ZYGO-
LOPHODON
N. Africa

Europe

?Europe

PRIMELEPHAS
E.& Central
Africa

STEGOTETRA-
BELODON
N.&.E.
Africa

Asia
Europe

Asia
Europe

Europe

Asia

GOMPHOTHERIUM
N.&.E. Africa

AMEBELODON
'PLATYBELODON'
N.&.E.
Africa

MAMMUT
E. Africa

Asia

*Gomphotherium
angustidens

*Primelephas
gomphotheroides

PALAEOMASTODON
N.Africa

*Palaeomastodon
beadnelli

PLEISTOCENE

PLIOCENE

MIOCENE

OLIGOCENE

EOCENE

Fig.29.9. Evolution of elephants in Africa, and the dentition from some representative animals. Major faunal communications with other continents (→); absence of African fossil records (-----). (Coppens, Maglio, Madden, and Beden 1978.)

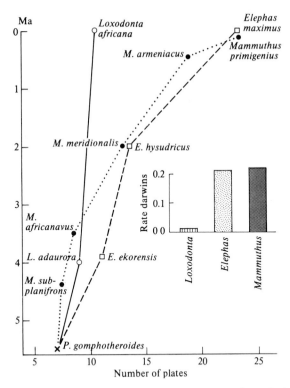

FIG.29.10. Evolutionary changes in plate numbers for molar[3] of three lineages of elephants. The absolute rate of change, measured in *darwins*, is given for the entire lineage as represented. (Maglio 1973.)

regular tooth row, with no enlargement of the incisors or canines and hypsodont molars.

9. Order Sirenia (= mermaids)

The sea cows are herbivorous creatures, cropping vegetation growing on the banks, along the coasts and in rivers. They are highly adapted to aquatic life but have reached this condition by modification of a basic ungulate type of organization, probably not very different from that of the early proboscidians. The two modern forms, the manatee of the Atlantic (Fig. 29.12) and dugong of the Pacific and Indian oceans, are different in many respects and represent lines that have been separate for a long time, probably since the Eocene. *Trichechus* (= *Manatus*) has three species on the tropical Atlantic coasts and in the rivers of Africa and America. *Dugong* (= *Halicore*) is a purely marine animal extending from the Red Sea throughout the Indian Ocean to Taiwan and Australia. *Rhytina (Steller's sea cow) was an Arctic form that became extinct in the eighteenth century.

Sea cows have a 'streamlined' body-form, with few hairs and thick 'blubber'. There are no hind limbs and

the pelvic girdle remains only as small rods to which the corpus cavernosum is attached in the male. The fore limbs are large, the digits joined to form paddles, with a full pentadactyl structure and no hyperphalangy or hyperdactyly. The caudal vertebrae are well developed and swimming is effected by the body and tail, the latter carrying a terminal horizontal fin. The vertebrae articulate with each other by flat surfaces, as in other aquatic forms, but there are zygapophyses, and the whole column is not quite reduced to the condition of a simple compression strut. The bones have a characteristic solid structure (pachyostosis) with little or no marrow. The manatee is the only mammal except some sloths with six cervical vertebrae. The ribs are round and the diaphragm is oblique, as in elephants and whales, allowing the lungs to reach far back. Respiration is probably mainly by means of the barrel-like ribs. The lungs contain large air sacs. Sea cows remain submerged only for relatively short periods (10 min). The blood system shows rete mirabile in the brain and elsewhere, as in other aquatic mammals (p. 501). The brain is small and the ventricles exceptionally large. The forebrain is rounded but the rhinencephalon less reduced than might be expected by comparison with whales. The neopallium is smaller and less folded than in almost any other mammal of comparable size. The eyes are small and protected by muscular lids; the animals do not see well. The external auditory meatus is reduced to a channel a few millimetres wide, as in whales. Little is known of the hearing but reports are that it is acute.

In the manatees, the upper lip is greatly developed to form a strong yet sensitive pad with fleshy lateral lobes, used for cropping. The front parts of the jaws carry no front teeth but have horny pads. The teeth form a series of pegs, with two transverse ridges as in early elephants; there may be up to twenty of them and those in front drop out when worn. It has been supposed that there is a continual replacement from behind, as in elephants, but this is doubtful. In the dugong the teeth are much reduced and the lower jaw carries a horny pad; the upper carries a pair of tusks in the male and the premaxillae are very large. The stomach is complex but not like that of either the whales or other ungulates. The intestine is very long. It is not known whether there is bacterial fermentation.

The reproductive system shows such primitive features as abdominal testes (with no signs that there was a descent in the ancestors) and a bicornuate uterus. The placenta shows a zonary arrangement and haemochorial structure, resembling that of elephants and conies. The young are born in the water and nursed at pectoral teats, which habit, with other features, may have produced some of the legends of mermaids.

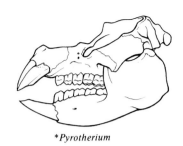

Pyrotherium

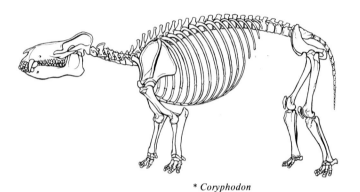

* *Coryphodon*

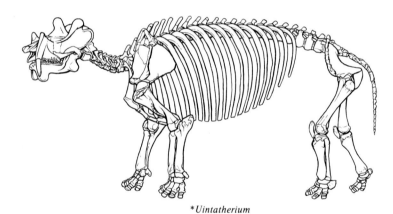

Uintatherium

FIG.29.11. Skull and skeletons of a pyrothere, a dinocerate, and a pantodont. (After Woodward 1898; Flower and Lydekker 1891; Romer 1945.)

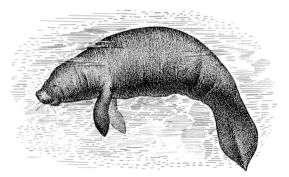

FIG.29.12. Manatee, *Manatus*. (From photographs.)

Sirenians are relatively common fossils, presumably because they live and die near estuaries. The Eocene *Protosiren*, while definitely a sirenian, shows distinct similarity to the paenungulates of those times such as *Moeritherium*. The nostrils were directed dorsally as in modern forms, but the tooth row was complete and small hind limbs were present.

Manatees and dugongs are much valued and hunted as sources of excellent meat and oil, but they are still common in some places, being abundant even in rivers in the centre of Miami. They are protected in the United States, Australia, and elsewhere. Unfortunately too little is known of their natural history to form a basis for policies of conservation.

30 Perissodactyls

1. Perissodactyl characteristics

THE protungulate and paenungulate herbivorous types achieved their chief radiation and greatest numbers early in the Tertiary period. Their organization was not profoundly different from that of the original eutherians and although a few of them, such as the elephants, have persisted to the present day, most have been supplanted by ungulates that appeared by later modification of the original type. Very roughly we may say that the protungulate is the chief Palaeocene mammalian herbivorous type and the paenungulate that of the Eocene. The Perissodactyla, including horses, rhinoceroses, tapirs, and certain early extinct types, then represent a third or Oligocene–Miocene development, supplanting the paenungulates and itself then largely replaced in Pliocene, Pleistocene, and Recent times by the Artiodactyla. This analysis must of course be taken only as a very rough approximation, especially as it is given unsupported by the quantitative data that it requires. It is subject to many exceptions, for example the large development of the elephants in post-Miocene times.

The early perissodactyls were much like all other early ungulates and it is not easy to characterize the group as a whole. The limb structure developed the mesaxonic condition, with the digits arranged around the third as the main weight-bearing member, the others being reduced. With the power of fast movement the lower part of the limbs became elongated and the upper segments shortened, with reduction of the ulna and fibula, but these are characters found also in artiodactyls. A distinctive feature of the perissodactyls was the plan of the carpus and tarsus (Fig. 28.2). One distal carpal, the capitate (magnum), became enlarged and interlocked with the proximal carpals, while in the foot the ectocuneiform developed into a large flat bone, transmitting the thrust through a flat navicular to the talus, which has a flat undersurface, not a pulley-like

one as in Artiodactyla. Modifications of the backbone for carrying great weight or for running were similar to those of other orders (elephants, Dinocerata), including increase in the number of ribs and the vertical position of the ilium (p. 514).

The feeding mechanism, though it has been the basis of the success of the perissodactyls, is in several ways less specialized than that of artiodactyls. The incisors are preserved and used for cropping, having a pit on the free surface, so that sharp edges are presented as the tooth wears away (incidentally allowing the age of the animal to be determined). The canine may be reduced or absent, and there is often a diastema. The molars of many of the earlier types remained bunodont and low-crowned, but those of the later rhinoceroses and horses developed an elaborate grinding surface. This was achieved by formation of a longitudinal ectoloph along the outer edge of the upper molar and parallel transverse ridges, the protoloph and metaloph (Fig. 28.3). Even with the secondary complications of the latest forms these teeth remain recognizably of quadritubercular pattern, and the same might be said of the lower molars. The premolars come to resemble the molars, giving a long battery of teeth. The gut shows less specialization than in artiodactyls, the stomach being undivided, but in horses there is a large cardiac area of non-glandular, oesophageal structure. Digestion of cellulose takes place by symbionts in the caecum and large intestine, which may be greatly developed. The brain of Perissodactyla is relatively small, especially in the earlier forms, such as the tapirs. It is of macrosmatic type and the sensory portion of the nose is highly developed.

The reproductive system also shows primitive features. The uterus is bicornuate and the placenta of the diffuse epitheliochorial type, with a large allantoic sac. The yolk sac grows to a large size and forms a yolk sac placenta during the early part of the development.

2. Classification

Superorder 4: Mesaxonia (= middle axis)
  Order Perissodactyla (= uneven toes). Lower Eocene–Recent
Suborder 1: Ceratomorpha (= horn forms)
  Superfamily 1: Tapiroidea. Tapirs (= thick skins). Eocene–Recent
    Homogalax; *Hyrachyus; Tapirus*, tapir, Asia, South America
  Superfamily 2: Rhinocerotoidea (nose horns). Rhinoceroses. Eocene–Recent
    Family *Hyracodontidae (running rhinoceroses). Eocene–Oligocene. Holarctic
      Hyracodon
    Family *Amynodontidae (= defence teeth). Eocene–Miocene. Holarctic
      Amynodon
    Family Rhinocerotidae. Oligocene–Recent
      Aceratherium; *Baluchitherium*, Asia; *Coelodonta*, woolly rhinoceros, N. Europe, Sub-Recent;
      Rhinoceros, Indian and Javan rhinoceros, Asia; *Dicerorhinus*, Sumatran rhinoceros, Sumatra;
      Diceros, African rhinoceros, Africa
Suborder 2: *Ancylopoda (= hook-footed). Eocene–Pliocene. Holarctic
    Eomoropus; *Chalicotherium*, Eurasia, Africa; *Moropus*, North America
Suborder 3: Hippomorpha (= horse-like). Eocene–Recent
  Family 1: *Palaeotheriidae (= ancient beasts). Eocene–Oligocene. Eurasia
    Palaeotherium
  Family 2: Equidae. Horses. Eocene–Recent
    Hyracotherium (=*Eohippus*), Lower Eocene, Holarctic; *Orohippus*, Eocene, North America;
    Epihippus, Upper Eocene, North America; *Mesohippus*, Oligocene, North America; *Miohippus*,
    Oligocene–Miocene, North America; *Anchitherium*, Miocene, Holarctic; *Parahippus*, Miocene,
    North America; *Merychippus*, Miocene, North America; *Hipparion*, Miocene-Pliocene, Holarctic,
    Africa; *Pliohippus*, Pliocene, North America; *Hippidion*, Pleistocene, South America; *Equus*,
    horses, asses, zebras, Pliocene–Recent, world-wide
  Family 3: *Brontotheriidae (= *Titanotheridae) (= thunder beasts). Eocene–Oligocene. Holarctic.
    Lambdotherium; *Eotitanops*; *Brontops*

3. Perissodactyl radiation

The fossil history of animals with the perissodactyl structure is perhaps better known than that of any other mammals; the type reached its peak during a period from which many fossils have been preserved. We have therefore a better opportunity to watch them develop, flourish, and decline than in the case of forms whose maximum development occurred either earlier or later. Here if anywhere we should be able to learn lessons about the nature of the evolutionary process and to study the forces that produce change in mammalian form. Because of the very abundance of the fossils it is necessary, however, to be cautious in interpretation and to recognize what can be proved from the evidence.

The specimens of fossil horses are divided into 350 'species', but only a small proportion of these can be confidently placed close to the direct line of evolution to *Equus*. Abundant though the material is, we have not, therefore, anything like a complete series of fossils to show every shade and grade of change of the populations throughout the 50 million years or so of their evolution. Our knowledge is based on a small sample of individuals, preserved at random at scattered intervals.

The remains often suggest evolutionary sequences and many accounts speak confidently of changes and trends. We shall try, even in this brief account, to describe the actual discoveries and to indicate clearly what evidence is available for evolutionary speculation. With all the mass of information we possess it must yet be realized that the study of the details of perissodactyl evolution has hardly begun, for example we have little quantitative information about the variability of the characters.

The earliest perissodactyls had departed but little from condylarth conditions. *Hyracotherium* (= *Eohippus*) from the Eocene of Europe and North America (Fig. 30.1) was the size of a terrier and resembled the condylarth *Phenacodus*. The tooth row was complete, with low crowned, square bunodont molars (Fig. 28.3). In addition to the four main upper molar cusps an anterior protoconule and posterior metaconule are suggested, and between these and their neighbours the dentine is partly built up to form two transverse ridges (lophs). The premolars were tritubercular. The lower molars had four cusps. The gait was digitigrade, with rather long metapodials and with hoofs, the front leg having four and the hind leg three toes, the central ones the longest. These animals were already distinctly horse-like, in spite of their small size, and they probably lived in forests, browsing on leaves.

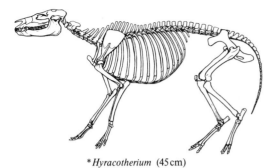

Hyracotherium (45 cm)

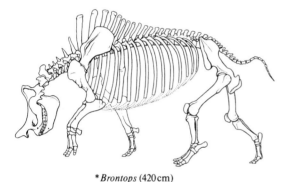

Brontops (420 cm)

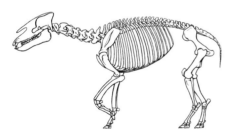

Palaeotherium (210 cm)

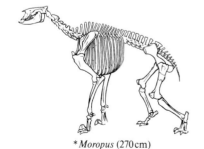

Moropus (270 cm)

FIG. 30.1. Skeletons of some early perissodactyls. Not drawn to scale; length of each shown in parentheses. (After Woodward 1898; Scott 1913.)

From a population of animals of this type there evolved a varied host of herbivores (Fig. 30.2), which may be divided into three suborders. The Ceratomorpha include the tapirs, remaining little changed, and the heavy-bodied rhinoceroses. The Ancylopoda (chalicotheres) acquired claws. Thirdly the Hippomorpha including the early line of large brontotheres (titanotheres), the palaeotheres, a horse-like stock, and finally the horses themselves. Of course, each of these lines had many subdivisions and branches, producing a most complex evolutionary bush.

4. Tapirs and rhinoceroses

The modern tapirs with several species in Central and South America, Malaya, and the East Indies (Fig. 30.3) are solitary, nocturnal creatures, mostly living in dense forests on damp, soft ground and eating lush vegetation and fruit. They have retained many of the conditions of the ancestors of all Perissodactyla. The legs and feet are short with four digits in the forefoot and three in the hind with small hoofs; the ulna and fibula are complete and distinct. The tooth row is complete (42 teeth in tapirs), but the premolars are molariform, though of a simple square pattern, with low crowns and no cement (Fig. 30.4). The nose has developed into a short trunk, with characteristic shortening of the nasal bones. The stomach is like that of horses and there is a large caecum. The placenta is diffuse. The brain is relatively smaller than in horses.

Fossil tapirs, very similar to modern forms, are found back to the Oligocene, and somewhat more primitive Eocene related forms (*Homogalax*) might have been close to the ancestors of *Hyracotherium*. We may say therefore with some confidence that the modern tapirs show us with little change the condition of the perissodactyl stock in late Eocene or Oligocene times, perhaps 40 million years ago. Fossils almost exactly similar to the existing genus occur back to the Miocene, say, 20 million years ago, and the tapirs were then of widespread distribution, which explains the modern discontinuous distribution. We may notice once again the important fact that, given suitable environments, types persist with little change even when their relatives, moving into other conditions, become greatly changed.

The rhinoceroses (Fig. 30.5) are the only surviving perissodactyls that show the graviportal type of body form that was adopted by many of the extinct forms (brontotheres, etc.) and also by many other large eutherians. Dinocerates, elephants, hippopotamuses, and rhinoceroses all show the same type of skeleton. The vertebral column (Fig. 30.6) has long neural spines above the fore legs, and there are many ribs, reaching back nearly to the pelvis. The whole column thus makes

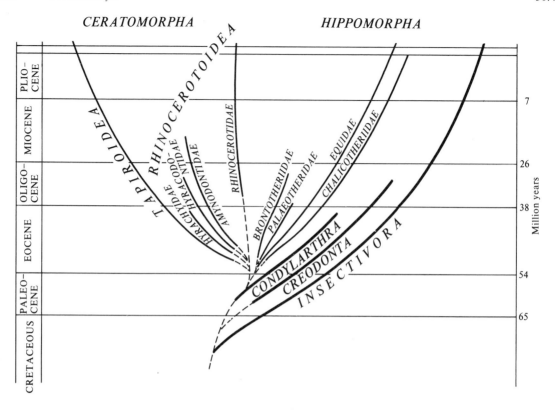

FIG. 30.2. Chart of the evolution of the perissodactyls.

FIG. 30.3. Malayan tapir, *Tapirus*. (From photographs.)

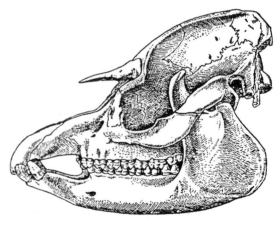

FIG. 30.4. Skull of the tapir. (After Reynolds 1897.)

FIG. 30.5. Indian rhinoceros (*Rhinoceros*).

a girder balanced on the fore legs, and the head, being very heavy, counterbalances the body weight. (Fig. 30.7). The molars are low crowned, with transverse lophs and little or no cement. The first upper and second lower incisors form enlarged cutting teeth. The hind legs provide the main locomotor thrust. It is characteristic of this 'single-girder' type of backbone that the ilia are wide and vertically placed. The feet are basically similar in all these groups in that several digits (three or four in the rhinoceros) are retained, making supports of large area, with nail-like hoofs. The brain of the rhinoceros is small and the chief receptors are those of smell and hearing; the eyes are mainly used in weak light. Like the tapirs they are essentially timid animals, mainly nocturnal, though defending themselves with a charge if attacked. African black rhinoceroses are solitary, bush animals, each holding a small compact territory, marked by shuffling their feet in their dung. The white (wide-mouthed) rhinoceros graze in small herds on the plains.

The earliest members of the rhinoceros group may have been *Hyrachyus* of the Eocene, now placed with the tapirs. They were very like other primitive perisso-

dactyls, mostly small and with a complete tooth row, in which the molars already show an ectoloph and the parallel transverse lophs characteristic of the group. The *Hyracodonts were an Eocene and Oligocene line specialized for swifter movement ('running rhino-ceroses') with long legs and three toes in each foot, much as in the earlier horses, who perhaps then supplanted them. The members of another early line, the *Amyn-odonts, were larger, probably semi-aquatic forms. The true rhinoceroses appear in the Oligocene, already as large creatures, fully terrestrial and hence with stout limbs and a good grinding battery, with molarized premolars. *Baluchitherium* of the Oligocene and early Miocene of Asia became enormous, 20 000 kg weight and 5.5 m high at the shoulders, the largest of all known land mammals. The skull, 1.4 m long, was set on a long neck for reaching high branches. Rhinoceroses became numerous in the Miocene and Pliocene of Eurasia (*Aceratherium*). Various types persisted through the Pliocene and Pleistocene, and the modern single-horned *Rhinoceros* of Asia and two-horned *Dicerorhinus* of Sumatra and *Diceros* of Africa are derived from some of these. Extinct types such as the woolly rhinoceros (*Coelodonta*) are known from Palaeolithic drawings and from specimens 'embalmed' in waxy material in an oil seep. Modern rhinoceroses all have a very thick, almost hairless skin, with characteristic folds. The horns of the rhinoceros, one, two, or occasionally three median outgrowths on the head, consist of a compact mass of coarse, keratinized threads which are a modified form of hair (Spearman 1966).

5. *Brontotheres (*Titanotheres)

These were early, heavily built ungulates, reaching large size in the later Eocene and Oligocene, in fact preceding

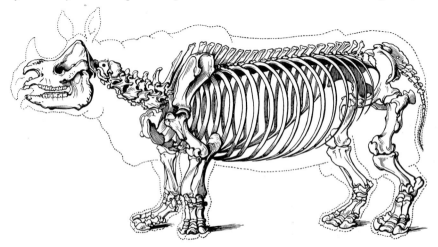

FIG. 30.6. Skeleton of the Indian rhinoceros. (Owen 1866.)

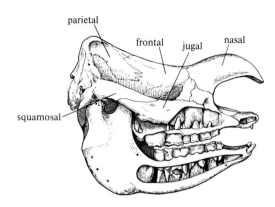

FIG. 30.7. Skull and teeth of a young Indian rhinoceros. The grinding surface is made up of four milk premolars and one adult molar on each jaw. The remaining permanent teeth have not erupted. (After Reynolds 1897.)

the rhinoceroses as large herbivores. The fully developed forms, such as *Brontops* (Fig. 30.1) of the Lower Oligocene, were of typical graviportal type, up to 2.4 m high, with high thoracic spines, numerous ribs, vertical and laterally expanded ilia, and rather short legs, with four digits in front and three behind. The tooth row was complete and the molars large but low-crowned, with a ridge along the outer side, but isolated cusps on the inner (hence 'bunolophodont'). They were not suited for eating hard vegetation or continuous wear. A single pair of large horns was carried on the front of the skull. The brain was even smaller than that of rhinoceroses.

The earliest fossils that can be referred to this type, *Lambdotherium* of the Lower Eocene of North America, were much smaller, and without horns; they could well have been derived from *Hyracotherium*. *Eotitanops* from the Middle Eocene was larger, but still hornless. From some such stage numerous lines probably diverged, each becoming larger and independently acquiring horns. For obvious reasons it is difficult to obtain a proper idea of the evolution of such giants, as Simpson points out, genera and perhaps even sub-families have probably been created on a basis of differences that may be only sexual or individual.

6. *Chalicotheres (= *Ancylopoda)

One remarkable side line of perissodactyl evolution, while becoming large and horse-like in some ways, acquired structures resembling claws instead of hoofs (*Moropus*, Fig. 30.8). These chalicotheres, all rather alike, existed from the Eocene to the Pleistocene and were therefore a successful group. The terminal phalanges of the three toes of each foot were cleft and undoubtedly carried a nail or claw of some sort, though

not necessarily one like that of true unguiculates. There is no doubt that in a sense this is a case of reversal of evolution, but we cannot assert much about its possible genetic implications unless we can find details of the nails.

When chalicothere digits were first discovered in 1823 Cuvier applied his 'law of correlation' and suggested that this was the remains of an ant-eater, 'un Pangolin gigantesque', while teeth and other bones found near by he referred to an ungulate. It was only when skeletons were found in such a position that the association of the bones could not be denied that the danger of this attempt to apply deductive principles in biology was exposed. The other parts of the skeleton are unambiguously perissodactyl (Fig. 30.1), the teeth rather like those of brontotheres. There may have been a short proboscis. The neck vertebrae of some forms show very strong zygapophyses, and it has been suggested that the snout and claws were used for digging for roots or water. It is more probable that the chalicotheres reared up on their hind legs and used the claws to cling to tree-trunks while reaching for leaves with their flexible necks, or perhaps to drag down branches. Like the toxodont *Homalodotherium* they had long front legs and large ischia. Moreover, their remains are found in association with those of forest dwellers. The attraction of speculating about these creatures has not diminished with the demonstration of its dangers.

7. *Palaeotheres

These animals, from the later Eocene and early Oligocene of Europe, were an early offshoot that paralleled in many ways the evolution of horses in North America. For example, they developed slim metapodials and three-toed feet, and the premolars became molarized. The teeth developed ridges on a similar plan to the horses, but differing in details. Some forms became hypsodont. Several lines of descent are included in the

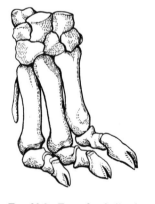

FIG. 30.8.   Feet of a chalicothere. (After Romer 1945.)

group. *Palaeotherium* became large, though not so large as the graviportal brontotheres and rhinoceroses. The shortness of its nasal bones suggests that it had a proboscis like a tapir. They acquired features of the teeth and legs not found in the true horses until the Miocene. They illustrate the importance of parallelism in evolution, and serve to warn us against the easy assumption that a character that is shown by two animals must have been present in their common ancestor.

8. Horses

The horse, besides its special interest as one of our oldest and most useful commensals, has provided a rather complete and convincing record of its origin. We shall therefore first describe its present structure and then analyse the fossil record to discover exactly what can be demonstrated about the evolution of the group. Existing horses, asses, and zebras, all referred to six species of the genus *Equus* (Figs. 30.9–30.11), are highly specialized for swift movement and grazing grasses (p. 514). The humerus and femur are short and attached to much longer radius and tibia, with ulna and fibula reduced (p. 515). Only the third digits are developed and covered with hoofs. These are elaborately organized pads, including several sorts of keratin, harder in front, more elastic behind. The metapodials of digits II and IV are present as small splint-bones. There is a horny callosity on the inner side of the fore limb in all species (also on the hind limbs in *E. caballus*, the domestic horse), representing the vestigial hoofs of lateral digits.

There are three incisors in each jaw, and a canine, or tush, which is present in the stallion only. Then there is a long diastema with the first premolar being vestigial in each jaw ('wolf-tooth'); whilst the remaining three premolars resemble the three molars. All the cheek teeth are hypsodont, square in cross-section, with ectoloph

FIG. 30.10. Racehorse (*Equus*). (From a photograph.)

and transverse protoloph and metaloph, joined by longitudinal ridges that give the tooth a certain resemblance to the selenodont molars of artiodactyls, hence 'selenolophodont'. The enamel, dentine, and cement wear at different rates, leaving a rough surface for grinding siliceous grasses. The skull is modified to allow space for the deep, continually growing teeth and for the large jaw muscles, and there is a complete post-orbital bar.

The hair is long all over the body and the tail is long, its hairs beginning close to the base in horses, half-way along in the others.

All horses in the native state live in migratory herds, as do so many herbivores that dwell on plains. Most zebras and horses are polygamous without being territorial. Each family consisting of a stallion and up to twelve mares, the excess stallions forming batchelor groups. Unmated stallions abduct young mares during oestrus, fighting the family stallion. Grevy's zebra and wild African asses have a somewhat different social system, with no permanent bonds, whilst each dominant

FIG. 30.9. Zebra (*Equus*). (From photographs.)

FIG. 30.11. Shire horse (*Equus*). (From a photograph.)

stallion holds and defends a large territory (Klingel 1974).

The brain is large, and although the organs of smell are well developed, the eyes are also large and the neopallium is extensive. Receptors for touch are well developed in the muzzle, in the skin beneath the hoofs, and elsewhere. Hearing is exceptionally acute. Besides the keen senses common to many herbivores the horse, with its large brain, also has considerable powers of learning and ability to vary and restrain its behaviour. There is an elaborate communication system, involving not only sounds but movements of the ears, tail, and lips. In these respects horses and elephants, and perhaps also modern artiodactyls, are probably very different from the small-brained herbivores of the Eocene, though, of course, we can only guess at the behaviour of these.

Modern equids show considerable genetical diversity (Figs. 30.10 and 30.11), but none of the 'species' are mutually sterile, though the F_1 resulting from the cross may be nearly so, as in the case of the mule, produced from the horse–ass cross. The domestic horse (*E. caballus*) is presumed to be descended from Przewalski's horse of central Asia. This probably survives only as a few small herds in captivity. The domestic horse has 64 chromosomes, Przewalski's 66, but they are fully interfertile. The ass (*E. africanus*) probably still exists in the wild and has 62 chromosomes, *E. hemionus*, the onager of Asia has 54. The several species of zebra all live in Africa, the commonest *E. burchelli* having 44 chromosomes. Grevy's zebra (*E. grevyi*) has 46.

The pattern of stripes of the zebra varies in a cline, in the north they are sharp and extend onto the legs, but are paler and fewer in the south. It is doubtful whether they serve for disruptive concealment. Zebras live in the open and rely on their senses to escape, moreover they frequently live together in herds with dark-coloured gnu.

Between the modern *Equus* and the lower Eocene *Hyracotherium* a great number of fossil stages can be recognized (Fig. 30.12) (Simpson 1951). The chief changes that can be followed may be listed as (1) increase of size, (2) lengthening of the distal portion of the legs, (3) reduction of accessory digits, (4) increase in the relative length of the front part of the skull, (5) increase of depth (hypsodonty) and of the grinding lophs of the molars, (6) approximation of premolars to molar structure, (7) completion of post-orbital bar. No doubt there has been change also in many other characteristics, for instance the brain and behaviour; these are difficult to follow in a fossil series, but study of cranial casts suggests that a rapid increase in size and folding of the cerebrum occurred relatively early in the evolution.

The fossil remains are not usually available in stratified series of vertical beds, such that we can be sure that one population has evolved into the next. However, the dating of the fossils can often be done with considerable accuracy by means of the associated animals, and a series can thus be produced such as would be expected in the progress from *Hyracotherium* to *Equus*. There are, however, many fossils that show special developments, and cannot be fitted into the direct series. These are presumed to be divergent lines; it must be emphasized that this is an arbitrary though probably justified procedure. These 'side-lines' are so numerous that they immediately throw doubt on the idea that there has been any single uniform 'trend' in horse evolution. At least twelve types sufficiently marked to be classified as genera are known, in addition to those directly on the line leading to *Equus*; of course there is a much larger number of shorter, independent, evolutionary lines within these genera. We have enough evidence to glimpse the extraordinary complexity that would be revealed by the complete evolutionary 'bush', even in this single family. A further complication is produced by migrations. It is at present believed that the main course of horse evolution went on in North America, with migration at various times to the Old World and South America. Certainly a more continuous series of forms has been revealed in North America than elsewhere, but it must be remembered that they have been looked for intensively, and brilliantly studied. It is not impossible that further study of Old World horses will produce still greater complications by revealing sequences of evolution within that area.

Throughout the Eocene epoch the horses all possessed four toes on the front feet. The fossils classed as *Orohippus* and *Epihippus* from the Middle and Upper North American Eocene are little different from *Hyracotherium*, except for molarization of the hinder premolars. The size remains small.

The Oligocene horses, *Mesohippus* and *Miohippus*, were the size of large dogs and walked with three toes on the ground, and all the premolars were molarized. The ectoloph was well formed but the inner cusps were still separate, and the teeth low-crowned. Some horses of this type (*Anchitherium* and its descendant *Hypohippus*) persisted into the Miocene and became larger, presumably surviving as browsers in forests, while other descendants took to the plains. These browsing forest horses migrated to the Old World in the Miocene, then died out there, as they did also in North America.

Parahippus of the American Lower Miocene shows the beginning of the adaptation for life on the plains. The lateral digits 2 and 4 still carried hoofs, but since the central proximal phalanx was much the longest and

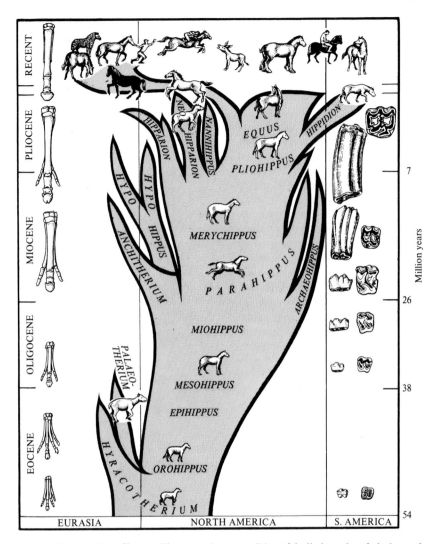

FIG. 30.12. Table to show the evolution of horses. The approximate condition of the limbs and teeth during each epoch are shown to the left and right. (Based on Stirton.)

strongest, it is probable that the lateral ones touched the ground only to maintain balance over uneven surfaces, or in soft conditions. The teeth were still rather low, but were beginning to be elongated and to show cement on the crowns. The protoloph and metaloph were connected by a narrow bridge. There was a partial postorbital bar.

Merychippus comes from later Miocene beds and could have been directly derived from *Parahippus* by increase in the depth of the teeth and reduction of the lateral digits to short stumps, still three-jointed and carrying hoofs, but vestigial in the sense of never touching the ground. Several mammalian lineages increased their hypsodonty and cursorial adaptations in

the Miocene, not as formerly supposed because of the increase in prairies but rather because a woodland savannah with broad interdigitations of riparian forest now existed in the Great Plains (Webb 1977).

Apparently the type was very successful and later in the Miocene it produced various populations. *Hipparion*, with the two lateral toes remaining as vestiges, spread through Eurasia in the Miocene. *Nannippus* was a small form that remained in America. *Pliohippus* was another American descendant from *Merychippus*, and here the lateral digits were lost altogether in the Pliocene, the metapodials remaining as long thin vestiges. When the land connection with South America became open this type of horse migrated there and produced a special development, *Hippidion of the Pleistocene, with rather short legs, perhaps correlated with a mountain habit.

Meanwhile in the late Pliocene the *Pliohippus stock of North America finally reduced the lateral metapodials to short splint bones and produced the Equus type, which rapidly spread thence over all the available land masses, becoming then extinct in North and South America until reintroduced by man.

9. Allometry in the evolution of horses

Although *Equus* is certainly a very different creature from *Hyracotherium, we are fortunate in that many of the differences are due to measurable changes in proportions. A beginning has been made with attempts to estimate the rate of evolutionary change, as a preliminary to study of the factors that influence it. Some of the changes in proportion seen during horse evolution are a consequence of the increase in size. If an organ grows relatively faster or slower than the body as a whole it is obvious that its proportions will differ in animals of differing adult size. The size of an organ, y, in relation to that of the body, x, is often expressed as $y = bx^k$, where the constant k describes the relative growth rate. If $k > 1$ the organ becomes larger in larger animals and is said to be positively allometric (see Huxley 1932). The demonstration that growth actually follows this law in particular cases is not easy, and the underlying assumptions have been questioned. It is probably true, however, that organs do sometimes differ in relative growth-rates, and the method provides a means of investigation of the proportions of an organ not only between adults of different sizes but throughout the growth period, and indeed also throughout an evolutionary sequence. Thus Figure 30.13 shows that the length of a horse's face increases between embryo and adult along a line similar to that found in the series *Hyracotherium to Equus, and that adult horses of different sizes vary similarly in face proportion. A nearly fitting line gives constants $b = 0.25$ and $k = 1.23$. Other methods of plotting, for instance face length against cranium length, give somewhat different results and it cannot be considered certain that no new genetic factors have been involved in the increase of face length throughout the whole evolutionary sequence.

10. Rate of evolution of horses

Study in the same way of other horse characters, for instance those involved in hypsodonty, shows that special genetic changes may be involved and that genetic change does not go on at a constant rate.

Estimates of rate of evolution for the whole animal have also been made. Assuming that the 'genus' is

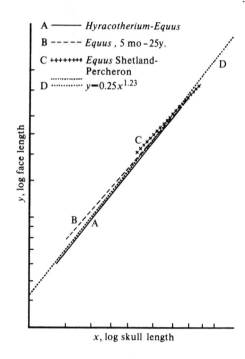

FIG. 30.13. Relative rates of growth in horses. The lines show regression of log skull length on log face length; in A, the line of horse phylogeny; B, the ontogeny of *Equus*; C, various races of *Equus* of different sizes; D, $y = 0.25 x^{1.23}$. (After Simpson, from the data of Robb.)

assessed as a comparable entity throughout (a large assumption, this!) and dividing the eight genera (excluding *Equus*) on the direct line into the time involved between *Hyracotherium and *Pliohippus (50 million years), we should have 6.3 million years per genus. However, there is reason to suppose that individual genera lasted for very different times, *Miohippus, for instance, less than a third of the time of *Merychippus. Therefore if the criterion of a genus is constant, the rate of evolution must vary.

Sufficient fossil horse material is available to allow consideration of whether known rates of mutation are likely to be adequate to account for the observed evolutionary changes. Simpson calculates that from *Hyracotherium to Equus there must have been at least 15 million generations, which, with a population in North America of 100 000 (a low estimate), gives a total of 1.5×10^{12} individuals in the 'real and potential ancestry of the modern horse'. One in a million is a moderate rate for large mutations at any locus in *Drosophila*, and this would give 1.5 million such mutations for a single locus in the horse ancestry. It would be safe to assume that one-fifth of these (300 000) were in the direction favoured by selection and that one-tenth of all such genes affect a structural change, such as ectoloph length. The actual increase in this length between *Hyracotherium* and *Equus* was from 8 to 40 mm which, divided into 300 steps, gives an increase per mutation of only 0.1 mm. This is a reasonable figure, and such calculations suggest that observed mutation rates are quite adequate to account for the evolutionary changes, even neglecting possible multiple actions and interactions of genes, by which the speed of evolution could be further increased.

11. Conclusions from the study of the evolution of horses

Careful consideration of the fossil horse material therefore suggests that their evolution could have proceeded by gradual change (see p. 534). As more and more evidence becomes available the series becomes more and more complete, and incidentally the nomenclature increasingly confusing. The data are still not sufficient to show whether the sequence included small jumps (saltation), especially when, as in the Old World horses, there have been successive migrations into one region from another. Early European palaeontologists, finding *Hyracotherium*, *Anchitherium*, *Hipparion*, and *Equus*, without intermediate forms, interpreted the evidence as showing evolution by saltation. This was indeed a reasonable conclusion from the facts, but was not the only possible one, as has since been shown by the discovery of the much fuller sequence in North America.

An outstanding conclusion from a study of horse evolution is that it is very difficult to describe the change as occurring in a single direction, as supposed by believers in 'orthogenesis'. Apart from the fact that many 'side-lines' can be detected in addition to the line that happens to have survived, it is important to remember that not every line evolves in the same direction. Thus in at least two genera of horses size became progressively *smaller* (*Nannippus* and *Archaeohippus*). However, it is certainly true that in some lines evolution may proceed for long periods in one direction. We have no clear evidence why this should be so, but it is reasonable to suppose that it is due to 'orthoselection', that is to say, the survival of animals that adopt a particular method of life for which they are suited by a particular make-up. The effect of this would be gradually to select all those genetic factors that make for success in one environment (say, grazing on grassy plains) and hence to produce evolutionary change in one direction.

It is interesting to consider whether we should be justified in describing modern horses as 'more advanced' than their Eocene ancestors or than the tapirs, which have remained at the Eocene level. The legs of the horse are highly specialized and very well adapted for rapid running on hard ground – but not for any other type of movement. Their brains are larger and senses perhaps keener than those of their ancestors, but with unadaptable legs and teeth even their good brains would probably not have allowed them to survive except in the service of man. Without domestication they would probably have become extinct in the Holocene, because of climatic change and the spread of forests.

31 Artiodactyls

1. Characteristics of artiodactyls

THE even-toed ungulates, though they can be traced as a distinct line back to the Eocene, may be considered as the latest mammalian herbivores, having radiated out chiefly in the Miocene and attained then a dominance that has persisted to the present day. Except for man and the horse all the large terrestrial mammals really well established and successful at the present time are artiodactyls. Any attempt to be dogmatic about the reasons for the success of a group of animals is apt to be superficial, but it is not unreasonable to suggest that in this case the result is due to swiftness of foot, combined with keenness of sense and brain, efficient cropping and grinding mechanisms, and especially a complex stomach, allowing the digestion of cellulose by symbionts. Two families of artiodactyls survive without these special features, the pigs and hippopotamuses, forest- and water-living remnants that show us approximately the condition of the group in the Eocene.

The characters of artiodactyls show a fascinating 'similarity with a difference' to those of perissodactyls. The common origin of the two groups (p. 514) was little above the insectivore level of organization and nearly every feature has been evolved independently; the general structural similarities and detailed differences therefore show the effect produced by similar ways of life on slightly differing populations. For example, a post-orbital bar developed in both groups, for attachment of the large masseter muscle, but whereas in the horses it is formed wholly of a process of the frontal bone, in ruminants there is a union of extensions of the jugal and frontal. Many such similarities and differences are seen throughout the body, and especially in the limbs.

The skull of later artiodactyls shows changes of shape to accommodate the very deep molars and to support the horns that are commonly found. It becomes very long and there is a sharp kink between the basisphenoid and presphenoid, so that the face slopes steeply downward. The frontal bones become large and the parietals restricted to the vertical posterior face of the skull, to which the powerful neck muscles are attached. In many ruminants there is a scent gland, lying in a pre-orbital fossa of the skull and opening on the side of the head. The pre-lachrymal fossa is a gap in the skull, where the nasal cavity is separated from the outside only by the skin.

The vertebral column shows the characteristics of other large mammals in the development of high thoracic spines. Some of the heavier types have a long rib series and graviportal 'single girder' structure, but the tendency has been to retain and develop the break in structure of the vertebral column behind the thoracic region, giving a long lumbar region with forwardly directed transverse processes. In rabbits and other mammals this division of the column is associated with the jumping habit, and this is also found, though in a different form, in ruminants. Associated with this method of progress is a fore-and-aft elongation of the pelvic girdle, the ischium being well developed for the attachment of the retractor muscles of the thigh. In making the jumping movements, which are common in all artiodactyls and are especially used by the mountain-living types, the extensor muscles of the back (sacrospinalis and multifidus) work with the retractors of the two hind limbs to give a powerful thrust.

The characteristic of the limbs is, of course, the equal development of digits 3 and 4, with reduction of the rest. The gait was at first plantigrade, then digitigrade; hoofs, differing from those of perissodactyls, have developed on the toes. The elongation of the lower segments of the limbs and shortening of the upper has been similar to that of perissodactyls, but the long metapodials have become united in later forms to make the 'cannon bone'. The ulna and fibula become reduced, as in horses. The presence of two digits has led to the retention of two bones in the distal row of carpals, the hamate (unciform) and fused magnum-trapezoid, and these articulate in interlocking fashion with the three proximal carpals (Fig. 28.2). Similarly in the hind foot the two cuneiforms are fused to thrust upon the third digit, while the fourth sends its thrusts to the cuboid and the latter is fused with the navicular. Between this compound bone and the talus (astragalus) there is a very characteristic joint, the under surface of the talus being grooved like its upper surface. These joints of the carpus and tarsus are evidently an important part of the apparatus of loco-motion. Probably in both limbs they serve to take strain when the animal is moving over uneven ground, and in the hind legs they are the seat of a considerable propulsive thrust from the calf muscles. In walking, the limb of artiodactyls is moved as a whole at the shoulder and hip, by action of the upper muscles. The wrist and ankle joints bend just enough to raise the feet off the ground, and the elbow and knee joints, lying so high as to be hardly visible externally, also bend little. The essence of artiodactyl locomotion is the use of the upper limb muscles; indeed the hinder part of the vertebral column has almost become part of the limb!

The dentition of artiodactyls is highly specialized. The upper incisors are lost in later types, which crop by means of the lower incisors biting against the hardened gum of the premaxilla. The canines may form tusks. Premolars are not molarized, but an efficient grinding battery is often provided by the very elongated, hyps-odont molars. These acquire a grinding surface by the development of each of the four original cusps into a longitudinal ridge – the selenodont ('moon-tooth') condition. The effect is similar to that arrived at, by parallel evolution, in horses, and the enamel, dentine, and cement, wearing at differing rates, provide a continually roughened surface. The temporo-mandibular joint is flattened, allowing rotary move-ments of the jaw, produced by the powerful pterygoid muscles.

The tongue is large and is an important part of the cropping and grinding mechanism; it is very mobile, protrusible and pointed, and the papillae covering it are often horny. Elaboration of the stomach is common to all artiodactyls. In the pigs and hippopotamuses there is a pocket close to the opening of the oesophagus and the whole cardiac side secretes only mucus, pepsin being produced on the right side. In the fully developed stomach of Ruminantia there are four chambers (Fig. 31.1), rumen (= paunch), reticulum, omasum (= psalterium or manyplies), and abomasum (= reed). The first three are lined by a stratified epithelium of oeso-phageal type, folded into muscular ridges. These are low in the rumen, form a network in the reticulum, and are overlapping leaves in the omasum. Food is first swal-lowed into the rumen, where it is mixed with mucus and acted upon by a fauna of anaerobic cellulose-splitting bacteria, whose enzymes break up the walls of the plant food and reduce the whole to pulp. Organic acids, from acetic acid upwards, are produced, absorbed into the

circulation, and metabolized. There are also protozoans and bacteria in the rumen, which digest cellulose and are themselves later digested.

The process of rumination depends upon an oeso-phageal groove running from the cardia to the opening of the omasum. When the lips of this are brought together food does not enter the reticulum and is returned from the rumen to the mouth. After chewing, the bolus is again swallowed, the groove opens, and the food passes to the reticulum and omasum. Here water is pressed out and absorbed and the remainder proceeds to the abomasum, the 'true' stomach, with peptic glands. This elaborate digestive mechanism has no doubt contributed largely to the success of the artiodactyls. The efficient cellulose-splitting system enables them to make use of hard grasses and other unpromising sources of nutrient.

The brain is moderately well developed in later artiodactyls, but even here the cerebral hemispheres only partly cover the cerebellum, and in the earlier forms the brain was relatively small, as it is today in hip-popotamuses and pigs. The olfactory organ and related parts of the brain are well developed and most artiodac-tyls also have large eyes, with a horizontal pupil, and long ears and an acute sense of hearing.

In many species there is an elaborate communication system of conventional signals, which may serve for attraction (as in courting), repulsion, submission, or warning (see Leuthold 1977). Signals may be visual, by posture or movement, auditory, olfactory, or tactile. The vocalizations of different species of bovids are often very similar. This may be due to selection for particular efficient features, such as ease of localization, but it may be advantageous that alarm calls are understood by all members of multi-species associations.

Olfaction plays a very important part in the in-teractions of ungulates with each other and with their environment. On mating the animals sniff each other's nose, mouth, and genital regions, gaining information from the pheromone secretions there about their iden-tity and physiological and sexual condition. Mothers and offspring recognize each other by smell. Social and territorial organization is probably served by scent-marking, the deposition of urine or faeces, often around the borders of a territory. The antorbital gland of many bovids is a pit between the eyes whose secretion is spread on selected sticks or twigs or on the ground. The marking is connected with social status and often with agonistic behaviour. The colour of the coat and es-pecially the form of the head and horns also play an important part in the communication system between individuals.

The reproductive system remains rather close to the

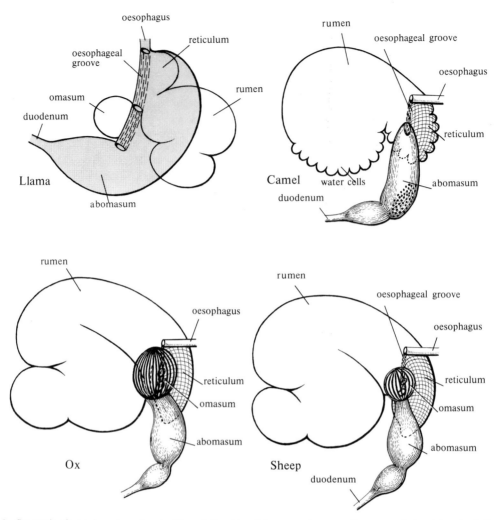

Fig. 31.1. Stomach of camels and ruminants. The relationship of the normal mammalian stomach (top left, stippled) can be compared with that of ruminants. The rumen represents the cardiac region, the reticulum the body. The oesophageal groove and omasum are derived from the lesser curvature as far as the incisura angularis and the abomasum represents the pyloric antrum. The omasum and abomasum are shown as if pulled downwards. In the camel are the water cells. The abomasum is mostly lined with stratified squamous epithelium; fundic glands are found only in the dotted area. (Material for figure kindly supplied by the late Professor A.T. Phillipson, partly after Pemkopf.)

presumed original eutherian condition. The uterus is bicornuate and in pigs the placenta is of the diffuse epitheliochorial type. In ruminants there is a cotyle-donary placenta, but the contact between maternal and foetal tissues is never very close (syndesmochorial) and the allantois is usually large.

2. Classification

Superorder 5: Paraxonia
   Order Artiodactyla
      Suborder 1: *Palaeodonta (ancient teeth). Lower Eocene–Oligocene
         Diacodexis, North America; *Homacodon*, North America
      Suborder 2: Suina (pigs). Eocene–Recent
         Entelodon, Holarctic; *Sus*, pig, Eurasia (then world-wide); *Phacochoerus*, wart-hog, Africa; *Tayassu*

(= *Dicotyles*), peccary, Central and South America; *Potamochoerus*, water-hog, Africa;
    *Anthracotherium, Eurasia; *Hippopotamus*, Eurasia, Africa
Suborder 3: Ruminantia (repeat chewers). Eocene–Recent
   Infraorder 1: Tylopoda (knob feet). Eocene–Recent
      *Merycoidodon (= **Oreodon*), North America; **Agriochoerus*, North America; **Anoplotherium*,
         Europe; **Cainotherium*, Europe; **Xiphodon*, Europe, **Amphimeryx*, Europe; **Protylopus*, North
         America; **Poebrotherium*, North America; **Alticamelus*, North America; *Lama*, alpaca, South
         America; *Camelus*, camel, dromedary, Asia
   Infraorder 2: Pecora (cattle). Oligocene–Recent
      Superfamily 1: Traguloidea (nibblers). Eocene–Recent. Holarctic, Africa
         *Archaeomeryx, Asia; *Tragulus*, chevrotain, Asia; *Hyemoschus*, water chevrotain, Africa
      Superfamily 2: Cervoidea (deer-like). Oligocene–Recent. Holarctic, South America
      Family 1: Palaeomerycidae. Oligocene–Recent
         *Blastomeryx, North America; **Palaeomeryx*, Europe; *Moschus*, musk deer, Asia
      Family 2: Cervidae (deer). Miocene–Recent
         Cervus, red deer, American elk, etc., Holarctic; *Dama*, fallow deer, Eurasia; *Rangifer*, reindeer,
            Holarctic; *Capreolus*, roe deer, Eurasia; *Alce*, moose, European elk, Holarctic
      Family 3: Giraffidae. Miocene–Recent. Eurasia, Africa
         Giraffa, giraffe, Asia, Africa; *Okapia*, okapi, Africa; **Palaeotragus*, Eurasia; **Sivatherium*, Asia
      Superfamily 3: Bovoidea (ox-like). Miocene–Recent
      Family 1: Antilocapridae (antelope-goat). Miocene–Recent. North America
         *Merycodus; *Antilocapra*, prong-horn antelope, North America
      Family 2: Bovidae (oxen). Miocene–Recent
         *Eotragus, Europe, Africa; *Gazella*, gazelles, Eurasia, Africa; *Taurotragus*, eland, Africa; *Aepyceros*,
            impala, Africa; *Bos*, cattle, yak, Eurasia and North America (now world-wide); *Bison*, buffalo,
            Holarctic; *Capra*, goat, Eurasia, Africa (now world-wide); *Ovis*, sheep, Holarctic, Africa (now
            world-wide)

3. The earliest artiodactyls

Although abundant fossil material is available, the lines of evolution within the artiodactyls are not altogether clear, and numerous classificatory arrangements have been suggested. We shall recognize three suborders. The suborder Palaeodonta contains the ancestral Eocene forms and some of their little-modified descendants. The Suina includes the pigs and hippopotamuses. Probably no members of this group developed the ruminating habit, and they are sometimes called 'non ruminantia'. The suborder Ruminantia includes all the other modern forms of artiodactyl. The Palaeodonta were close to the ancestral stock of all placentals. In **Diacodexis* from the North American Lower Eocene there were tritubercular molars and it was probably a small, running, omnivorous form, with four toes on each foot. These animals could indeed almost equally well be classified as insectivores or creodonts, and the only reason for placing them as artiodactyls is that the talus had the typical pulley-like lower surface. Later Eocene and early Oligocene forms developed a bunodont condition, with sometimes six cusps in the upper molars, protocone, paracone, metacone, hypocone, protoconule, and metaconule. In some later forms these cusps show a selenodont condition (Fig. 31.2).

4. Suborder Suiformes. Pigs and hippopotamuses

The pigs of the Old World and peccaries of America have remained essentially in the Eocene condition; Simpson (1945) recognizes this by classifying them with the Eocene forms in one infraorder Palaeodonta, distinct from the amphibious Ancodonta (hippopotamuses and anthracotheres) and the **Oreodonta. Close re-latives of the pigs are found from the Eocene, with bunodont molars. Several lines can be recognized, including the entelodonts, giant pigs of the Oligocene, over 1.5 m high and 3.6 m long and of graviportal structure (Fig. 31.3). The modern pigs (Fig. 31.4) show a nearly complete dentition, with persistently growing canine tusks in the male (Leuthold 1977). The tusks are curved upwards in pigs, downwards in peccaries and are used for combat between males, for defence, and for digging roots. The orbit is continuous with the temporal fossa. There are no horns. There are four toes, but only two reach the ground. The brain is small. With the exception of the African wart-hog they mostly live in marshy, forest conditions and are omnivorous, digging with the robust but sensitive snout for food detected by smell. Large domestic boars have been known to break up a concrete floor. The neck muscles are very large.

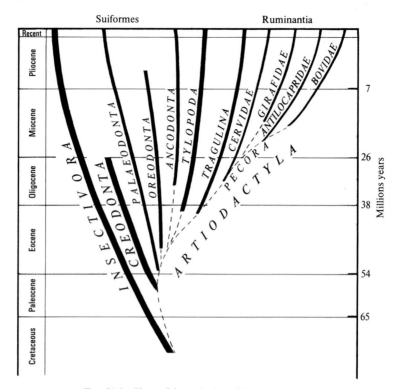

Fig. 31.2. Chart of the evolution of the Artiodactyla.

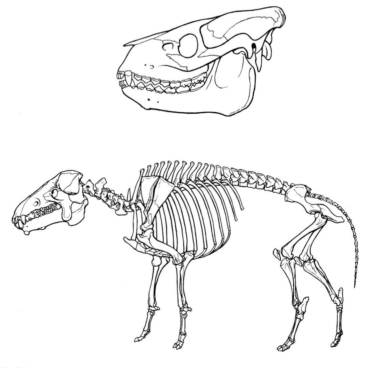

Fig. 31.3. Above, skull of the oreodont *Merycoidodon*. (After Scott 1898.) Below, skeleton of *Entelodon*. (After Woodward.)

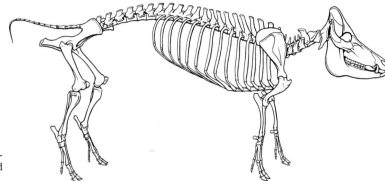

Fig. 31.4. Skeleton of the pig. (Modified after Ellenberger, from Sisson and Grossman 1938.)

Pigs live in families or small troops. The male produces a great volume of semen and there are a large number of young.

Sus is an Old World genus, found from the Miocene onwards and now extends from Europe to Japan. It was represented in Great Britain by the wild boar, until the sixteenth century. The African wart-hog (*Phacochoerus*), is diurnal, crops grass, and does not dig (Fig. 31.5). *Potamochoerus* is the red river-hog of Africa. The peccaries of Central and South America are similar to the pigs, but have been distinct since the Oligocene (Fig. 31.6). There are two small accessory digits in the fore foot and one in the hind. A large scent gland opening on the back resembles a second navel.

An offshoot of the palaeodont line in the Eocene led to the development of a race of large amphibious animals, the anthracotheres and hippopotamuses (Figs. 31.7 and 31.8). They have an enormous barrel-like thorax and large lungs, short, relatively thin legs with four digits having nail-like hooves, and a complete dentition with tusk-like canines and low-crowned bunodont molars, wearing to a foliage pattern. The stomach is enormous and partly divided into three chambers, only the posterior part of the third produces gastric juices. The rest contain symbiotic bacteria and

Fig. 31.6. The collared peccary, *Tayassu*. (From photographs.)

ciliates, more varied and less specialized than those of the ruminants but able to digest cellulose and produce fatty acids. There are many specializations for life in the water, including eyes, ears, and nose on the top of the head, muscles for closing them, and a broad muzzle. They can remain submerged for five minutes. Hippopotamuses produce a secretion, the so-called 'bloody sweat' in which the skin appears pink. There are no red blood cells in the sweat and its composition is as yet unknown although a bacteriostatic agent is present, perhaps to give some protection against the polluted waters.

Fig. 31.5. The wart-hog, *Phacochoerus*. (From photographs.)

Fig. 31.7. Hippopotamus, *Hippopotamus*. (From photographs.)

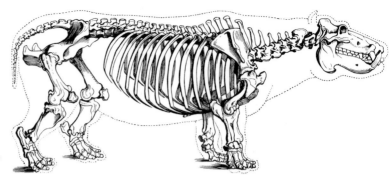

FIG. 31.8. Skeleton of the hippopotamus. (From Owen 1866.)

Modern hippopotamuses are found only in Africa, but they were widespread throughout the Old World until Recent times. They live in schools of up to 100, comprising cows, juveniles, and a few bulls. They graze mainly on land and at night. The males maintain territory by threat and fighting, often with serious wounding and sometimes death.

5. Suborder Ruminantia

Several further lines developed from palaeodont ancestors and are all classed together. Most or all of them probably had a complex stomach and some form of rumination. The molars are selenodont and the canines reduced. They fall into two main groups the more primitive infraorder Tylopoda including the camels and several early lines, and the Pecora for all the other forms. However this is a rather rough 'horizontal' division, and obscures the many lines of artiodactyls evolving in parallel (Fig. 31.2).

6. Infraorder Tylopoda

The oreodonts were abundant and successful herbivores, living in North America from the Eocene to the early Pliocene (Fig. 31.3). They had long bodies and short legs and perhaps somewhat resembled pigs. There were four functional digits in each foot, and a complete tooth row, including molars whose cusps were selenodont and in some later forms quite high-crowned. Although Eocene intermediate forms have not been found, it may be presumed that the oreodonts arose from a basal palaeodont ancestor. They pursued an independent evolution in North America parallel in some ways to that of the ruminants in the Old World. Unlike the latter, they were at a disadvantage in the changed conditions of the Pliocene and then died out.

Agriochoerus and its relatives were oreodonts that acquired claws and therefore represent a parallel to ancylopod perissodactyls, with which they were for a long time confused. It has been guessed by some that the claws were used for digging roots, by others that they

were for climbing. Several other lines of ruminants developed in the Eocene and disappeared by the end of the Oligocene. *Anoplotherium*, common in Europe, was 90 cm high at the shoulders. *Cainotherium* and its relatives in Europe were small creatures like hares with long slender hind legs. Two further European types were *Xiphodon* perhaps near to the ancestry of the camels and *Amphimeryx* possibly to the Pecora.

7. Camels

The camels have been common animals since the late Eocene, flourishing especially in North America although, like the horses, they died out there very recently and survive today in the wild only as remnants, which migrated from North America in the Pleistocene, the camels to the Old World and the llamas to South America. The one-humped camel (*Camelus dromedarius*) and the two-humped (*C. bactrianus*) are closely related and cross readily. The one-humped exists only in domestication, the two-humped lives mainly in central Asia with a small population of possibly wild animals existing in the Gobi desert. The llamas (Fig. 31.9), though similar in basic structure to the camels, differ in the smaller size, long hair, and lack of hump. The vicuña

FIG. 31.9. Llama, *Lama*. (From photographs.)

(*Lama vicugna*) are mountain dwellers and have very numerous, small red cells (14 million per mm³, against 5.4 million in man). The guanaco (*L. glama*) is larger and lives in deserts. They form small herds of up to 10 females and a male. Each troop holds a territory. The llama and the alpaca are domesticated camelids bred as beasts of burden and for wool and meat. The spitting of the llama is a protective device performed by blowing out saliva and nasal mucus.

The Tylopoda show some features retained from the Eocene condition, some developments parallel to those found in the Pecora, and various special features of their own, the latter mainly in characters that suit them for life in sandy desert conditions. In the limbs (Fig. 31.10) there has been complete loss of the lateral digits and of some carpals and tarsals, but not the fusion of navicular and cuboid that is so characteristic of the Pecora. A specialized feature is the loss of the hoofs. They were present in early camels but are replaced in modern forms by a nail and large pad. The toes thus spread sideways and enable the animals to walk on soft or sandy ground. The walk of a camel is a 'rock' like that of a giraffe, in which the fore and hind legs of one side leave the ground together.

The large hump of fat on the back provides calories when metabolized. Camels can lose up to 25 per cent of their body weight by desiccation and they conserve water by reduced evaporation and a concentrated urine. They store heat during the day to avoid sweat loss and if short of water can tolerate a body temperature of 41 °C (Schmidt-Nielsen 1979). This is equivalent to a saving of 5 litres of water and also reduces the gain of heat from the environment. They can survive for 6–8 days in the desert without drinking in conditions where one day would be fatal for a man. When water becomes available they can drink up to 9 litres and re-hydrate their tissues.

The ruminating mechanism is different from that of the Pecora and simpler (Fig. 31.1). The wall of the rumen contains a number of pockets separated by muscular walls. These are usually called water pockets and have been supposed to have a storage function, with sphincters. However their walls are glandular and their function may be digestive. There is no external separation of omasum and abomasum, which form a single tubular organ with glandular lining. These differences suggest that the ruminating habit may have been evolved separately in camels and Pecora (Bohlken 1960).

In the head there are many features of similarity to the Pecora. Cropping is by means of procumbent lower incisors, working against specialized premaxillary gums; but an upper incisor and canine are still present (three incisors in the young). The molars have a typical selenodont pattern, and the structure of the skull shows the developments so commonly seen as a result of herbivorous life, such as closure of the post-orbital bar. The lips and tongue are tough and able to chew spiny desert plants. A peculiar feature of camels is that the red blood corpuscles are oval, as in no other mammal. A nuchal gland on the back of the neck of both sexes provides a sexual stimulant when rubbed against that of the partner. The placenta is of a diffuse (non-cotyledonary) syndesmochorial type.

There is no doubt that many of the skeletal features have developed independently in camels and Pecora; the Eocene camel-ancestors did not show them. At the stage

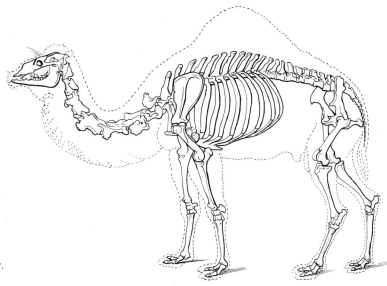

Fig. 31.10. Skeleton of dromedary. (From Owen 1866.)

of *Protylopus* the camels were small and had short limbs, with separate radius and ulna and four digits in the manus. Throughout the Oligocene many primitive features still remained. *Poebrotherium* was about 90 cm high, with a complete dentition, and orbit only partly closed behind. However, the lateral toes had been lost and the digits began to diverge distally, though probably still carrying hoofs. The remaining increase of size, and the development of other special camel features can be traced slowly through such types as *Procamelus* of the Miocene and Pliocene. During the dry Miocene times the type was very successful in North America and developed various lines, such as *Alticamelus*, the giraffe-camels, with long necks and legs. Camels persisted in the south-west of North America almost to modern times.

8. Infraorder Pecora

The most successful modern artiodactyls, the deer (Cervidae) and cattle, sheep, and antelopes (Bovidae), have flourished only since the Miocene and are thus the most recent ungulate group, largely replacing the tylopods, oreodonts, and still earlier protungulate and paenungulate types. They have always been mainly an Old World group and this remains their headquarters, though some have reached other parts of the world. The early ancestors of the ruminants can be traced to Oligocene and late Eocene forms very like the early camels, oreodonts, and other primitive artiodactyls; the modern chevrotains (*Tragulus*) retain some of the features of this early stage.

The characteristic features of the ruminants are the full development of the feeding system described on page 545, with loss of the upper incisors and often also of the canines, development of selenodont molars, and of a four-chambered stomach. In the legs only two digits are functional, though traces of others may be found. Besides loss of the extra carpal and tarsal bones there is fusion of those that remain and in particular of the navicular and cuboid. Protection is afforded mainly by swift running, kicking, and keen senses. Nearly all ruminants also have antlers or horns on the head, which are used mainly though not exclusively for intraspecific fighting, especially among males. In the more primitive species the fighting organs are enlarged teeth, in the Tragulidae large upper canines. Short sharp horns are the most primitive type, useful for defence but in social fighting they are apt to cause injury and species possessing them may seldom fight but have ritual displays (Fig. 31.11). In the more evolved ruminants the horns of the males become elaborate organs to allow ritualized head-on contests in which there is little damage. The shapes become complex to allow interlock-

FIG. 31.11 Grant's gazelle. Intimidation display showing a combination of lateral display and facing away. (From Walther (1974). *Calgary Symposium* 56–106.)

ing (Fig. 31.12). The cervids and bovids have independently evolved this method of social organization.

The ruminants have been an actively expanding group since the Miocene and are now the most successful of the ungulates, existing in vast herds in Africa and to a lesser extent in Asia and North America, and several species even penetrating to South America. The modern deer (Cervidae), are mostly browsing creatures with bony deciduous antlers, still close to the central stock. From this have been derived the giraffes and the bovoids, the great group of grazing ruminants including the primitive prong-horn antelope and the host of sheep, goats, cattle, and antelopes.

9. Superfamily Traguloidea. Chevrotains

The deerlets or mouse-deer of Africa and Asia (Fig. 31.13) keep so many primitive features that al-

FIG. 31.12. Impala. Two males showing grooming behaviour during an agonistic encounter. (Leuthold 1977.)

FIG. 31.13. Chevrotain, *Tragulus*. (After Beddard, *Cambridge Natural History*. Macmillan, London.)

though they are certainly related to the ancestors of the other ruminants it would be almost as easy to class them with the camels. This is another case where it is difficult to decide whether to make horizontal or vertical classificatory divisions; any system is bound to be arbitrary. They are peculiar little creatures, only 30 cm high, with more external resemblance to a rodent than to modern deer. In some features they show suggestions of similarity to the pigs. There are no horns, but the upper canines are large and tusk-like. The upper incisors have been lost and the molars are selenodont, but the stomach has only three chambers. The fibula is complete. The feet have four hoofed digits in each limb, and although the two main metatarsals are fused to form a cannon bone the metacarpals are still partly separate. The navicular and cuboid make a single bone, this being a 'diagnostic' ruminant character. The placenta is diffuse, as in camels. Chevrotains are nocturnal and the individuals live alone in the forests, pairing only for breeding. *Tragulus* has three species in India, Malaysia, and the East Indies. *Hyemoschus*, the water chevrotain, lives in tropical West Africa.

In many of these characteristics the chevrotains show signs of retention of an ancient organization, and fossil forms similar to them are found in the Miocene. Animals not very different (*Archaeomeryx*) are found back to the Eocene. They had a full set of upper incisors and the metapodials were separate in both limbs – but the fibula was incomplete. They were very like the early camels or palaeodonts and may well be close to the ancestry of all ruminants. Several similar types and lines can be recognized. Evidently the group represents a population persisting with rather little change since the Eocene, and this status is represented by recognizing a superfamily Traguloidea of the Pecora, contrasting with the other superfamilies of higher ruminants.

10. Superfamily Cervoidea, Family Palaeomerycidae. Ancestral deer

The ancestral population from which the deer arose must have been quite similar to the Eocene traguline *Archaeomeryx*, but the group does not become distinctly recognizable until the Oligocene. The early members mostly had large canines, low-crowned molars for browsing, and no antlers, as in *Blastomeryx* of the Miocene of North America. Others such as *Palaeomeryx* of the Miocene of Europe possessed a bony outgrowth covered with skin and not shed. In *Moschus*, the musk-deer (Fig. 31.14) of central Asia, the males have large permanently growing upper canine tusks and no antlers and the animal is probably a survivor of this Miocene stage of evolution. It is intermediate in many respects between *Tragulus* and the Cervidae and some classify it with the former others with the latter. We shall include it among the palaeomerycids, following our plan to call attention to surviving types. Musk-deer are about 60 cm high and the individuals live alone in mountain forests. The much-valued musk of the male is secreted by a preputial gland, in the form of a sac.

11. Family Cervidae. Deer

The true deer appeared in the Miocene of Eurasia. The early forms were small with simple antlers and large canine tusks, such as are found today in the males of the muntjacs (*Muntiacus*) of south-east Asia, which are solitary nocturnal animals. From the Pliocene onwards larger forms with complex antlers appeared, presumably with social habits. These bony growths are shed each year and form progressively more branches as the animal grows older (Figs. 31.15 and 31.16). In the reindeer the females also have antlers. A sign of the rather primitive nature of the deer is the retention in the manus of definite rudiments of the first two phalanges of the lateral digits (dew claws). The molars are brachydont, but the placenta cotyledonary as in bovidae.

FIG. 31.14. Musk-deer, *Moschus*. (After Beddard, *Cambridge Natural History*. Macmillan, London.)

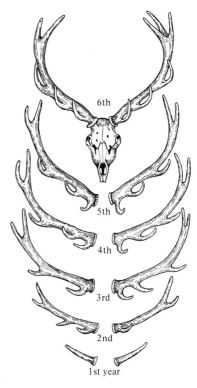

FIG. 31.15. Growth of the antlers of a male deer. The bony outgrowth is covered with very vascular skin (velvet) which is shed when growth is complete, shortly before the rut. The antlers are then shed. (From Hamilton 1939.)

FIG. 31.16. Series of antlers in the British Museum (Natural History) showing the increasing number of tines in successive years. (After Romanes.)

The deer (Fig. 31.17) have been common since the Pliocene, as browsing animals of the forests of the Holarctic region and South America, but not Africa. They live in herds with an elaborate social organization, based on the supremacy of a leading male, maintained by a succession of 'fights' with his rivals. These fights are very fierce, but seldom result in death, and indeed the complicated antlers interlock in such a way as to mitigate their danger to the challenger. Red deer (*Cervus elaphus*) are still wild in Britain in Scotland, the Lake

FIG. 31.17. Deer, *Cervus*. (From life.)

District, Exmoor, and the New Forest. The antlers have six or more points. Roe deer (*Caproelus*) have smaller antlers (three points) and are not gregarious. They are also indigenous in Great Britain. Fallow deer (*Dama*) have been introduced, and are usually spotted, with palmate antlers.

12. Family Giraffidae

The giraffes (Fig. 31.18), like the Cervidae, from which they diverged in the Miocene, are browsing animals, now restricted to tropical Africa. The teeth are low-crowned and the head bears up to five simple skin-covered bony prongs in both sexes. The horns develop in ossification centres below the skin, fusing later with the skull. Extra bone is laid down as the animal grows older, giving extra weight for use in the necking exercises. In the okapi there is a rudimentary keratinous horn at the tip. In giraffes the lateral digits are completely absent and the legs are very long; the whole structure is specialized to carry the great bulk on the fore legs, the head and neck balancing the weight of the body and the hind legs being used mainly for propulsion. The walk of a giraffe is a 'rock' in which the fore and hind legs of one side move together. Since the weight is balanced on the fore legs there is no use of the tripodal method of movement that is usual in quadrupeds. This is an extreme development of the type of vertebral organi-zation in which the weight-carrying beam ends in the middle of the back. There is only a small number of ribs, so that the hinder part of the column functions as an upper segment of the hind limbs and the extensor muscles of the back aid in propulsion. The long neck makes it possible to balance the great weight on the fore legs. It is of use not only for reaching high branches but also as a look-out among the long grasses. Special mechanisms are present to prevent changes of blood flow when the head is raised from 2.13 m below the heart to 2.74 m above it.

Giraffes live in local associations of mixed herds. There are also herds of unmated bulls, in which dominance is established by a form of sexual play by swinging blows and rubbing of the heads, known as necking behaviour. Giraffes make few sounds, indeed have little communication beyond visual danger signals.

The rare okapi (*Okapia*) (Fig. 31.19), discovered in 1900 in west central Africa, is a ruminant with shorter legs and neck, almost indistinguishable from *Palaeotragus* of the Pliocene, also possessing small horns. Other lines (*Sivatherium*) acquired a pair of large, branched, non-deciduous horns.

13. Superfamily Bovoidea

The remaining Pecora are all rather alike and are often placed in a single family, Bovidae. However, the prong-horn 'antelope' *Antilocapra* (Fig. 31.20) of the North American west, and its numerous fossil allies, all New World forms, have been distinct since before the Miocene from the true bovids, evolving in the Old World. The origin of the two groups is obscure and we have no Eocene or Oligocene fossils that are certainly on the bovid line of evolution. As already mentioned *Archaeomeryx* and other Eocene tragulines show us a type of population from which the Pecora could all have been evolved, but the stages of the transformation have not been found.

FIG. 31.18. Giraffe (*Giraffa*). (From photographs.)

FIG. 31.19. Okapi, *Okapia*. (From photographs.)

FIG. 31.20. Pronghorn, *Antilocapra*. (From a photograph.)

Antilocaprids and bovids are alike in living in herds and in their grazing habits, with which are associated deeply hypsodont molars. The side toes have been almost or completely lost, a development occurring parallel to that of the cervids, since the common ancestry almost certainly possessed lateral toes, which are indeed present in rudimentary form in some bovids. The upper canines are always reduced and the molars have high crowns and selenodont cusps. In *Antilocapra* (Fig. 31.20) the horns have two branches in the bucks, usually simple in the does but sometimes with a small prong. In both sexes they have a bony core and rather soft keratinous covering, the latter but not the former being shed. This therefore suggests how a skin-covered antler, such as that of the Cervidae, may have become converted into the bovid horn. In earlier antilocaprids, such as the Miocene *Merycodus*, the horns were more elaborately branched; evidently the group has developed a horn structure parallel to that of the Cervidae.

In all true bovids the horns are permanent coverings for the bony core. They are unbranched, though curved and twisted in various geometrically interesting ways. Moreover, they are usually borne in both sexes (though often larger in the male) and their function is definitely defensive, as well as social and sexual. Correspondingly the social organization is often into large herds, rather than into the small family groups under a dominant male such as are found among Cervidae. Grazing on open plains and mountains has presumably led to the formation of the larger herds, smaller and more closely knit groups being more suitable for forest life. The social organization varies. As an example in the impala (*Aepyceros*) of Africa some 100 females and their young are herded throughout the year by a dominant male, while the bachelors form separate herds. In other antelopes the males only dominate seasonally.

The diversification of bovids into many types may have advantages for the whole group and indeed for all ungulates. The ranges of the species often overlap. This does not lead to interspecific competition since each sympatric species occupies its own niche. The seasonal plant growth is used by a sequence first of heavier then lighter animals. Thus in Tanzania after the rains the elephants trample the vegetation and are followed by buffaloes which tread the vegetation down to lawns and these are then grazed by a succession of antelopes to the end of the dry season.

The Bovidae, with 50 genera and 115 species, is much the largest ungulate family. The original centre of evolution of the family was in Eurasia, where they are now less common, whereas in Africa they are particularly successful at present. A few types, such as the bison, reached North America, but none entered South America until man showed that they can flourish there, and indeed also in Australia. The fact that we possess numerous fossil remains and that the group is still at the height of its development makes classification very difficult. This is the situation that we should expect, if evolution consists largely in the slow change of the characteristics of populations. At first thought it may seem paradoxical that in a group so recently evolved and of which we know so much it should be exceptionally difficult to trace affinities and lines of descent. The fact is that the numerous remains of fossil bovids from the Pliocene and Miocene are still quite insufficient to enable us to reconstruct the changes in the populations. It is not really to be expected that the relatively few specimens of these large animals that can be collected and studied should show us the detailed changes, extending over 20 million years or more, by which a population of perhaps a million small creatures such as *Eotragus* of the Eurasian Miocene, developed into the present bovid population of, say, a thousand million animals, divisible into hundreds of non-interbreeding populations that range in structure and habits from the gazelle to the bison. An imaginative look at the details of evolutionary change reveals a terrifyingly complicated system, which we can hardly hope to follow in detail. The geological information can surely never be sufficient to show us the necessary facts about the variation of such great populations, and their gradual changes, at least in the case of animals as large and rarely preserved as Bovidae. We know hardly anything about variation and heredity in our own cattle, so how can we hope to follow the genetics of their ancestors? Yet nothing less than a full view of the gradual population changes will show us how the evolution of a group has proceeded.

Following types of organization over long geological periods gives a deceptively simplified idea of the stages traversed. We recognize the 'stages' because, fortunately for us, only tiny remnants of the populations have been preserved, and perhaps some 'primitive' types remain to the present day. Thus in the long history of the perissodactyls we can refer all our modern and fossil forms to some 160 genera. The tapirs are there to show us a very ancient condition, and we know just enough fossil horses to arrange them in a number of series with side branches, so that we feel that we can imagine the whole evolution of the group. It is interesting that the sequence of evolution used as a type specimen for students is so often that of the horses and not of the bovids, although we have very much more material for the latter, at least in the later stages. It would be wise to study the two together and to learn from the difficulty of recognizing clear-cut boundaries among the millions in the herds of intergrading sheep, goats, oxen, and antelopes that the most important result of the discovery of evolutionary change was the realization that our logic and use of words can no longer depend, as the ancients thought and many backward-lookers still wish today, on the recognition of a certain number of 'species', to one of which every individual can be referred.

In trying to classify the Bovidae we may perhaps recognize a central group of 'antelopes', but the term is vague and certainly includes several diverse lines; it is not even possible to find criteria for saying 'this is an antelope, that a cow, and this other a sheep'. 'Typical' antelopes (Fig. 31.21) live in Eurasia and Africa, especially the latter. They are rather tall and slender, with smooth hair and backward-curving horns, living mostly on warm or tropical plains. The gazelles may be taken as an example among many. The oxen are heavier animals, often almost of graviportal structure, but with very high thoracic spines; they live in cooler conditions on more northern plains and have more and shaggier hair. The untwisted horns usually curve forwards. They originated in Eurasia. The domestic cattle and yaks, *Bos* (Fig. 31.22), are good examples, and the *Bison* (Fig. 31.23) are related creatures, now almost restricted to North America. There are animals, however, that,

FIG. 31.21. Impala antelope, *Aepyceros*. (From life.)

with the criteria used, cannot be classed as either 'antelopes' or 'oxen'; for instance, the elands (*Taurotragus* (Fig. 31.24) of Africa are large and cow-like, but have backwardly directed and twisted horns. Similarly the caprine (sheep and goat) section of the family is not clearly marked off from the antelopine. The goats

FIG. 31.22. Central Asian Yak, *Bos*. (From photographs.)

FIG. 31.23. American bison, *Bison*. (From photographs.)

Fig. 31.24. Eland, *Taurotragus*. (From photographs.)

(*Capra*) are characteristically mountain-living animals of the Holarctic region, with backwardly curved but not twisted horns, and the sheep, *Ovis* (Fig. 31.25), are closely related, but addicted to less mountainous country and with a slight spiral on the horns. The beard and scent glands of the male are signs that there are great differences in social and sexual organization between sheep and goat life, in spite of the similarity in structure. The small Soay sheep on the island of St. Kilda is perhaps the most primitive domestic form in Europe and resembles the original wild species and the domesticated form of it that was brought to Britain about 5000 BC and persisted on the mainland until medieval times (Jewell, Milner, and Boyd 1974).

Thus there is very great variety of life and structure in the modern bovids, and one of the most striking conclusions about the group is that the specialists have not yet succeeded in finding an agreed system of classification. Of more practical importance than these logical problems is the fact that with their fine cropping mechanism, grinding battery of teeth and stomach symbionts the bovids provide a satisfactory intermediary by which grass can be used as a contributor to human life. This, with their peaceful and gregarious disposition, has made them our most important commensal mammal. If there were no Bovidae there would be fewer human beings in the world, and our social organization would be very different. The obverse is also true; it is probable that their lives and ours will continue to evolve together and mutually to modify each other.

Fig. 31.25. Barbary sheep, *Ammotragus*. (From life.)

32 The efficiency of mammals

1. What is efficiency?

MAMMALS have undoubtedly taken the place of reptiles in many niches. Is it possible to show that this success has been due to their better organization? Again, since many of the earlier types of mammal have been replaced by later descendants can we recognize this as the result of further improvements? It has been emphasized in Chapter 1 and throughout the book that questions of the efficiency of living organization can only be considered in relation to the difficulty that each particular environment presents for the maintenance of life. Yet it is very hard to give quantitative estimates of the 'problems' that different organisms need to overcome. We tend to fall back on comparisons of the complexity of the animals themselves, or their inherited control systems or behaviour.

It is indeed difficult to arrange mammals, or for that matter any animals, in an order of 'efficiency' (see Schmidt-Nielsen, Bolis, and Taylor 1980). Each species considered alone is adequate so long as its members continue to survive in their particular surroundings. Fishes, usually considered to be 'lower' vertebrates, are still the dominant aquatic vertebrates. Snakes are more 'efficient' than any mammal for that particular way of life. Yet there is good evidence that there have been repeated replacements in the history of many types of animal, for instance fishes, or large terrestrial herbivores (p. 581). We should very much like to know what was less efficient about the unsuccessful types, but as they are extinct we cannot study them directly. Two methods are open to us, we can use the fossil remains, say of teeth or limbs, to deduce how the earlier types functioned. Or we can compare survivors that retain earlier types of organization with their modern descendants. This method has great dangers, however, because the very fact of survival of, say, *Latimeria* or *Ornithorhynchus* means that their organization is efficient for the life that they live. We shall see some examples of how such survivors have actually taken advantage of the earlier type of organization. It is therefore unwise to label them as 'primitive' in contrast

to 'advanced'. Rather we shall speak of their conservative (or plesiomorphic) characters, contrasting with those that are derived (or apomorphic).

To make our judgements, therefore, we can compare the way in which given environmental challenges are met. An alternative criterion of efficiency would be to estimate the variety of conditions in which each animal type can survive. Such assessments of adaptability are often made, but are seldom quantitative. Another alternative is to make judgements of efficiency by comparisons between the performance of animals and that of human artefacts, but the value of such comparison varies. Energy exchanges and mechanisms of locomotion are easier to compare than hormonal or nervous control systems. Yet some useful comparisons can be made even of such matters, and of reproductive strategies, which we might call the techniques of capital replacement, although they are utterly different from those of manufacturing industry.

2. Efficiency of temperature control

Endothermic temperature regulation is one of the main features in which mammals show an 'advance' over reptiles. There can be little doubt that a high and constant temperature has allowed them to displace cold-blooded, or ectothermic, animals from many niches and to occupy others not previously open, for instance activity at night or in cold climates. Homeothermy involves many special devices such as hot and cold sensors, hair, shivering, high metabolism, control systems, and, of course, the consumption of much food. Therefore in one sense reptiles are the *more* efficient since they regulate without having to use all this expensive equipment, simply by basking in the sun or retiring to the shade. In order to do this, however, they already have sensors and even hypothalamic temperature-sensitive centres (Whittow 1970). They lack special cooling devices for the whole body and can only keep a degree or two below ambient temperature, but they have arrangements that selectively cool (or heat) the brain (p. 276).

The state of homeothermy might be considered as 'primitive' or 'inefficient' if the body temperature is low, or if it is unstable, or if the system is unable to meet cold or hot conditions. Many conservative mammals among the monotremes, marsupials and edentates do, in fact, have lower temperatures than most eutherians, when measurements are made in the unstressed state (which is not easy to do for a small animal) (Table 32.1). There may, however, be an *advantage* in regulation of temperature about a rather low set point, as an echidna does between 28 and 30 °C. Small animals cannot afford to cool themselves in environments that are more than 2 or 3 °C above their body temperature, because of excessive water loss. It is efficient, therefore, for them to have a low body temperature and a rather broad band of regulation. This allows temperature to rise during a period of activity, followed by retreat for cooling (as does the chipmunk, *Eutomias*).

This is one reason why so many small mammals are nocturnal. Perhaps the earliest use of homeothermy was to invade the nocturnal niche and for this a lower temperature is adequate and economical. Higher temperatures may have evolved later when mammals invaded diurnal habitats (see Schimdt-Nielsen *et al.*

1980). Sweating and panting may have evolved independently in several groups. Some small modern animals such as dasyurids and rodents do not pant or sweat. Some insectivores seem to have remained in the nocturnal niche since their appearance some 75 Ma ago and they are efficient for this, as fishes are in water. Thus *Tenrec* in Madagascar (p. 436) has a body temperature of 29.5 °C in ambient temperatures of 24 °C and their metabolic rates during exercise are similar to those of reptiles. Failure to maintain a high temperature can also be 'efficient' for a small mammal with a large surface area. Thus microchiropteran bats regularly become cool when they cease flying (p. 440). But this habit limits their life pattern because to remain safe they must retire to a protected place.

Many of these conservative mammals are now known to be able to maintain a relatively constant temperature (even monotremes) (see Whittow 1970). They may have a lesser capacity to respond to cold than eutherians, perhaps owing to their lower basal heat production (see later). Most of them have some response to high temperatures, marsupials mainly by panting and licking; they tend to store heat during exercise and lose it later. A low resting temperature is of course an advan-

TABLE 32.1

	Standard metabolism (W kg$^{-0.75}$) (ml O$_2$ kg$^{-0.75}$ h^{-1})	Body temperature (°C)	Normalized (38 °C) standard metabolism (W kg$^{-0.75}$)
Reptiles			
Tuatara	0.19	30	0.39
Lizards	0.40	30	1.02
Snakes	0.48	30	1.02
Crocodiles	0.29	23	1.06
Turtles	0.15	20	0.58
Mean			0.81
Mammals			
Monotremes,			
Echidnas	0.92	31	1.55
Platypus	2.21	32	3.45
Marsupials	2.37	35	3.00
Eutherians			
Advanced	3.34	38	3.34
Insectivores	2.76	35	3.63
Edentates	1.69	33	2.66
Mean			2.94

TABLE 32.1. Comparison of the standard metabolism of reptiles and mammals. The metabolism is given as oxygen consumption in ml W kg$^{-0.75}$ h^{-1}; this is the minimum resting oxygen consumption or standard metabolic rate (SMR) at 25 °C. (Hulbert 1980.)

tage for heat storage. Sweating is rare in marsupials but occurs in large kangaroos, though at a rate less than half that of cattle or horses.

3. Energy metabolism

In order to compare the energy consumption in different animals we must eliminate the effects of size and temperature (Table 32.1). Size can be allowed for since energy metabolism has been found to vary as the 0.75 power of body weight in all animals (though the reason for this relation is not agreed). It is found to vary with temperature by a Q_{10} between 2.1 and 3.3. The standard metabolism can therefore be normalized to 38 °C and $Wkg^{-0.75}$ and is then found to be four to five times higher in mammals and birds than in reptiles. The difference is due to several factors. The brain is metabolically very active and is larger in mammals; the cytochrome oxidase is more active in mammals, and there are more mitochondria. The thyroid gland has a powerful influence on the metabolism of a mammal, but after thyroidectomy reptiles show no fall in its level. These features must depend upon new genetic factors introduced by selection at some point in the origin of mammals and giving them advantages over reptiles by allowing faster contraction of muscles, conduction of nerves and activity generally.

There are differences of metabolic level within mammals. The standard level for marsupials is about 70 per cent of that for eutherians, while in echidnas it is lower still. These values are, of course, associated with the low body temperatures and mostly disappear when normalized to 38 °C (Table 32.1). However, the rate for echidnas and edentates is low even when normalized, and these perhaps retain the earlier condition. Thyroidectomy of an echidna is not followed by lower metabolism. Some desert animals such as camels and bandicoots (*Macrotis*) also have low metabolic rates, presumably to minimize expenditure on scarce food.

The changes of metabolic level during exercise differs among mammals and the energy cost of locomotion varies. It is lower ($\frac{1}{3}-\frac{1}{5}$) in insectivores than in more advanced mammals (Taylor 1980). Brachiation is one of the most energy 'expensive' means of progression, but such comparisons illustrate the problem of defining 'efficiency'. The brachiating gibbon may use a lot of energy, but is able to survive in a difficult 'habitat' that is inaccessible to the hedgehog. The hopping of a kangaroo incurs a high cost at low speeds but above 18 km/h it moves more cheaply than quadrupedal eutherians of similar size. This economy at the high speeds at which they normally move (25–30 km/h) enables them quickly to avoid fires or to cover the long distances needed to reach freshly sprouted grass after rain. The heart rate at rest or during exercise is generally lower in marsupials than in eutherians but the adjustments of cardiac output and stroke volume are similar in the two groups.

4. The mechanics of locomotion

The efficiency of locomotion depends upon such factors as the fuel economy of the engine, the materials used and design that applies power effectively and minimizes resistance. The basic materials, bones, muscle, and tendons, seem to have similar properties in all mammals. Muscles display a spectrum of properties of speed of contraction and relaxation and oxidative capacities between fast-twitch high-oxidative (FH), fast-twitch low-oxidative (FL), and slow-twitch high-oxidative (S) fibres. The FH-fibres generate high-peak tensions for movement but fatigue quickly. S-fibres contract and relax slowly but are resistant to fatigue and are thus effective for postural movement. These types appear in all mammals, but mixed in various proportions in the muscles according to the work they have to do. Thus for obvious reasons the insectivorous bats have no S-fibres. Conversely the slow loris (*Nycticebus*), which often hangs suspended, has many of them in both flexor and extensor muscles.

There are great differences in the way the muscles and bones are arranged. Cursorial animals such as horses have mostly evolved elongated distal limb segments. This reduces the mass of the limbs and hence the energy needed to accelerate and decelerate them. However, it is doubtful whether this is really a major part of the power requirement. The long limbs give other advantages, for instance in the use of the elasticity of the long tendons of the gastrocnemius and plantaris muscles as springs to save energy. For this reason it is advantageous to have short fibres in these muscles, which have a pennate structure. The kangaroo has relatively longer fibres in the plantaris than either dogs or gazelles while in the camel the plantaris has no muscle fibres at all but is reduced to a tendon running from the femur to the phalanges (Alexander 1980). This may explain the fact that the kangaroo requires more weight-specific power to move its centre of mass than does a fast-moving dog although a galloping quadruped can be compared mechanically to two kangaroos! However, we cannot take this to show greater 'efficiency' by the eutherians. Muscles store more elastic energy than tendons at high speeds, allowing the kangaroo to bounce along cheaply, as we have seen (see Cavagna, Heglund, and Taylor 1977).

5. The efficiency of tooth replacement

One of the earliest advantages obtained by the mammalian type of organization was that nutrition by milk dispenses with the need for teeth until a substantial part of the growth of the skull has been completed. A single replacement of premolars and the addition of three or four molars is all that is necessary to allow growth after weaning. This was a great economy, achieved in even the earliest prototherian mammals such as *Morganucodon*. In the mammal-like reptiles there were as many as six generations of teeth. Even more important than economy of teeth was the fact that *occlusion* became possible when new teeth no longer erupted between the old. The capacity to grind the food may have been essential for homeothermy and of course it made many new diets possible. The limited replacement also had the consequence of setting a fixed adult size for mammals, in contrast to the gradual termination of growth in reptiles.

The capacity to masticate opened the possibility for a whole new series of methods of life and in this sense represented a definite advance of the earliest mammals. This possibility was very quickly exploited and all the main mammalian orders became differentiated within a few million years in the late Palaeocene and early Eocene (p. 402). However, the earliest members of each group, whether herbivore, carnivore, or insectivore, still showed the original trituincular pattern of cusps. As animals with each way of life developed, the tooth patterns became increasingly specialized. Herbivores gradually came to have deeper and deeper hypsodont teeth with grinding surfaces of enamel, dentine, and cement. The culmination of this pattern only appeared quite recently in the Pleistocene bovids, horses, and elephants. In each case the masticatory mechanisms required special muscular arrangements. The rodents developed a different mechanism, again with gradual 'improvement' along several lines.

Carnivore teeth show the gradual development of tearing and cutting mechanisms, with special developments in some for aquatic life. The earliest whales appeared already in the Eocene but the specialized dentitions for consumption of fish, squids, or plankton only appeared later.

The Insectivora retain what was probably the original mammalian diet with little change of the teeth, even in bats which are so highly specialized in other ways. The adoption of a diet of ants or termites has led to the appearance of the same set of characteristics several times, in a marsupial, among edentates, in *Manis* and in the aardvark.

The primates show perhaps the greatest variety of diets of any order and some have developed specialized teeth, such as the molars of herbivorous monkeys. A detailed study would show that the teeth of each genus and species have features related to their different forms

of life. The heavy molars of *Australopithecus* were adapted for a different way of life from those of *Homo*.

Thus the achievement of the mammalian form of dentition has undoubtedly allowed the development of a great many ways of life, perhaps more than among reptiles. The teeth of reptiles mostly serve only to capture or retain food, though there are obvious specializations in snakes, and some mollusc-eating lizards have teeth modified for crushing. There were also specialized herbivores and carnivores among the dinosaurs, but without such dental refinements as are found in mammals.

It seems that there has been a gradual improvement in the dental mechanism within each mammalian order. Teeth are especially suitable for study of the change of efficiency, assuming the consumption of a particular diet. The grinding mechanisms of modern horses and elephants are certainly able to handle larger amounts of material than did those of *Eohippus* or *Moeritherium*. Even so we must be careful. The cusped molars of the earlier horses or the bunodont and mastodont elephants were probably suited to diets *different* from those of their descendants.

6. The efficiency of digestion

The earliest mammals were probably insectivorous and the gut was a rather simple tube as it is today in many lizards, monotremes, insectivores, and bats (Fig. 32.1). Among marsupials the carnivorous and insectivorous species also have simple guts. In the herbivores there are often complications of the stomach or caecum (as in the koala). In kangaroos the stomach is divided but never as completely as in ruminant artiodactyls. Microbial fermentation is as active as in the sheep but food is retained for a shorter time in the stomach. There is no true rumination though food may be regurgitated. Production of volatile fatty acids is similar. The process of digestion of grass is thus about as efficient in kangaroos as in ruminants.

It is equally difficult to compare the efficiency of digestion among eutherians. Those with herbivorous diets usually have a complex digestive tract, but fermentative digestion is not limited to animals with chambers specialized for it. Microbial production of organic acids occurs in hedgehogs with a simple gut and bushbabies with a caecum but rapid passage through the gut. The nutritional value of these and other digestive processes is difficult to assess. Complex mechanisms such as the stomach of ruminants, or the caecum of rabbits or horses, or colon of vervet monkeys have been evolved in association with particular ways of life. We have no fossil evidence of their gradual appearance as we have with teeth. It seems likely that during evolution within

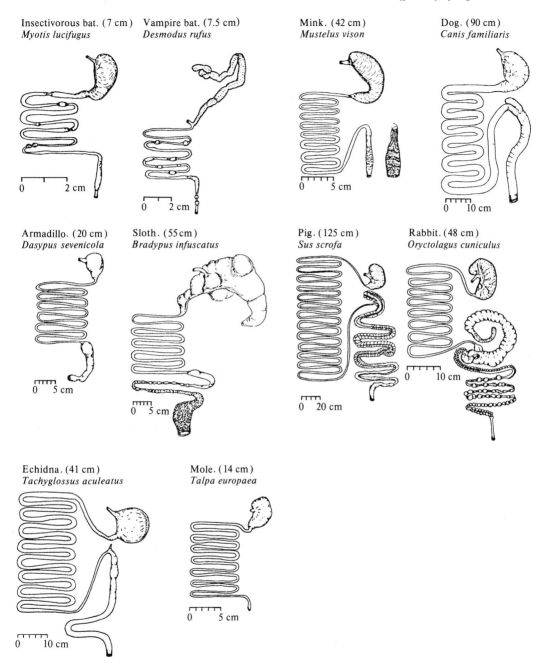

Insectivorous bat. (7 cm)
Myotis lucifugus

Vampire bat. (7.5 cm)
Desmodus rufus

Mink. (42 cm)
Mustelus vison

Dog. (90 cm)
Canis familiaris

Armadillo. (20 cm)
Dasypus sevenicola

Sloth. (55 cm)
Bradypus infuscatus

Pig. (125 cm)
Sus scrofa

Rabbit. (48 cm)
Oryctolagus cuniculus

Echidna. (41 cm)
Tachyglossus aculeatus

Mole. (14 cm)
Talpa europaea

FIG. 32.1. The digestive tract of several mammals to show various modifications. The body length is shown in parenthesis. (After Schmidt-Nielsen 1980, drawings by E. Maleck.)

each order there has been an increase in efficiency and capacity to digest materials not available before.

7. Changes in endocrine systems

Handicapped by the absence of fossil evidence we can only compare animals considered to be conservative or

advanced for other reasons. Data are scarce, especially for endocrine functions in the wild. The morphology of the various endocrine systems and the hormones they secrete are in general similar in all mammals. There is some suggestion that the more conservative mammals lack some special endocrine responses. The echidna has an unusually low rate of secretion of glucocorticoids and aldosterone; its glucocorticoids have little diabetogenic effect. Echidnas can survive loss of either the thyroid or the adrenal, but only if not stressed. This may be an example of how 'progress' in evolution has allowed for increased opportunity for survival. However, other mammals vary in these respects and it is not clear that the condition in echidna is primitive.

The enzyme 11 β-hydroxylase is very much less specific in echidna and in marsupials than in eutherians. This is an enzyme that responds readily to ACTH for the production of cortisol under stress. Eutherians make this response much more rapidly than the others and they reach a higher level. The evolution of specific 11 β-hydroxylase in eutherians may thus have given them finer control of the synthesis of cortisol, especially in response to ACTH, allowing quicker response to stress.

8. Efficiency of the brain

There is no sense in which the overall capacity of the brain of different animals can be compared. There is no such entity as 'general intelligence' nor methods by which it can be assessed. 'Brain power' (whatever that may be) presumably increases with the number of neurons, which is related in a complex way to brain size. The neurons of larger brains are more numerous but also more widely separated. This may allow more intricate connections in the 'neuropil' between them, but there is no detailed evidence about this.

Brain size increases with body size (Fig. 32.2) and various methods have been suggested to compensate for this in order to allow comparison. Jerison (1973) suggests that this can be done if all the points indicating brain and body size of a given group of animals are included in a polygon, with a principal axis fitted by eye using a slope of $\frac{2}{3}$ (Fig. 32.3). This Jerison considers to be the usual exponent for the relationship of brain weight (E) to body weight (P) where $E = kP^{\alpha}$. From Fig. 32.4 he concludes that if $\alpha = \frac{2}{3}$ there has been an increase of ten times in the proportionality constant in higher vertebrates. This constant (k) has been used as an 'index of encephalization.' But k has no biological meaning. It represents the brain weight of a hypothetical adult animal weighing 1 g. Moreover it depends upon an

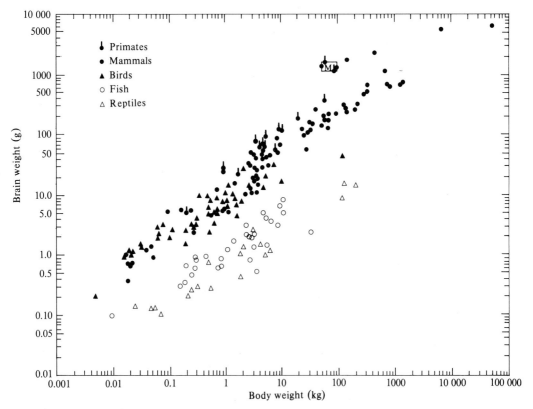

FIG. 32.2. Brain and body weights of the largest specimens of 198 vertebrates. The rectangle contains the range of data on living men. Logarithmic scale. Note that no elasmobranchs are included. (After Jerison 1973.)

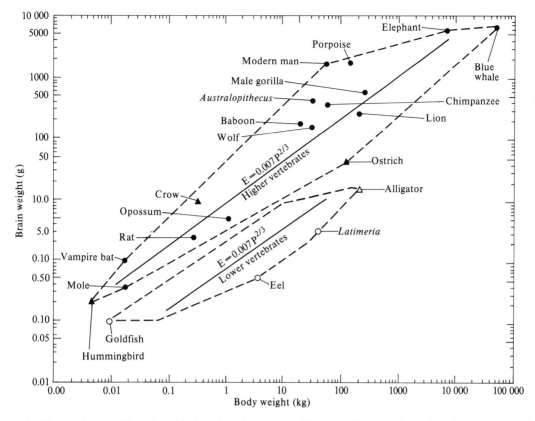

FIG. 32.3. The brain:body weight on logarithmic scale to show relationship in some living vertebrates including *Latimeria* and the fossil hominid *Australopithecus*. Minimum convex polygons are shown by dashed lines enclosing visually fitted lines with slopes of 2/3. (After Jerison 1973.)

arbitrary assumption about the slope of the line. The $\frac{2}{3}$ exponent does approximately show the relationship when numerous types of mammals are compared (Fig. 32.4), though other data show much flatter curves. Comparison of brains is often made using an 'encephalization quotient' that assumes $k = 0.12$ and $\alpha = 0.75$ (Jerison 1973). Direct comparisons of mammals of similar body weight as in Fig. 32.4 show that there has indeed been a general increase in brain size throughout the Tertiary period and moreover within several individual orders. Of course in many orders the increase of brain weight follows the greater body size, but the bigger brains are no less useful for that. It is not obvious that the basic functioning of a larger animal requires a larger brain, except possibly to handle input from the surface. So the larger brains of larger animals provide what Jerison has called 'extra neurons', giving greater power of processing information and this is shown by the cerebral powers of elephants and whales, let alone man.

It is probably justifiable to think of the mammalian

brain as the most complex of all devices for the regulation of homeostasis, although there are few measures for such comparisons. The weight of the brain in comparison with that of the body is higher in mammals and birds than other vertebrates. From this we can argue that their whole system of information transfer and control is more complex, even with the qualifications already discussed. The cerebral cortex, with its columns and layers is unique to mammals, and is present already in monotremes. However, we should not consider the cortex alone but rather the whole brain organization which includes reciprocal activity between the receptor systems, midbrain, thalamus, cerebral cortex, and cerebellum. This most elaborate system somehow gives to mammals capacities for such actions as exploration, learning, manipulation of the environment and social behaviour that are greater than those of other vertebrates, even the birds. It is curious and sad that we cannot state specifically what neural factors give these capacities.

The basic equipment that provides the standards for

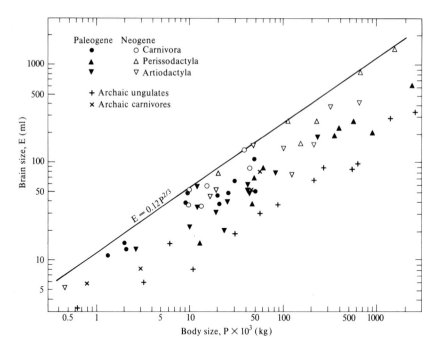

Fig. 32.4. The endocast volume, to give measure of brain size, as a function of body size in 69 fossil ungulates and carnivores; logarithmic scale. The line is the 'average' for living mammals fitted to a large and diverse set of species. (After Jerison 1973.)

homeostasis lies in the hypothalamus. It must surely be different and more elaborate in birds and mammals to allow the maintenance of set points for values that vary in lower vertebrates. But the stability of the higher animals depends upon regulating a whole complex of behaviour and not merely adjustment of the internal environment. The secret of their success may be in the mechanisms that allow for such activities as exploration and social behaviour. The frontal lobes and cingulate cortex are probably connected with the maintenance of proper levels of activity of these more elaborate mammalian brain programs. It is significant that the cingulate cortex receives inputs from all the main sensory areas and sends part of its output to the hippocampus. This is in turn connected with the basal forebrain regions concerned both with regulating homeostasis and with recording in the memory. Further understanding of these basal systems is likely to show that much that is characteristic of mammalian behaviour is to be found here as well as in the complexities of the sensory analysers of the neocortex.

We conclude therefore that the brain more than any other part of the body allows animals to find new means of life. Its relative size and complexity has certainly

increased along many lines of mammalian evolution and this has helped the animals to adjust to changes of climate and other conditions and to invade new niches.

9. Life strategies

The 'efficiency' of reproduction is a concept that includes the effectiveness of all other aspects of life and is therefore impossible to define precisely. We can in principle measure the energy expenditure used to produce enough offspring to maintain a stable population. But how can we compare the efforts needed by say cod with their millions of eggs and a fulmar guarding its single one? Each is an effective way of exploiting a particular environment. Yet there has certainly been a change during vertebrate evolution in the direction of producing fewer offspring and guarding them as they develop. Amphibia produce quite a lot of eggs but many modern species have few and nourish them well. No reptiles produce more than a few hundred and no bird or mammal produces more than 20. Such a change has all sorts of implications for genetics, energy balance, processing of information, and evolutionary potential. The condition with fewer young has in general prevailed among terrestrial vertebrates and is presumably therefore more efficient for these animals. The same comparison can also be made for reproductive strategies among mammals.

There is a complex relationship between use of the brain and the general life pattern of a species. Where the environment is unstable species may adjust by rapid fluctuation of population involving large litters, short life-times, and small brains, for instance among rodents (this is a form of 'r selection'). A more stable environment allows a slower rate of reproduction, larger brain and hence more time for learning from experience ('K selection'). Yet it is doubtful how far this distinction can be pressed. The rat uses its brain to meet the changing conditions, and the environment of the elephant can hardly be called stable. Social life is often supposed to be more complex in animals with larger brains but it may be simply that these animals are easier for us to observe because they are more like ourselves. It is hard to make an adequate ethogram of small animals communicating by smells or by high-frequency sounds that are beyond our range. Mother–young interactions may be complex even if they are brief. Social life in mammals is not like that of the pre-programmed behaviour of the neuter castes of Hymenoptera and termites, whose genetic systems allow extreme 'altruism'. The mammalian patterns depend upon pre-programming to pay attention to the signals emitted by relatives and then to adapt behaviour in the way that seems best for survival in the light of experience. This produces the combination of altruism and selfishness that has been so successful in man.

10. Efficiency of reproduction

The very name of the class shows that it is in reproduction that we consider that the mammals excel. The evolution of the mammary glands was indeed a fundamental advance over the reptilian condition (see Lillegraven 1979). It made possible the biting dentition with all its varieties and, of course, homeothermy and its many consequences (p. 402). Monotremes secrete milk and it was probably produced by Mesozoic mammals. It may perhaps have been necessary for some reason before placentation became possible. Some species in all classes of vertebrates have interchanges between mother and embryo, but none as complex as those of a mammal.

Marsupial reproduction is often considered to be less advanced than that of eutherians but in fact is a distinct strategy with its own advantages (p. 424). The birth of small young who are then fed in a pouch produces by the end of lactation an individual as well equipped as one retained for a long time and nourished by a complex placenta.

Within the eutherians there are large differences in the

extent to which foetal and maternal blood are separated. In the epitheliochorial placenta of prosimians, artiodactyls, and perissodactyls, where several layers remain, there is also a large allantois for storage of excreta. This seems to be a less efficient arrangement than the endotheliochorial placenta of carnivores or the haemochorial placenta of higher primates, where fewer layers separate the bloodstreams and excreta are returned to the maternal circulation. However, the haemochorial condition is also found in insectivores and other conservative mammals and there is some doubt as to which way the series should be read! (see p. 431 and *Life of Mammals*, Chapter 53).

11. Summary of mammalian advantages and improvements

We may conclude that it is possible to see some of the reasons for the success of the mammals and their subsequent evolutionary changes. The advantages that mammals have over reptiles are clear in nearly all the aspects of life that have been considered. Homeothermy has allowed continuation of many species in face of the fluctuations of temperature that occur with both short and long periods of time. The earlier mammals probably had rather low and variable temperatures, which were advantageous for their niches. Progressive improvement of food-collecting systems allowed higher temperatures to be maintained and new habitats occupied. The high and constant temperature as well as constancy of other internal factors may have allowed the power of information processing and memorizing, which has dominated later mammalian evolution.

Reptiles and mammals differ more in their cerebral hemispheres than in any other parts of the body. No really intermediate forms of cortex are known. If the mammalian condition appeared rapidly it must have been before the monotremes evolved as they have well developed cerebral cortices (unless of course these have evolved in parallel). However, the mammalian brain remained small for an immense period of over 100 Ma throughout the Mesozoic. It is not clear what selective factor then favoured its sudden increase in several lines. The behavioural plasticity that the mammalian brain provides is corelated with their reproductive strategy. Placentation limits the number of young that can be produced but good brains ensure that an adequate number survive.

The changes that have occurred in the skeleton and teeth provide good evidence that the replacements during evolution were a result of later improvements. It is very likely that this was true also of the soft parts that we cannot study, and especially the brain. There has been an increase in body size in several lines living in conditions where this is an advantage. The correspondingly larger brains have allowed still further development of the K selection strategy of producing few young and providing care and protection for them.

Information about the environment is acquired much more rapidly by the brain than by natural selection and this may be the main reason for the success of the later mammals. The power to pass such information to succeeding generations gives such an outstanding advantage that man is in danger of exterminating all the other mammals.

33 Conclusion. Evolutionary changes of the life of vertebrates

1. The life of the earliest chordates

WE set out to try to define the features that are characteristic of vertebrate life, hoping then to show how these features have changed during evolutionary history. We may now summarize the evidence collected and see how far it is possible to make general statements about vertebrate life and the factors that change it.

The vertebrate type of organization has proved capable of supporting life under a wide variety of circumstances; most of its modern forms operate under conditions very different from those in which the type first appeared. According to the most probable theory chordate life began at the sea surface, as the ciliated larvae of some creatures rather like sessile echinoderms (p. 69). The first fish-like animals, with the characteristic chordate organization, appeared when such larvae acquired powers of rhythmic metachronal muscular movement, in order to allow support by swimming. We can still see approximately this stage today in amphioxus.

The earliest chordates showed a rather low level of metazoan organization, with a relatively small number of distinct cell types and few special organs. Nitrogen and other raw materials were obtained in the form of minute plants, collected by ciliary action of the pharynx and gill-slits. The food was broken down by a system of enzymes working in alkaline solution, and absorbed through the walls of a simple intestine. There was probably no specialization of cells of the walls of the gut to produce enzymes or to perform particular operations of conversion or storage; at least no special liver, pancreas, or other organs were present for these purposes. There were no special respiratory surfaces and the oxygen was carried to the tissues in simple solution in colourless blood. The circulatory system perhaps at first involved little more than an irregular set of spaces among the cells, but quite early there must have appeared the distinct contractile vessels, containing a blood with composition distinct from that of the surrounding lymphatic or tissue spaces. The method of excretion of the earliest chordates is not clearly known;

it perhaps involved no highly specialized cells, but occurred all over the body surface. Since the animals were marine there were no serious osmotic problems. Movement was by the contraction of a series of blocks of longitudinally arranged muscle fibres and this serial repetition of the muscles and their attendant nerves and blood vessels has left a large mark on the chordate plan of life.

The nervous organization was at first based on a system of nerve cells and fibres lying spread out below the epidermis, but then became concentrated dorsally in the walls of a neural tube. The special receptor organs were probably simple and lay either in the skin or within the tube, perhaps along its whole length, with little concentration at the front end and no definite anterior enlargement or brain. The system functioned as a series of more or less separate reflex arcs, activation coming from the stimulation of receptor organs by changes in the world around. There were no large masses of nervous tissue and little possibility of sustained independent action by the creatures, which probably showed little flexibility of behaviour or variation of action with experience. The only endocrine influences were the effects of cell by-products on neighbouring tissues; there were no specialized glands of internal secretion.

Reproduction was presumably sexual (perhaps also by budding) and development followed the pattern of radial (indeterminate) cleavage and gastrulation by invagination, with chordomesoderm separating from the endoderm. The young were provided with yolk for their development, but were probably not otherwise cared for by their parents.

This gives a rough picture of chordate organization in the Cambrian period when it probably first arose, 550 Ma ago, after a paedomorphic change by which a previously larval creature became sexually mature. It was an organization that had already proceeded far from the aggregation of similar cells that presumably characterized the first metazoans. Its embryological processes were already sufficiently elaborate to produce a creature with well marked organ systems, though these did not have the numerous cell types and anatomically separate parts that are found later.

2. Comparison of the life of early chordates with that of mammals

Such an early chordate is immensely complicated when considered as a chemical system, yet it lacks the specializations that later became so characteristic of vertebrates. The difference appears very clearly if we contrast the organization that maintains homeostasis and life in some amphioxus-like chordate with that of a mammal. In almost every part of the body of the latter we find cell types and organs that are not yet differentiated in the former.

For illustration of this difference we can look at almost any tissue, say the skin, the blood, the gut, or the brain. In a mammal the skin contains far more types of cell than are present in amphioxus; there are hairs and these are different in various parts of the body; there are many types of gland and of receptor organ. The blood, again, circulates with great rapidity and in two circuits; it maintains a constant composition and allows rapid flow of materials to the tissues; it contains haemoglobin in special corpuscles. In addition there are numerous types of cell able to be used for defence, and a system of antibodies for the same purpose that is almost certainly also far beyond anything found in the earlier creature. Digestion in a mammal involves an elaborate arrangement of mouth, oesophagus, stomach, and intestine, each with a controlled pH and special masses of cells aggregated into groups, such as the salivary glands, pancreas, and liver, the latter a most elaborate chemical workshop.

Finally the nervous system possesses an enormous number of cells and elaborate receptor organs. It gives the power to react to many aspects of environmental change that cannot be discriminated by the simpler organism. Nervous conduction is rapid, allowing these large creatures to be well coordinated. The nervous system, working through the many contractile parts that are provided by the muscular system, enables the performance of numerous elaborate acts, helpful in obtaining food, escaping enemies, and perhaps particularly in providing for the care of young, which is another characteristic mammalian feature. The pattern of behaviour does not always follow one single course, but is adaptable and suited to the conditions that are likely to be encountered in view of past experience. There is an elaborate system of chemical signalling by many endocrine glands.

No doubt this greater complexity found in mammalian organs reflects a similar complication of the metabolic processes throughout the body. Moreover, an organism with so many diverse parts presumably depends for its replication on a very elaborate genetical system containing much more information than in a protochordate. Such a comparison can leave no doubt that the life of later vertebrates is in many respects more complicated than that of its Pre-Cambrian ancestors. We now have to ask whether in any sense it can be said that this change indicates an 'advance' or 'progress'.

3. Progress in vertebrate evolution

In the first chapter we discussed some 'eternal questions' that have been asked about the change of organic form. Study of the sequences of vertebrate lives should enable us to see at least some answers to these questions about the direction and purpose of life, which are of interest to everyone. After a long study of these problems the ornithologist Mayr had no doubt that 'If we study the record of life on earth from its very beginning, we do note a progression' (1976). But what are the 'causes' behind this progress? Can it be related to environmental changes or has it been the result of random stochastic processes such as were discussed in Chapter 1? Probably no biologist would be prepared to agree that evolution has been wholly random. Discussion of the problem of the causes of evolution is so extensive and technical that a summary is bound to be inadequate. Fortunately a set of essays collected by Hallam (1977) shows how a group of seventeen palaeontologists would answer some of these questions about the extent and nature of change during the evolution of vertebrates. From their statements we can extract some answers to the main problem.

Rather surprisingly most of the authors hold that there has been an approximately constant diversity of habitats, at least for vertebrates: 'At the end of the Devonian, or the beginning of the Carboniferous, fishes had come to occupy all available carnivorous niches in the aquatic environment' (Thomson 1977). The corollary of this would be that all evolution consists of replacement. But obviously this is not the whole story. Vertebrates have invaded new habitats, for instance the land and the air.

Most of these authors detect some directionality in the changes that have occurred. 'We have direct evidence of such a spiral (of efficiency) within the Cenozoic Mammalia, both for locomotory adaptations and for intelligence' (Stanley 1977). Such directional tendencies are now generally considered to be the result of the continued pressure of competition, ensuring increasing 'efficiency' of structure through time. All the authors agree that *some* structures have become more complex and efficient. The question of whether there have been directional trends in a *majority* of structures is technically unanswerable since we can only sample the trends of a few hard parts (Schopf 1977). Nevertheless there seem to be *beliefs* among these authors that some such general increase in efficiency has occurred even if it cannot be quantified. Before trying to reach conclusions on this fundamental question we may survey how the different authors describe the processes of replacement during evolution. This can be done quantitatively by recording the changes through time in the numbers of genera, families, and orders in the different groups of vertebrates.

4. Succession among fishes

The fishes have shown two major periods of diversification, in the Devonian and the Tertiary, the latter being mainly among teleosts (Fig. 33.1). These differences are not artefacts of sampling since they remain if the numbers of the various taxa are related to the relative volumes of deposit laid down during each geological period. The agnathan fishes of the Ordovician and early Silurian showed little diversity, probably they were in competition with invertebrates in confined adaptive zones. Then in the mid-Silurian the fishes achieved superiority over some benthic invertebrates and began to diversify into the major gnathostome groups, the Acanthodii, Actinopterygii, Sarcopterygii, and Placodermi. In the full diversification of the Devonian new niches were filled both by active predators in open water and bottom-livers and scavengers. No further increase

in diversity occurred until the teleostean outburst in the late Cretaceous and Tertiary.

During their history the fishes, amphibians and reptiles have shown a certain uniformity of pattern of evolution. Each group (at ordinal level) gives a simple curve of increase in diversity to a maximum and then decrease, all within about 100 Ma (Fig. 33.2). This pattern of advance and decline of each group, giving peaks of diversity, seems to have occurred in a regular manner throughout the Phanerozoic (Fig. 33.3). Each group, having acquired an advantage by some particular set of adaptations diversifies very rapidly, reaches a peak and then declines. Moreover maxima occur at about the same times in fishes, amphibia, and reptiles (Fig. 33.4). Thomson therefore suggests that some factor has produced maxima of diversification at intervals of 62 Ma throughout the whole period of history of the vertebrates (1977). There were also similar peaks of diversification among some invertebrates. Certainly however the periodicity was not entirely regular. As we shall see later the amphibia and reptiles both showed two early peaks at intervals of 30 Ma. Periods for the mammals were even shorter (p. 577).

These apparent regularities in the appearance of new genera are subject to many qualifications, depending as they do on arbitrary and subjective classifications and the hazards of an incomplete fossil record. Many of the best palaeontologists therefore distrust such generalizations and comparisons. Moreover, the extent of

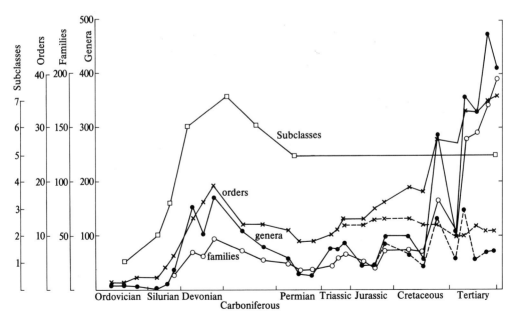

Fig. 33.1. Variety of types of fishes throughout their history as shown by the numbers of various taxa. Dotted lines show counts exclusive of teleosts. (After Thomson 1977, data from Romer 1966.)

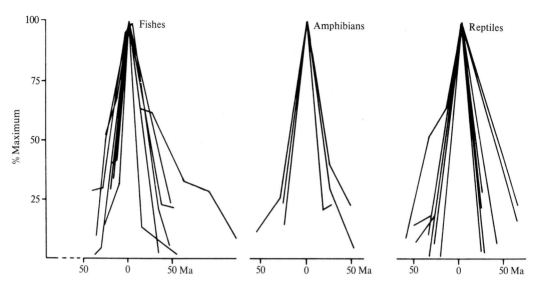

FIG. 33.2. Diversification and disappearance among orders of the lower vertebrates. The percentage of genera in each order found before and after its maximum. (After Thomson 1977.)

supposed diversification of genera may vary with the morphological complexity of each group and with other factors. Nevertheless the pattern can hardly be wholly an artefact and it suggests that some oscillatory mechanism is at work. Various suggestions for the causes of mass extinction have been discussed in Chapter 1. They

include sea floor spreading, plate motions, geomagnetic reversals, changes in sea level, marine transgressions and regressions, and changes in provinciality (isolation). No one of these factors seems to provide a clear explanation, but several of them may have worked together. The curve of the area of marine transgressions

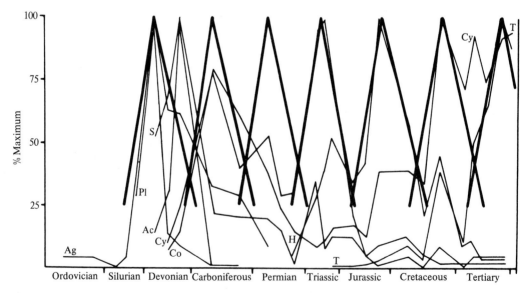

FIG. 33.3. Recurrent peaks of diversification in fishes. The numbers of genera in each major group of fishes plotted as a percentage of the maximum in the group. Heavy lines are superimposed to show the apparent regularity of peaks of diversification. See also Fig. 33.4.
Ac., Acanthodii; Ag., Agnatha; Cy., Chondrichthyes; Co., Chondrostei; H., Holostei; Pl., Placodermi; S., Sarcopterygii; T., Teleostei. (Thomson 1977.)

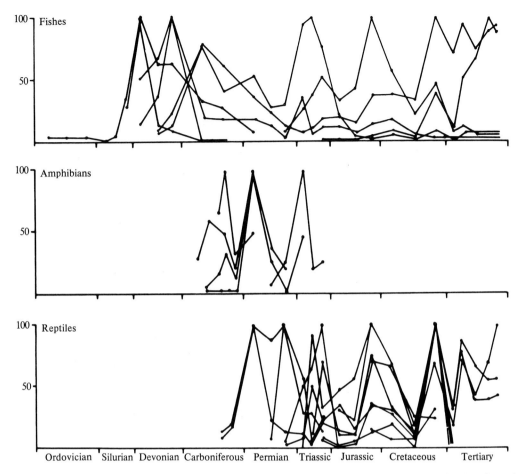

FIG. 33.4. Peaks of diversity occur at similar times in various groups. Data for fishes as in 33.3. Similar data for orders of amphibians and reptiles. Ordinate shows percentage of maximum diversity of order. (From Thomson 1977.)

shows peaks somewhat similar to those of diversification (Fig. 33.5), but the peaks of rates of geomagnetic reversal are at rather different points (Fig. 33.6). We can thus only as yet glimpse the *possibility* of a general explanation for the succession of vertebrate forms. Knowledge of the subject has advanced greatly in recent years and future findings may give us a fuller understanding and so a more general science of Zoology.

5. Succession among Amphibia

The anamniote tetrapods have shown a curious bimodal pattern of diversity (Fig. 33.7). After their appearance in the late Devonian they had the land to themselves and radiated rapidly into numerous types in the Carboniferous. Some were fully terrestrial, others partially so and some returned to the water. With the emergence of the amniote reptiles in the Permian the labyrinthodont amphibia rapidly declined and none survived beyond

the end of the Triassic. Modern amphibia appeared at the end of the Mesozoic but were rare (at least in the fossil record) until the end of the Cretaceous. The majority of living families were already present in the Eocene and several modern species of frogs and salamanders are known from the Miocene. These are further examples of the strong forces resisting change.

6. The successive dynasties of reptiles

The sequence of evolution of the reptiles shows evidence like that of fishes of cycles of diversification and extinction. Bakker recognizes 6 or 7 of these between the Permian and Cretaceous, each of 20–60 Ma and he calls them 'dynasties' (1977). Each begins with a sudden mass extinction of most of the families of big terrestrial and marine aquatic tetrapods. Families of small terrestrial and freshwater aquatic tetrapods show no such clear pattern of mass extinctions and may increase in number

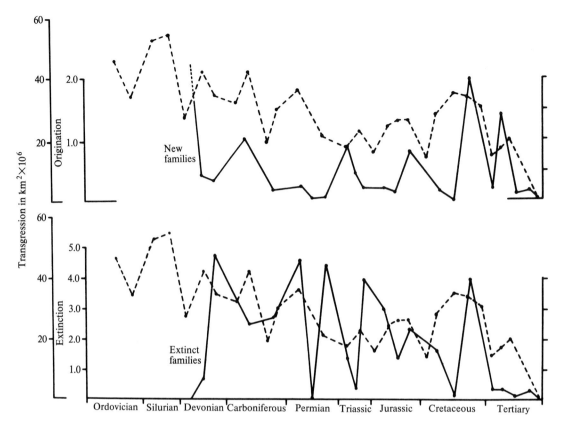

FIG. 33.5. Originations and extinctions of fishes plotted against area of marine transgression (dotted lines). (Data from Hallam 1971.) New families at each interval are shown as ratio of number of families in preceding interval. Extinct families as those absent divided by those present in preceding interval. (From Thomson 1977.)

during the decreases of big terrestrial and marine forms. After the extinctions terrestrial diversity is gradually built up by radiation from the surviving large tetrapods and from small-bodied groups. Bakker does not find evidence that during the mass extinctions climatic change was greater than the usual fluctuations; but there is coincidence between the extinctions and major regressions of shallow seas and decrease in orogeny (Fig. 33.8). Such a change obviously limits the habitat area of marine forms and increases that of those living in fresh water. The effect on land animals depends on the fact that new species tend to be formed in mountainous regions. These provide greater diversity of habitats and geographic barriers, which are major influences on the rate of speciation. It has been shown that among recent mammals, birds, angiosperms, and other groups the diversity is greater in mountains and high plains than in lowland basins. Small animals are isolated by small barriers and so are less affected by more uniform condition of the land. This theory (known as the Haug

effect; see Fig. 33.8) explains extinction on the assumption that new species formation is essential for survival of a particular type, say at family level. There is indeed palaeontological evidence that the usual fate of a species in a basin is extinction in about a million years.

The tetrapods of the first dynasty in the late Carboniferous, were probably all ectothermic. Since the temperature gradient towards the poles was then steep they were restricted to the warm tropics of Laurasia and northern Gondwanaland. Then in the Permian, with intense orogeny and an arid climate, the ectothermic families died out and were replaced by the second dynasty, comprising the first radiation of large and small therapsids (p. 406). These were probably endothermic and produced the extinction of the ectotherms through competition and predation.

All the large families of herbivores and carnivores of the second dynasty disappeared abruptly from the Karoo sequence in southern Africa in the middle of the Permian. The new large herbivores of the third dynasty

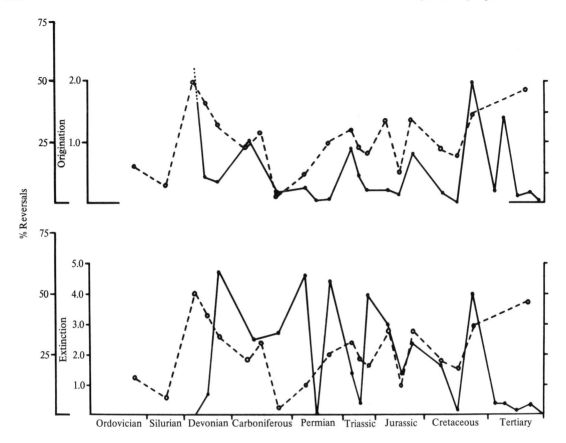

Fig. 33.6. As 33.5 but with rates of geomagnetic reversal (dotted line) after McElhinny's data (1971). (From Thomson 1977.)

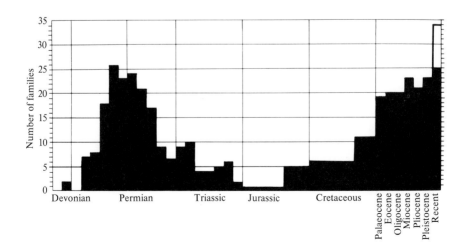

Fig. 33.7. Total number of families of Amphibia at various geological horizons. (After Carroll 1977.)

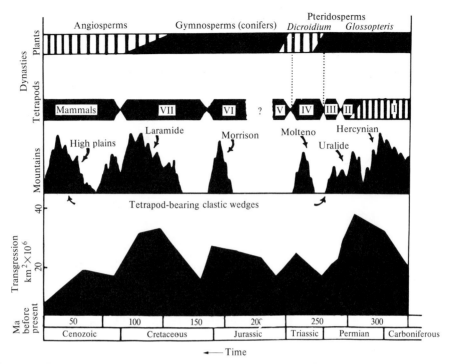

FIG. 33.8. The Haug Effect and terrestrial dynasties. World-wide marine transgression is shown at the bottom. The tetrapods show mass extinctions of families, except between first and second dynasties. In contrast the floral transitions appear gradual. (After Bakker 1977.)

were the several families of dicynodonts and pareiasaurs and the carnivores were the gorgonopsids and thecodonts.

Then at the Permian–Triassic boundary all of these disappeared and were replaced by a new dynasty lasting throughout most of the Triassic. The herbivores included great numbers of new dicynodont forms such as *Lystrosaurus* but, curiously, no predators above 1 kg weight. Later in this dynasty the lystrosaurids were replaced by other dicynodonts, the kannemeyerids and big cynodonts and, later, rhynchosaurs, in all four families of large herbivores. Large carnivores also appeared, such as *Cynognathus* and the thecodonts.

In their turn all of these big forms disappeared at the end of the Triassic and were replaced by the fifth dynasty including two families of prosauropod dinosaurs. The top predators of this time were still thecodonts, with theropod dinosaurs entering at the Triassic–Jurassic boundary. There are few specimens from the early Jurassic to follow the transition to the sixth dynasty at the end of that period, when there was a great diversity of herbivores, the giant sauropods and stegosaurs. By the earliest Cretaceous these too were extinct and replaced by the ornithischian families, such as iguanodonts and later hadrosaurs, forming dynasty seven.

New families continued to appear right up to the end of the Cretaceous. The carnivores of this period were the tyrannosaurs.

Then in a very short interval as judged by the rocks, a few million years or less, all dinosaurs became extinct (see Desmond 1975; Russell 1979) leaving no animals over 10 kg, but only small mammals and lizards. The next group of large herbivores, the eighth and last of Bakker's dynasties, was mammalian but the large forms did not appear at first. Nearly all of the many mammals of the Palaeocene and Eocene were small insectivores or omnivores. Only in the Oligocene did the first mammals who possessed batteries of continuously replaced teeth appear, the successors of the Cretaceous hadrosaurs and ceratopsids.

Many causes have been suggested for the faunal change at the Cretaceous–Tertiary boundary (p. 19). The poleward temperature gradient probably became rather steep, but there were crocodiles in Saskatchewan in the early Palaeocene. Dinosaurs seem to have flourished in many climates and to have survived successive fluctuations throughout the Cretaceous. Probably they were buffered from climatic change by endothermy (p. 277). The dinosaurs evolved with the increasing diversity of angiosperms in the Cretaceous and there is no evidence of a marked floral change when they became extinct.

The explanation for these successive extinctions that Bakker adopts is the major regressions of seas, which were marked at the ends of the Permian, Triassic, Jurassic, and Cretaceous periods. The figure he provides (Fig. 33.8) does not, however, seem to show very close correlation between the regressions and extinctions. We are left with the important conclusion that there has been a succession of faunas of large tetrapods, separated by mass extinctions. However the periodicity of the change is not very regular and the cause of the phenomenon remains obscure.

7. Successive extinctions among mammals

The succession of periods of extinction followed by new radiations has continued throughout the evolution of the mammals. If we include the first Eocene radiation four distinct cycles appear, with intervals of about 20 Ma between the peaks of diversification (Figs. 33.9 and 33.10). The early progress of the mammalian stock was slow and gradual. For the first two-thirds of their 200 Ma history they were a small and little diversified component of the vertebrate fauna (Fig. 33.11). They began to form numerous types in the later Cretaceous, and throughout the Tertiary the number of genera recognized increased in an exponential manner. This is of course partly an artefact of sampling and classifi-

cation but the curve emphasizes that mammals are at present at the peak of their development, perhaps they are already declining.

The radiation at the end of the Cretaceous led to the Palaeocene fauna of multituberculates, archaic primates, proteutherian insectivores, and Condylartha. In the late Palaeocene other placentals appeared, including Carnivora, Pantodonta, and several south American orders (p. 529). There was a climatic cooling at the end of the Palaeocene, restricting the ranges of mammals as the subtropical zone moved southwards. This was followed by warmer conditions in the early Eocene and these changes may have been the cause of the almost complete disappearance of the Palaeocene fauna and the sudden arrival of modern types. These included rodents, modern types of primates (adapids and omomyids), bats, miacine carnivores, artiodactyls, and perissodactyls. The mammals also invaded the sea in the Eocene (Sirenia and Cetacea).

It is certainly tempting to look for some environmental change to explain the appearance of so many new types. The early Eocene was a time of active mountain building with several changes in ocean levels, part of the long-drawn out Laramide orogenesis. The warmer Eocene conditions would have extended the possibilities for migration along the high latitude connections between the Holarctic continents. This was followed by the final opening of the north Atlantic later in the

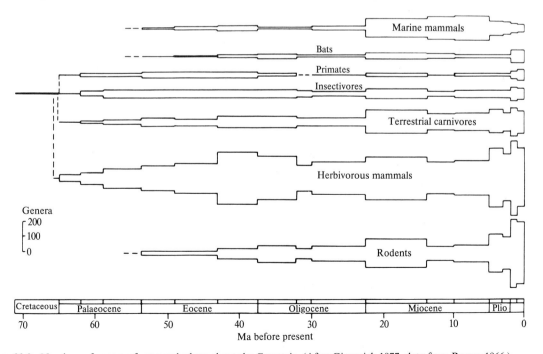

FIG. 33.9. Numbers of genera of mammals throughout the Cenozoic. (After Gingerich 1977, data from Romer 1966.)

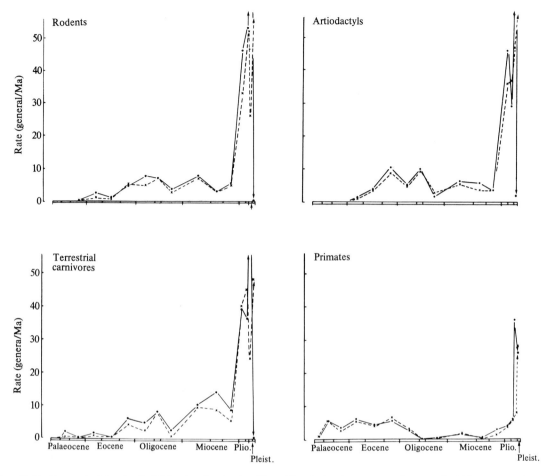

FIG. 33.10. Rates of origination (solid lines) and extinction (dotted lines) of various groups of mammals. (After Gingerich 1977, data from Romer 1966.)

Eocene, separating Europe and North America. The climatic changes and these expansions and contractions of habitats and formation and breaking of connections were probably the basis for the divergent evolution of mammals without any special factor producing high rates of diversification ('cladogenesis'). There is indeed evidence sometimes interpreted as showing that during the early Eocene there was gradual change of one species of mammals into several (anagenesis).

Some of these new Eocene mammalian types possessed obvious advantages. Thus the continually growing and self-sharpening rodent incisors and specialized jaw muscles arrived very early and allowed their possessors to displace the multituberculates and archaic primates. They probably also entered new adaptive zones and have continued to diversify ever since (Fig. 33.9). Similarly the perissodactyls replaced the

condylarths in the Eocene and Oligocene and were in turn replaced by more efficient herbivores, the artiodactyls, in the Miocene. Again the early creodont carnivores were replaced by the modern Carnivora mainly between the Oligocene and Miocene.

There are obvious functional reasons for these replacements, but it is very striking that they tend to occur at the same times in many orders of mammals, occupying very different habitats (e.g. whales and rodents) (Fig. 33.9). Extinctions and new originations of genera were numerous throughout the class at the beginning of the Eocene, Oligocene, Miocene, and Pliocene. These were times of climatic change, mainly warming. There were especially large changes, with numerous new genera and loss of old ones, in the early Pleistocene, this time as the climate became colder. Extinctions were higher in the late Pleistocene than at any other time in the Tertiary. 67 genera of rodents, 72 of artiodactyls, 13

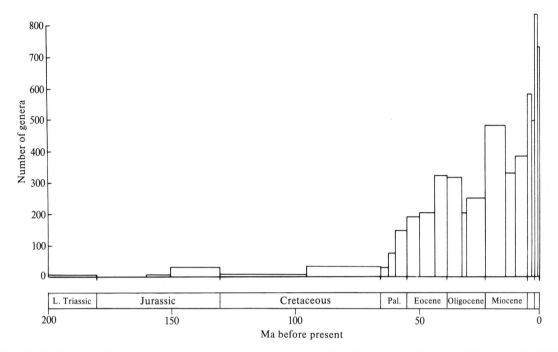

FIG. 33.11. Numbers of genera of mammals present in each subdivision of the Mesozoic and Cenozoic. (After Gingerich 1977, data from Romer 1966.)

of primates, and 23 of land carnivores disappeared. The Pleistocene glaciations were presumably a main cause and *Homo sapiens* may have been a factor. More likely the extinctions were due to competition from the even more numerous *new* genera that had appeared in the early Pleistocene (Fig. 33.9). This must surely have been the result of environmental changes due to the climatic fluctuations and glaciations. Man himself diversified into several lineages early in the Pleistocene, but only one survived to the end.

The conclusions we can draw from these facts are limited. They show that change has been very prevalent in mammalian evolution and that it occurs in waves, presumably conditioned by external events. Times of change of sea level and of mountain building and hence of habitat diversity seem to have accentuated evolution in mammals as they did among reptiles. In some situations we can see that the later types are more efficient than those they replaced and selection by competition is certainly a major factor inducing change, although selection also acts conservatively in the opposite direction. The degree of stability of the environment may be the factor that decides in which way selection operates at any particular time. Where conditions remain more or less constant selection rejects new types. Rapid changes in the environment favour the survival of new ones.

8. The increasing complexity and variety of vertebrates

In each of the successions that have been recorded in the previous sections there has been a series of increasing and decreasing variety as one type replaced another. We still have to enquire whether it is possible to discern any overall change in the type of life or its variety throughout the evolution of vertebrates. The comparison between early and late chordates at the beginning of this chapter indicated that the difference between the two is in the greater number of diverse parts and actions found in the later type. At every stage of the life cycle there are more alternative possible actions available and better methods for selecting the appropriate behaviour. In other words in higher organisms there are larger repertoires of possible actions, giving greater possibilities of choice. This greater complexity of the higher animals enables their life to be carried on under conditions that would have been impossible for the simpler ancestors. Many of the later types are even more improbable systems than the earlier and they continue to live by virtue of the greater quantities of information that they process. The survey of the evolution of chordates has certainly shown that since the Cambrian the chordate organization has invaded situations very different from the sea surface in which it probably arose.

It would not be profitable now to recapitulate all the stages of this change – they have already been described throughout the book. If we consider ecological niches in detail the number of fresh situations invaded by vertebrate life is almost as great as that of the species in the group. Among the earlier changes were the transfer from the sea surface to other waters and to the sea bottom. The entrance into fresh water must have called into play many special mechanisms of adjustment. Development of jaws, perhaps 350 Ma ago, probably from the anterior branchial arches, gave the possibility not only of eating new types of food (including fellow fishes) but also of performing simple acts of 'handling' of the environment. The heavy armour of the early types was given up and the body form was then greatly improved from a hydrodynamic point of view and with development of the swim bladder into a hydrostatic organ the fishes achieved their full mastery of the water.

Meanwhile other fishes left the water, probably in the Devonian period, rather less than 380 Ma ago. At first they operated with little modification of the method of life they had used in the water, but they later developed all sorts of devices to meet the new conditions, the earlier types dying out as the later developed. This process has been going on ever since, to produce the modern amphibia, reptiles, birds, and mammals, inhabiting a great variety of situations.

In the water tetrapods are found at all levels, including great depths and in perpetually dark caverns. They live in the most varied situations on the land and also by burrowing beneath its surface. Not a few are able to move in the air, some even to feed there. It is hardly possible to overestimate the great variety of vertebrate life; at each new examination of any phase one is amazed at the extraordinary number of special modes of life that are adopted by variants of each type.

There are no sure means of telling the number of types or of individuals constituting the biomass that was present in past times, but it is probable that by means of the above special devices the vertebrate stock has increased and colonized new regions, though not perhaps continuously or at a uniform rate. It is not unlikely that today there are more and more varied vertebrates than at any previous period. It has been pointed out that the number of species described from deposits tends to increase geometrically with time and this is not an artefact due to poor preservation.

9. Vertebrates that have evolved slowly

We may accept then the concept of bursts of rapid evolution, followed by slower change. In many lines after the rapid change there is a period over which many genera become extinct. However, a few linger on for

times longer than would be expected. This phenomenon of bradytely produces phylogenetic relics, of which there seem to be so many that some general attempt to explain them is desirable. *Neoceratodus* has a good claim to be considered the 'oldest' living vertebrate; it is very similar to fossils found in the Triassic, 200 Ma ago. Even in this case, however, there have been slight changes and the Triassic form is placed in a distinct genus **Ceratodus*. *Latimeria* provides us with an example of survival with little change for nearly 100 Ma, as well as the humbling thought that no fossil relatives are known throughout that time. *Heterodontus*, the Port Jackson shark, is another very ancient fish; it is closely similar to fossils found in the Triassic; indeed, all sharks are in many ways like their Palaeozoic ancestors.

Sphenodon has changed little since Permian times and hardly at all since the Jurassic, perhaps 150 Ma ago. Among mammals, the opossums and hedgehogs are quite similar to those of Cretaceous times, nearly 100 Ma ago, and there are several mammals that have survived with little change for the 50 Ma since the Eocene, for instance dogs, pigs, and lemurs. In none of these, however, has a form survived absolutely without change; they are examples of the persistence of a type of organization rather than of superficial details.

It is interesting to consider possible reasons for these very slowly evolving (bradytelic) populations (Simpson 1953). (1) It might be low mutation rate or low variability, but there is no evidence that opossums (say) are less variable than other mammals and indeed they have undergone much speciation. (2) It is often implied that survival is assisted by some special habit, such as being nocturnal or abyssal, but others with the same habits evolve fast. (3) Survival sometimes seems to be assisted by isolation (e.g. lemurs) but there is no evidence that this is necessarily a factor (*Latimeria*). (4) Low rate of evolutionary change is not a function of 'primitive' organization as such. In any case *Sphenodon* and *Crocodylus* were not especially 'primitive' when they stopped evolving in the Triassic and Jurassic. (5) Long survival must depend upon some special relationship between the genetical and information-carrying powers of the species, the risks imposed by the environment, and the stability of the latter. (6) A narrow adaptive zone is unlikely to persist in 'difficult' habitats such as deserts, or impermanent ones (salt lakes), or variable ones, such as Alps. Long survival is perhaps more to be expected in a broad adaptive zone such as the ocean or shore, lowland rivers or forest belts, especially in the tropics. Such environments present, however, many niches that can be considered as 'corridors', leading to diversification, and it is surprising that forms nevertheless remain stable in them. Thus opossum-like creatures gave rise to various offshoots in South America but themselves changed little. (7) Bradytelic populations must be genetically so integrated that any deviation is subject to counter-selection (though in that case it is hard to see how the offshoots have arisen). (8) Simpson concludes that these bradytelic organisms 'have run the whole repertory of baffles and ... persist indefinitely'. Most organisms are turned off into one or other of the corridors presented by the environment; when a group has met and passed them all it persists.

The discussion of animals that have evolved very slowly is thus a stimulus to considering the whole balance of factors by which a population of organisms maintains its homeostasis. Evidently there are some circumstances in which it can do this with little genetic change. In the great majority, however, change of the genes and hence of the structure and physiology is a part of the very mechanism by which the living system continues to survive in spite of changes around it.

10. Successive replacement among aquatic vertebrates

Does the examination of the sequences of types enable us to say anything about the nature of these evolutionary changes? Can we record any sense in which it represents a 'progress' or 'advance'? One striking feature that we have noticed is that often one type of organism seems to replace another. There is always a difficulty in establishing that this has occurred, because the fossil record does not leave us sufficient information to show for certain that the two types occupied identical 'niches'. However, if we take broad 'habitats', and particularly those that change relatively little, such as the waters, we cannot but be impressed with the succession of tenants that appears, each replacing the one before. Thus, among fish-like vertebrates we can recognize ostracoderms, placoderms, crossopterygians, palaeoniscids, holosteans, and teleosteans, each almost completely replacing the one before.

Again, there has been an astonishing series of tetrapods returning from the land to water, developing characters suitable for aquatic life and then becoming extinct, apparently displaced by later migrants, also returning from the land. To name only a few of these returners we have among amphibians the phyllospondyls, lepospondyls, branchiosaurs, and some urodeles; among reptiles the phytosaurs, crocodiles, plesiosaurs, ichthyosaurs, mesosaurs, mosasaurs, aigialosaurs, dolichosaurs, and snakes. Finally of the mammals there are the basilosaurs, modern whales, seals, and sea-cows, as well as some less completely aquatic types.

However much we make allowance for the fact that the sea itself may be changing, it is difficult not to find in these facts a suggestion that the later types are replacing the earlier by their greater 'efficiency'. These returned aquatics are especially interesting because each type when it first re-enters the water seems to be not very well suited to that medium – because of its shape for instance – and would therefore be expected to be at a disadvantage in relation to the 'streamlined' creatures that were already there.

11. Successive replacement among land vertebrates

It is equally easy to trace out successions of types occupying habitats on land, though here it is even more difficult to be sure that the successive animals are occupying identical niches. There has been a long series of large land herbivores, including the labyrinthodonts, pareiasaurs, herbivorous synapsids, various dinosaurs, multituberculates, condylarths, dinocerates, pantodonts, brontotheres, horses, pigs, rhinoceroses, elephants, and artiodactyls. Clearly not all of these lived in similar surroundings (and there were, of course, other herbivores), but the succession is impressive. As with the aquatic animals we have the curious phenomenon that the earlier members of each group seem to be clumsy creatures, no better fitted for their life than those they are replacing. The early mammalian herbivores, with their large limbs and small brains, do not seem greatly superior to the stegosaurs and ceratopsians of the Cretaceous. It is, of course, exceedingly difficult to know enough to settle such questions, for instance to assess the value of warm-bloodedness.

If we look at other ecological niches we see the same picture of continued replacement. Thus there has been a succession of land carnivores, first synapsid reptiles, then archosaurian reptiles, followed in the Tertiary period by the creodonts, which were replaced by modern carnivores and carnivorous birds.

12. Convergent and parallel evolution

A remarkable fact that has appeared many times in our survey is that during these radiations similar features repeatedly occur in distinct lines (Fig. 33.12). It is as if the vertebrate organization produced time after time slightly different variations on a series of themes. Thus some of the animals that feed on fishes acquire long jaws and numerous teeth. We have examples of these characteristics among the fishes themselves (garpike, *Belone*) and in the crocodiles, phytosaurs, ichthyosaurs, plesiosaurs, birds, and mammals. Yet, of course, many piscivorous fishes do not have long jaws and many have small teeth. One could continue with endless examples of the same sort, for instance, the large mouth of insect-eating animals – the frogs, swallows, swifts, and bats.

Marsupials

Placental mammals

Pouched mouse
(*Sminthopsis*)

Harvest mouse
(*Mus*)

Marsupial mole
(*Notoryctes*)

Common mole
(*Talpa*)

Flying opossum
(*Petaurus*)

Flying squirrel
(*Petaurista*)

Wombat (*Vombatus*)

Marmot (*Marmota*)

Banded ant-eater
(*Myrmecobius*)

Great ant-eater
(*Myrmecophaga*)

Eastern native cat
(*Dasyurus*)

Serval (*Felis*)

Tasmanian wolf
(*Thylacinus*)

Wolf (*Canis*)

FIG. 33.12. To show parallel evolution of placental and Australasian marsupial mammals. (Figure kindly suggested by G.G. Simpson.)

The five sorts of ant-eater found among mammals provide another remarkable example of this 'convergence'; all have an elongated snout, long sticky tongue, and large salivary glands. The hands of the moles show how a similar result may be arrived at by various slightly differing means (Fig. 33.13).

There can be no doubt that vertebrates adopting a given mode of life tend to acquire a particular structure. There is also evidence of what we might call the converse, namely that animals with a particular form tend to develop in certain directions. This is one of the forms of the situation described by some as 'pre-adaptation'. The elasmobranch fishes maintain equilibrium in the horizontal plane by a heterocercal tail, driving the head downwards, and horizontal pectoral fins and flattened front end of the body, having the opposite effect. From this type of organization creatures of ray-like type have several times developed, flattened dorsoventrally and obtaining their propulsive thrust from the pectoral fins. Conversely among teleosts, where the compression is in a transverse plane, many bottom-living types are flattened laterally, as in the sole or plaice. Where the swim-bladder has been lost, however, there may be dorsoventral flattening (e.g. angler fishes).

The more closely we examine the evolution of populations the more signs of these similar tendencies in evolution appear. Parallel evolution is so common that it is almost a rule that detailed study of any group produces a confused taxonomy. Investigators are unable to distinguish populations that are parallel new developments from those truly descended from each other. Examples we have noticed from study of modern populations are the various types of tree-living and burrowing anurans (p. 273). The 'tree-frog' and 'burrowing' animals have been evolved both from true frogs and from toads, probably several times in each case. Again, the habit of burrowing underground, with loss of the limbs, has appeared in a number of squamate reptiles; the slow-worms, amphisbaenas, and subterranean skinks are certainly distinct lines, possibly each contains more than one, and the snakes are probably derived from another group that went underground.

13. Some tendencies in vertebrate evolution

These are only a few scattered examples, but they already suggest that the evolutionary changes of vertebrate populations follow certain patterns. Although in the history of each type there is no doubt much that is unique, yet most types tend to follow along certain lines, according to the situations they have reached. It is not possible yet for the systematic morphologist to make any complete classification or analysis of these tendencies. He can only point to a few of them, already so familiar as to be almost banal. Thus vertebrates that move in the water tend to have the fish form, with streamlined body, vertebrae with flat articulations, paddle-like limbs and tail-flukes. We are so used to this that perhaps its interest is often overlooked, especially the fact that the pectoral limb may revert from an elongated pentadactyl structure to a paddle, sometimes with increased number of digits and of phalanges. Evidently this form of limb is suitable for the uses demanded in the water and tends to be developed again when needed.

Similar examples that can be called in a sense a reversal of evolution are interesting not only from genetic and embryological points of view, but especially because they show strikingly that under given conditions vertebrate populations tend to react alike, giving us the possibility of developing a reasonable science of morphology. Thus vertebrates that take to the air develop light and thin bones, those that burrow underground often lose their limbs (but among mammals these are often the main digging agents). Similar general

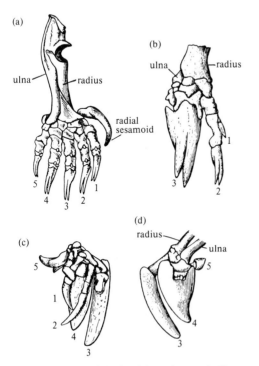

FIG. 33.13. Skeleton of the hand in various mole-like mammals. (a) European mole, *Talpa* (Insectivora); (b) Cape golden mole, *Chrysochloris* (Insectivora); (c) and (d) Marsupial mole, *Notoryctes* (Marsupialia). Palmar and dorsal views respectively. (Lull 1917.)

evolutionary principles could probably be developed for each organ system. Thus the eyes develop cone-like structures in diurnal, rod-like in nocturnal species, and they become buried under the skin or lost altogether in populations that live in the dark.

Examples of complete reversal are the re-appearance of limbs, which occurs occasionally in whales and of the lost digits in horses, dogs and guinea pigs. It is suggested that the genes responsible for these structures have other (pleiotropic) effects (see Lande 1978).

14. Summary. The direction of evolutionary change

Some very simple facts that have emerged from the evidence about vertebrate evolution are (1) that populations tend to be replaced by others of different form, often themselves descended from the first. (2) That the later populations often have more complicated organizations than the earlier ones; we can hardly be said to have established this last clearly as a rule in vertebrates, but it has repeatedly been suggested that the evidence supports some such thesis. (3) Some later populations invade habitats not previously occupied by vertebrates (e.g. the land).

Facts of this sort have led us repeatedly to the suspicion that (4) the later types are often in some way better able to carry on the self-maintenance of life than their predecessors, and that vertebrate life is continually invading new situations. To do this the organisms have needed fresh genetic information with which to regulate the building of new parts of the body. (5) Often the new habitat makes demands for the processing of more information by the nervous system. (6) New sets of standards for the regulation of the aims of the systems are set up, homeostatic mechanisms are improved. With this goes the suspicion that (7) the total biomass of vertebrate life (perhaps of all life) has been increasing and the energy flux becoming more rapid, though we have no exact estimates to verify this.

For organizing our knowledge about the evolution of vertebrates or other animals the above facts about the progress or direction of change are of great value. The invasion of new habitats is of particular interest, because it is usually made possible by the increase of complexity, which provides a system of elaborate adjustments that maintains the internal conditions nearly constant in face of fluctuations in the environment. This internal constancy or homeostasis is, of course, only a development of the general tendency to self-maintenance, that characterizes all living organisms; but it is carried to its extreme in the higher vertebrates, giving them the power to maintain life under varied and unpropitious conditions.

15. Evolutionary progress

If we have interpreted the situation correctly there can be said to be an evolutionary path that is progressive, in the sense of enabling life to be lived effectively under wider and wider conditions. This involves increasing turn-over of energy and more and more complicated mechanisms for ensuring homeostasis in spite of external changes. These mechanisms in turn depend upon an increasing store of instructions in the genotype. If this view is correct there is a tendency for organisms to come to represent more and more features of the environment, which is another way of saying that they have more information about it. More simply still we may say that they come to be able to live under ever more 'difficult' conditions, gathering and expending more energy in order to keep alive (Young 1938). It seems reasonable that this increase of complexity should be progressive; as the organisms acquire more information about the environment they also gain in possibility of acquiring still more. In particular those that develop mechanisms for learning directly with the nervous system will be successful and will evolve fast.. Adaptation during the lifetime of the individuals leads to increasingly intense competition. Perhaps such 'runaway evolution' has been at work in man for the last million years.

During the present study we have not been able to show rigorously that all evolution follows such principles, but the data are not inconsistent with this. It may be that vertebrates have proceeded farther along the path suggested than any other animals. The birds and mammals probably have more capacity for alternative action at all levels of metabolism and behaviour than other vertebrates. They process more information to make these decisions and turn over energy faster to implement them. The point that must not be overlooked is that they do this in order to provide the means by which they can remain constant, in spite of fluctuations in the external conditions of an environment that is different in composition from the living system.

Even if we cannot at present discern with certainty the principles that determine the change of animal form, we can see something of the influences that have modified the original vertebrate type to produce the great variety of creatures that has existed and remains today. We cannot give tables of numbers to show how the variables have operated to keep *Ceratodus* nearly constant for 200 Ma while related descendants have gone on to produce the whole variety of tetrapod life. But we can suggest that it is worth while pursuing the study as a means of building a truly general science of zoology. We begin to see signs of the principles according to which the vast populations of animals interact with each other, with the plants, and with the inorganic world. The changes that these interactions produce seem to be, if not constantly in one direction, at least often such as to allow the appearance of more varied and complicated forms of life, ever more able to maintain themselves constant and apart from the environment, and hence to exist in a wider range of conditions. The effect of this evolutionary change has been to increase the amount of information available to ensure survival, and hence the quantity of material organized into living things, and the total energy flux through the system.

This finding is not a mere rehabilitation of the complacent anthropocentric prejudices of the nineteenth century. The conclusion that there is a sense in which the mammals and man are among the highest animals can be based on objective analysis of the behaviour of the populations of vertebrates and the flow of information and energy through them. However, it has been repeatedly emphasized that we still have a long way to go before we have the knowledge necessary to understand these very slow changes. It has only been possible in this volume to suggest certain principles that may one day make possible a satisfactory scientific study of the life of vertebrates.

References

General works

ALEXANDER, R. McN. (1975). *The chordates*. Cambridge University Press, London.

AUSTIN, C. R. and SHORT, R. V. (1972–continuing). *Reproduction in mammals*. Cambridge University Press.

BARRINGTON, E. J. W. (ed.) (1979). *Hormones and evolution*, 2 Vols. Academic Press, New York.

BELLAIRS, A. d'A. (1969). *The life of reptiles*, 2 Vols. Weidenfeld and Nicolson, London.

BOLK, L., GÖPPERT, E., KALLIUS, E., and LUBOSCH, W. (1931–9). *Handbuch der vergleichenden Anatomie der Wirbeltiere*, 6 Vols. Urban and Schwarzenberg, Berlin.

CORBET, G. B. and HILL, J. E. (1980) *World list of mammalian species*. Cornell University Press.

DAY, M. H. (ed.) (1981). Vertebrate locomotion. *Symposia of the zoological Society of London* No. 48 (in press).

DOBZHANSKY, T., AYALA, F. J., STEBBINS, G. L. and VALENTINE, J. W. (1977). *Evolution*. Freeman, San Francisco.

FARNER, D. S. and KING, J. R. (1971–5). *Avian biology*, 5 Vols. Academic Press, London.

FLORKIN, M. and SCHEER, B. T. (1967–79). *Chemical zoology*, 11 Vols. Academic Press, New York.

GANS, C. (general ed.) (1969–continuing). *Biology of the Reptilia*. Academic Press, London.

GOODRICH, E. S. (1930). *Studies on the structure and development of vertebrates*. Macmillan, London.

GRASSÉ, P.-P. (1953–continuing). *Traité de zoologie, anatomie, systematique, biologie*. Masson, Paris.

GRZIMEK, B. (ed.) (1972–5). *Grimek's animal encyclopedia*. Van Nostrand Reinhold, New York.

HALLAM, A. (1977). *Patterns of evolution as illustrated by the fossil record. (Developments in palaeontology and stratigraphy*, Vol. 5.*)* Elsevier, Amsterdam.

—— (1981). *Facies interpretation and the stratigraphic record*. Freeman, Oxford.

Handbook of physiology: a critical, comprehensive presentation of physiological knowledge and concepts. (1959–continuing). American Physiological Society, Washington, DC.

Handbook of sensory physiology (1971–continuing). Edited by H. Autrum, R. Jung, W. R. Loewenstein, D. M. MacKay and H. L. Teuber. Springer, Berlin.

HARMER, S. F. and SHIPLEY, A. E. (eds.) (1895–1902). *The Cambridge natural history*. Macmillan, London.

HOAR, W. S. and RANDALL, D. J. (eds.) (1969–continuing). *Fish physiology*. Academic Press, New York.

JARVIK, E. (1980, 1981). *Basic structure and evolution of vertebrates*. 2 vols. Academic Press, London.

KAPPERS, C. U. A., HUBER, G. C. and CROSBY, E. C. (1936) *The comparative anatomy of the nervous system of vertebrates, including man*, 2 Vols. Macmillan, New York.

KING, A. S. and McLELLAND, J. (1980). *Form and function in birds*, Vol. 1. Academic Press, London.

KÜKENTHAL, W. and KRUMBACH, T. (1923–continuing). *Handbuch der Zoologie*. de Gruyter, Berlin.

MARSHALL, N. B. (1965). *The life of fishes*. Weidenfeld and Nicolson, London.

Marshall's physiology of reproduction (1956–66). Vols. 1–3. (3rd edn, ed. A. S. Parkes.) Longmans, London.

MATTHEWS, L. H. (1969) *The life of mammals*, 2 Vols. Weidenfeld and Nicolson, London.

MAYR, E. (1969). *Principles of systematic zoology*. McGraw Hill, New York.

MOORE, J. A. (ed.) (1964). *Physiology of the Amphibia*, Vol. 1. [Vols. 2 and 3 edited by B. Lofts.] Academic Press, New York.

PANCHEN, A. L. (ed.) (1980). *The terrestrial environment and the origin of land vertebrates*. Academic Press, London.

PARKER, T. J. and HASWELL, W. A. (1972). *Textbook of zoology*. (7th edn, ed. A. J. Marshall and W. D. Williams.) 2 Vols. Macmillan, London.

PIELOU, E. C. (1979). *Biogeography*. Wiley, New York.

RANSON, S. W. and CLARKE, S. L. (1959). *The anatomy of the nervous system: its development and function*, 10th edn. Saunders, Philadelphia.

ROMER, A. S. (1966). *Vertebrate paleontology*, 3rd edn.

University of Chicago Press.

SCHMIDT-NIELSEN, K. (1979). *Animal physiology, adaptation and environment*. Cambridge University Press.

SEDGWICK, A. (1898–1909). *A student's text-book of zoology*, 3 Vols. Sonnenschein, London.

SHERRINGTON, C. S. (1947). *The integrative action of the nervous system*, (revised edn). Cambridge University Press.

SIMPSON, G. G. (1944). *Tempo and mode of evolution*. Columbia University Press, New York.

——(1953). *The major features of evolution*. Columbia University Press, New York.

THOMPSON, D'ARCY W. (1961). *On growth and form*. (Abridged edition by J. T. Bonner.) Cambridge University Press.

WALKER, E. P. (1964). *Mammals of the world*, 3 Vols. Johns Hopkins Press, Baltimore.

WHITTOW, G. C. (1970–3). *Comparative physiology of thermoregulation*, 3 Vols. Academic Press, New York.

YOUNG, J. Z. (1951) *Doubt and certainty in science*. Clarendon Press, Oxford.

—— (1971). *An introduction to the study of man*. Clarendon Press, Oxford.

—— (1975). *The life of mammals, their anatomy and physiology*. Clarendon Press, Oxford.

—— (1978). *Programs of the brain*. Oxford University Press.

Chapter 1 Evolution, climate and geology

CAIN, A. J. (1963). *Animal species and their evolution*, 2nd edn. Hutchinson, London.

COOPE, G. R. (1975). Climatic fluctuations in northwest Europe since the Last Interglacial, indicated by fossil assemblages of Coleoptera. In *Ice ages: ancient and modern*. (ed. A. E. Wright and F. Moseley) pp. 153–168. (*Geological Journal* Special Issue No. 6.)

DANSGAARD, W., JOHNSEN, S. J., CLAUSEN, H. B., and LANGWAY, Jr. C. C. (1971). Climatic record revealed by the Camp Century ice core. In *The Late Cenozoic glacial ages*. (ed. K. K. Turekian pp. 37–56. Yale University Press, New Haven.

DAYHOFF, M. O. (ed.) (1969). *Atlas of protein sequence and structure*, Vol. 4. National Biomedical Research Foundation, Silver Spring, Md.

——(ed.) (1972). *Atlas of protein sequence and structure*, Vol. 5. National Biomedical Research Foundation, Silver Spring, Md.

DOBZHANSKY, T. and SPASSKY, B. (1947). Evolutionary changes in laboratory cultures of *D. pseudoobscura*. *Evolution, Lancaster, Pa.* **1**, 191–216.

—— AYALA, F. J., STEBBINS, G. L., and VALENTINE, J. W. (1977). *Evolution*. Freeman, San Francisco.

DONOVAN, D. T. and JONES, E. J. (1979). Causes of world-wide changes in sea level. *Journal of the Geological Society, London* **136**, 187–92.

FELLER, W. (1968). *An introduction to probability theory and its application*. Wiley, New York.

FITCH, W. M. (1976). Molecular evolutionary clocks. In *Molecular evolution* (ed. F. J. Ayala) pp. 160–78. Sinauer, Sunderland, Mass.

—— and MARGOLIASH, E. (1967). Construction of phylogenetic trees. *Science, New York* **155**, 279–84.

FRAKES, L. A. (1979). *Climates throughout geologic time*. Elsevier, Amsterdam.

GINGERICH, P. D. (1977). Patterns of evolution in the mammalian fossil record. In *Patterns of evolution as illustrated by the fossil record* (ed. A. Hallam) pp. 469–500. Elsevier, Amsterdam.

GOULD, S. J. (1977). Eternal metaphors of palaeontology. In *Patterns of evolution as illustrated by the fossil record* (ed. A. Hallam) pp. 1–26. Elsevier, Amsterdam.

—— and ELDREDGE, N. (1977). Punctuated equilibria: the tempo and mode of evolution reconsidered. *Paleobiology* **3**, 115–51.

—— and LEWONTIN, R. C. (1979). The spandrels of San Marco and the Panglossian paradigm: a critique of the adaptationist programme. *Proceedings of the Royal Society of London* **B205**, 581–98.

GREENWOOD, P. H. (1965). The cichlid fishes of Lake Nabugabo, Uganda. *British Museum Natural History Bulletin (Zoology)* **12**, 315–57.

HAECKEL, E. (1866). *Allgemeine Anatomie der Organismen*. Reimer, Berlin.

HALDANE, J. B. S. (1949). Suggestions as to quantitative measurements of rates of evolution. *Evolution, Lancaster, Pa.* **3**, 51–6.

HALLAM, A. (1975). *Jurassic environments*. Cambridge University Press.

—— (ed.) (1977a). *Patterns of evolution as illustrated by the fossil record. (Developments in Palaeontology and Stratigraphy*, Vol. 5.) Elsevier, Amsterdam.

—— (1977b). Secular changes in marine inundation of USSR and North America through the Phanerozoic. *Nature, London* **269**, 769–72.

—— (1981). *Facies interpretation and the stratigraphic record*. Freeman, Oxford.

HARTLEY, B. S. (1979). Evolution of enzyme structure. *Proceedings of the Royal Society of London* **B205**, 443–52.

HENNIG, W. (1966). *Phylogenetic systematics*. University of Illinois Press, Urbana, Ill. (2nd edn. 1979.)

IRVING, E. (1977). Drift of the major continental blocks since the Devonian. *Nature, London* **270**, 304–9.

KAVANAUGH, D. H. (1972). Hennig's principles and

methods of phylogenetic systematics. *The Biologist* **54**, 115–27.

KIMURA, M. (1968). Evolutionary rate at the molecular level. *Nature, London* **217**, 624–6.

—— (1977). Causes of evolution and polymorphism at the molecular level. In *Molecular evolution and polymorphism* (ed. M. Kimura) pp. 1–28. National Institute of Genetics, Mishima, Japan.

KOHNE, D. E., CHISCON, J. A., and HOYER, B. H. (1972). Evolution of primate DNA sequences. *Journal of Human Evolution* **1**, 627–44.

LAPORTE, L. F. (1979). *Ancient environments*, 2nd edn. Prentice-Hall, Englewood Cliffs, New Jersey.

LOCKWOOD, J. G. (1979). *Causes of climate.* Arnold, London.

MAYR, R. (1976). *Evolution and the diversity of life: selected essays.* Belknap, Cambridge, Mass.

NEWELL, N. D. (1967). Revolutions in the history of life. *Geological Society of America Special Papers*, No. 89, pp. 63–91.

OLSON, E. C. (1944). Origin of mammals based upon cranial morphology of the therapsid suborders. *Geological Society of America Special Papers*, No. 55.

OXNARD, C. E. (1975). *Uniqueness and diversity in human evolution: morphometric studies of Australopithecines.* University of Chicago Press.

PAULSON, D. R. (1973). Predator polymorphism and stochastatic selection. *Evolution, Lancaster, Pa.* **27**, 269–77.

PITRAT, C. W. (1973). Vertebrates and the Permo-Triassic extinction. *Paleogeography, Paleoclimate and Paleoecology* **14**, 249–64.

POWERS, D. A. and PLACE, A. R. (1978). Biochemical genetics of *Fundulus heteroclitus* (L.). 1. Temporal and spatial variation in gene frequencies of Ldh-B, Mdh-A, Gpi-B, and Pgm-A. *Biochemical Genetics* **16**, 593–607.

RAUP, D. M. (1977). Stochastic models in evolutionary palaeontology. In *Patterns of evolution as illustrated by the fossil record* (ed. A. Hallam) pp. 59–78. Elsevier, Amsterdam.

RIEDL, R. (1975). *Die Ordnung des Lebendigen.* Parey, Hamburg.

—— (1979). *Order in living organisms: a systems analysis of evolution.* Wiley, London.

ROMER, A. S. (1966). *Vertebrate paleontology.* University of Chicago Press.

RUSSELL, D. A. (1979). The enigma of the extinction of the dinosaurs. *Annual Review of Earth and Planetary Sciences* **7**, 163–82.

SAVAGE, J. M. (1977). *Evolution*, 3rd edn. Holt, Rinehart & Winston, New York.

SCHOPF, T. J. M. (1974). Permo-Triassic extinctions:

relations to sea floor spreading. *Journal of Geology* **82**, 129–43.

SEYFERT, C. and SIRKIN, L. (1979). *Earth history and plate tectonics.* Harper & Row, London.

SIMBERLOFF, D. S. (1974). Permo-Triassic extinctions: effects of area on biotic equilibrium. *Journal of Geology* **82**, 267–74.

SIMPSON, G. G. (1944). *Tempo and mode in evolution.* Columbia University Press, New York.

—— (1953). *The major features of evolution.* Columbia University Press, New York.

SMITH, A. G. and BRIDEN, J. C. (1977). *Mesozoic and Cenozoic paleocontinental maps.* Cambridge University Press.

STANLEY, S. M. (1975). A theory of evolution above the species level. *Proceedings of the National Academy of Sciences U.S.A.* **72**, 646–50.

—— (1978). Chronospecies' longevities, the origin of genera, and the punctuational model of evolution. *Paleobiology* **4**, 26–40.

—— (1980). *Macroevolution: pattern and process.* Freeman, San Francisco.

TARLING, D. H. and TARLING, M. P. (1977). *Continental drift: a study of the earth's moving surface*, 2nd edn. Bell, London.

TASHIAN, R. E., GOODMAN, M., FERRELL, R. E., and TANIS, R. J. (1976). Evolution of carbonic anhydrase in primates and other mammals. In *Molecular anthropology* (ed. M. Goodman and R. E. Tashian). Plenum, New York.

VAN VALEN, L. (1973). A new evolutionary law. *Evolutionary Theory* **1**, 1–30.

VON BAER, K. E. (1828–88). *Über Entwicklungsgeschichte der Thiere.* Bornträger, Königsberg.

WEGENER, A. (1912). Die enstehung der Kontinente. *Geologische Rundschau* **3**, 276–92.

—— (1966). *The origin of continents and oceans.* [Translated from the 4th revised edition in German (1929) by J. Biram, with an introduction by B. C. King.] Methuen, London.

WESTOLL, T. S. (1949). On the evolution of the Dipnoi. In *Genetics, paleontology and evolution* (ed. G. L. Jepsen, E. Mayr, and G. G. Simpson) pp. 121–88. Princeton University Press.

WRIGHT, S. (1931). Evolution in Mendelian populations. *Genetics, Princeton* **16**, 97–159.

—— (1967). Comments on the preliminary working papers of Eden and Waddington. In *Mathematical challenges to the Neo-Darwinian Theory of Evolution* (ed. P. S. Moorehead and M. M. Kaplan). *Wistar Institute Symposium*, Vol. 5, pp. 117–20. Philadelphia.

—— (1968–78). *Evolution and the genetics of popu-*

lations. *A treatise in four volumes*. University of Chicago Press.

YOUNG, J. Z. (1938). The evolution of the nervous system and of the relationship of organism and environment. In *Evolution: essays on aspects of evolutionary biology, presented to Professor E. S. Goodrich on his 70th birthday* (ed. G. R. de Beer) pp. 179–204. Clarendon Press, Oxford.

—— (1971). *An introduction to the study of man*. Oxford University Press.

Chapters 2 and 3 The chordates

BARRINGTON, E. J. W. (1938). VI – The digestive system of *Amphioxus (Branchiostoma) lanceolatus. Philosophical Transactions of the Royal Society, London* **B228**, 269–311.

—— (1940). Observations on feeding and digestion in *Glossobalanus minutus. Quarterly Journal of microscopical Science* **82**, 227–60.

—— (1959). Some endocrinological aspects of the Protochordata. In *Comparative endocrinology*. Proceedings of the Columbia University Symposium on Comparative Endocrinology (ed. A. Gorbman). Wiley, New York.

—— (1965). *The biology of Hemichordata and Protochordata*. Oliver and Boyd, Edinburgh.

—— (1974). Biochemistry of primitive deuterostomians. In *Chemical zoology*, Vol. VIII (ed. M. Florkin and B. T. Scheer) pp. 61–95. Academic Press, London.

BATESON, W. (1886). The ancestry of the Chordata. *Quarterly Journal of microscopical Science* **26**, 535–71.

BERRILL, N. J. (1950). *The Tunicata with an account of the British species*. The Ray Society, No. 133.

—— (1955). *The origin of vertebrates*. Clarendon Press, Oxford.

—— (1975). Chordata: Tunicata. In *Reproduction of marine invertebrates II. Ectoprocts and lesser Coelomates* (ed. A. C. Giese and J. S. Pearse) pp. 241–82. Academic Press, New York.

BOEKE, J. (1935). The autonomic (enteric) nervous system of *Amphioxus lanceolatus. Quarterly Journal of microscopical Science* **77**, 623–58.

BONE, Q. (1958). The central nervous system in larval acraniates. *Quarterly Journal of microscopical Science* **100**, 509–27.

—— (1960). The central nervous system in amphioxus. *Journal of comparative Neurology* **115**, 27–64.

—— (1961). The organisation of the atrial nervous system of Amphioxus (*Branchiostoma lanceolatum* (Pallas)). *Philosophical Transactions of the Royal Society, London* **B243**, 241–69.

—— (1979). *The origin of chordates*, 2nd edn. Carolina Biological Supply, Burlington, NC.

—— and RYAN, K. P. (1978). Cupular sense organs in *Ciona* (Tunicata: Ascidiacea). *Journal of Zoology, London* **186**, 417–29.

—— (1979). The Langerhans receptor of *Oikopleura* (Tunicata: Larvacea). *Journal of the marine biological Association of the United Kingdom* **59**, 69–75.

—— ANDERSON, P. A. V., and PULSFORD, A. (1980). The communication between individuals in salp chains 1. Morphology of the system. *Proceedings of the Royal Society of London* B, **210**, 549–58.

BRIEN, P. (1974). General characteristics and evolution of Craniata or vertebrates. In *Chemical zoology*, Vol. VIII (ed. M. Florkin and B. T. Scheer) pp. 99–146. Academic Press, London.

BULLOCK, T. H. (1940). The functional organization of the nervous system of Enteropneusta. *Biological Bulletin. Marine biological Laboratory, Woods Hole, Mass.* **79**, 91–113.

—— (1944). The giant nerve fibre system in balanoglossids. *Journal of Comparative Neurology* **80**, 355–68.

BURDON-JONES, C. (1953). Development and biology of the larva of *Saccoglossus horsti* (Enteropneusta). *Philosophical Transactions of the Royal Society, London* **B236**, 553–89.

CARLISLE, D. B. (1968). Vanadium and other metals in ascidians. *Proceedings of the Royal Society, London* **B171**, 31–42.

CLONEY, R. A. (1978). Ascidian metamorphosis: review and analysis. In *Settlement and metamorphosis of marine invertebrate larvae* (ed. Fu-Shiang Chia and M. E. Rice) pp. 255–82. Elsevier, New York.

DILLY, P. N. (1973). The larva of *Rhabdopleura compacta* (Hemichordata). *Marine Biology* **18**, 69–86.

—— (1975). The pterobranch *Rhabdopleura compacta*: its nervous system and phylogenetic position. *Symposium of the zoological Society of London* No. 36, 1–16.

DODD, J. M. and DODD, M. H. I. (1966). An experimental investigation of the supposed pituitary affinities of the ascidian neural complex. In *Some contemporary studies in marine science* (ed. H. Barnes) Allen and Unwin, London.

EAKIN, R. M. and KUDA, A. (1971). Ultrastructure of sensory receptors in ascidian tadpoles. *Zeitschrift für Zellforschung und mikro-skopische Anatomie* **112**, 287–312.

FLOREY, E. (1951). Reizphysiologie Untersuchungen an der Ascidie *Ciona intestinatis*, L. *Biologisches Zentralblatt* **70**, 523–30.

FLOOD, PER R. (1968). Structure of the segmental trunk

muscle in amphioxus: with notes on the course and "endings" of the so-called ventral root fibres. *Zeitschrift für Zellforschung und mikroskopische Anatomie* **84**, 389–416.

——(1978). Filter characteristics of appendicularian food catching nets. *Experientia* **34**, 173–5.

FLORKIN, M. and SCHEER, B. T. (eds.) (1974). *Chemical zoology*, Vol. VIII. *Deuterostomians, cyclostomes and fishes*. Academic Press, London.

GARSTANG, W. (1894). Preliminary note on a New Theory of the Phylogeny of the Chordata. *Zoologischer Anzeiger* **17**, 122–5.

—— (1928). The morphology of the Tunicata and its bearings on the phylogeny of the Chordata. *Quarterly Journal of microscopical Science* **72**, 51–187.

GASKELL, W. H. (1908). *The origin of vertebrates*. Longman, London.

GODEAUX, J. E. A. (1974). Primitive deuterostomians. In *Chemical zoology*, Vol. VIII (ed. M. Florkin and B. T. Scheer) pp. 4–60. Academic Press, London.

GOODBODY, I. (1974). The physiology of ascidians. *Advances in Marine Biology* **12**, 1–149.

GOODRICH, E. S. (1902). On the structure of the excretory organs of Amphioxus, Part 1. *Quarterly Journal of microscopical Science* **45**, 493–501.

HATSCHEK, B. (1893). *The amphioxus and its development* (trans. and ed. J. Tuckey). Sonnenschein, London.

JEFFERIES, R. P. S. (1973). The Ordovician fossil *Lagynocystis pyramidalis* (Barrande) and the ancestry of amphioxus. *Philosophical Transactions of the Royal Society, London* **B265**, 409–69.

—— (1975). Fossil evidence concerning the origin of the chordates. *Symposia of the zoological society of London* No. 36, 253–318.

MACKIE, G. O. (1974). Behaviour of a compound ascidian. *Canadian Journal of Zoology* **52**, 23–7.

MACKIE, G. O., PAUL, D. H., SINGLA, C. M., SLEIGH, M. A. and WILLIAMS, D. E. (1974). Branchial innervation and ciliary control in the ascidian *Corella*. *Proceedings of the Royal Society, London* **B187**, 1–35.

MARCHALONIS, J. J. (1977). *Immunity in evolution*. Arnold, London.

MILLAR, R. H. (1953). *Ciona*. Liverpool Marine Biological Committee Memoirs on typical British marine Plants and Animals **35**, 122 pp.

MONNIOT, C. and MONNIOT, F. (1978). Recent work on the deep-sea tunicates. *Oceanography and marine Biology Annual Review* **16**, 181–228.

ORTON, J. H. (1913). The ciliary mechanisms on the gill and the mode of feeding in Amphioxus, Ascidians and *Solenomyo togato*. *Journal of marine biological Association of the United Kingdom* **10**, 19–49.

PATTEN, W. (1912). *The evolution of the vertebrates and their kin*. Churchill, London.

RÄHR, H. (1979). The circulatory system of Amphioxus (*Branchiostoma lanceolatum* (Pallas)). A light-microscopic investigation based on intravascular injection technique. *Acta zoologica, Stockholm* **60**, 1–18.

RAMÓN Y CAJAL, S. (1906). The structure and connexions of neurons. Nobel Lecture, December 12, 1906. In *Nobel Lectures including presentation speeches and Laureates' Biographies, Physiology or Medicine 1901–1921*. Elsevier, Amsterdam.

SHERRINGTON, C. S. (1906). *The integrative action of the nervous system*. Yale University Press, New Haven, Conn.

SOUTHWARD, E. C. (1975). Fine structure and phylogeny of the Pogonophora. *Symposia of the zoological Society of London* No. 36, 235–51.

WATTS, D. C. (1975). Evolution of phosphagen kinases in the chordate line. *Symposia of the zoological Society of London* No. 36, 105–27.

WEBB, J. E. (1973). The role of the notochord in forward and reverse swimming and burrowing in the amphioxus *Branchiostoma lanceolatum*. *Journal of Zoology, London* **170**, 325–38.

—— (1976). A review of swimming in amphioxus. In *Perspectives in experimental biology*, Vol. 1, *Zoology*. *Proceedings of the Fiftieth Anniversary Meeting of the Society of Experimental Biology* (ed. P. Spencer Davies) pp. 447–54. Pergamon, Oxford.

WELSCH, U. (1975). The fine structure of the pharynx, cyrtopodocytes and digestive caecum of amphioxus (*Branchiostoma lanceolatum*). *Symposia of the zoological Society of London* No. 36, 17–41.

WICKSTEAD, J. H. (1975). Chordata: Acrania (Cephelochordata). In *Reproduction of marine invertebrates II, Entoprocts and lesser Coelomates* (ed. A. C. Giese and J. S. Pearse) pp. 283–319. Academic Press, New York.

WILLMER, E. N. (1975). The possible contribution of nemertines to the problem of the phylogeny of the protochordates. *Symposia of the zoological Society of London* No. 36, 319–45.

YOUNG, J. Z. (1974). The George Bidder Lecture 1973. Brains and worlds: the cerebral cosmologies. *Journal of experimental Biology* **61**, 5–17.

Chapter 4 Vertebrates without jaws. Lampreys

BARDACK, D. and LANGERL, R. (1971). Lampreys in the fossil record. In *The biology of lampreys* (ed. M. W. Hardisty and I. C. Potter) **1**, 67–84. Academic Press, London.

BARRINGTON, E. J. W. (1936). Proteolytic digestion and

the problem of the pancreas in *Lampetra*. *Proceedings of the Royal Society, London* **B121**, 221–32.

—— (1942). Blood sugar and the follicles of Langerhans in the ammocoete larva. *Journal of experimental Biology* **19**, 45–55.

DEAN, B. (1895). *Fishes, living and fossil: an outline of their forms and probable relationships*. Macmillan, New York.

DOBZHANSKY, T. (1951). *Genetics and the origin of species*, 3rd edn. Columbia University Press, New York.

EBBESSON, S. O. and NORTHCUTT, R. G. (1976). Neurology of anamniotic vertebrates. In *Evolution of brain and behavior in vertebrates* (ed. R. B. Masterton, C. B. G. Campbell, M. E. Bitterman, and N. Hotton). Erlbaum, Hillsdale, New Jersey.

EDDY, J. M. P. (1972). The pineal complex. In *The biology of lampreys* (ed. M. W. Hardisty and I. C. Potter) **2**, 91–103. Academic Press, London.

GRAY, J. E. (1851). *List of the specimens in the collection of the British Museum*, Part 1. *Chondropterygii*. British Museum of Natural History, London.

HALSTEAD, L. B. (1973). The heterostracan fishes. *Biological Reviews* **48**, 279–332.

—— (1979). Internal anatomy of the polybranchiaspids (Agnatha, Galeaspida). *Nature, London* **282**, 833–6.

—— LIU, Y.-H. and P'AN, K. (1979). Agnathans from the Devonian of China. *Nature, London* **282**, 831–3.

HARDISTY, M. W. (1979). *Biology of cyclostomes*. Chapman and Hall, London.

—— and POTTER, I. C. (1971, 1972). *The biology of lampreys*, Vols. 1 and 2. Academic Press, London.

—— (1971). The general biology of adult lampreys. In *The biology of lampreys* (ed. M. W. Hardisty and I. C. Potter) **1**, 127–206. Academic Press, London.

HOLMES R. L. and BALL, J. N. (1974). *The pituitary gland: a comparative account*. Cambridge University Press.

HUBBS, C. L. and POTTER, I. C. (1971). Distribution, phylogeny and taxonomy. In *The biology of lampreys* (ed. M. W. Hardisty and I. C. Potter) **1**, 1–65. Academic Press, London.

JERISON, H. J. (1973). *Evolution of the brain and intelligence*. Academic Press, London.

KLEEREKOPER, H. and SIBAKIN, K. (1956). Spike potentials produced by the sea lamprey (*Petromyzon marinus*) in the water surrounding the head region. *Nature, London* **178**, 490–1.

LARSEN, L. O. and ROTHWELL, B. (1972). Adenohypophysis. In *The biology of lampreys* (ed. M. W. Hardisty and I. C. Potter) **2**, 1–67. Academic Press, London.

LOWENSTEIN, O., OSBORNE, M. P., and THORNHILL, R.

A. (1968). The anatomy and ultrastructure of the labyrinth of the lamprey (*Lampetra fluviatilis* L.). *Proceedings of the Royal Society, London* **B170**, 113–34.

MORMAN, R. H. (1979). Distribution and ecology of lampreys in the lower peninsula of Michigan, 1957–75. Great Lakes Fishery Commission, Technical Report No. 33, 59pp.

MOY-THOMAS, J. A. and MILES, R. S. (1971). *Palaeozoic fishes*, 2nd edn. Chapman and Hall, London.

ØRVIG, T. (1967). Phylogeny of tooth tissues: evolution of some calcified tissues in early vertebrates. In *Structural and chemical organization of teeth* (ed. A. E. W. Miles). Academic Press, London.

ROVAINEN, C. M. (1974). Synaptic interactions of identified nerve cells in the spinal cord of the sea lamprey. *Journal of comparative Neurology* **154**, 189–206.

—— (1976). Vestibulo-ocular reflexes in the adult sea lamprey. *Journal of comparative Physiology* **112**, 159–64.

—— (1979). Neurobiology of lampreys. *Physiological Reviews* **59**, 1007–77.

STENSIÖ, E. (1958). Les cyclostomes fossiles ou Ostracodermes. In *Traité de zoologie*, 1st fasc., Vol. 13 (ed. P. P. Grassé). Masson et Cie, Paris.

STERBA, G. (1972). Neuro- and gliasecretion. In *The biology of lampreys* **2**, 69–89. Academic Press, London.

STEVEN, D. M. (1950). Some properties of the photoreceptors of the brook lamprey. *Journal of experimental Biology* **27**, 350–64.

—— (1963). The dermal light sense. *Biological Reviews* **38**, 204–39.

TARLO, L. B. H. (1969). Calcified tissues in the earliest vertebrates. *Calcified Tissue Research* **3**, 107–24.

WALLS, G. L. (1942). *The vertebrate eye and its adaptive radiation*. The Cranbrook Institute of Science, Bloomfield Hills, Mich.

WATSON, D. M. S. (1954). A consideration of ostracoderms. *Philosophical Transactions of the Royal Society, London* **B238**, 1–25.

WHITING, H. P. (1977). Cranial anatomy of the ostracoderms in relation to the organisation of larval lampreys. In *Problems in vertebrate evolution*: Essays presented to Professor T. S. Westoll, F.R.S., F.L.S. (ed. S. M. Andrews, R. S. Miles, and A. D. Walker). Academic Press, London. [*Linnean Society Symposium Series* No. 4, 1–23.]

Chapters 5–10 Fishes

ADEY, W. R. and BAWIN, S. M. (1977). Brain interactions with weak electric and magnetic fields.

Neurosciences Research Program Bulletin **15**, No. 1, 129 pp.

ADRIAN, E. D. and BUYTENDYK, F. J. (1931). Potential changes in the isolated brain stem of the goldfish. *Journal of Physiology, London* **71**, 121–35.

ARONSON, L. R. and KAPLAN, H. (1968). Function of the teleostean forebrain. In *The central nervous system and fish behavior* (ed. D. Ingle) pp. 107–25. University of Chicago Press.

—— ARONSON, F. R., and CLARK, E. (1967). Instrumental conditioning and light–dark discrimination in young nurse sharks. *Bulletin of marine Science of the Gulf and Caribbean* **17**, 249–56.

ATTENBOROUGH, D. (1979). *Life on earth, a natural history*. BBC/Collins, London.

BALFOUR, F. M. (1878). *A monograph on the development of elasmobranch fishes*. Macmillan, London.

BAINBRIDGE, R. (1961). Problems of fish locomotion. *Symposia of the zoological Society of London* No. 5, 13–32.

BARDACH, J. E. and VILLARS, T. (1974). The chemical senses of fishes. In *Chemoreception in marine organisms* (ed. P. T. Grant and A. M. Mackie) pp. 49–104. Academic Press, London.

BENNETT, M. V. L. (1968). Neural control of electric organs. In *The central nervous system and fish behavior* (ed. D. Ingle) pp. 147–69. University of Chicago Press.

—— (1971). Electric organs. In *Fish physiology*, Vol. V (ed. W. S. Hoar and D. J. Randall) pp. 347–491. Academic Press, New York.

BENTLEY, P. J. (1971). *Endocrines and osmoregulation. A comparative account of regulation of water and salt in vertebrates*. Springer, Berlin.

BLIGHT, A. R. (1976). Undulatory swimming with and without waves of contraction. *Nature, London* **264**, 352–4.

—— (1977). The muscular control of vertebrate swimming movements. *Biological Reviews* **52**, 181–218.

BONE, Q. (1978). Locomotor muscle. In *Fish physiology*, Vol. VII (ed. W. S. Hoar and D. J. Randall). Academic Press, New York.

BOORD, R. L. and CAMPBELL, C. B. G. (1977). Structural and functional organization of the lateral line system of sharks. *American Zoologist* **17**, 431–41.

BREDER, C. M. and ROSEN, D. E. (1966). *Modes of reproduction in fishes: how fishes breed*. Natural History Press, New York.

BULL, H. O. (1957). Behavior: conditioned responses. In *The physiology of fishes*, Vol. II (ed. M. E. Brown) pp. 211–28. Academic Press, New York.

COLLIS, C. S. (1979). Melanophore potentials of the chromatically intact spinal stoneloach (*Neomacheilus barbatulus* L.) following adaptation to varying backgrounds. *Journal of comparative Physiology A* **131**, 13–21.

COMPAGNO, L. J. V. (1977). Phyletic relationships of living sharks and rays. *American Zoologist* **17**, 303–22.

DAY, M. H. (ed.) (1981). Vertebrate locomotion. *Symposia of the zoological Society of London* (in press).

DEAN, B. (1895). *Fishes, living and fossil: an outline of their forms and probable relationships*. Macmillan, New York.

DENTON, E. J. (1970). Review lecture on the organization of reflecting surfaces in some marine animals. *Philosophical Transactions of the Royal Society of London* **B258**, 285–313.

—— and NICOL, J. A. C. (1964). The chorioidal tapeta of some cartilaginous fishes (Chondrichthyes). *Journal of the marine biological Association of the United Kingdom* **44**, 219–58.

—— GILPIN-BROWN, J. B., and WRIGHT, P. G. (1970). On the "filters" in the photophores of mesopelagic fish and on a fish emitting red light and especially sensitive to red light. *Journal of Physiology, London* **208**, 72P.

—— (1972). The angular distribution of the light produced by some mesopelagic fish in relation to their camouflage. *Proceedings of the Royal Society, London* **B182**, 145–58.

—— GRAY, J. A. B., and BLAXTER, J. H. S. (1979). The mechanics of the clupeid acoustico-lateralis system: frequency responses. *Journal of the marine biological Association of the United Kingdom* **59**, 27–47.

EBBESSON, S. O. E. (1972a). New insights into the organization of the shark brain. *Comparative Biochemistry and Physiology* **42**, 121–9.

—— (1972b). A proposal for a common nomenclature for some optic nuclei in vertebrates and the evidence for a common origin of two such cell groups. *Brain, Behavior and Evolution* **6**, 75–91.

—— (ed.) (1980). *Comparative neurology of the telencephalon*. Plenum Press, New York.

—— and NORTHCUTT, R. G. (1976). Neurology of anamniotic vertebrates. In *Evolution of brain and behavior in vertebrates* (ed. R. B. Masterton, C. B. G. Campbell, M. E. Bitterman, and N. Hotton). Erlbaum, Hillsdale, New Jersey.

ECCLES, J. C., TÁBOŘÍKOVÁ, H., and TSUKAHARA, N. (1970). Responses of the Purkyně cells of a selachian cerebellum (*Mustellus canis*). *Brain Research* **17**, 57–86.

FLORKIN, M. and SCHEER, B. T. (eds.) (1974). *Chemical Zoology*, Vol. VIII. *Deuterostomians, cyclostomes and fishes*. Academic Press, London.

GILBERT, P. W., MATHEWSON, R. F., and RALL, D. P. (eds.) (1967). *Sharks, skates and rays*. The Johns Hopkins Press, Baltimore.

GOLDSTEIN, S. (ed.) (1938). *Modern developments in fluid dynamics*. Oxford University Press.

GOODRICH, E. S. (1909). *Vertebrata Craniata* (First Fascicle: Cyclostomes and fishes). Part IX of *A treatise on zoology* (ed. Sir Ray Lankester). Black, London.

—— (1930). *Studies on the structure and development of vertebrates*. Macmillan, London.

GRAY, J. (1936). Studies in animal locomotion VI. The propulsive powers of the dolphin. *Journal of experimental Biology* **13**, 192–9.

—— (1957). How fishes swim. *Scientific American* **197**, 48–54.

—— (1968). *Animal locomotion*. Weidenfeld and Nicolson, London.

GREENWOOD, P. H., MILES, R. S., and PATTERSON, C. (eds.) (1973). *Interrelationships of fishes*. Academic Press, London. [Supplement No. 1 to the *Zoological Journal of the Linnean Society* **53**.]

—— ROSEN, D. E., WEITZMAN, S. H., and MYERS, G. S. (1966). Phyletic studies of teleostean fishes, with a provisional classification of living forms. *Bulletin of the American Museum of natural History* **131**, Art 4, 339–456.

GRIFFITH, R. W., UMMINGER, B. L., GRANT, B. F., PANG, P. K. T., and PICKFORD, G. E. (1974). Serum composition of the coelacanth *Latimeria chalumnae* Smith. *Journal of experimental Zoology* **187**, 87–102.

GRUBER, S. H. (1977). The visual system of sharks, adaptations and capability. *American Zoologist* **17**, 453–69.

—— and MYRBERG, Jr, A. A. (1977). Approaches to the study of the behavior of sharks. *American Zoologist* **17**, 471–86.

HARRIS, J. E. (1936). The role of the fins in the equilibrium of the swimming fish I. Wind-tunnel tests on a model of *Mustelus canis* (Mitchill). *Journal of experimental biology* **13**, 476–93.

—— (1938). The role of the fins in the equilibrium of the swimming fish II. The role of the pelvic fins. *Journal of experimental Biology* **15**, 32–47.

HASLER, A. D. (1971). Orientation and fish migration. In *Fish physiology*, Vol. VI (ed. W. S. Hoar and D. J. Randall) pp. 429–510. Academic Press, New York.

HERRING, P. J. (ed.) (1978). *Bioluminescence in action*. Academic Press, London.

HOAR, W. S. and RANDALL, D. J. (eds.) (1969–79). *Fish physiology*, Vols. I–VIII. Academic Press, New York.

HOLMES, R. L. and BALL, J. N. (1974). *The pituitary gland: a comparative account*. Cambridge University Press.

HULET, W. H., FISCHER, J., and REITBERG, B. J. (1972). Electrolyte composition of *Anguilliform leptocephali* from the Straits of Florida. *Bulletin of marine Science of the Gulf and Caribbean* **22**, 432–48.

JARVIK, E. (1980, 1981). *Basic structure and evolution of vertebrates*. 2 vols. Academic Press, London.

JERISON, H. J. (1973). *Evolution of the brain and intelligence*. Academic Press, New York.

JOHANSEN, K., LENFANT, C., and GRIGG, G. C. (1967). Respiratory control in the lungfish. *Comparative Biochemistry and Physiology* **20**, 835–54.

JOLLIE, M. T. (1977). Segmentation of the vertebrate head. *American Zoologist* **17**, 323–33.

JONES, G. M. (1974). The functional significance of semicircular canal size. In *Vestibular system part 1: Basic mechanisms* (ed. H. H. Kornhuber) pp. 171–84. Part VI/1 of *Handbook of sensory physiology*. Springer, Berlin.

—— SPELLS, K. E. (1963). A theoretical and comparative study of the functional dependence of the semicircular canal upon its physical dimensions. *Proceedings of the Royal Society of London* **B157**, 403–19.

KAPPERS, C. U. A., HUBER, G. C., and CROSBY, E. C. (1936). *The comparative anatomy of the nervous system of vertebrates, including man*. Macmillan, New York.

KLITGAARD, T. (1978). Morphology and histology of the heart of the Australian lungfish, *Neoceratodus forsteri* (Krefft). *Acta zoologica, Stockholm* **59**, 135–47.

KUCHNOW, K. P. (1971). The elasmobranch pupillary response. *Vision Research* **11**, 1395–406.

LAUDER, Jr, G. V. (1979). Feeding mechanics in primitive teleosts and in the halecomorph fish *Amia calva*. *Journal of Zoology, London* **187**, 543–78.

—— (1980). Evolution of the feeding mechanism in primitive actinopterygian fishes: a functional anatomical analysis of *Polypterus, Lepisosteus*, and *Amia*. *Journal of Morphology* **163**, 283–317.

LAURENT, P. (1974). Pseudobranchial receptors in teleosts. In *Electroreceptors and other specialized receptors in lower vertebrates* (ed. A. Fessard) pp. 279–96. Part III/3 of *Handbook of sensory physiology*. Springer, Berlin.

LIGHTHILL, Sir J. (1975). *Mathematical biofluiddynamics*. Society for Industrial and Applied Mathematics, Philadelphia.

LOCKETT, N. A. (1977). Adaptations to the deep-sea environment. In *The visual system in vertebrates* (ed. F. Crescitelli). Part III/5 of *Handbook of sensory physiology*. Springer, Berlin.

—— (1980). Review Lecture. Some advances in coelacanth biology. *Proceedings of the Royal Society of London* **B208**, 265–384.

McNulty, J. A. (1979). A comparative light and electron microscope study of the pineal complex in the deep-sea fishes *Cyclothone signata* and *C. acclinidens*. *Journal of Morphology* **162**, 1–16.

Marshall, N. B. (1960). Swimbladder structure of deep-sea fishes in relation to their systematics and biology. *Discovery Report* **31**, 1–122.

—— (1965). *The life of fishes*. Weidenfeld and Nicolson, London.

—— (1971). *Explorations in the life of fishes*. Harvard University Press, Cambridge, Mass.

—— (1979). *Developments in deep-sea biology*. Blandford, Poole, Dorset.

Miles, R. S. (1973). Relationships of acanthodians. In *Interrelationships of fishes* (ed. P. H. Greenwood, R. S. Miles, and C. Patterson). Academic Press, London. [Supplement No. 1 to the *Zoological Journal of the Linnean Society* **53**.]

Millar, P. J. (1980). Fish phenology: anabolic adaptiveness in teleost fishes. *Symposia of the zoological Society of London* No. 44.

Mittal, A. K. and Whitear, M. (1979). Keratinization of fish skin with special reference to the catfish *Bagurius bagurius*. *Cell and Tissue Research* **202**, 213–30.

Moss, M. L. (1977). Skeletal tissues in sharks. *American Zoologist* **17**, 335–42.

Nicol, J. A. G. (1969). Bioluminescence. In *Fish physiology*, Vol. III (ed. W. S. Hoar and D. J. Randall) pp. 355–400. Academic Press, New York.

Nielsen, J. G. and Munk, O. (1964). A hadal fish (*Bassogigas profundisomus*) with a functional swimbladder. *Nature, London* **204**, 594–5.

Nieuwenhuys, R. (1966). The interpretation of the cell masses in the teleostean forebrain. In *Evolution of the forebrain: phylogenesis and ontogenesis of the forebrain* (ed. R. Hassler and H. Stephan) pp. 32–9. Georg Thieme, Stuttgart.

—— (1967). Comparative anatomy of olfactory centres and tracts. *Progress in Brain Research* **23**, 1–64.

Norman, J. R. (1931). *A history of fishes*. Benn, London. [3rd edn 1975 revised by P. H. Greenwood. Benn, London.]

Northcutt, R. G. (1977). Elasmobranch central nervous system organization and its possible evolutionary significance. *American Zoologist* **17**, 411–29.

Packard, A. (1960). Electrophysiological observations on a sound-producing fish. *Nature, London* **187**, 63–4.

—— and Wainwright, A. W. (1973). Brain growth of young herring and trout. In *The early life history of fish* (ed. J. H. S. Blaxter). Springer, Berlin.

Pang, P. K. T., Griffith, R. W., and Atz, J. W. (1977). Osmoregulation in elasmobranchs. *American Zoologist* **17**, 365–77.

Patterson, C. (1973). Interrelationships of holosteans. In *Interrelationships of fishes* (ed. P. H. Greenwood, R. S. Miles, and C. Patterson). Academic Press, London. [Supplement No. 1 to the *Zoological Journal of the Linnean Society* **53**.]

Pietsch, T. W. (1975). Precocious sexual parasitism in the deep sea ceratioid anglerfish, *Cryptopsaras conesi* Gill. *Nature, London* **256**, 38–40.

Popper, A. N. and Fay, R. R. (1977). Structure and function of the elasmobranch auditory system. *American Zoologist* **17**, 443–52.

Pye, J. D. (1964). Nervous control of chromatophores in teleost fishes. I. Electrical stimulation in the minnow (*Phoxinus phoxinus* (L.)). *Journal of experimental Biology* **41**, 525–34.

Retzius, M. C. (1881–4). *Das Gehörgan der Wirbelthiere. Morphologisch-histologische Studien*. Kongel. Boktr., Stockholm.

Roberts, B. L. (1969). The response of a proprioceptor to the undulatory movements of dogfish. *Journal of experimental Biology* **51**, 775–85.

Russell, I. J. and Roberts, B. L. (1972). Inhibition of spontaneous lateral-line activity by efferent nerve stimulation. *Journal of experimental Biology* **57**, 77–82.

Sand, A. (1937). The mechanism of the lateral sense organs of fishes. *Proceedings of the Royal Society, London* **B123**, 472–95.

Satchell, G. H. (1971). *Circulation in fishes*. Cambridge University Press.

Schadé, J. P. and Weiler, I. J. (1959). Electroencephalographic patterns of the goldfish (*Carassius auratus*). *Journal of experimental Biology* **36**, 435–52.

Schaeffer, B. (1952). The Triassic Coelacanth fish *Diplurus*, with observations on the evolution of the Coelacanthini. *Bulletin of the American Museum of natural History* **99**, 31–78.

—— (1967). Comments on elasmobranch evolution. In *Sharks, skates and rays* (ed. P. W. Gilbert, R. G. Mathewson, and D. P. Rall) pp. 3–35. Johns Hopkins Press, Baltimore.

Scheich, H. and Bullock, T. H. (1974). The detection of electric fields from electric organs. In *Electroreceptors and other specialized receptors in lower vertebrates* (ed. A. Fessard) pp. 201–56. Part III/3 of *Handbook of sensory physiology*. Springer, Berlin.

Schmidt-Nielsen, K. (1979). *Animal physiology: adaptation and environment*, 2nd edn. Cambridge University Press.

Schnitzlein, H. N. (1968). Introductory remarks on the telencephalon of fish. In *The central nervous system and fish behavior* (ed. D. Ingle) pp. 97–105. University of Chicago Press.

SHARP, G. D. and DIZON, A. E. (1978). *The physiological ecology of tunas*. Academic Press, London.

SHERRINGTON, Sir C. S. (1947). *The integrative action of the nervous system*, 2nd edn. Cambridge University Press.

SMITH, C. L., RAND, C. S., SCHAEFFER, B., and ATZ, J. W. (1975). *Latimeria*, the living coelacanth, is ovoviviparous. *Science, New York* **190**, 1105–6.

SPONDER, D. L. and LAUDER, G. V. (1981). Terrestrial feeding in the mudskipper *Periophthalmus* (Pisces: Teleostei): a cineradiographic analysis. *Journal of Zoology, London* **193**, 517–30.

SUTHERLAND, N. S. (1968). Shape discrimination in the goldfish. In *The central nervous system and fish behavior* (ed. D. Ingle) pp. 35–50. University of Chicago Press.

TAVOLGA, W. N. (1971). Sound production and detection. In *Fish physiology*, Vol. V (ed. W. S. Hoar and D. J. Randall) pp. 135–205. Academic Press, New York.

THOMSON, K. S. (1966). Intracranial mobility in the coelacanth. *Science, New York* **153**, 999–1000.

—— (1969). The biology of the lobe-finned fishes. *Biological Reviews* **44**, 91–154.

—— (1975). On the biology of cosmine. *Peabody Museum of Natural History, Yale University, Bulletin* **40**, 58 pp.

—— (1976). On the heterocercal tail in sharks. *Paleobiology* **2**, 19–38.

WAINWRIGHT, S. A., VOSBURGH, F., and HEBRANK, J. H. (1978). Shark skin: function in locomotion. *Science, N.Y.* **202**, 747–9.

WALLS, G. L. (1942). *The vertebrate eye and its adaptive radiation*. The Cranbrook Institute of Science, Bloomfield Hills, Michigan.

WAXMAN, S. G. (1975). Integrative properties and design principles of axons. *International Review of Neurobiology* **18**, 1–40.

WESTOLL, T. S. (1949). On the evolution of the Dipnoi. In *Genetics, paleontology and evolution* (ed. G. L. Jepsen, E. Mayr, and G. G. Simpson) pp. 121–84. Princeton University Press.

WHITING, H. P. and BONE, Q. (1980). Ciliary cells in the epidermis of the larval Australian dipnoan, *Neoceratodus*. *Zoological Journal of the Linnean Society* **68**, 125–37.

YOUNG, J. Z. (1933). The autonomic nervous systems of selachians. *Quarterly Journal of microscopical Science* **75**, 571–624.

—— (1978). *Programs of the brain*. Oxford University Press.

—— (1980). Nervous control of movements of the gut of fishes: 1. The stomach of dogfishes and rays. 2. *Lophius*. *Journal of the marine biological Association of the United Kingdom* **60**, 1–17.

ZANGERL, R. (1973). Interrelationships of early chondrichthyans. In *Interrelationships of fishes* (ed. P. H. Greenwood, R. S. Miles, and C. Patterson). Academic Press, London. [*Zoological Journal of the Linnean Society*, Suppl. No. 1, **53**, 1–14.]

Chapters 11 and 12 Amphibia

BAGNARA, J. T. (1976). Color change. In *Physiology of the Amphibia*, Vol. III (ed. B. Lofts) pp. 1–52. Academic Press, New York.

BARRINGTON, E. J. W. (1975). *An introduction to general and comparative endocrinology*. Clarendon Press, Oxford.

CAMPENHAUSEN, C. V. (1963). Quantitative Beziehungen zwischen Lichtreiz und Kontraktion des Musculus sphincter pupillae vom Scheibenzüngler (*Discoglossus pictus*). *Kybernetik* **1**, 249–67.

COX, C. B. (1966). The Amphibia – an evolutionary backwater. In *Looking at animals again: contributions to a further understanding and investigation of some common animals* (ed. D. R. Arthur) pp. 97–118. Freeman, London.

EVANS, F. G. (1944). The morphological status of the modern Amphibia among the Tetrapoda. *Journal of Morphology* **74**, 43–100.

—— (1946). The anatomy and function of the foreleg in salamander locomotion. *Anatomical Record* **95**, 257–81.

EWERT, J.-P. and BORCHERS, H.-W. (1971). Reaktions charakteristik von Neuronen aus dem Tectum opticum und subtectum der Erdkröte *Bufo bufo* (L). *Zeitschrift für vergleichende Physiologie* **71**, 165–89.

FOXON, G. E. H. (1955). Problems of the double circulation in vertebrates. *Biological Reviews* **30**, 196–228.

—— (1964). Blood and respiration. In *Physiology of the Amphibia* (ed. J. A. Moore) pp. 151–209. Academic Press, London.

FRANCIS, E. T. B. (1961). *The anatomy of the salamander*. Clarendon Press, Oxford.

GAUPP, E. (1896–1904). In A. Ecker's and R. Wiederheim's *Anatomie des Frosches auf Grund eigener untersuchungen Durchaus neu Bearbeitet*, Vols. I–III. Vieweg, Braunschweig.

GAZE, R. M. and STRAZNICKY, C. (1980). Regeneration of optic nerve fibres from a compound eye to both tecta in *Xenopus*: evidence relating to the state of specification of the eye and tectum. *Journal of embryology and experimental morphology* **60**, 125–40.

GOODRICH, E. S. (1930). *Studies on the structure and development of vertebrates*. Macmillan, London.

GORDON, M. S., SCHMIDT-NIELSEN, K., and KELLY, H. M. (1961). Osmotic regulation in the crab-eating frog (*Rana crancrivora*). *Journal of experimental Biology* **38**, 659–78.

GOUDER, B. Y. M. and DESAI, R. N. (1966). Studies on the carotid body in the frog *Rana tigrina* Daud. *Naturwissenschaften* **53**, 535–6.

GRAY, J. and LISSMANN, H. W. (1947). The co-ordination of limb movements in the Amphibia. *Journal of experimental Biology* **23**, 133–42.

GREGORY, W. K. and RAVEN, H. C. (1942). Studies on the origin and early evolution of paired fins and limbs. *Annals of the New York Academy of Sciences* **42**, 273–360.

GRIFFITHS, I. (1954). On the nature of the fronto-parietal in Amphibia, Salientia. *Proceedings of the zoological Society of London* **123**, 781–92.

—— (1963). The phylogeny of the Salientia. *Biological Reviews* **38**, 241–92.

HABERMEHL, G. G. (1974). Venoms of Amphibia. In *Chemical zoology*, IX (ed. M. Florkin and B. T. Scheer) pp. 161–83. Academic Press, London.

HALLIDAY, T. R. (1977). The courtship of European newts: an evolutionary perspective. In *The reproductive biology of amphibians* (ed. D. H. Taylor and S. I. Guttman). Plenum, New York.

HANKE, W. (1974). Endocrinology of amphibia. In *Chemical zoology*, Vol. IX (ed. M. Florkin and B. T. Scheer) pp. 123–159. Academic Press, London.

HORRIDGE, G. A. (1968). *Interneurons: their origin, action, specificity growth and plasticity*. Freeman, London.

JAEGER, C. B. and HILLMAN, D. E. (1976). Morphology of gustatory organs. In *Frog neurobiology* (ed. R. Llinas and W. Precht) pp. 588–606. Springer, Berlin.

JARVIK, E. (1980, 1981). *Basic structure and evolution of vertebrates*. 2 vols. Academic Press, London.

JASIŃSKI, A. and MIODOŃSKI, A. (1978). Model of skin vascularization in *Rana esculenta* L.: scanning electron microscopy of microcorrosion casts. *Cell and Tissue Research* **191**, 539–48.

KEMALI, M. and BRAITENBERG, V. (1969). *Atlas of the frog's brain*. Springer, Heidelberg.

LLINÁS, R. (1976). Cerebellar physiology. In *Frog neurobiology* (ed. R. Llinás and W. Precht) pp. 892–923. Springer, Berlin.

—— and PRECHT, W. (eds.) (1976). *Frog neurobiology*. Springer, Berlin.

LOFTS, B. (ed.) (1974 and 1976). *Physiology of the Amphibia*, Vols. II and III. Academic Press, New York. [Vol. I edited by J. A. Moore.]

LOMBARD, R. E. and BOLT, J. R. (1979). Evolution of the tetrapod ear: an analysis and re-interpretation. *Biological Journal of the Linnean Society* **11**, 19–76.

—— and STRAUGHAN, I. R. (1974). Functional aspects of anuran middle ear structure. *Journal of experimental Biology* **61**, 71–93.

MARSHALL, A. M. (1920). *The frog: an introduction to anatomy, histology and embryology*, 11th edn. Macmillan, London.

MOORE, J. A. (ed.) (1964). *Physiology of the Amphibia*, Vol. 1. Academic Press, New York. [Vols. II and III edited by B. Lofts.]

MUNTZ, W. R. A. (1962). Effectiveness of different colors of light in releasing the positive phototactic behavior of frogs, and a possible function of the retinal projection to the direncephalon. *Journal of Neurophysiology* **25**, 712–20.

OKSCHE, A. and UECK, M. (1976). The nervous system. In *Physiology of the Amphibia*, Vol. III (ed. B. Lofts) pp. 314–419. Academic Press, New York.

PAPEZ, J. W. (1929). *Comparative neurology: a manual and text for the study of the nervous system of vertebrates*. Crowell, New York.

SALTHE, S. N. and MECHAM, J. S. (1974). Reproductive and courtship patterns. In *Physiology of the Amphibia*, Vol. II (ed. B. Lofts). Academic Press, New York.

WALLS, G. L. (1942). *The eye and its adaptive radiation*. The Cranbrook Institute of Science, Blomfield Hills, Michigan.

WATSON, D. M. S. (1925). Croonian Lecture – The evolution and origin of the Amphibia. *Philosophical Transactions of the Royal Society of London* **B214**, 189–257.

WELLS, K. D. (1977). The courtship of frogs. In *The reproductive biology of amphibians* (ed. D. H. Taylor and S. I. Guttman). Plenum, New York.

WESTOLL, T. S. (1943). The origin of the tetrapods. *Biological Reviews* **18**, 78–98.

WHITING, H. P. (1961). Pelvic girdle in amphibian locomotion. *Symposia of the zoological Society of London* No. 5, 43–57.

WILLISTON, S. W. (1925). *The osteology of the reptiles*. Harvard University Press, Cambridge, Mass.

Chapters 13 and 14 The reptiles

BAKKER, R. T. (1977). Tetrapod mass extinctions – a model of the regulation of speciation rates and immigration by cycles of topographic diversity. In *Patterns of evolution as illustrated by the fossil record* (ed. A. Hallam) pp. 439–68. (*Developments in Palaeontology and Stratigraphy*, Vol. 5.) Elsevier, Amsterdam.

BARRETT, R. (1970). The pit organ of snakes. In *Biology of the Reptilia*, Vol. 2 (ed. C. Gans) pp. 277–300. Academic Press, London.

BELEKHOVA, M. G. (1979). Neurophysiology of the forebrain. In *Biology of the Reptilia* (ed. C. Gans) **10**, pp. 287–359. Academic Press, London.

BELLAIRS, A. d'A. and ATTRIDGE, J. (1975). *Reptiles*, 4th edn. Hutchinson, London.

—— and COX, C. G. (eds.) (1976). *Morphology and biology of reptiles. Linnean Society of London Symposium Series* No. 3. Academic Press, London.

BENNET, A. F. and DAWSON, W. R. (1976). Metabolism. In *Biology of the Reptilia*, Vol. 5 (ed. C. Gans) pp. 127–223. Academic Press, London.

BIRD, R. T. (1954). We captured a 'live' brontosaur. *National Geographic Magazine* **105**, 707–722.

BOLTT, R. E. and EWER, R. F. (1964). The functional anatomy of the head of the puff adder, *Bitis arietans* (Merr.). *Journal of Morphology* **114**, 83–106.

BRAMWELL, C. D. (1971). Aerodynamics of *Pteranodon. Biological Journal of the Linnean Society* **3**, 313–28.

—— and WHITFIELD, G. R. (1974). Biomechanics of *Pteranodon. Philosophical Transactions of the Royal Society* **B267**, 503–81.

BULLOCK, T. H. and FOX, W. (1957). The anatomy of the infrared sense organ in the facial pit of pit vipers. *Quarterly Journal of microscopical Science* **98**, 219–234.

BURGHARDT, G. M. (1977). Of iguanas and dinosaurs: social behavior and communication in neonate reptiles. *American Zoologist* **17**, 177–90.

CARPENTER, C. C. (1977). Communication and displays of snakes. *American Zoologist* **17**, 217–23.

CHARIG, A. (1979). *A new look at the dinosaurs.* Heinemann, London.

CORDIER, R. (1964). Sensory cells. In *The cell*, Vol. 6 (ed. J. Brachet and A. Mirsky) pp. 313–86. Academic Press, New York.

DESMOND, A. (1975). *The hot-blooded dinosaurs. A revolution in palaeontology.* Blond & Briggs, London.

FLORKIN, M. and SCHEER, B. T. (eds.) (1974). *Chemical zoology*, Vol. IX. *Amphibia and Reptilia.* Academic Press, London.

FOX, H. (1977). The urinogenital system of reptiles. In *Biology of the Reptilia*, Vol. 6 (ed. C. Gans) pp. 1–157. Academic Press, London.

GANS, C. (general ed.) (1969–79). *Biology of the Reptilia*, Vols. 1–9. (continuing series). Academic Press, London.

—— (1978). Reptilian venoms: some evolutionary considerations. In *Biology of the Reptilia*, Vol. 8 (ed. C. Gans) pp. 1–42. Academic Press, London.

GOODRICH, E. S. (1930). *Studies on the structure and development of vertebrates.* Macmillan, London.

—— (1919). Note on the reptilian heart. *Journal of Anatomy, London* **53**, 298–304.

GREENBERG, N. (1977). A neuroethological study of display behavior in the lizard *Anolis carolinensis* (Reptilia, Cacertilia, Iguanidae). *American Zoologist* **17**, 191–201.

HARKNESS, L. (1977). Chameleons use accommodation cues to judge distance. *Nature, London* **267**, 346–9.

HARLESS, M. and MORLOCK, H. (1979). *Turtles: perspectives and research.* Wiley, New York.

HARTLINE, P. H. (1974). Thermoreception in snakes. In *Handbook of sensory physiology*, Part III/3 (ed. A. Fessard). pp. 297–312. Springer, Berlin.

—— KASS, L., and LOOP, M. (1978). Merging of modalities in the optic tectum: infrared and visual integration in rattlesnakes. *Science, New York* **199**, 1225–9.

HEATON, M. J. (1980). The Cotylosauria: a reconsideration of a group of archaic tetrapods. In *The terrestrial environment and the origin of land vertebrates* (ed. A. L. Panchen) pp. 497–551. Academic Press, London.

HOLMES, E. B. (1975). A reconsideration of the phylogeny of the tetrapod heart. *Journal of Morphology* **147**, 209–28.

HORNER, J. R. and MAKELA, R. (1979). Nest of juveniles provides evidence of family structure among dinosaurs. *Nature, London* **282**, 296–8.

IHLE, J. E. W., KAMPEN, P. N. VAN, NIERSTRASZ, H. F., and VERSLUYS, J. (1927). *Vergleichende Anatomie der Wirbeltiere.* Springer, Berlin.

JOHANSEN, K. (1977). In *Chordate structure and function*, 2nd edn (ed. A. G. Klume). Macmillan, London.

JOHNSTON, P. A. (1979). Growth rings in dinosaur teeth. *Nature, London* **278**, 635–6.

NOPCSA, F. (1907). Ideas on the origin of flight. *Proceedings of the zoological Society of London* 223–36.

—— (1923). On the origin of flight in birds. *Proceedings of the zoological Society of London* 463–77.

PORTER, K. R. (1972) *Herpetology.* Saunders, Philadelphia.

RETZIUS, M. C. (1881–84). *Das Gehörgan der Wirbelthiere. Morphologisch-histologische Studien.* Kongel. Boktr., Stockholm.

ROMER, A. S. (1956). *Osteology of the reptiles.* University of Chicago Press.

—— (1966). *Vertebrate paleontology.* University of Chicago Press.

SOLOMON, S. E. and BAIRD, T. (1979). Aspects of the biology of *Chelonia mydas* L. *Oceanography and marine Biology Annual Review* **17**, 347–61.

THOMSON, J. A. (1923). *The biology of birds*. Sidgwick & Jackson, London.

TUCKER, V. A. (1977). Scaling and avian flight. In *Scale effects in animal locomotion* (ed. T. J. Pedley) pp. 497–509. Academic Press, London.

UNDERWOOD, G. (1970). The eye. In *Biology of the Reptilia*, Vol. 2 (ed. C. Gans) pp. 1–97. Academic Press, London.

WALLS, G. L. (1942). *The vertebrate eye and its adaptive radiation*. Cranbrook Institute of Science, Bloomfield Hills, Mich.

WEVER, E. G. (1978). *The reptile ear: its structure and function*. Princeton University Press.

WHEELER, P. E. (1978). Elaborate CNS cooling structures in large dinosaurs. *Nature, London* **275**, 441–2.

WHITE, F. N. (1968). Functional anatomy of the heart of reptiles. *American Zoologist* **8**, 211–19.

—— (1976). Circulation. In *Biology of the Reptilia*, Vol. 5 (ed. C. Gans) pp. 275–334. Academic Press, London.

WILLISTON, S. W. (1925). *The osteology of the reptiles*. Harvard University Press, Cambridge, Mass.

YOUNG, J. Z. (1978). *Programs of the brain*. Oxford University Press.

Chapters 15–17 The birds

ATTENBOROUGH, D. (1979). *Life on earth*. Collins and BBC, London.

AYMAR, G. C. (1935). *Bird flight*. Bodley Head, London.

BELLAIRS, M. R. (1971). *Developmental processes in higher vertebrates*. Logos, London.

BERTHOLD, P. (1972). Migration: control and metabolic physiology. In *Avian biology*, Vol. V (ed. D. S. Farner and J. R. King) pp. 77–128. Academic Press, New York.

BOCK, W. J. (1964). Kinetics of the avian skull. *Journal of Morphology* **114**, 1–41.

BOWMAKER, J. K. (1979). Visual pigments and oil droplets in the pigeon retina, as measured by microspectrophotometry and their relationship to spectral sensitivity. In *Neural mechanisms of behavior in the pigeon* (ed. A. M. Granda and J. H. Maxwell) pp. 287–306. Plenum Press, New York.

BRAMWELL, C. D. and WHITFIELD, G. R. (1974). Biomechanics of *Pteranodon*. *Philosophical Transactions of the Royal Society* **B267**, 503–81.

BRODKORB, P. (1971). Origin and evolution of birds. In *Avian biology*, Vol. I (ed. D. S. Farner and J. R. King) pp. 19–55. Academic Press, New York.

BROWN, R. H. J. (1953). The flight of birds. II Wing function in relation to flight speed. *Journal of experimental Biology* **30**, 90–103.

CROSSLAND, W. J. (1979). Identification of tectal synaptic terminals in the avian isthmotopic nucleus. In *Neural mechanisms of behavior in the pigeon* (ed. A. M. Granda and J. H. Maxwell) pp. 267–86. Plenum Press, New York.

DE BEER, G. (1956). Evolution of ratites. *Bulletin of the British Museum (Natural History), Zoology* **4**, 57–70.

DELIUS, J. D. and EMMERTON, J. (1979). Visual performance of pigeons. In *Neural mechanisms of behavior in the pigeon* (ed. A. M. Granda and J. H. Maxwell) pp. 51–70. Plenum Press, New York.

DOBZHANSKY, T., AYALA, F. J., STEBBINS, G. L., and VALENTINE, J. W. (1977). *Evolution*. Freeman, San Francisco.

DUNNET, G. M. and OLLASON, J. C. (1979). The fulmar. *Biologist* **26**, 117–22.

EMLEN, S. T. (1975). Migration: orientation and navigation. In *Avian biology*, Vol. V (ed. D. S. Farner and J. R. King) pp. 129–219. Academic Press, New York.

FARNER, D. S. and KING, J. R. (eds.) (1971–75). *Avian biology*, Vols. I–V. Academic Press, New York.

GRIFFIN, D. R. (1955). Bird navigation. In *Recent Studies in avian biology* (ed. A. Wolfson) pp. 154–97. University of Illinois Press, Urbana, Ill.

HEILMANN, G. (1926). *The origin of birds*. Witherby, London.

HINDE, R. A. (1959). Behaviour and speciation in birds and lower vertebrates. *Biological Reviews* **34**, 85–128.

—— (1973). Behavior. In *Avian biology*, Vol. III (ed. J. S. Farner and J. R. King) pp. 479–535. Academic Press, New York.

HORN, G., McCABE, B. J., and BATESON, P. P. G. (1979). An autoradiographic study of the chick brain after imprinting. *Brain Research* **168**, 361–3.

HOWARD, H. E. (1920). *Territory in bird life*. Murray, London.

HUXLEY, J. S. (1914). The courtship of the great crested grebe (*Podiceps cristatus*); with an addition to the theory of sexual selection. *Proceedings of the zoological Society, London* 491–562.

JONES, D. R. and JOHANSEN, K. (1972). The blood vascular system of birds. In *Avian biology*, Vol. II (ed. D. S. Farner and J. R. King) pp. 157–285. Academic Press, New York.

JUVIK, J. O. and AUSTRING, A. P. (1979). The Hawaiian avifauna: biogeographic theory in evolutionary time. *Journal of Biogeography* **6**, 205–24.

KARTEN, H. J. (1963). Ascending pathways from spinal cord in the pigeon *Columba livia*. *International Congress of Zoology* **16**, 23.

—— (1979). Visual lemniscal pathways in birds. In *Neural mechanisms of behavior in the pigeon* (ed. A. M. Granda and J. H. Maxwell) pp. 409–30. Plenum Press, New York.

KEETON, W. T. (1971). Magnets interfere with pigeon homing. *Proceedings of the National Academy of Sciences, U.S.A.* **68**, 102–6.

—— (1972). Effects of magnets on pigeon homing. NASA Special Publication NASA SP-262, 579–94.

KING, A. S. (1966). Structural and functional aspects of the avian lungs and air sacs. *International Review of general and experimental Zoology* **2**, 171–267.

—— and COWIE, A. F. (1969). The functional anatomy of the bronchial muscle of the bird. *Journal of Anatomy, London* **105**, 323–36.

—— and MCLELLAND, J. (1980). *Form and function in birds*. Academic Press, London.

KOBAYASHI, H. and WADA, M. (1973). Neuroendocrinology in birds. In *Avian biology*, Vol. III (ed. D. S. Farner and J. R. King) pp. 287–347. Academic Press, New York.

KRAMER, G. (1951). Eine neue Methode zur Erforschung der Zugorientierung und die bisher damit erzielten Ergebnisse. *Proceedings of the 10th International Ornithological Congress*, 1950, 269–80.

LACK, D. (1933). Habitat selection in birds with special reference to the effects of afforestation on the Breckland avifauna. *Journal of animal Ecology* **2**, 239–62.

—— (1947). *Darwin's finches*. Cambridge University Press.

—— (1968). *Ecological adaptations for breeding in birds*. Methuen, London.

—— (1971). *Ecological isolation in birds*. Blackwell, Oxford.

—— (1976). *Island biology: illustrated by the land birds of Jamaica*. Blackwell, Oxford.

LASIEWSKI, R. C. (1972). Respiratory function in birds. In *Avian biology*, Vol. II (ed. D. S. Farner and J. R. King) pp. 287–342. Academic Press, New York.

LIGHTHILL, Sir James (1975). *Mathematical biofluiddynamics*. Society for Industrial and Applied Mathematics, Philadelphia, Pennsylvania.

MASTERTON, R. B., CAMPBELL, C. B. G. BITTERMAN, M. E., and HOTTON, N. (eds.) (1976). *Evolution of brain and behavior in vertebrates*. Erlbaum, Hillsdale, New Jersey.

MATTHEWS, G. V. T. (1955). *Bird navigation*. Cambridge University Press.

—— (1963). The astronomical bases of 'nonsense' orientation. *Proceedings of the 13th International Ornithological Congress*, 1962. 415–29.

MATURANA, H. R. (1962). Functional organisation of the pigeon retina. *Proceedings of the International Union of Physiological Sciences. 22nd International Congress* Vol. 3, pp. 170–8.

MAXWELL, J. H. and GRANDA, A. M. (1979). Receptive fields movement-sensitive cells in the pigeon thalamus. In *Neural mechanisms of behavior in the pigeon* (ed. A. M. Granda and J. H. Maxwell) pp. 177–98. Plenum Press, New York.

MOREAU, R. E. (1972). *The Palaearctic–African bird migration system*. Academic Press, New York.

OSTROM, J. H. (1979). Bird flight: how did it begin? *American Scientist* **67**, 46–56.

PEARSON, R. (1972). *The avian brain*. Academic Press, London.

PENNYCUICK, C. J. (1968). A wind-tunnel study of gliding flight in the pigeon *Columba livia*. *Journal of experimental Biology* **49**, 509–26.

—— (1972). *Animal flight*. Arnold, London.

—— (1975). Mechanics of flight. In *Avian biology*, Vol. V (ed. D. S. Farner and J. R. King) pp. 1–75. Academic Press, New York.

PERDECK, A. C. (1958). Two types of orientation in migrating starlings *Sturnus vulgaris* L., and chaffinches *Fringilla welebs* L., as revealed by displacement experiments. *Ardea* **46**, 1–37.

PETERS, J. L. (1931–70). *Check-list of birds of the world* I–XV. Harvard, Cambridge, Mass.

PRESTI, D. and PETTIGREW, D. (1980). Ferromagnetic coupling to muscle receptors as a basis for geomagnetic field sensitivity in animals. *Nature, London* **285**, 99–100.

PUMPHREY, R. J. (1948a). The sense organs of birds. *Ibis* **90**, 171–90.

—— (1948b). The theory of the fovea. *Journal of experimental Biology* **25**, 299–312.

PYCRAFT, W. P. (1910). *A history of birds*. Methuen, London.

ROWAN, W. (1925). Relation of light to bird migration and developmental changes. *Nature, London* **115**, 494–5.

—— (1931). *The riddle of migration*. Williams and Williams, Baltimore.

SCHMIDT-KOENIG, K. (1979). *Avian orientation and navigation*. Academic Press, London.

SCHMIDT-NIELSEN, K. (1979). *Animal physiology: adaptation and environment*. Cambridge University Press.

—— HAINSWORTH, F. R. and MURRISH, D. E. (1970). Countercurrent heat exchange in the respiratory passages: effect on water and heat balance. *Respiratory Physiology* **9**, 263–76.

SCHWARTZKOPF, J. (1973). Mechanoreception. In *Avian biology*, Vol. III, (ed. D. S. Farner and J. R. King) pp. 417–77. Academic Press, New York.

SKINNER, B. F. (1938). *The behavior of organisms; an experimental analysis*. Appleton, New York.

SLUCKIN, W. (1972). *Imprinting and early learning*, 2nd. edn. Methuen, London.

STETTENHEIM, P. (1972). The integument of birds. In

Avian biology, Vol. II (ed. D. S. Farner and J. R. King) pp. 1–63. Academic Press, New York.

STRESEMANN, E. (1934). *Handbuch der Zoologie eine naturgeschichte der Stämme des Tierreiches*, Vol. 7, Part 2. Sauropsida: Aves (ed. W. Kükenthal). de Gruyter, Berlin.

THOMSON, J. A. (1923). *The biology of birds*. Sidgwick & Jackson, London.

THORPE, W. H. (1961). *Bird song*. Cambridge University Press.

—— (1963). *Learning and instinct in animals*, 2nd edn. Methuen, London.

VON HOLST, E. (1973). *The behavioural physiology of animals and man: the selected papers of E. von Holst* (transl. R. Martin). Methuen, London.

WALLMAN, J. (1979). Role of the retinal oil droplets in the color vision of Japanese quail. In *Neural mechanisms of behavior in the pigeon* (ed. A. M. Granda and J. H. Maxwell) pp. 327–52. Plenum Press, New York.

WALLS, G. L. (1937). Significance of the foveal depression. *Archives of Ophthalmology* **18**, 912–19.

—— (1942). *The vertebrate eye and its adaptive radiation*. Cranbrook Institute of Science, Bloomfield Hills, Mich.

WEBSTER, K. (1980). Some aspects of the comparative study of the corpus striatum. In *The neostriatum*, EBBS Workshop (ed. I. Divac). Pergamon, Oxford.

WEIS-FOGH, T. (1973). Quick estimates of flight fitness in hovering animals, including novel mechanisms for lift production. *Journal of experimental Biology* **59**, 169–230.

WILTSCHKO, W. (1972). The influence of magnetic total intensity and inclination on directions preferred by migrating European robins (*Erithacus rubecula*). NASA Special Publication NASA SP-262, pp. 115–28.

WRIGHT, S. (1968–78). *Evolution and the genetics of populations: a treatise*, Vols. 1–4. Chicago University Press.

WYNNE-EDWARDS, V. C. (1962). *Animal dispersion in relation to social behaviour*. Oliver & Boyd, Edinburgh.

YOUNG, J. Z. (1978). *Programs of the brain*. Oxford University Press.

ZAHAVI, A. (1977). Reliability in communication systems and the evolution of altruism. In *Evolutionary ecology* (ed. B. Stonehouse and C. Perrins) pp. 253–9. Macmillan, London.

Chapter 18 Origin of mammals. Monotremes

BARCROFT, J. (1932). 'La fixité du milieu intérieur est la condition de la vie libre.' *Biological Reviews* **7**, 24–87.

BERNARD, C. (1878). *Leçons sur les phénomènes de la vie communs aux animaux et aux végétaux*. Baillière, Paris.

BROOM, R. (1932). *The mammal-like reptiles of South Africa*. Witherby, London.

BUTLER, P. M. (1939). Studies of the mammalian dentition. Differentiation of the postcanine dentition. *Proceedings of the zoological Society of London* **109**, 1–36.

—— (1972). Some functional aspects of molar evolution. *Evolution* **26**, 474–8.

—— (1978). Molar cusp nomenclature and homology. In *Development, function and evolution of teeth*, (ed. P. M. Butler and K. A. Joysey) pp. 439–454. Academic Press, London.

CROMPTON, A. W. (1980). Biology of the earliest mammals. In *Comparative physiology: primitive mammals* (ed. K. Schmidt-Nielson, L. Bolis, and C. R. Taylor) pp. 1–12. Cambridge University Press.

—— and KIELAN-JAWOROWSKA, Z. (1978). Molar structure and occlusion in Cretaceous therian mammals. In *Development, function and evolution of teeth* (edited by P. M. Butler & K. A. Joysey) pp. 249–87. Academic Press, London.

GRIFFITHS, M. (1968). *Echidnas*. Pergamon, Oxford.

—— (1978). *The biology of the monotremes*. Academic Press, New York.

HIIEMAE, K. M. (1978). Mammalian mastication: a review of the activity of the jaw muscles and the movements they produce in chewing. In *Development, function and evolution of teeth* (ed. P. M. Butler and K. A. Joysey) pp. 359–98. Academic Press, London.

HILL, J. P. (1910). The early development of the Marsupialia, with special reference to the native cat (*Dasyurus viverrinus*). *Quarterly Journal of microscopical Science* **56**, 1–134.

HOPSON, J. A. (1970). The classification of non-therian mammals. *Journal of Mammalogy* **51**, 1–9.

HOTTON, N. (1959). The pelycosaur tympanum and early evolution of the middle ear. *Evolution* **13**, 99–121.

JENKINS, F. A. Jr and PARRINGTON, F. R. (1976). The postcranial skeletons of the Triassic mammals *Eozostrodon, Megazostrodon* and *Erythrotherium*. *Philosophical Transactions of the Royal Society of London* **B273**, 387–431.

KAY, R. F. and HIIEMAE, K. M. (1974). Jaw movements and tooth use in recent and fossil primates. *American Journal of physical Antropology* **40**, 227–56.

KERMACK, K. A. (1967). The interrelations of early mammals. *Journal of the Linnean Society, Zoology* **47**, 241–9.

—— (1972). The origin of mammals and the evolution

of the temperomandibular joint. *Proceedings of the Royal Society of Medicine* **65**, 389–92.

KIELAN-JAWOROWSKA, Z. (1971). Results of the Polish–Mongolian Palaeontological Expeditions – Part III. Skull structure and affinities of the Multituberculata. *Palaeontologia Polonica* **25**, 5–41.

—— (1979). Pelvic structure and nature of reproduction in Multituberculata. *Nature, London* **277**, 402–3.

LENDE, R. A. (1964). Representation in the cerebral cortex of a primitive mammal. Sensorimotor, visual, and auditory fields in the echidna (*Tachyglossus aculeatus*). *Journal of Neurophysiology* **27**, 37–48.

LILLEGRAVEN, J. A., KIELAN-JAWOROWSKA, Z., and CLEMENS, W. A. (1979). *Mesozoic mammals: the first two-thirds of mammalian history.* University of California Press, Berkeley.

OSBORN, H. F. (1888). The nomenclature of the mammalian molar cusps. *American Naturalist* **22**, 926–8.

—— (1907). *Evolution of mammalian molar teeth to and from the triangular type.* Macmillan, London.

OSBORN, J. W. (1978). Morphogenetic gradients: fields versus clones. In *Development, function and evolution of teeth* (ed. P. M. Butler and K. A. Joysey) pp. 171–201. Academic Press, London.

PARRINGTON, F. R. (1978). A further account of the Triassic mammals. *Philosophical Transactions of the Royal Society of London* **B282**, 177–204.

SELIGSOHN, D. and SZALAY, F. S. (1978). Relationship between natural selection and dental morphology. In *Development, function and evolution of teeth* (ed. P. M. Butler and K. A. Joysey) pp. 289–308. Academic Press, London.

SIMPSON, G. G. (1959). Mesozoic mammals and the polyphyletic origin of mammals. *Evolution* **13**, 404–14.

—— (1961). Evolution of Mesozoic mammals. International colloquium on the evolution of lower and non-specialised mammals. *Koninklijke Vlaamsche Academie voor Wettenschappen, Letteren en schoone kunsten van België*, Part 1, pp. 57–95. Brussels.

YOUNG, J. Z. (1978). *Programs of the brain.* Oxford University Press.

Chapter 19 Marsupials

BAILEY, L. F. and LEMON, M. (1966). Specific mille protein associated with resumption of development by the quiescent blastocyst of the lactating red kangaroo. *Journal of Reproduction and Fertility* **II**, 473–5.

FLOWER, W. H. and LYDEKKER, R. (1891). *An introduction to the study of mammals, living and extinct.* Blade, London.

HEARN, J. P. (1975a). Hypophysectomy of the tammar wallaby, *Macropus eugenii:* surgical approach and general effects. *Journal of Endocrinology* **64**, 403–16.

—— (1975b). The role of the pituitary in the reproduction of the male tammar wallaby, *Macropus eugenii. Journal of Reproduction and Fertility* **42**, 399–402.

JONES, F. W. (1923). The mammals of South Australia. *Handbook of the Flora and Fauna of South Australia. Issued by the British Science Guild (South Australian Branch).*

KIRKBY, R. J. (1977). Learning and problem-solving behaviour in marsupials. In *The biology of marsupials* (Ed. B. Stonehouse and D. Gilmore) pp. 193–220. University Park Press, Baltimore, Md.

KIRSCH, J. A. W. and CALABY, J. H. (1977). The species of living marsupials: an annotated list. In *The biology of marsupials* (Ed. B. Stonehouse and D. Gilmore) pp. 9–26. University Park Press, Baltimore, Md.

LEMON, M. and BARKER, S. (1967). Changes in mille composition of the red kangaroo, *Megaleia rufa* (Desmarest) during lactation. *Australian Journal of experimental Biology and medical Science* **45**, 213–19.

RIDE, W. D. L. (1970). *A guide to the native mammals of Australia.* Oxford University Press, Melbourne.

ROMER, A. S. (1966). *Vertebrate paleontology.* University of Chicago Press.

SIMPSON, G. G. (1945). The principles of classification and a classification of mammals. *Bulletin of the American Museum of natural History* **85**, 1–350.

STONEHOUSE, B. and GILMORE, D. (1977). *The biology of marsupials.* University Park Press, Baltimore, Md.

TYNDALE-BISCOE, C. H. (1973). *Life of marsupials.* Arnold, London.

—— HEARN, J. P., and RENFREE, M. B. (1974). Review. Control of reproduction in macropodid marsupials. *Journal of Endocrinology* **63**, 589–614.

Chapters 20–31 Mammals

ALLEN, K. R. (1980). *Conservation and management of whales.* University of Washington Press, Seattle.

ATTENBOROUGH, D. (1979). *Life on earth, a natural history.* Collins/BBC, London.

BERRY, R. J., JAKOBSON, M. E., and PETERS, J. (1978). The house mice of the Faroe Islands: a study in microdifferentiation. *Journal of Zoology, London* **185**, 73–92.

BOHLKEN, H. (1960). Remarks on the stomach and the systematic position of the Tylopoda. *Proceedings of the zoological Society of London* **134**, 207–14.

BRODIE, E. D. (1977). Hedgehogs use toad venom in their own defence. *Nature, London* **268**, 627–8.

BRODMANN, K. (1909). *Vergleichende Lokalisationslehre*

der Grosshirnrinde in Prinzipien dargesellt auf Grunddes Zellenbause. Barth, Leipzig.

BUETTNER-JANUSCH, J. (1966). *Origins of man.* Wiley, New York.

BUTLER, P. M. and MILLS, J. R. E. (1959). A contribution to the odontology of *Oreopithecus. Bulletin of the British Museum (Natural History), Geology* **4**, 1–26.

CAMPBELL, B. G. (1964). Quantitative taxonomy and human evolution. In *Classification and human evolution* (Ed. S. L. Washburn). Methuen, London.

CAMPBELL, C. B. G. (1975). The central nervous system: its uses and limitations in assessing phylogenetic relationships. In *Phylogeny of the primates* (Ed. W. P. Luckett and F. S. Szalay) pp. 183–97. Plenum Press, New York.

CARTMILL, M. (1974). *Daubentonia, Dactylopsila,* woodpeckers and klinorhynchy. In *Prosimian biology* (Ed. R. D. Martin, G. A. Doyle, and A. C. Walker) pp. 655–70. Duckworth, London.

CLARK, W. E. LE GROS (1934). *Early forerunners of man. A morphological study of the evolutionary origin of the primates.* Ballière, Tindall & Cox, London.

CLARKE, M. R. (1978). Buoyancy control as function of the spermaceti organ in the sperm whale. *Journal of the marine biological Association of the United Kingdom* **58**, 27–71.

—— (1979). The head of the Sperm whale. *Scientific American* **240**, 128–41.

COPPENS, Y., MAGLIO, V. J., MADDEN, C. T., and BEDEN, M. (1978). Proboscidae. In *Evolution of African mammals* (Ed. V. J. Maglio and H. B. S. Cooke) pp. 336–67. Harvard University Press, Cambridge, Mass.

CORBET, G. B. and HILL, J. E. (1980). *World list of mammalian species.* Cornell University Press.

CURTIS, G. H. (1981). Establishing a relevant time scale in anthropological and archaeological research. *Philosophical Transactions of the Royal Society of London* **B292**, 7–20.

DART, R. A. (1925). *Australopithecus africanus*: the man-ape of South Africa. *Nature, London* **115**, 195–9.

DAY, M. H. (1965). Guide to fossil man. *A handbook of human palaeontology.* Cassell, London.

—— (1973). Locomotor features of the lower limb in hominids. *Symposium of the zoological Society of London* No. 33, 29–51.

—— (1974). The interpolation of isolated fossil footbones into a discriminant analysis – or reply. *American Journal of physical Anthropology* **41**, 233–6.

DELANY, M. J. and HAPPOLD, D. C. D. (1979). *Ecology of African mammals.* Longman, London.

ELTRINGHAM, S. K. (1979). *The ecology and conservation of large African mammals.* Macmillan, London.

EWER, R. F. (1973). *The carnivores.* Weidenfeld & Nicolson, London.

FENNER, F. and RATCLIFFE, F. N. (1965). *Myxomatosis.* Cambridge University Press.

FLOWER, W. H. and LYDEKKER, R. (1891). *An introduction to the study of mammals living and extinct.* Black, London.

FOX, M. W. (ed.) (1975). *The wild canids: their systematics, behavioral ecology and evolution.* Van Nostrand Reinhold, New York.

FRASER, F. C. and PURVES, P. E. (1960). Hearing in cetaceans. Evolution of the accessory air sacs and the structure and function of the outer and middle ear in recent cetaceans. *Bulletin of the British Museum (Natural History) Zoology* **7**, 1–140.

GASKIN, D. E. (1978). Form and function in the digestive tract and associated organs in Cetacea, with a consideration of metabolic rates and specific energy budgets. *Oceanography and Marine Biology* **16**, 313–45.

GOULD, E., NEGUS, N. C., and NOVICK, A. (1964). Evidence for echolocation in shrews. *Journal of experimental Zoology* **156**, 19–38.

GOULD, S. J. and ELDREDGE, N. (1977). Punctuated equilibria: the tempo and mode of evolution reconsidered. *Paleobiology* **3**, 115–51.

GRASSÉ, P.-P. (ed.) (1955). *Traité de zoologie. Anatomie, systematique, biologie.* Tome XVII. *Mammifères. Les ordres: Anatomie ethologie, systematique*, second fasicule. Masson, Paris.

GRIFFIN, D. R. (1958). *Listening in the dark.* Yale University Press, New Haven, Conn.

HALDANE, J. B. S. (1949). Suggestions as to quantitative measurements of rates of evolution. *Evolution* **3**, 51–6.

HAMILTON, W. J. (1902). *American mammals; their lives, habits, and economic relations.* McGraw-Hill, New York.

HARTRIDGE, H. (1945). Acoustical control in the flight of bats. *Nature, London* **156**, 490–4.

HARTRIDGE, H. (1920). The avoidance of objects by bats in their flight. *Journal of Physiology, London* **54**, 54–7.

HENNIG, W. (1950). *Grundzüge einer Theorie der Phylogenetischen Systematik.* Deutscher Zentralverlag, Berlin. (English translation 1966, *Phylogenetic systematics.* University of Illinois, Urbana.)

HENSON, O. W. (1970). The ear and audition. In *Biology of bats*, Vol. II (ed. W. A. Wimsatt) pp. 181–263. Academic Press, New York.

HORRIDGE, G. A. (1968). *Interneurons.* Freeman, London.

HOWELL, F. C. (1978). Hominidae. In *Evolution of*

African mammals (ed. V. J. Maglio and H. B. S. Cooke) Harvard University Press, Cambridge, Mass.

HOWELLS, W. (1944). *Mankind so far.* Doubleday, Doran, New York. [*American Museum of Natural History, Science Series* No. 5.]

HUXLEY, J. S. (1932). *Problems of relative growth.* Methuen, London.

JEWELL, P. R., MILNER, C., and BOYD, J. M. (1974). *Island survivors: the ecology of the Soay sheep of St. Kilda.* Athlone Press, London.

JOHANSON, D. C. and EDEY, M. A. (1981). *Lucy: the beginnings of humankind.* Granada, London.

KAPLAN, H. M. and TIMMONS, E. H. (1979). *The rabbit: a model for the principles of mammalian physiology and surgery.* Academic Press, New York.

KELSO, J. (1970). *Physical anthropology.* Lippincott, Philadelphia.

KINGDON, J. (1971–9). *East African mammals: an atlas of evolution in Africa.* 3 vols. Academic Press, London.

KLINGEL, H. (1974). A comparison of the social behaviour of the Equidae. In *The behaviour of ungulates and its relation to management,* Vol. I (ed. V. Geist and F. Walther). *International Union for Conservation of Nature and Natural Resources,* Morges, Switzerland. New Series No. 24, **1**, 124–32.

LAWICK-GOODALL, J. VAN (1971). *In the shadow of man.* Collins, London.

LAWS, R. M. (1966). Age criteria for the African elephant. *East African Wildlife Journal* **4**, 1–37.

—— (1969). Aspects of reproduction in the African elephant, *Loxondonta africana. Journal of Reproduction and Fertility* Suppl. **6**, 193–218.

—— PARKER, I. S. C., and JOHNSTONE, R. C. B. (1975). *Elephants and their habits: the ecology of elephants in North Bunyoro, Uganda.* Clarendon Press, Oxford.

LEAKEY, M. (1981). Tracks and tools. *Philosophical Transactions of the Royal Society of London,* **B292**, 95–102.

LEAKEY, R. E. F. (1973). Australopithecines and hominines: a summary on the evidence from the early Pleistocene of eastern Africa. *Symposium of the zoological Society of London* No. 33, 53–69.

——(1981). *The making of mankind.* Michael Joseph, London.

LEUTHOLD, W. (1977). *African ungulates: a comparative review of their ethology and behavioural ecology.* Springer, Berlin.

LUCKETT, W. P. and SZALAY, F. S. (eds.) (1975). *Phylogeny of the primates: a multidisciplinary approach.* Plenum, New York. [*Proceedings of the Wenner Gren Symposium* No. 61 in Burg Wartenstein, Austria, 6–14 July 1974.]

LULL, R. S. (1917). *Organic evolution.* (1929, 2nd edn.) Macmillan, London.

McKENNA, M. C. (1975). Toward a phylogenetic classification of the Mammalia. In *Phylogeny of the primates* (Edited by W. P. Luckett and F. S. Szalay) pp. 21–46. Plenum Press, New York.

MAGLIO, V. J. (1973). Origin and evolution of the Elephantidae. *Transactions of the American Philosophical Society (Philadelphia) New Series* **63**, Pt 3, 149pp.

—— and COOKE, H. B. S. (edn.) (1978). *Evolution of African mammals.* Harvard University Press, Cambridge, Mass.

—— and RICCA, A. B. (1977). Dental and skeletal morphology of the earliest elephants. *Verhandelingen der Koninklijke Nederlandse Akademie van Wetenschappen, Afd. Natuurkunde,* Eerste Reeks, Deel 29.

MARTIN, R. D., DOYLE, G. A., and WALKER, A. C. (eds.) (1974). *Prosimian biology.* Duckworth, London.

MATTHEWS, L. H. (1969). *Life of mammals,* Vols. I and II. Weidenfeld & Nicolson, London.

—— (1978). *The natural history of the whale.* Columbia University Press, New York.

MEAD-BRIGGS, A. R. (1977). The European rabbit, the European rabbit flea and myxomatosis. *Applied Biology* **2**, 184–262.

NAPIER, J. R. and NAPIER, P. H. (1967). *A handbook of living primates.* Academic Press, London.

NOVICK, A. (1977). Acoustic orientation. In *Biology of bats,* Vol. III (ed. W. A. Wimsatt) pp. 73–287. Academic Press, New York.

OSBORN, H. F. (1936, 1942). *Proboscidae: a monograph of the discovery, evolution, migration and extinction of the mastodonts and elephants of the world,* Vols I and II. American Museum of Natural History, New York.

OWEN, R. (1866–68). *On the anatomy of vertebrates,* Vols. 1–3. Longmans Green, London.

OXNARD, C. E. (1975). *Uniqueness and diversity in human evolution: morphometric studies of Australopithecines.* University of Chicago Press.

PAYNE, R. and WEBB, D. (1971). Orientation by means of long range acoustic signaling in baleen whales. *Annals of the New York Academy of Sciences* **188**, 110–41.

POIRIER, F. E. (1977). *Fossil evidence. The human evolutionary journey,* 2nd edn. Mosby, Saint Louis, Mo.

PYE, J. D. (1960). A theory of echolocation by bats. *Journal of Laryngology and Otology* **74**, 718–29.

REYNOLDS, S. H. (1897). *The vertebrate skeleton.* Cambridge University Press. (2 edn 1913)

ROMER, A. S. (1957). *Man and the vertebrates.* University of Chicago Press.

—— (1966). *Vertebrate paleontology*, 3rd edn. University of Chicago Press.

RONALD, K. and MANSFIELD, A. W. (eds.) (1975). *Biology of the seal. Proceedings of a symposium held in Guelph 14–17 August 1972*. Conseil International pour l'Exploration de la Mer, Charlottenlund Swt. [*Rapport et procès-verbaux des réunions* 69, 1–557.]

SALES, G. D. and PYE, J. D. (1974). *Ultrasonic communication by animals*. Chapman & Hall, London.

SCHALLER, G. B. (1963). *The mountain gorilla: ecology and behavior*. Chicago University Press.

SCHMIDT-NIELSEN, K. (1964). *Desert animals. Physiological problems of heat and water*. Oxford University Press. [1979 edn. Dover Publications, New York.]

—— (1979). *Animal physiology*, 2nd edn. Cambridge University Press.

——, BOLIS, L. and TAYLOR, C. R. (1980). *Comparative physiology: primitive mammals*. Cambridge University Press.

SCHULTZ, A. H. (1944). Age changes and variability in gibbons. A morphological study on a population sample of a man-like ape. *American Journal of physical Anthropology* New Series 2, 1–129.

—— (1950). The physical distinctions of man. *Proceedings of the American Philosophical Society* 94, 428–49.

—— (1963). Age changes, sex differences, and variability as factors in the classification of primates. In *Classification and human evolution* (ed. S. L. Washburn) pp. 85–115. Aldine, Chicago.

SCOTT, W. B. (1913). *A history of land mammals in the Western hemisphere*. Macmillan, New York.

SIKES, S. K. (1971). *The natural history of the African elephant*. Weidenfeld & Nicolson, London.

SILMAN, R. E., LEONE, R. M., HOOPER, R. J. L., and PREECE, M. A. (1979). Melatonin, the pineal gland and human puberty. *Nature, London* 282, 301–3.

SIMONS, E. L. (1972). *Primate evolution: an introduction to man's place in nature*. Macmillan, New York.

—— (1978). Diversity among the early hominids: a vertebrate palaeontologist's viewpoint. In *Early hominids of Africa* (ed. C. J. Jolly) pp. 543–66. Duckworth, London.

—— and DELSON, E. (1978). Cercopithecidae and Parapithecidae. In *Evolution of African mammals* (ed. V. J. Maglio and H. B. S. Cooke) pp. 100–119. Harvard University Press, Cambridge, Mass.

—— and PILBEAM, D. R. (1978). *Ramapithecus* (Hominidae, Hominoidea). In *Evolution of African mammals* (ed. V. J. Maglio and H. B. S. Cooke) pp. 147–53. Harvard University Press, Cambridge, Mass.

—— ANDREWS, P., and PILBEAM, D. R. (1978). Cenozoic apes. In *Evolution of African mammals* (ed. V. J. Maglio and H. B. S. Cooke) pp. 120–46. Harvard University Press, Cambridge, Mass.

SIMPSON, G. G. (1945). The principles of classification and a classification of mammals. *Bulletin of the American Museum of Natural History* 85, 1–350.

—— (1951). *Horses: the story of the horse family in the modern world and through sixty million years of history* Oxford University Press.

—— (1975). Recent advances in methods of phylogenetic inference. In *Phylogeny of the primates* (ed. W. P. Luckett and F. S. Szalay) pp. 3–19. Plenum Press, New York.

—— (1978). Early mammals in South America: fact, controversy, and mystery. *Proceedings of the American Philosophical Society* 122, 318–28.

—— (1980). *Splendid isolation: the curious history of South American mammals*. Yale University Press, New Haven.

SISSON, S. and GROSSMAN, J. D. (1938). *The anatomy of the domestic animals*. (1975, R. Getty, 5th edn. 2 Vols.) Saunders, Philadelphia.

SPEARMAN, R. I. C. (1966). The keratinization of epidermal scales, feathers and hairs. *Biological Reviews* 41, 59–96.

STANLEY, S. M. (1980). *Macroevolution pattern and process*. Freeman, San Francisco.

TATTERSALL, I. (1975). *The evolutionary significance of Ramapithecus*. Burgess, Minnesota.

—— and SUSSMAN, R. W. (1975). *Lemur biology*. Plenum Press, New York.

WALKER, E. P. (1964). *Mammals of the world*, 3 vols. Johns Hopkins Press, Baltimore.

WEBB, S. D. (1977). A history of savanna vertebrates in the New World. Part I: North America. *Annual Review of ecological Systematics* 8, 355–88.

WIMSATT, W. A. (ed.) (1970–77). *Biology of bats*, Vols. I–III. Academic Press, New York.

WINN, H. E. and WINN, L. K. (1978). the song of the humpback whale *Megaptera novaengliae* in the West Indies. *Marine Biology* 47, 97–114.

WOOD, A. E. (1974). The evolution of the Old World and New World hystricomorphs. *Symposia of the zoological Society of London* No. 34, 21–60.

WOOD, B. A. (1978). *Human evolution*. Chapman & Hall, London.

WOOD JONES, F. (1916). *Arboreal man*. Arnold, London.

WOODWARD, A. S. (1898). *Outlines of vertebrate palaeontology for students of zoology*. Cambridge University Press.

YALDEN, D. W. and MORRIS, P. A. (1975). *The lives of bats*. David & Charles, Newton Abbot, Devon.

YOUNG, J. Z. (1971). *An introduction to the study of man*.

Clarendon Press, Oxford.

——(1975). *The life of mammals*. Clarendon Press, Oxford.

——, JOPE, E. M. and OAKLEY, K. P. (eds.) (1981). Emergence of man. *Philosophical Transactions of the Royal Society of London*, **B292**.

Chapter 32 Efficiency of mammals

ALEXANDER, R. McN. (1980). Elasticity in the locomotion of mammals. In *Comparative physiology: primitive mammals* (eds. K. Schmidt-Nielsen, L. Bolis, and C. R. Taylor) pp. 220–30. Cambridge University Press.

CAVAGNA, G. A., HEGLUND, N. C., and TAYLOR, C. R. (1977). Mechanical work in terrestrial locomotion: two basic mechanisms for minimizing energy expenditure. *American Journal of Physiology* **233**, R243–61.

DOBZHANSKY, T., AYALA, F. J., STEBBINS, G. L., and VALENTINE, J. W. (1977). *Evolution*. Freeman, San Francisco.

GRANT, T. R. and DAWSON, T. J. (1978). Temperature regulation in the platypus, *Ornithorhynchus anatinus*: production and loss of metabolic heat in air and water. *Physiological Zoology* **51**, 315–32.

HULBERT, A. J. (1980). The evolution of energy metabolism in mammals. In *Comparative physiology: primitive mammals* (ed. K. Schmidt-Nielsen, L. Bolis, and C. R. Taylor) pp. 130–9. Cambridge University Press.

JERISON, H. J. (1973). *Evolution of the brain and intelligence*. Academic Press. New York.

LILLEGRAVEN, J. A. (1979). Reproduction in Mesozoic mammals. In *Mesozoic mammals the first two-thirds of mammalian history* (ed. J. A. Lillegraven, Z. Kielan-Jaworowska, and W. A. Clemens) pp. 259–76. University of California Press, Berkeley.

SCHMIDT-NIELSEN, K., BOLIS, L., and TAYLOR, C. R. (eds.) (1980). *Comparative physiology: primitive mammals*. Cambridge University Press.

SIMPSON, G. G. (1953). *The major features of evolution*. Columbia University Press, New York.

TAYLOR, C. R. (1980). Energetics of locomotion: primitive and advanced mammals. In *Comparative physiology: primitive mammals* (ed. K. Schmidt-Nielsen, L. Bolis, and C. R. Taylor) pp. 192–9. Cambridge University Press.

VAN VALEN, L. (1973). A new evolutionary law. *Evolutionary Theory* **1**, 1–30.

——and SLOAN, R. E. (1977) Ecology and the extinction of the dinosaurs. *Evolutionary Theory* **2**, 37–64.

WHITTOW, G. C. (1970–73). *Comparative physiology of thermoregulation*. Academic Press, New York.

Chapter 33 Evolutionary changes

BAKKER, R. T. (1977). Tetrapod mass extinctions – a model of the regulation of speciation rates and immigration by cycles of topographic diversity. In *Patterns of evolution as illustrated by the fossil record* (ed. A. Hallam) pp. 439–68. Elsevier, Amsterdam.

CARROLL, R. L. (1977). Patterns of amphibian evolution: an extended example of the incompleteness of the fossil record. In *Patterns of evolution as illustrated by the fossil record* (ed. A. Hallam) pp. 405–38. Elsevier, Amsterdam.

DESMOND, A. J. (1975). *The hot-blooded dinosaurs: a revolution in palaeontology*. Blond & Briggs, London.

GINGERICH, P. D. (1977). Patterns of evolution in the mammalian fossil record. In *Patterns of evolution as illustrated by the fossil record* (ed. A. Hallam) pp. 469–500. Elsevier, Amsterdam.

HALLAM, A. (1963). Major epeirogenic and eustatic changes, since the Cretaceous, and their possible relationship to crustal structure. *American Journal of Science* **261**, 397–423.

—— (1971). Re-evaluation of the palaeographic argument for an expanding earth. *Nature, London* **232**, 180–2.

——(ed.) (1977). *Patterns of evolution as illustrated by the fossil record*. Elsevier, Amsterdam. [*Developments in Palaeontology and Stratigraphy*, Vol. 5.]

HAYS, J. D. and PITMAN, W. C. (1973). Lithospheric motion, sea level changes and climatic and ecological consequences. *Nature, London* **246**, 18–22.

LANDE, R. (1978). Evolutionary mechanism of limb loss in tetrapods. *Evolution* **32**, 73–92.

LULL, R. S. (1917). *Organic evolution*. Macmillan, London.

McELHINNY, M. W. (1971). Geomagnetic reversals during the Phanerozoic. *Science, New York* **172**, 157–9.

MAYR, E. (1976). *Evolution and the diversity of life: selected essays*. Belknap, Cambridge, Mass.

ROMER, A. S. (1966). *Vertebrate paleontology*. University of Chicago Press.

RUSSELL, D. A. (1979). The enigma of the extinction of the dinosaurs. *Annual Review of Earth and Planetary Sciences* **7**, 163–82.

SCHOPF, T. J. M. (1974). Permo–Triassic extinctions: relation to sea floor spreading. *Journal of Geology* **82**, 129–43.

—— (1977). Patterns of evolution: a summary and discussion. In *Patterns of evolution as illustrated by the fossil record* (ed. A. Hallam) pp. 547–62. Elsevier, Amsterdam.

SIMPSON, G. G. (1953). *The major features of evolution*. Columbia University Press, New York.

STANLEY, S. M. (1977). Trends, rates, and patterns of evolution in the Bivalvia. In *Patterns of evolution as illustrated by the fossil record* (ed. A. Hallam) pp. 209–250. Elsevier, Amsterdam.

THOMSON, K. S. (1977). The pattern of diversification in fishes. In *Patterns of evolution as illustrated by the fossil record* (ed. A. Hallam) pp. 377–404. Elsevier, Amsterdam.

YOUNG, J. Z. (1938). The evolution of the nervous system and of the relationship of organism and environment. In *Evolution: essays on aspects of evolutionary biology presented to Professor E. S. Goodrich on his 70th birthday* (ed. G. R. de Beer) pp. 179–204. Clarendon Press, Oxford.

Author Index

Subject Index